VOLUME ONE HUNDRED AND TWELVE

Current Topics in DEVELOPMENTAL BIOLOGY

Cellular Adhesion in Development and Disease

CURRENT TOPICS IN DEVELOPMENTAL BIOLOGY

"A meeting-ground for critical review and discussion of developmental processes"
A.A. Moscona and Alberto Monroy (Volume 1, 1966)

VOLUME ONE HUNDRED AND TWELVE

CURRENT TOPICS IN DEVELOPMENTAL BIOLOGY

Cellular Adhesion in Development and Disease

Edited by

ALPHA S. YAP

Division of Cell Biology and Molecular Medicine,
Institute for Molecular Bioscience,
The University of Queensland,
Brisbane, Queensland, Australia

ELSEVIER

AMSTERDAM • BOSTON • HEIDELBERG • LONDON
NEW YORK • OXFORD • PARIS • SAN DIEGO
SAN FRANCISCO • SINGAPORE • SYDNEY • TOKYO

Academic Press is an imprint of Elsevier

Academic Press is an imprint of Elsevier
225 Wyman Street, Waltham, MA 02451, USA
525 B Street, Suite 1800, San Diego, CA 92101-4495, USA
125 London Wall, London, EC2Y 5AS, UK
The Boulevard, Langford Lane, Kidlington, Oxford OX5 1GB, UK

First edition 2015

ISBN: 978-0-12-407758-4
ISSN: 0070-2153

For information on all Academic Press publications visit our website at store.elsevier.com

CONTENTS

CONTRIBUTORS

Elias H. Barriga
Cell and Developmental Biology Department, University College London, London, United Kingdom

Deanna L. Benson
Fishberg Department of Neuroscience, Friedman Brain Institute and the Graduate School of Biomedical Sciences, Icahn School of Medicine at Mount Sinai, New York, USA

Nicholas H. Brown
Department of Physiology, Development and Neuroscience, The Gurdon Institute, University of Cambridge, Cambridge, United Kingdom

Alexander N. Combes
Institute for Molecular Bioscience, The University of Queensland, St. Lucia, Brisbane, Queensland, Australia

Jamie A. Davies
Centre for Integrative Physiology, University of Edinburgh, Edinburgh, United Kingdom

Andrew J. Ewald
Department of Cell Biology, Center for Cell Dynamics, and Department of Oncology, Johns Hopkins University School of Medicine, Baltimore, Maryland, USA

François Fagotto
Department of Biology, McGill University, Montréal, Québec, Canada

Lauren G. Friedman
Fishberg Department of Neuroscience, Friedman Brain Institute and the Graduate School of Biomedical Sciences, Icahn School of Medicine at Mount Sinai, New York, USA

Cara J. Gottardi
Cellular and Molecular Biology, and Medicine, Northwestern University Feinberg School of Medicine, Chicago, Illinois, USA

Benjamin M. Hogan
Institute for Molecular Bioscience, The University of Queensland, Brisbane, Queensland, Australia

George W. Huntley
Fishberg Department of Neuroscience, Friedman Brain Institute and the Graduate School of Biomedical Sciences, Icahn School of Medicine at Mount Sinai, New York, USA

Anne Karine Lagendijk
Institute for Molecular Bioscience, The University of Queensland, Brisbane, Queensland, Australia

Terry Lechler
Department of Dermatology, and Department of Cell Biology, Duke University Medical Center, Durham, North Carolina, USA

Melissa H. Little
Institute for Molecular Bioscience, The University of Queensland, St. Lucia, Brisbane, Queensland, Australia

Aidan P. Maartens
Department of Physiology, Development and Neuroscience, The Gurdon Institute, University of Cambridge, Cambridge, United Kingdom

Meghan T. Maher
Department of Biology, Washington University in St. Louis, St. Louis, Missouri, USA

Kenji Mandai
Division of Pathogenetic Signaling, Kobe University Graduate School of Medicine, and CREST, Japan Science and Technology Agency, Kobe, Japan

Roberto Mayor
Cell and Developmental Biology Department, University College London, London, United Kingdom

Pierre D. McCrea
Department of Genetics, University of Texas MD Anderson Cancer Center; Program in Genes & Development, Graduate School in Biomedical Sciences, Houston, Texas, USA

Masahiro Mori
CREST, Japan Science and Technology Agency; Division of Neurophysiology, Department of Physiology and Cell Biology, Kobe University Graduate School of Medicine, and Faculty of Health Sciences, Kobe University Graduate School of Health Sciences, Kobe, Japan

Nicolas Plachta
European Molecular Biology Laboratory (EMBL) Australia, Australian Regenerative Medicine Institute, Monash University, Clayton, Victoria, Australia

Rashmi Priya
Division of Cell Biology and Molecular Medicine, Institute for Molecular Bioscience, The University of Queensland, Brisbane, Queensland, Australia

Yoshiyuki Rikitake
CREST, Japan Science and Technology Agency; Division of Signal Transduction, Department of Biochemistry and Molecular Biology, and Division of Cardiovascular Medicine, Department of Internal Medicine, Kobe University Graduate School of Medicine, Kobe, Japan

Katja Röper
MRC-Laboratory of Molecular Biology, Cambridge Biomedical Campus, Cambridge, United Kingdom

Pierre Savagner
IRCM, Institut de Recherche en Cancérologie de Montpellier, INSERM U896, Institut régional du Cancer Université Montpellier1, Montpellier, France

Eliah R. Shamir
Department of Cell Biology, Center for Cell Dynamics, and Department of Oncology, Johns Hopkins University School of Medicine, Baltimore, Maryland, USA

Kaelyn D. Sumigray
Department of Dermatology, and Department of Cell Biology, Duke University Medical Center, Durham, North Carolina, USA

Yoshimi Takai
Division of Pathogenetic Signaling, Kobe University Graduate School of Medicine, and CREST, Japan Science and Technology Agency, Kobe, Japan

Melanie D. White
European Molecular Biology Laboratory (EMBL) Australia, Australian Regenerative Medicine Institute, Monash University, Clayton, Victoria, Australia

Alpha S. Yap
Division of Cell Biology and Molecular Medicine, Institute for Molecular Bioscience, The University of Queensland, Brisbane, Queensland, Australia

PREFACE

Cell adhesion is a fundamental determinant of development in metazoan organisms. For over a century—from the early observations of HV Wilson, through the seminal studies of Townes and Holtfreter, and since—we have endeavored to understand how adhesion helps make multicellular organisms more than just the sum of their parts. We now know that physical interactions between cells and their environment (other cells and components of the extracellular matrix) influence critical parameters of development, including tissue cohesion, cellular patterning, differentiation, and population control. These diverse functional effects reflect the complex ways in which distinct adhesion systems interact with cellular processes such as signaling, the cytoskeleton, and membrane trafficking. In this volume, we aim to survey recent developments in understanding how the cellular and molecular mechanisms of adhesion determine the development of organisms and their constituent organs.

The early chapters in this volume endeavor to define some of the key processes that allow adhesion to influence development. Melanie White and Nicolas Plachta review how adhesion cooperates with the cytoskeleton to drive the earliest cellular events in the preimplantation mouse embryo: compaction, change in cell shape, polarity, and cell fate. François Fagotto then addresses one of the long-standing problems in developmental biology: understanding how boundaries are formed in the embryo. Building on the long-standing realization that boundaries reflect physical differences between populations of cells, Fagotto outlines how different cell–cell adhesion systems may cooperate with the cytoskeleton to segregate cell populations at boundaries.

We then have a series of chapters that focus on the mechanisms by which cadherin cell adhesion molecules influence animal development. Here, a major advance has come from the realization that cadherins cooperate with the contractile apparatus, that is, the actomyosin cytoskeleton. Accordingly, Rashmi Priya and Alpha Yap discuss the molecular and cellular mechanisms that allow cadherin adhesion systems to physically interact with, and also regulate, the actomyosin cytoskeleton. Katja Röper then addresses how cooperation between cell–cell adhesion and contractility determines morphogenesis in the early *Drosophila* embryo. In their chapter, Pierre McCrea, Meghan Maher, and Cara Gottardi broaden the discussion to review how cadherins and their associated proteins signal to the nucleus, a paradigm that underlies canonical Wnt signaling and also impinges on other fundamental developmental pathways, such as the Hippo signaling pathway.

Of course, cadherins are not the only adhesion systems that influence development. Another large family of cell–cell adhesion molecules are the nectins and nectin-like proteins. Kenji Mandai, Yoshimi Takai, and their colleagues discuss the fundamental cell biology of nectins and review how these molecules affect the development of many organs in the body. Aidan Maartens and Nicholas Brown then outline developments in understanding how integrin cell–matrix adhesion molecules contribute to *Drosophila* development, including notable developments in how integrins influence cell fate, cell migration, and cell polarity.

The two subsequent chapters focus on developmental processes that integrate adhesion, signaling, and the cytoskeleton. Pierre Savagner discusses the concept of epithelial-to-mesenchymal transition, providing a historical and conceptual framework for this complex phenomenon, with its often controversial mechanistic underpinnings. Elias Barriga and Roberto Mayor then take the example of neural crest migration to consider how adhesive events generate collective patterns of cell migration.

Finally, we examine how cell adhesion influences the development of individual organs. Anne Lagendijk and Benjamin Hogan review how cell signaling and cell–cell adhesion cooperate during vascular development. Eliah Shamir and Andrew Ewald focus on how individual and collective cell migration are regulated by cell–cell adhesion to drive epithelial morphogenesis of the mammary gland. Kaelyn Sumigray and Terry Lechler review how multiple junctions (adherens, tight, and desmosomes) contribute to development of the epidermis as a fundamental biological barrier in the body. Lauren Friedman, Deanna Benson, and George Huntley consider the role that cadherins play in the nervous system, with a particular focus on understanding their role in synapse formation and the generation of synaptic networks, the bases of neural activity. And in the final chapter of this volume, Alexander Combes, Jamie Davies, and Melissa Little discuss how cell adhesion drives self-organization in the embryonic kidney, providing insights relevant to tissue engineering and regenerative medicine.

We hope that the contributions in this volume illustrate some of the different perspectives that are now being used to understand how cell adhesion contributes to development. A final perspective lies in the relationship between development and disease. Many of the cellular mechanisms and biological processes that we consider are also implicated in disease. Thus, we also sought, where possible, to highlight how basic biology illuminates our understanding of disease and vice versa. We hope that these reviews will then be a useful guide to students of fundamental biology and pathology. And we will be well pleased if they prompt further research at the interface between these disciplines.

Alpha S. Yap

CHAPTER ONE

How Adhesion Forms the Early Mammalian Embryo

Melanie D. White, Nicolas Plachta[1]
European Molecular Biology Laboratory (EMBL) Australia, Australian Regenerative Medicine Institute, Monash University, Clayton, Victoria, Australia
[1]Corresponding author: e-mail address: nicolas.plachta@emblaustralia.org

Contents

Abstract

The early mouse embryo is an excellent system to study how a small group of initially rounded cells start to change shape and establish the first forms of adhesion-based cell–cell interactions in mammals *in vivo*. In addition to its critical role in the structural integrity of the embryo, we discuss here how adhesion is important to regulate cell polarity and cell fate. Recent evidence suggests that adherens junctions participate in signaling pathways by localizing key proteins to subcellular microdomains. E-cadherin has been identified as the main player required for the establishment of adhesion but other mechanisms involving additional proteins or physical forces acting in the embryo may also contribute. Application of new technologies that enable high-resolution quantitative imaging of adhesion protein dynamics and measurements of biomechanical forces will provide a greater understanding of how adhesion patterns the early mammalian embryo.

1. THE MOUSE PREIMPLANTATION EMBRYO AS A MODEL OF ADHESION IN MAMMALIAN DEVELOPMENT

Most research on adhesion has been performed on cells in tissue culture due to their availability and ease of manipulation. However, it is only

Current Topics in Developmental Biology, Volume 112
ISSN 0070-2153
http://dx.doi.org/10.1016/bs.ctdb.2014.11.022

during true cellular differentiation within an embryo that the contribution of adhesion to development can be examined directly. The mouse preimplantation embryo provides an ideal system to study adhesion mechanisms that are based exclusively on cell–cell interactions. A glycoprotein membrane, the *zona pellucida*, encloses the preimplantation embryo so cell–cell adhesion can be studied in the complete absence of extracellular matrix interactions.

Preimplantation development naturally occurs within the oviduct, but it can be recapitulated *in vitro* without adversely affecting the developmental potential of embryos (Summers & Biggers, 2003). Mouse embryos can be easily removed from the maternal oviducts and cultured in simple media conditions. Under these *ex utero* conditions, the embryos develop almost as rapidly as they do *in utero* and if transferred back to the uterus they can implant and continue developing to produce viable offspring.

During the first 2 days of development, the fertilized mouse egg undergoes three cleavage divisions to produce an 8-cell embryo (Fig. 1A). At this stage, the cells are round and visibly indistinguishable. The first major cell morphological changes begin as the 8-cell embryo undergoes compaction. Concomitant with a rise in intercellular adhesion, the cells flatten their membranes against each other, maximizing contact and forming a highly packed mass. This process of increased adhesion and embryo compaction occurs ubiquitously during preimplantation

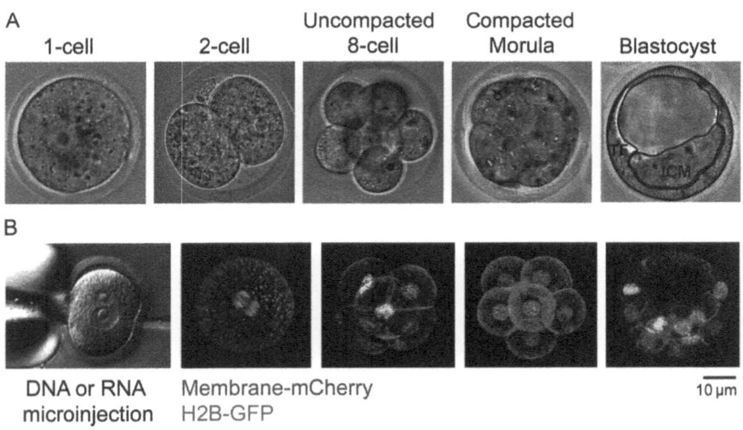

Figure 1 Imaging preimplantation development in the mouse embryo. (A) DIC images showing development of mouse embryo from 1-cell to blastocyst stage. (B) Microinjection of mRNA or DNA into the pronucleus allows visualization of proteins of interest throughout preimplantation development. In the example shown, the membrane is labeled with mCherry and the nucleus is labeled with H2B-GFP. ICM, inner cell mass; TE, trophectoderm. (See the color plate.)

development in different mammalian species and is an absolute requirement for embryo viability. This process is very similar in mouse and humans, making the preimplantation mouse embryo an ideal model to study the role of adhesion in cell shape, position, and fate in early mammalian development. In addition, the cells of the mouse embryo are relatively large, facilitating imaging of subcellular processes. Furthermore, there are many available genetic tools that are applicable in the mouse for manipulation of proteins of interest. Pronuclear microinjection of mRNA or DNA is a well-established technique for expression of exogenous proteins and mouse embryos are resilient enough to withstand this process with high efficiency (Fig. 1B). Moreover, thousands of genetically modified animals are now available carrying targeted endogenous genes or expressing various transgenic constructs.

1.1. Adhesion molecules in the preimplantation mouse embryo

Early studies identified a critical role for calcium-dependent adhesion in embryo compaction, subsequent spatial segregation of the inner cell mass (ICM) and formation of the first differentiated tissue, the trophectoderm (Fleming, Sheth, & Fesenko, 2001). Interfering with adhesion by chelating calcium ions or using antibodies targeting a cell surface glycoprotein decompacted embryos and prevented blastocyst formation (Ducibella & Anderson, 1975; Wales, 1970; Whitten, 1971). In 1981, this cell surface glycoprotein was identified as uvomorulin, now more commonly known as E-cadherin (Hyafil, Babinet, & Jacob, 1981).

Although usually expressed in epithelial cell layers, E-cadherin is also expressed from the very early stages of development. It is initially maternally derived in the oocyte and at the 2-cell stage, *de novo* E-cadherin zygotic synthesis starts (Vestweber, Gossler, Boller, & Kemler, 1987). Embryos lacking zygotic E-cadherin are preimplantation lethal. They do undergo compaction due to residual maternal E-cadherin but fail to form a blastocyst (Larue, Ohsugi, Hirchenhain, & Kemler, 1994). Using siRNAs to knockdown E-cadherin expression in just half of the embryo prevents those cells from integrating into the compacting embryo (Fig. 2) (Fierro-Gonzalez, White, Silva, & Plachta, 2013). Embryos lacking both maternal and zygotic E-cadherin are unable to compact or form a blastocyst and they appear as loose aggregates of cells (Stephenson, Yamanaka, & Rossant, 2010). Deleting maternal E-cadherin alone delays compaction until the late morula stage but embryos then develop normally due to zygotic E-cadherin expression

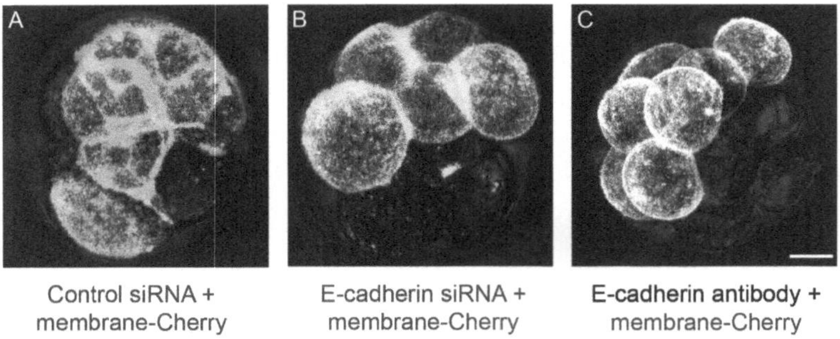

Figure 2 E-cadherin is required for cell–cell adhesion and embryo compaction. (A) Microinjection of one cell at the 2-cell stage results in an embryo expressing a control siRNA and a membrane-Cherry marker in half of its cells. The transgenic cells have normal morphology and integrate into the compacting embryo mass. (B) An siRNA targeting E-cadherin reduces cell–cell adhesion in the transgenic half of the embryo. The nontransgenic cells compact normally but the E-cadherin knockdown cells are very spherical and do not integrate into the embryo mass. (C) Treating the embryo with the DECMA-1 E-cadherin function-blocking antibody reduces adhesion and causes all cells to become very spherical. The embryo does not compact. Scale bar = 10 μm.

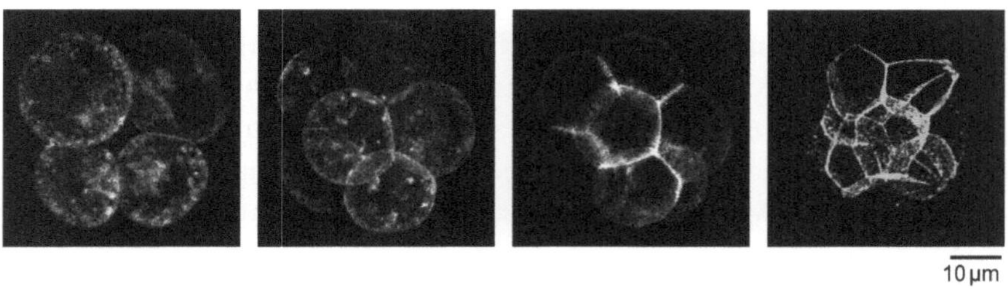

Figure 3 E-cadherin localization changes during preimplantation development. E-cadherin is distributed throughout the membrane until the late 8-cell stage. Then, it begins to accumulate in cell–cell junctions and is predominantly localized to basolateral regions by the 16-cell stage.

(De Vries et al., 2004). Adhesion does not develop until the late morula stage in these embryos indicating that although E-cadherin and its binding partners are expressed (Ohsugi et al., 1996; Sefton, Johnson, & Clayton, 1992; Vestweber et al., 1987), they are not required to form functional adhesion complexes at very early stages.

E-cadherin is uniformly distributed in the cell membrane until the 8-cell stage when PKC-α-mediated phosphorylation of β-catenin, a key protein regulating E-cadherin intracellular signaling, is thought to activate the compaction process (Fig. 3; Pauken & Capco, 1999). Compaction can be blocked by inhibition of PKC-α and induced early at the 2- and 4-cell stages by PKC-α activation (Ohsugi, Ohsawa, & Semba, 1993; Winkel, Ferguson, Takeichi, & Nuccitelli, 1990). E-cadherin accumulates basolaterally, forming adherens junctions between cells and connecting to the actin cytoskeleton via catenin proteins (Ozawa, Ringwald, & Kemler, 1990). As adhesion initiates, the actin cytoskeleton is reorganized to define the orientation of the first cellular polarity in the embryo (Stephenson et al., 2010).

After compaction has occurred, tight junctions begin to assemble at apicolateral cell–cell junctions in a process that requires prior activation of E-cadherin-mediated adhesion (Fleming, McConnell, Johnson, & Stevenson, 1989; Ohsugi, Larue, Schwarz, & Kemler, 1997). The timing of tight junction formation is tightly regulated by staggered expression of the constituent proteins from the 8- to 32-cell stage (Sheth et al., 1997). E-cadherin-mediated adhesion may also stabilize tight junction proteins, preventing their turnover once they are assembled at the membrane (Javed, Fleming, Hay, & Citi, 1993). The close intercellular adhesion at tight junctions then forms a permeability seal between adjacent epithelial cells. This allows the formation of the blastocoel cavity and the generation of the blastocyst. During blastocyst expansion, small, strongly adhesive junctions called desmosomes are assembled between adjacent cells in the trophectoderm. These junctions are multiprotein complexes containing the desmosomal cadherins, desmocollins, and desmogleins (Fleming, Garrod, & Elsmore, 1991). Adherens junctions and desmosomes functionally synergize to maintain epithelial polarity and structure. Thus, in the sequence of developmental events generating adhesion in the embryo, E-cadherin is a central player in a pathway that enables the progressive formation and organization of different types of anchoring junctions and the establishment of the first forms of tissue-like organization during development.

2. ADHESION REGULATES CELL SHAPE

There are many studies demonstrating the importance of adhesion for morphogenesis in other experimental models (Lecuit, Lenne, & Munro, 2011). However, relatively little work has been performed in the preimplantation mouse embryo to elucidate how adhesion controls cell shape. The first discernible changes accompanying the increased intercellular adhesion during compaction are morphological. From the late 8-cell stage, cells lose their previously spherical shape and flatten into a tightly packed mass with indistinguishable membranes (Fig. 1). The formation of adherens junctions was widely believed to be responsible for this global change in morphology since perturbations of E-cadherin using antibodies or gene deletion approaches resulted in cells displaying a rounder morphology. However, recent work has demonstrated that in addition to its localization to adherens junctions, E-cadherin also accumulates in long cellular protrusions identified as filopodia, which appear specifically during the stages of embryo compaction (Fierro-Gonzalez et al., 2013). Cells extend these filopodia from their junctions onto the apical membrane of neighboring cells and adhere to them via E-cadherin *trans* interactions (Fig. 4). The filopodia then maintain tension to elongate the cell's membrane over its neighbors, bringing them into close apposition, and facilitating compaction (Fierro-Gonzalez et al., 2013). Whereas E-cadherin localized at adherens junctions play an important role in keeping cells together, the pool of E-cadherin in the recently discovered filopodia helps cells establish tight interactions at their apical membranes in order to compact the embryo.

Physically disrupting these filopodia with laser ablation causes cells to revert to a more spherical shape. Moreover, interfering with the molecular components present in these filopodia, which include E-cadherin, α-catenin, β-catenin, or Myo10, prevents compaction. Expressing a mutant form of E-cadherin lacking the extracellular domain or treating embryos with an E-cadherin function-blocking antibody also disrupts filopodia formation and embryo compaction, suggesting that the adhesive function of

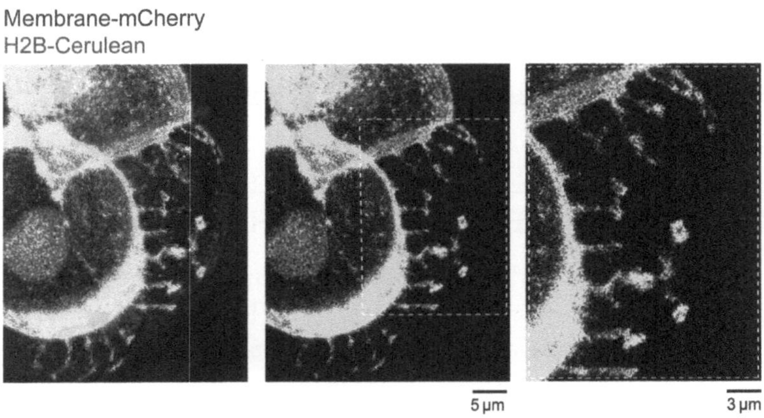

Figure 4 E-cadherin-dependent filopodia control embryo compaction. Two cells labeled with a mCherry protein targeted to the cell membrane extend filopodia over their nonlabeled neighbors during compaction at the late 8-cell stage. Dashed box shows filopodia at higher magnification. (See the color plate.)

E-cadherin is critical. Only around 60% of cells in each embryo were found to form filopodia and there is no direct association with subsequent cell fate so it remains to be revealed what determines whether a cell will employ this newly described mechanism of adhesion during preimplantation development.

3. ADHESION CONTROLS CELL POLARITY

The initiation of adhesion at the 8-cell stage also directs establishment of the first forms of cellular polarity in the developing mouse embryo. During this time, the actin cytoskeleton is reorganized to establish an apicobasal polarity together with the formation of an apical microvillous pole (Handyside, 1980) and a polarized distribution of cytoplasmic organelles and cytoskeletal elements (Fleming & Pickering, 1985; Houliston, Pickering, & Maro, 1987; Johnson & Maro, 1984). Some classical cell polarity proteins such as Ezrin, Pard6b, and the aPKCs (PKCζ and PKCλ) are then localized to the apical domain while Par-1, Jam-1, and Na/K ATPase are found at basolateral cell–cell contacts (Barcroft, Moseley, Lingrel, & Watson, 2004; Louvet, Aghion, Santa-Maria, Mangeat, & Maro, 1996; Pauken & Capco, 2000; Thomas et al., 2004; Vinot et al., 2005; Wang, Ojakian, & Nelson, 1990). Polar cells remain on the outside of the embryo and differentiate into trophectoderm, whereas apolar cells are enclosed inside the embryo and form the pluripotent ICM (Dyce, George, Goodall, & Fleming, 1987). Adhesion does not appear to be required for the initiation of polarization as cells isolated from early mouse embryos can polarize in the absence of cell–cell contacts (Johnson, Maro, & Takeichi, 1986; Ziomek & Johnson, 1980) or following inhibition of E-cadherin function with antibodies (Houliston, Pickering, & Maro, 1989; Johnson et al., 1986). However, the polarization of these disaggregated cells is delayed and also random in orientation. Deleting both maternal and zygotic E-cadherin confirms that E-cadherin is required to restrict the area of the apical domain and confine basolateral proteins, ensuring correct segregation of apical and basolateral domains (Stephenson et al., 2010). Together, these studies reveal that E-cadherin-mediated adhesion controls the timing and axis of polarization in the preimplantation mouse embryo.

4. ADHESION DETERMINES CELL FATE

Polarization has been closely linked to cell fate. Disruption of the apical proteins Par3 and PKCλ preferentially directs cells in the embryo toward an

ICM fate (Plusa et al., 2005). By determining the orientation of polarity, intercellular adhesion has a critical role in specifying cell fate. The maternal and zygotic deletion of E-cadherin demonstrated that E-cadherin-mediated adhesion is important to restrict trophectoderm fate (Stephenson et al., 2010). More cells in the mutant embryos express the trophectoderm-specific marker Cdx2, and the normal spatial allocation of trophectoderm cells to the outside of the embryo and ICM cells to the inside is disrupted. Recent work has now established how adhesion affects cell fate indirectly through polarity and shown that it also has an additional direct role in fate determination (Fig. 5).

Differentiation of the outer cells of the morula into trophectoderm requires expression of the transcription factors Cdx2 and Gata3, which in turn is driven

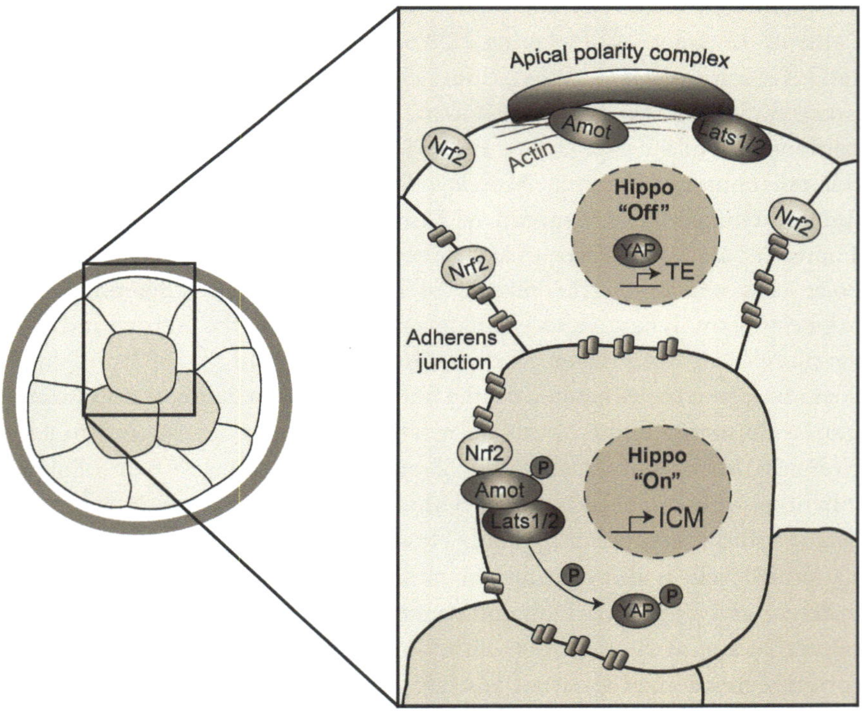

Figure 5 Adhesion and polarity determine cell fate in the preimplantation mouse embryo. The apical polarity complex in outer cells sequesters components of the Hippo signaling pathway preventing its activation. Unphosphorylated Yap can enter the nucleus and drive expression of trophectoderm-specific genes. In inner cells, Amot localizes to adherens junctions where it binds to Lats1/2 and the E-cadherin adhesion complex via Nrf2. Lats 1/2 phosphorylates Amot, activating it and this complex phosphorylates Yap. Phosphorylated Yap is excluded from the nucleus and the Hippo pathway is activated, allowing transcription of inner cell mass specific genes.

by the transcription factor Tead4 and its coactivator, Yap1 (Ralston et al., 2010). Tead4 knockout mice can specify an ICM but do not form the trophectoderm (Nishioka et al., 2008; Yagi et al., 2007). The Hippo signaling pathway regulates Yap1 subcellular localization by phosphorylation (Nishioka et al., 2009). When Hippo signaling is active, the kinase Lats1/2 phosphorylates Yap1, excluding it from the nucleus, promoting transcription of ICM-specific genes, and repressing trophectoderm fate. In the absence of sufficient Hippo signaling, Yap1 is free to enter the cell nucleus and induce transcription of the trophectoderm-specific genes Cdx2 and Gata3. However, Yap1, Tead4, and the members of the Hippo signaling pathway are expressed in all cells of the preimplantation mouse embryo. So, how is Hippo signaling suppressed in outer cells and activated in inner cells? The answer lies in sequestration of key members of the pathway by proteins involved in polarity and adhesion.

Angiomotin (Amot) and Nf2 are required to activate Lats1/2 and switch on the Hippo signaling pathway (Hirate et al., 2013). Nf2 is uniformly distributed through the membrane but in outer cells, Amot is sequestered by components of the apical polarity complex and localized to the apical domain (Hirate et al., 2013). Here, it is bound to actin and held in an inactive state. Lats1/2 may also be sequestered by the polarity complex as it too has an apical localization in outer cells (Cockburn, Biechele, Garner, & Rossant, 2013). The apical localization of Amot and Lats1/2 prevents activation of the Hippo signaling pathway, unphosphorylated Yap1 enters the nucleus and the cell reverts to a trophectoderm fate.

In inner cells that lack apical polarity, Amot can interact with Lats1/2 and the E-cadherin adhesion complex at adherens junctions via Nf2 (Hirate et al., 2013). This interaction activates Amot and stabilizes it at adherens junctions where it switches the Hippo signaling pathway on. Phosphorylated Yap1 is excluded from the nucleus and ICM genes are transcribed.

When Nf2 is removed by maternal and zygotic deletion, the Hippo pathway cannot be activated and the mutant embryos fail to generate an ICM (Cockburn et al., 2013). Instead, inner cells express trophectoderm-specific genes demonstrating that, regardless of their position in the embryo, cells revert to a trophectoderm fate when the components of the Hippo pathway are not correctly localized to adherens junctions.

5. EMERGING TECHNOLOGIES TO STUDY ADHESION

In vivo imaging of developing mouse embryos has recently been used to track cell progeny and fate (Bischoff, Parfitt, & Zernicka-Goetz, 2008;

Kurotaki, Hatta, Nakao, Nabeshima, & Fujimori, 2007; Morris et al., 2010; Plachta, Bollenbach, Pease, Fraser, & Pantazis, 2011). In coming years, imaging technologies that allow tracking of the levels, stability and distribution of proteins controlling adhesion will enable visualization of the establishment of adhesion and how this influences downstream processes such as polarity and fate. For example, photobleaching methods like fluorescence recovery after photobleaching (FRAP) allow measurement of the overall stability of membrane and cytoplasmic proteins regulating adhesion. FRAP has been useful to elucidate how E-cadherin is positioned and maintained at adherens junctions in different cellular contexts (de Beco, Gueudry, Amblard, & Coscoy, 2009; Huang et al., 2011). However, it has not been applied to study adhesion yet in the early mouse embryo. Methods like fluorescence correlation spectroscopy (FCS), based on the study of the fluorescent fluctuations in a small volume, can provide even more detailed information about the biophysical properties of proteins (Digman & Gratton, 2011). FCS has been utilized to measure the dynamics of proteins related to cell adhesion, such as actin (Gowrishankar et al., 2012). More recently, FCS has been applied to living preimplantation mouse embryos to study the dynamics of nuclear gene-regulatory proteins (Kaur et al., 2013). In the future, it will be important to establish its use for cytoplasmic and membrane-bound proteins involved in adhesion to reveal their dynamic behavior throughout development in different subcellular contexts. Super-resolution microscopy approaches are already starting to yield impressively detailed information about adhesion complexes in fixed cells (Guillaume et al., 2013; Indra, Hong, Troyanovsky, Kormos, & Troyanovsky, 2013). So far, these methods are restricted to thin specimens or sliced tissues, but they may soon be applicable in more complex three-dimensional structures such as the mouse embryo and may even allow data to be gathered at different time points, as has been achieved in living brain slices (Berning, Willig, Steffens, Dibaj, & Hell, 2012).

Finally, the study of adhesion will be greatly advanced by the development of engineered proteins. E-cadherin has been engineered to contain a stretchable Förster resonance energy transfer probe that can be used to directly measure the tensile forces transmitted through the cytoplasmic domain of E-cadherin (Borghi et al., 2012). So far, this approach has only been used in cultured epithelial cells, but it will be highly informative to visualize tension across adherens junctions in the developing mouse embryo. Optogenetic control of Rho-family GTPases engineered to be light-sensitive has been used to remodel actin and alter cell shape *in vitro* (Leung, Otomo, Chory, & Rosen, 2008; Levskaya, Weiner, Lim, & Voigt, 2009). Chromophore assisted light inactivation can be used to selectively inactivate proteins (reviewed in Sano, Watanabe, & Matsunaga, 2014). It will be revealing to use these technologies to express light-sensitive adhesion proteins in the preimplantation mouse embryo, and specifically alter their activity through targeting of laser stimulation to defined subcellular microdomains.

6. QUESTIONS FOR THE FUTURE

Recent studies have just begun to explain the connection between cell adhesion, polarity, and fate in the early mammalian embryo. However, many important questions remain unanswered and the development of several new technologies will enable them to be addressed in future work.

For example, it remains unclear whether, aside from its role in activating the Hippo pathway, adhesion has any other signaling functions in the mouse preimplantation embryo. E-cadherin has been shown to negatively regulate the receptor tyrosine kinases EGFR, IGF-1R, and c-Met and inhibit cell growth *in vitro* in an adhesion-dependent manner (Qian, Karpova, Sheppard, McNally, & Lowy, 2004). The small GTPases Rho, Rac, and Cdc42 are activated upon E-cadherin-mediated cell contact formation *in vitro* (Betson, Lozano, Zhang, & Braga, 2002; Calautti et al., 2002; Kim, Li, & Sacks, 2000; Kovacs, Ali, McCormack, & Yap, 2002; Noren, Niessen, Gumbiner, & Burridge, 2001). Little is known about potential signaling functions for adhesion in the preimplantation mouse embryo. However, E-cadherin has recently been shown to be required to activate Igf1r signaling at adherens junctions for trophectoderm formation (Bedzhov, Liszewska, Kanzler, & Stemmler, 2012). It remains to be established whether it also influences signaling from other receptor tyrosine kinases. Given the *in vitro* evidence and the recent role for adherens junctions in modulating the Hippo signaling pathway, it seems likely that adherens junctions may serve as scaffolds where signaling proteins are recruited and regulated. Careful imaging studies will be required to determine if this is the case in the embryo.

In addition, it will be important to establish if adhesion is involved in signaling through the E-cadherin-dependent filopodia that control compaction. A similar mechanism has been reported in *Drosophila* where filopodia-like structures known as cytonemes are involved in spatial patterning of the embryo (Roy, Hsiung, & Kornberg, 2011). Further research will ascertain whether adhesion contributes to spatiotemporally regulated signaling outside of adherens junctions in the preimplantation mouse embryo.

Most of our current knowledge about adhesion in the mouse embryo relates to E-cadherin. However, it is likely that other proteins and mechanisms also play a role. When E-cadherin is disrupted, cells become rounder but still maintain some adhesive properties (Fierro-Gonzalez et al., 2013; Shirayoshi, Okada, & Takeichi, 1983; Stephenson et al., 2010). While some of this remaining adhesion may be explained by residual expression of other less known cadherin members (for example, P-Cadherin; Stephenson et al., 2010), other proteins, such as GalTase (Bayna, Shaper, & Shur, 1988) may also participate.

Adhesion is one of the main physical forces acting between cells of the embryo. However, it remains largely unknown how the adhesion force interacts with some of the other main forces acting in the embryo. In particular, it is now well recognized that cells behave like viscoelastic fluids and the opposing forces of adhesion and cortical tension determine the degree of cell–cell contact (Lecuit & Lenne, 2007). Cells have been shown to behave like fluid objects, with a tendency to maximize their intercellular adhesion in the same way that liquids maximize their intermolecular attraction and simultaneously minimize their free surface energy (Foty, Forgacs, Pfleger, & Steinberg, 1994; Heintzelman, Phillips, & Davis, 1978). E-cadherin is likely to initiate this adhesive process by bringing cell surfaces into contact and providing the first anchoring point. The adhesion may then be reinforced by cells minimizing their cell–liquid interfacial tension as recently proposed in the zebrafish embryo (Maitre et al., 2012). Research addressing the overriding question of how adhesion and tension interact to control cell–cell interactions and embryo patterning has begun in other nonmammalian systems like the zebrafish (Maitre et al., 2012) and *Drosophila* (Rauzi, Verant, Lecuit, & Lenne, 2008) embryos using laser ablations of the actomyosin cortex or indirect measurements of tensile forces. Similar work will likely be focused in the mammalian embryo in the future.

In summary, future work should provide a better understanding of how adhesion is integrated with other key processes patterning the early mammalian embryo and controlling cell–cell interactions *in vivo*. The application of new technologies permitting real time and quantitative studies will help reveal how adhesion controls cell shape, fate, and position and vice versa in the mammalian embryo.

REFERENCES

Barcroft, L. C., Moseley, A. E., Lingrel, J. B., & Watson, A. J. (2004). Deletion of the Na/K-ATPase alpha1-subunit gene (Atp1a1) does not prevent cavitation of the preimplantation mouse embryo. *Mechanisms of Development, 121*(5), 417–426.

Bayna, E. M., Shaper, J. H., & Shur, B. D. (1988). Temporally specific involvement of cell surface beta-1,4 galactosyltransferase during mouse embryo morula compaction. *Cell, 53*(1), 145–157.

Bedzhov, I., Liszewska, E., Kanzler, B., & Stemmler, M. P. (2012). Igf1r signaling is indispensable for preimplantation development and is activated via a novel function of E-cadherin. *PLoS Genetics, 8*(3), e1002609.

Berning, S., Willig, K. I., Steffens, H., Dibaj, P., & Hell, S. W. (2012). Nanoscopy in a living mouse brain. *Science, 335*(6068), 551.

Betson, M., Lozano, E., Zhang, J., & Braga, V. M. (2002). Rac activation upon cell-cell contact formation is dependent on signaling from the epidermal growth factor receptor. *Journal of Biological Chemistry, 277*(40), 36962–36969.

Bischoff, M., Parfitt, D. E., & Zernicka-Goetz, M. (2008). Formation of the embryonic-abembryonic axis of the mouse blastocyst: Relationships between orientation of early cleavage divisions and pattern of symmetric/asymmetric divisions. *Development, 135*(5), 953–962.

Borghi, N., Sorokina, M., Shcherbakova, O. G., Weis, W. I., Pruitt, B. L., Nelson, W. J., et al. (2012). E-cadherin is under constitutive actomyosin-generated tension that is increased at cell-cell contacts upon externally applied stretch. *Proceedings of the National Academy of Sciences of the United States of America, 109*(31), 12568–12573.

Calautti, E., Grossi, M., Mammucari, C., Aoyama, Y., Pirro, M., Ono, Y., et al. (2002). Fyn tyrosine kinase is a downstream mediator of Rho/PRK2 function in keratinocyte cell-cell adhesion. *Journal of Cell Biology, 156*(1), 137–148.

Cockburn, K., Biechele, S., Garner, J., & Rossant, J. (2013). The Hippo pathway member Nf2 is required for inner cell mass specification. *Current Biology, 23*(13), 1195–1201.

de Beco, S., Gueudry, C., Amblard, F., & Coscoy, S. (2009). Endocytosis is required for E-cadherin redistribution at mature adherens junctions. *Proceedings of the National Academy of Sciences of the United States of America, 106*(17), 7010–7015.

De Vries, W. N., Evsikov, A. V., Haac, B. E., Fancher, K. S., Holbrook, A. E., Kemler, R., et al. (2004). Maternal beta-catenin and E-cadherin in mouse development. *Development, 131*(18), 4435–4445.

Digman, M. A., & Gratton, E. (2011). Lessons in fluctuation correlation spectroscopy. *Annual Review of Physical Chemistry, 62*, 645–668.

Ducibella, T., & Anderson, E. (1975). Cell shape and membrane changes in the eight-cell mouse embryo: Prerequisites for morphogenesis of the blastocyst. *Developmental Biology, 47*(1), 45–58.

Dyce, J., George, M., Goodall, H., & Fleming, T. P. (1987). Do trophectoderm and inner cell mass cells in the mouse blastocyst maintain discrete lineages? *Development, 100*(4), 685–698.

Fierro-Gonzalez, J. C., White, M. R., Silva, J., & Plachta, N. (2013). Cadherin-dependent filopodia control preimplantation embryo compaction. *Nature Cell Biology, 15*, 1424–1433.

Fleming, T. P., Garrod, D. R., & Elsmore, A. J. (1991). Desmosome biogenesis in the mouse preimplantation embryo. *Development, 112*(2), 527–539.

Fleming, T. P., McConnell, J., Johnson, M. H., & Stevenson, B. R. (1989). Development of tight junctions de novo in the mouse early embryo: Control of assembly of the tight junction-specific protein, ZO-1. *Journal of Cell Biology, 108*(4), 1407–1418.

Fleming, T. P., & Pickering, S. J. (1985). Maturation and polarization of the endocytotic system in outside blastomeres during mouse preimplantation development. *Journal of Embryology and Experimental Morphology, 89*, 175–208.

Fleming, T. P., Sheth, B., & Fesenko, I. (2001). Cell adhesion in the preimplantation mammalian embryo and its role in trophectoderm differentiation and blastocyst morphogenesis. *Frontiers in Bioscience, 6*, D1000–D1007.
Foty, R. A., Forgacs, G., Pfleger, C. M., & Steinberg, M. S. (1994). Liquid properties of embryonic tissues: Measurement of interfacial tensions. *Physical Review Letters, 72*(14), 2298–2301.
Gowrishankar, K., Ghosh, S., Saha, S., Rumamol, C., Mayor, S., & Rao, M. (2012). Active remodeling of cortical actin regulates spatiotemporal organization of cell surface molecules. *Cell, 149*(6), 1353–1367.
Guillaume, E., Comunale, F., Do Khoa, N., Planchon, D., Bodin, S., & Gauthier-Rouviere, C. (2013). Flotillin microdomains stabilize cadherins at cell-cell junctions. *Journal of Cell Science, 126*(Pt 22), 5293–5304.
Handyside, A. H. (1980). Distribution of antibody- and lectin-binding sites on dissociated blastomeres from mouse morulae: Evidence for polarization at compaction. *Journal of Embryology and Experimental Morphology, 60*, 99–116.
Heintzelman, K. F., Phillips, H. M., & Davis, G. S. (1978). Liquid-tissue behavior and differential cohesiveness during chick limb budding. *Journal of Embryology and Experimental Morphology, 47*, 1–15.
Hirate, Y., Hirahara, S., Inoue, K., Suzuki, A., Alarcon, V. B., Akimoto, K., et al. (2013). Polarity-dependent distribution of angiomotin localizes Hippo signaling in preimplantation embryos. *Current Biology, 23*(13), 1181–1194.
Houliston, E., Pickering, S. J., & Maro, B. (1987). Redistribution of microtubules and pericentriolar material during the development of polarity in mouse blastomeres. *Journal of Cell Biology, 104*(5), 1299–1308.
Houliston, E., Pickering, S. J., & Maro, B. (1989). Alternative routes for the establishment of surface polarity during compaction of the mouse embryo. *Developmental Biology, 134*(2), 342–350.
Huang, J., Huang, L., Chen, Y. J., Austin, E., Devor, C. E., Roegiers, F., et al. (2011). Differential regulation of adherens junction dynamics during apical-basal polarization. *Journal of Cell Science, 124*(Pt 23), 4001–4013.
Hyafil, F., Babinet, C., & Jacob, F. (1981). Cell-cell interactions in early embryogenesis: A molecular approach to the role of calcium. *Cell, 26*(3 Pt 1), 447–454.
Indra, I., Hong, S., Troyanovsky, R., Kormos, B., & Troyanovsky, S. (2013). The adherens junction: A mosaic of cadherin and nectin clusters bundled by actin filaments. *Journal of Investigative Dermatology, 133*(11), 2546–2554.
Javed, Q., Fleming, T. P., Hay, M., & Citi, S. (1993). Tight junction protein cingulin is expressed by maternal and embryonic genomes during early mouse development. *Development, 117*(3), 1145–1151.
Johnson, M. H., & Maro, B. (1984). The distribution of cytoplasmic actin in mouse 8-cell blastomeres. *Journal of Embryology and Experimental Morphology, 82*, 97–117.
Johnson, M. H., Maro, B., & Takeichi, M. (1986). The role of cell adhesion in the synchronization and orientation of polarization in 8-cell mouse blastomeres. *Journal of Embryology and Experimental Morphology, 93*, 239–255.
Kaur, G., Costa, M. W., Nefzger, C. M., Silva, J., Fierro-Gonzalez, J. C., Polo, J. M., et al. (2013). Probing transcription factor diffusion dynamics in the living mammalian embryo with photoactivatable fluorescence correlation spectroscopy. *Nature Communications, 4*, 1637.
Kim, S. H., Li, Z., & Sacks, D. B. (2000). E-cadherin-mediated cell-cell attachment activates Cdc42. *Journal of Biological Chemistry, 275*(47), 36999–37005.
Kovacs, E. M., Ali, R. G., McCormack, A. J., & Yap, A. S. (2002). E-cadherin homophilic ligation directly signals through Rac and phosphatidylinositol 3-kinase to regulate adhesive contacts. *Journal of Biological Chemistry, 277*(8), 6708–6718.

Kurotaki, Y., Hatta, K., Nakao, K., Nabeshima, Y., & Fujimori, T. (2007). Blastocyst axis is specified independently of early cell lineage but aligns with the ZP shape. *Science*, *316*(5825), 719–723.

Larue, L., Ohsugi, M., Hirchenhain, J., & Kemler, R. (1994). E-cadherin null mutant embryos fail to form a trophectoderm epithelium. *Proceedings of the National Academy of Sciences of the United States of America*, *91*(17), 8263–8267.

Lecuit, T., & Lenne, P. F. (2007). Cell surface mechanics and the control of cell shape, tissue patterns and morphogenesis. *Nature Reviews. Molecular Cell Biology*, *8*(8), 633–644.

Lecuit, T., Lenne, P. F., & Munro, E. (2011). Force generation, transmission, and integration during cell and tissue morphogenesis. *Annual Review of Cell and Developmental Biology*, *27*, 157–184.

Leung, D. W., Otomo, C., Chory, J., & Rosen, M. K. (2008). Genetically encoded photoswitching of actin assembly through the Cdc42-WASP-Arp2/3 complex pathway. *Proceedings of the National Academy of Sciences of the United States of America*, *105*(35), 12797–12802.

Levskaya, A., Weiner, O. D., Lim, W. A., & Voigt, C. A. (2009). Spatiotemporal control of cell signalling using a light-switchable protein interaction. *Nature*, *461*(7266), 997–1001.

Louvet, S., Aghion, J., Santa-Maria, A., Mangeat, P., & Maro, B. (1996). Ezrin becomes restricted to outer cells following asymmetrical division in the preimplantation mouse embryo. *Developmental Biology*, *177*(2), 568–579.

Maitre, J. L., Berthoumieux, H., Krens, S. F., Salbreux, G., Julicher, F., Paluch, E., et al. (2012). Adhesion functions in cell sorting by mechanically coupling the cortices of adhering cells. *Science*, *338*(6104), 253–256.

Morris, S. A., Teo, R. T., Li, H., Robson, P., Glover, D. M., & Zernicka-Goetz, M. (2010). Origin and formation of the first two distinct cell types of the inner cell mass in the mouse embryo. *Proceedings of the National Academy of Sciences of the United States of America*, *107*(14), 6364–6369.

Nishioka, N., Inoue, K., Adachi, K., Kiyonari, H., Ota, M., Ralston, A., et al. (2009). The Hippo signaling pathway components Lats and Yap pattern Tead4 activity to distinguish mouse trophectoderm from inner cell mass. *Developmental Cell*, *16*(3), 398–410.

Nishioka, N., Yamamoto, S., Kiyonari, H., Sato, H., Sawada, A., Ota, M., et al. (2008). Tead4 is required for specification of trophectoderm in pre-implantation mouse embryos. *Mechanisms of Development*, *125*(3–4), 270–283.

Noren, N. K., Niessen, C. M., Gumbiner, B. M., & Burridge, K. (2001). Cadherin engagement regulates Rho family GTPases. *Journal of Biological Chemistry*, *276*(36), 33305–33308.

Ohsugi, M., Hwang, S. Y., Butz, S., Knowles, B. B., Solter, D., & Kemler, R. (1996). Expression and cell membrane localization of catenins during mouse preimplantation development. *Developmental Dynamics*, *206*(4), 391–402.

Ohsugi, M., Larue, L., Schwarz, H., & Kemler, R. (1997). Cell-junctional and cytoskeletal organization in mouse blastocysts lacking E-cadherin. *Developmental Biology*, *185*(2), 261–271.

Ohsugi, M., Ohsawa, T., & Semba, R. (1993). Similar responses to pharmacological agents of 1,2-OAG-induced compaction-like adhesion of two-cell mouse embryo to physiological compaction. *Journal of Experimental Zoology*, *265*(5), 604–608.

Ozawa, M., Ringwald, M., & Kemler, R. (1990). Uvomorulin-catenin complex formation is regulated by a specific domain in the cytoplasmic region of the cell adhesion molecule. *Proceedings of the National Academy of Sciences of the United States of America*, *87*(11), 4246–4250.

Pauken, C. M., & Capco, D. G. (1999). Regulation of cell adhesion during embryonic compaction of mammalian embryos: Roles for PKC and beta-catenin. *Molecular Reproduction and Development*, *54*(2), 135–144.

Pauken, C. M., & Capco, D. G. (2000). The expression and stage-specific localization of protein kinase C isotypes during mouse preimplantation development. *Developmental Biology*, *223*(2), 411–421.

Plachta, N., Bollenbach, T., Pease, S., Fraser, S. E., & Pantazis, P. (2011). Oct4 kinetics predict cell lineage patterning in the early mammalian embryo. *Nature Cell Biology*, *13*(2), 117–123.

Plusa, B., Frankenberg, S., Chalmers, A., Hadjantonakis, A. K., Moore, C. A., Papalopulu, N., et al. (2005). Downregulation of Par3 and aPKC function directs cells towards the ICM in the preimplantation mouse embryo. *Journal of Cell Science*, *118*(Pt 3), 505–515.

Qian, X., Karpova, T., Sheppard, A. M., McNally, J., & Lowy, D. R. (2004). E-cadherin-mediated adhesion inhibits ligand-dependent activation of diverse receptor tyrosine kinases. *EMBO Journal*, *23*(8), 1739–1748.

Ralston, A., Cox, B. J., Nishioka, N., Sasaki, H., Chea, E., Rugg-Gunn, P., et al. (2010). Gata3 regulates trophoblast development downstream of Tead4 and in parallel to Cdx2. *Development*, *137*(3), 395–403.

Rauzi, M., Verant, P., Lecuit, T., & Lenne, P. F. (2008). Nature and anisotropy of cortical forces orienting Drosophila tissue morphogenesis. *Nature Cell Biology*, *10*(12), 1401–1410.

Roy, S., Hsiung, F., & Kornberg, T. B. (2011). Specificity of Drosophila cytonemes for distinct signaling pathways. *Science*, *332*(6027), 354–358.

Sano, Y., Watanabe, W., & Matsunaga, S. (2014). Chromophore-assisted laser inactivation—Towards a spatiotemporal-functional analysis of proteins, and the ablation of chromatin, organelle and cell function. *Journal of Cell Science*, *127*(Pt 8), 1621–1629.

Sefton, M., Johnson, M. H., & Clayton, L. (1992). Synthesis and phosphorylation of uvomorulin during mouse early development. *Development*, *115*(1), 313–318.

Sheth, B., Fesenko, I., Collins, J. E., Moran, B., Wild, A. E., Anderson, J. M., et al. (1997). Tight junction assembly during mouse blastocyst formation is regulated by late expression of ZO-1 alpha + isoform. *Development*, *124*(10), 2027–2037.

Shirayoshi, Y., Okada, T. S., & Takeichi, M. (1983). The calcium-dependent cell-cell adhesion system regulates inner cell mass formation and cell surface polarization in early mouse development. *Cell*, *35*(3 Pt. 2), 631–638.

Stephenson, R. O., Yamanaka, Y., & Rossant, J. (2010). Disorganized epithelial polarity and excess trophectoderm cell fate in preimplantation embryos lacking E-cadherin. *Development*, *137*(20), 3383–3391.

Summers, M. C., & Biggers, J. D. (2003). Chemically defined media and the culture of mammalian preimplantation embryos: Historical perspective and current issues. *Human Reproduction Update*, *9*(6), 557–582.

Thomas, F. C., Sheth, B., Eckert, J. J., Bazzoni, G., Dejana, E., & Fleming, T. P. (2004). Contribution of JAM-1 to epithelial differentiation and tight-junction biogenesis in the mouse preimplantation embryo. *Journal of Cell Science*, *117*(Pt 23), 5599–5608.

Vestweber, D., Gossler, A., Boller, K., & Kemler, R. (1987). Expression and distribution of cell adhesion molecule uvomorulin in mouse preimplantation embryos. *Developmental Biology*, *124*(2), 451–456.

Vinot, S., Le, T., Ohno, S., Pawson, T., Maro, B., & Louvet-Vallee, S. (2005). Asymmetric distribution of PAR proteins in the mouse embryo begins at the 8-cell stage during compaction. *Developmental Biology*, *282*(2), 307–319.

Wales, R. G. (1970). Effects of ions on the development of the preimplantation mouse embryo in vitro. *Australian Journal of Biological Sciences*, *23*, 421–429.

Wang, A. Z., Ojakian, G. K., & Nelson, W. J. (1990). Steps in the morphogenesis of a polarized epithelium. I. Uncoupling the roles of cell-cell and cell-substratum contact in

establishing plasma membrane polarity in multicellular epithelial (MDCK) cysts. *Journal of Cell Science*, *95*(Pt 1), 137–151.
Whitten, W. K. (1971). Nutritional requirements for the culture of preimplantation embryos in vitro. *Advances in Bioscience*, *6*, 129–139.
Winkel, G. K., Ferguson, J. E., Takeichi, M., & Nuccitelli, R. (1990). Activation of protein kinase C triggers premature compaction in the four-cell stage mouse embryo. *Developmental Biology*, *138*(1), 1–15.
Yagi, R., Kohn, M. J., Karavanova, I., Kaneko, K. J., Vullhorst, D., DePamphilis, M. L., et al. (2007). Transcription factor TEAD4 specifies the trophectoderm lineage at the beginning of mammalian development. *Development*, *134*(21), 3827–3836.
Ziomek, C. A., & Johnson, M. H. (1980). Cell surface interaction induces polarization of mouse 8-cell blastomeres at compaction. *Cell*, *21*(3), 935–942.

CHAPTER TWO

Regulation of Cell Adhesion and Cell Sorting at Embryonic Boundaries

François Fagotto[1]
Department of Biology, McGill University, Montréal, Québec, Canada
[1]Corresponding author: e-mail address: francois.fagotto@mcgill.ca

Contents

Current Topics in Developmental Biology, Volume 112
ISSN 0070-2153
http://dx.doi.org/10.1016/bs.ctdb.2014.11.026

Abstract

Embryonic boundaries are sharp delimitations that prevent intermingling between different cell populations. They are essential for the development of well-organized structures and ultimately a functional organism. It has been long believed that this process was driven by global differences in cell adhesion strength, or expression of different types of adhesion molecules. The actual picture turns out to be quite different: Boundaries should be viewed as abrupt discontinuities, where cortical contractility is acutely upregulated in response to specific cell surface contact receptors which act as repulsive cues. Cell adhesion is also modulated along the interface, in different ways depending on the type of boundary, but in all cases the process is subordinated to the function of the cortical actomyosin cytoskeleton.

1. INTRODUCTION

Development proceeds by subdivision of a single mass of cells into progressively smaller regions, which will eventually give rise to the tissues and organs of the adult organism. The position and size of these regions are determined by the interplay between patterning signals and gene regulatory networks, which have been characterized in detail and show amazing degrees of precision and sophistication. It is perhaps less widely appreciated that the newly determined regions become rapidly physically separated by embryonic boundaries, which impede any future exchange of cells. We will see that the property to separate from an adjacent population is acquired as an inheritable, cell-autonomous property. Without this physical separation, embryonic cells, which divide frequently and are generally highly motile, would be constantly at risk of ending up in the wrong territory. Short term, patterning signals are capable of reprogramming misplaced cells. To constantly maintain a developing structure based on patterning information, however, would be a challenging task, in particular for regions undergoing intense proliferation, such as the insect imaginal discs. It would certainly be close to impossible during large scale movements such as gastrulation. Boundaries largely relieve development from these constraints and allow each separate region to further evolve into complex structures. Consistently, experimental interference with boundary formation causes distortion of the insect wings (e.g., Janody, Martirosyan, Benlali, & Treisman, 2003) and catastrophic defects in the general body plan when it targets the initial vertebrate ectoderm–mesoderm separation (e.g., Rohani, Canty, Luu, Fagotto, & Winklbauer, 2011; Winklbauer, Medina, Swain,

& Steinbeisser, 2001). The capacity of cells to sort into different populations is thus a fundamental property indissociable with the multicellularity of metazoans, which can be traced to the deepest roots of animal evolution (Wilson, 1907).

The phenomenon has fascinated scientists for decades, and despite recent significant advances in understanding its cellular basis, it continues to withhold many mysteries. Regulation of cell–cell adhesion is clearly at the core of the problem, but different views have been opposed as to whether it constituted the main force that caused cell sorting and tissue separation, or whether other parameters were driving the process, such as the contractility of the cell cortex, or regulated repulsive reactions. One of the goals of this review is to provide an updated compilation of the scattered information gathered over the past years on tissue and boundary properties and to confront them to the theoretical models. This exercise will highlight the paucity of evidence for global differences between separating tissues, opposed to a strong case for local high tension as hallmark of all boundaries. I will then focus on the role of ephrin–Eph signaling in building this tension at vertebrate boundaries, discussing the different possible mechanisms through which these repulsive cues may control the tension and adhesive properties of the boundaries. A significant part of this essay will be devoted to the many remaining open issues in the field. One should note in particular that the prototypic compartment boundaries in *Drosophila* are still in want of upstream cues, which so far have remained inexplicably elusive. As for vertebrates, tissue separation clearly depends on an ephrin–Eph-mediated reaction that resembles classical contact inhibition, but a description of the process only based on this mechanism is likely to be a coarse oversimplification. We will see in particular that ephrin–Eph signaling may have multiple effects beyond simple repulsion. We will also see that even its main target, i.e., stimulation of actomyosin contractility, can have different effects on cell adhesion and boundary properties. Furthermore, several other molecules have been implicated in separation, which bear no obvious direct connection to ephrins and Eph receptors, except for having myosin as common target. Although far less understood, these other components must be taken into account, and I will propose some ideas for their integration in a general model. The most exciting next challenge in my opinion is to move from a coarse description of a generic boundary to the more subtle regulations that likely provide each boundary with the right properties required for each different morphogenetic process.

2. A SHORT HISTORY OF TISSUE SEPARATION

I summarize here the original discoveries of "tissue affinities" and compartment boundaries, and briefly review the major hypotheses that were proposed to explain these phenomena.

2.1. Cell sorting and "affinities"

The field was founded by the discovery of the phenomenon of cell sorting: When cells dissociated from different embryonic regions are mixed and left to reaggregate, they initially form a mixed aggregate, but then gradually sort into distinct populations. Remarkably, these cell clusters develop into organized structures that bear the histological signatures of the tissues that would normally derive from the regions from which the cells originated. First observed in sponges (Wilson, 1907), the phenomenon was systematically analyzed in frog embryos (Holtfreter, 1939; Townes & Holtfreter, 1955), and its generality was confirmed in chicken embryos (Moscona & Moscona, 1952). The key characteristics conceptualized by Holtfreter constitute fundamental principles of metazoan organization and continue to bear deep implications for our understanding of the process: First, the fact that mixed aggregates can be produced implies that all cells of an embryos share a common adhesive mechanism. We now know that the main actors are cadherin adhesion molecules. Second, each cell, once determined, acquires an autonomous tissue identity, which can be maintained after cell dissociation and isolation, and even when the cell finds itself surrounded by cells of another type. This identity translates into the capacity to discriminate between neighbors and react by adopting a specific cell behavior, i.e., to group with cells of the same type. Holtfreter named this property "tissue affinity" (Holtfreter, 1939). The nature of the mechanisms that mediate the recognition of self (homotypic contacts) and nonself (heterotypic contacts) and drive the appropriate response remains the central question in the field.

2.2. Compartments

A second key discovery was made by *Drosophila* geneticists, who observed that the expansion of proliferating clones in the embryo blastoderm and in the larval imaginal discs was restricted by invisible yet sharp and fully impermeable partitions. Thus, these epithelial sheets were subdivided into "compartments," which were delimited by stably inherited "boundaries"

(Garcia-Bellido, Ripoll, & Morata, 1973; Lawrence & Green, 1975; Lawrence, Green, & Johnston, 1978; Morata & Lawrence, 1978; Martinez-Arias & Lawrence, 1985; reviewed in Dahmann & Basler, 1999). Similar compartment boundaries were subsequently found in the vertebrate embryo, e.g., in the brain and limb buds (Altabef, Clarke, & Tickle, 1997; Dahmann, Oates, & Brand, 2011; Fraser, Keynes, & Lumsden, 1990; Zervas, Millet, Ahn, & Joyner, 2004).

2.3. The differential adhesion hypothesis

Steinberg had the revolutionary idea to consider cell–cell adhesion from a physical point of view. He noticed that tissue explants behaved very much like liquids, from which he conceived the following analogy: Individual cells would correspond to the molecules of a liquid and cell adhesion to the cohesive bonds between these molecules. The principle of liquid surface tension predicted with astonishing accuracy many of the configurations adopted by cells and tissues. For instance, single cells and pieces of tissues in isolation invariably round up, thus minimizing the surface exposed to the medium, just as a drop of oil in water. When placed against an adhesive surface, whether matrix or cells, they spread, or in biophysical terms they "wet" the surface. When two groups of cells are put into contact, they either coalesce or, on the contrary, they remain fully separated, again similar to the behavior of immiscible liquids. Based on this analogy, Steinberg proposed that quantitative differences in cell adhesion were sufficient to explain cell sorting and thus tissue separation, a model that was named the differential adhesion hypothesis (DAH) (Davis, Phillips, & Steinberg, 1997; Steinberg, 1970, 1978). Sorting in mixed aggregates based on DAH was demonstrated *in vitro*, using cells expressing different cadherin levels (Foty & Steinberg, 2004, 2005). We will see that DAH in its original form does not seem to explain tissue separation in the embryo. However, the concept of representing morphogenesis based on a simple combination of adhesive and tensile forces exerted on the cell surface (i.e., cell membrane and its actin cortex) arguably constituted the most influential ideas in the field of morphogenesis. The analogy to liquid surface tension remains a very successful way to simulate the behavior of cells and tissues (Foty & Steinberg, 2004; Lecuit & Lenne, 2007; Manning, Foty, Steinberg, & Schoetz, 2010).

2.4. Differential CAM expression

As one entered the cloning era, and cellular functions could finally start to be assigned to particular gene products, biophysical considerations on cell sorting were temporarily left aside, and attention shifted naturally to a resolutely molecular perspective. The two first cell adhesion molecules (CAMs), N-CAM and E-cadherin (Edelman, 1986), soon joined by many other CAMs, were found to preferentially bind to themselves (Inuzuka, Miyatani, & Takeichi, 1991; Matsuzaki et al., 1990; Nose, Nagafuchi, & Takeichi, 1988; Steinberg & Takeichi, 1994). This homophilic adhesive property and the striking tissue-specific expression of most CAMs seemed to provide a perfect explanation for tissue segregation (Takeichi, 1995). According to this model, individualization of each tissue would rely on the expression of a particular kit of CAMs (reviewed in Oda & Takeichi, 2011). This hypothesis received a broad acceptance among developmental biologists, entered the textbooks, and reigned almost undisputed until recently.

Evidence supporting this hypothesis however remained scarce (Inoue et al., 2001; Price, De Marco Garcia, Ranscht, & Jessell, 2002). More accurate methods led to a reevaluation of the concept of homophilic binding. Notably, two studies showed that cells which expressed two different cadherins—at similar levels—failed to sort (Duguay, Foty, & Steinberg, 2003; Niessen & Gumbiner, 2002). *In vitro* measurements confirmed that binding is not strictly homophilic (Katsamba et al., 2009; Ounkomol, Yamada, & Heinrich, 2010; Prakasam, Maruthamuthu, & Leckband, 2006; Shi, Chien, & Leckband, 2008; Shimoyama, Tsujimoto, Kitajima, & Natori, 2000), and functional N-/E-cadherin heterotypic interactions were detected *in vivo* (Straub et al., 2011). A role for differential expression of homophilic CAMs in separation remains uncertain. Note that several studies hint at specific functions for different cadherin cytoplasmic tails in regulating signaling pathways (e.g., Schafer, Narasimha, Vogelsang, & Leptin, 2014; Seidel, Braeg, Adler, Wedlich, & Menke, 2004; Wheelock, Shintani, Maeda, Fukumoto, & Johnson, 2008), which could provide an alternative explanation for the multiplicity of cadherins and for their mosaic expression.

2.5. Contact inhibition

Contemporarily to the discovery of CAMs, the identification of ephrins and Eph receptors as cell surface repulsive cues (Flanagan & Vanderhaeghen, 1998) pointed to a completely different model for tissue separation, based on contact inhibition of migration, a phenomenon that had been originally proposed based on observations on cell lines (Abercrombie, 1967). Although ephrins and Ephs have mostly been studied for their role in development of neuronal networks and in angiogenesis, they are also widely expressed in early embryos (see Fagotto, Winklbauer, & Rohani, 2014 for review). The striking complementarity of ephrin/Eph expression patterns in the developing hindbrain suggested a role in segmentation, which was supported by functional data (Xu, Alldus, Holder, & Wilkinson, 1995). Similar observations were then made for segmentation of the somites (Durbin et al., 1998). This model, although rather straightforward, remained less popular than the cadherin-based models, perhaps because it was until recently limited to the two vertebrate segmentation processes. However, ephrins and Ephs are now known to control other types of boundaries, including between germ layers (Fagotto, Rohani, Touret, & Li, 2013; Park, Cho, Kim, Choi, & Han, 2011; Rohani et al., 2011), and emerge as major regulators of separation in vertebrates.

2.6. Differential interfacial tension

The turn of the century witnessed strong revival of biophysical approaches to developmental processes. Myosin II-mediated contractility was known to be involved in most aspects of cell adhesion and motility, and it became apparent that this parameter could explain quite a few aspects of morphogenesis. The importance of cell cortex contractility had been already highlighted in a seminal critical analysis of DAH by Harris (1976). Adhesion and contractility were formally integrated into a broader theory, named differential interfacial tension hypothesis (DITH) (Brodland, 2002; Brodland, Yang, & Sweny, 2009). DITH was used to interpret sorting of zebrafish germ layers, and it was concluded that the system was dominated by differences in cortical tension (Krieg et al., 2008; Maitre et al., 2012). Note that an old puzzle remains: When tested in reaggregation experiments, ectoderm cells always sorted toward the center, endoderm to the periphery, and mesoderm in between. This configuration fitted perfectly with DITH, but was opposite to the normal organization of embryos. This discrepancy may at least partly be due to the fact that the dissected explants are artificially exposed to the medium, which causes a strong surface tension, whereas in the embryo, the tissues are wrapped in an outer polarized epithelial layer (outer ectoderm in *Xenopus*, enveloping layer in zebrafish), which alleviates internal tensions (Ninomiya & Winklbauer, 2008).

2.7. Direct investigation of basic properties: Embryonic boundaries are not stable physical structures, but the dynamic product of cell–cell interactions

What about the extracellular matrix (ECM)? It is a robust isolator of adult organs. Would not it be an obvious candidate for the separation of embryonic tissues? ECM is indeed deposited soon after separation and contributes to consolidate the boundaries (Julich, Mould, Koper, & Holley, 2009; Koshida et al., 2005), but it does not appear to participate to the initial boundaries, which are clearly not physical fences. A boundary is in fact only impermeable insofar as it keeps cells of the two populations within their respective territories. Missorted cells, on the contrary, can freely cross it to reintegrate the proper tissue. This has been unambiguously demonstrated by following migration of single cells in mosaic notochords (Fagotto et al., 2013; Reintsch, Habring-Mueller, Wang, Schohl, & Fagotto, 2005) and rhombomeres (Calzolari, Terriente, & Pujades, 2014).

Further support came from *in vitro* reconstitution of the ectoderm–mesoderm boundary. This assay, based on the simple juxtaposition of tissue explants (Wacker, Grimm, Joos, & Winklbauer, 2000), played a significant role in the recent progress in uncovering the mechanism of tissue separation (Ibrahim & Winklbauer, 2001; Medina, Swain, Kuerner, & Steinbeisser, 2004; Rohani et al., 2011; Rohani, Parmeggiani, Winklbauer, & Fagotto, 2014; Wacker et al., 2000; Winklbauer et al., 2001). It thus became clear that cells react almost instantaneously to contacts with a tissue of the same or of the other cell type, "melting" selectively in the former within minutes, while remaining stably separated from the latter. This separation behavior (Wacker et al., 2000) can be observed for single dissociated cells laid on an explant (Wacker et al., 2000) and even between two single dissociated cells (Rohani et al., 2014).

These observations are important because they firmly validate Holtfreter's original interpretation that tissue separation relies on cell-autonomous properties. A key experimental consequence of this property is the possibility to study the mechanisms underlying tissue separation *in vitro* using isolated cells (Rohani et al., 2014).

What is then the mechanism that allows single cells to find their way and eventually gather with cells of the same type? Classical DAH and DITH state that boundaries result from the juxtaposition of two cell populations displaying global differences in cell–cell adhesion, cortical contractility, or both. Alternatively, cell populations may express different types of adhesion molecules. Contact inhibition does not presume of such differences, but predicts local effects at heterotypic contacts. One would thus expect that the analysis of the tissue properties would provide support for one or the other model.

3. ADHESION AND CONTRACTILITY OF EMBRYONIC TISSUES

There are four main models of boundaries in vertebrates: The mesoderm is first separated from the ectoderm (Fig. 1A). Its most axial portion then splits from the paraxial (or presomitic) mesoderm (PSM) to form the notochord (Fig. 1B). The PSM eventually segments into somites (Fig. 1D). For the developing neural tissue, the best studied process is the segmentation of the hindbrain into seven rhombomeres (Fig. 1C). In *Drosophila*, the two main models are the parasegment boundaries and the compartment boundaries of imaginal discs (Fig. 1F). I have also included the particular case of the egg appendages, which has been studied for the function of the Echinoid protein (Fig. 1E, see below).

I have compiled in this section the available information on adhesion and contractility, which is fragmentary and heteroclite, making comparison between models still rather rash. A brief overview of these methods will be useful to define what has been actually measured.

3.1. Methodology

Cell adhesion can be estimated *in vitro* by determining the degree of reaggregation of dissociated cells, resistance of tissues to mechanical dissociation or adhesion of single cells plated on immobilized recombinant CAMs (e.g., Yap, Niessen, & Gumbiner, 1998). Adhesion can also be directly measured as the force necessary to pull apart a pair of cells using AFM or dual aspiration pipette (Krieg et al., 2008; Maitre et al., 2012). These two methods can also be used to measure cortical stiffness of single cells (Krieg et al., 2008; Maitre et al., 2012). The advantage of *in vitro* measurements is a better controlled environment and reduced parameter complexity. A caveat of these single cell measurements is the fact that cells are bound to actively react to the artificial environment, including the large free surface exposed to the medium, and in the case of AFM, the type of substrate used to hold the cells (inert or adherent, chemical or biological). Thus, it is important to validate these measurements *in vivo*. Cortical tension can be estimated by laser ablation (e.g., Landsberg et al., 2009). Biosensors are being developed for direct force

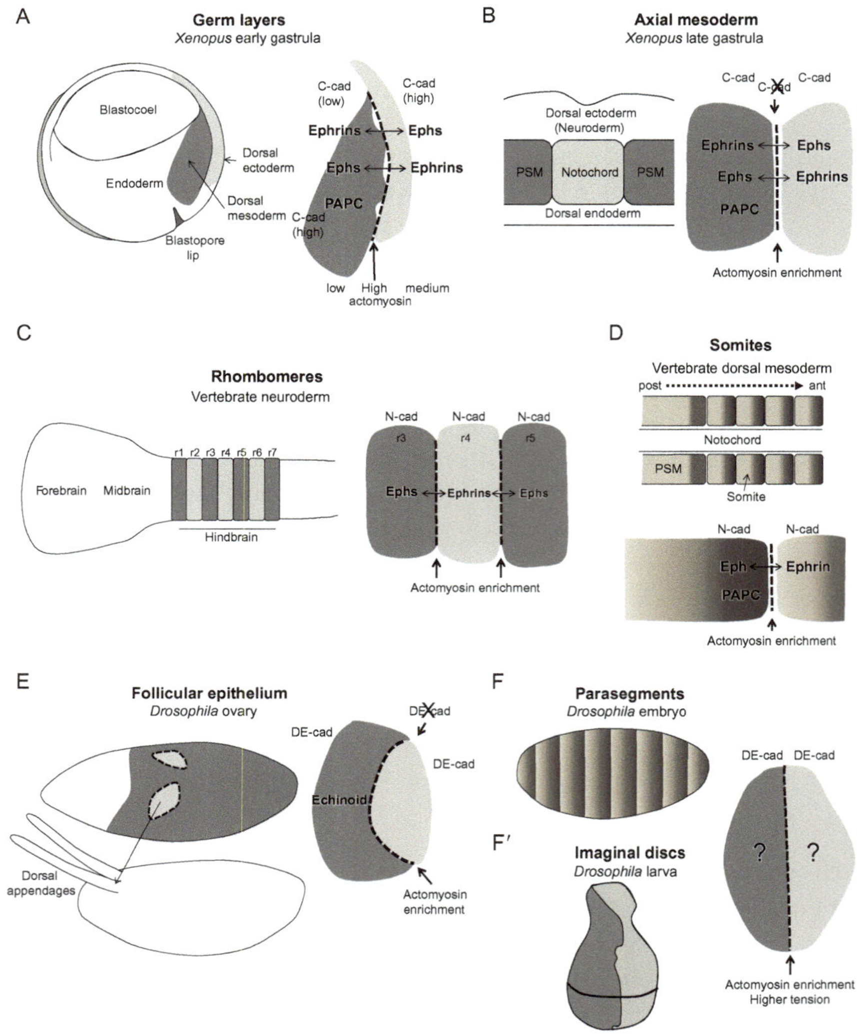

Figure 1 Embryonic boundary models. The enlarged view of each boundary provides information about cadherin distribution (C-cad, N-cad, and DE-Cad), actomyosin structures the cell surface, and the identity and localization of cell surface cues functionally implicated in separation (Ephrins, Eph receptors, PAPC, and Echinoid). The boundary is represented as a dashed line. (A) Separation of the dorsal ectoderm and mesoderm in the early *Xenopus* gastrula. The process depends on a complex network of partially selective ephrin–Eph pairs setting bidirectional signals across the boundary. PAPC also participate in a parallel and less understood pathway. Differences in C-cadherin levels and actomyosin activity are indicated. Gaps between the two tissues represent the occurrence of dynamic cycles of detachments and reattachments. (B) Separation of the most axial mesoderm (notochord) from the paraxial or presomitic mesoderm (PSM). A similar ephrin–Eph network controls separation. PAPC is restricted to the

measurement in live tissues (Borghi et al., 2012; Kuriyama et al., 2014) but they have not yet been used in the context of tissue separation.

Indirect information can also be obtained from the myosin distribution (e.g., Calzolari et al., 2014; Landsberg et al., 2009; Rohani et al., 2014) and from the relative cadherin levels and their degree clustering (Fagotto et al., 2013). Another simple but quite informative criterion is the cell geometry, which is predicted to directly reflect the strength and direction of forces exerted on the cell: At the two extremes, cells with high adhesion tend to maximize their contacts and adopt a hexagonal shape, while highly contractile/low adhering cells are close to round. In principle, a rather precise map of local tensions may be drawn using refined morphological criteria, including membrane curvature, angles at cell vertices, or cell elongation (e.g., Brodland et al., 2014, 2009; Lynch, Veldhuis, Brodland, & Hutson, 2014; Manning et al., 2010). Finally, cohesion is a general physical property integrating adhesion and tension, which can be measured for whole tissue explants (David et al., 2014; David, Ninomiya, Winklbauer, & Neumann, 2009; Kalantarian et al., 2009; Luu, David, Ninomiya, & Winklbauer, 2011; Ninomiya & Winklbauer, 2008).

PSM, but a role in formation of this boundary has not yet been demonstrated. Myosin activation and C-cadherin levels are similar on both sides. Myosin is hyperactivated along the boundary and cadherin adhesion is inhibited, which is indicated by the space separating the two tissues. Eventually, the gap is permanently stabilized by secretion of a thick layer of extracellular matrix. (C) Hindbrain segmentation. The hindbrain becomes segmented in seven rhombomeres r1–7. The process depends on several ephrins and Eph receptors. The central segments r3–5 presented here express complementary sets of multiple ligands and receptors. Expression in the other segments is more complex. All segments express N-cadherin homogenously. Actin and myosin are enriched along the boundaries. (D) Somitogenesis. The PSM becomes progressively segmented, starting anteriorly, under the control of ephrinB2, expressed in the posterior half of the newly forming somite, and EphA4, complementary expressed in the most anterior portion of the unsegmented PSM. PAPC, also expressed anteriorly, contributes to the process. Eventually, somites become completely isolated from each other (empty space), through deposition of extracellular matrix. (E) *Drosophila* dorsal appendages. These two cuticular structures of the *Drosophila* egg are secreted by extensions of the follicular epithelium. Their primordia are delimited as two regions devoid of Echinoid. The juxtaposition of Echinoid-positive and -negative cells produces a smooth interface with high actomyosin and low DE-cadherin levels. (F and F′) *Drosophila* compartment boundaries. The embryonic blastoderm (F) and the wing imaginal disc epithelium (F′) are two examples of epithelia partitioned by sharp compartment boundaries. Increased cortical tension is consistent with accumulation of actomyosin fibers along the interface, but upstream cues have not yet been identified (?).

3.2. Germ layers

The best studied systems are the germ layers of lower vertebrates, fish and amphibians. There are discrepancies between studies, which partly reflect intrinsic specificities of the two models, but mostly differences in the type of approach and in the interpretation of the data. Thus, there is the need to discuss the two cases in some detail.

In zebrafish, adhesion, measured by AFM, was found to be lowest for ectoderm, intermediate for endoderm, and highest for mesoderm cells (Krieg et al., 2008). Consistently, E-cadherin levels were reported to be higher in the mesoderm (Montero et al., 2005; Ulrich et al., 2005). For cortical stiffness, AFM measurements gave the highest value for ectoderm, intermediate for mesoderm, and lowest for endoderm (Krieg et al., 2008). Inference of the adhesive and tensile components of interfacial tension highlighted the predominant influence of cortical tension (Maitre et al., 2012). Results from cell sorting experiments also yielded configurations consistent with cortical tension serving as the main driving force (Krieg et al., 2008). The caveat of these conclusions is that the interfacial properties of heterotypic contacts were estimated based on DITH, which assumes that these properties should be somewhat intermediate between those of the two tissues. Yet heterotypic adhesion had been in fact measured in the original report, although not commented, and values were lower than for homotypic adhesion (Krieg et al., 2008), a result that did not fit with DAH/DITH. Other parameters of the boundary interface have not yet been explicitly studied, but some information can be extracted from published images: Cadherin staining showed no particularity at the boundary (Krieg et al., 2008), but there seems to be some myosin enrichment (Maitre et al., 2012). The angles formed by the cell edges appear close to 90° along the boundary (Fig. 1F in Krieg et al., 2008), which is a typical configuration that reflects high interfacial tension. Based on these various criteria, interfacial tension at heterotypic contacts must be significantly higher than in each of the two tissues. Although this conclusion awaits confirmation, it would be fully consistent with the properties of the other boundaries, including the *Xenopus* ectoderm–mesoderm boundary.

Cortical properties seem to be shared between zebrafish and *Xenopus* embryos: *Xenopus* ectoderm cells are also stiffer than mesoderm cells (AFM unpublished data, Canty and Fagotto), consistent with significantly higher levels of activated myosin (Rohani et al., 2014). The parallel with zebrafish does not hold for adhesion, though cadherin levels were found to be higher in the ectoderm (Angres, Müller, Kellermann, & Hausen, 1991; Ogata et al., 2007), and ectoderm performed better than mesoderm in all adhesion assays, reaggregation (Brieher & Gumbiner, 1994), adhesion to immobilized recombinant cadherin extracellular domain (Zhong, Brieher, & Gumbiner, 1999), and dissociation assay (Canty and Fagotto, unpublished data). However, a more detailed analysis of different mesoderm subregions showed that this tissue was not homogenous, and that the presumed lower adhesion of this layer only applied to the anterior region, while the posterior chordomesoderm appeared similar to the ectoderm in all these aspects (Chen, Koh, Yoder, & Gumbiner, 2009; Fagotto, unpublished observations; Winklbauer, 2009). This was corroborated by measurements of surface tension of ectoderm, anterior and posterior mesoderm explants (Kalantarian et al., 2009; Ninomiya & Winklbauer, 2008; Winklbauer, 2009). The lack of correlation between tissue properties and separation behavior is incompatible with a role of DAH/DITH. The distinct characteristics of each region of the embryo probably reflect specific requirement for other aspects of gastrulation, which also explains differences between zebrafish and *Xenopus*. As for the boundary, it showed unique properties: heterotypic contacts across the interface were constantly disrupted by transient but dramatic repulsive reactions followed by a phase of relaxation and reattachment (Rohani et al., 2011). This behavior, which was never observed within the tissues, suggested a mechanism of contact inhibition. We demonstrated that it was indeed directly controlled by ephrins and Eph receptors interacting across the boundary (Rohani et al., 2011). Detachments were found to correlate with bursts of Rho activation at heterotypic contacts (Rohani et al., 2011), consistent with myosin accumulation along the boundary (Rohani et al., 2014).

3.3. Notochord–presomitic mesoderm boundary

Both tissues showed identical cadherin and myosin staining and similar cell shapes (Fagotto et al., 2013). The boundary, however, was characterized by strong accumulation of actomyosin structures and intense membrane blebbing, indicative of extreme cortical tension, and most interestingly, almost complete lack of cadherin clusters (see below). All these particularities depended on myosin activity, which, similar to the ectoderm–mesoderm case, was activated downstream of ephrin–Eph signaling (Fagotto et al., 2013).

3.4. Somite and hindbrain segmentation

Somitogenesis is a particularly complex process, where segmentation is accompanied by other morphogenetic movements, including compaction and epithelization, a 90° rotation of cell alignment in the lower vertebrates, and differentiation of somite subregions. This complexity makes it difficult to distinguish those parameters that are directly involved in separation *sensus stricto* from those reflecting other events. N-cadherin and N-CAM have distinct distributions in the forming somites (Duband et al., 1987), but these patterns appear to relate to epithelization. Another classical cadherin, cadherin 11 is specifically expressed in somites, starting first in the posterior half of the newly formed somite (Kimura et al., 1995), but its appearance is also probably more relevant for subsequent somite cohesion, not for the initial segmentation. Loss of N-cadherin in mice did not impair segmentation, but on the contrary led to somite fragmentation, a phenotype that was enhanced in double N-cadherin/cadherin 11 knockouts (loss of cadherin 11 alone had no effect) (Horikawa, Radice, Takeichi, & Chisaka, 1999). Somite segmentation is actually resistant to general interference with type I cad (Giacomello et al., 2002). Actin and myosin distribution was examined in zebrafish. Levels were higher along the somatic boundaries, but homogenous within the forming somites (Julich et al., 2009).

Less is known about the properties of the rhombomeres. The hindbrain expresses N-cadherin homogenously. Cadherin 6, however, is expressed specifically in rhombomeres 6 and 7, but its function is not known (Inoue, Asami, & Inoue, 2008; Inoue, Chisaka, Matsunami, & Takeichi, 1997). Similar to the somites, actin and myosin are enriched at the boundaries but homogeneous within the rhombomeres (Calzolari et al., 2014).

As already mentioned, both segmentation processes are known to depend on ephrin–Eph signaling, which readily explains the accumulation of actomyosin along the boundary (Calzolari et al., 2014).

3.5. *Drosophila* tissues

In *Drosophila*, all criteria examined so far argue that tissue properties are similar on both sides of compartment boundaries, whereas cortical tension is higher along the boundaries: Cells have identical polygonal shapes on both sides, but vertex angles approach 90° along the smooth boundary interface (Aliee et al., 2012). Myosin levels are indistinguishable between the tissues, but significantly higher at the boundary (Aliee et al., 2012; Landsberg et al., 2009; Laplante & Nilson, 2006; Major & Irvine, 2006; Monier,

Pelissier-Monier, Brand, & Sanson, 2010). Laser ablation in the wing imaginal disc confirmed that cortical tension is identical in anterior and posterior compartments, but higher along the boundary (Landsberg et al., 2009). Cell adhesion has not yet been examined.

3.6. Boundaries reflect abrupt discontinuities in tissue properties

We have seen that, in most cases, the cell populations show no obvious differences in terms of adhesive or tensile properties. The only exception are the ectoderm and the mesoderm, but even this case hardly supports DAH/DITH, since there is no consistency between fish and frog tissues and high heterogeneity between mesoderm subregions. According to DAH/DITH, tissue interfaces should represent some kind of middle point (~average) between tissue properties, which certainly does not predict for the remarkable characteristics of the actual boundaries. In fact, simply considering the stereotypical cell alignment and smooth interface of all boundaries from a biophysical point of view leads to the inescapable conclusion that boundaries must be sharp discontinuities.

Although direct evidence for high interfacial tension so far has only been obtained on the wing compartment boundary (Landsberg et al., 2009), we have seen numerous indications that this is a general property of embryonic boundaries, including actomyosin enrichment (all boundaries), repulsive behavior between ectoderm and mesoderm cells in *Xenopus* (Rohani et al., 2011, 2014), and blebbing along the notochord–PSM boundary (Fagotto et al., 2013) (see below and Fig. 5).

Functional evidence for the importance of contractility has been demonstrated in all cases, either by biochemical inhibition of myosin function (*Drosophila* imaginal discs, Landsberg et al., 2009; vertebrate ectoderm–mesoderm, Rohani et al., 2011, 2014; notochord, Fagotto et al., 2013; hindbrain, Calzolari et al., 2014) or by myosin optical inactivation (*Drosophila* parasegments, Monier et al., 2010).

Retrospectively, it makes sense that boundary formation must be driven by a local and robust mechanism. In a DAH/DITH situation, sorting would rely on probing differences between different neighbors, which, in order to produce fast and sharp separation, would need to be extreme. In other words, such a mechanism would work only if one of the tissues is extremely compact and the other one extremely loose. These conditions would be cripplingly limiting for embryonic development. They would be incompatible, for instance, with morphogenesis of the axial mesoderm, since both the notochord and the paraxial mesoderm undergo simultaneously convergent extension, for which they need essentially identical biophysical properties. Creation of a local discontinuity uncouples separation from other morphogenetic requirements.

The next exciting question is about the mechanisms which create this discontinuity. We have seen that boundaries form as a result of local reactions at heterotypic contacts, they rely on cell-autonomous components, and myosin is systematically overactivated. This is highly reminiscent of the established repulsive mechanisms active in neuronal guidance. The potential cues responsible for setting *Drosophila* compartment boundaries have still not been identified. The case is however largely elucidated in vertebrates, where ephrin–Eph receptors play a major role. These molecules are well known to activate myosin through Rho, but we will see that the consequences of this activity on cell adhesion are potentially complex. Before discussing these reactions, it may be useful to remember some of the general factors that are thought to regulate adhesion, in particular those related to the actin cytoskeleton.

3.6.1 Principles of regulation of cell adhesion

This brief overview is restricted to classical cadherins, which account for most adhesion in all metazoans and are far better understood than any other CAM. The simplest and most common experimental strategy to modify cell–cell adhesion is to change cadherin levels (e.g., Dahmann & Basler, 2000; Foty & Steinberg, 2005). Yet, it is not clear that expression levels constitute a major mode of regulation *in vivo*. In the embryo, manipulation of cadherin levels has surprisingly mild effects on gastrulation movements, cell sorting being particularly resilient (Fagotto et al., 2013; Ninomiya et al., 2012; Reintsch et al., 2005).

Similar to integrins, cell–cell (or "*trans*") interactions between cadherins are weak, and efficient adhesion involves clustering (Brasch, Harrison, Honig, & Shapiro, 2012). Clustering is due to the intrinsic property of the cadherin extracellular domains to form both *cis* and *trans* interactions, with additional contributions from the transmembrane domain and from the cytoplasmic tail. The size of the clusters is considered to relate to adhesive strength, and may be regulatable, for instance, through p120catenin (Yap et al., 1998).

Cadherin stability at the cell surface is an important determinant of adhesion, which has been implicated in a variety of morphogenetic processes (e.g., Kuriyama et al., 2014; Levayer, Pelissier-Monier, & Lecuit, 2011).

Several factors have been identified that control the rate of cadherin internalization: Cadherin endocytosis is antagonized by p120catenin binding to cadherin cytoplasmic tail (Ishiyama et al., 2010; Nanes et al., 2012). Surface localization is also strongly stabilized by *trans* cell–cell adhesive interactions. However, the largest clusters are also the favored targets for internalization (Truong Quang, Mani, Markova, Lecuit, & Lenne, 2013). Cadherins can also be inactivated by shedding of the extracellular domain, performed by ADAM metalloproteases (Maretzky et al., 2005; Reiss et al., 2005).

The actin cytoskeleton has a predominant influence on adhesion. Its effects are diverse and complex, and here I will only touch upon a few aspects, focusing on actomyosin contractility: Mechanical coupling between cadherin adhesions and actin structures is considered essential to cell adhesion, similar to the link between focal contacts and stress fibers for cell-matrix adhesion. Like integrin-matrix adhesion, cadherin adhesion involves reinforcement of the actin connection, which can occur via the mechanosensing molecules α-catenin and vinculin (Huveneers et al., 2012; Ladoux et al., 2010; le Duc et al., 2010; Yonemura, Wada, Watanabe, Nagafuchi, & Shibata, 2010), or by myosin II-dependent regulation of actin dynamics (Engl, Arasi, Yap, Thiery, & Viasnoff, 2014). On the other hand, the establishment of new adhesive bonds requires cell protrusive activity, a process that is countered by Rho. More generally, contractility of the actomyosin cell cortex represents a force that works toward minimizing the cell contact surface, thus antagonizing cell adhesion (Fig. 2A). There are two ways to consider these two antagonistic activities, which I will call here proadhesive and antiadhesive (Fig. 2). In one model, two distinct actomyosin pools would be in charge of the tension on the cell cortex and of cadherin anchoring. Optimal adhesion would result from the balance of these two opposite activities (Fig. 2B). There is experimental evidence supporting the existence of these two pools (e.g., Cavey, Rauzi, Lenne, & Lecuit, 2008) (Fig. 2A and B), but details on their composition, distribution, and interplay remains unclear. In the second model, the cadherin–actomyosin structures are integrated with the cortical cytoskeleton in a single tensile network (Fig. 2A). The existence of such continuum between various actin structures was recently visualized by high-resolution electron microscopy (Hoelzle & Svitkina, 2012). The pro- and antiadhesive roles could then correspond to two different regimes of a single, intrinsically bimodal, system (Fig. 2B′). The mechanosensing α-catenin–vinculin interaction could provide an underlying molecular mechanism: Weak tension is needed to expose the vinculin binding site of α-catenin, but too high tension

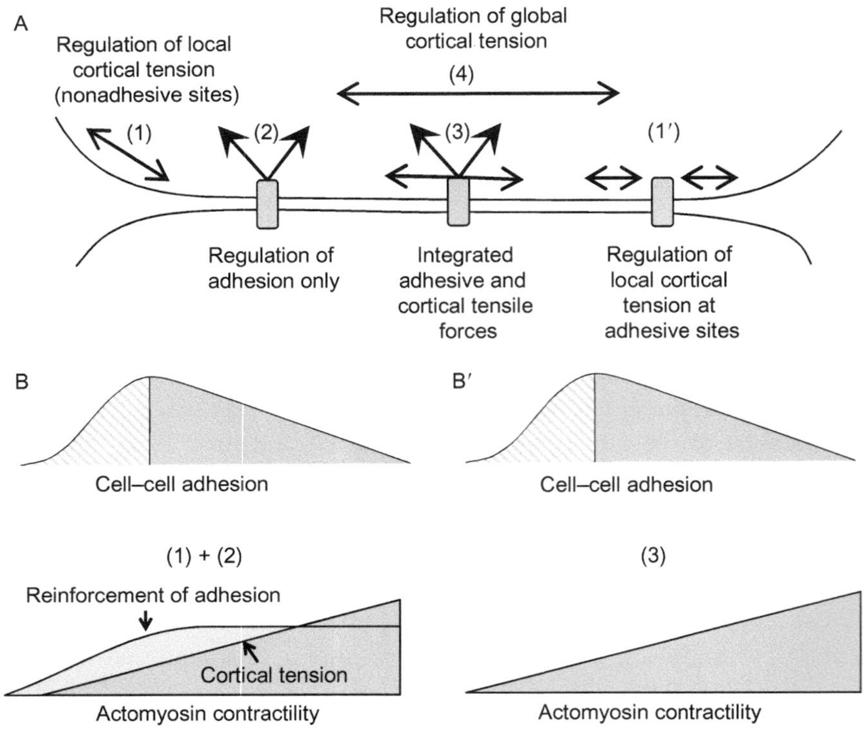

Figure 2 Models for integration of actomyosin contractile structures and cell–cell adhesion. (A) Independent versus integrated systems. Actomyosin may form two independent types of tensile structures responsible for contractility of the cell cortex (1) or reinforcement of adhesions (2). The former may be locally regulated, for instance, along nonadhesive membrane domains (1), or on the contrary in the proximity of adhesive sites (1′). Alternatively, mechanical connections between the adhesive sites and the cytoskeleton may be integrated with the cell cortex (3). Global changes in actomyosin contractility may propagate to multiple pools (4). (B) Bimodal dependence of adhesion on tension. Available data argue for an optimal tension, beyond which increasing tension leads to decreased adhesion. This optimum may be defined by the overlap of the two opposite activities of cortical contractility (1) and adhesion reinforcement (2) (B). Alternatively, a single contractile system (3) may be intrinsically bimodal (B′).

unfolds it (Yao et al., 2014). The actual situation is likely to be a combination of these two models. We will see below that both pro- and antiadhesive effects must be taken into account to explain the properties of embryonic boundaries.

Cell–cell adhesion also feeds back on cortical tension: Indeed, cadherin engagement induces local downregulation of Rho activity and decreased contractility along the cell contacts (Yamada & Nelson, 2007). This regulation is essential, as the tension of the cortex along a free cell surface would be too high to be overcome by the strength of cadherin–cadherin bonds (David et al., 2014; Maitre et al., 2012).

Cortical actin can also influence adhesion more indirectly, for instance, by restricting cadherin lateral diffusion by the process of corralling or by cadherin tethering, which, even in the absence of cadherin–cadherin engagement, contribute to cadherin stabilization and clustering (Hong, Troyanovsky, & Troyanovsky, 2013; Sako, Nagafuchi, Tsukita, Takeichi, & Kusumi, 1998). Furthermore, actin dynamics are also involved in the process of endocytosis, and thus can also control cadherin cell surface distribution. The complexity of these multiple functions explains why interference with general cytoskeleton regulators such as RhoGTPases has often yielded inconsistent results. Note that most of what we know about cadherin adhesion comes from studies on epithelial cells. The fact that these cells adhere through highly specialized adherens junctions adds extra layers of complexity. The study of simpler adhesion clusters (spot junctions or puncta) present in nonepithelial cells, including embryonic mesoderm, should help clarify the basic regulatory mechanisms of cadherin-mediated adhesion.

4. MOLECULAR BASE OF SEPARATION IN VERTEBRATES: EPHRINS–EPH SIGNALING

Ephrins and Eph receptors are widely expressed in all embryonic tissues, including early germ layers. The fact that highly localized patterns delineated specific structures pointed to a role in segmentation (Xu et al., 1995; Xu & Wilkinson, 1997). It turns out that ephrin–Eph signaling plays a major role for all vertebrate boundaries examined so far, ectoderm–mesoderm (Park et al., 2011; Rohani et al., 2011, 2014), notochord–paraxial mesoderm (Fagotto et al., 2013), somites (Davy & Soriano, 2007; Durbin et al., 1998; Watanabe, Sato, Saito, Tadokoro, & Takahashi, 2009), hindbrain rhombomeres (Calzolari et al., 2014; Xu et al., 1995), and eye field (Cavodeassi, Ivanovitch, & Wilson, 2013) (see Batlle & Wilkinson, 2012; Fagotto, 2014; Fagotto et al., 2014 for reviews). The identity, distribution, and function of various ephrins and Eph receptors were recently reviewed in detail (Fagotto et al., 2014). I summarize here the key points and focus the discussion on the effect of ephrin–Eph signaling on the adhesive properties of the boundary.

Gain- and loss-of-function experiments have solidly established a requirement for these molecules in all of the above-mentioned models.

Their interaction across the boundary is expected to trigger contraction of the actomyosin cortices, leading to mutual cell retraction/repulsion. There is evidence that these events do occur at the boundary (e.g., Calzolari et al., 2014; Rohani et al., 2011), and this simple mechanism explains satisfactorily how cell mixing is inhibited. However, there is far more complexity in these systems, which must be addressed in order to understand how separation may truly work and, moving one step further, start to distinguish mechanistic similarities and differences between different types of boundaries.

Ephrins and Eph receptors represent large families, subdivided into A and B subfamilies. It was long assumed that A ligands bind promiscuously to A receptors and B ligands to B receptors. There are, however, strong differences in affinity between individual pairs, even within subfamilies (Blits-Huizinga, Nelersa, Malhotra, & Liebl, 2004; Pabbisetty et al., 2007). Thus, only some of the possible ephrin–Eph combinations may pair efficiently enough to generate physiologically significant signals. We have recently confirmed such partial selectivity in the context of embryonic tissues (Rohani et al., 2014).

Ephrin and Eph expression patterns in early embryos show an unexpected sophistication, comparable to what is found at later stages in the developing brain. The simplest case, where one tissue expresses one ephrin and the adjacent tissue a cognate Eph receptor, is rare. Instead, most tissues express more than one ephrin and/or one Eph, creating various degree of complexity: Segregation of the eye field is a relatively simple example, where the eye field itself expresses several ephrins and the surrounding neural tissue expresses a set of complementary to Eph receptors, all likely to contribute to strong repulsion at the interface (Cavodeassi et al., 2013). In most other cases, however, both ligands and receptors are coexpressed (reviewed in Fagotto et al., 2014). The output is then difficult to predict, since functional ephrin–Eph interactions can occur both within the tissues and across the boundaries.

The logic of these networks was elucidated for the *Xenopus* dorsal ectoderm–mesoderm boundary (Rohani et al., 2014). Here, the system could be satisfactorily explained based on two parameters, partial selectivity of ephrin–Eph pairing and partial complementary of expression (Fig. 3): Despite widespread ephrin and Eph expression, more ephrin–Eph interactions could form at the tissue interface than within each tissue, due to the asymmetric expression of key pairs, such as ephrinB3 and EphA4. The resulting global signal is low in the ectoderm, moderate in the mesoderm, but strongest across the boundary, where it overcomes cell–cell adhesion

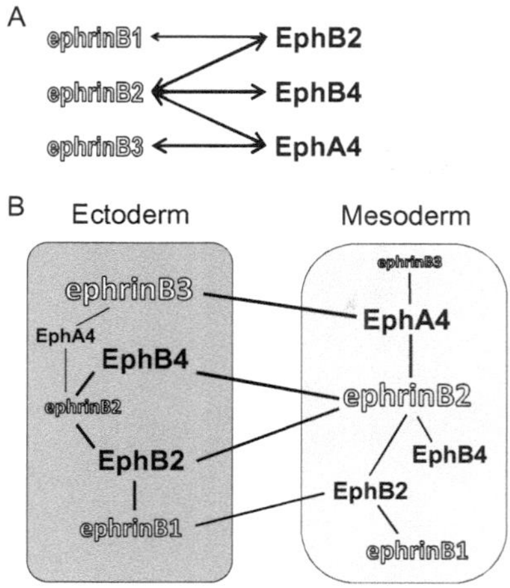

Figure 3 Partial selectivity in ephrin–Eph pairing serves as an identity code for vertebrate embryonic tissues. (A) Selective interactions between ephrinB ligands and cognate receptors. (B) Ephrin and Eph network in the *Xenopus* dorsal ectoderm and mesoderm. Tissues are represented by single boxes. Each cell of a tissue expresses the full set of ligands and receptors. The relative abundance of each component is represented by the font size. The lines represent ephrin–Eph interactions between cells of the same tissue and across the boundary. The thickness of the lines corresponds to the relative signal intensity, which depends on the ligand–receptor binding affinity and of the abundance of the two partners. Multiple interactions can form in the mesoderm. On the contrary, few combinations are available in the ectoderm, despite high expression of several components. The strongest global output is generated at the tissue boundary, and only there it is sufficient to cause overt cell–cell deadhesion.

and results in overt repulsion. The occurrence of weaker repulsive reactions in the mesoderm was revealed when adhesion was experimentally weakened by partial cadherin depletion (Rohani et al., 2014). Similar principles appear to account for ventral ectoderm–mesoderm and notochord–paraxial mesoderm separation (Rohani et al., 2014) and are likely to apply to other boundaries, such as somites and rhombomeres, which also show some partial coexpression of ligands and receptors.

Ephrin–Eph signaling is usually defined as repulsive, based on the typical reaction of neurites contacting a negative cue. Ephrin–Eph signaling is however complex and its effects on cell behavior diverse, which will be here briefly discussed in the context of tissue separation (Fig. 4).

The classical reaction, often defined as "collapse" for neuron growth cones, typically involves RhoA activation as the major target. The process of collapse is still relatively poorly characterized. It may involve increased

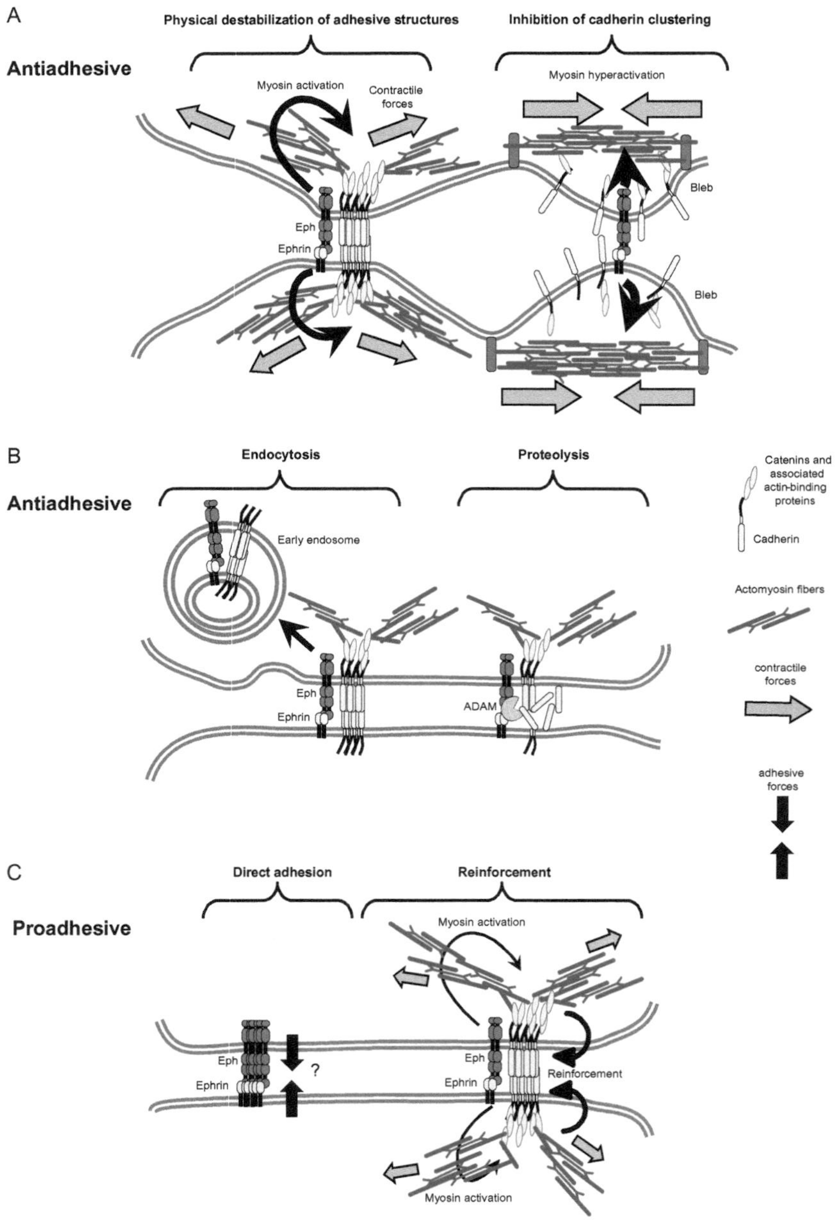

Figure 4 Regulation of cell–cell adhesion by ephrin–Eph signaling. (A) Myosin-dependent antiadhesive mechanisms. Ephrin–Eph signaling generates high cortical tension through activation of the Rho pathway. Contraction of the cell cortex pulls cells apart, exerting a physical force antagonistic to cell–cell adhesion. Exacerbated cortical tension can further lead to transient rupture of the cortex–membrane connections,

actomyosin contractility and/or stimulation of actin depolymerization in the lamellipodium. Within a compact tissue, a simple action of ephrin–Eph-dependent boost in cortical actomyosin contractility is to pull on the adhesive contacts, eventually disrupting them (Fig. 4A). Time lapses of the *Xenopus* ectoderm–mesoderm boundary are consistent with this mechanism (Rohani et al., 2011). The observed cycles of alternating detachments–reattachments may be described as follows (Fig. 5B): Tight cadherin contacts favor ephrin–Eph interactions. Rho activation builds up, and eventually triggers repulsion and deadhesion. Once cells apart, the repulsive signal decays, contractility decreases, and protrusions are reemitted until adhesive contacts are reestablished. This system is well suited for the ectoderm–mesoderm boundary, which must accommodate tissue separation with the need of the mesoderm to migrate using the ectoderm as a cell substrate. Note that the details of the process may be more sophisticated than described here: Ephrin–Eph signaling may contribute to cell detachment not only by increasing cortical tension, but could also involve cointernalization of cadherin clusters with ephrin–Eph complexes or ephrin–Eph-induced shedding of cadherin extracellular domains by ADAM metalloproteases (Fig. 4B) (Solanas, Cortina, Sevillano, & Batlle, 2011).

In other contexts, ephrin–Eph-dependent Rho and myosin activation can create boundary interfaces with different properties. In the case of the notochord boundary, contractility seems to be particularly strong. It causes rupture of the links between the actomyosin cortical structures and the plasma membrane, leading to membrane blebbing (Fagotto et al., 2013) (Fig. 4A). Under this peculiar situation, cadherins fail to cluster, even though the two membranes remain closely apposed due to the compaction of the tissues (Fagotto et al., 2013). The mechanism is not understood, but may depend on the composition and mechanical properties of the blebbing

---

leading to membrane blebbing. Under these conditions, cadherin clustering appears to be inhibited through an unknown mechanism. (B) Other potential antiadhesive mechanisms. Repulsion requires removal of ephrin–Eph clusters, which can occur either by endocytosis or by proteolytic cleavage by ADAM proteinases. Cadherins could conceivably be internalized together with ephrin–Eph clusters. Ephrin–Eph-dependent ADAM activation can also cause cadherin shedding. (C). Ephrins and Eph receptors are also known to stimulate cell-cell adhesion. It has been proposed that under certain conditions, ephrin-Eph clusters may fail to be internalized and that their trans-interactions may then contribute to adhesion (?). Another potential mechanism could rely on mild myosin activation, which could reinforce cell adhesion.

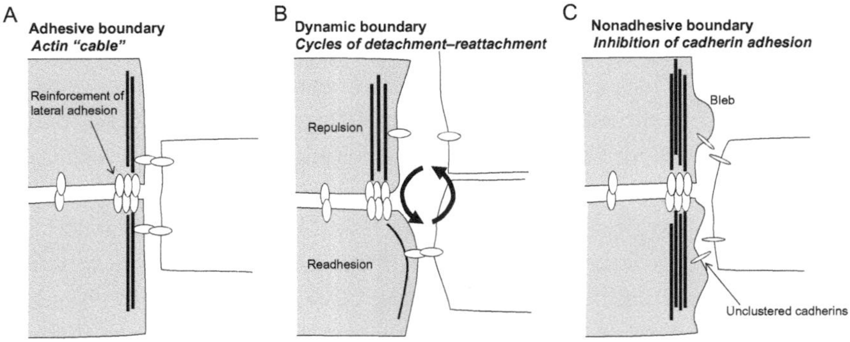

Figure 5 Potential impact of cortical tension on boundary properties. (A) Adhesive boundary. Mild cortical contractility reinforces lateral cadherin adhesions. The mechanical coupling between adjacent cells through "actin cables" connected by cadherin adhesion can be viewed as a physical seal that prevents incursion of cells from the other tissue, while maintaining stable adhesion between the tissues. This mechanism may apply to insect compartment boundaries. (B) Dynamic boundary. Ephrin–Eph-dependent myosin activation along the ectoderm–mesoderm boundary in *Xenopus* leads to transient cell detachment. The repulsive signal then decays, cells extend protrusions, and cadherin bonds are reestablished. Cell–cell contact stimulates a new burst of ephrin–Eph-dependent contraction, initiating a new cycle of detachment–reattachment (symbolized by the two curved arrows). (C) Nonadhesive boundary. Maximal tension can produce a fully not adhesive boundary interface by inhibiting cadherin adhesion (see Fig. 3A).

membrane and/or of the associated actin layer. An important future question will be to determine whether inhibition of cadherin clustering is a reaction that only occurs when tension reaches extreme intensities, or if it may constitute a more pervasive effect, the importance of which would progressively increase as tension increases.

Paradoxically, ephrins and Eph receptors are also known to promote cell adhesion under some circumstances (Halloran & Wolman, 2006) (Fig. 4C). The decrease in tissue cohesion observed upon ephrin/Eph loss-of-function in zebrafish rhombomeres (Cooke, Kemp, & Moens, 2005) and in the *Xenopus* ectoderm (Rohani et al., 2011) suggests that this phenomenon is relevant for tissue separation. Ephrins and Eph receptors could then control separation by two parallel mechanisms, simultaneously decreasing adhesion at the boundary and increasing it within the tissues. The molecular mechanism responsible for this proadhesive activity is unclear. In one model, ephrin–Eph complexes would directly function as adhesive bonds. In fact, it is thought that these complexes must be removed from the cell surface to allow repulsion, which can occur via endocytosis or proteolytic cleavage (Janes et al., 2005; Marston, Dickinson, & Nobes, 2003; Zimmer, Palmer, Kohler, & Klein, 2003). Although other types of regulations could be involved, one possible parameter could simply be intensity of receptor activation, which may need to reach a given threshold to trigger internalization/cleavage of these molecules, thus shifting from an adhesive to a repulsive mode (Halloran & Wolman, 2006). I would argue that the ability of mild actomyosin contractility to reinforce cell adhesion could constitute another explanation for the ephrin–Eph proadhesive activity (Fig. 4C). This simple model is generally not discussed in the context of ephrin–Eph signaling, but would be compatible with observations of the formation of the notochord boundary, where transient strong cadherin clusters are observed across the boundary, likely reflecting cell–cell contact attempting to resist the increasing ephrin–Eph-dependent tension (Fagotto et al., 2013).

Extrapolating these observations to different separation processes, one may postulate that different types of boundaries could be built simply based on different intensities of contractile reactions along the interface (Fig. 5) (Fagotto, 2014). Positive feedback between cadherins and moderate tension could explain the presence of the so-called "actin cables" along many boundaries. Under these conditions, all cells would remain adherent, but cell mixing would be prevented by a lateral compaction of the row of cells lining the boundary (Fig. 5A). This scenario would correspond to the situation of *Drosophila* compartments. Higher tension will disrupt cell contacts, producing either a "dynamic" or a "nonadhesive" boundary (Fig. 5B and C). The *Xenopus* ectoderm–mesoderm boundary is the prototype of the former type, where local bursts of repulsion temporarily disrupt adhesion (Fig. 5B) (Rohani et al., 2011). The notochord represents a nonadhesive boundary, seemingly maintained under a stronger interfacial tension (Fig. 5C) (Fagotto et al., 2013). This tentative classification is undoubtedly an oversimplification, as we are only at the very beginning of the exploration of boundary properties. An important piece of the puzzle will be to determine the strength and orientation of the various forces exerted on lateral and interfacial contacts, and how adhesion bonds are affected by local manipulation of these forces. An ephrin–Eph-based model is valid for vertebrates, but the determinants setting interfacial tension at *Drosophila* boundaries are not known. Even in vertebrates, the effects of ephrin–Eph signaling are likely to be complex and diverse, and additional mechanisms are certainly involved. The next section gives a glimpse of other interesting surface molecules which were shown to induce separation.

5. HOMOPHILIC CONTACT MOLECULES AT EMBRYONIC BOUNDARIES

The identification of additional cell surface cues came essentially from studies in *Drosophila* and in *Xenopus*. These proteins can all form direct homophilic or heterophilic cell to cell interactions, consistent with the importance of direct cell–cell contact at tissue interfaces. Most of them were originally classified as CAMs, but a bona fide direct adhesive function has generally not been validated, and their effect on tissue separation seems to be rather attributable to intracellular activities. The downstream pathways remain generally poorly understood, but, quite strikingly, they all seem to target myosin (Fig. 6). The four best studied examples, Echinoid in *Drosophila*, FLRT3, paraxial protocadherin (PAPC), and EpCAM in *Xenopus*, belong to four different families of membrane proteins and illustrate well the diversity of strategies of regulation of cell contractility and adhesion.

5.1. Immunoglobulin CAMs, Echinoid

The immunoglobulin superfamily includes classical CAMs such as N-CAM and L1CAM, as well as a variety of other cell surface proteins that can undergo homophilic or heterophilic interactions (Shimono, Rikitake, Mandai, Mori, & Takai, 2012). In vertebrate models, most of them have been studied in the context of the nervous system or other specialized tissues. One pair, Nephrin and of its heterophilic partner Neph1 is responsible for forming the filtering slits of the kidney glomerula (Heikkila et al., 2011). In *Drosophila*, Nephrin-like Hibris and Sticks-and-Stones (Sns), and Neph-like Kirre and Roughest control the organization of the omnatidia, which can be considered a cell sorting process (Bao & Cagan, 2005; Bao, Fischbach, Corbin, & Cagan, 2010; Shimono et al., 2012).

The best studied member is Echinoid: Natural and experimental juxtaposition of Echinoid-expressing and nonexpressing cells create smooth boundaries in the ovarian follicular epithelium and in the wing imaginal disc of the larvae, marked by an accumulation of actomyosin cortical structures along the heterotypic contacts (Chang et al., 2011; Laplante & Nilson, 2006, 2011). A similar structure also forms at the edge of the ventral epidermis as it closes over the amnioserosa in the embryo. This actin "cable" requires the presence of Echinoid in the epidermis and its disappearance from the amnioserosa (Laplante & Nilson, 2011). Based on sequence alignment of the extracellular domain, Echinoid is most similar to *Drosophila* Sns, to

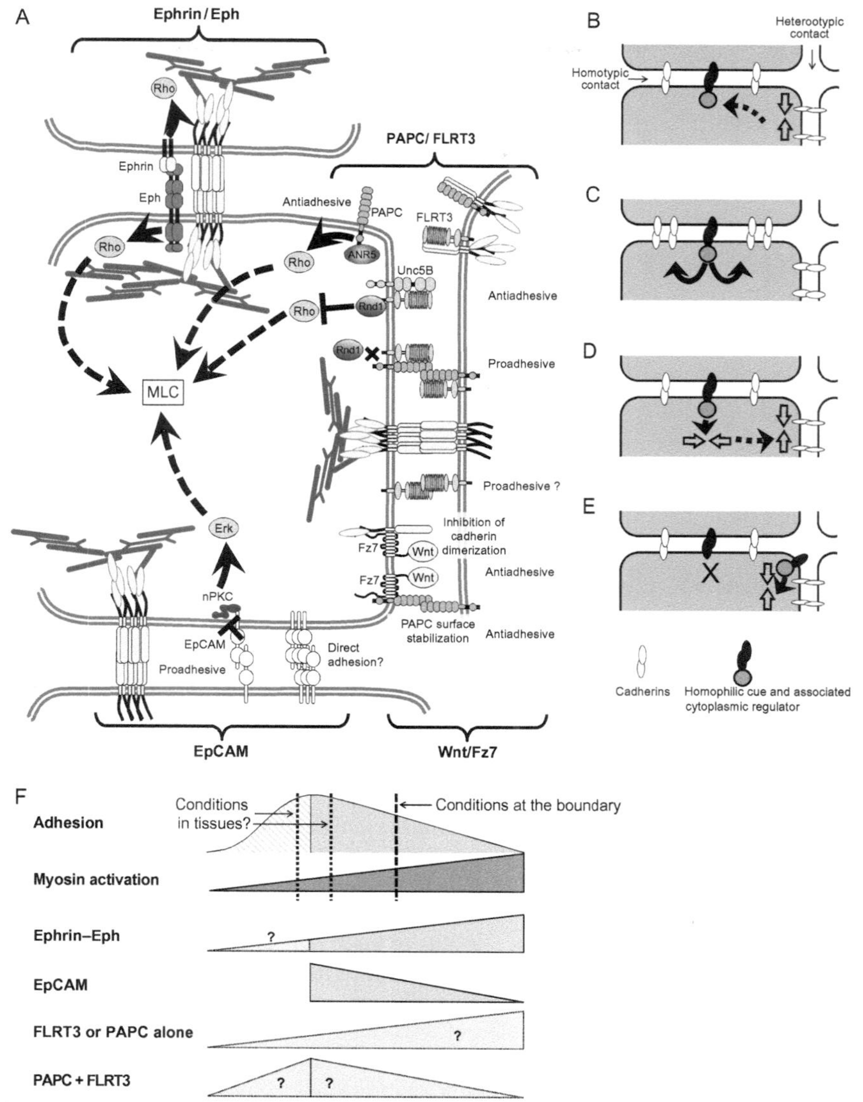

Figure 6 Models of regulation of separation by repulsive and proadhesive cues. (A) Contact cues involved in *Xenopus* ectoderm–mesoderm separation. The diagram represents the cell surface cues, their direct molecular interactions, and their major downstream targets, which all regulate myosin: PAPC activates Rho via ANR5. FLRT3 (together via Unc5B) inhibits it via Rnd1. EpCAM represses Erk activity through direct inhibition of novel PKCs. PAPC and FLRT3 also directly interact, which inhibits FLRT3–Rnd1 interaction. Additional reported interactions include direct binding to cadherins (PAPC and FLRT3) or indirect cadherin regulation by competition with the Wnt receptor Frizzled 7 (PAPC). The function of these heterophilic interactions remains unclear. Note

(Continued)

vertebrate ECM proteins hemicentrin and fibulin, and to nephrin. The intracellular domain of Ed does not show any conservation with other sequences. Functionally, Echinoid appears quite comparable to another subfamily of vertebrate IgG CAMs, the Nectin and nectin-like molecules, which in vertebrates establish the initial contacts between epithelial cells, prior to recruitment of cadherins and various other components that build mature adherens and tight junctions (Takai, Miyoshi, Ikeda, & Ogita, 2008). Both Nectin and Echinoid recruit the junctional soluble proteins Afadin/Canoe and Par3/Bazooka (Takai et al., 2008; Wei et al., 2005). Echinoid is known to be localized to homotypic contacts, presumably through homophilic binding but absent (or less stable) from heterotypic contacts. As a consequence, Bazooka is also asymmetrically distributed, and its absence from heterotypic contacts leads to the accumulation of actomyosin along the boundary (Laplante & Nilson, 2011). Whether sequestration of Bazooka is sufficient to account for Echinoid function is unclear: From data on the process of convergence extension, where Bazooka and myosin II are also complementarily distributed, myosin localization does not depend on Bazooka (Simoes Sde et al., 2010). It is then likely that another pathway is responsible for myosin II regulation downstream of Echinoid. An alternative mechanism based on differential adhesion has also been proposed (Chang et al., 2011).

Figure 6—Cont'd that Frizzled 7 was reported to inhibit cadherin dimerization, but the data did not discriminate between dimerization and higher order clustering. The effect of *cis* and *trans* homophilic interactions is still unknown. (B–E) Models for regulation of boundary by homophilic cues. Homophilic cell surface proteins such as PAPC and Echinoid are differentially expressed on one side of the boundary. *Trans* interactions stabilize them at homotypic contacts. Various mechanisms may explain their ability to induce separation: (B) sequestration of a negative regulator of cortical tension away from the heterotypic contact. (C) Strengthening of cell adhesion on one side of the boundary. (D) Global increase in tension, indirectly boosting tension at the interface. (E) A free pool could also directly activate tension (and/or inhibit adhesion) at the boundary interface, in the absence of *trans* interactions. (F) Putative range of regulation of tension and adhesion by ephrins/Eph receptors, EpCAM, PAPC, and FLRT3. The top curve represents the dependence of adhesion on increasing myosin activation (tension), as in Fig. 2B′. For each surface cue, the corresponding triangles symbolize levels/activity, aligned to the hypothetical effect on myosin activity and adhesion. Ephrin–Eph signaling typically produces cell retraction and deadhesion, but a proadhesive function is also observed, perhaps restricted to the lowest range of activity. EpCAM was found to function exclusively as a proadhesive cue. PAPC and FLRT3 seem to be antiadhesive when expressed alone, but proadhesive when coexpressed, but the actual relationship to tension levels is unknown.

5.2. Leucine-rich repeat proteins, FLRT3

Several leucine-rich repeat proteins function as guidance cues for neurons, either as ligands (Slit is one example) or receptors (Trks; de Wit, Hong, Luo, & Ghosh, 2011). Two members, Tartan and Capricious, play a role at the dorsoventral boundary of the *Drosophila* wing imaginal disc and at the leg segment boundaries (Milan, Weihe, Perez, & Cohen, 2001; Milan, Perez, & Cohen, 2005; Sakurai, Kojima, Aigaki, & Hayashi, 2007; Tepass, Godt, & Winklbauer, 2002). Note that they are not absolutely required for boundary formation, and may rather contribute to its refinement (Sakurai et al., 2007). Neither the cellular mechanism involved nor the identity of their partners in adjacent cells is known.

Vertebrate "fibronectin and leucine-rich repeat" proteins FLRT1–3 are potent regulators of cell–cell adhesion, and, although a role for endogenous boundary formation has not yet been demonstrated, they are quite remarkable for their ability to induce cell sorting in the early *Xenopus* embryo (Chen et al., 2009; Karaulanov et al., 2009). The most studied is FLRT3, which is particularly strongly expressed in the early *Xenopus* mesoderm and plays a role in gastrulation movements (Ogata et al., 2007). FLRT3 can interact with several transmembrane proteins, including itself (Karaulanov, Bottcher, & Niehrs, 2006), the FGF receptor (Bottcher, Pollet, Delius, & Niehrs, 2004), the Netrin receptor Unc5B (Karaulanov et al., 2009), PAPC (Chen et al., 2009), and classical cadherins (Chen et al., 2009) (Fig. 6A). FLRT3 and Unc5B were originally proposed to bind in *cis* on the same membrane, but *trans* interactions should also be considered (Yamagishi et al., 2011). Data aiming at determining the function of FLRT proteins gave inconsistent results: FLRT3 homophilic binding was originally proposed to mediate cell–cell adhesion (Karaulanov et al., 2006), but since this effect was observed in cells expressing endogenous cadherins and was calcium-dependent, it could be due to indirect positive regulation of cadherin-mediated adhesion. On the contrary, gain and loss-of-function experiments in *Xenopus* indicated that FLRT3 regulated cadherin adhesion negatively (Karaulanov et al., 2009; Ogata et al., 2007). The effect appeared to involve FLRT3 interaction with Rnd1 (Ogata et al., 2007) (Fig. 6A), an atypical RhoGTPase that is thought to antagonize Rho. The deadhesive activity of FLRT3 and Rnd1 was originally attributed to stimulation of cadherin endocytosis and consequently decrease in cadherin levels at the cell surface (Ogata et al., 2007). However, Chen et al. (2009) showed evidence that the inhibition of adhesion preceded and was probably the cause for the drop in cadherin levels.

A similar conclusion was obtained for PAPC (see below). These data are again consistent with an indirect regulation of adhesion through the actin cytoskeleton.

5.3. Protocadherins, PAPC

The discovery of the large protocadherin subfamily had originally brought the promise of definitive explanation of cell sorting based on differential CAM expression. However, protocadherins appear to be unable to promote efficient cell–cell adhesion (Chen & Gumbiner, 2006), most likely due to the absence of a β-catenin binding site. Although a potential direct role in cell–cell adhesion cannot be excluded, the current consensus is that they rather function as cell contact receptors. Consistently, their cytoplasmic tails, which are quite diverse in sequence, regulate a variety of intracellular pathways (e.g., Nap1/WAVE for protocadherin 10, Wnt for protocadherin 11Y, Rho for protocadherin 8, Kim, Yasuda, Tanaka, Yamagata, & Kim, 2011).

Protocadherin 8, also called paraxial protocadherin (PAPC) has been implicated in early vertebrate morphogenesis, including in tissue separation. It induces efficient cell sorting, producing exceptionally sharp boundaries when ectopically expressed (Chen & Gumbiner, 2006; Chen et al., 2009). During early embryonic development it is systematically expressed on one side of a forming boundary (Fig. 1A, B, and D): It first appears in the dorsal mesoderm as this layer separates from the ectoderm, later it disappears from the axial mesoderm (future notochord), and remains restricted to the paraxial mesoderm when both structures split. The same process of restriction occurs once more during somitogenesis, with PAPC remaining in the anterior half of the forming somites (Kim, Jen, De Robertis, & Kintner, 2000; Kim, Yamamoto, Bouwmeester, Agius, & Robertis, 1998; Rhee, Takahashi, Saga, Wilson-Rawls, & Rawls, 2003). PAPC loss-of-function impairs ectoderm–mesoderm separation in *Xenopus* (Medina et al., 2004) and somite segmentation in *Xenopus* and the mouse (Kim et al., 2000; Rhee et al., 2003). A role for formation of the notochord boundary has not yet been confirmed.

Chen and Gumbiner showed that PAPC was unable to provide cell–cell adhesion on its own. On the contrary, it negatively regulates the function of classical cadherins (Chen & Gumbiner, 2006). PAPC was found to upregulate Rho activity (Medina et al., 2004) through an unknown mechanism that involves the ankyrin repeat protein ANR5 (Chung, Yamamoto, & Ueno, 2007) (Fig. 6A). As explained above, Rho overactivation and cortical hypercontractility could be sufficient to explain the observed decrease in adhesion. The actual function of PAPC is however more complicated. First, PAPC cytoplasmic domain, and thus presumably its interaction with ANR5, is dispensable for the capacity to induce sorting (Chen & Gumbiner, 2006; Chen et al., 2009). PAPC was also reported to laterally interact with FLRT3 (Chen et al., 2009), with classical cadherins (Chen et al., 2009) (also observed for another protocadherin, Pcdh19, Biswas et al., 2014), and with the Wnt receptor Fz7 (Kraft, Berger, Wallkamm, Steinbeisser, & Wedlich, 2012) (Fig. 6A). The consequences of these interactions are incompletely understood. In one model, PAPC mitigates the antiadhesive activity of FLRT3 by preventing recruitment of Rnd1 (Chen et al., 2009). The interaction with Fz7 seems to favor PAPC stabilization at the cell membrane (Kraft et al., 2012). Note that Fz7 also forms a complex with classical cadherins that inhibits cadherin dimerization (Kraft et al., 2012). These authors failed to detect a PAPC–cadherin interaction. There is clearly a need to reconcile and integrate these various data into a coherent picture.

Note that another protocadherin, axial protocadherin (AxPC of Pcadh1) replaces PAPC in the notochord at the time of its separation from the paraxial mesoderm (Kuroda, Inui, Sugimoto, Hayata, & Asashima, 2002). AxPC depletion disrupted the corresponding boundary (Yoder & Gumbiner, 2011). The underlying mechanism has not yet been investigated.

5.4. EpCAM, inducer of tissue mixing

Although its loss-of-function in early *Xenopus* and zebrafish does not lead to any obvious separation phenotype, EpCAM is included here as prototypic example of positive regulator of cell migration and adhesion, which must be taken into account to understand how tissue properties influence tissue boundaries. Its case is quite informative about the relationship between regulation of cortical actomyosin, cell adhesion, and cell motility.

EpCAM was mostly known as a tumor-associated protein highly expressed in human carcinomas, but very little is known about its potential role in metastasis. EpCAM can bind homophilically (Balzar et al., 1999), but its classification as CAM remains to be validated. It turns out that EpCAM is also abundantly expressed in early vertebrate embryos. We originally identified *Xenopus* EpCAM in a gain-of-function screen through its remarkable capacity to induce mixing between ectoderm and mesoderm (Maghzal, Vogt, Reintsch, Fraser, & Fagotto, 2010). As mentioned, EpCAM depletion does not affect ectoderm–mesoderm separation, but has other effects on morphogenesis: it first impairs the thinning and spreading of the ectoderm during gastrulation, a process called epiboly (Maghzal et al., 2010; Slanchev et al., 2009). After gastrulation, EpCAM loss yields a dramatic phenotype, with cells losing adhesion, leading to complete tissue disaggregation and embryo death (Maghzal, Kayali, Rohani, Kajava, & Fagotto, 2013). We will see that, although diverse in appearance, all these phenotypes have a single common molecular cause. Tissue mixing upon EpCAM overexpression and epiboly defect upon EpCAM depletions are two opposite results of modulating intercellular motility, i.e., the capacity of cells to move among other cells. We found that EpCAM controls this process by repressing myosin activation (Maghzal et al., 2010). When EpCAM levels are lowered, high myosin activity makes the cells poorly motile and the tissue stiff, when EpCAM is high, low myosin activity allows cells to be more motile and tissues more "fluid." Cell motility is boosted enough to bypass boundary repulsion and cause tissue mixing. EpCAM-mediated repression of myosin activity has also a positive effect on cadherin adhesion. Note that unlike commonly assumed, decreasing myosin activation stimulates both adhesion and migration. We will discuss below possible ways to reconcile this phenomenon with the assumed requirement for myosin in cell–cell adhesion.

Cell dissociation in EpCAM depleted postgastrula embryos is an extreme result of myosin overactivation (Maghzal et al., 2013). High contractility prevents maintaining cell–cell adhesion. This in turn leads to cadherin internalization and degradation, which further feeds back into accelerating loss of adhesion. At the molecular level, EpCAM acts by directly binding and inhibiting novel PKCs, thus blocking the downstream PKD–Raf–Erk cascade (Maghzal et al., 2013). Although several targets are certainly affected, all the developmental defects result from myosin deregulation (Maghzal et al., 2013).

6. REGULATION OF TENSION AND ADHESION BY CONTACT CUES

The identification of these various cell surface molecules has given us some idea of the type of regulations that control tissue and boundary properties, and has highlighted the central role of myosin. Yet the preliminary study of these molecules, far from solving the problem, has brought additional puzzles. I discuss here general conceptual issues raised by the implication of homophilic molecules in regulating heterotypic contacts and by the coexistence of pro- and antiadhesive activities.

6.1. The action of homophilic regulators

PAPC and Echinoid share the ability to bind homophilically and the particularity of being expressed only on one side of the boundary. As a consequence, they both accumulate at homotypic contacts and are low or absent from the membranes abutting the boundary (Laplante & Nilson, 2006 and unpublished observations), which, by analogy with classical cadherins, is probably determined by internalization of the nonengaged molecules and stabilization by homophilic interactions. PAPC and Echinoid can be thus considered as "homophilic cues," by opposition to "heterophilic cues" such as ephrins and Eph receptors.

How can then homophilic cues control the properties of an interface from which they are precisely absent? Among several possible scenarios, the one favored for Echinoid is that it sequesters some regulator of myosin contractility, resulting in its depletion from the boundary interface (Laplante & Nilson, 2011) (Fig. 6B). Alternatively, PAPC and Echinoid may favor cell–cell adhesion within the tissue, either functioning directly as CAMs, or indirectly, via regulation of cadherins or of contractility. Separation would then be stimulated by the building of differential adhesion (and/or tension) between homotypic and heterotypic contacts (Fig. 6C). Changes in adhesion or contractility along homophilic contacts may also affect global cellular forces, which could propagate to the heterotypic contact (Fig. 6D). These possible scenarios are based on the reasonable assumption that the functional pool is the one engaged in homophilic interactions. This, however, has not yet been demonstrated, and it remains formally possible that what counts is the free pool, which, despite its presumed instability, may directly affect the boundary properties (Fig. 6E). A better characterization of homophilic and heterophilic interactions, and of their impact on intracellular activities should clarify the mode of action of these interesting molecules.

6.2. Putting pro- and antiadhesive activities together

So far, the function of the various cell surface cues can be largely explained by their action on actomyosin contractility. Although our knowledge is still too incomplete to draw more specific conclusions, studies on the best known molecules, ephrins/Eph receptors and EpCAM (Fig. 6F) give us some general ideas about how multiple pathways may be integrated. Ephrin–Eph signaling and EpCAM work essentially antagonistically: the former stimulates myosin activation, contractility, and decreases adhesion. EpCAM has the exact inverse action. Since both are simultaneously active in the early embryo, one first conclusion is that the local properties must be influenced by their balance. This explains, for instance, why EpCAM overexpression causes ectoderm to mix with mesoderm (Maghzal et al., 2010): by inhibiting myosin, it decreases basal cortical tension, such that the bursts induced by ephrin–Eph signaling are no longer sufficient to overcome cell–cell adhesion.

Assuming this simple scenario is however not necessarily sufficient to predict the combined outcome of multiple cues. It is indeed important to remember that myosin has multiple functions, which can be pro- or antiadhesive, which could be differently regulated by different cues, either because a specific pool may be preferentially targeted or because the regulation would be restricted to a given range of tensile strength. The case of EpCAM seems relatively straightforward: when EpCAM levels were manipulated, myosin activation and of adhesion always consistently responded in opposite directions. Taking a simple linear model (Fig. 6F), we may conclude that EpCAM activity spans essentially the "antiadhesive" range of contractility. The case of ephrins and Ephs is complicated by the ill-defined proadhesive activity, which may occur within a weak tensile mode, even though the system is probably antiadhesive over a broad range of ephrin–Eph signaling (Fig. 6F). Predictions are currently more risky for PAPC and FLRT3 (Fig. 6F), due to outstanding inconsistencies in the reported effects on adhesion and cell sorting and their mutual regulations (Chen et al., 2009). It remains that PAPC and FLRT proteins are remarkably potent regulators of adhesion and inducers of cell sorting, and one should consider them as potential heavy weights in setting the adhesive and tensile conditions. Defining their mode and range of regulation should be a priority if one wants to build a coherent picture of separation in vertebrates.

6.2.1 Note about complexity and redundancy

Studies in fish and *Xenopus* have implicated several components in tissue separation, including multiple ephrins and Eph receptors, PAPC, and FLRT3. Quite strikingly, depletion of a single component, including one out of several ephrins/Ephs, is sufficient to affect separation rather severely (Cooke et al., 2005; Fagotto et al., 2013; Rohani et al., 2011, 2014; Watanabe et al., 2009). On the contrary, the corresponding knockouts in mice gave no obvious phenotype (e.g., Dottori et al., 1998; Yamamoto, Kemp, Bachiller, Geissert, & De Robertis, 2000). An extreme case is the removal of all ephrinBs, which did not impair any of the major early developmental processes (Senturk, Pfennig, Weiss, Burk, & Acker-Palmer, 2011). Yet the expression patterns of all these molecules is very conserved throughout vertebrates, which makes it difficult to imagine that the function would not be similarly conserved. The lack of phenotype in mammals is probably due to a higher degree of redundancy, particularly between the ephrins and PAPC, two systems that are active in parallel for at least three boundaries. This apparent redundancy is a serious obstacle for studying boundaries and probably other morphogenetic processes in mammals.

Note that a certain degree of redundancy must also occur in fish and frogs, suggested by the fact that none of the depletions completely abolished separation. Complexity could then serve as buffer for small variations in levels/activity of individual components. The existence of multiple pathways may also be related to additional functions in other morphogenetic processes. This is clearly the case for PAPC, which is involved in mesoderm convergence extension (Medina et al., 2004; Unterseher et al., 2004). Redundancy could similarly explain the current lack of strong candidates for setting high contractility at *Drosophila* compartment boundaries. Mutants for Tartan and Capricious, for instance, do yield a wiggling boundary (Milan et al., 2005), suggesting that they do play a role in separation. Using a sensitized mutant background could reveal a second parallel pathway.

6.2.2 After separation: Relationship between separation and epithelization

Formation of *Drosophila* parasegments, imaginal disc boundaries and vertebrate hindbrain segmentation, all correspond to the transverse partition of an epithelial layer, with little direct impact on the tissue structure. In other cases, however, the separating tissues undergo important reorganization. A well-known example is somitogenesis, where segmentation is immediately followed by compaction of the new somite into a closed epithelial layer that constitutes an independent structure. The *de novo* appearance of a polarized organization in a mesoderm tissue is a particularly intriguing phenomenon, which is not restricted to somitogenesis. By this stage, it has already occurred twice during mesoderm morphogenesis, each time as a consequence of a separation event. This first happens after the separation of the mesoderm from the ectoderm (and from the endoderm, an event that has not yet been studied): Mesoderm organizes then as two parallel layers, which correspond to the somatic mesoderm (facing the ectoderm) and the splanchnic mesoderm (facing the endoderm), two key elements of the coelomate body structure. When the dorsal section of this mesoderm splits into the notochord and the presomitic mesoderm, a similar reorganization takes place along the new boundary. The alignment and polarization of the cells abutting the boundary results in connecting the dorsal and ventral mesoderm layers on both sides of the boundary into two continuous monolayers, as summarized in Fig. 7. When somites segment out of the PSM, the process is repeated, virtually identical except for the 90° rotation of the plane of separation.

In the case of somitogenesis, epithelization had been sometimes considered as a step toward separation (e.g., Duband et al., 1987). However, a boundary interface is unambiguously detected before any morphological sign of epithelization (e.g., Youn & Malacinski, 1981), although this latter process contributes to the appearance of a visible gap (Fig. 7D). Comparison of the three consecutive separations suggests a common mechanism, initiated by local ephrin–Eph-mediated repulsion (we will ignore for simplicity the poorly understood contribution from PAPC). The resulting decrease or loss in cell adhesion creates a nonadhesive interface, which is interpreted by the cells as a "basal" side, where extracellular matrix is deposited. It is likely that the matrix serves then as a basal cue to reorganize and polarize the cells along the boundary.

7. CONCLUSIONS

High actomyosin cortical tension appears to be a fundamental property of embryonic boundaries, and ephrin–Eph-dependent repulsion provides a satisfactory mechanism to set this local tension, at least in vertebrates. With the generalization of these concepts, our view of morphogenesis has been radically modified: Rather than selecting their neighbors based on shared properties, cells appear to stick together by default and to specifically detect those other cell types that they should repulse. In other words, development does not proceed by increasing "affinities," but by creating "identities." This is one more example where behind a phenomenon that seemed to be explainable based on simple physical principles, is actually controlled by active cellular signals.

How much did we solve so far of this fascinating process? We have then learnt how embryonic tissues are being cut into parts, which answers an old

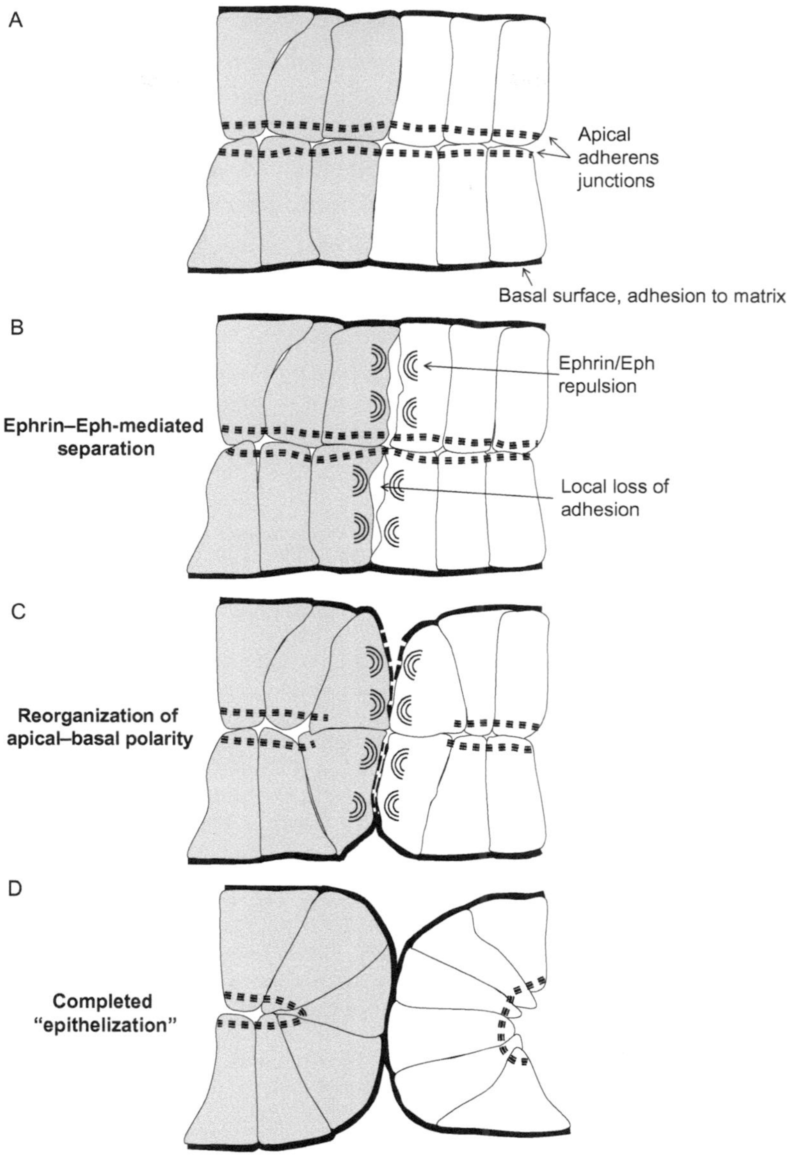

Figure 7 Local epithelization during separation of vertebrate mesoderm tissues. Once separated from the ectoderm and the mesoderm, the mesoderm layer becomes organized as two parallel layers of polarized cells. The basal domain faces the outside, and adherens junction-like structures are in the center (A). During the two successive separations that first produce the notochord and the presomitic mesoderm, then the somites, a similar pattern of changes in tissue organization is observed: A boundary forms due to ephrin–Eph-dependent increase in cortical tension and a corresponding drop in cell–cell adhesion (B). The boundary interface becomes equivalent to a basal domain, where extracellular matrix is deposited (C). Cells are reoriented, the continuity of the polarized layer is reestablished on each side of the boundary, and the two new regions become permanently isolated by extracellular matrix (D).

and fundamental question. We are however still far from being able to build a realistic picture of tissue separation. We have in particular only vague hints for what confers different boundaries with specific properties, such that tissues can at times slide relative to each other, crawl, or remain tightly connected. There is certainly a whole set of regulations to uncover, a mine of new excitement for cell biologists.

ACKNOWLEDGMENTS

F. F. research is supported by the Canadian Institute for Health Research, the Canadian Cancer Society Research Institute, and the Cancer Research Society.

REFERENCES

Abercrombie, M. (1967). Contact inhibition: The phenomenon and its biological implications. *In Vitro, 26*, 246–261.

Aliee, M., Roper, J. C., Landsberg, K. P., Pentzold, C., Widmann, T. J., Julicher, F., et al. (2012). Physical mechanisms shaping the Drosophila dorsoventral compartment boundary. *Current Biology, 22*(11), 967–976.

Altabef, M., Clarke, J. D., & Tickle, C. (1997). Dorso-ventral ectodermal compartments and origin of apical ectodermal ridge in developing chick limb. *Development, 124*, 4547–4556.

Angres, B., Müller, A. H. J., Kellermann, J., & Hausen, P. (1991). Differential expression of two cadherins in *Xenopus laevis*. *Development, 111*, 8229–8244.

Balzar, M., Prins, F. A., Bakker, H. A., Fleuren, G. J., Warnaar, S. O., & Litvinov, S. V. (1999). The structural analysis of adhesions mediated by Ep-CAM. *Experimental Cell Research, 246*, 108–121.

Bao, S., & Cagan, R. (2005). Preferential adhesion mediated by Hibris and Roughest regulates morphogenesis and patterning in the Drosophila eye. *Developmental Cell, 8*, 925–935.

Bao, S., Fischbach, K. F., Corbin, V., & Cagan, R. L. (2010). Preferential adhesion maintains separation of ommatidia in the Drosophila eye. *Developmental Biology, 344*, 948–956.

Batlle, E., & Wilkinson, D. G. (2012). Molecular mechanisms of cell segregation and boundary formation in development and tumorigenesis. *Cold Spring Harbor Perspectives in Biology, 4*, a008227.

Biswas, S., Emond, M. R., Duy, P. Q., Hao le, T., Beattie, C. E., & Jontes, J. D. (2014). Protocadherin-18b interacts with Nap1 to control motor axon growth and arborization in zebrafish. *Molecular Biology of the Cell, 25*, 633–642.

Blits-Huizinga, C. T., Nelersa, C. M., Malhotra, A., & Liebl, D. J. (2004). Ephrins and their receptors: Binding versus biology. *IUBMB Life, 56*, 257–265.

Borghi, N., Sorokina, M., Shcherbakova, O. G., Weis, W. I., Pruitt, B. L., Nelson, W. J., et al. (2012). E-cadherin is under constitutive actomyosin-generated tension that is increased at cell-cell contacts upon externally applied stretch. *Proceedings of the National Academy of Sciences of the United States of America, 109*, 12568–12573.

Bottcher, R. T., Pollet, N., Delius, H., & Niehrs, C. (2004). The transmembrane protein XFLRT3 forms a complex with FGF receptors and promotes FGF signalling. *Nature Cell Biology, 6*, 38–44.

Brasch, J., Harrison, O. J., Honig, B., & Shapiro, L. (2012). Thinking outside the cell: How cadherins drive adhesion. *Trends in Cell Biology, 22*, 299–310.

Brieher, W. M., & Gumbiner, B. M. (1994). Regulation of C-cadherin function during activin induced morphogenesis of Xenopus animal caps. *The Journal of Cell Biology*, *126*, 519–527.
Brodland, G. W. (2002). The Differential Interfacial Tension Hypothesis (DITH): A comprehensive theory for the self-rearrangement of embryonic cells and tissues. *Journal of Biomechanical Engineering*, *124*, 188–197.
Brodland, G. W., Veldhuis, J. H., Kim, S., Perrone, M., Mashburn, D., & Hutson, M. S. (2014). CellFIT: A cellular force-inference toolkit using curvilinear cell boundaries. *PLoS One*, *9*, e99116.
Brodland, G. W., Yang, J., & Sweny, J. (2009). Cellular interfacial and surface tensions determined from aggregate compression tests using a finite element model. *Human Frontier Science Program*, *3*, 273–281.
Calzolari, S., Terriente, J., & Pujades, C. (2014). Cell segregation in the vertebrate hindbrain relies on actomyosin cables located at the interhombomeric boundaries. *The EMBO Journal*, *33*, 686–701.
Cavey, M., Rauzi, M., Lenne, P. F., & Lecuit, T. (2008). A two-tiered mechanism for stabilization and immobilization of E-cadherin. *Nature*, *453*, 751–756.
Cavodeassi, F., Ivanovitch, K., & Wilson, S. W. (2013). Eph/Ephrin signalling maintains eye field segregation from adjacent neural plate territories during forebrain morphogenesis. *Development*, *140*, 4193–4202.
Chang, L. H., Chen, P., Lien, M. T., Ho, Y. H., Lin, C. M., Pan, Y. T., et al. (2011). Differential adhesion and actomyosin cable collaborate to drive Echinoid-mediated cell sorting. *Development*, *138*, 3803–3812.
Chen, X., & Gumbiner, B. M. (2006). Paraxial protocadherin mediates cell sorting and tissue morphogenesis by regulating C-cadherin adhesion activity. *The Journal of Cell Biology*, *174*, 301–313.
Chen, X., Koh, E., Yoder, M., & Gumbiner, B. M. (2009). A protocadherin-cadherin-FLRT3 complex controls cell adhesion and morphogenesis. *PLoS One*, *4*, e8411.
Chung, H. A., Yamamoto, T. S., & Ueno, N. (2007). ANR5, an FGF target gene product, regulates gastrulation in Xenopus. *Current Biology*, *17*, 932–939.
Cooke, J. E., Kemp, H. A., & Moens, C. B. (2005). EphA4 is required for cell adhesion and rhombomere boundary formation in the zebrafish. *Current Biology*, *15*, 536–542.
Dahmann, C., & Basler, K. (1999). Compartment boundaries at the edge of development. *Trends in Genetics*, *15*, 320–326.
Dahmann, C., & Basler, K. (2000). Opposing transcriptional outputs of Hedgehog signaling and engrailed control compartmental cell sorting at the Drosophila A/P boundary. *Cell*, *100*, 411–422.
Dahmann, C., Oates, A. C., & Brand, M. (2011). Boundary formation and maintenance in tissue development. *Nature Reviews Genetics*, *12*, 43–55.
David, R., Luu, O., Damm, E. W., Wen, J. W. H., Nagel, M., & Winklbauer, R. (2014). Tissue cohesion and the mechanics of cell rearrangement. *Development*, *141*, 3672–3682.
David, R., Ninomiya, H., Winklbauer, R., & Neumann, A. W. (2009). Tissue surface tension measurement by rigorous axisymmetric drop shape analysis. *Colloids and Surfaces B: Biointerfaces*, *72*, 236–240.
Davis, G. S., Phillips, H. M., & Steinberg, M. S. (1997). Germ-layer surface tensions and "tissue affinities" in *Rana pipiens* gastrulae: Quantitative measurements. *Developmental Biology*, *192*, 630–644.
Davy, A., & Soriano, P. (2007). Ephrin-B2 forward signaling regulates somite patterning and neural crest cell development. *Developmental Biology*, *304*, 182–193.
de Wit, J., Hong, W., Luo, L., & Ghosh, A. (2011). Role of leucine-rich repeat proteins in the development and function of neural circuits. *Annual Review of Cell and Developmental Biology*, *27*, 697–729.

Dottori, M., Hartley, L., Galea, M., Paxinos, G., Polizzotto, M., Kilpatrick, T., et al. (1998). EphA4 (Sek1) receptor tyrosine kinase is required for the development of the corticospinal tract. *Proceedings of the National Academy of Sciences of the United States of America, 95*, 13248–13253.

Duband, J. L., Dufour, S., Hatta, K., Takeichi, M., Edelman, G. M., & Thiery, J. P. (1987). Adhesion molecules during somitogenesis in the avian embryo. *The Journal of Cell Biology, 104*, 1361–1374.

Duguay, D., Foty, R. A., & Steinberg, M. S. (2003). Cadherin-mediated cell adhesion and tissue segregation: Qualitative and quantitative determinants. *Developmental Biology, 253*, 309–323.

Durbin, L., Brennan, C., Shiomi, K., Cooke, J., Barrios, A., Shanmugalingam, S., et al. (1998). Eph signaling is required for segmentation and differentiation of the somites. *Genes & Development, 12*(19), 3096–3109.

Edelman, G. M. (1986). Cell adhesion molecules in the regulation of animal form and tissue pattern. *Annual Review of Cell Biology, 8*(4), 81–116.

Engl, W., Arasi, B., Yap, L. L., Thiery, J. P., & Viasnoff, V. (2014). Actin dynamics modulate mechanosensitive immobilization of E-cadherin at adherens junctions. *Nature Cell Biology, 16*, 587–594.

Fagotto, F. (2014). The cellular basis of tissue separation. *Development, 141*, 3303–3318.

Fagotto, F., Rohani, N., Touret, A. S., & Li, R. (2013). A molecular base for cell sorting at embryonic boundaries: Contact inhibition of cadherin adhesion by ephrin/Eph-dependent contractility. *Developmental Cell, 27*, 72–87.

Fagotto, F., Winklbauer, R., & Rohani, N. (2014). Ephrin-Eph signaling in embryonic tissue separation. *Cell Adhesion & Migration, 8*, 308–326.

Flanagan, J. G., & Vanderhaeghen, P. (1998). The ephrins and Eph receptors in neural development. *Annual Review of Neuroscience, 21*, 309–345.

Foty, R. A., & Steinberg, M. S. (2004). Cadherin-mediated cell-cell adhesion and tissue segregation in relation to malignancy. *The International Journal of Developmental Biology, 48*, 397–409.

Foty, R. A., & Steinberg, M. S. (2005). The differential adhesion hypothesis: A direct evaluation. *Developmental Biology, 278*, 255–263.

Fraser, S., Keynes, R., & Lumsden, A. (1990). Segmentation in the chick embryo hindbrain is defined by cell lineage restrictions. *Nature, 344*, 431–435.

Garcia-Bellido, A., Ripoll, P., & Morata, G. (1973). Developmental compartmentalisation of the wing disk of Drosophila. *Nature: New Biology, 245*, 251–253.

Giacomello, E., Vallin, J., Morali, O., Coulter, I. S., Boulekbache, H., Thiery, J. P., et al. (2002). Type I cadherins are required for differentiation and coordinated rotation in *Xenopus laevis* somitogenesis. *The International Journal of Developmental Biology, 46*, 785–792.

Halloran, M. C., & Wolman, M. A. (2006). Repulsion or adhesion: Receptors make the call. *Current Opinion in Cell Biology, 18*, 533–540.

Harris, A. K. (1976). Is cell sorting caused by differences in the work of intercellular adhesion? A critique of the Steinberg hypothesis. *Journal of Theoretical Biology, 61*, 267–285.

Heikkila, E., Ristola, M., Havana, M., Jones, N., Holthofer, H., & Lehtonen, S. (2011). Trans-interaction of nephrin and Neph1/Neph3 induces cell adhesion that associates with decreased tyrosine phosphorylation of nephrin. *The Biochemical Journal, 435*, 619–628.

Hoelzle, M. K., & Svitkina, T. (2012). The cytoskeletal mechanisms of cell-cell junction formation in endothelial cells. *Molecular Biology of the Cell, 23*, 310–323.

Holtfreter, J. (1939). Gewebsaffinität, ein Mittel der embryonalen Formbildung. *Archiv für experimentelle Zellforschung, besonders Gewebezüchtung, 23*, 169–209.

Hong, S., Troyanovsky, R. B., & Troyanovsky, S. M. (2013). Binding to F-actin guides cadherin cluster assembly, stability, and movement. *The Journal of Cell Biology*, *201*, 131–143.
Horikawa, K., Radice, G., Takeichi, M., & Chisaka, O. (1999). Adhesive subdivisions intrinsic to the epithelial somites. *Developmental Biology*, *215*, 182–189.
Huveneers, S., Oldenburg, J., Spanjaard, E., van der Krogt, G., Grigoriev, I., Akhmanova, A., et al. (2012). Vinculin associates with endothelial VE-cadherin junctions to control force-dependent remodeling. *The Journal of Cell Biology*, *196*, 641–652.
Ibrahim, H., & Winklbauer, R. (2001). Mechanisms of mesendoderm internalization in the Xenopus gastrula: Lessons from the ventral side. *Developmental Biology*, *240*, 108–122.
Inoue, Y. U., Asami, J., & Inoue, T. (2008). Cadherin-6 gene regulatory patterns in the postnatal mouse brain. *Molecular and Cellular Neurosciences*, *39*, 95–104.
Inoue, T., Chisaka, O., Matsunami, H., & Takeichi, M. (1997). Cadherin-6 expression transiently delineates specific rhombomeres, other neural tube subdivisions, and neural crest subpopulations in mouse embryos. *Developmental Biology*, *183*, 183–194.
Inoue, T., Tanaka, T., Takeichi, M., Chisaka, O., Nakamura, S., & Noriko, O. (2001). Role of cadherins in maintaining the compartment boundary between the cortex and striatum during development. *Development*, *128*, 561–569.
Inuzuka, H., Miyatani, S., & Takeichi, M. (1991). R-cadherin: A novel Ca(2+)-dependent cell-cell adhesion molecule expressed in the retina. *Neuron*, *7*, 69–79.
Ishiyama, N., Lee, S. H., Liu, S., Li, G. Y., Smith, M. J., Reichardt, L. F., et al. (2010). Dynamic and static interactions between p120 catenin and E-cadherin regulate the stability of cell-cell adhesion. *Cell*, *141*, 117–128.
Janes, P. W., Saha, N., Barton, W. A., Kolev, M. V., Wimmer-Kleikamp, S. H., Nievergall, E., et al. (2005). Adam meets Eph: An ADAM substrate recognition module acts as a molecular switch for ephrin cleavage in trans. *Cell*, *123*, 291–304.
Janody, F., Martirosyan, Z., Benlali, A., & Treisman, J. E. (2003). Two subunits of the Drosophila mediator complex act together to control cell affinity. *Development*, *130*, 3691–3701.
Julich, D., Mould, A. P., Koper, E., & Holley, S. A. (2009). Control of extracellular matrix assembly along tissue boundaries via integrin and Eph/Ephrin signaling. *Development*, *136*, 2913–2921.
Kalantarian, A., Ninomiya, H., Saad, S. M., David, R., Winklbauer, R., & Neumann, A. W. (2009). Axisymmetric drop shape analysis for estimating the surface tension of cell aggregates by centrifugation. *Biophysical Journal*, *96*, 1606–1616.
Karaulanov, E. E., Bottcher, R. T., & Niehrs, C. (2006). A role for fibronectin-leucine-rich transmembrane cell-surface proteins in homotypic cell adhesion. *EMBO Reports*, *7*, 283–290.
Karaulanov, E., Bottcher, R. T., Stannek, P., Wu, W., Rau, M., Ogata, S., et al. (2009). Unc5B interacts with FLRT3 and Rnd1 to modulate cell adhesion in Xenopus embryos. *PLoS One*, *4*, e5742.
Katsamba, P., Carroll, K., Ahlsen, G., Bahna, F., Vendome, J., Posy, S., et al. (2009). Linking molecular affinity and cellular specificity in cadherin-mediated adhesion. *Proceedings of the National Academy of Sciences of the United States of America*, *106*, 11594–11599.
Kim, S. H., Jen, W. C., De Robertis, E. M., & Kintner, C. (2000). The protocadherin PAPC establishes segmental boundaries during somitogenesis in Xenopus embryos. *Current Biology*, *10*, 821–830.
Kim, S. H., Yamamoto, A., Bouwmeester, T., Agius, E., & Robertis, E. M. (1998). The role of paraxial protocadherin in selective adhesion and cell movements of the mesoderm during Xenopus gastrulation. *Development*, *125*, 4681–4690.
Kim, S. Y., Yasuda, S., Tanaka, H., Yamagata, K., & Kim, H. (2011). Non-clustered protocadherin. *Cell Adhesion & Migration*, *5*, 97–105.

Kimura, Y., Matsunami, H., Inoue, T., Shimamura, K., Uchida, N., Ueno, T., et al. (1995). Cadherin-11 expressed in association with mesenchymal morphogenesis in the head, somite, and limb bud of early mouse embryos. *Developmental Biology, 169*, 347–358.
Koshida, S., Kishimoto, Y., Ustumi, H., Shimizu, T., Furutani-Seiki, M., Kondoh, H., et al. (2005). Integrinalpha5-dependent fibronectin accumulation for maintenance of somite boundaries in zebrafish embryos. *Developmental Cell, 8*, 587–598.
Kraft, B., Berger, C. D., Wallkamm, V., Steinbeisser, H., & Wedlich, D. (2012). Wnt-11 and Fz7 reduce cell adhesion in convergent extension by sequestration of PAPC and C-cadherin. *The Journal of Cell Biology, 198*, 695–709.
Krieg, M., Arboleda-Estudillo, Y., Puech, P. H., Kafer, J., Graner, F., Muller, D. J., et al. (2008). Tensile forces govern germ-layer organization in zebrafish. *Nature Cell Biology, 10*, 429–436.
Kuriyama, S., Theveneau, E., Benedetto, A., Parsons, M., Tanaka, M., Charras, G., et al. (2014). In vivo collective cell migration requires an LPAR2-dependent increase in tissue fluidity. *The Journal of Cell Biology, 206*, 113–127.
Kuroda, H., Inui, M., Sugimoto, K., Hayata, T., & Asashima, M. (2002). Axial protocadherin is a mediator of prenotochord cell sorting. *Developmental Biology, 244*, 267–277.
Ladoux, B., Anon, E., Lambert, M., Rabodzey, A., Hersen, P., Buguin, A., et al. (2010). Strength dependence of cadherin-mediated adhesions. *Biophysical Journal, 98*, 534–542.
Landsberg, K. P., Farhadifar, R., Ranft, J., Umetsu, D., Widmann, T. J., Bittig, T., et al. (2009). Increased cell bond tension governs cell sorting at the Drosophila anteroposterior compartment boundary. *Current Biology, 19*, 1950–1955.
Laplante, C., & Nilson, L. A. (2006). Differential expression of the adhesion molecule Echinoid drives epithelial morphogenesis in Drosophila. *Development, 133*, 3255–3264.
Laplante, C., & Nilson, L. A. (2011). Asymmetric distribution of Echinoid defines the epidermal leading edge during Drosophila dorsal closure. *The Journal of Cell Biology, 192*, 335–348.
Lawrence, P. A., & Green, S. M. (1975). The anatomy of a compartment border. The intersegmental boundary in Oncopeltus. *The Journal of Cell Biology, 65*, 373–382.
Lawrence, P. A., Green, S. M., & Johnston, P. (1978). Compartmentalization and growth of the Drosophila abdomen. *Journal of Embryology and Experimental Morphology, 43*, 233–245.
le Duc, Q., Shi, Q., Blonk, I., Sonnenberg, A., Wang, N., Leckband, D., et al. (2010). Vinculin potentiates E-cadherin mechanosensing and is recruited to actin-anchored sites within adherens junctions in a myosin II-dependent manner. *The Journal of Cell Biology, 189*, 1107–1115.
Lecuit, T., & Lenne, P. F. (2007). Cell surface mechanics and the control of cell shape, tissue patterns and morphogenesis. *Nature Reviews Molecular Cell Biology, 8*, 633–644.
Levayer, R., Pelissier-Monier, A., & Lecuit, T. (2011). Spatial regulation of Dia and Myosin-II by RhoGEF2 controls initiation of E-cadherin endocytosis during epithelial morphogenesis. *Nature Cell Biology, 13*, 529–540.
Luu, O., David, R., Ninomiya, H., & Winklbauer, R. (2011). Large-scale mechanical properties of Xenopus embryonic epithelium. *Proceedings of the National Academy of Sciences of the United States of America, 108*, 4000–4005.
Lynch, H. E., Veldhuis, J., Brodland, G. W., & Hutson, M. S. (2014). Modeling cell elongation during germ band retraction: Cell autonomy versus applied anisotropic stress. *New Journal of Physics, 16*, 055003.
Maghzal, N., Kayali, H. A., Rohani, N., Kajava, A. V., & Fagotto, F. (2013). EpCAM controls actomyosin contractility and cell adhesion by direct inhibition of PKC. *Developmental Cell, 27*(3), 263–277. http://dx.doi.org/10.1016/j.devcel.2013.10.003.

Maghzal, N., Vogt, E., Reintsch, W., Fraser, J. S., & Fagotto, F. (2010). The tumor-associated EpCAM regulates morphogenetic movements through intracellular signaling. *The Journal of Cell Biology, 191*, 645–659.
Maitre, J. L., Berthoumieux, H., Krens, S. F., Salbreux, G., Julicher, F., Paluch, E., et al. (2012). Adhesion functions in cell sorting by mechanically coupling the cortices of adhering cells. *Science, 338*, 253–256.
Major, R. J., & Irvine, K. D. (2006). Localization and requirement for Myosin II at the dorsal-ventral compartment boundary of the Drosophila wing. *Developmental Dynamics, 235*, 3051–3058.
Manning, M. L., Foty, R. A., Steinberg, M. S., & Schoetz, E. M. (2010). Coaction of intercellular adhesion and cortical tension specifies tissue surface tension. *Proceedings of the National Academy of Sciences of the United States of America, 107*, 12517–12522.
Maretzky, T., Reiss, K., Ludwig, A., Buchholz, J., Scholz, F., Proksch, E., et al. (2005). ADAM10 mediates E-cadherin shedding and regulates epithelial cell-cell adhesion, migration, and beta-catenin translocation. *Proceedings of the National Academy of Sciences of the United States of America, 102*, 9182–9187.
Marston, D. J., Dickinson, S., & Nobes, C. D. (2003). Rac-dependent trans-endocytosis of ephrinBs regulates Eph-ephrin contact repulsion. *Nature Cell Biology, 5*, 879–888.
Martinez-Arias, A., & Lawrence, P. A. (1985). Parasegments and compartments in the Drosophila embryo. *Nature, 313*, 639–642.
Matsuzaki, F., Mege, R. M., Jaffe, S. H., Friedlander, D. R., Gallin, W. J., Goldberg, J. I., et al. (1990). cDNAs of cell adhesion molecules of different specificity induce changes in cell shape and border formation in cultured S180 cells. *The Journal of Cell Biology, 110*, 1239–1252.
Medina, A., Swain, R. K., Kuerner, K. M., & Steinbeisser, H. (2004). Xenopus paraxial protocadherin has signaling functions and is involved in tissue separation. *The EMBO Journal, 23*, 3249–3258.
Milan, M., Perez, L., & Cohen, S. M. (2005). Boundary formation in the Drosophila wing: Functional dissection of Capricious and Tartan. *Developmental Dynamics, 233*, 804–810.
Milan, M., Weihe, U., Perez, L., & Cohen, S. M. (2001). The LRR proteins capricious and Tartan mediate cell interactions during DV boundary formation in the *Drosophila* wing. *Cell, 106*, 785–794.
Monier, B., Pelissier-Monier, A., Brand, A. H., & Sanson, B. (2010). An actomyosin-based barrier inhibits cell mixing at compartmental boundaries in Drosophila embryos. *Nature Cell Biology, 12*, 60–65.
Montero, J. A., Carvalho, L., Wilsch-Brauninger, M., Kilian, B., Mustafa, C., & Heisenberg, C. P. (2005). Shield formation at the onset of zebrafish gastrulation. *Development, 132*, 1187–1198.
Morata, G., & Lawrence, P. A. (1978). Anterior and posterior compartments in the head of Drosophila. *Nature, 274*, 473–474.
Moscona, A., & Moscona, H. (1952). The dissociation and aggregation of cells from organ rudiments of the early chick embryo. *Journal of Anatomy, 86*, 287–301.
Nanes, B. A., Chiasson-MacKenzie, C., Lowery, A. M., Ishiyama, N., Faundez, V., Ikura, M., et al. (2012). p120-catenin binding masks an endocytic signal conserved in classical cadherins. *The Journal of Cell Biology, 199*, 365–380.
Niessen, C. M., & Gumbiner, B. M. (2002). Cadherin-mediated cell sorting not determined by binding or adhesion specificity. *The Journal of Cell Biology, 156*, 389–399.
Ninomiya, H., David, R., Damm, E. W., Fagotto, F., Niessen, C. M., & Winklbauer, R. (2012). Cadherin-dependent differential cell adhesion in Xenopus causes cell sorting in vitro but not in the embryo. *Journal of Cell Science, 125*, 1877–1883.

Ninomiya, H., & Winklbauer, R. (2008). Epithelial coating controls mesenchymal shape change through tissue-positioning effects and reduction of surface-minimizing tension. *Nature Cell Biology, 10*, 61–69.

Nose, A., Nagafuchi, A., & Takeichi, M. (1988). Expressed recombinant cadherins mediate cell sorting in model systems. *Cell, 54*, 993–1001.

Oda, H., & Takeichi, M. (2011). Evolution: Structural and functional diversity of cadherin at the adherens junction. *The Journal of Cell Biology, 193*, 1137–1146.

Ogata, S., Morokuma, J., Hayata, T., Kolle, G., Niehrs, C., Ueno, N., et al. (2007). TGF-beta signaling-mediated morphogenesis: Modulation of cell adhesion via cadherin endocytosis. *Genes & Development, 21*, 1817–1831.

Ounkomol, C., Yamada, S., & Heinrich, V. (2010). Single-cell adhesion tests against functionalized microspheres arrayed on AFM cantilevers confirm heterophilic E- and N-cadherin binding. *Biophysical Journal, 99*, L100–L102.

Pabbisetty, K. B., Yue, X., Li, C., Himanen, J. P., Zhou, R., Nikolov, D. B., et al. (2007). Kinetic analysis of the binding of monomeric and dimeric ephrins to Eph receptors: Correlation to function in a growth cone collapse assay. *Protein Science, 16*, 355–361.

Park, E. C., Cho, G. S., Kim, G. H., Choi, S. C., & Han, J. K. (2011). The involvement of Eph-Ephrin signaling in tissue separation and convergence during Xenopus gastrulation movements. *Developmental Biology, 350*, 441–450.

Prakasam, A. K., Maruthamuthu, V., & Leckband, D. E. (2006). Similarities between heterophilic and homophilic cadherin adhesion. *Proceedings of the National Academy of Sciences of the United States of America, 103*, 15434–15439.

Price, S. R., De Marco Garcia, N. V., Ranscht, B., & Jessell, T. M. (2002). Regulation of motor neuron pool sorting by differential expression of type II cadherins. *Cell, 109*, 205–216.

Reintsch, W. E., Habring-Mueller, A., Wang, R. W., Schohl, A., & Fagotto, F. (2005). beta-Catenin controls cell sorting at the notochord-somite boundary independently of cadherin-mediated adhesion. *The Journal of Cell Biology, 170*, 675–686.

Reiss, K., Maretzky, T., Ludwig, A., Tousseyn, T., de Strooper, B., Hartmann, D., et al. (2005). ADAM10 cleavage of N-cadherin and regulation of cell-cell adhesion and beta-catenin nuclear signalling. *The EMBO Journal, 24*, 742–752.

Rhee, J., Takahashi, Y., Saga, Y., Wilson-Rawls, J., & Rawls, A. (2003). The protocadherin PAPC is involved in the organization of the epithelium along the segmental border during mouse somitogenesis. *Developmental Biology, 254*, 248–261.

Rohani, N., Canty, L., Luu, O., Fagotto, F., & Winklbauer, R. (2011). EphrinB/EphB signaling controls embryonic germ layer separation by contact-induced cell detachment. *PLoS Biology, 9*, e1000597.

Rohani, N., Parmeggiani, A., Winklbauer, R., & Fagotto, F. (2014). Variable combinations of specific ephrin ligand/Eph receptor pairs control embryonic tissue separation. *PLoS Biology, 12*, e1001955.

Sako, Y., Nagafuchi, A., Tsukita, S., Takeichi, M., & Kusumi, A. (1998). Cytoplasmic regulation of the movement of E-cadherin on the free cell surface as studied by optical tweezers and single particle tracking: Corralling and tethering by the membrane skeleton. *The Journal of Cell Biology, 140*, 1227–1240.

Sakurai, K. T., Kojima, T., Aigaki, T., & Hayashi, S. (2007). Differential control of cell affinity required for progression and refinement of cell boundary during Drosophila leg segmentation. *Developmental Biology, 309*, 126–136.

Schafer, G., Narasimha, M., Vogelsang, E., & Leptin, M. (2014). Cadherin switching during the formation and differentiation of the Drosophila mesoderm—Implications for epithelial-to-mesenchymal transitions. *Journal of Cell Science, 127*, 1511–1522.

Seidel, B., Braeg, S., Adler, G., Wedlich, D., & Menke, A. (2004). E- and N-cadherin differ with respect to their associated p120ctn isoforms and their ability to suppress invasive growth in pancreatic cancer cells. *Oncogene, 23*, 5532–5542.

Senturk, A., Pfennig, S., Weiss, A., Burk, K., & Acker-Palmer, A. (2011). Ephrin Bs are essential components of the Reelin pathway to regulate neuronal migration. *Nature*, *472*, 356–360.

Shi, Q., Chien, Y. H., & Leckband, D. (2008). Biophysical properties of cadherin bonds do not predict cell sorting. *The Journal of Biological Chemistry*, *283*, 28454–28463.

Shimono, Y., Rikitake, Y., Mandai, K., Mori, M., & Takai, Y. (2012). Immunoglobulin superfamily receptors and adherens junctions. *Sub-Cellular Biochemistry*, *60*, 137–170.

Shimoyama, Y., Tsujimoto, G., Kitajima, M., & Natori, M. (2000). Identification of three human type-II classic cadherins and frequent heterophilic interactions between different subclasses of type-II classic cadherins. *The Biochemical Journal*, *349*, 159–167.

Simoes Sde, M., Blankenship, J. T., Weitz, O., Farrell, D. L., Tamada, M., Fernandez-Gonzalez, R., et al. (2010). Rho-kinase directs Bazooka/Par-3 planar polarity during Drosophila axis elongation. *Developmental Cell*, *19*, 377–388.

Slanchev, K., Carney, T. J., Stemmler, M. P., Koschorz, B., Amsterdam, A., Schwarz, H., et al. (2009). The epithelial cell adhesion molecule EpCAM is required for epithelial morphogenesis and integrity during zebrafish epiboly and skin development. *PLoS Genetics*, *5*, e1000563.

Solanas, G., Cortina, C., Sevillano, M., & Batlle, E. (2011). Cleavage of E-cadherin by ADAM10 mediates epithelial cell sorting downstream of EphB signalling. *Nature Cell Biology*, *13*, 1100–1107.

Steinberg, M. S. (1970). Does differential adhesion govern self-assembly processes in histogenesis? Equilibrium configurations and the emergence of a hierarchy among populations of embryonic cells. *The Journal of Experimental Zoology*, *173*, 395–434.

Steinberg, M. S. (1978). Cell–cell recognition in multicellular assembly: Levels of specificity. *Symposia of the Society for Experimental Biology*, *32*, 25–49.

Steinberg, M. S., & Takeichi, M. (1994). Experimental specification of cell sorting, tissue spreading, and specific spatial patterning by quantitative differences in cadherin expression. *Proceedings of the National Academy of Sciences of the United States of America*, *91*, 206–209.

Straub, B. K., Rickelt, S., Zimbelmann, R., Grund, C., Kuhn, C., Iken, M., et al. (2011). E-N-cadherin heterodimers define novel adherens junctions connecting endoderm-derived cells. *The Journal of Cell Biology*, *195*, 873–887.

Takai, Y., Miyoshi, J., Ikeda, W., & Ogita, H. (2008). Nectins and nectin-like molecules: Roles in contact inhibition of cell movement and proliferation. *Nature Reviews Molecular Cell Biology*, *9*, 603–615.

Takeichi, M. (1995). Morphogenetic roles of classic cadherins. *Current Opinion in Cell Biology*, *7*, 619–627.

Tepass, U., Godt, D., & Winklbauer, R. (2002). Cell sorting in animal development: Signalling and adhesive mechanisms in the formation of tissue boundaries. *Current Opinion in Genetics & Development*, *12*, 572–582.

Townes, P. L., & Holtfreter, J. (1955). Directed movements and selective adhesion of embryonic amphibian cells. *The Journal of Experimental Zoology*, *128*, 53–120.

Truong Quang, B. A., Mani, M., Markova, O., Lecuit, T., & Lenne, P. F. (2013). Principles of E-cadherin supramolecular organization in vivo. *Current Biology*, *23*, 2197–2207.

Ulrich, F., Krieg, M., Schotz, E. M., Link, V., Castanon, I., Schnabel, V., et al. (2005). Wnt11 functions in gastrulation by controlling cell cohesion through Rab5c and E-cadherin. *Developmental Cell*, *9*, 555–564.

Unterseher, F., Hefele, J. A., Giehl, K., De Robertis, E. M., Wedlich, D., & Schambony, A. (2004). Paraxial protocadherin coordinates cell polarity during convergent extension via Rho A and JNK. *The EMBO Journal*, *23*, 3259–3269.

Wacker, S., Grimm, K., Joos, T., & Winklbauer, R. (2000). Development and control of tissue separation at gastrulation in Xenopus. *Developmental Biology*, *224*, 428–439.

Watanabe, T., Sato, Y., Saito, D., Tadokoro, R., & Takahashi, Y. (2009). EphrinB2 coordinates the formation of a morphological boundary and cell epithelialization during somite segmentation. *Proceedings of the National Academy of Sciences of the United States of America, 106*, 7467–7472.

Wei, S. Y., Escudero, L. M., Yu, F., Chang, L. H., Chen, L. Y., Ho, Y. H., et al. (2005). Echinoid is a component of adherens junctions that cooperates with DE-Cadherin to mediate cell adhesion. *Developmental Cell, 8*, 493–504.

Wheelock, M. J., Shintani, Y., Maeda, M., Fukumoto, Y., & Johnson, K. R. (2008). Cadherin switching. *Journal of Cell Science, 121*, 727–735.

Wilson, H. V. (1907). On some phenomena of coalescence and regeneration in sponges. *The Journal of Experimental Zoology, 5*, 235–278.

Winklbauer, R. (2009). Cell adhesion in amphibian gastrulation. *International Review of Cell and Molecular Biology, 278*, 215–275.

Winklbauer, R., Medina, A., Swain, R. K., & Steinbeisser, H. (2001). Frizzled-7 signalling controls tissue separation during Xenopus gastrulation. *Nature, 413*, 856–860.

Xu, Q., Alldus, G., Holder, N., & Wilkinson, D. G. (1995). Expression of truncated Sek-1 receptor tyrosine kinase disrupts the segmental restriction of gene expression in the Xenopus and zebrafish hindbrain. *Development, 121*, 4005–4016.

Xu, Q., & Wilkinson, D. G. (1997). Eph-related receptors and their ligands: Mediators of contact dependent cell interactions. *Journal of Molecular Medicine, 75*, 576–586.

Yamada, S., & Nelson, W. J. (2007). Localized zones of Rho and Rac activities drive initiation and expansion of epithelial cell-cell adhesion. *The Journal of Cell Biology, 178*, 517–527.

Yamagishi, S., Hampel, F., Hata, K., Del Toro, D., Schwark, M., Kvachnina, E., et al. (2011). FLRT2 and FLRT3 act as repulsive guidance cues for Unc5-positive neurons. *The EMBO Journal, 30*, 2920–2933.

Yamamoto, A., Kemp, C., Bachiller, D., Geissert, D., & De Robertis, E. M. (2000). Mouse paraxial protocadherin is expressed in trunk mesoderm and is not essential for mouse development. *Genesis, 27*, 49–57.

Yao, M., Qiu, W., Liu, R., Efremov, A. K., Cong, P., Seddiki, R., et al. (2014). Force-dependent conformational switch of alpha-catenin controls vinculin binding. *Nature Communications, 5*, 4525.

Yap, A. S., Niessen, C. M., & Gumbiner, B. M. (1998). The juxtamembrane region of the cadherin cytoplasmic tail supports lateral clustering, adhesive strengthening, and interaction with p120ctn. *The Journal of Cell Biology, 141*, 779–789.

Yoder, M. D., & Gumbiner, B. M. (2011). Axial protocadherin (AXPC) regulates cell fate during notochordal morphogenesis. *Developmental Dynamics, 240*, 2495–2504.

Yonemura, S., Wada, Y., Watanabe, T., Nagafuchi, A., & Shibata, M. (2010). alpha-Catenin as a tension transducer that induces adherens junction development. *Nature Cell Biology, 12*, 533–542.

Youn, B. W., & Malacinski, G. M. (1981). Comparative analysis of amphibian somite morphogenesis: Cell rearrangement patterns during rosette formation and myoblast fusion. *Journal of Embryology and Experimental Morphology, 66*, 1–26.

Zervas, M., Millet, S., Ahn, S., & Joyner, A. L. (2004). Cell behaviors and genetic lineages of the mesencephalon and rhombomere 1. *Neuron, 43*, 345–357.

Zhong, Y., Brieher, W. M., & Gumbiner, B. M. (1999). Analysis of C-cadherin regulation during tissue morphogenesis with an activating antibody. *The Journal of Cell Biology, 144*, 351–359.

Zimmer, M., Palmer, A., Kohler, J., & Klein, R. (2003). EphB-ephrinB bi-directional endocytosis terminates adhesion allowing contact mediated repulsion. *Nature Cell Biology, 5*, 869–878.

CHAPTER THREE

Active Tension: The Role of Cadherin Adhesion and Signaling in Generating Junctional Contractility

Rashmi Priya, Alpha S. Yap[1]
Division of Cell Biology and Molecular Medicine, Institute for Molecular Bioscience, The University of Queensland, Brisbane, Queensland, Australia
[1]Corresponding author: e-mail address: a.yap@uq.edu.au

Contents

Abstract

In this chapter, we discuss the cell biology of contractility at cell–cell junctions. As discussed elsewhere in this volume, contractile forces play key roles in development and tissue homeostasis. Here, we review our understanding of the cellular mechanisms that functionally and physically link cadherin adhesion to the actomyosin contractile apparatus of the cell. Focusing on epithelia, we argue that E-cadherin junctions can be

Current Topics in Developmental Biology, Volume 112
ISSN 0070-2153
http://dx.doi.org/10.1016/bs.ctdb.2014.11.016

considered as active mechanical agents, which contribute to the assembly of actomyosin at the junctional cortex itself. This reflects cortical signaling, notably that regulated by the Rho GTPase, coordinated with actin regulation at junctions. The product, contractile tension at junctions, can then be regarded as an emergent property of a complex dynamical system that integrates adhesion with the cytoskeleton.

1. INTRODUCTION

Cell–cell interactions have long been recognized to participate in the genesis and homeostasis of tissue organization. Adhesive interactions between cells influence patterning in the early embryo and counteract forces that tend to disrupt tissue integrity (Levine, Lee, Kintner, & Gumbiner, 1994; Takeichi, 2014). These adhesive interactions have commonly been envisaged as passive resistance elements that respond to large-scale forces acting on organisms and tissues. Increasingly, we have come to realize that many of the forces that act upon cell–cell adhesions are generated by the neighboring cells themselves (Gomez, McLachlan, & Yap, 2011; Luo et al., 2013). Effectively, cells pull upon each other and cell–cell adhesion systems serve to mechanically couple the contractile apparatuses of neighboring cells together (Maitre et al., 2012). This is best understood for epithelia where a dominant adhesion system is mediated by E-cadherin. Mechanical coupling of contractility to E-cadherin adhesion is reflected in contractile tension within junctions that scales-up to tissue level patterns of tension (Fernandez-Gonzalez, Simoes Sde, Roper, Eaton, & Zallen, 2009; Martin, Gelbart, Fernandez-Gonzalez, Kaschube, & Wieschaus, 2010; Rauzi, Verant, Lecuit, & Lenne, 2008). Adhesion can effectively convert the isolated contractile apparatus of a single cell into a supracellular contractile network. This is implicated in both local cellular rearrangements and tissue-level morphogenesis (Martin, 2010).

At the cellular level, contractility is predominantly generated by the actomyosin system, involving the interaction of F-actin microfilaments with myosin II motors (Vicente-Manzanares, Ma, Adelstein, & Horwitz, 2009). Mechanistically, a key question is how cadherin molecular systems physically interact with actomyosin to couple adhesion and contractility. However, cadherin adhesion also contributes to the biogenesis of actomyosin at cell–cell junctions (Ladoux et al., 2010; Shewan et al., 2005). This entails the recruitment of cytoskeletal regulators to the junctional cortex, responding to cortical signals that are often influenced by cadherin adhesion itself (Ratheesh & Yap, 2012). Thus, the integration of adhesion and contractility not only reflects the passive association of cadherins with the cytoskeleton but also involves an *active* role of adhesion in establishing the junctional cytoskeleton (Kovacs, Goodwin, Ali, Paterson, & Yap, 2002; Kovacs et al., 2011; Priya, Yap, & Gomez, 2013; Shewan et al., 2005; Smutny et al., 2011). In this chapter, we review our growing understanding of how this comes about. A recurring theme in this discussion is the emerging role that contractile tension itself can play in regulating actomyosin at cell–cell junctions.

2. THE CONTRACTILE APPARATUS: ACTOMYOSIN

Actomyosin constitutes the principal contractile apparatus in nonmuscle cells, as well as in muscle (Fernandez-Gonzalez & Zallen, 2009; Kasza & Zallen, 2011; Fig. 1). Contractility is generated by the physical interaction of the myosin II head domains with actin filaments. In the best-understood model, force is generated by the ability of myosin II to slide actin filaments once the motor is activated by phosphorylation of its regulatory light chain (RLC) (Craig, Smith, & Kendrick-Jones, 1983; Matsumura, 2005). myosin II activation also entails the assembly of Myosin motors into antiparallel minifilaments that can each consist of 10–30 individual myosins (Heissler & Manstein, 2013; Niederman & Pollard, 1975). This antiparallel orientation allows minifilaments to slide F-actin inward, thereby generating contractility (Fig. 1B). Importantly, the organization of the actin filaments with which myosin interacts can influence contractile output (Reymann et al., 2012). This is exemplified by the alignment of myosin minifilaments with parallel actin filaments that is found in the sarcomeres of muscle cells. Something similar has also been observed at epithelial cell–cell junctions (Ebrahim et al., 2013). However, Myosin minifilaments can also act upon actin filaments that are organized into less well organized, nonparallel networks; under these latter circumstances, contractility can aggregate F-actin and even induce filament turnover (as will be discussed further below) (Haviv, Gillo, Backouche, & Bernheim-Groswasser, 2008; Reymann et al., 2012).

The functional impact of actomyosin is demonstrated by the observation that morphogenetic events, such as germ-band extension during *Drosophila* development, are perturbed when components of myosin II are depleted from embryos (Bertet, Sulak, & Lecuit, 2004) This is thought to reflect local

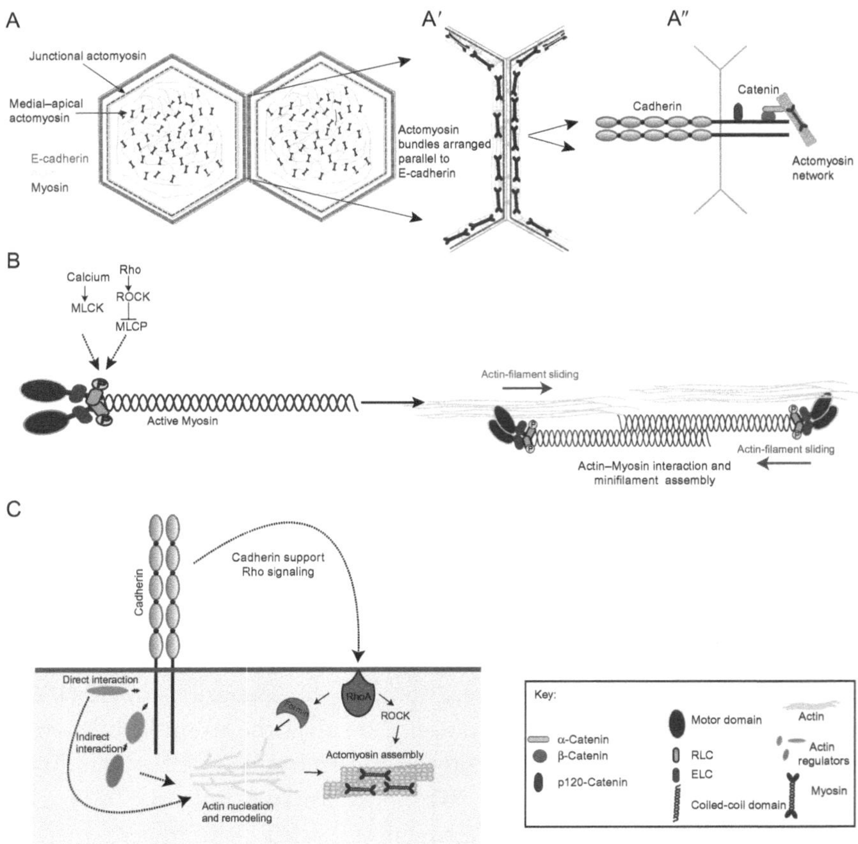

Figure 1 The contractile apparatus at junctions and its biogenesis. (A) Spatial relation of actomyosin network with cadherin junctions. F-actin and myosin form a mesh at the apical pole of the cells (medial–apical network) and also orient themselves parallel to the junctions (junctional actomyosin). (A′) The contractile cortices of neighboring cells are mechanically coupled through association with the cadherin–catenin complex (A″). (For clarity, this coupling is illustrated only for junctional actomyosin.) (B) Kinases (ROCK and MLCK) activate myosin by phosphorylating its RLC. This leads to association of myosin with F-actin and minifilament assembly (which is also controlled by heavy chain phosphorylation). (C) Cadherin ligation recruits actin regulators directly or indirectly via adaptor proteins to modulate local actin assembly and remodeling. Cadherin can also scaffold upstream Rho activators and thus support Rho signaling, which further mediates actomyosin assembly via its effectors. (See the color plate.)

cellular rearrangements that are driven by contractile coupling to cadherin cell junctions. Consistent with this, a number of patterns of myosin II networks are found to mechanically associate with junctions in tissues and cultured epithelial cells. One pattern involves myosin II networks located at the apical poles of epithelial cells (often described as medial–apical networks; Martin, Kaschube, & Wieschaus, 2009); these commonly display pulsatile contractions which are transmitted to junctions. As well, myosin II accumulates at cell–cell junctions themselves, typically decorating F-actin bundles that circumscribe the apical junctions (Shewan et al., 2005; Smutny et al., 2010; Wu et al., 2014; Fig. 1A). These also contribute to contractile tension in junctions. But the latter appear to manifest as a line tension in the apical junctions rather than the pulsatile fluctuations seen with medial–apical networks. In both these systems, active myosin must physically interact with the cadherin adhesion system to allow force to be transmitted to, and through, junctions. This involves both cell-signaling pathways that regulate the activity and localization of myosin II and association of myosin-bound actin filaments with the cadherin complex (Fig. 1C).

2.1. The role of cadherin signaling in myosin II regulation

As noted in the preceding text, myosin II is regulated by a variety of cellular signals. In the best-characterized model, phosphorylation of the RLC (at Thr18 and Ser19) promotes F-actin binding and actin-induced ATPase activity (Heissler & Manstein, 2013); it also allows myosins to assemble into minifilaments (Sellers, 1991). Monophosphorylation of Ser19 is sufficient to stimulate myosin ATPase activity, motor activity, and minifilament assembly, although these can be further enhanced by additional phosphorylation of Thr18 (Julian & Olson, 2014). These combined effects promote actin filament sliding and therefore contractility. RLC phosphorylation is, in turn, determined by the balance between the action of kinases and myosin light-chain phosphatase (MLCP; Julian & Olson, 2014; Schofield & Bernard, 2013). A number of kinases can directly phosphorylate RLC, including myosin light-chain kinase (MLCK), which commonly responds to changes in intracellular Ca^{2+} levels that can activate contractility (Heissler & Manstein, 2013). Rho kinase (ROCK) is another Ser/Thr kinase that mediates signaling downstream of the small GTPase, Rho, a major regulator of cellular contractility (Riento & Ridley, 2003). Although ROCK can directly phosphorylate myosin *in vitro*, within cells its major impact may occur indirectly. In contrast to MLCK, ROCK can also inhibit the action of MLCP, thereby preventing RLC dephosphorylation (Amano et al., 1996; Kimura et al., 1996; Totsukawa et al., 2000).

Both MLCK and ROCK participate in building the contractile cytoskeleton at cadherin junctions (Smutny et al., 2010). In the case of ROCK, this appears to reflect an E-cadherin-based Rho signaling pathway (which will be discussed in greater detail below). Thus, myosin II can be recruited to sites of adhesion when E-cadherin alone is ligated, which is isolated from other juxtacrine pathways that may come into effect when cell surfaces come into contact with one another (Shewan et al., 2005). This implies that cadherin adhesion itself can provide an instructive cue to recruit myosin II. Myosin recruitment to cadherin adhesions is disrupted when ROCK is inhibited, implying that ROCK serves in a cadherin-dependent signaling pathway to activate myosin II (Shewan et al., 2005). Similarly, junctional myosin II levels are reduced when ROCK is inhibited (Smutny et al., 2010). Consistent with the role of ROCK as a major mediator of Rho signaling (Riento & Ridley, 2003), junctional myosin II and contractile tension are also reduced when Rho signaling itself is blocked or the molecular mechanisms that activate Rho at junctions are disrupted (Ratheesh et al., 2012). The ability of myosin activation to promote junctional localization of the motor likely reflects the increased F-actin binding that accompanies RLC phosphorylation, as junctional myosin II is lost when junctional F-actin is reduced (Leerberg et al., 2014; Verma et al., 2012). MLCK is also necessary for E-cadherin adhesion to recruit myosin II (Shewan et al., 2005; Smutny et al., 2010), and intracellular calcium signaling is increased at junctions when cells make contacts with one another (Nigam, Rodriguez-Boulan, & Silver, 1992). However, what relationship this has to cadherin signaling has yet to be determined. As well, it is likely that many other signaling pathways impinge to influence the biogenesis and contractile activity of actomyosin at junctions.

Interestingly, mammalian cells have three myosin II paralogs (A–C) (Vicente-Manzanares et al., 2009), in contrast to the single myosin II possessed by invertebrates, and different sets of signaling pathways may influence individual myosin II paralogs at cadherin junctions in mammalian cells. For example, although both myosin IIA and myosin IIB were found at the zonulae adherente (ZA) in cultured MCF7 cells, a breast epithelial cell line, inhibition of Rho-ROCK signaling reduced the junctional accumulation of myosin IIA to a significantly greater extent than it did myosin IIB accumulation (Smutny et al., 2010). In contrast, myosin IIB localization was preferentially sensitive to Src and Rap1 activity (Smutny et al., 2010). Thus, multiple signaling pathways may determine the ability of different myosin paralogs to recruit to, and exert contractile force upon, cadherin-based cell–cell junctions.

2.2. F-actin mediates the association of myosin motors with cadherin adhesion

For activated myosin to generate contractile force upon junctions, molecular mechanisms are necessary to physically link myosin II to the cadherin adhesion complex. To date, there is no convincing evidence that myosin II interacts directly with cadherins or their associated proteins. Instead, it is probable that the mechanical coupling is achieved through the actin filaments with which the myosins interact. Junctional myosin is reduced, and contractile tension compromised, when the assembly of actin filaments is decreased at the junctions, either by drugs that disrupt F-actin or when regulators of F-actin dynamics are perturbed (Leerberg et al., 2014; Verma et al., 2012; Wu et al., 2014). This implies that cortical F-actin also contributes to localizing myosin II to the junctions.

How cadherin complexes associate with F-actin is an active and controversial area which has been reviewed in detail elsewhere (Ratheesh & Yap, 2012). For the purposes of the present discussion, it is sufficient to emphasize the following points. A number of actin-binding proteins can associate with E-cadherin and potentially serve to couple adhesion complexes to cortical F-actin (Fig. 2A). These include α-catenin, undoubtedly the most extensively discussed candidate, and also proteins such as vinculin, cortactin, and myosin VI. The reason(s) for this variety have yet to be resolved. Importantly, there is increasing evidence that contractile tension may influence the activity of some of these proteins. For example, α-catenin can directly associate with F-actin. Despite this, early efforts failed to reconstitute a minimal quaternary complex consisting of the cadherin cytoplasmic tail, β- and α-catenin bound to purified actin filaments (Drees, Pokutta, Yamada, Nelson, & Weis, 2005; Yamada, Pokutta, Drees, Weis, & Nelson, 2005). Those studies were performed in solution using purified proteins. However, it has recently been reported that the minimal cadherin–catenin complex will bind stably to actin filament when force is applied. Kinetic analysis of bond dissociation indicated that the interaction could be explained by a catch-bond model, where force strengthened the bond between cadherin/catenin and F-actin (Buckley et al., 2014).

α-Catenin can also interact indirectly with F-actin, through the recruitment of proteins such as vinculin (Choi et al., 2012; Rangarajan & Izard, 2012; Watabe-Uchida et al., 1998). Although other proteins may contribute (Peng, Cuff, Lawton, & DeMali, 2010), the junctional recruitment of vinculin is dominantly mediated by α-catenin (Huveneers et al., 2012; Leerberg et al., 2014). This effect is also force dependent, being decreased when myosin is

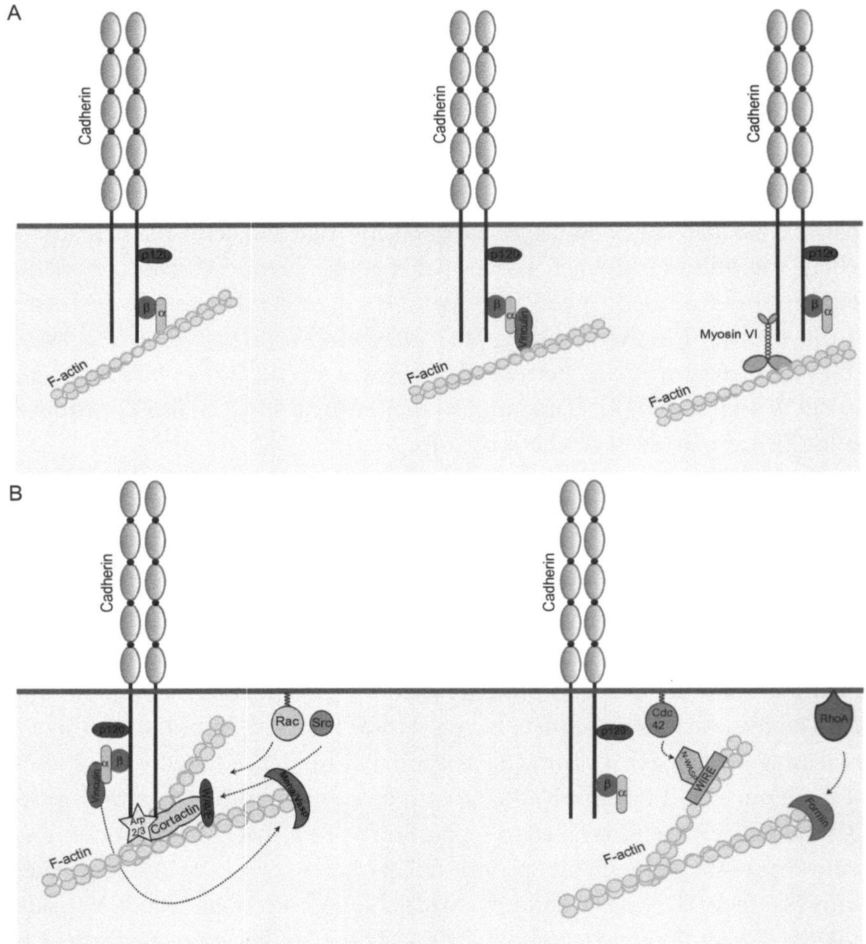

Figure 2 Cadherin and actin cooperativity. (A) The core cadherin–catenin complex is composed of cadherin, β-catenin, α-catenin, and p120-catenin. This core complex associates with actin in various ways: direct binding of α-catenin to F-actin; α-catenin binds to proteins such as vinculin which can interact with F-actin; myosin VI can directly associate with the cytoplasmic domain of cadherin and thus mediates the interaction of cadherin and F-actin. (For clarity, one of the cadherin dimers is drawn without its associated core catenins.) (B) Actin assembly and remodeling at junctions is an outcome of multiple overlapping pathways. WAVE-Arp2/3 responds to Rac signaling and nucleates actin assembly at junctions. Cortactin can directly bind WAVE and Arp2/3 and thus promote actin nucleation by scaffolding these regulators at cadherin junctions. Cortactin can also be phosphorylated by Src kinases. Further, Mena-VASP is recruited to cadherin junctions via vinculin to promote postnucleation elongation of actin filaments. N-WASP acts downstream of Cdc42 and binds to F-actin via the WIP family protein, WIRE. This promotes actin stability and organization. Rho-mediated activation of formins promotes elongation of preexisting filaments and also contributes to actin assembly by nucleation.

inhibited (Leerberg et al., 2014; Yonemura, Wada, Watanabe, Nagafuchi, & Shibata, 2010). An attractive notion is that contractile tension alters the molecular conformation of α-catenin to reveal or promote its vinculin-binding capacity. This concept was first suggested by evidence that an epitope near the vinculin-binding domain of α-catenin is revealed in a force-dependent fashion within cells (Yonemura et al., 2010). Recently, it was demonstrated that the vinculin-binding domain of α-catenin became unfolded when isolated molecules were stretched using magnetic tweezers. This promoted binding of vinculin, which then stabilized α-catenin in its open conformation (Yao et al., 2014). Thus, it is possible that the number of sites for cadherin complexes to interact with F-actin may be modulated in response to the force acting upon those adhesion complexes. Similar, myosin VI, which can bind directly to the E-cadherin cytoplasmic tail and contribute to the organization of F-actin at junctions (Mangold, Norwood, Yap, & Collins, 2012; Mangold et al., 2011), can function as a tension-sensitive actin-based anchor (Altman, Sweeney, & Spudich, 2004; Chuan, Spudich, & Dunn, 2011). Overall, this implies that cadherin–actin binding may not be a static process, but one that can be tuned in response to the forces that act upon junctions.

The association of actin with myosin II not only influences the junctional recruitment of myosin but also affects cortical F-actin itself. Myosin can control F-actin organization both by crosslinking filaments and by aggregating filaments as they slide. Furthermore, contractile stresses can induce F-actin severing, which leads to actin disassembly and turnover (Reymann et al., 2012). Effectively, stress-induced turnover would constitute a mechanism that inherently limits the force that can be generated by actomyosin, as it would turn over the F-actin scaffolds that myosin II requires to generate force. This has been demonstrated to occur in polarized cultured epithelial cells where stress-induced F-actin turnover limits the tension that is generated in the lateral cell–cell junctions found below the apical ZA (Wu et al., 2014). What function might be served by limiting the contractile stresses that junctional actomyosin can generate? One possibility is that it limits the forces that lateral membranes can exert on the cytoplasm, thereby allowing hydrodynamic behavior to be coupled across cell boundaries to drive cellular flows during morphogenesis (He, Doubrovinski, Polyakov, & Wieschaus, 2014). Additionally, control of contractility at lateral junctions contributes to integrating cells within epithelia (Wu et al., 2014). This further implies that the morphogenetic impact of actomyosin may depend not only on the myosin II and its regulation but also on the F-actin networks with which it interacts.

3. CADHERINS AND BIOGENESIS OF THE JUNCTIONAL ACTIN CYTOSKELETON

As well as influencing signaling pathways that impinge on myosin, cadherin adhesion can influence the biogenesis of junctional actomyosin by regulating multiple aspects of the actin cytoskeleton itself. These include actin filament dynamics and also potentially actin filament organization. At the mechanistic level, they are likely to reflect both intermolecular interactions that recruit actin regulators to the junctional cortex and cell signaling at cadherin junctions.

3.1. The dynamic junctional actin cytoskeleton

F-actin at the cortex of cell–cell junctions is dynamic. FRAP studies suggest that actin filaments undergo turnover on time scales of tens of seconds (Kovacs et al., 2011; Smutny et al., 2010; Yamada et al., 2005). It is also likely that multiple pools of F-actin exist at junctions that have different degrees of stability (Cavey, Rauzi, Lenne, & Lecuit, 2008; Zhang et al., 2005). Nonetheless, the fact that filaments are highly dynamic implies that the maintenance of the overall F-actin content at cell–cell junctions involves ongoing assembly of filaments to replenish those that are being lost. Moreover, the disruption of actin homeostasis affects junctional integrity and contractile tension (Verma et al., 2012), as will be discussed further. This indicates that, although filament pools may differ in their stability, the regulation of filament dynamics is functionally important.

3.2. Actin assembly

The process of assembling actin filaments can be thought of as consisting of two stages. The first is actin filament nucleation, the energetically rate-limiting step of building filaments from monomers to trimers (Pollard, Blanchoin, & Mullins, 2000). The subsequent growth of filaments from trimers can occur by actin self-assembly, but this is relatively slow. Therefore, within cells, the postnucleation assembly (growth) of filaments is commonly accelerated by proteins such as Ena/VASP proteins and formins (Chesarone & Goode, 2009; Krause, Dent, Bear, Loureiro, & Gertler, 2003).

Consistent with the notion that actin assembly may contribute to the homeostasis of dynamic filaments, actin assembly, including nucleation itself, can be identified at E-cadherin-based cell–cell junctions themselves

(Kovacs et al., 2011; Tang & Brieher, 2012; Vasioukhin, Bauer, Yin, & Fuchs, 2000; Yamada et al., 2005). A number of proteins can mediate actin filament nucleation, and several of these have been identified at E-cadherin junctions (Fig. 2B). These include the Arp2/3 complex, which mediates the assembly of branched networks of actin filaments (Pollard et al., 2000; Ydenberg, Smith, Breitsprecher, Gelles, & Goode, 2011). Arp2/3 concentrates at E-cadherin junctions (Tang & Brieher, 2012; Verma et al., 2012), through an interaction with the scaffolding protein, cortactin, which can directly associate with Arp2/3 (Han et al., 2014; Weed & Parsons, 2001). In cultured epithelial cells, depletion of Arp2/3 or cortactin reduces actin assembly at junctions, decreases steady-state levels of F-actin, and leads to fragmentation of the apical ZA (Han et al., 2014; Verma et al., 2012). Of note, Arp2/3 depletion reduces contractile tension at junctions, as well as the junctional accumulation of myosin II (Verma et al., 2012). Therefore, the local generation of F-actin by Arp2/3 appears to contribute to junctional contractility, by generating actin filaments that promote junctional accumulation of myosin and which are themselves necessary for active myosin to generate force. Arp2/3 has little intrinsic activity and instead is stimulated by cell-signaling pathways that are mediated by proteins, particularly those of the WASP/WAVE family, which transduce signals from Cdc42 and Rac1, respectively (Insall & Machesky, 2009). Indeed, both N-WASP and WAVE2 are found at cell–cell junctions, and WAVE2 is necessary for actin assembly to occur there (Kovacs et al., 2011; Verma et al., 2012). Consistent with this, Rac1 can promote actin assembly at E-cadherin adhesions (Kraemer, Goodwin, Verma, Yap, & Ali, 2007). Therefore, Arp2/3 may serve to mediate Rac-dependent actin assembly at cadherin-based junctions.

In contrast to Arp2/3, members of the formin family promote nucleation of parallel actin filaments (Chesarone & Goode, 2009), which would be predicted to be favored for the generation of F-actin bundles. Two formin family members have been identified at cell–cell junctions, formin-1 in keratinocytes and mDia1 in human mammary epithelial cells. Formin-1 was reported to interact with α-catenin, which may be responsible for its junctional recruitment (Kobielak, Pasolli, & Fuchs, 2004), whereas how mDia1 localizes to junctions remains poorly understood, although it is likely to respond to Rho (Carramusa, Ballestrem, Zilberman, & Bershadsky, 2007). Formin activity is also tightly regulated in response to cellular context (Chesarone & Goode, 2009). Interestingly, formins can also respond to changes in cellular tension, potentially independent of other known signals

(Higashida et al., 2013), although whether this is relevant at junctions remains to be determined.

A growing list of other proteins also cooperates to promote actin nucleation at junctions. α-Actinin-4 appears to facilitate Arp2/3-based actin nucleation at E-cadherin adhesion in cultured mammalian epithelial cells (Tang & Brieher, 2012), and cortactin also has a weak capacity to nucleate filaments *in vitro* (Weed & Parsons, 2001). The major contribution of cortactin to junctional nucleation in cells may, however, be to recruit Arp2/3 and WAVE2 to the junctions (Han et al., 2014). The plethora of molecules that can support actin nucleation attests to the functional importance of actin filament homeostasis at junctions, although it is possible that these diverse nucleators serve different functions depending on cellular context.

Postnucleation actin assembly also plays an important part in F-actin homeostasis at cadherin junctions. This is exemplified by members of the Ena/VASP protein family, which vary in their number from *Drosophila* (where Enabled [Ena] is the only member) to mammalian cells (which have three: Mena, VASP, and EVL; Krause et al., 2003). *In vitro*, Ena/VASP proteins have been reported to have weak nucleating capacity, but in cells their primary function appears to be to promote postnucleation elongation at F-actin barbed ends (Barzik et al., 2005). They may achieve this by antagonizing capping protein (Bear et al., 2002), which inhibits F-actin elongation at the barbed ends, and/or by acting as processive actin polymerases (Hansen & Mullins, 2010). In mammalian epithelia, Mena/VASP proteins localize to E-cadherin junctions where they contribute to actin assembly (Scott et al., 2006; Vasioukhin et al., 2000). Interestingly, their recruitment to junctions appears to constitute a mechanism for tension-dependent regulation of actin assembly (Leerberg et al., 2014). Thus, actin assembly at adherens junctions was sensitive to tension, increasing when contractility was stimulated and decreasing when myosin II was inhibited. This entailed the tension-sensitive recruitment of Mena/VASP downstream of vinculin (Leerberg et al., 2014; Peng et al., 2010). Of note, this tension-sensitive actin assembly apparatus was necessary for sustained contractile stress to be generated at the apical ZA (Leerberg et al., 2014).

Together, these findings argue that both nucleation and postnucleation mechanisms may play important roles in actin assembly at cell–cell junctions. This is further attested by the diverse impacts of Arp2/3 and Mena/VASP proteins during development. Arp2/3 and its activator, WASP, were reported to promote E-cadherin endocytosis for junctional remodeling during *Drosophila* morphogenesis (Georgiou, Marinari, Burden, & Baum, 2008;

Leibfried, Fricke, Morgan, Bogdan, & Bellaiche, 2008). Arp2/3 also contributes to junctional maturation and maintenance in the *Caenorhabditis elegans* gastrointestinal epithelium, although in that system other adhesion molecules may play a greater role in their recruitment than cadherin (Bernadskaya, Patel, Hsu, & Soto, 2011). Although Mena/VASP proteins do not appear to be essential during development, their impact becomes more apparent in tissues that are under mechanical stress. For example, mouse embryos lacking all Mena/VASP proteins have a defective endothelial barrier, a tissue that experiences shear forces (Furman et al., 2007), while during *Drosophila* development the contribution of Ena is most evident during dynamic morphogenetic events where cell–cell junctions experience strong forces (Gates et al., 2007, 2009). Together, these observations imply that Ena/VASP proteins may especially contribute to strengthening junctions against stress, potentially by mediating tension-sensitive actin assembly (Leerberg et al., 2014).

3.3. Actin filament stability and junctional contractility

Conversely, what are the mechanisms that challenge junctional actin homeostasis by inducing actin filament turnover? Isolated actin filaments have an intrinsic tendency toward disassembly, which is favored at the pointed (minus) end compared with the barbed end. However, this intrinsic process is often quite slow compared with the turnover rates of actin filaments within cells. Thus, other molecular mechanisms must exist within cells to accelerate actin filament turnover and promote the remodeling of actin networks (Brieher & Yap, 2013). One of these is contractile stress driven by myosin II (Haviv et al., 2008; Medeiros, Burnette, & Forscher, 2006; Reymann et al., 2012). As noted earlier, in epithelial cells, stress-induced filament turnover can limit the contractile tension that is generated at the lateral junctions located below the ZA (Wu et al., 2014). This carries the important implication that mechanisms that allow filaments to resist, or compensate for, stress-induced turnover can influence the active stresses that cells generate at junctions.

Indeed, several such compensatory mechanisms have recently been identified. One is tension-sensitive actin assembly itself; of note, targeting of Mena/VASP to junctions could restore contractile tension at the ZA when vinculin was depleted (Leerberg et al., 2014). A second is N-WASP, a member of the WASP/WAVE protein family that is selectively localized at the ZA in mammalian epithelial cells (Kovacs et al., 2011; Wu et al., 2014).

Although best understood as a regulator of Arp2/3-mediated actin nucleation (Padrick & Rosen, 2010), N-WASP at the ZA appears to function to stabilize actin filaments (Kovacs et al., 2011). This contributes to sustaining contractile tension at this apical junction (Wu et al., 2014). The precise molecular mechanism that allows N-WASP to stabilize F-actin is less clear. It involves the WIP family protein, WIRE, which associates directly with the N-terminal of N-WASP (Kovacs et al., 2011). WIRE can bundle actin filaments (Kato & Takenawa, 2005), an observation that is consistent with increasing evidence that network organization critically influences the sensitivity of F-actin to stress-induced turnover (Reymann et al., 2012). Branched networks appear to be especially sensitive to myosin-induced turnover (Medeiros et al., 2006), perhaps because contractility is more prone to generate compressive forces that buckle F-actin and increase its sensitivity to severing by cofilin. Conversely, actin filaments that are tense, as occur in bundles such as stress fibers, are actually resistant to cofilin (Hayakawa, Tatsumi, & Sokabe, 2011). Whether N-WASP and WIRE stabilize filaments by promoting bundling has yet to be determined. However, it is noteworthy that other bundling proteins, such as α-actinin (Knudsen, Soler, Johnson, & Wheelock, 1995) are recruited to junctions and associate with E-cadherin complexes.

4. REGULATION OF CORTICAL SIGNALING BY CADHERIN ADHESION

In this section, we will consider how cortical signals are regulated at cadherin junctions to influence junctional contractility. We focus on the Rho family of small GTPases, which have diverse and profound contributions to cytoskeletal regulation (Etienne-Manneville & Hall, 2002). Indeed, the best-known members of this family—Rho, Rac, and Cdc42—have all been identified as active at cadherin junctions (Kim, Li, & Sacks, 2000; Ratheesh et al., 2012). For the purposes of the present discussion, we will concentrate on Rho, given its central role in regulating contractility, cadherin biology and in development.

4.1. Rho signaling and morphogenesis

Morphogenesis involves the coordinated self-assembly of cells into multilayered tissues that undergo deformations such as bending, folding, and tube formation to generate specialized three-dimensional structures (Guillot & Lecuit, 2013; Mammoto, Mammoto, & Ingber, 2013).

Rho contributes to many morphogenetic processes during development. Its role has often been identified by experiments that targeted its upstream regulators, thereby identifying Rho as a common effector in many developmental signaling pathways. For example, *DRhoGEF2* (an activator of Rho) is important for cell-shape changes required for gastrulation: deletion of DRhoGEF2 or expression of dominant-negative Rho caused gastrulation to fail in *Drosophila* embryos (Barrett, Leptin, & Settleman, 1997). The junctional localization of DRhoGEF2 protein also appears to be important for its developmental impact in *Drosophila*, as the transcription factor Twist induces the expression of a transmembrane protein, T48, that serves to concentrate RhoGEF2 at cell junctions. The subsequent stimulation of Rho signaling was necessary to initiate mesoderm invagination (Kolsch, Seher, Fernandez-Ballester, Serrano, & Leptin, 2007). This emphasizes the important role that spatial coordination of signaling plays in the developmental impact of Rho. Similarly, the transcription factor Snail acts through its target gene *folded gastrulation* (Fog) to promote Rho-ROCK activity for apical constriction during fly gastrulation (Mammoto et al., 2013).

Rho is also implicated in developmental processes where junction remodeling is coupled to morphogenesis. For example, during the process of germ-band extension in *Drosophila*, cells intercalate to extend the body axis (Guillot & Lecuit, 2013; Mammoto et al., 2013). Here, Rho-mediated activation of ROCK leads to polarized localization of myosin filaments at the junctional interface undergoing shrinkage and at the same time regulates the preferential localization and stability of adherens junction components at the growing interface (Bertet et al., 2004).

The close link among Rho signaling, contractility, and cell–cell junctions is also evident in vertebrates. In chick embryos, Rho, along with its downstream effector myosin, becomes concentrated at the apical surface of neural plate cells, and drugs that inhibited this cascade caused gross disruptions in neural plate morphogenesis (Kinoshita, Sasai, Misaki, & Yonemura, 2008). ROCK has been implicated in neural tube closure, as the neural folds failed to bend and close normally when ROCK-1 signaling was diminished at the apical surfaces of the neural plates (Nishimura & Takeichi, 2008; Wei et al., 2001). Although RhoA null-knockout mice are embryonic lethal (Thumkeo, Watanabe, & Narumiya, 2013), conditional gene targeting has revealed its role in a variety of developing tissues. For example, ROCK-mediated phosphorylation of myosin RLC is required for eyelid closure as gross defects in actin-filament organization, and loss of pRLC was observed in the eyelid epithelium of ROCK-1-knockout mice

(Shimizu et al., 2005). Similarly, using conditional knockout in the lens epithelium, Chauhan, Lou, Zheng, and Lang (2011) reported that mutual antagonism of Rho and Rac modulates apical constriction and the changes in cell width required for epithelial invagination. RhoA, via ROCK, generates active myosin and contractility, which is needed for apical constriction. Thus, when RhoA was deleted, lens pits were open in shape (Chauhan et al., 2011). In the neuroepithelium of mice, ablation of RhoA disrupts adherens junctions and apical–basal cell polarity, causing aberrant proliferation of neuronal progenitors and defects in organization of ventricular region (Herzog et al., 2011; Katayama et al., 2011). Similar defects were seen when the Rho effector, mDia, was disrupted (Thumkeo et al., 2011).

4.2. Rho and cadherin biology

Rho is also essential for the stability and integrity of cadherin-based junctions. Although three Rho family genes are found in mammals, much work has focused on RhoA, as disruption of RhoA signaling perturbs the integrity of cadherin-based adhesions (Braga, 2000; Braga, Machesky, Hall, & Hotchin, 1997; Priya et al., 2013; Ratheesh et al., 2012; Ratheesh & Yap, 2012; Shewan et al., 2005; Smith, Dohn, Brown, & Reynolds, 2012; Takaishi, Sasaki, Kotani, Nishioka, & Takai, 1997). Rho signaling can contribute to adherens junction integrity by stabilizing cadherin receptors, cytoskeleton organization, generation of tension, and possibly facilitating E-cadherin clustering (Braga, Betson, Li, & Lamarche-Vane, 2000; Priya et al., 2013; Ratheesh et al., 2012; Ratheesh, Priya, & Yap, 2013; Shewan et al., 2005; Smutny et al., 2010, 2011).

In the context of our present discussion, myosin II is a major downstream target for Rho to promote cadherin biology. The recruitment and activity of myosin II at junctions responded to the Rho pathway. Inhibition of Rho kinase signaling not only led to the loss of myosin II from junctions but also abolished its activity as suggested by the loss of MLC phosphorylation and decreased junctional staining of ppRLC (Shewan et al., 2005). Moreover, the phenotype observed after ROCK inhibition was very similar to that of myosin inhibition, namely, an inability to concentrate E-cadherin in junctions and reduced adhesion to cadherin-coated substrata (Shewan et al., 2005). These observations placed myosin II as a potent downstream effector that allows Rho to modulate cadherin adhesion (Shewan et al., 2005). Interestingly, as noted earlier, although both myosin IIA and myosin IIB can be found at apical ZA, the junctional concentration of myosin IIA was more dependent on the Rho/ROCK pathway than that of myosin IIB and served to promote cadherin clustering and the concentration of E-cadherin at the apical ZA (Smutny et al., 2010). Thus, a Rho-ROCK-Myosin II pathway can play an integral role in supporting the integrity of the ZA.

However, to perform its biological functions, Rho must undergo a regulated activation/inactivation cycle facilitated by ancillary stimulatory and inhibitory proteins (Jaffe & Hall, 2005). The spatiotemporal activity of these regulatory molecules decides the site and timing of Rho signaling (Bement, Miller, & von Dassow, 2006). An important emerging theme here is the ability of cells to localize contiguous elements of signaling pathways to cadherin junctions. This is well illustrated by the Rho-ROCK pathway. Both active, GTP-loaded Rho and its immediate downstream mediator, Rho-activated kinase (ROCK), are found at cadherin junctions (Priya et al., 2013; Ratheesh et al., 2012; Smith et al., 2012; Figs. 3B and 4A). Inhibition of either Rho or ROCK decreases junctional contractility. Moreover, disruption of mechanisms that activate or recruit Rho or ROCK to junctions also compromises junctional contractility (Ratheesh et al., 2012). This implies that the local regulation of this signaling module plays an important role in regulating junctional actomyosin. Indeed, this notion is teleologically attractive, as it would provide a mechanism to focus contractility at junctions. The important mechanistic issue is how such junctional localization of Rho-ROCK signaling might be achieved.

4.3. Focusing Rho signaling at cell–cell junctions

One major paradigm for the spatial control of Rho signaling lies in the proteins that control its activation and inactivation at the plasma membrane (Etienne-Manneville & Hall, 2002). This reflects two broad processes: the control of Rho nucleotide status and the localization of active Rho at the plasma membrane.

Like other GTPases, the ability of Rho to mediate cellular regulation depends on its nucleotide-loaded status. Binding of GTP induces conformational changes that allow Rho to interact with, and regulate the activity of, downstream effectors (Bustelo, Sauzeau, & Berenjeno, 2007). In contrast, conversion of bound GTP to GDP, mediated by the intrinsic GTPase activity of Rho, disables its ability to interact with its effectors, thereby terminating its signaling. However, the intrinsic nucleotide turnover rate for Rho is slow (half-life of ~30 min), relative to many of the cellular processes that it

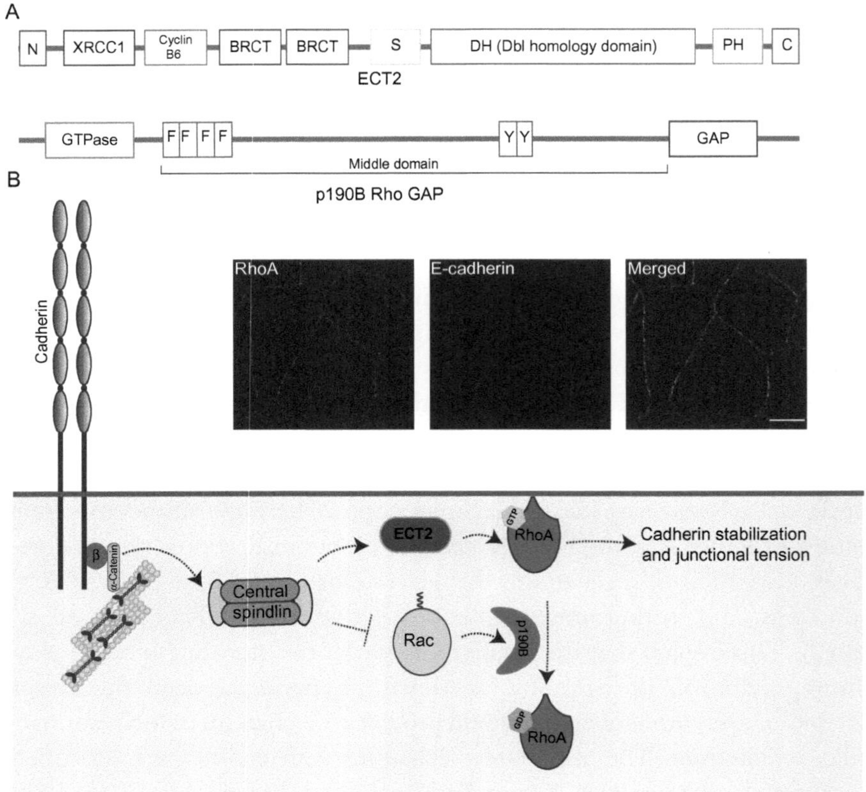

Figure 3 Junctional Rho regulation. (A) Domain structure of ECT2: N, amino-terminal; BRCT, BRAC1 C-terminal domain; XRCC1, X-ray repair complementing defective repair in Chinese hamster cells 1 domain; PH, pleckstrin-homology domain; C, carboxyl-terminal region; S, small central region. Domain structure of p190B Rho GAP: The N-terminus consists of a GTPase-like domain, followed by a relatively less well-characterized middle domain and a GAP domain at the C-terminus; FF, phosphorylation-regulated protein–protein interacting domain; Y, SH2 domain. (B) Rho signaling at the epithelial cadherin junctions. RhoA localizes to E-cadherin-based junctions as illustrated by immunofluorescence in MCF-7 cells. ECT2 acts as a GEF, which can activate Rho at epithelial cadherin junctions, while p190B Rho GAP is the GAP that can inhibit Rho activity. Junctional localization of ECT2 relies on the centralspindlin complex, which can also inhibit Rac and thus prevents p190B GAP localization. The GTP-Rho further supports cadherin stabilization and junctional tension via myosin-ROCK pathway. Scale bar = 10 μm.

controls (Zhang & Zheng, 1998). Instead, the nucleotide status of Rho is further regulated by two sets of proteins. Guanine nucleotide-exchange factors (GEFs) catalyze the exchange of GDP for GTP, thereby activating Rho, and GTPase-activating proteins (GAPs) stimulate nucleotide hydrolysis to

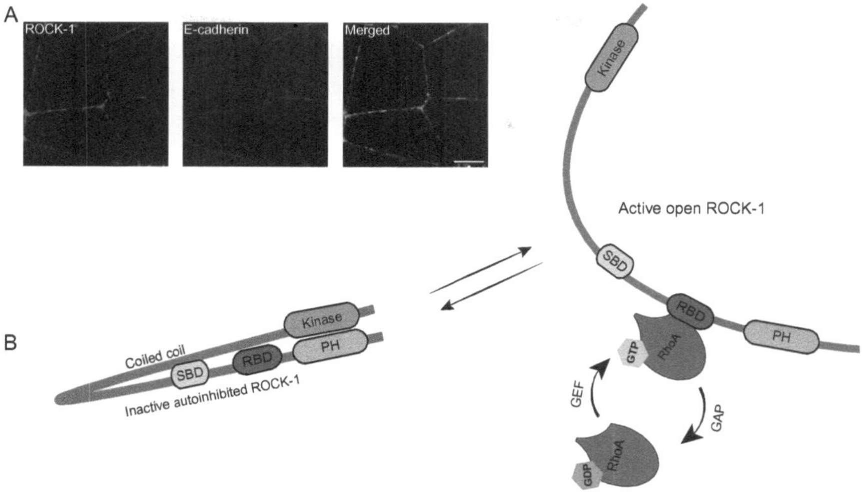

Figure 4 Rho kinase and cell–cell junctions. (A) ROCK-1 localizes to cadherin-based junctions as illustrated by immunofluorescence in MCF-7 cells. Scale bar = 10 μm. (B) The C-terminus of ROCK interacts with the kinase domain at the N-terminus leading to an autoinhibited conformation, which is inactive. Binding of GTP-Rho to the RBD (Rho-binding domain) disrupts this autoinhibition, resulting in an active ROCK. SBD, Shroom-binding domain; RBD, Rho-binding domain; PH, pleckstrin-homology domain.

GDP to inactivate Rho (Bos, Rehmann, & Wittinghofer, 2007; Etienne-Manneville & Hall, 2002).

In addition, Rho signaling typically occurs at cellular membranes. In the context of contractility, this allows GTP-Rho to ultimately activate the contractile apparatus at the cell cortex where it can interact with membrane proteins to control cell shape and motility. Accordingly, Rho bears a C-terminal CAAX box that serves as a lipid-binding membrane anchor. Conversely, inactivation of Rho signaling also involves its removal from the membrane. This is mediated by Guanine nucleotide dissociation inhibitors (GDIs) that extract Rho from membranes and sequester it in its GDP-bound state within the cytosol (Ridley, 2012).

Ultimately, the localized expression of Rho signaling at specific sites within cells is influenced by the action of these three sets of proteins. While it is likely that GDI function will also contribute, coordination of GEF and GAP activity is currently the best-understood way in which localized zones of Rho signaling can be established. Formally, one might envisage that a localized zone of GTP-Rho might be achieved at sites where GEF activity predominates over GAP activity. Evidence for such a model has been identified during cell division, where Rho signaling is focused to support contractility at the cytokinetic furrow (Glotzer, 2005; Piekny, Werner, & Glotzer, 2005). To date, the GEFs and GAPs that regulate Rho signaling at cell–cell junctions are poorly understood. Surprisingly, however, one set involves proteins that are already implicated in cytokinesis, the GEF ECT2 and the GAP p190B RhoGAP (Fig. 3A).

4.4. ECT2

ECT2 belongs to the Dbl family of Rho GEFs and was first identified as a proto-oncogene (Chan et al., 1993). ECT2 is highly conserved across evolution, human ECT2 sharing a significant degree of similarity in its coding region with murine ECT2, *Let-21* (the ECT2 ortholog in *C. elegans*), and *XECT2* (*Xenopus* ECT2) (Fields & Justilien, 2010). ECT2 is composed of various structural domains, each with a distinct role (Fig. 3A). The N-terminal half contains many domains which are common to cell-cycle regulators while the C-terminal half is mainly responsible for the Rho-GEF catalytic activity (Tatsumoto, Xie, Blumenthal, Okamoto, & Miki, 1999). At the extreme N-terminus lies a XRCC1 domain, which shows sequence homology to human XRCC1, a protein involved in DNA repair. The major structural motif in the N-terminal half is a tandem array of BRCT repeats, which are highly conserved in proteins involved in DNA repair and cell-cycle checkpoint responses (Fields & Justilien, 2010). This BRCT motif can bind phosphorylated peptides (Manke, Lowery, Nguyen, & Yaffe, 2003), and it has been suggested that it can also bind to the C-terminal half of ECT2, leading to autoinhibition (discussed further below). The C-terminal half is the catalytic core of the protein and contains a tandem array of Dbl-homology (DH) and pleckstrin-homology (PH) domains (Tatsumoto et al., 1999). The extreme C-terminus region of ECT2 does not exhibit significant homology to any known protein domains or motifs (Fields & Justilien, 2010). The N- and C-terminal domains of ECT2 are separated by a small central S domain, which harbors two nuclear localization sequences that regulate the intracellular localization of ECT2 (Fields & Justilien, 2010).

ECT2 is best understood for its role in cytokinesis (Glotzer, 2005; Piekny et al., 2005; Wolfe & Glotzer, 2009; Wolfe, Takaki, Petronczki, & Glotzer, 2009; Yuce, Piekny, & Glotzer, 2005). RhoA plays the role of chief modulator in cytokinesis by initiating the formation of the contractile ring (Werner & Glotzer, 2008). This involves the Rho-mediated activation of the actin assembly factor, formin, and phosphorylation of the RLC of myosin II. ECT2 acts as a primary GEF for Rho during cytokinesis. It ensures the correct localization and activation of Rho during cell division and is a crucial regulator for cytokinesis in all the metazoans analyzed to date (Werner & Glotzer, 2008; Yuce et al., 2005).

The profound contribution of ECT2 to cytokinesis has tended to obscure any potential extramitotic function that it may have. This, taken with its predominant localization in the nucleus of isolated interphase cells (the common system to study cytokinesis), has led to the notion that ECT2 may principally serve to regulate cytokinesis. Nonetheless, extramitotic functions for ECT2 are beginning to emerge. *Pebbles*, the *Drosophila* homolog of ECT2, contribute to mesoderm migration in the fly embryo (Schumacher, Gryzik, Tannebaum, & Muller, 2004; Smallhorn, Murray, & Saint, 2004), while ECT2 has been identified at the apical junctions of cultured mammalian epithelia when they are grown as monolayers (Liu, Ishida, Raziuddin, & Miki, 2004; Ratheesh et al., 2012). In the latter situation, ECT2 can interact with the Par6 apical polarity complex to influence apical–basal polarity (Liu, Ohno, & Miki, 2006). Additionally, it contributes to stimulating Rho signaling at the ZA. Depletion of ECT2 reduced GTP-Rho signaling at apical cadherin junctions. This was accompanied by loss of myosin II and a decrease in contractile tension at the junctions (Ratheesh et al., 2012). Thus, ECT2 represents one GEF that can support Rho signaling at cell–cell junctions.

4.5. p190 RhoGAP

The p190RhoGAP family comprises two members: p190A and p190B, which are widely expressed in human, rat, fly, and mouse (Chakravarty et al., 2000; Settleman, Narasimhan, Foster, & Weinberg, 1992). These proteins are encoded by separate genes and share 50% sequence homology (Ponik, Trier, Wozniak, Eliceiri, & Keely, 2013). Both contain three major structural features: an N-terminal GTP-binding domain (GBD), a large-middle domain responsible for various protein–protein interactions, and a C-terminal GAP domain (Tcherkezian & Lamarche-Vane, 2007; Fig. 3A). They show GAP activity toward RhoA, Rac 1, and Cdc42, with highest activity for RhoA (Settleman et al., 1992). Of note for this discussion, both p190 RhoGAPs have been identified at cadherin-based cell–cell junctions. p190A RhoGAP was found at N-cadherin-mediated cell–cell junctions, where its recruitment involved an association p120-ctn (Wildenberg et al., 2006). p190B RhoGAP can also be found at epithelial cell–cell junctions (Ratheesh et al., 2012).

Its junctional localization is especially interesting, as p190B has been implicated in regulating epithelial architecture both during morphogenesis and in disease. p190B is highly expressed in the terminal end buds of the mouse mammary gland and its deletion impairs ductal morphogenesis (Chakravarty, Hadsell, Buitrago, Settleman, & Rosen, 2003). On the other hand, overexpression of p190B caused complete disruption of mammary epithelial architecture (Vargo-Gogola, Heckman, Gunther, Chodosh, & Rosen, 2006). The precise expression of p190B activity thus seems to be critical for its impact on tissue organization. p190B is also overexpressed in mammary carcinomas, and its expression can influence mammary tumor progression in mouse models. Overexpression of p190B accelerated tumor development in an MMTV-Neu model system, while haploinsufficency for p190B retarded tumor development in this system (Heckman-Stoddard et al., 2009). Thus, p190B was necessary for epithelial morphogenesis but its overactivity might contribute to tumorigenesis. Given the close relationship between E-cadherin function and dysfunction during epithelial morphogenesis and tumorigenesis, respectively, this raises the interesting possibility that p190 might influence Rho signaling at junctions.

Indeed, both p190A and p190B have been implicated in regulating Rho signaling at junctions. Early studies demonstrated that acute engagement of Xenopus C-cadherin induced a transient decrease in Rho signaling through p190, although the precise paralog responsible was not identified (Noren, Arthur, & Burridge, 2003; Noren, Liu, Burridge, & Kreft, 2000). Subsequently, Wildenberg et al. (2006) reported that p190A RhoGAP could be recruited to N-cadherin-based adhesions through an association with p120-ctn, leading to local inactivation of the Rho-ROCK pathway. In epithelial cells, recruitment of p190B to apical cadherin junctions also contributes to inactivating Rho (Ratheesh et al., 2012). Therefore, both p190A and p190B RhoGAP may serve to inactivate Rho at cadherin junctions, perhaps with different contributions in different cellular contexts.

4.6. Coordinating GEF and GAP activity

Formally, it might be predicted that the generation of a Rho signaling zone would necessitate coordinated regulation of both GEF and GAP activities. Recruitment and/or activation of a GEF(s) would stimulate Rho signaling, while GAP activity would need to be simultaneously antagonized to prevent the premature inactivation of Rho. This paradigm has already been developed for the Rho zone of the cytokinetic furrow (Mikawa, Su, & Parsons, 2008). Interestingly, a common mechanism—the centralspindlin complex—is also implicated in supporting the Rho zone at ZA (Fig. 3B).

Centralspindlin is an evolutionarily conserved protein complex, which was first identified at the central spindle of the mitotic apparatus. It is a tetrameric complex comprising a dimer of MKLP-1/ZEN-4, a member of the kinesin family of microtubule-binding proteins, and a dimer of hCYK-4/MgcRacGAP (Mishima, Kaitna, & Glotzer, 2002). It concentrates at the contractile furrow during cytokinesis, a process that reflects multiple molecular mechanisms (Glotzer, 2005; Somers & Saint, 2003). These include association with the microtubules of the central spindle, likely mediated by MKLP-1/ZEN-4, and binding of hCYK-4/MgcRacGAP to phosphoinositides at the inner leaflet of the plasma membrane (Glotzer, 2009). The junctional localization of centralspindlin at the ZA also requires dynamic microtubules and binding of centralspindlin to the N-terminus of α-catenin (Ratheesh et al., 2012). Importantly, depletion of centralspindlin reduced Rho signaling at the ZA and decreased junctional contractility (Ratheesh et al., 2012). Similar effects occurred when the mechanisms responsible for recruiting centralspindlin to junctions were perturbed. Thus, centralspindlin controls Rho signaling to support contractility at interphase cadherin junctions as it does to support the contractile furrow during cytokinesis. At the ZA, as well as in cytokinesis, localization of centralspindlin appears to be regulated through coincidence detection (Carlton & Cullen, 2005), where multiple mechanisms must coincide to ensure that centralspindlin is localized to the correct place with the cell with both spatial and temporal fidelity.

Once it is correctly localized to the cortex, centralspindlin can regulate Rho through ECT2. CYK-4/MgcRacGAP can bind directly to the N-terminal BRCT domain of ECT2, an interaction, which is responsible for recruiting ECT2 to the central spindle during cytokinesis (Yuce et al., 2005). In the absence of centralspindlin, ECT2 does not localize either to the central spindle during cell division or to the ZA of interphase epithelial monolayers (Ratheesh et al., 2012; Yuce et al., 2005). In cell division, this association is regulated by protein phosphorylation, as Polo-like kinase 1 (PLK1) phosphorylates CYK-4 to generate a phophoepitope that is recognized by the BRCT domain of ECT2 (Wolfe et al., 2009). This interaction may also serve to activate ECT2, by relieving its autoinhibitory conformation. The C-terminus of ECT2 may also contribute to its cortical localization during cytokinesis (Su, Takaki, & Petronczki, 2011), but this also appears to be downstream of the influence of centralspindlin and protein phosphorylation by CDK1. Therefore, centralspindlin may lie at the top of a signaling pathway that activates Rho through ECT2 to support junctional contractility. However, what protein kinases may regulate centralspindlin and ECT2 at junctions remain to be identified.

In addition to promoting Rho activation at the ZA, centralspindlin can also antagonize the junctional accumulation of p190B RhoGAP, which increased when either MLKP1 or hCYK-4/MgcRacGAP was depleted. This accumulation of p190B contributed to the fall in junctional Rho signaling when centralspindlin was perturbed, as the change in Rho could be ameliorated by simultaneous depletion of p190B. Therefore, centralspindlin can support junctional Rho signaling by both recruiting a GEF and antagonizing an inactivator (Ratheesh et al., 2012).

How, then, does centralspindlin antagonize junctional p190B? One possibility is that centralspindlin blocks the mechanism responsible for recruiting p190B to junctions. One such recruitment mechanism is signaling by another Rho family GTPase, Rac, which can bind directly to p190B and promote its cortical localization (Bustos, Forget, Settleman, & Hansen, 2008; Wildenberg et al., 2006). Indeed, depletion of centralspindlin increased Rac signaling at junctions and inhibition of Rac blocked the junctional accumulation of p190B. Presumably, then, centralspindlin limits the junctional accumulation of p190B by dampening Rac signaling. Exactly how centralspindlin may inhibit junctional Rac signaling is not understood; however, CYK-4 itself is a candidate, as it possesses Rac GAP activity (Jantsch-Plunger et al., 2000). Interestingly, Rac signaling also promoted the recruitment of p190A RhoGAP to N-cadherin junctions (Wildenberg et al., 2006). However, this did not appear to involve direct binding of GTP-Rac to p190A. Instead, Rac-induced generation of reactive oxygen species promoted tyrosine phosphorylation of p190A and its associated protein, p120-RhoGAP, by inhibiting a protein tyrosine phosphatase (Noren et al., 2003). Whether centralspindlin may also participate in regulating the junctional recruitment of p190A RhoGAP has yet to be investigated.

4.7. Other mechanisms

While current experience with ECT2 and p190B illustrates the capacity for coordination of GEFS and GAPs, it is likely that other combinations of GEFs and GAPs contribute to regulating junctional Rho signaling in other circumstances. Nor is control of the Rho GTPase cycle the only way in which its spatial expression can be influenced. Potentially, other Rho-binding proteins may contribute by limiting the diffusion of GTP-Rho and/or promoting its localization with GEFs and effectors. An interesting example is provided by anillin, a scaffolding protein that is found at the cytokinetic furrow and was also recently identified at cell–cell junctions (Piekny & Glotzer, 2008; Reyes et al., 2014). Anillin can bind directly to several proteins that regulate contractility, including Rho itself, ECT2, and myosin II (Piekny & Maddox, 2010). Depletion of anillin in Xenopus embryos perturbed the integrity of cadherin-based cell–cell junctions, associated with unstable bursts of GTP-Rho signaling at those junctions (Reyes et al., 2014). Whether that aberrant Rho signaling was responsible for the loss of junctional integrity remains to be determined.

4.8. Localizing ROCK to junctions

Spatial regulation of signaling is not confined to the upstream elements of the Rho signaling pathway. Instead, it is increasingly evident that it can extend to mediators of Rho signaling, including the ROCK itself. Consistent with the notion that coincidence detection can define the spatial rigor of Rho expression (Carlton & Cullen, 2005) ROCK localization is also determined by multiple mechanisms.

ROCK was among the first Rho effectors to have been identified. Two ROCK isoforms (ROCK-1 and ROCK-2) are found in mammalian cells (Amano, Nakayama, & Kaibuchi, 2010; Julian & Olson, 2014; Riento & Ridley, 2003). These share a broadly similar primary structure consisting of an N-terminal kinase domain, a central coiled-coil domain, and a C-terminal region which bears both the GTP-Rho-binding domain (RBD), and a lipid-binding PH domain (Fig. 4B). The kinase domain of both isoforms is highly conserved at the amino-acid level (92% identity), while overall they share 62% identity. GTP-bound Rho activates ROCK by altering its intramolecular conformation. In the absence of GTP-Rho, ROCK exists in an autoinhibited state where an intramolecular association between the C-terminal region and the N-terminal kinase domain inhibits ROCK activity. Binding of GTP-Rho to the RBD then releases this autoinhibitory conformation, thus activating the kinase (Fig. 4B; Amano et al., 2010; Riento & Ridley, 2003). As noted earlier, ROCK plays a canonical role in mediating the regulation of myosin II by Rho, including that at cadherin junctions.

Since the kinase domain is highly conserved between these two isoforms, it has often been assumed that they phosphorylate a similar range of substrates. However, it is increasingly apparent that they differ in their expression and protein–protein interactions, suggesting that they may have different biological functions (Amano et al., 2010; Sandquist & Means, 2008; Sandquist, Swenson, Demali, Burridge, & Means, 2006; Schofield & Bernard, 2013). Interestingly, these differences extend to their subcellular localization. ROCK-2 has been reported to localize in the cytoplasm, globally at the plasma membrane and also at the cleavage furrow during cytokinesis (Schofield & Bernard, 2013). In contrast, recent studies have reported the selective concentration of ROCK-1 at the apical junctions of epithelial cells (Smith et al., 2012). This suggests that mechanisms exist to selectively colocalize GTP-Rho with this key effector at cadherin junctions, thereby establishing a Rho-ROCK-signaling module at the junctions.

ROCK bears multiple binding domains that can influence its cortical localization. The PH domain appears to be necessary for its association with the membrane, but was not sufficient to confer concentration at cadherin junctions (Schofield & Bernard, 2013). Other coincident factors therefore appear to collaborate to concentrate ROCK at junctions. One of these is likely to be GTP-Rho itself, as deletion of the RBD generated a mutant ROCK that could associate with the membrane but was unable to concentrate at cadherin junctions (Simoes Sde, Mainieri, & Zallen, 2014).

There is also emerging evidence that junctional proteins themselves can contribute to localizing ROCK-1. One interesting role is played by the adaptor protein Shroom, which was first identified in a screen for mouse mutants that disrupted neural tube closure (Hildebrand & Soriano, 1999). Subsequent characterization identified a single Shroom gene in *Drosophila* and four Shroom isoforms (Shroom 1–4) in mammalian cells. Despite their number, many of these mammalian Shroom genes participate in regulating processes that involve contractility at junctions, including apical constriction in embryonic epithelia and junctional contractility in endothelia (Bolinger, Zasadil, Rizaldy, & Hildebrand, 2010; Hildebrand, 2005; Simoes Sde et al., 2014). Shroom is an actin-binding protein (Hildebrand & Soriano, 1999) that localizes to cell–cell junctions, including cadherin junctions through an association with the cadherin-associated protein, p120-catenin (p120-ctn) (Lang, Herman, Reynolds, Hildebrand, & Plageman, 2014). Its contribution to junctional contractility may reflect in part its ability to recruit both F-actin and myosin II to junctions (Hildebrand, 2005). However, an independent contribution to contractility comes from its additional ability to interact directly with ROCK. The junctional accumulation of ROCK could be promoted or reduced by the expression or depletion of Shroom, respectively (Nishimura & Takeichi, 2008). Further, dominant-negative mutants of Shroom that disrupt the interaction between ROCK and endogenous Shroom perturb junctional tension and apical constriction (Nishimura & Takeichi, 2008).

Shroom may then collaborate with Rho in recruiting ROCK to junctions, an influence that appears to be closely tied to cell polarity (Simoes Sde et al., 2014). Not only does Shroom contribute to localizing ROCK at the ZA in epithelial cells that have undergone apical–basal polarization, but also it participates in regulating junctional signaling during planar polarization. This is demonstrated when *Drosophila* embryos undergo cellular rearrangements during axis elongation. At this stage, ROCK is not only accumulated at apical adherens junctions but also distributed in a planar polarized pattern to promote the remodeling of cell–cell junctions. Rho is necessary for this process, because of its ability to recruit ROCK to junctions, but Shroom also contributes by conferring planar polarized distribution on junctional ROCK (Simoes Sde et al., 2014).

Thus, the coordinated spatial localization of ROCK and Rho is a characteristic feature that contributes to contractility at cadherin junctions. It is likely that we have only begun to unravel the ways in which this coordination is established and how it affects contractility. For example, it remains to be determined whether the ability of p120-ctn to bind ROCK (Smith et al., 2012) is solely mediated by Shroom, or whether other interactions also participate. Furthermore, ROCK can also potentially feedback to influence Rho signaling by regulating p190 RhoGAPs. Both p190A and p190B are recruited to the plasma membrane by association with members of a family of atypical GTPases, Rnd1 and Rnd3, respectively (Oinuma, Kawada, Tsukagoshi, & Negishi, 2012; Wennerberg et al., 2003). These recruitment mechanisms are disrupted by ROCK, albeit by different molecular mechanisms (Riento et al., 2005). ROCK can phosphorylate p190A directly, thereby disrupting its interaction with Rnd1 (Mori et al., 2009). In contrast, ROCK does not appear to directly target p190B but, instead, phosphorylates Rnd3 to sequester it in the cytosol, preventing it from associating with the plasma membrane (Riento et al., 2005; Riou et al., 2013). By either pathway, ROCK antagonizes the membrane accumulation of p190, thereby potentially sustaining Rho signaling. However, whether these affect Rho signaling at cadherin junctions remains to be explored.

5. CLOSING COMMENTS

In conclusion, cell–cell junctions can be considered as active mechanical agents, where many molecular pathways are coordinated to ultimately integrate adhesion with the actomyosin contractile apparatus. The product, contractile tension at junctions, may then be regarded as the emergent property of an extremely complex system. An interesting challenge for the future will be to develop approaches to make sense of this complexity: defining the key architecture(s) of the system and parsing the rules that regulate it. One issue will be to understand whether (and if so, how) levels of tension may be regulated. It might be supposed, for example, that established tissues at homeostasis would display relatively stable levels of tension, compared with morphogenetically active tissues that are undergoing neighbor exchange or apical constriction. We will need to develop approaches to quantitatively characterize contractile tension in these different contexts as the basis for understanding their regulation. Finally, although we have focused on the basic cell biology of junctional contractility, what we learn is likely to have implications for understanding disease as well as development. It is interesting to note that a regulator of junctional Rho, such as p190B, is implicated in cancer. Whether this impact occurs through dysregulation of junctional contractility is an interesting, open question. Our growing understanding of the complex mechanisms that make junctions active mechanical agents then holds the promise of informing our understanding of cell biology in disease.

ACKNOWLEDGMENTS

Our work was supported by the National Health and Medical Research Council of Australia (1044041, 1037320) and an ANZ Trustees Ph.D. Scholarship in Medical Research to R. P.

REFERENCES

Altman, D., Sweeney, H. L., & Spudich, J. A. (2004). The mechanism of myosin VI translocation and its load-induced anchoring. *Cell*, *116*, 737–749.

Amano, M., Ito, M., Kimura, K., Fukata, Y., Chihara, K., Nakano, T., et al. (1996). Phosphorylation and activation of myosin by Rho-associated kinase (Rho-kinase). *The Journal of Biological Chemistry*, *271*(34), 20246–20249.

Amano, M., Nakayama, M., & Kaibuchi, K. (2010). Rho-kinase/ROCK: A key regulator of the cytoskeleton and cell polarity. *Cytoskeleton (Hoboken)*, *67*(9), 545–554. http://dx.doi.org/10.1002/cm.20472.

Barrett, K., Leptin, M., & Settleman, J. (1997). The rho GTPase and a putative RhoGEF mediate a signaling pathway for the cell shape changes in Drosophila gastrulation. *Cell*, *91*(7), 905–915. http://dx.doi.org/10.1016/S0092-8674(00)80482-1.

Barzik, M., Kotova, T. I., Higgs, H. N., Hazelwood, L., Hanein, D., Gertler, F. B., et al. (2005). Ena/VASP proteins enhance actin polymerization in the presence of barbed end capping proteins. *The Journal of Biological Chemistry*, *280*(31), 28653–28662.

Bear, J. E., Svitkina, T. M., Krause, M., Schafer, D. A., Maly, I. V., Chaga, O. Y., et al. (2002). Antagonism between Ena/VASP proteins and actin filament capping regulates fibroblast motility. *Cell*, *109*, 509–521.

Bement, W. M., Miller, A. L., & von Dassow, G. (2006). Rho GTPase activity zones and transient contractile arrays. *Bioessays*, *28*(10), 983–993.

Bernadskaya, Y. Y., Patel, F. B., Hsu, H. T., & Soto, M. C. (2011). Arp2/3 promotes junction formation and maintenance in the Caenorhabditis elegans intestine by regulating membrane association of apical proteins. *Molecular Biology of the Cell*, *22*(16), 2886–2899. http://dx.doi.org/10.1091/Mbc.E10-10-0862.

Bertet, C., Sulak, L., & Lecuit, T. (2004). Myosin-dependent junction remodelling controls planar cell intercalation and axis elongation. *Nature*, *429*(6992), 667–671. http://dx.doi.org/10.1038/Nature02590.

Bolinger, C., Zasadil, L., Rizaldy, R., & Hildebrand, J. D. (2010). Specific isoforms of Drosophila shroom define spatial requirements for the induction of apical constriction. *Developmental Dynamics*, *239*(7), 2078–2093. http://dx.doi.org/10.1002/dvdy.22326.

Bos, J. L., Rehmann, H., & Wittinghofer, A. (2007). GEFs and GAPs: Critical elements in the control of small G proteins. *Cell*, *129*(5), 865–877. http://dx.doi.org/10.1016/j.cell.2007.05.018.

Braga, V. (2000). Epithelial cell shape: Cadherins and small GTPases. *Experimental Cell Research*, *261*(1), 83–90. http://dx.doi.org/10.1006/excr.2000.5050, S0014-4827(00)95050-X [pii].

Braga, V. M., Betson, M., Li, X., & Lamarche-Vane, N. (2000). Activation of the small GTPase Rac is sufficient to disrupt cadherin-dependent cell-cell adhesion in normal human keratinocytes. *Molecular Biology of the Cell*, *11*(11), 3703–3721.

Braga, V. M., Machesky, L. M., Hall, A., & Hotchin, N. A. (1997). The small GTPases Rho and Rac are required for the establishment of cadherin-dependent cell-cell contacts. *The Journal of Cell Biology*, *137*(6), 1421–1431.

Brieher, W. M., & Yap, A. S. (2013). Cadherin junctions and their cytoskeleton(s). *Current Opinion in Cell Biology*, *25*(1), 39–46. http://dx.doi.org/10.1016/j.ceb.2012.10.010.

Buckley, C. D., Tan, J. L., Anderson, K. L., Hanein, D., Volkmann, N., Weis, W. I., et al. (2014). The minimal cadherin-catenin complex binds to actin filaments under force. *Science*, *346*, 1254211–1254218.

Bustelo, X. R., Sauzeau, V., & Berenjeno, I. M. (2007). GTP-binding proteins of the Rho/Rac family: Regulation, effectors and functions in vivo. *Bioessays*, *29*(4), 356–370. http://dx.doi.org/10.1002/bies.20558.

Bustos, R. I., Forget, M. A., Settleman, J. E., & Hansen, S. H. (2008). Coordination of Rho and Rac GTPase function via p190B RhoGAP. *Current Biology*, *18*(20), 1606–1611. http://dx.doi.org/10.1016/J.Cub.2008.09.019.

Carlton, J. G., & Cullen, P. J. (2005). Coincidence detection in phosphoinositide signaling. *Trends in Cell Biology*, *15*(10), 540–547. http://dx.doi.org/10.1016/j.tcb.2005.08.005.

Carramusa, L., Ballestrem, C., Zilberman, Y., & Bershadsky, A. D. (2007). Mammalian diaphanous-related formin Dia1 controls the organization of E-cadherin-mediated cell-cell junctions. *Journal of Cell Science*, *120*(21), 3870–3882. http://dx.doi.org/10.1242/Jcs.014365.

Cavey, M., Rauzi, M., Lenne, P. F., & Lecuit, T. (2008). A two-tiered mechanism for stabilization and immobilization of E-cadherin. *Nature*, *453*(7196), 751–756. http://dx.doi.org/10.1038/nature06953.

Chakravarty, G., Hadsell, D., Buitrago, W., Settleman, J., & Rosen, J. M. (2003). p190-B RhoGAP regulates mammary ductal morphogenesis. *Molecular Endocrinology, 17*(6), 1054–1065. http://dx.doi.org/10.1210/me.2002-0428.

Chakravarty, G., Roy, D., Gonzales, M., Gay, J., Contreras, A., & Rosen, J. M. (2000). p190-B, a Rho-GTPase-activating protein, is differentially expressed in terminal end buds and breast cancer. *Cell Growth & Differentiation, 11*(7), 343–354.

Chan, A. M., Fleming, T. P., McGovern, E. S., Chedid, M., Miki, T., & Aaronson, S. A. (1993). Expression cDNA cloning of a transforming gene encoding the wild-type G alpha 12 gene product. *Molecular and Cellular Biology, 13*(2), 762–768.

Chauhan, B. K., Lou, M., Zheng, Y., & Lang, R. A. (2011). Balanced Rac1 and RhoA activities regulate cell shape and drive invagination morphogenesis in epithelia. *Proceedings of the National Academy of Sciences of the United States of America, 108*(45), 18289–18294. http://dx.doi.org/10.1073/pnas.1108993108.

Chesarone, M. A., & Goode, B. L. (2009). Actin nucleation and elongation factors: Mechanisms and interplay. *Current Opinion in Cell Biology, 21*(1), 28–37. http://dx.doi.org/10.1016/j.ceb.2008.12.001, S0955-0674(08)00198-1 [pii].

Choi, H. J., Pokutta, S., Cadwell, G. W., Bobkov, A. A., Bankston, L. A., Liddington, R. C., et al. (2012). αE-catenin is an autoinhibited molecule that coactivates vinculin. *Proceedings of the National Academy of Sciences of the United States of America, 109*, 8576–8581. http://dx.doi.org/10.1073/pnas.1203906109.

Chuan, P., Spudich, J. A., & Dunn, A. R. (2011). Robust mechanosensing and tension generation by myosin VI. *Journal of Molecular Biology, 405*(1), 105–112. http://dx.doi.org/10.1016/j.jmb.2010.10.010.

Craig, R., Smith, R., & Kendrick-Jones, J. (1983). Light-chain phosphorylation controls the conformation of vertebrate non-muscle and smooth muscle myosin molecules. *Nature, 302*(5907), 436–439.

Drees, F., Pokutta, S., Yamada, S., Nelson, W. J., & Weis, W. I. (2005). Alpha-catenin is a molecular switch that binds E-cadherin-beta-catenin and regulates actin-filament assembly. *Cell, 123*(5), 903–915.

Ebrahim, S., Fujita, T., Millis, B. A., Kozin, E., Ma, X., Kawamoto, S., et al. (2013). NMII forms a contractile transcellular sarcomeric network to regulate apical cell junctions and tissue geometry. *Current Biology, 23*(8), 731–736. http://dx.doi.org/10.1016/j.cub.2013.03.039.

Etienne-Manneville, S., & Hall, A. (2002). Rho GTPases in cell biology. *Nature, 420*(6916), 629–635. http://dx.doi.org/10.1038/nature01148.

Fernandez-Gonzalez, R., Simoes Sde, M., Roper, J. C., Eaton, S., & Zallen, J. A. (2009). myosin II dynamics are regulated by tension in intercalating cells. *Developmental Cell, 17*(5), 736–743. http://dx.doi.org/10.1016/j.devcel.2009.09.003.

Fernandez-Gonzalez, R., & Zallen, J. A. (2009). Cell mechanics and feedback regulation of actomyosin networks. *Science Signaling, 2*(101), pe78. http://dx.doi.org/10.1126/scisignal.2101pe78.

Fields, A. P., & Justilien, V. (2010). The guanine nucleotide exchange factor (GEF) ECT2 is an oncogene in human cancer. *Advances in Enzyme Regulation, 50*(1), 190–200. http://dx.doi.org/10.1016/j.advenzreg.2009.10.010, S0065-2571(09)00067-3 [pii].

Furman, C., Sieminski, A. L., Kwiatkowski, A. V., Rubinson, D. A., Vasile, E., Bronson, R. T., et al. (2007). Ena/VASP is required for endothelial barrier function in vivo. *Journal of Cell Biology, 179*(4), 761–775. http://dx.doi.org/10.1083/Jcb.200705002.

Gates, J., Mahaffey, J. P., Rogers, S. L., Emerson, M., Rogers, E. M., Sottile, S. L., et al. (2007). Enabled plays key roles in embryonic epithelial morphogenesis in Drosophila. *Development, 134*(11), 2027–2039. http://dx.doi.org/10.1242/dev.02849.

Gates, J., Nowotarski, S. H., Yin, H., Mahaffey, J. P., Bridges, T., Herrera, C., et al. (2009). Enabled and Capping protein play important roles in shaping cell behavior during Drosophila oogenesis. *Developmental Biology*, *333*(1), 90–107. http://dx.doi.org/10.1016/j.ydbio.2009.06.030.

Georgiou, M., Marinari, E., Burden, J., & Baum, B. (2008). Cdc42, Par6, and aPKC regulate Arp2/3-mediated endocytosis to control local adherens junction stability. *Current Biology*, *18*(21), 1631–1638. http://dx.doi.org/10.1016/j.cub.2008.09.029.

Glotzer, M. (2005). The molecular requirements for cytokinesis. *Science*, *307*(5716), 1735–1739. http://dx.doi.org/10.1126/science.1096896, 307/5716/1735 [pii].

Glotzer, M. (2009). The 3Ms of central spindle assembly: Microtubules, motors and MAPs. *Nature Reviews Molecular Cell Biology*, *10*(1), 9–20. http://dx.doi.org/10.1038/nrm2609, nrm2609 [pii].

Gomez, G. A., McLachlan, R. W., & Yap, A. S. (2011). Productive tension: Force-sensing and homeostasis of cell-cell junctions. *Trends in Cell Biology*, *21*(9), 499–505. http://dx.doi.org/10.1016/j.tcb.2011.05.006.

Guillot, C., & Lecuit, T. (2013). Mechanics of epithelial tissue homeostasis and morphogenesis. *Science*, *340*(6137), 1185–1189. http://dx.doi.org/10.1126/science.1235249.

Han, S. P., Gambin, Y., Gomez, G. A., Verma, S., Giles, N., Michael, M., et al. (2014). Cortactin scaffolds Arp2/3 and WAVE2 at the epithelial zonula adherens. *Journal of Biological Chemistry*, *289*(11), 7764–7775. http://dx.doi.org/10.1074/Jbc.M113.544478.

Hansen, S. D., & Mullins, R. D. (2010). VASP is a processive actin polymerase that requires monomeric actin for barbed end association. *The Journal of Cell Biology*, *191*(3), 571–584. http://dx.doi.org/10.1083/jcb.201003014.

Haviv, L., Gillo, D., Backouche, F., & Bernheim-Groswasser, A. (2008). A cytoskeletal demolition worker: myosin II acts as an actin depolymerization agent. *Journal of Molecular Biology*, *375*(2), 325–330. http://dx.doi.org/10.1016/J.Jmb.2007.09.066.

Hayakawa, K., Tatsumi, H., & Sokabe, M. (2011). Actin filaments function as a tension sensor by tension-dependent binding of cofilin to the filament. *Journal of Cell Biology*, *195*(5), 721–727. http://dx.doi.org/10.1083/Jcb.201102039.

He, B., Doubrovinski, K., Polyakov, O., & Wieschaus, E. (2014). Apical constriction drives tissue-scale hydrodynamic flow to mediate cell elongation. *Nature*, *508*(7496), 392–396. http://dx.doi.org/10.1038/nature13070.

Heckman-Stoddard, B. M., Vargo-Gogola, T., McHenry, P. R., Jiang, V., Herrick, M. P., Hilsenbeck, S. G., et al. (2009). Haploinsufficiency for p190B RhoGAP inhibits MMTV-Neu tumor progression. *Breast Cancer Research*, *11*(4), R61. http://dx.doi.org/10.1186/bcr2352.

Heissler, S. M., & Manstein, D. J. (2013). Nonmuscle myosin-2: Mix and match. *Cellular and Molecular Life Sciences*, *70*(1), 1–21. http://dx.doi.org/10.1007/s00018-012-1002-9.

Herzog, D., Loetscher, P., van Hengel, J., Knusel, S., Brakebusch, C., Taylor, V., et al. (2011). The small GTPase RhoA is required to maintain spinal cord neuroepithelium organization and the neural stem cell pool. *Journal of Neuroscience*, *31*(13), 5120–5130. http://dx.doi.org/10.1523/Jneurosci. 4807-10.2011.

Higashida, C., Kiuchi, T., Akiba, Y., Mizuno, H., Maruoka, M., Narumiya, S., et al. (2013). F- and G-actin homeostasis regulates mechanosensitive actin nucleation by formins. *Nature Cell Biology*, *15*(4), 395–405. http://dx.doi.org/10.1038/ncb2693.

Hildebrand, J. D. (2005). Shroom regulates epithelial cell shape via the apical positioning of an actomyosin network. *Journal of Cell Science*, *118*(Pt 22), 5191–5203. http://dx.doi.org/10.1242/jcs.02626.

Hildebrand, J. D., & Soriano, P. (1999). Shroom, a PDZ domain-containing actin-binding protein, is required for neural tube morphogenesis in mice. *Cell*, *99*(5), 485–497. http://dx.doi.org/10.1016/S0092-8674(00)81537-8.

Huveneers, S., Oldenburg, J., Spanjaard, E., van der Krogt, G., Grigoriev, I., Akhmanova, A., et al. (2012). Vinculin associates with endothelial VE-cadherin junctions to control force-dependent remodeling. *The Journal of Cell Biology*, *196*(5), 641–652. http://dx.doi.org/10.1083/jcb.201108120.

Insall, R. H., & Machesky, L. M. (2009). Actin dynamics at the leading edge: From simple machinery to complex networks. *Developmental Cell*, *17*(3), 310–322. http://dx.doi.org/10.1016/J.Devcel.2009.08.012.

Jaffe, A. B., & Hall, A. (2005). Rho GTPases: Biochemistry and biology. *Annual Review of Cell and Developmental Biology*, *21*, 247–269. http://dx.doi.org/10.1146/annurev.cellbio.21.020604.150721.

Jantsch-Plunger, V., Gonczy, P., Romano, A., Schnabel, H., Hamill, D., Schnabel, R., et al. (2000). CYK-4: A Rho family gtpase activating protein (GAP) required for central spindle formation and cytokinesis. *The Journal of Cell Biology*, *149*(7), 1391–1404.

Julian, L., & Olson, M. F. (2014). Rho-associated coiled-coil containing kinases (ROCK): Structure, regulation, and functions. *Small GTPases*, *5*, e29846. http://dx.doi.org/10.4161/sgtp.29846.

Kasza, K. E., & Zallen, J. A. (2011). Dynamics and regulation of contractile actin-myosin networks in morphogenesis. *Current Opinion in Cell Biology*, *23*(1), 30–38. http://dx.doi.org/10.1016/j.ceb.2010.10.014.

Katayama, K. I., Melendez, J., Baumann, J. M., Leslie, J. R., Chauhan, B. K., Nemkul, N., et al. (2011). Loss of RhoA in neural progenitor cells causes the disruption of adherens junctions and hyperproliferation. *Proceedings of the National Academy of Sciences of the United States of America*, *108*(18), 7607–7612. http://dx.doi.org/10.1073/Pnas.1101347108.

Kato, M., & Takenawa, T. (2005). WICH, a member of WASP-interacting protein family, cross-links actin filaments. *Biochemical and Biophysical Research Communications*, *328*(4), 1058–1066. http://dx.doi.org/10.1016/j.bbrc.2005.01.058.

Kim, S. H., Li, Z., & Sacks, D. B. (2000). E-cadherin-mediated cell-cell attachment activates Cdc42. *The Journal of Biological Chemistry*, *275*, 36999–37005.

Kimura, K., Ito, M., Amano, M., Chihara, K., Fukata, Y., Nakafuku, M., et al. (1996). Regulation of myosin phosphatase by Rho and Rho-associated kinase (Rho-kinase). *Science*, *273*(5272), 245–248.

Kinoshita, N., Sasai, N., Misaki, K., & Yonemura, S. (2008). Apical accumulation of rho in the neural plate is important for neural plate cell shape change and neural tube formation. *Molecular Biology of the Cell*, *19*(5), 2289–2299. http://dx.doi.org/10.1091/Mbc.E07-12-1286.

Knudsen, K. A., Soler, A. P., Johnson, K. R., & Wheelock, M. J. (1995). Interaction of alpha-actinin with the cadherin/catenin cell-cell adhesion complex via alpha-catenin. *The Journal of Cell Biology*, *130*(1), 67–77.

Kobielak, A., Pasolli, H. A., & Fuchs, E. (2004). Mammalian formin-1 participates in adherens junctions and polymerization of linear actin cables. *Nature Cell Biology*, *6*(1), 21–30. http://dx.doi.org/10.1038/ncb1075.

Kolsch, V., Seher, T., Fernandez-Ballester, G. J., Serrano, L., & Leptin, M. (2007). Control of Drosophila gastrulation by apical localization of adherens junctions and RhoGEF2. *Science*, *315*(5810), 384–386. http://dx.doi.org/10.1126/Science.1134833.

Kovacs, E. M., Goodwin, M., Ali, R. G., Paterson, A. D., & Yap, A. S. (2002). Cadherin-directed actin assembly: E-cadherin physically associates with the Arp2/3 complex to direct actin assembly in nascent adhesive contacts. *Current Biology*, *12*(5), 379–382.

Kovacs, E. M., Verma, S., Ali, R. G., Ratheesh, A., Hamilton, N. A., Akhmanova, A., et al. (2011). N-WASP regulates the epithelial junctional actin cytoskeleton through a non-canonical post-nucleation pathway. *Nature Cell Biology*, *13*(8), 934–943. http://dx.doi.org/10.1038/Ncb2290.

Kraemer, A., Goodwin, M., Verma, S., Yap, A. S., & Ali, R. G. (2007). Rac is a dominant regulator of cadherin-directed actin assembly that is activated by adhesive ligation independently of Tiam1. *American Journal of Physiology. Cell Physiology, 292*(3), C1061–C1069.

Krause, M., Dent, E. W., Bear, J. E., Loureiro, J. J., & Gertler, F. B. (2003). ENA/VASP proteins: Regulators of the actin cytoskeleton and cell migration. *Annual Review of Cell and Developmental Biology, 19*, 541–564. http://dx.doi.org/10.1146/Annurev.Cellbio.19.050103.103356.

Ladoux, B., Anon, E., Lambert, M., Rabodzey, A., Hersen, P., Buguin, A., et al. (2010). Strength dependence of cadherin-mediated adhesions. *Biophysical Journal, 98*(4), 534–542. http://dx.doi.org/10.1016/j.bpj.2009.10.044.

Lang, R. A., Herman, K., Reynolds, A. B., Hildebrand, J. D., & Plageman, T. F. (2014). p120-Catenin-dependent junctional recruitment of Shroom3 is required for apical constriction during lens pit morphogenesis. *Development, 141*(16), 3177–3187. http://dx.doi.org/10.1242/Dev.107433.

Leerberg, J. M., Gomez, G. A., Verma, S., Moussa, E. J., Wu, S. K., Priya, R., et al. (2014). Tension-sensitive actin assembly supports contractility at the epithelial zonula adherens. *Current Biology, 24*(15), 1689–1699. http://dx.doi.org/10.1016/J.Cub.2014.06.028.

Leibfried, A., Fricke, R., Morgan, M. J., Bogdan, S., & Bellaiche, Y. (2008). Drosophila Cip4 and WASp define a branch of the Cdc42-Par6-aPKC pathway regulating E-cadherin endocytosis. *Current Biology, 18*(21), 1639–1648. http://dx.doi.org/10.1016/j.cub.2008.09.063.

Levine, E., Lee, C. H., Kintner, C., & Gumbiner, B. M. (1994). Selective disruption of E-cadherin function in early Xenopus embryos by a dominant negative mutant. *Development, 120*(4), 901–909.

Liu, X. F., Ishida, H., Raziuddin, R., & Miki, T. (2004). Nucleotide exchange factor ECT2 interacts with the polarity protein complex Par6/Par3/protein kinase Czeta (PKCzeta) and regulates PKCzeta activity. *Molecular and Cellular Biology, 24*(15), 6665–6675.

Liu, X. F., Ohno, S., & Miki, T. (2006). Nucleotide exchange factor ECT2 regulates epithelial cell polarity. *Cellular Signalling, 18*(10), 1604–1615. http://dx.doi.org/10.1016/j.cellsig.2006.01.007.

Luo, W., Yu, C. H., Lieu, Z. Z., Allard, J., Mogilner, A., Sheetz, M. P., et al. (2013). Analysis of the local organization and dynamics of cellular actin networks. *The Journal of Cell Biology, 202*(7), 1057–1073. http://dx.doi.org/10.1083/jcb.201210123.

Maitre, J. L., Berthoumieux, H., Krens, S. F., Salbreux, G., Julicher, F., Paluch, E., et al. (2012). Adhesion functions in cell sorting by mechanically coupling the cortices of adhering cells. *Science, 338*(6104), 253–256. http://dx.doi.org/10.1126/science.1225399.

Mammoto, T., Mammoto, A., & Ingber, D. E. (2013). Mechanobiology and developmental control. *Annual Review of Cell and Developmental Biology, 29*(29), 27–61. http://dx.doi.org/10.1146/Annurev-Cellbio-101512-122340.

Mangold, S., Norwood, S. J., Yap, A. S., & Collins, B. M. (2012). The juxtamembrane domain of the E-cadherin cytoplasmic tail contributes to its interaction with Myosin VI. *Bioarchitecture, 2*(5), 185–188. http://dx.doi.org/10.4161/bioa.22082.

Mangold, S., Wu, S. K., Norwood, S. J., Collins, B. M., Hamilton, N. A., Thorn, P., et al. (2011). Hepatocyte growth factor acutely perturbs actin filament anchorage at the epithelial zonula adherens. *Current Biology, 21*(6), 503–507. http://dx.doi.org/10.1016/J.Cub.2011.02.018.

Manke, I. A., Lowery, D. M., Nguyen, A., & Yaffe, M. B. (2003). BRCT repeats as phosphopeptide-binding modules involved in protein targeting. *Science, 302*(5645), 636–639.

Martin, A. C. (2010). Pulsation and stabilization: Contractile forces that underlie morphogenesis. *Developmental Biology, 341*(1), 114–125. http://dx.doi.org/10.1016/J.Ydbio.2009.10.031.

Martin, A. C., Gelbart, M., Fernandez-Gonzalez, R., Kaschube, M., & Wieschaus, E. (2010). Integration of contractile forces during tissue invagination. *The Journal of Cell Biology, 188*, 735–749.

Martin, A. C., Kaschube, M., & Wieschaus, E. F. (2009). Pulsed contractions of an actin-myosin network drive apical constriction. *Nature, 457*(7228), 495–499. http://dx.doi.org/10.1038/nature07522.

Matsumura, F. (2005). Regulation of myosin II during cytokinesis in higher eukaryotes. *Trends in Cell Biology, 15*(7), 371–377. http://dx.doi.org/10.1016/J.Tcb.2005.05.004.

Medeiros, N. A., Burnette, D. T., & Forscher, P. (2006). myosin II functions in actin-bundle turnover in neuronal growth cones. *Nature Cell Biology, 8*(3), 215–226.

Mikawa, M., Su, L., & Parsons, S. J. (2008). Opposing roles of p190RhoGAP and ECT2 RhoGEF in regulating cytokinesis. *Cell Cycle*, 7(13), 2003–2012.

Mishima, M., Kaitna, S., & Glotzer, M. (2002). Central spindle assembly and cytokinesis require a kinesin-like protein/RhoGAP complex with microtubule bundling activity. *Developmental Cell, 2*(1), 41–54, doi: S1534580701001101 [pii].

Mori, K., Amano, M., Takefuji, M., Kato, K., Morita, Y., Nishioka, T., et al. (2009). Rho-kinase contributes to sustained RhoA activation through phosphorylation of p190A RhoGAP. *Journal of Biological Chemistry, 284*(8), 5067–5076. http://dx.doi.org/10.1074/Jbc.M806853200.

Niederman, R., & Pollard, T. D. (1975). Human platelet myosin. II. In vitro assembly and structure of myosin filaments. *The Journal of Cell Biology, 67*(1), 72–92.

Nigam, S. K., Rodriguez-Boulan, E., & Silver, R. B. (1992). Changes in intracellular calcium during the development of epithelial polarity and junctions. *Proceedings of the National Academy of Sciences of the United States of America, 89*(13), 6162–6166.

Nishimura, T., & Takeichi, M. (2008). Shroom3-mediated recruitment of Rho kinases to the apical cell junctions regulates epithelial and neuroepithelial planar remodeling. *Development, 135*(8), 1493–1502. http://dx.doi.org/10.1242/Dev.019646.

Noren, N. K., Arthur, W. T., & Burridge, K. (2003). Cadherin engagement inhibits RhoA via p190RhoGAP. *The Journal of Biological Chemistry, 278*(16), 13615–13618. http://dx.doi.org/10.1074/jbc.C200657200.

Noren, N. K., Liu, B. P., Burridge, K., & Kreft, B. (2000). p120 Catenin regulates the actin cytoskeleton via rho family GTPases. *The Journal of Cell Biology, 150*, 567–579.

Oinuma, I., Kawada, K., Tsukagoshi, K., & Negishi, M. (2012). Rnd1 and Rnd3 targeting to lipid raft is required for p190 RhoGAP activation. *Molecular Biology of the Cell, 23*(8), 1593–1604. http://dx.doi.org/10.1091/Mbc.E11-11-0900.

Padrick, S. B., & Rosen, M. K. (2010). Physical mechanisms of signal integration by WASP family proteins. *Annual Review of Biochemistry, 79*, 707–735. http://dx.doi.org/10.1146/annurev.biochem.77.060407.135452.

Peng, X., Cuff, L. E., Lawton, C. D., & DeMali, K. A. (2010). Vinculin regulates cell-surface E-cadherin expression by binding to beta-catenin. *Journal of Cell Science, 123*(Pt 4), 567–577. http://dx.doi.org/10.1242/jcs.056432.

Piekny, A. J., & Glotzer, M. (2008). Anillin is a scaffold protein that links RhoA, actin, and myosin during cytokinesis. *Current Biology, 18*(1), 30–36. http://dx.doi.org/10.1016/J.Cub.2007.11.068.

Piekny, A. J., & Maddox, A. S. (2010). The myriad roles of Anillin during cytokinesis. *Seminars in Cell and Developmental Biology, 21*(9), 881–891. http://dx.doi.org/10.1016/j.semcdb.2010.08.002.

Piekny, A., Werner, M., & Glotzer, M. (2005). Cytokinesis: Welcome to the Rho zone. *Trends in Cell Biology, 15*(12), 651–658. http://dx.doi.org/10.1016/j.tcb.2005.10.006, S0962-8924(05)00250-3 [pii].

Pollard, T. D., Blanchoin, L., & Mullins, R. D. (2000). Molecular mechanisms controlling actin filament dynamics in nonmuscle cells. *Annual Review of Biophysics and Biomolecular Structure, 29*, 545–576.

Ponik, S. M., Trier, S. M., Wozniak, M. A., Eliceiri, K. W., & Keely, P. J. (2013). RhoA is down-regulated at cell-cell contacts via p190RhoGAP-B in response to tensional homeostasis. *Molecular Biology of the Cell*, *24*(11), 1688–1699. http://dx.doi.org/10.1091/Mbc.E12-05-0386.

Priya, R., Yap, A. S., & Gomez, G. A. (2013). E-cadherin supports steady-state Rho signaling at the epithelial zonula adherens. *Differentiation*, *86*(3), 133–140. http://dx.doi.org/10.1016/j.diff.2013.01.002.

Rangarajan, E. S., & Izard, T. (2012). The cytoskeletal protein α-Catenin unfurls upon binding to vinculin. *The Journal of Biological Chemistry*, *287*(22), 18492–18499. http://dx.doi.org/10.1074/jbc.M112.351023.

Ratheesh, A., Gomez, G. A., Priya, R., Verma, S., Kovacs, E. M., Jiang, K., et al. (2012). Centralspindlin and alpha-catenin regulate Rho signalling at the epithelial zonula adherens. *Nature Cell Biology*, *14*(8), 818–828. http://dx.doi.org/10.1038/ncb2532.

Ratheesh, A., Priya, R., & Yap, A. S. (2013). Coordinating Rho and Rac: The regulation of Rho GTPase signaling and cadherin junctions. *Progress in Molecular Biology and Translational Science*, *116*, 49–68. http://dx.doi.org/10.1016/B978-0-12-394311-8.00003-0.

Ratheesh, A., & Yap, A. S. (2012). A bigger picture: Classical cadherins and the dynamic actin cytoskeleton. *Nature Reviews. Molecular Cell Biology*, *13*(10), 673–679. http://dx.doi.org/10.1038/Nrm3431.

Rauzi, M., Verant, P., Lecuit, T., & Lenne, P. F. (2008). Nature and anisotropy of cortical forces orienting Drosophila tissue morphogenesis. *Nature Cell Biology*, *10*(12), 1401–1410. http://dx.doi.org/10.1038/ncb1798.

Reyes, C. C., Jin, M., Breznau, E. B., Espino, R., Delgado-Gonzalo, R., Goryachev, A. B., et al. (2014). Anillin regulates cell-cell junction integrity by organizing junctional accumulation of Rho-GTP and actomyosin. *Current Biology*, *24*(11), 1263–1270. http://dx.doi.org/10.1016/j.cub.2014.04.021.

Reymann, A. C., Boujemaa-Paterski, R., Martiel, J. L., Guerin, C., Cao, W. X., Chin, H. F., et al. (2012). Actin network architecture Can determine myosin motor activity. *Science*, *336*(6086), 1310–1314. http://dx.doi.org/10.1126/Science.1221708.

Ridley, A. J. (2012). Historical overview of Rho GTPases. *Rho GTPases: Methods and Protocols*, *827*, 3–12. http://dx.doi.org/10.1007/978-1-61779-442-1_1.

Riento, K., & Ridley, A. J. (2003). Rocks: Multifunctional kinases in cell behaviour. *Nature Reviews. Molecular Cell Biology*, *4*(6), 446–456. http://dx.doi.org/10.1038/nrm1128.

Riento, K., Totty, N., Villalonga, P., Garg, R., Guasch, R., & Ridley, A. J. (2005). RhoE function is regulated by ROCK I-mediated phosphorylation. *The EMBO Journal*, *24*(6), 1170–1180. http://dx.doi.org/10.1038/Sj.Emboj.7600612.

Riou, P., Kjaer, S., Garg, R., Purkiss, A., George, R., Cain, R. J., et al. (2013). 14-3-3 Proteins interact with a hybrid prenyl-phosphorylation motif to inhibit G proteins. *Cell*, *153*(3), 640–653. http://dx.doi.org/10.1016/j.cell.2013.03.044.

Sandquist, J. C., & Means, A. R. (2008). The C-terminal tail region of nonmuscle myosin II directs isoform-specific distribution in migrating cells. *Molecular Biology of the Cell*, *19*(12), 5156–5167. http://dx.doi.org/10.1091/mbc.E08-05-0533, E08-05-0533 [pii].

Sandquist, J. C., Swenson, K. I., Demali, K. A., Burridge, K., & Means, A. R. (2006). Rho kinase differentially regulates phosphorylation of nonmuscle myosin II isoforms A and B during cell rounding and migration. *The Journal of Biological Chemistry*, *281*(47), 35873–35883.

Schofield, A. V., & Bernard, O. (2013). Rho-associated coiled-coil kinase (ROCK) signaling and disease. *Critical Reviews in Biochemistry and Molecular Biology*, *48*(4), 301–316. http://dx.doi.org/10.3109/10409238.2013.786671.

Schumacher, S., Gryzik, T., Tannebaum, S., & Muller, H. A. J. (2004). The RhoGEF Pebble is required for cell shape changes during cell migration triggered by the Drosophila FGF receptor Heartless. *Development, 131*(11), 2631–2640. http://dx.doi.org/10.1242/Dev.01149.

Scott, J. A., Shewan, A. M., den Elzen, N. R., Loureiro, J. J., Gertler, F. B., & Yap, A. S. (2006). Ena/VASP proteins can regulate distinct modes of actin organization at cadherin-adhesive contacts. *Molecular Biology of the Cell, 17*(3), 1085–1095. http://dx.doi.org/10.1191/Mbc.E05-07-0644.

Sellers, J. R. (1991). Regulation of cytoplasmic and smooth muscle myosin. *Current Opinion in Cell Biology, 3*(1), 98–104.

Settleman, J., Narasimhan, V., Foster, L. C., & Weinberg, R. A. (1992). Molecular cloning of cDNAs encoding the GAP-associated protein p190: Implications for a signaling pathway from ras to the nucleus. *Cell, 69*(3), 539–549.

Shewan, A. M., Maddugoda, M., Kraemer, A., Stehbens, S. J., Verma, S., Kovacs, E. M., et al. (2005). Myosin 2 is a key Rho kinase target necessary for the local concentration of E-cadherin at cell-cell contacts. *Molecular Biology of the Cell, 16*(10), 4531–4542. http://dx.doi.org/10.1091/mbc.E05-04-0330.

Shimizu, Y., Thumkeo, D., Keel, J., Ishizaki, T., Oshima, H., Oshima, M., et al. (2005). ROCK-I regulates closure of the eyelids and ventral body wall by inducing assembly of actomyosin bundles. *The Journal of Cell Biology, 168*(6), 941–953. http://dx.doi.org/10.1083/jcb.200411179.

Simoes Sde, M., Mainieri, A., & Zallen, J. A. (2014). Rho GTPase and Shroom direct planar polarized actomyosin contractility during convergent extension. *The Journal of Cell Biology, 204*(4), 575–589. http://dx.doi.org/10.1083/jcb.201307070.

Smallhorn, M., Murray, M. J., & Saint, R. (2004). The epithelial-mesenchymal transition of the Drosophila mesoderm requires the Rho GTP exchange factor Pebble. *Development, 131*(11), 2641–2651. http://dx.doi.org/10.1242/dev.01150, dev.01150 [pii].

Smith, A. L., Dohn, M. R., Brown, M. V., & Reynolds, A. B. (2012). Association of Rho-associated protein kinase 1 with E-cadherin complexes is mediated by p120-catenin. *Molecular Biology of the Cell, 23*(1), 99–110. http://dx.doi.org/10.1091/Mbc.E11-06-0497.

Smutny, M., Cox, H. L., Leerberg, J. M., Kovacs, E. M., Conti, M. A., Ferguson, C., et al. (2010). myosin II isoforms identify distinct functional modules that support integrity of the epithelial zonula adherens. *Nature Cell Biology, 12*(7), 696–702. http://dx.doi.org/10.1038/ncb2072.

Smutny, M., Wu, S. K., Gomez, G. A., Mangold, S., Yap, A. S., & Hamilton, N. A. (2011). Multicomponent analysis of junctional movements regulated by myosin II isoforms at the epithelial zonula adherens. *PLoS One, 6*(7), e22458. http://dx.doi.org/10.1371/journal.pone.0022458.

Somers, W. G., & Saint, R. (2003). A RhoGEF and Rho family GTPase-activating protein complex link the contractile ring to cortical microtubules at the onset of cytokinesis. *Developmental Cell, 4*, 29–39.

Su, K. C., Takaki, T., & Petronczki, M. (2011). Targeting of the RhoGEF ECT2 to the equatorial membrane controls cleavage furrow formation during cytokinesis. *Developmental Cell, 21*(6), 1104–1115. http://dx.doi.org/10.1016/j.devcel.2011.11.003.

Takaishi, K., Sasaki, T., Kotani, H., Nishioka, H., & Takai, Y. (1997). Regulation of cell-cell adhesion by rac and rho small G proteins in MDCK cells. *The Journal of Cell Biology, 139*(4), 1047–1059.

Takeichi, M. (2014). Dynamic contacts: Rearranging adherens junctions to drive epithelial remodelling. *Nature Reviews. Molecular Cell Biology, 15*(6), 397–410. http://dx.doi.org/10.1038/nrm3802.

Tang, V. W., & Brieher, W. M. (2012). alpha-Actinin-4/FSGS1 is required for Arp2/3-dependent actin assembly at the adherens junction. *Journal of Cell Biology, 196*(1), 115–130. http://dx.doi.org/10.1083/Jcb.201103116.

Tatsumoto, T., Xie, X. Z., Blumenthal, R., Okamoto, I., & Miki, T. (1999). Human ECT2 is an exchange factor for Rho GTPases, phosphorylated in G2/M phases, and involved in cytokinesis. *Journal of Cell Biology*, *147*(5), 921–927.

Tcherkezian, J., & Lamarche-Vane, N. (2007). Current knowledge of the large RhoGAP family of proteins. *Biology of the Cell*, *99*(2), 67–86. http://dx.doi.org/10.1042/BC20060086.

Thumkeo, D., Shinohara, R., Watanabe, K., Takebayashi, H., Toyoda, Y., Tohyama, K., et al. (2011). Deficiency of mDia, an actin nucleator, disrupts integrity of neuroepithelium and causes periventricular dysplasia. *PLoS One*, *6*(9), e25465. http://dx.doi.org/10.1371/journal.pone.0025465.

Thumkeo, D., Watanabe, S., & Narumiya, S. (2013). Physiological roles of Rho and Rho effectors in mammals. *European Journal of Cell Biology*, *92*(10–11), 303–315. http://dx.doi.org/10.1016/j.ejcb.2013.09.002.

Totsukawa, G., Yamakita, Y., Yamashiro, S., Hartshorne, D. J., Sasaki, Y., & Matsumura, F. (2000). Distinct roles of ROCK (Rho-kinase) and MLCK in spatial regulation of MLC phosphorylation for assembly of stress fibers and focal adhesions in 3T3 fibroblasts. *The Journal of Cell Biology*, *150*(4), 797–806.

Vargo-Gogola, T., Heckman, B. M., Gunther, E. J., Chodosh, L. A., & Rosen, J. M. (2006). P190-B Rho GTPase-activating protein overexpression disrupts ductal morphogenesis and induces hyperplastic lesions in the developing mammary gland. *Molecular Endocrinology*, *20*(6), 1391–1405. http://dx.doi.org/10.1210/me.2005-0426.

Vasioukhin, V., Bauer, C., Yin, M., & Fuchs, E. (2000). Directed actin polymerization is the driving force for epithelial cell-cell adhesion. *Cell*, *100*(2), 209–219. http://dx.doi.org/10.1016/S0092-8674(00)81559-7.

Verma, S., Han, S. P., Michael, M., Gomez, G. A., Yang, Z., Teasdale, R. D., et al. (2012). A WAVE2-Arp2/3 actin nucleator apparatus supports junctional tension at the epithelial zonula adherens. *Molecular Biology of the Cell*, *23*(23), 4601–4610. http://dx.doi.org/10.1091/Mbc.E12-08-0574.

Vicente-Manzanares, M., Ma, X., Adelstein, R. S., & Horwitz, A. R. (2009). Non-muscle myosin II takes centre stage in cell adhesion and migration. *Nature Reviews Molecular Cell Biology*, *10*(11), 778–790. http://dx.doi.org/10.1038/nrm2786, nrm2786 [pii].

Watabe-Uchida, M., Uchida, N., Imamura, Y., Nagafuchi, A., Fujimoto, K., Uemura, T., et al. (1998). alpha-Catenin-vinculin interaction functions to organize the apical junctional complex in epithelial cells. *The Journal of Cell Biology*, *142*(3), 847–857.

Weed, S. A., & Parsons, J. T. (2001). Cortactin: Coupling membrane dynamics to cortical actin assembly. *Oncogene*, *20*(44), 6418–6434.

Wei, L., Roberts, W., Wang, L., Yamada, M., Zhang, S. X., Zhao, Z. Y., et al. (2001). Rho kinases play an obligatory role in vertebrate embryonic organogenesis. *Development*, *128*(15), 2953–2962.

Wennerberg, K., Forget, M. A., Ellerbroek, S. M., Arthur, W. T., Burridge, K., Settleman, J., et al. (2003). Rnd proteins function as RhoA antagonists by activating p190 RhoGAP. *Current Biology*, *13*(13), 1106–1115. http://dx.doi.org/10.1016/S0960-9822(03)00418-4.

Werner, M., & Glotzer, M. (2008). Control of cortical contractility during cytokinesis. *Biochemical Society Transactions*, *36*(Pt 3), 371–377. http://dx.doi.org/10.1042/BST0360371, BST0360371 [pii].

Wildenberg, G. A., Dohn, M. R., Carnahan, R. H., Davis, M. A., Lobdell, N. A., Settleman, J., et al. (2006). p120-catenin and p190RhoGAP regulate cell-cell adhesion by coordinating antagonism between Rac and Rho. *Cell*, *127*(5), 1027–1039. http://dx.doi.org/10.1016/j.cell.2006.09.046.

Wolfe, B. A., & Glotzer, M. (2009). Single cells (put a ring on it). *Genes & Development*, *23*(8), 896–901. http://dx.doi.org/10.1101/gad.1801209, 23/8/896 [pii].

Wolfe, B. A., Takaki, T., Petronczki, M., & Glotzer, M. (2009). Polo-like kinase 1 directs assembly of the HsCyk-4 RhoGAP/ECT2 RhoGEF complex to initiate cleavage furrow formation. *PLoS Biology*, 7(5), e1000110. http://dx.doi.org/10.1371/journal.pbio.1000110.

Wu, S. K., Gomez, G. A., Michael, M., Verma, S., Cox, H. L., Lefevre, J. G., et al. (2014). Cortical F-actin stabilization generates apical-lateral patterns of junctional contractility that integrate cells into epithelia. *Nature Cell Biology*, *16*(2), 167–178. http://dx.doi.org/10.1038/ncb2900.

Yamada, S., Pokutta, S., Drees, F., Weis, W. I., & Nelson, W. J. (2005). Deconstructing the cadherin-catenin-actin complex. *Cell*, *123*(5), 889–901. http://dx.doi.org/10.1016/j.cell.2005.09.020.

Yao, M., Qiu, W., Liu, R., Efremov, A. K., Cong, P., Seddiki, R., et al. (2014). Force-dependent conformational switch of alpha-catenin controls vinculin binding. *Nature Communications*, *5*, 4525. http://dx.doi.org/10.1038/ncomms5525.

Ydenberg, C. A., Smith, B. A., Breitsprecher, D., Gelles, J., & Goode, B. L. (2011). Cease-fire at the leading edge: New perspectives on actin filament branching, debranching, and cross-linking. *Cytoskeleton (Hoboken)*, *68*(11), 596–602. http://dx.doi.org/10.1002/cm.20543.

Yonemura, S., Wada, Y., Watanabe, T., Nagafuchi, A., & Shibata, M. (2010). alpha-Catenin as a tension transducer that induces adherens junction development. *Nature Cell Biology*, *12*(6), 533–542. http://dx.doi.org/10.1038/Ncb2055.

Yuce, O., Piekny, A., & Glotzer, M. (2005). An ECT2-centralspindlin complex regulates the localization and function of RhoA. *Journal of Cell Biology*, *170*(4), 571–582.

Zhang, J., Betson, M., Erasmus, J., Zeikos, K., Bailly, M., Cramer, L. P., et al. (2005). Actin at cell-cell junctions is composed of two dynamic and functional populations. *Journal of Cell Science*, *118*(Pt 23), 5549–5562.

Zhang, B., & Zheng, Y. (1998). Regulation of RhoA GTP hydrolysis by the GTPase-activating proteins p190, p50RhoGAP, Bcr, and 3BP-1. *Biochemistry*, *37*(15), 5249–5257. http://dx.doi.org/10.1021/bi9718447.

CHAPTER FOUR

Integration of Cell–Cell Adhesion and Contractile Actomyosin Activity During Morphogenesis

Katja Röper[1]
MRC-Laboratory of Molecular Biology, Cambridge Biomedical Campus, Cambridge, United Kingdom
[1]Corresponding author: e-mail address: kroeper@mrc-lmb.cam.ac.uk

Contents

Abstract

During embryonic development, cells become organized into complex tissues. Cells need to adhere and communicate with their immediate and remote neighbors to allow morphogenesis to take place in a coordinated way. Cell–cell adhesion, mediated by transmembrane adhesion receptors such as Cadherins and their intracellular interaction partners, is intimately linked to cell contractility that drives cell shape changes. Research in recent years has revealed that the contractile machinery responsible for cell shape

Current Topics in Developmental Biology, Volume 112
ISSN 0070-2153
http://dx.doi.org/10.1016/bs.ctdb.2014.11.017

changes, actomyosin, can in fact be organized into a number of different functional assemblies such as cortical-junctional actomyosin, apical–medial actomyosin, supracellular actomyosin cables as well as basal actomyosin networks. During coordinated shape changes of a tissue, these assemblies have to be functionally and mechanically linked between cells through cell–cell junctions. Although many actin-binding proteins associated with adherens junctions have been identified, which specific factors are required for the linkage of particular actomyosin assemblies to junctions is not well understood. This review will summarize our current knowledge, based mainly on the *in vivo* study of morphogenesis in the fruit fly *Drosophila melanogaster*.

1. INTRODUCTION

A crucial step during embryonic development and a prerequisite for all morphogenetic events that shape the embryo is the establishment of cell–cell adhesion. Cell–cell adhesion allows for mechanical as well as chemical communication between cells, so that morphogenetic movements are coordinated across large assemblies of cells.

Many morphogenetic processes take place in epithelial sheets of cells. A hallmark of epithelial cells is their apical–basal polarity, with adhesive cell–cell junctions, so-called adherens junctions (AJ), concentrated near the apical end of lateral contact sites, and lateral tight, or septate junctions that seal the contact and provide a diffusion barrier between external and internal environments. As such, epithelia form the protective barriers between the embryo's internal space and the outside world.

Adhesions arise at different times in the development of different model organisms frequently used to study morphogenesis. In *Drosophila*, cell–cell adhesions are only formed during cellularization of the blastoderm embryo, which at this stage is a syncytium containing several thousand nuclei localized near the surface of the embryo (Harris, Sawyer, & Peifer, 2009). While membrane is being pulled down and assembled between neighboring nuclei, early spot AJ are being formed in a microtubule/dynein- and bazooka/Par3-dependent way near the new apical surface at the outside of the embryo (Harris & Peifer, 2005). During the major morphogenetic events of gastrulation, AJ mature into more belt-like structures surrounding the apical circumference (zonula adherens, ZA). This latter process depends on actin, as well as the apical polarity regulator Crumbs, a transmembrane protein, and its associated proteins (Harris et al., 2009; Tepass, 2012). Thus, AJ are in close contact with the cytoskeleton already during their assembly. In contrast to flies, AJ in mouse embryos begin to develop during the compaction of the 8-cell embryo and appear functional and mature at the 32-cell stage (Fleming, Sheth, & Fesenko, 2001; Sheth et al., 2000).

In this review, I will discuss how the molecular composition of the Cadherin adhesion complex and its link to the actin cytoskeleton allow AJ to be the prime communicator of forces and tension during morphogenesis, thereby coordinating cell shape changes and allowing for cell rearrangements in large assemblies of epithelial cells. As the *in vivo* study of these processes is most advanced in *Drosophila* I will mostly draw on examples from this organism.

2. MOLECULAR INTERPLAY BETWEEN CADHERIN ADHESION RECEPTORS AND THE ACTOMYOSIN CYTOSKELETON

2.1. The Cadherin complex

E-Cadherin, epithelial Cadherin, is the major homophilic interactor in AJ, and is a representative of the classical Cadherin family that appeared with the advent of metazoans (for review, see Harris & Tepass, 2010). Extracellular Cadherin (EC) domains mediate homophilic interactions in *trans* between molecules on neighboring cells (for review, see Pokutta & Weis, 2007; Fig. 1A). On the cytoplasmic site, the Cadherin intracellular domain mediates interactions with a range of partners that confer functionality on E-Cadherin (Fig. 1A). The Cadherin tail interacts with the cortical actin cytoskeleton (see below), it binds factors mediating Cadherin turnover and recycling at the plasma membrane, and it ties Cadherin and AJ establishment into the apical–basal polarity machinery (Baum & Georgiou, 2011).

Individual Cadherins are able to associate in *trans* between neighboring cells, but increased density of Cadherin monomers cooperatively increases the likelihood of binding in *trans* (Zhang, Sivasankar, Nelson, & Chu, 2009). Whether actual *cis*-dimers are a prerequisite to the formation of interactions in *trans* is still under debate (Wu, Vendome, Shapiro, Ben-Shaul, & Honig, 2011; Zhang et al., 2009). In either case though, controlling the clustering state of Cadherins is a likely mechanism to control adhesive strength (Kovacs & Yap, 2008).

An essential aspect of the role of AJ during development is that they need to be dynamic and to be able to accommodate morphogenetic processes while at the same time maintaining epithelial integrity and structure. Cadherin endocytosis can drive junction disassembly by promoting adhesive

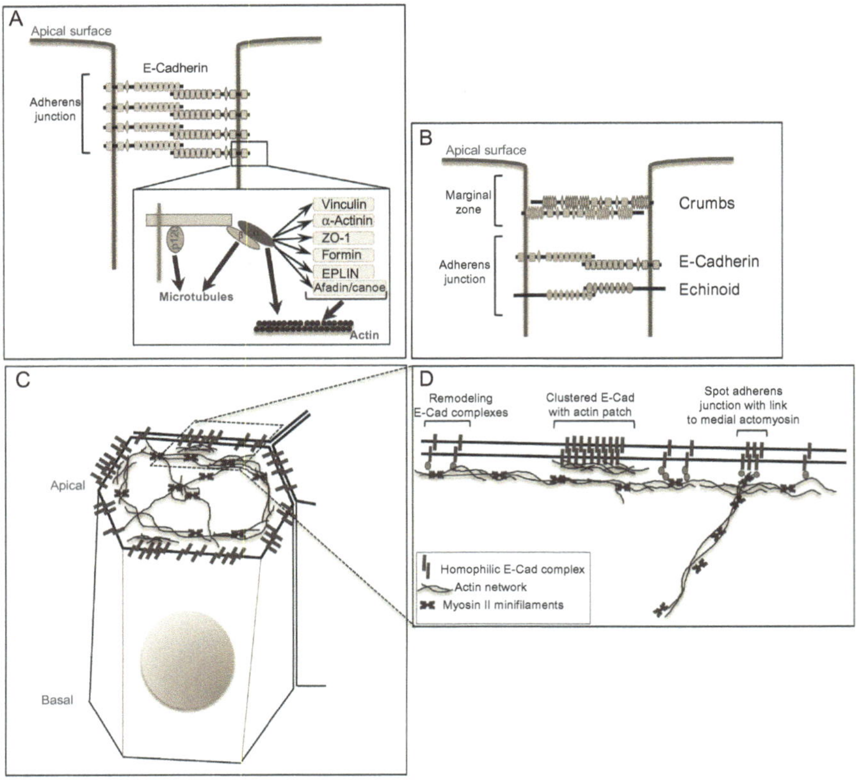

Figure 1 Components and organization of adherens junctions in epithelial cells. (A) Schematic illustrating the position of adherens junctions (AJ) formed mainly by homophilic E-Cadherin interactions between neighboring cells at the apical-most end of the lateral sides of epithelial cells. Intracellularly, E-Cadherin interacts directly with α-, β-, and p120-catenin. α-Catenin can directly bind to actin but can also recruit further factors that mediate actin association. (B) At least three classes of transmembrane proteins with the ability to interact homophilically in *trans* are found in the apical-lateral domain of epithelial cells. E-Cadherin is joined in the AJ by the Ig-family molecule Echinoid in *Drosophila*, a nectin. Just above the AJ, in the marginal zone, the apical polarity regulator Crumbs mediates homophilic interactions between cells. All three classes of proteins have the ability to impinge on the organization and function of the apical actin cytoskeleton (Thompson, Pichaud, & Röper, 2013). (C), (D) Surface view of the apical actomyosin cytoskeleton with regards to AJ in a typical epithelial cell, the dashed box in (C) is magnified in (D). E-Cadherin can be found associated with F-actin in different ways: as homophilic *trans*-clusters associated with actin patches that do not contain myosin II, as homophilic E-Cadherin complexes likely involved in remodeling that interact with a more wide-spread membrane-near actomyosin network (Cavey, Rauzi, Lenne, & Lecuit, 2008), and in form of spot AJ that in some cases associate with actomyosin bundles that run perpendicular to the plasma membrane. These spot adherens junctions might link medial actomyosin networks across neighboring cells (Martin, Kaschube, & Wieschaus, 2009).

dimer dissociation (Troyanovsky, Sokolov, & Troyanovsky, 2006). Endo- and exocytosis of Cadherins is carefully controlled by a variety of downstream factors (for a review, see Baum & Georgiou, 2011), and in particular the local interaction with the actomyosin cytoskeleton during morphogenesis can determine the stability of E-Cadherin clusters at specific sites within the plasma membrane (Levayer & Lecuit, 2013; Levayer, Pelissier-Monier, & Lecuit, 2011; Rauzi, Lenne, & Lecuit, 2010).

Positioning of AJ at the apical end of the lateral domains of epithelial cells is tightly controlled by factors that also control apicobasal polarity *per se* (Tepass, 2012). In particular, apical polarity regulators serve a crucial role in triggering segregation of the apical membrane from the AJ domain, a process that relies on aPKC-dependent phosphorylation of Bazooka/Par3 (Morais-de-Sa, Mirouse, & St Johnston, 2010; Tepass, 2012; Walther & Pichaud, 2010).

2.2. The Cadherin–actin link

Intracellularly, E-Cadherin associates with members of the catenin protein family. β-Catenin/armadillo and p120-catenin bind directly to the Cadherin tail, with β-catenin providing the link to α-catenin (Fig. 1A). β-Catenin itself is required for biosynthetic transport of E-Cadherin to the plasma membrane and might have a regulatory role in E-Cadherin endocytosis (Chen, Stewart, & Nelson, 1999; Delva & Kowalczyk, 2009; Langevin et al., 2005). p120-Catenin also functions in regulating E-Cadherin endocytosis and turnover by masking an endocytic signal and thus blocking the association with factors that promote endocytosis (Nanes et al., 2012). α-Catenin can directly bind to actin (Rimm, Koslov, Kebriaei, Cianci, & Morrow, 1995) but in addition it binds a multitude of proteins that can themselves also bind actin (Fig. 1A; for a review, see Harris & Tepass, 2010).

Whether a quaternary complex of E-Cadherin/β-catenin/α-catenin/actin is ever formed and functional in cells has been the cause of much debate over the last 10 years. Genetic evidence mostly from *Drosophila* suggested this to be the case whereas data from mammalian cells indicated that for mammalian αE-catenin binding to β-catenin (in form of a dimer) and binding to actin was mutually exclusive (Benjamin et al., 2010; Pokutta & Weis, 2000; Yamada, Pokutta, Drees, Weis, & Nelson, 2005). Recent evidence from *Drosophila* strongly supports the view that in flies a monomeric α-catenin provides as an essential physical link between the E-Cadherin/β-catenin complex and actin (Desai et al., 2013). Further support for a direct interaction comes from recent studies of mammalian E-Cadherin and αE-catenin: insertion of a FRET-based tension-sensing module into the E-Cadherin cytoplasmic tail revealed that E-Cadherin in the plasma membrane of tissue culture cells is under constitutive stretch that can be abolished equally by disrupting actin, myosin, or αE-catenin function (Borghi et al., 2012). Furthermore, recent force-measurements using optical traps also support that under force a quaternary complex of E-Cadherin/ß-catenin/α-catenin/actin exists (Buckley et al., 2014).

The link of E-Cadherin into the actin network can even be subdivided into different functional entities (Fig. 1C and D). Work from flies reveals that homophilic *trans*-interacting E-Cadherin complexes partition into very stable clusters that depend on a subpool of cortical actin, so-called actin patches, for their stability, though not their mobility (Fig. 1D; Cavey et al., 2008). Mobility of these clusters as well as of unclustered, potentially remodeling, E-Cadherin homophilic complexes depends on a much more dynamic and contractile actin network underneath the plasma membrane (Cavey et al., 2008). This differential interaction with submembraneous actin pools is likely influencing E-Cadherin turnover and recycling (Baum & Georgiou, 2011).

α-Catenin is a promiscuous molecule that in addition to binding directly to actin has also been shown to bind a number of other proteins, including α-actinin, vinculin, ZO-1, afadin/canoe, formin, and EPLIN that all in turn can also associate with the actin cytoskeleton (Fig. 1A; for a review, see Kobielak & Fuchs, 2004). Interaction of the E-Cadherin complex with actin through some of these proteins could determine the spatial arrangement of the actin network linked to the membrane. In established epithelial cells with mature junctions, most junction-associated actin appears arranged parallel to the membrane, forming a circumferential apical actin belt (Fig. 1C). Epithelial cells in culture that contact and initiate AJ formation, as well as epithelial cells during sheet closure in model organisms such as *Drosophila* or *C. elegans* show actin bundles emanating from newly formed spot AJ at right angles to the plasma membrane (reviewed in Baum & Georgiou, 2011; Harris & Tepass, 2010). During this establishment of new AJ, a number of actin regulators associate with the E-Cadherin-catenin cluster, including Rac, WAVE, RhoGAP, and Arp2/3 as an actin-nucleator at early stages, followed by Rho, myosin II, and formins coming in at a later stage when the initial-branched actin network resolves into the circumferential belt while the AJ expand (Braga, Betson, Li, & Lamarche-Vane, 2000; Delanoe-Ayari, Al Kurdi, Vallade, Gulino-Debrac, & Riveline, 2004; Sahai & Marshall, 2002; Vaezi, Bauer, Vasioukhin, & Fuchs, 2002; Yamada & Nelson, 2007; Zandy, Playford, & Pendergast, 2007). The α-catenin-binding mammalian protein EPLIN has a role in maintaining the actin network near mature AJ organized into arrays lying parallel to the plasma membrane. Without EPLIN, actin in epithelial tissue culture cells organizes into bundles that contact the plasma membrane at right angle (Abe & Takeichi, 2008).

Apart from the earliest forming spot AJ during cell contact formation or processes such as mesenchymal–epithelial transition (Baum, Settleman, & Quinlan, 2008), the actin cytoskeleton associated with AJ is a contractile structure of actin and nonmuscle myosin II (Fig. 1C). Rho and Rho-kinase (Rok) association with maturing junctions leads to recruitment of actomyosin to AJ, and this in turn promotes AJ clustering (Delanoe-Ayari et al., 2004). During germband extension (GBE) in the fly embryo, the recruitment of myosin II and the formin Diaphanous (Dia) to junctions in a planar polarized manner triggers selective clathrin-dependent endocytosis of clustered E-Cadherin from these junctions (Levayer et al., 2011), indicating a complex relationship between myosin activity and E-Cadherin function that might change depending on the tissue context.

2.3. Other apical adhesion and homophilic receptors acting on the cytoskeleton

Although Cadherins are the major homophilic cell–cell adhesion receptors in AJ and in general in epithelial cell, they are not the only ones (Fig. 1B). Ig-Domain containing proteins of the nectin family are also found in AJ (Rikitake, Mandai, & Takai, 2012). In addition, in a domain just apical to where AJ are located, in the so-called subapical domain, the transmembrane protein and apical polarity regulator Crumbs has recently been shown to undergo homophilic interactions and might also mediate cell–cell adhesion (Thompson et al., 2013). Below I will briefly outline how these two classes of homophilic interactors affect the actomyosin cytoskeleton.

2.3.1 Nectins/Echinoid and the actin cytoskeleton

Nectins comprise a Ca^{2+}-independent class of cell-adhesion molecules. In contrast to Ca^{2+}-sensitive E-Cadherin in AJ, they can display heterophilic *trans*-interactions between different nectins on neighboring cells, as well as homophilic *trans*-associations between the same nectin species. There are four nectin paralogs in vertebrates, and the Ig-domain containing surface receptor Echinoid is the single closest relative in flies (Chang et al., 2011; Indra, Hong, Troyanovsky, Kormos, & Troyanovsky, 2013). Both nectin-1 and-3 and Echinoid localize to AJ where they recruit intracellular factors through their C-terminal PDZ-binding domains, including Par3 and afadin/canoe (Rikitake et al., 2012; Fig. 1B). Afadin/canoe can bind directly to α-catenin, and it can also bind to actin (for review, see Takai, Ikeda, Ogita, & Rikitake, 2008). In fly imaginal wing discs, clones of Echinoid mutant cells show increased apical actin, narrowed apical surfaces as well as loss of E-Cadherin from the boundary of the clone. In fact, a paracellular actomyosin cable forms at the boundary of the clones (see below and Wei et al., 2005). This effect on actomyosin is also important during morphogenesis, where absence of Echinoid from some tissues is involved in triggering the assembly of an actomyosin cable at the interface between Echinoid-expressing and nonexpressing cells, such as in dorsal closure (see below) and formation of the dorsal appendages in the fly (Laplante & Nilson, 2006, 2011).

2.3.2 Crumbs and the organization of actomyosin

Crumbs is a crucial apical-polarity regulator in flies and vertebrates involved in maintaining apicobasal polarization of epithelial cells. In both vertebrates and *Drosophila*, Crumbs localizes just apical to the AJ in the subapical region/marginal zone of epithelial cells (Fig. 1B). Here, it recruits further important polarity regulators, such as Par6 and aPKC, through a membrane-proximal FERM-binding domain and a C-terminal PDZ-binding ERLI motif (for review, see Tepass, 2012). Work in the last couple of years has revealed that the large extracellular domain of Crumbs, containing EGF and Lamin AG repeats, can undergo homophilic interactions in *trans*, and that these interactions are important for functions of Crumbs beyond apical–basal polarity regulation. In zebrafish, these interactions are important to establish the correct mosaic arrangement of photoreceptors in the developing retina (Röper, 2012; Thompson et al., 2013; Zou, Wang, & Wei, 2012). In fly embryos, step changes in Crumbs protein levels between different groups of cells lead to the anisotropic localization of Crumbs at the boundary between the groups, mediated by the homophilic interactions of extracellular domains. Downstream of Crumbs this anisotropy then triggers the assembly of supracellular actomyosin cables at these boundaries (Röper, 2012; Thompson et al., 2013; Zou et al., 2012).

3. INTEGRATION OF ADHESION AND CONTRACTILITY DURING MORPHOGENESIS

The textbook view of the interaction of actomyosin and E-Cadherin in AJ is that of a circumferential actomyosin belt that helps to maintain apical shape and junctional integrity. However, the analysis of many morphogenetic processes mainly in *Drosophila* that depend on changes in apical shape of epithelial cells over the last 5 years have revealed that this view of apical actomyosin organization and function is too simplistic. In addition to actomyosin associated with the AJ, large and very dynamic pools of actomyosin can be found that are located within the apical–medial region of epithelial cells (Fig. 2A, C, and D; details below and for recent reviews, see Gorfinkiel & Blanchard, 2011; Levayer & Lecuit, 2012). In addition, as already mentioned above, myosin can assemble into larger scale assemblies, in particular so-called actomyosin cables, that seemingly span many cell diameters (Fig. 2A and B; Röper, 2013).

Thus, a recent challenge in understanding how contractility and adhesion are integrated during morphogenesis is to dissect how the different pools of actomyosin are coupled with AJ and their components, and what their respective functions are.

3.1. Medial myosin during mesoderm invagination in *Drosophila*

A major morphogenetic event during gastrulation in the fly embryo is the invagination of the presumptive mesoderm from the surface of the embryo, a very fast process that only takes about 15 min. It was known for many years that apical constriction of the mesodermal cells, fueled by the apical accumulation of nonmuscle myosin II, was key to correct tissue bending and invagination (Nikolaidou & Barrett, 2004; Young, Pesacreta, & Kiehart, 1991). Live imaging of GFP-tagged nonmuscle myosin II regulatory light chain (MRLC) revealed only a few years ago that an unexpected pool of apically localized myosin is responsible for this fast cell constriction. This pool of myosin is not localized near the apical AJ, but instead sits in a central, medial position within the apical-most region of the cells and was thus termed apical–medial myosin (Martin et al., 2009; Figs. 2D and 3A). The apical–medial myosin in the fly mesoderm increases and decreases in intensity over a time frame of minutes in a pulsatile manner, and increases in myosin precede and drive apical area constriction (Martin, Gelbart, Fernandez-Gonzalez, Kaschube, & Wieschaus, 2010; Martin et al., 2009). Careful analysis revealed that the apical–medial actomyosin is in contact with spot AJ at the periphery through long fibers that span the apical cell surface all the way to the junction (Martin et al., 2010). Reducing cell–cell adhesion using embryos with impaired β-catenin/arm function leads to failure in cell constriction, tissue relaxation, and an intracellular aggregation of

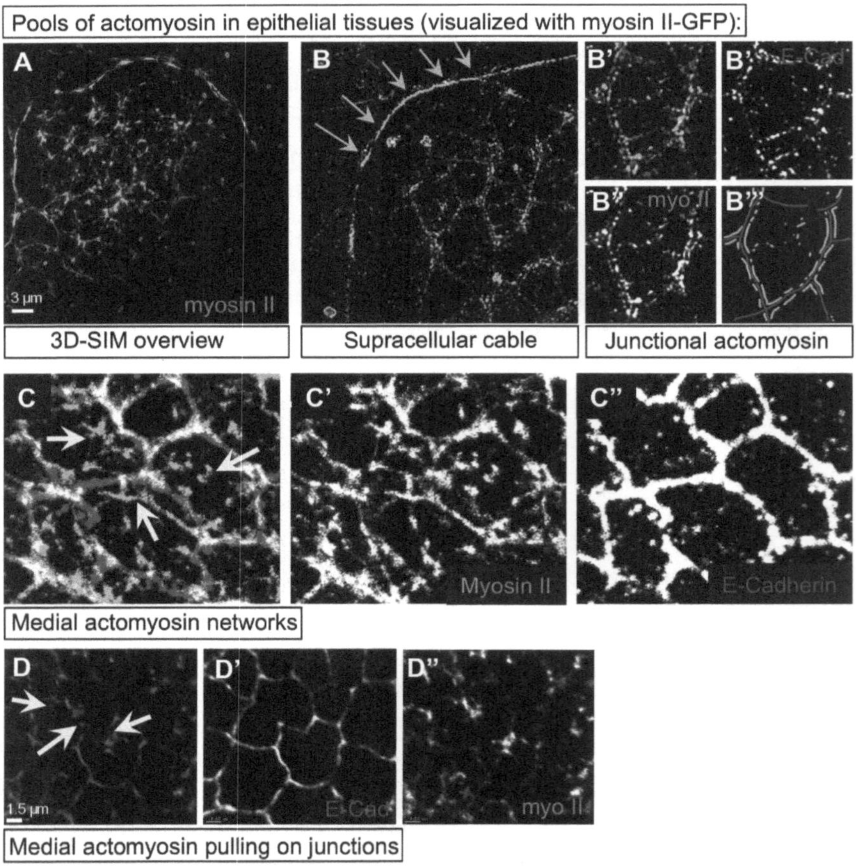

Figure 2 Apical actomyosin assemblies in epithelial cells. (A) Overview of apical myosin II organization in a patch of epithelial cells in a *Drosophila* embryo. The patch of cells shown here form the so-called salivary gland placode that will undergo the morphogenetic process of tubulogenesis through a process of regulated apical constriction and convergent extension (for review, see Girdler & Röper, 2014). This image was acquired using structured illumination microscopy (3D-SIM) on an embryo expressing myosin II-GFP. Extensive networks of myosin II, some of which are spanning many cell diameters can be seen (the cells in the placode have an apical diameter of 2–5 µm). (B) 3D-SIM image of myosin II incomparison to E-Cadherin labeling at the boundary of the placode shown in (A). A supracellular actomyosin cable surrounds the placode (arrows). The cable is perfectly aligned between many cells. (B′)–(B″) Junctional myosin II is in close apposition with E-Cadherin on either side of the junction. (B″) is a schematic of (B′). (C)–(C‴). In many tissues undergoing morphogenesis, actomyosin forms an apical–medial meshwork that connects between neighboring cells through spot AJ. Arrows point to medial pools of myosin II. (D)–(D‴) AJ labeled by E-Cadherin appear deformed near clusters of apical–medial actomyosin, indicating pulling forces (arrows). Myosin II is in green and E-Cadherin in magenta. Scale bar in (A) is 3 µm, in (D) 1.5 µm. *Panels (A) and (B) were adapted from Fig. 3B and H in Röper (2012).* (See the color plate.)

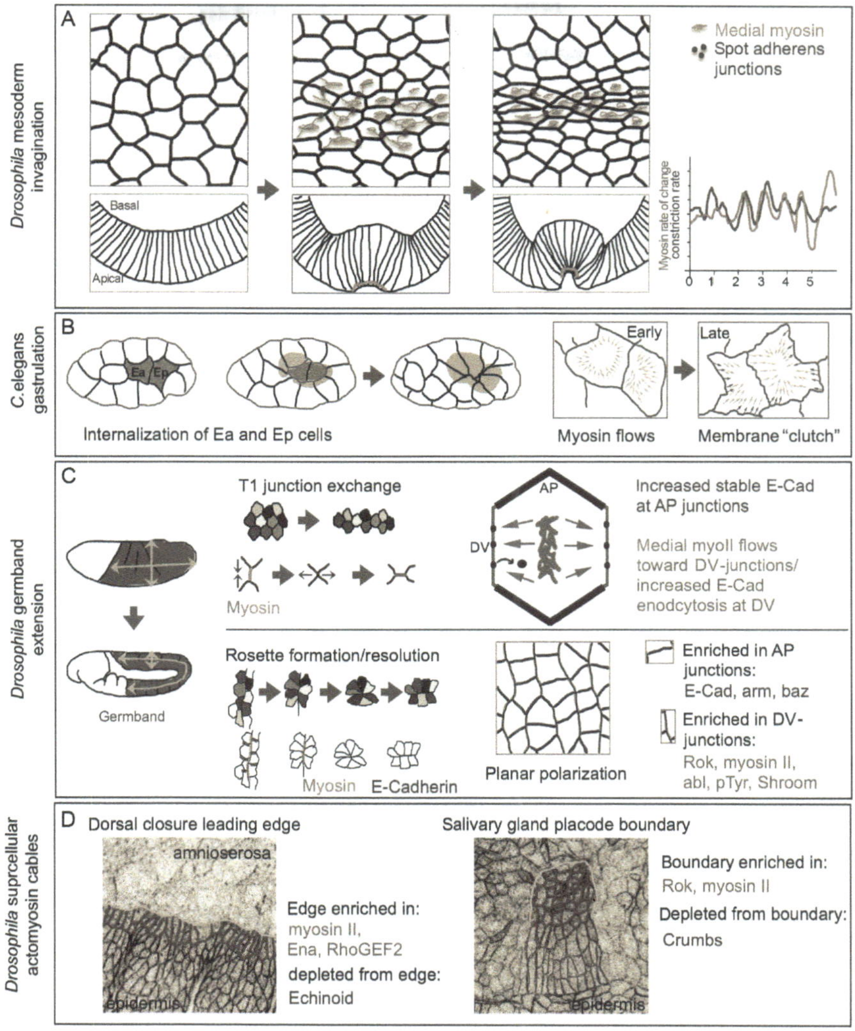

Figure 3 Adherens junction–actomyosin interactions in morphogenesis. (A) During mesoderm invagination in the fly embryo, apical–medial actomyosin networks develop that show a pulsatile increase/decrease in intensity (see graph on right). These myosin II assemblies (green) pull on spot AJ (magenta) to decrease apical area (Martin et al., 2009; Martin & Wieschaus, 2010). Top panels show surface views, bottom panels show cross sections of the presumptive mesoderm. (B) During gastrulation and internalization of the Ea and Ep cells (green) on the ventral side of the *C. elegans* embryo, pulsatile centripetal myosin II flows can already be observed prior to apical constriction. A clutch mechanism that depends on Cadherin and Rac activity allows the flows to be harnessed for productive constriction at a later stage (Roh-Johnson et al., 2012). (C) During *Drosophila*

(Continued)

contracted myosin that is not efficiently linked to the membrane. Also, in embryos completely lacking afadin/canoe, the medial actomyosin cytoskeleton loses connection to the cortex and again forms aggregates. These mutant embryos start to constrict apices but cannot maintain the constriction and the mesoderm remains on the outside of the embryo (Sawyer, Harris, Slep, Gaul, & Peifer, 2009). These experiments illustrate how correct integration of the amount of adhesion and contractility at any given moment is crucial for morphogenesis.

3.2. Myosin flows during gastrulation in *C. elegans*

Similar to *Drosophila* gastrulation, during gastrulation in *C. elegans* pulsatile apicomedial actomyosin plays an important role in driving cell shape changes. Mesoderm invagination in the worm consists of two cells, the endodermal precursors Ea and Ep, being internalized through apical constriction (Fig. 3B). Initially, apical–medial myosin coalesces into central spots but myosin flow is not coupled to apical cortex movement and constriction. Later on, the centripetal movement of myosin particles is clearly

Figure 3—Cont'd germband extension, the germband shrinks in width by half and doubles in length within the confinement of the egg (schematic on the left). This is achieved through a process termed convergent extension, that in the germband seems to be fueled by two main mechanism. T1 junctional exchanges involve groups of four cells where a vertical or DV junction shrinks, followed by expansion of a perpendicular horizontal or AP junction, leading to exchange in neighbors (Bertet, Sulak, & Lecuit, 2004). Flows of medial myosin II clusters flow toward DV-junctions and help shrink them (tope panel diagram), with the flows being triggered by oscillating amounts of E-Cadherin-mediated tension from those junction. There is also an increase in E-Cadherin endocytosis from DV-junctions (Levayer & Lecuit, 2013; Rauzi et al., 2010). The second mechanisms affect larger groups of cells. Myosin II localization and activity is overall planar polarized within the germband, with myosin II being enriched in DV-junctions together with Rok, phospho-Tyrosine (pTyr), abelson kinase (abl), and Shroom (that can recruit Rok) (Blankenship, Backovic, Sanny, Weitz, & Zallen, 2006; Simoes Sde et al., 2010; Simoes Sde, Mainieri, & Zallen, 2014; Tamada, Farrell, & Zallen, 2012). Myosin II in DV-junctions leads to the pulling of larger groups of cells into a single vertex, forming a rosette structure than then resolves in a perpendicular fashion (schematic in lower panel). (D) During dorsal closure in the fly embryo, an actomyosin cable form at the dorsal-most edge of the leading edge cells, where Ena and RhoGEF2 are also enriched. This enrichment depends on the anisotropic localization of Echinoid in the leading edge cells, with loss of Echinoid from the edge surface triggered by the absence of Echinoid expression in the amnioserosa (Laplante & Nilson, 2011). An actomyosin cable also forms around the salivary gland placode in the fly embryo, where instead of Echinoid the anisotropic localization of Crumbs triggers Rok and myosin II enrichment at the boundary (Röper, 2012). (See the color plate.)

correlated and coupled to cortical displacement (Roh-Johnson et al., 2012; Fig. 3B, right panels). This implies a mechanisms whereby at a certain time point a "clutch" is engaged, leading to an effective linking of pre-existing myosin activity and cell shape change. The *C. elegans* classical Cadherin as well as the GTPase Rac are required for this linking, but not for the actomyosin contraction *per se*. The same appears to be true for fly mesoderm invagination, where myosin fluctuations can also already be observed for about 2 min prior to the start of actual constriction (Roh-Johnson et al., 2012).

The existence of apical myosin activity in form of medial actomyosin pulses prior to any morphogenetic activity suggests that such pulsatile myosin behavior could be a more general feature of cells that is actively exploited during morphogenesis.

3.3. Planar polarized myosin and myosin flows during germband extension in *Drosophila*

An important morphogenetic process that drives the coordinated elongation of tissues in many animals is convergent extension. In fly embryos, convergent extension leads to the doubling of the embryos length while halving its width in a process termed GBE (Fig. 3C).

Elongation and perpendicular narrowing of the tissue is achieved through a combination of cell rearrangements and cell shape changes (Butler et al., 2009). The cell rearrangements are planar polarized to achieve the extension, and two main mechanisms with a related molecular basis have been identified that drive these, T1 junctional exchanges and rosette formation/resolution (Bertet et al., 2004; Blankenship et al., 2006; Fig. 3C). Both processes rely on the planar polarized accumulation of myosin II in the so-called vertical or DV (dorsoventrally oriented) junctions, an accumulation that eventually triggers the selective shrinkage of these junctions. In the case of T1 exchanges this is followed by the extension of a single perpendicular junction (Bertet et al., 2004), whereas in the case of rosettes, linked DV-junctions shrink, pulling many cells into a multicellular rosette with a central vertex, followed by resolution and relaxation of the rosette in the perpendicular orientation (Blankenship et al., 2006).

At the onset of cell intercalation, and in a pattern that is complementary to the myosin accumulation, junctional proteins such as E-Cadherin and β-catenin become enriched at horizontal or AP (anterior–posterior oriented) junctions (Blankenship et al., 2006; Fig. 3C). The junctional planar polarity of myosin II as well as β-catenin depends on Bazooka/Par3 enrichment in AP junctions, with Bazooka planar polarity in turn being directed by Rok and early AP-patterning genes (Simoes Sde et al., 2010). Laser ablation experiments as well as computational modeling of the localization and effects of known components such as myosin II revealed that the anisotropy in cortical tension between DV and AP boundaries generated in this way is sufficient to drive tissue elongation through predictive steps of relaxations generated through junction remodeling and neighbor exchanges (Rauzi, Verant, Lecuit, & Lenne, 2008).

Although myosin II is clearly enriched at DV-junctions during GBE and perturbance of this accumulation blocks proper intercalation and junction shrinkage (Blankenship et al., 2006; Rauzi et al., 2008; Simoes Sde et al., 2010), the actual junctional shrinkage during GBE in the fly was recently shown to be driven not by this junctional pool of myosin but rather by dynamic apical myosin II pulses that flow specifically toward both DV-oriented junctions of a germband cell (Rauzi et al., 2010). This orientation of flows toward DV-junctions critically depends on the planar polarized E-Cadherin distribution, with myosin II flowing toward junctions with less E-Cadherin, probably reflecting some force balance across the cell (Rauzi et al., 2010). In fact, it appears that subtle fluctuations in E-Cadherin local concentration in these DV-junctions, that are caused by fluctuating levels of endocytosis of E-Cadherin, trigger the direction of flows (Levayer & Lecuit, 2013; Fig. 3C). The endocytosis of E-Cadherin from DV-junctions is also triggered by the accumulation of abl kinase in these junctions, leading to β-catenin phosphorylation and destabilization (Tamada et al., 2012).

Thus, medial myosin II is a crucial driver of morphogenetic changes in the germband of the *Drosophila* embryo. The main difference in comparison to the mesoderm appears to be the anisotropic distribution of E-Cadherin, leading to an anisotropy in linkage of apical–medial myosin II to the AJ or spot AJ, that hence triggers directional flows of myosin II pulses during GBE that are not observed during mesoderm invagination (Levayer & Lecuit, 2013).

Interestingly, the mechanism of rosette formation/resolution during convergent extension is not restricted to flies. Recent work in mouse and zebrafish kidney has identified cells arranged in rosettes as a key feature during the period of mediolateral intercalation in these vertebrate kidney tubules. Live imaging of the zebrafish embryonic pronephron shows that rosettes form in a myosin-dependent fashion and preferably resolve in a perpendicular way, with the shift in orientation being controlled by the planar cell polarity machinery (Lienkamp et al., 2012). During this process, myosin II is enriched at circumferential boundaries, similar to the organization in the fly germband. This supports the notion that dissecting morphogenetic principles in the fly is a highly valuable tool to better understand morphogenesis of vertebrate tissues in systems where live analysis is either very complex or impossible to achieve.

3.4. Medial myosin and a myosin cable during dorsal closure in *Drosophila*

During mid-embryogenesis in the fly embryo, a patch of extraembryonic tissue, the so-called amnioserosa, contracts and disappears from the embryo surface while concomitantly the surrounding dorsal epidermis moves in and zippers across the patch previously occupied by the amnioserosa. During contraction, amnioserosa cells also display highly dynamic pulsatile myosin, with pulses flowing across the apical surface of these cells, driving constriction (Blanchard, Murugesu, Adams, Martinez-Arias, & Gorfinkiel, 2010; David, Tishkina, & Harris, 2010; David, Wang, Feng, & Harris, 2013). In contrast to both mesoderm and germband, amnioserosa cells show an isotropically distributed population of junctional/cortical actomyosin that increases over time (Blanchard et al., 2010). In this tissue, medial actomyosin drives pulsatile cell behavior, with low-frequency pulses leading to area fluctuation and higher frequency pulses driving active cell constriction. In contrast, the amount of junctional actomyosin correlates with the straightness of the overall cell shape at the level of the AJ.

During dorsal closure, in addition to the contraction of the amnioserosa, the surrounding epidermis has to move together and fuse over the dorsal midline of the embryo to complete the process. It has long been appreciated that during this dorsal-ward movement of the epidermis a surpracellular actomyosin cable forms within the dorsal-most epidermal cells, the so-called leading edge cells (Blake, Myette, & Jack, 1998; Harden, Ricos, Ong, Chia, & Lim, 1999; Martin & Parkhurst, 2004; Fig. 3D). The contact-zone between leading edge and amnioserosa constitutes a modified AJ zone (Kaltschmidt et al., 2002). E-Cadherin is present at the amnioserosa-leading edge interphase, but other components such as Echinoid, canoe/afadin, and Crumbs are lost from the leading edge. Loss of Echinoid and Crumbs localization in the leading edge is due to loss of Echinoid and Crumbs expression in the amnioserosa prior to the start of dorsal closure, leading to an anisotropic distribution of both Echinoid and Crumbs in the leading edge cells (Fig. 3D). Loss of Echinoid anisotropy, in mutant embryos or through overexpression of Echinoid in the amnioserosa itself, leads to loss of actomyosin accumulation at the leading edge (Laplante & Nilson, 2006, 2011).

3.5. An actomyosin cable during tube morphogenesis in *Drosophila*

The ability of actomyosin to form assemblies embracing many cells, in particular actomyosin cables, is a common feature observed in different tissues and organisms. Differing functional classes of cables exist: (1) dynamic circumferential cables that pull in the surrounding epidermis (Fig. 2A and B), (2) long static cables at compartment boundaries that separate cells of different fate, and (3) short dynamic cables that assist morphogenesis, such as during germband extension discussed above (Röper, 2013). The actomyosin cable at the leading edge during dorsal closure is one example of a dynamic circumferential cable. A second example of this type is the circumferential cable surrounding the salivary gland placode in the fly embryo (Fig. 2A and B) that forms as a consequence of an anisotropic localization of Crumbs at the boundary of the placode (Fig. 3D). High levels of Crumbs prevent accumulation of Rok and myosin II, and these instead accumulate at the Crumbs-free boundary to form the cable (Röper, 2012). Interestingly, none of the other upstream junctional components that can induce anisotropic myosin accumulation into cables in either the germband or at compartment boundaries seem to be at work in this process. E-Cadherin, Bazooka/Par3, Echinoid, and Canoe are all also found at the interface where Crumbs is depleted (Röper, 2012). Thus, the junctional trigger for actomyosin cable assembly can be very tissue- and context-specific, whereas the downstream activating machinery that actually drives cable assembly is likely to be more conserved.

More static cables assist morphogenesis in that they keep populations of cells with different developmental programs separated, such as at parasegmental boundaries in the fly embryo (Monier, Pelissier-Monier, Brand, & Sanson, 2010) and at the DV boundary in the wing disc (Major & Irvine, 2006). Short dynamically contracting cables such as during GBE in the fly embryo are not restricted to flies, but have been shown to be important for convergent extension of vertebrate kidney tubules (see above; Lienkamp et al., 2012) and for the folding of the chicken neural tube, where cables that run perpendicular to the long axis of the tube constrict and assist bending at the hinge points (Nishimura, Honda, & Takeichi, 2012).

Supracellular myosin cables are the cytoskeletal structure that appears most continuous over many cell diameters (Fig. 2A and B). At the moment, it is completely unclear, though, how this continuity is achieved. It is assumed that a modified type of AJ will link the ends of cables between adjacent cells and transmit the forces generated by contraction of a cable section within a cell. Which intracellular E-Cadherin-binding partners would specifically mediate such a link is currently unclear. In addition, actomyosin cables might require a special type of lateral tethering to the apical domain they run along to. This lateral interaction, in topology, resembles the interaction of the cytokinetic ring, a unique type of intracellular actomyosin cable, with the plasma membrane (Fededa & Gerlich, 2012). In the case of lateral membrane interactions of multicellular actomyosin cables during morphogenetic processes, no specific components have yet been identified that would tie a cable to the lateral membrane. Components mediating the membrane interaction of the cytokinetic ring with the plasma membrane could be involved. Of particular interest is here that the centralspindlin complex, known for its role in Rho activation in the cytokinetic furrow, is also localized to AJ in mammalian cells in culture. Recruited through interactions with α-catenin, centralspindlin helps to promote Rho activity at junctions through recruitment of the RhoGEF Ect2/pebble and inhibition of recruitment of a RhoGAP, and it thereby activates myosin IIA that is needed to stabilize the AJ in the ZA (Ratheesh et al., 2012). Thus, the centralspindlin complex or further mitotic components that promote actomyosin cytokinetic cable assembly and attachment to the plasma membrane might well be involved in the lateral tethering of supracellular actomyosin cables during morphogenesis.

4. ROLE OF TENSION-SENSING IN ADHERENS JUNCTIONS IN REGULATING CELL SHAPE AND MORPHOGENESIS

One of the most exciting developments over the last few years is the realization and growing understanding that AJ are not only important to couple contractile actomyosin networks or fibers to the plasma membrane and connect them in a multicellular tissue-level network, but that AJ have active mechanisms to determine the amount of force impinging on a junction. Several proteins associated with AJ have recently been shown to undergo conformational changes in response to stretch or tension (Fig. 4; for a recent review, see Huveneers & de Rooij, 2013).

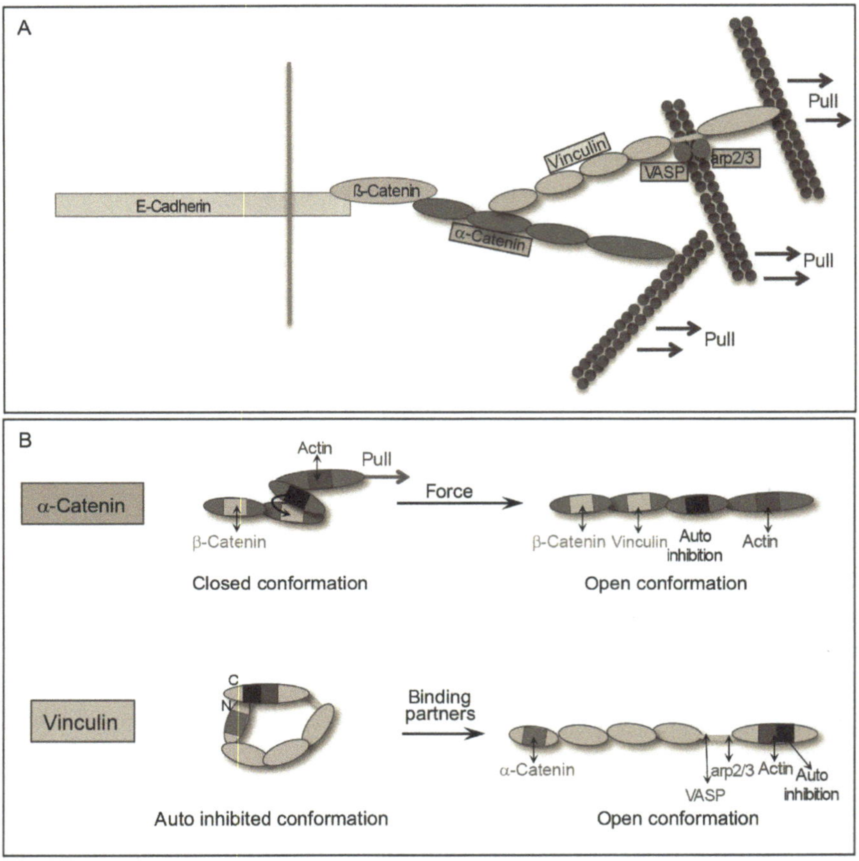

Figure 4 Tension-sensing at adherens junctions. (A) Schematic of E-Cadherin interacting with the two main tension-sensing proteins found in AJ: α-catenin and vinculin. (B) α-Catenin is found in a closed conformation with a masked vinculin-binding site, but already associated with β-catenin/E-Cadherin and actin prior to application of force. Pulling through actomyosin on the molecule leads to a conformational change into an open conformation, in which α-catenin can now bind to vinculin (Choi et al., 2012; Yonemura et al., 2010). Vinculin itself exist in solution in an autoinhibited conformation, but binding of two or more of its interactors will trigger change into and open conformation that can bind further downstream partners as well as actin (Bakolitsa et al., 2004).

4.1. Tension-sensing roles for α-catenin and vinculin in Adherens Junctions

In a central position is α-catenin itself. Recent evidence from epithelial tissue culture suggests that vinculin binding to α-catenin is force-dependent, and that the vinculin-binding site within α-catenin is masked prior to application of actomyosin-mediated tension (Yonemura, Wada, Watanabe, Nagafuchi, & Shibata, 2010). The predicted open and stretched conformation of α-catenin recruits vinculin, that itself is required for the reinforcement of AJ and resulting cell stiffening in culture (Huveneers et al., 2012; Twiss et al., 2012). The AJ strengthening could be the effect of vinculin now providing two further binding sites for F-actin (Huttelmaier, Bubeck, Rudiger, & Jockusch, 1997). In addition, vinculin can bind further downstream factors that influence the actin cytoskeleton, including VASP and Arp2/3 (Brindle, Holt, Davies, Price, & Critchley, 1996; DeMali, Barlow, & Burridge, 2002), further strengthening and modulating the E-Cadherin–actin link. Free vinculin exists in a folded autoinhibited conformation that requires activation, through interaction with two or more of its binding partners, to allow interaction with actin and further binding partners (Bakolitsa et al., 2004). αE-catenin is one of the vinculin-activating factors, but as discussed above, it itself requires activation to be able to do so (Choi et al., 2012). The exact molecular nature of the steps leading to activation of α-catenin at the junction is currently unclear, but is likely to involve α-catenin bound to junctional F-actin already in a closed conformation that is then opened by increased actomyosin-generated stretch (Borghi et al., 2012; Ishiyama et al., 2013).

In mammalian tissue culture, EPLIN is another α-catenin-binding partner that cooperates with vinculin to mediate a mechanosensitive response, in particular the remodeling of spot AJ into a mature AJ in form of a ZA (Taguchi, Ishiuchi, & Takeichi, 2011).

4.2. Using tension sensors to study contractility and tension during morphogenesis

The discovery of mechanosensitive proteins associated with AJ opened up the possibility to utilize engineered tension sensors as *in vivo* read-outs of the amount of tension that a given junctional complex is under. The best tool in fixed mammalian tissues thus far is an antibody that recognizes only the opened conformation of α-catenin that is under stretch (Yonemura et al., 2010). This antibody, α18, at least in epithelial cells seems to faithfully label only junctions that are known to be or artificially induced to be under stretch.

A sensor that has recently been developed and proven to work in both tissue culture as well as an *in vivo* (*Drosophila*) context is a modified version of E-Cadherin itself. In this E-Cadherin tension sensor, called ECad-TSMod, a

FRET module with a spring that can unfold under stretch is inserted into the intracellular domain of E-Cadherin before the β-catenin-binding domain (Borghi et al., 2012; Cai et al., 2014; Grashoff et al., 2010). Analysis of this sensor in comparison to the FRET-spring module alone and an ECad-TSMod where the tensor module is inserted beyond the β-catenin-binding site and that is thus insensitive to stretch, revealed the continuous stretch applied on E-Cadherin inside and outside of junctions in cultured cells, as discussed above (Borghi et al., 2012). Use of the ECad-TSMod sensor in analyzing collective migration of the border cells in the fly ovary revealed that stretch-sensing through E-Cadherin assists the chemoattractant-based direction sensing in this process (Cai et al., 2014).

Tools such as these will in the future be essential to guide our understanding of the role of force in AJ assembly, maturation, and turnover.

5. SUMMARY

Over the last few years, our understanding of the molecular mechanisms that guide the interplay between the contractile actomyosin cytoskeleton and components of the AJ has increased dramatically. Many key players have been identified, many structures have been determined, and the central debate on whether there is a ternary complex of E-Cadherin/β-catenin/α-catenin/actin has been laid to rest with many recent experiments showing that this is likely to be the case. What is still fairly poorly understood is how different types of actin fibers or actomyosin assemblies are linked differentially into the junctions. What are the discriminating interactors that favor one over the other? And how are very specialized structures such as supracellular actomyosin cables linked across many cells?

With respect to understanding epithelial morphogenesis, the advent of functioning *in vivo* tension sensors will likely revolutionize our understanding of the forces and strains impinging on junctions, cells, and tissues during morphogenetic events.

ACKNOWLEDGMENTS

The author apologizes to colleagues whose work could not be cited or discussed in sufficient depth owing to space limitations.

The author would like to thank member of the Röper lab for comments on the chapter. Work in the lab is supported by the Medical Research Council (MRC file reference number U105178780).

REFERENCES

Abe, K., & Takeichi, M. (2008). EPLIN mediates linkage of the cadherin catenin complex to F-actin and stabilizes the circumferential actin belt. *Proceedings of the National Academy of Sciences of the United States of America*, *105*(1), 13–19.

Bakolitsa, C., Cohen, D. M., Bankston, L. A., Bobkov, A. A., Cadwell, G. W., Jennings, L., et al. (2004). Structural basis for vinculin activation at sites of cell adhesion. *Nature*, *430*(6999), 583–586.

Baum, B., & Georgiou, M. (2011). Dynamics of adherens junctions in epithelial establishment, maintenance, and remodeling. *The Journal of Cell Biology*, *192*(6), 907–917.

Baum, B., Settleman, J., & Quinlan, M. P. (2008). Transitions between epithelial and mesenchymal states in development and disease. *Seminars in Cell & Developmental Biology*, *19*(3), 294–308.

Benjamin, J. M., Kwiatkowski, A. V., Yang, C., Korobova, F., Pokutta, S., Svitkina, T., et al. (2010). AlphaE-catenin regulates actin dynamics independently of cadherin-mediated cell-cell adhesion. *The Journal of Cell Biology*, *189*(2), 339–352.

Bertet, C., Sulak, L., & Lecuit, T. (2004). Myosin-dependent junction remodelling controls planar cell intercalation and axis elongation. *Nature*, *429*(6992), 667–671.

Blake, K. J., Myette, G., & Jack, J. (1998). The products of ribbon and raw are necessary for proper cell shape and cellular localization of nonmuscle myosin in Drosophila. *Developmental Biology*, *203*(1), 177–188.

Blanchard, G. B., Murugesu, S., Adams, R. J., Martinez-Arias, A., & Gorfinkiel, N. (2010). Cytoskeletal dynamics and supracellular organisation of cell shape fluctuations during dorsal closure. *Development*, *137*(16), 2743–2752.

Blankenship, J. T., Backovic, S. T., Sanny, J. S., Weitz, O., & Zallen, J. A. (2006). Multicellular rosette formation links planar cell polarity to tissue morphogenesis. *Developmental Cell*, *11*(4), 459–470.

Borghi, N., Sorokina, M., Shcherbakova, O. G., Weis, W. I., Pruitt, B. L., Nelson, W. J., et al. (2012). E-cadherin is under constitutive actomyosin-generated tension that is increased at cell-cell contacts upon externally applied stretch. *Proceedings of the National Academy of Sciences of the United States of America*, *109*(31), 12568–12573.

Braga, V. M., Betson, M., Li, X., & Lamarche-Vane, N. (2000). Activation of the small GTPase Rac is sufficient to disrupt cadherin-dependent cell-cell adhesion in normal human keratinocytes. *Molecular Biology of the Cell*, *11*(11), 3703–3721.

Brindle, N. P., Holt, M. R., Davies, J. E., Price, C. J., & Critchley, D. R. (1996). The focal-adhesion vasodilator-stimulated phosphoprotein (VASP) binds to the proline-rich domain in vinculin. *The Biochemical Journal*, *318*(Pt. 3), 753–757.

Buckley, C. D., Tan, J., Anderson, K. L., Hanein, D., Volkmann, N., Weis, W. I., et al. (2014). Cell adhesion. The minimal cadherin-catenin complex binds to actin filaments under force. *Science*, *346*(6209), 1254211. http://dx.doi.org/10.1126/science.1254211.

Butler, L. C., Blanchard, G. B., Kabla, A. J., Lawrence, N. J., Welchman, D. P., Mahadevan, L., et al. (2009). Cell shape changes indicate a role for extrinsic tensile forces in Drosophila germ-band extension. *Nature Cell Biology*, *11*(7), 859–864.

Cai, D., Chen, S. C., Prasad, M., He, L., Wang, X., Choesmel-Cadamuro, V., et al. (2014). Mechanical feedback through E-cadherin promotes direction sensing during collective cell migration. *Cell*, *157*(5), 1146–1159.

Cavey, M., Rauzi, M., Lenne, P. F., & Lecuit, T. (2008). A two-tiered mechanism for stabilization and immobilization of E-cadherin. *Nature*, *453*(7196), 751–756.

Chang, L. H., Chen, P., Lien, M. T., Ho, Y. H., Lin, C. M., Pan, Y. T., et al. (2011). Differential adhesion and actomyosin cable collaborate to drive Echinoid-mediated cell sorting. *Development*, *138*(17), 3803–3812.

Chen, Y. T., Stewart, D. B., & Nelson, W. J. (1999). Coupling assembly of the E-cadherin/beta-catenin complex to efficient endoplasmic reticulum exit and basal-lateral membrane targeting of E-cadherin in polarized MDCK cells. *The Journal of Cell Biology*, *144*(4), 687–699.

Choi, H. J., Pokutta, S., Cadwell, G. W., Bobkov, A. A., Bankston, L. A., Liddington, R. C., et al. (2012). alphaE-catenin is an autoinhibited molecule that coactivates vinculin. *Proceedings of the National Academy of Sciences of the United States of America*, *109*(22), 8576–8581.

David, D. J., Tishkina, A., & Harris, T. J. (2010). The PAR complex regulates pulsed actomyosin contractions during amnioserosa apical constriction in Drosophila. *Development*, *137*(10), 1645–1655.

David, D. J., Wang, Q., Feng, J. J., & Harris, T. J. (2013). Bazooka inhibits aPKC to limit antagonism of actomyosin networks during amnioserosa apical constriction. *Development*, *140*(23), 4719–4729.

Delanoe-Ayari, H., Al Kurdi, R., Vallade, M., Gulino-Debrac, D., & Riveline, D. (2004). Membrane and acto-myosin tension promote clustering of adhesion proteins. *Proceedings of the National Academy of Sciences of the United States of America*, *101*(8), 2229–2234.

Delva, E., & Kowalczyk, A. P. (2009). Regulation of cadherin trafficking. *Traffic*, *10*(3), 259–267.

DeMali, K. A., Barlow, C. A., & Burridge, K. (2002). Recruitment of the Arp2/3 complex to vinculin: Coupling membrane protrusion to matrix adhesion. *The Journal of Cell Biology*, *159*(5), 881–891.

Desai, R., Sarpal, R., Ishiyama, N., Pellikka, M., Ikura, M., & Tepass, U. (2013). Monomeric alpha-catenin links cadherin to the actin cytoskeleton. *Nature Cell Biology*, *15*(3), 261–273.

Fededa, J. P., & Gerlich, D. W. (2012). Molecular control of animal cell cytokinesis. *Nature Cell Biology*, *14*(5), 440–447.

Fleming, T. P., Sheth, B., & Fesenko, I. (2001). Cell adhesion in the preimplantation mammalian embryo and its role in trophectoderm differentiation and blastocyst morphogenesis. *Frontiers in Bioscience*, *6*, D1000–D1007.

Girdler, G. C., & Röper, K. (2014). Controlling cell shape changes during salivary gland tube formation in Drosophila. *Seminars in Cell & Developmental Biology*, *31*, 74–81.

Gorfinkiel, N., & Blanchard, G. B. (2011). Dynamics of actomyosin contractile activity during epithelial morphogenesis. *Current Opinion in Cell Biology*, *23*(5), 531–539.

Grashoff, C., Hoffman, B. D., Brenner, M. D., Zhou, R., Parsons, M., Yang, M. T., et al. (2010). Measuring mechanical tension across vinculin reveals regulation of focal adhesion dynamics. *Nature*, *466*, 263–266.

Harden, N., Ricos, M., Ong, Y. M., Chia, W., & Lim, L. (1999). Participation of small GTPases in dorsal closure of the Drosophila embryo: Distinct roles for Rho subfamily proteins in epithelial morphogenesis. *Journal of Cell Science*, *112*(Pt. 3), 273–284.

Harris, T. J., & Peifer, M. (2005). The positioning and segregation of apical cues during epithelial polarity establishment in Drosophila. *The Journal of Cell Biology*, *170*(5), 813–823.

Harris, T. J., Sawyer, J. K., & Peifer, M. (2009). How the cytoskeleton helps build the embryonic body plan: Models of morphogenesis from Drosophila. *Current Topics in Developmental Biology*, *89*, 55–85.

Harris, T. J., & Tepass, U. (2010). Adherens junctions: From molecules to morphogenesis. *Nature Reviews Molecular Cell Biology*, *11*(7), 502–514.

Huttelmaier, S., Bubeck, P., Rudiger, M., & Jockusch, B. M. (1997). Characterization of two F-actin-binding and oligomerization sites in the cell-contact protein vinculin. *European Journal of Biochemistry*, *247*(3), 1136–1142.

Huveneers, S., & de Rooij, J. (2013). Mechanosensitive systems at the cadherin-F-actin interface. *Journal of Cell Science*, *126*(Pt. 2), 403–413.

Huveneers, S., Oldenburg, J., Spanjaard, E., van der Krogt, G., Grigoriev, I., Akhmanova, A., et al. (2012). Vinculin associates with endothelial VE-cadherin junctions to control force-dependent remodeling. *The Journal of Cell Biology, 196*(5), 641–652.

Indra, I., Hong, S., Troyanovsky, R., Kormos, B., & Troyanovsky, S. (2013). The adherens junction: A mosaic of cadherin and nectin clusters bundled by actin filaments. *The Journal of Investigative Dermatology, 133*(11), 2546–2554.

Ishiyama, N., Tanaka, N., Abe, K., Yang, Y. J., Abbas, Y. M., Umitsu, M., et al. (2013). An autoinhibited structure of alpha-catenin and its implications for vinculin recruitment to adherens junctions. *The Journal of Biological Chemistry, 288*(22), 15913–15925.

Kaltschmidt, J. A., Lawrence, N., Morel, V., Balayo, T., Fernandez, B. G., Pelissier, A., et al. (2002). Planar polarity and actin dynamics in the epidermis of Drosophila. *Nature Cell Biology, 4*(12), 937–944. http://dx.doi.org/10.1038/ncb882.

Kobielak, A., & Fuchs, E. (2004). Alpha-catenin: At the junction of intercellular adhesion and actin dynamics. *Nature Reviews Molecular Cell Biology, 5*(8), 614–625.

Kovacs, E. M., & Yap, A. S. (2008). Cell-cell contact: Cooperating clusters of actin and cadherin. *Current Biology, 18*(15), R667–R669.

Langevin, J., Morgan, M. J., Sibarita, J. B., Aresta, S., Murthy, M., Schwarz, T., et al. (2005). Drosophila exocyst components Sec5, Sec6, and Sec15 regulate DE-Cadherin trafficking from recycling endosomes to the plasma membrane. *Developmental Cell, 9*(3), 365–376.

Laplante, C., & Nilson, L. A. (2006). Differential expression of the adhesion molecule Echinoid drives epithelial morphogenesis in Drosophila. *Development, 133*(16), 3255–3264.

Laplante, C., & Nilson, L. A. (2011). Asymmetric distribution of Echinoid defines the epidermal leading edge during Drosophila dorsal closure. *The Journal of Cell Biology, 192*(2), 335–348.

Levayer, R., & Lecuit, T. (2012). Biomechanical regulation of contractility: Spatial control and dynamics. *Trends in Cell Biology, 22*(2), 61–81.

Levayer, R., & Lecuit, T. (2013). Oscillation and polarity of E-cadherin asymmetries control actomyosin flow patterns during morphogenesis. *Developmental Cell, 26*(2), 162–175.

Levayer, R., Pelissier-Monier, A., & Lecuit, T. (2011). Spatial regulation of Dia and Myosin-II by RhoGEF2 controls initiation of E-cadherin endocytosis during epithelial morphogenesis. *Nature Cell Biology, 13*(5), 529–540.

Lienkamp, S. S., Liu, K., Karner, C. M., Carroll, T. J., Ronneberger, O., Wallingford, J. B., et al. (2012). Vertebrate kidney tubules elongate using a planar cell polarity-dependent, rosette-based mechanism of convergent extension. *Nature Genetics, 44*(12), 1382–1387.

Major, R. J., & Irvine, K. D. (2006). Localization and requirement for myosin II at the dorsal-ventral compartment boundary of the Drosophila wing. *Developmental Dynamics, 235*(11), 3051–3058.

Martin, A. C., Gelbart, M., Fernandez-Gonzalez, R., Kaschube, M., & Wieschaus, E. F. (2010). Integration of contractile forces during tissue invagination. *The Journal of Cell Biology, 188*(5), 735–749.

Martin, A. C., Kaschube, M., & Wieschaus, E. F. (2009). Pulsed contractions of an actin-myosin network drive apical constriction. *Nature, 457*(7228), 495–499.

Martin, P., & Parkhurst, S. M. (2004). Parallels between tissue repair and embryo morphogenesis. *Development, 131*(13), 3021–3034.

Martin, A. C., & Wieschaus, E. F. (2010). Tensions divide. *Nature Cell Biology, 12*(1), 5–7.

Monier, B., Pelissier-Monier, A., Brand, A. H., & Sanson, B. (2010). An actomyosin-based barrier inhibits cell mixing at compartmental boundaries in Drosophila embryos. *Nature Cell Biology, 12*(1), 60–65, sup pp. 61–69.

Morais-de-Sa, E., Mirouse, V., & St Johnston, D. (2010). aPKC phosphorylation of Bazooka defines the apical/lateral border in Drosophila epithelial cells. *Cell, 141*(3), 509–523.

Nanes, B. A., Chiasson-MacKenzie, C., Lowery, A. M., Ishiyama, N., Faundez, V., Ikura, M., et al. (2012). p120-catenin binding masks an endocytic signal conserved in classical cadherins. *The Journal of Cell Biology*, *199*(2), 365–380.

Nikolaidou, K. K., & Barrett, K. (2004). A Rho GTPase signaling pathway is used reiteratively in epithelial folding and potentially selects the outcome of Rho activation. *Current Biology*, *14*(20), 1822–1826.

Nishimura, T., Honda, H., & Takeichi, M. (2012). Planar cell polarity links axes of spatial dynamics in neural-tube closure. *Cell*, *149*(5), 1084–1097.

Pokutta, S., & Weis, W. I. (2000). Structure of the dimerization and beta-catenin-binding region of alpha-catenin. *Molecular Cell*, *5*(3), 533–543.

Pokutta, S., & Weis, W. I. (2007). Structure and mechanism of cadherins and catenins in cell-cell contacts. *Annual Review of Cell and Developmental Biology*, *23*, 237–261.

Ratheesh, A., Gomez, G. A., Priya, R., Verma, S., Kovacs, E. M., Jiang, K., et al. (2012). Centralspindlin and alpha-catenin regulate Rho signalling at the epithelial zonula adherens. *Nature Cell Biology*, *14*(8), 818–828.

Rauzi, M., Lenne, P. F., & Lecuit, T. (2010). Planar polarized actomyosin contractile flows control epithelial junction remodelling. *Nature*, *468*(7327), 1110–1114.

Rauzi, M., Verant, P., Lecuit, T., & Lenne, P. F. (2008). Nature and anisotropy of cortical forces orienting Drosophila tissue morphogenesis. *Nature Cell Biology*, *10*(12), 1401–1410.

Rikitake, Y., Mandai, K., & Takai, Y. (2012). The role of nectins in different types of cell-cell adhesion. *Journal of Cell Science*, *125*(Pt. 16), 3713–3722.

Rimm, D. L., Koslov, E. R., Kebriaei, P., Cianci, C. D., & Morrow, J. S. (1995). Alpha 1(E)-catenin is an actin-binding and -bundling protein mediating the attachment of F-actin to the membrane adhesion complex. *Proceedings of the National Academy of Sciences of the United States of America*, *92*(19), 8813–8817.

Roh-Johnson, M., Shemer, G., Higgins, C. D., McClellan, J. H., Werts, A. D., Tulu, U. S., et al. (2012). Triggering a cell shape change by exploiting preexisting actomyosin contractions. *Science*, *335*(6073), 1232–1235.

Röper, K. (2012). Anisotropy of crumbs and aPKC drives myosin cable assembly during tube formation. *Developmental Cell*, *23*(5), 939–953.

Röper, K. (2013). Supracellular actomyosin assemblies during development. *Bioarchitecture*, *3*(2), 45–49.

Sahai, E., & Marshall, C. J. (2002). ROCK and Dia have opposing effects on adherens junctions downstream of Rho. *Nature Cell Biology*, *4*(6), 408–415.

Sawyer, J. K., Harris, N. J., Slep, K. C., Gaul, U., & Peifer, M. (2009). The Drosophila afadin homologue Canoe regulates linkage of the actin cytoskeleton to adherens junctions during apical constriction. *The Journal of Cell Biology*, *186*(1), 57–73.

Sheth, B., Fontaine, J. J., Ponza, E., McCallum, A., Page, A., Citi, S., et al. (2000). Differentiation of the epithelial apical junctional complex during mouse preimplantation development: A role for rab13 in the early maturation of the tight junction. *Mechanisms of Development*, *97*(1–2), 93–104.

Simoes Sde, M., Blankenship, J. T., Weitz, O., Farrell, D. L., Tamada, M., Fernandez-Gonzalez, R., et al. (2010). Rho-kinase directs Bazooka/Par-3 planar polarity during Drosophila axis elongation. *Developmental Cell*, *19*(3), 377–388.

Simoes Sde, M., Mainieri, A., & Zallen, J. A. (2014). Rho GTPase and Shroom direct planar polarized actomyosin contractility during convergent extension. *The Journal of Cell Biology*, *204*(4), 575–589.

Taguchi, K., Ishiuchi, T., & Takeichi, M. (2011). Mechanosensitive EPLIN-dependent remodeling of adherens junctions regulates epithelial reshaping. *The Journal of Cell Biology*, *194*(4), 643–656.

Takai, Y., Ikeda, W., Ogita, H., & Rikitake, Y. (2008). The immunoglobulin-like cell adhesion molecule nectin and its associated protein afadin. *Annual Review of Cell and Developmental Biology, 24*, 309–342.

Tamada, M., Farrell, D. L., & Zallen, J. A. (2012). Abl regulates planar polarized junctional dynamics through beta-catenin tyrosine phosphorylation. *Developmental Cell, 22*(2), 309–319.

Tepass, U. (2012). The apical polarity protein network in Drosophila epithelial cells: Regulation of polarity, junctions, morphogenesis, cell growth, and survival. *Annual Review of Cell and Developmental Biology, 28*, 655–685.

Thompson, B. J., Pichaud, F., & Röper, K. (2013). Sticking together the Crumbs—An unexpected function for an old friend. *Nature Reviews Molecular Cell Biology, 14*(5), 307–314.

Troyanovsky, R. B., Sokolov, E. P., & Troyanovsky, S. M. (2006). Endocytosis of cadherin from intracellular junctions is the driving force for cadherin adhesive dimer disassembly. *Molecular Biology of the Cell, 17*(8), 3484–3493.

Twiss, F., Le Duc, Q., Van Der Horst, S., Tabdili, H., Van Der Krogt, G., Wang, N., et al. (2012). Vinculin-dependent Cadherin mechanosensing regulates efficient epithelial barrier formation. *Biology Open, 1*(11), 1128–1140.

Vaezi, A., Bauer, C., Vasioukhin, V., & Fuchs, E. (2002). Actin cable dynamics and Rho/Rock orchestrate a polarized cytoskeletal architecture in the early steps of assembling a stratified epithelium. *Developmental Cell, 3*(3), 367–381.

Walther, R. F., & Pichaud, F. (2010). Crumbs/DaPKC-dependent apical exclusion of Bazooka promotes photoreceptor polarity remodeling. *Current Biology, 20*(12), 1065–1074.

Wei, S. Y., Escudero, L. M., Yu, F., Chang, L. H., Chen, L. Y., Ho, Y. H., et al. (2005). Echinoid is a component of adherens junctions that cooperates with DE-Cadherin to mediate cell adhesion. *Developmental Cell, 8*(4), 493–504.

Wu, Y., Vendome, J., Shapiro, L., Ben-Shaul, A., & Honig, B. (2011). Transforming binding affinities from three dimensions to two with application to cadherin clustering. *Nature, 475*(7357), 510–513.

Yamada, S., & Nelson, W. J. (2007). Localized zones of Rho and Rac activities drive initiation and expansion of epithelial cell-cell adhesion. *The Journal of Cell Biology, 178*(3), 517–527.

Yamada, S., Pokutta, S., Drees, F., Weis, W. I., & Nelson, W. J. (2005). Deconstructing the cadherin-catenin-actin complex. *Cell, 123*(5), 889–901.

Yonemura, S., Wada, Y., Watanabe, T., Nagafuchi, A., & Shibata, M. (2010). alpha-Catenin as a tension transducer that induces adherens junction development. *Nature Cell Biology, 12*(6), 533–542.

Young, P. E., Pesacreta, T. C., & Kiehart, D. P. (1991). Dynamic changes in the distribution of cytoplasmic myosin during Drosophila embryogenesis. *Development, 111*(1), 1–14.

Zandy, N. L., Playford, M., & Pendergast, A. M. (2007). Abl tyrosine kinases regulate cell-cell adhesion through Rho GTPases. *Proceedings of the National Academy of Sciences of the United States of America, 104*(45), 17686–17691.

Zhang, Y., Sivasankar, S., Nelson, W. J., & Chu, S. (2009). Resolving cadherin interactions and binding cooperativity at the single-molecule level. *Proceedings of the National Academy of Sciences of the United States of America, 106*(1), 109–114.

Zou, J., Wang, X., & Wei, X. (2012). Crb apical polarity proteins maintain zebrafish retinal cone mosaics via intercellular binding of their extracellular domains. *Developmental Cell, 22*(6), 1261–1274.

CHAPTER FIVE

Nuclear Signaling from Cadherin Adhesion Complexes

Pierre D. McCrea*[,1], Meghan T. Maher[†], Cara J. Gottardi[‡,§,1]
*Department of Genetics, University of Texas MD Anderson Cancer Center; Program in Genes & Development, Graduate School in Biomedical Sciences, Houston, Texas, USA
[†]Department of Biology, Washington University in St. Louis, St. Louis, Missouri, USA
[‡]Cellular and Molecular Biology, Northwestern University Feinberg School of Medicine, Chicago, Illinois, USA
[§]Medicine, Northwestern University Feinberg School of Medicine, Chicago, Illinois, USA
[1]Corresponding authors: e-mail address: pdmccrea@mdanderson.org; c-gottardi@northwestern.edu

Contents

Abstract

The arrival of multicellularity in evolution facilitated cell–cell signaling in conjunction with adhesion. As the ectodomains of cadherins interact with each other directly in *trans* (as well as in *cis*), spanning the plasma membrane and associating with multiple

Current Topics in Developmental Biology, Volume 112
ISSN 0070-2153
http://dx.doi.org/10.1016/bs.ctdb.2014.11.018

other entities, cadherins enable the transduction of "outside-in" or "inside-out" signals. We focus this review on signals that originate from the larger family of cadherins that are inwardly directed to the nucleus, and thus have roles in gene control or nuclear structure–function. The nature of cadherin complexes varies considerably depending on the type of cadherin and its context, and we will address some of these variables for classical cadherins versus other family members. Substantial but still fragmentary progress has been made in understanding the signaling mediators used by varied cadherin complexes to coordinate the state of cell–cell adhesion with gene expression. Evidence that cadherin intracellular binding partners also localize to the nucleus is a major point of interest. In some models, catenins show reduced binding to cadherin cytoplasmic tails favoring their engagement in gene control. When bound, cadherins may serve as stoichiometric competitors of nuclear signals. Cadherins also directly or indirectly affect numerous signaling pathways (e.g., Wnt, receptor tyrosine kinase, Hippo, NFκB, and JAK/STAT), enabling cell–cell contacts to touch upon multiple biological outcomes in embryonic development and tissue homeostasis.

The ability of cells to associate and organize themselves in response to one another the extracellular matrix or other cues to generate a variety of tissue shapes and forms has captivated biologists for centuries. Here, we discuss the broad principles of how the state of intercellular adhesion is communicated to a cell's interior, with a focus on changes in nuclear gene activity. We will examine signals that originate from or involve the classic cadherin–catenin complex present at cell–cell contacts or more distantly related cadherin family members. Over 100 types of cadherins are present in the larger superfamily, with a variety of structural characteristics and functions (Hulpiau & van Roy, 2009; Oda & Takeichi, 2011).

Related aspects of this topic have been addressed in a number of excellent reviews, to which the reader is directed (Cavallaro & Christofori, 2004; Cavallaro & Dejana, 2011; Daugherty & Gottardi, 2007; Gavard, 2013; Geletu, Guy, Arulanandam, Feracci, & Raptis, 2013; Heuberger & Birchmeier, 2010; McEwen, Escobar, & Gottardi, 2012; Nelson & Nusse, 2004; Paulson, Prasad, Thuringer, & Manzerra, 2014; Philippova et al., 2009; Stepniak, Radice, & Vasioukhin, 2009; Sun, Parrish, Hill, & Meininger, 2014; van Roy, 2014). In addition to providing a current analysis, we aim here to provide a perspective on cadherins that addresses a cross-section of the superfamily, especially in vertebrates. Some recent findings include evidence, for example, of the intersection of classic cadherin–catenin complex components or functions with the Wnt or Hippo signaling pathways, each of which has powerful roles in development and human pathology. Signaling trajectories to the nucleus are initiated from additional types of cadherins as well, including desmosomal cadherins, or from cadherins that are more distantly related and comprise part of the planar cell polarity network, used in establishing proximal–distal identities within tissues. While less has been uncovered in these latter cases with regard to nuclear signaling, intriguing possibilities have arisen or are postulated.

Much of what we address here is within the sphere of "outside-in" signaling, that is, a form of signaling where a cue that originates from a cell's exterior is propagated to the interior cytoplasm and nucleus. In the context of cadherin family members, this normally includes involvement in cytoskeletal control, such as the modulation of small GTPases. However, while such small GTPases and cytoskeletal effects are likely to contribute to downstream effects upon gene activity, we will give this area relatively brief mention and refer the reader to other reviews (Anastasiadis, 2007; Hatzfeld, 2005; Kourtidis, Ngok, & Anastasiadis, 2013; Pieters, van Roy, & van Hengel, 2012; Ratheesh, Priya, & Yap, 2013). In a similar vein, cadherins can mechanically transduce outside pulling forces originating from adjacent cells into intracellular signals that regulate gene activity. Since the nuclear skeleton and the cytoskeletons are interconnected, mechanical changes at cell contacts can be rapidly propagated through the nuclear cytoskeleton with consequent but little understood effects upon the organization and activity of chromatin. Concerning this and other modes of mechanotransduction, we again direct the reader to other expert reviews (Brieher & Yap, 2013; Cavey & Lecuit, 2009; DuFort, Paszek, & Weaver, 2011; Gomez, McLachlan, & Yap, 2011; Huveneers & de Rooij, 2013; Maitre & Heisenberg, 2013; Smutny & Yap, 2010; Twiss & de Rooij, 2013).

1. β-CATENIN, A DUAL-FUNCTION ADHESION/TRANSCRIPTIONAL COACTIVATOR PROTEIN

β-Catenin was first isolated in association with the cytoplasmic domains of classical cadherins at cell–cell junctions and later was recognized to be the prime signal transducer of the canonical Wnt pathway, passing through nuclear pores en route to gene control regions (reviewed in Fagotto, 2013). In response to upstream Wnt pathway activation, β-catenin enters the nucleus to derepress (activate) genes that are otherwise repressed by T cell factor (TCF)/lymphoid enhancer factor (LEF). Since β-catenin is essential to most vertebrate developmental processes, with pathologic activity contributing to multiple human diseases, there have been innumerable attempts to unravel the highly involved set of molecular participants in what is commonly referred to as the "canonical Wnt pathway."

How is the nuclear pool of β-catenin generated? Under normal cellular conditions, β-catenin that is not already participating in cadherin-based adhesions is rapidly destroyed. Wnt signals effectively block this elaborate destruction mechanism, which allows β-catenin protein to accumulate in the cytosol, translocate to the nucleus, bind TCF/LEF DNA-binding factors, and induce transcription. Although conceptually simple, a dizzying array of proteins contribute to the targeted destruction of β-catenin and its inhibition by Wnts, which implies that organisms go to great trouble to control the magnitude, duration, and cell-specific expression of genes imposed by Wnt signals in the name of normal development. Thus, rather than being comprehensive, we will summarize key steps in the Wnt pathway where cross talk with cadherins has been observed.

Briefly, the canonical Wnt pathway is initiated by an extracellular Wnt-ligand binding to a transmembrane coreceptor complex comprised of seven-membrane pass Frizzled (Fz) protein and Low-density lipoprotein Receptor-related Protein 5 or 6 (Lrp5/6). After engaging these receptors, a canonical Wnt acts to stabilize a pool of β-catenin that would otherwise be destroyed by a macromolecular assembly termed the "destruction complex." This complex is organized by a key scaffold component, Axin. Through its ability to bind β-catenin, its regulatory kinases, and other modifiers, Axin favors sequential phosphorylations by casein kinase 1 alpha (CK1α) and glycogen synthase kinase 3β (GSK3β), events that ultimately "flag" β-catenin for E3-ligase (β-TrCP) phosphorecognition, ubiquitylation, and proteosomal destruction. The adenomatous polyposis coli (APC) tumor suppressor protein is a scaffold component with binding sites for β-catenin and Axin (among other proteins) and is particularly noteworthy because loss-of-function mutations in APC cause both heritable and sporadic forms of colon cancer through excessive β-catenin nuclear signaling (reviewed in Heinen, 2010; Polakis, 2000). Despite the clear importance of APC, molecular details as to how this protein leads to the destruction of β-catenin are just coming into view. APC may facilitate β-catenin "flux" through the destruction complex by its higher affinity binding to β-catenin than Axin (Ha, Tonozuka, Stamos, Choi, & Weis, 2004) as well as shield β-catenin's N-terminal phosphate groups (required for β-TrCP-mediated ubiquitylation and destruction) from dephosphorylation by the phosphatase PP2 (Su et al., 2008), possibly through a region in APC that can bind the

F-actin-binding protein α-catenin (Choi, Estaras, Moresco, Yates, & Jones, 2013). Wnt binding to cell surface Fz and Lrp5/6 receptors (Chu et al., 2013) favors phosphorylations along the cytoplasmic tail of Lrp5/6, leading to the recruitment of Axin and the inhibition of GSK-3β within the Axin complex (Cselenyi et al., 2008; Mi, Dolan, & Johnson, 2006; Piao et al., 2008; Stamos & Weis, 2013; Wu, Huang, Garcia Abreu, & He, 2009), in addition to β-catenin hypophosphorylation and its ultimate stabilization and nuclear signaling. Wnt signaling can also target GSK3 to multivesicular endosomes, which may be important for sustained GSK3 inhibition required for long-term β-catenin signaling activity (Taelman et al., 2010). While any point along this pathway may be subject to regulation, evidence (discussed below) indicates that cadherins largely affect β-catenin signaling at three levels: (1) direct, high-affinity binding to β-catenin that competes for its interactions with other binding partners (TCF, Axin, APC) and thereby prevents nuclear import (Fagotto, Gluck, & Gumbiner, 1998; Suh & Gumbiner, 2003) (Fig. 1); (2) increasing the rate of β-catenin flux through the destruction complex (Hay et al., 2009; Maher, Flozak, Stocker, Chenn, & Gottardi, 2009) (Figs. 2A and 3); and (3) facilitating activation of the Wnt receptor complex at the level of Lrp5/6 phosphorylation, endocytosis, and "signalosome" formation (Bilic et al., 2007; Blitzer & Nusse, 2006) (Figs. 2B and 3). We reason that these distinct inputs arose during the course of evolution so that the state of cadherin adhesive interactions could inform the magnitude and duration of the nuclear differentiation programs dictated by β-catenin/TCF and perhaps additional gene regulatory complexes.

2. CADHERINS AS STOICHIOMETRIC INHIBITORS OF WNT/β-CATENIN SIGNALING

Evidence that β-catenin is both a critical component of the cadherin/catenin adhesive complex, serving as an important linker between the cadherin and the actin-binding protein, α-catenin (Dartsch, Schulte, Hagerling, Kiefer, & Vestweber, 2014), and an indispensible mediator of Wnt signaling raises intriguing questions as to whether or not adhesion and Wnt signaling are coordinated through the use of this common component, β-catenin. Indeed, experimental manipulations in cells and simple model organisms reveal that cadherin expression and β-catenin signaling are interrelated. For example, forced expression of the cadherin can antagonize β-catenin signaling activity in a number of systems (Fagotto, Funayama, Gluck, & Gumbiner, 1996; Heasman et al., 1994; Orsulic,

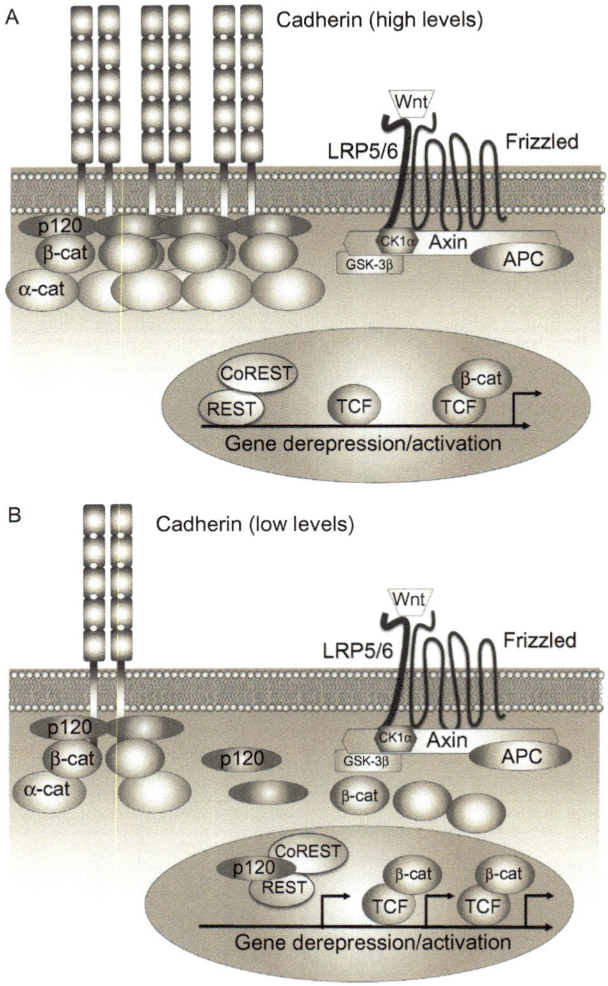

Figure 1 Cadherins as competitive stoichiometric inhibitors of p120^{ctn} and β-catenin signaling. The model presented in (A) and (B) reflects evidence that cells with greater cadherin abundance (A = high levels of cadherin expression; B = lower cadherin expression) can sequester and thereby inhibit the ability of p120^{ctn} and β-catenin to derepress activities of their respective DNA-binding factors, Rest/CoRest and TCF/LEF.

Huber, Aberle, Arnold, & Kemler, 1999; Sanson, White, & Vincent, 1996). Conversely, lowering cadherin levels in developing tissues can lead to enhanced β-catenin signaling (Ciruna & Rossant, 2001; Cox, Kirkpatrick, & Peifer, 1996). Since biochemical and crystallographic evidence show that β-catenin binds cadherins or TCFs through an overlapping

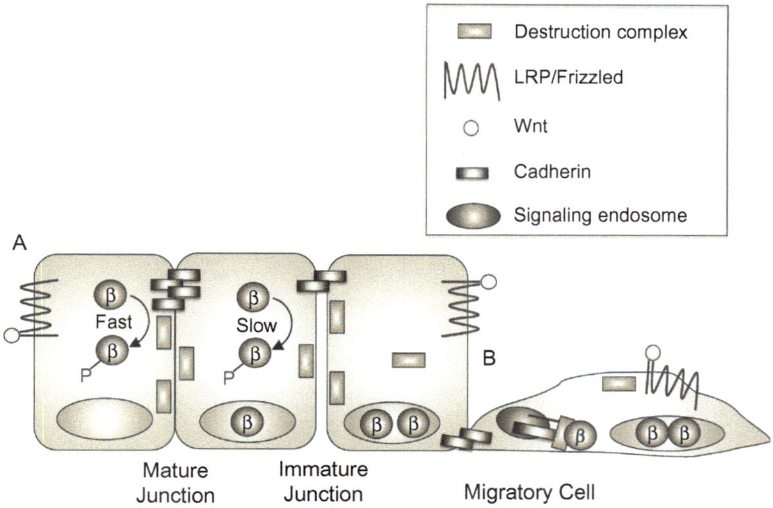

Figure 2 Cadherin-based adhesion as an enhancer of β-catenin destruction. In densely confluent cells with mature junctions, cadherins promote a faster turnover of β-catenin than in less adhesive (subconfluent) cells with immature junctions (A). This may explain why cells migrating adjacent to a wound appear sensitized to Wnt signals (Howard, Deroo, Fujita, & Itasaki, 2011; Maher et al., 2009). Mechanistically, the latter study suggests that enhanced β-catenin signaling is due to an activation step that depends on cadherins and a signaling endosome (B), while the former suggests that these cells show reduced capacity to degrade β-catenin (A). (See the color plate.)

binding interface (Graham, Weaver, Mao, Kimelman, & Xu, 2000; Huber & Weis, 2001), where the cadherin:β-catenin interaction is of higher affinity than that of TCF:β-catenin (Choi, Huber, & Weis, 2006), these data collectively show how cadherins can, in principle, function as stoichiometric inhibitors of β-catenin/TCF (etc.) signaling. Importantly, the ability of lowered E-cadherin levels to lead to β-catenin signaling activation is not universal, as epithelial cancers that have lost E-cadherin expression by various means often fail to exhibit constitutive β-catenin signaling activation (Caca et al., 1999; Herzig, Savarese, Novatchkova, Semb, & Christofori, 2007; van de Wetering et al., 2001). Indeed, targeted loss of E-cadherin can be associated with downregulation of β-catenin (Hendriksen et al., 2008), presumably because there are no other β-catenin-binding cadherins in these systems (e.g., N-cadherin or P-cadherin), and loss of this major high-affinity β-catenin-binding partner leads to β-catenin elimination by the phosphodestruction complex. Thus, the ability of a cadherin to limit

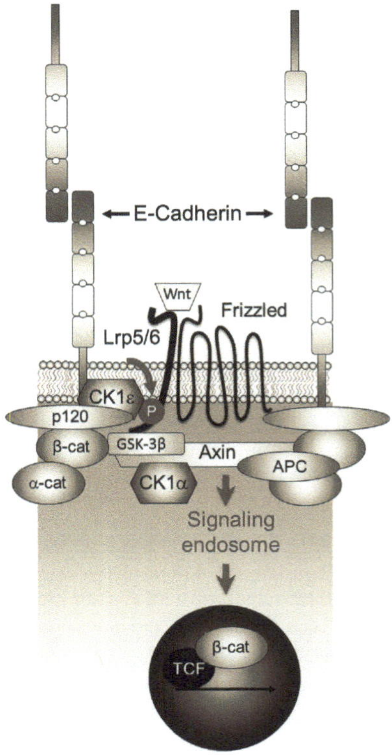

Figure 3 Cadherins as facilitators of Wnt/β-catenin signaling. A number of studies show that E- and N-cadherins are required for efficient Wnt/β-catenin signaling, working at the level of the Wnt receptor complex. Work from the Dunach group shows that $p120^{ctn}$ facilitates Lrp5/6 phosphorylation and signaling by functioning as a proximal scaffold for CK1ε. $p120^{ctn}$ also contributes to the inhibition of GSK3 by multivesicular bodies leading to enhanced signaling from an endosome.

β-catenin signaling is contextual and appears to occur only when cells are actively engaged in Wnt signaling (Fig. 1).

While the aforementioned gain- and loss-of-function examples clearly show that cadherins levels *can* set thresholds for β-catenin signaling, we have few examples where cadherin levels are modulated *in vivo* to dampen or elevate β-catenin signaling. The best early example of this involves an FGF-induced epithelial–mesenchymal transition (EMT) that leads to an E-cadherin downregulation that is important for full β-catenin signaling required for primitive streak formation in the mouse (Ciruna & Rossant, 2001). It is also clear that stem cells downregulate E-cadherin as they begin transitioning away from pluripotency (discussed below) (Bao et al., 2009;

del Valle et al., 2013; Lee, Ji, Furuta, Park, & McCrea, 2014; Li, Zhou, et al., 2010; Li, Wang, et al., 2010; reviewed in Pieters & van Roy, 2014). Conversely, while a few signals have been shown to increase E-cadherin transcription in both cell culture and developmental models (Hosono et al., 2000; Montell, Rorth, & Spradling, 1992; Niewiadomska, Godt, & Tepass, 1999; Ohira et al., 2003; Shimamura & Takeichi, 1992), it is unclear whether the observed increases in cadherin synthesis are used to dampen endogenous Wnt/β-catenin signals, in addition to enhancing cell–cell adhesiveness required in certain morphogenetic events. Interestingly, TCF-binding sites have been identified in the E-cadherin promoter (Huber et al., 1996), suggesting the existence of a negative feedback loop whereby Wnt/β-catenin signaling could increase E-cadherin levels, in turn dampening the signaling pool of β-catenin. While activation of β-catenin signaling has been associated with the upregulation of E-cadherin in mouse intestine (Wong, Rubinfeld, & Gordon, 1998) and a *Drosophila* cell line (Yanagawa et al., 1997), the universality of this feedback mechanism is unclear given evidence that a TCF site in the E-cadherin promoter interacts with other factors to inhibit E-cadherin transcription during hair follicle development (Jamora, DasGupta, Kocieniewski, & Fuchs, 2003).

3. THE CADHERIN–CATENIN COMPLEX AS BOTH A POTENTIATOR AND AN ATTENUATOR OF WNT/β-CATENIN SIGNALING

3.1. Cadherin-based adhesion and destruction complex activity

If it is clear that cadherin protein levels can impact the nuclear signaling capacity of β-catenin, it has been less clear whether changes in the state of cell–cell adhesion might impact β-catenin signaling. Intuitively, the pairing of a cell surface adhesion receptor (i.e., cadherin) with a transcriptional coactivator (i.e., β-catenin) for the purposes of cell–cell adhesion long suggested such a relationship. Recently, two labs independently uncovered proximal relationships between the cadherin/catenin and β-catenin destruction complexes (Hay et al., 2009; Maher et al., 2009). One group found that cadherin-based cell–cell adhesion across a number of epithelial cell lines could increase the amino-terminal phosphorylation of β-catenin and its subsequent rate of destruction (Maher et al., 2009). This ability of cadherins to promote β-catenin turnover was supported by evidence that the N-terminally phosphorylated forms of β-catenin colocalized at E-cadherin-based cell–cell contacts with its known destruction complex components, Axin, APC2, and GSK3β. Although this complex appeared to be molecularly distinct from the E-cadherin complex by sucrose density gradient analysis, the membrane proximity of the phosphodestruction complex suggests that it spatially poised to receive cadherin-contact-dependent signals. In osteoblasts, N-cadherin showed a similar capacity to promote the phosphorylation-dependent ubiquitylation (and ultimate destruction) of β-catenin (Hay et al., 2009). But in the latter study, the molecular relationship between N-cadherin, Lrp5/6, and Axin appears closer, as they were found to coassociate by immunoprecipitation (Hay et al., 2009). Differences aside, these results suggest a model where greater cadherin-mediated adhesion, as seen most in sedentary cells, disfavors canonical Wnt signaling by enhancing destruction complex activity (Fig. 2A). This model is consistent with earlier work demonstrating that less adhesive, motile cells display increased Wnt-reporter activity in *Zebrafish* embryos (Dorsky, Sheldahl, & Moon, 2002), as well as the more recent identification of a novel membrane-proximal inhibitor of β-catenin signaling, WTX/Amer1 (Major et al., 2007), which can impact the activity of the phosphodestruction complex (Tanneberger et al., 2011). An appealing feature of this model is that the activity of the β-catenin phosphodestruction complex can be "tuned" by the local adhesive environment, despite a uniform presence of Wnt ligand, which may be relevant to cell fate decisions that occur in various developmental contexts.

3.2. Cadherins as facilitators of Wnt/β-catenin signaling

For the two aforementioned examples where cadherin-based adhesion *limits* the activity of the phosphodestruction complex, recent studies highlight more numerous examples demonstrating a *requirement* of the cadherin complex for Wnt/β-catenin signaling. For example, N-cadherin:N-cadherin associations between cortical precursor cells in the ventricular neurogenic niche help to maintain β-catenin signaling in these cells, thus inhibiting their differentiation into mature neurons (Zhang et al., 2010). Later findings from the same group (Zhang et al., 2013) used conditional knockout approaches in mice together with *in vivo* electroporation of embryos *in utero* to induce focal loss of N-cadherin, as well as knockdown and antibody function-blocking approaches in cell lines to show that N-cadherin enhanced the phosphorylation of Lrp6 in the presence of Wnt signals in a cell-autonomous fashion (Zhang et al., 2013). Since Lrp5/6 phosphorylation leads to the recruitment of Axin and inhibition of GSK3 activity (Cselenyi et al., 2008; Mi et al., 2006; Piao et al., 2008; Stamos & Weis, 2013; Wu et al.,

2009), these data suggest that N-cadherin:N-cadherin engagement facilitates an obligate early step in Wnt signal transduction (Fig. 3).

While the mechanism of N-cadherin's requirement for Wnt signaling was not established in the above studies, it is important to bear in mind that, as the major cell-cell adhesion system in cells, the cadherin/β-catenin/α-catenin complex is critical not only for basic intercellular adhesion but also for establishing the close contacts that so many other juxtacrine signaling molecules depend upon, from membrane-anchored signaling pairs like Notch/Delta (Ferreira et al., 2011), Ephrins and Eph receptors (Zantek et al., 1999) to death receptor activation (Lu et al., 2014) and insulin secretion (Parnaud et al., 2014). Thus, via this fundamental mechanism, N-cadherin:N-cadherin engagement may enhance the interaction of Wnt ligands with Wnt receptor components to facilitate Wnt/β-catenin signaling. Relatedly, cadherin engagement is known to promote phosphatidylinositol 3-kinase activity (PI3K) (Tran, Adams, Vaillancourt, & Heimark, 2002), which through subsequent activation of AKT may facilitate β-catenin signaling through phosphoinhibition of GSK3β (via serine 9 phosphorylation; Fang et al., 2000), as well as phosphorylating a site in β-catenin that appears to enhance its transcriptional activity (Fang et al., 2007; Tian et al., 2004; Zhang et al., 2013).

Other evidence, however, suggests that the cadherin/catenin complex may play a more direct role in Wnt receptor complex activation. For example, the Dunach group has published a number of reports centered upon the role of the cadherin–catenin complex in recruiting kinases needed for the execution of canonical Wnt signals (Casagolda et al., 2010; Del Valle-Perez, Arques, Vinyoles, de Herreros, & Dunach, 2011; Del Valle-Perez, Casagolda, et al., 2011). The relationships uncovered are largely operating at the level of cadherin/catenin associations and their dissociations from Wnt–receptor complex components, Lrp5/6, CK1 family members, Dishevelled and Axin, as assessed by coimmunoprecipitation analysis in recombinant Wnt-treated cell lines. Indeed, a proximal relationship between a cadherin and Wnt receptor may be inferred from the existence of the Flamingo (Celsr in humans) planar cell polarity (PCP) protein, which appears to be a fusion between a nonclassical cadherin protein and a Frizzled receptor (Berger-Muller & Suzuki, 2011; Formstone, 2010).

As discussed above, Lrp5/6 phosphorylation is a key early step in Wnt signal transduction, leading to the recruitment of Dishevelled and Axin, inhibition of GSK3 activity, and β-catenin stabilization (MacDonald, Tamai, & He, 2009). While CK1γ and GSK3 can modify Lrp5/6 (Davidson et al., 2005; Mi et al., 2006; Zeng et al., 2005), what controls the local activation or accessibility of Lrp5/6 to these kinases has been unclear. Dunach and colleagues present evidence that p120-catenin ($p120^{ctn}$) is required for Wnt-mediated phosphoactivation of Lrp6 (Casagolda et al., 2010; Del Valle-Perez, Arques, et al., 2011) (Fig. 3). A subsequent step involves the recruitment of Axin bound to another CK1 family member, CK1α. CK1α in turn phosphorylates $p120^{ctn}$ (Serines 268 and 269), dissociating it from the cadherin complex (Del Valle-Perez, Casagolda, et al., 2011; Vinyoles et al., 2014). Once dissociated from the cadherin, $p120^{ctn}$ is able to transduce signals to the nucleus in a manner that partially resembles β-catenin (discussed below) (Del Valle-Perez, Casagolda, et al., 2011). Cadherin-free $p120^{ctn}$ can also associate with the small-GTPase modulator Vav2 (Valls et al., 2012) and activate JNK kinases to promote β-catenin's nuclear entry and activity (Wu et al., 2008). Lastly, evidence that endocytosis appears to be an obligate step for Wnt signal transduction (Bilic et al., 2007; Blitzer & Nusse, 2006), as is known for many other signaling pathways (Goh & Sorkin, 2013; Haglund & Dikic, 2012), raises the possibility that cadherins may impact Wnt signaling by controlling Wnt receptor complex internalization into signalosomes (Bilic et al., 2007). In this regard, recent studies show that Wnt receptor complex activation can promote the sequestration of GSK3 into multivesicular bodies (Taelman et al., 2010) (although perhaps not in flies; Gagliardi, Hernandez, McGough, & Vincent, 2014), leading to β-catenin stabilization and signaling, where the cadherin and $p120^{ctn}$ appear required for this process (Vinyoles et al., 2014). Given that the cadherin/catenin complex is fundamental not only for cell–cell adhesion but for basic membrane dynamics (Abreu-Blanco, Verboon, Liu, Watts, & Parkhurst, 2012; Schepis, Sepich, & Nelson, 2012), it may not be surprising to find these examples of the cadherin/catenin complex being both required for and inhibitory to Wnt/β-catenin signaling. A similar phenomenon has been described for the role of GSK3 in Wnt signaling, where it plays both activating and inhibitory roles at distinct steps of the pathway (Zeng et al., 2005). Nonetheless, we look forward to future studies that parse the activating and inhibitory contributions of the cadherin/catenin complex to Wnt signaling.

4. CADHERIN SIGNALING AND STEM CELL BEHAVIOR

One important context where cadherin levels are endogenously regulated during development occurs in mouse embryonic stem cells (mESCs), as they begin transitioning away from pluripotency (reviewed in Bhatt, Rizvi, Batta, Kataria, & Jamora, 2013; Farahani et al., 2014; Li,

Bennett, & Wang, 2012; Pieters & van Roy, 2014). E-Cadherin levels are reduced as this transition begins, with cadherin levels being restored in subsequent differentiating lineages. The types of cadherin(s) that subsequently become expressed are lineage based. That is, cells moving toward a neural lineage will begin to express N-cadherin, and such alterations in E-, N-, or other cadherin levels likely impact β-catenin signaling as described above. For example, N-cadherin expression in cortical neural progenitors favors stemness through maintaining robust β-catenin signaling, while reductions in N-cadherin levels led to premature differentiation of neurons and their migration away from the stem cell niche (Woodhead, Mutch, Olson, & Chenn, 2006; Zhang et al., 2010). Other work suggests that N-cadherin's ability to promote β-catenin signaling can favor differentiation of multipotential mesenchymal progenitor cells (murine C3H10T1/2 cells) toward the bone lineage, as opposed to maintaining stemness (Arnsdorf, Tummala, & Jacobs, 2009), fitting with long established evidence that Wnt signals can both promote stemness as well as differentiated states (Clevers, Loh, & Nusse, 2014; Lien & Fuchs, 2014; Serio, 2014).

5. β-CATENIN "RELEASE" FROM CORTICAL ENDOSOMES AND NUCLEAR SIGNALING

Two recent studies support the intriguing idea that a cadherin-dependent, membrane-proximal event may "prime" β-catenin for nuclear signaling. Similar to studies by the Chenn group (Zhang et al., 2013, 2010), Howard et al. found that N-cadherin is required for robust Wnt signaling in ingressing cells during mouse embryo gastrulation (Howard et al., 2011). Using a cell culture EMT model (meant to mimic aspects of ingression), this group demonstrated that an established phosphomimetic form of β-catenin previously shown to exhibit reduced binding to cadherin (Y654E; Roura, Miravet, Piedra, Garcia de Herreros, & Dunach, 1999; van Veelen et al., 2011) manifested reduced transcriptional activity compared to wild type or Y654F forms of β-catenin in the context of Wnt3 and HGF signals. Together with evidence that HGF-induced β-catenin signaling is sensitive to E-cadherin silencing and the endocytosis inhibitor dynasore, it appears that a cadherin-dependent priming event, from a signaling endosome, is required for efficient Wnt signaling. Additional evidence that HGF-induced β-catenin signaling is insensitive to the protein synthesis inhibitor, cycloheximide (despite an earlier report to the contrary; Willert, Epping, Pollack, Brown, & Nusse, 2002), raises the possibility that β-catenin signaling activity

Plakoglobin has been implicated in nuclear signaling by a number of groups (reviewed in Aktary & Pasdar, 2012; Swope, Li, & Radice, 2013), with some reports suggesting a signaling axis relating to the extent of its dissociation from classic or desmosomal cadherins present at desmosomal or adherens junctions (Hu, Berkowitz, O'Keefe, & Rubenstein, 2003; Hu, O'Keefe, & Rubenstein, 2001; Miravet et al., 2003; Williamson et al., 2006). In addition to TCF/LEF, plakoglobin is reported to associate with p53 and coregulate some p53 target genes (Aktary, Kulak, Mackey, Jahroudi, & Pasdar, 2013), as well as to associate with other proteins having nuclear activity (Aktary & Pasdar, 2013; Lam et al., 2012). Certain mutations in human desmosomal proteins, including desmosomal cadherins (e.g., desmoglein2 and desmocollin2), reduce plakoglobin's junctional association, enhance its nuclear localization, and thereby contribute to the heart pathology arrhythmogenic right ventricular cardiomyopathy (ARVC) (Garcia-Gras et al., 2006). The condition can be mimicked by transgenic plakoglobin expression in mice, where plakoglobin shows enhanced nuclear localization and suppresses canonical Wnt signals (Lombardi et al., 2011). Conversely, the conditional removal of plakoglobin from the mouse heart leads to a decrease in desmosomal proteins and a coincident increase in β-catenin stabilization and Wnt signaling (Li et al., 2011). Possibly reflecting in part such a relationship of plakoglobin's antagonism of Wnt/β-catenin signaling in some contexts, plakoglobin's removal in keratinocytes protects keratinocytes from apoptosis (Dusek et al., 2007), whereas plakoglobin expression in mouse epidermis and hair follicles results in stunted proliferative potential and hair growth, and premature differentiation (Charpentier, Lavker, Acquista, & Cowin, 2000).

In addition to β-catenin and plakoglobin, at adherens junctions the cytodomains of classic cadherins associate with another armadillo-domain protein group known as the $p120^{ctn}$ subfamily. Members of the $p120^{ctn}$ family bind to the cadherin directly, but at a more membrane-proximal site than β-catenin or plakoglobin. The armadillo domains of the $p120^{ctn}$-subfamily are similar in structure to β-catenin, even as their homology is not especially high. Members include $p120^{ctn}$, ARVCF (armadillo-repeat gene deleted in velocardiofacial syndrome), p0071, and δ-catenin, and they bind to classic cadherins in a mutually exclusive manner. α-Catenin, on the other hand, is more structurally similar to vinculin, binds cadherin only indirectly (via β-catenin), and lacks an armadillo domain.

$p120^{ctn}$, as well as other subfamily members, promotes cadherin stability, as best studied for E-, N-, and VE-cadherin (Chiasson, Wittich, Vincent,

Faundez, & Kowalczyk, 2009; Davis, Ireton, & Reynolds, 2003; Ferreri, Minnear, Yin, Kowalczyk, & Vincent, 2008; Hatanaka, Simons, & Murakami, 2011; Iyer, Ferreri, DeCocco, Minnear, & Vincent, 2004; Mege, Gavard, & Lambert, 2006; Reynolds & Carnahan, 2004; Xiao et al., 2005). Additionally, $p120^{ctn}$ exhibits key roles in modulating the functions of small GTPases, as will be discussed later (reviewed in Anastasiadis, 2007; Kourtidis et al., 2013; Ratheesh et al., 2013). While the $p120^{ctn}$ subfamily is far less studied than β-catenin, each catenin also has nuclear roles. We will mention such functions briefly, while directing the reader to additional papers (reviewed in Bass-Zubek, Godsel, Delmar, & Green, 2009; Daniel, 2007; Hatzfeld, 2007; Kourtidis et al., 2013; McCrea & Park, 2007; Schackmann, Tenhagen, van de Ven, & Derksen, 2013; Schmidt & Jager, 2005; Stepniak et al., 2009).

$p120^{ctn}$ subfamily nuclear functions bear similar features to some β-catenin activities. For example, despite some interesting alternative views (Blattler et al., 2013; Ruzov, Hackett, et al., 2009; Ruzov, Savitskaya, et al., 2009), a number of $p120^{ctn}$ gene targets are shared with β-catenin. Further, the stability of $p120^{ctn}$-isoform1, which determines the size of its signaling pool, is regulated by canonical Wnt signals in much the same way as β-catenin (Hong et al., 2010, 2012; Kim et al., 2004; Park et al., 2005; Spring et al., 2005). Even with a number of shared gene targets, each catenin is likely to have distinct nuclear roles. For example, in contrast to β-catenin, the expression of $p120^{ctn}$ or ARVCF alone cannot generate an ectopic dorsal axis when expressed on the ventral side of early *Xenopus* embryos (Geis, Aberle, Kuhl, Kemler, & Wedlich, 1998; Paulson, Fang, Ji, Reynolds, & McCrea, 1999). Future genome-wide analyses should assist in discerning the shared versus distinct gene targets and nuclear outputs of these catenins. At least three distinct transcription factors have thus far been found in direct protein–protein and functional association with $p120^{ctn}$, with some evidence indicating that p120's sequestration by cadherins or the cytoskeleton has an impact upon such nuclear signaling (Daniel & Reynolds, 1999; Hosking et al., 2007; Kim et al., 2004; Lee et al., 2014). Additional transcription factors bind to δ-catenin (Gu et al., 2011), ARVCF (unpublished), and plakophilin3-catenin (Munoz et al., 2014). Thus, just as β-catenin binds to multiple types of transcription factors (with the HMG-box containing TCF/LEF being the most prominent), each $p120^{ctn}$ subfamily member is likely to bind more than one DNA-binding factor, with some relationships being unique to that catenin. Future studies that uncover this larger network of

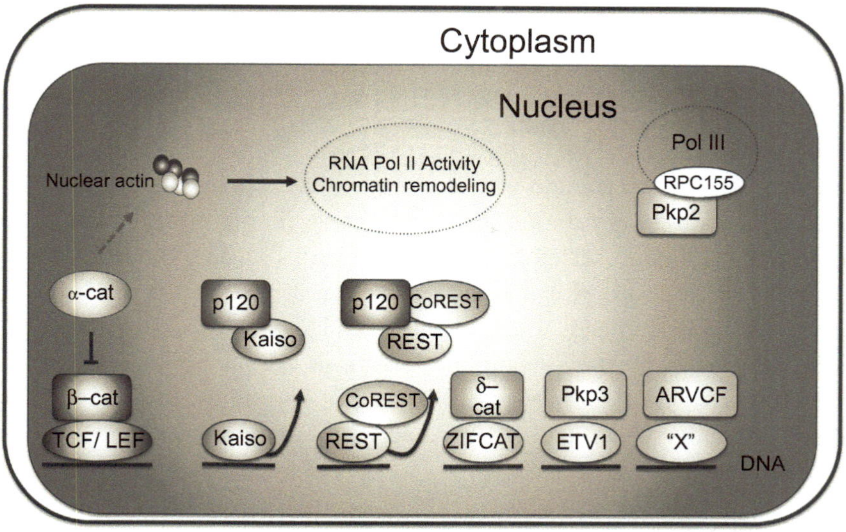

Figure 4 Nuclear roles for catenins. The armadillo family of catenin proteins (β-catenin, p120^{ctn}, δ-catenin, ARVCF, and plakophilins (PKP) 2 and 3 (shades of purple; light gray in the print version)) modulate gene expression via association with cognate DNA-binding factors (shades of blue; gray in the print version). In most instances characterized to date, derepression occurs to activate the target genes. However, Pkp3 appears to further activate gene targets already positively modulated by ETV1, and δ-catenin's action upon ZIFCAT is not yet known. Lacking an armadillo domain and homology to the other catenins, α-catenin (green; gray in the print version) is structurally related to vinculin and may impact transcription through its actin-binding function, inhibiting RNA polymerase II (RNA Pol II). PKP2 can interact with a component of the RNA Pol III complex RPC155.

catenin nuclear functions will likely provide additional opportunities to learn how the cadherin–catenin complex signals to the nucleus in varied contexts (Fig. 4).

Finally, it is worth noting that armadillo-repeat catenin proteins like β-catenin, plakoglobin, and p120^{ctn} family members appear to have arisen from the same evolutionary predecessor as the importins (Andrade, Petosa, O'Donoghue, Muller, & Bork, 2001), which shuttle substrates across the nuclear membrane. Thus, in metazoans, catenin nuclear roles were likely selected for in concert with those relating to cadherins. In this scenario, catenins would have enhanced the functionality of cell–cell junctions on a structural level, and additionally on a signaling level, given their predisposed capacity to enter the nucleus (reviewed in (Fagotto, 2013; Hulpiau & van Roy, 2011)).

7. NUCLEAR SIGNALING FUNCTIONS OF THE ACTIN-BINDING PROTEIN, α-CATENIN

α-Catenin is an F-actin-binding protein and vinculin homologue and has been long appreciated to play an indispensible role in cell–cell adhesion by forming an essential link between the cadherin/β-catenin and the underlying actin cytoskeleton (Maiden & Hardin, 2011). As cells contain a substantial amount of cadherin-free α-catenin (Benjamin et al., 2010; Schneider, Herrenknecht, Butz, Kemler, & Hausen, 1993), there has been long-standing interest in the unique functions of this pool of α-catenin. Indeed, recent studies show that this pool of α-catenin can impact actin dynamics, such as limiting Arp2/3-based actin polymerization and severing by cofilin (Benjamin et al., 2010; Drees, Pokutta, Yamada, Nelson, & Weis, 2005; Hansen et al., 2013), which ultimately promotes the assembly and stabilization of unbranched filament bundles. This ability of α-catenin to impact F-actin dynamics might explain how α-catenin loss has been linked to the context-dependent activation of numerous pathways, such as Ras (Vasioukhin, Bauer, Degenstein, Wise, & Fuchs, 2001), NFκB (Kobielak & Fuchs, 2006; Piao et al., 2014), Hedgehog (Lien, Klezovitch, Fernandez, Delrow, & Vasioukhin, 2006; Rhee, Ryu, Kim, Chun, & Chun, 2012), and Hippo/Warts (Schlegelmilch et al., 2011; Silvis et al., 2011). While in some of these cases, α-catenin can be co-immunoprecipitated with various components of these pathways (e.g., the NFκB negative regulator, IκBα (Piao et al., 2014); the Hippo/Warts effector, Yap (Schlegelmilch et al., 2011; Silvis et al., 2011); and the Hedgehog nuclear effector, Gli3 (Rhee et al., 2012)), the molecular details of these interactions and the extent to which they are mediated by the cadherin-free pool of α-catenin remain unknown. In this regard, it is important to bear in mind that as the major cell–cell adhesion system in cells, the cadherin/β-catenin/α-catenin complex is critical not only for intercellular adhesion but also for establishing the close cell contacts that so many other junction and juxtacrine signaling molecules depend upon, from tight junctions in epithelia to gap junctions and desmosomes in cardiac tissue to membrane-anchored signaling pairs like Notch/Delta or Ephrins and Eph receptors, as mentioned above (Fagotto & Gumbiner, 1996; Ferreira et al., 2011; Zantek et al., 1999). Thus, it may be useful to consider some of the pleiotrophic signaling effects of α-catenin loss in the context of its broader role in cell–cell adhesion.

Regardless of this issue, it is clear that one important target of the cytosolic pool of α-catenin is its major stoichiometric binding partner, β-catenin.

While a number of forced-expression studies showed that α-catenin could inhibit the nuclear accumulation (Takahashi, Ishihara, Takada, Tsukita, & Nagafuchi, 2000) and signaling of β-catenin (Giannini, Vivanco, & Kypta, 2000; Hwang et al., 2005; Merdek, Nguyen, & Toksoz, 2004; Sehgal, Gumbiner, & Reichardt, 1997), it was not until recently that α-catenin silencing studies were carried out in a way that revealed its capacity to function as a *bona fide* negative regulator of β-catenin signaling. In one study, α-catenin was found to limit β-catenin signaling by promoting its ubiquitylation and proteolysis by the "destruction complex" through an ability to bind a distinct region of APC (Choi et al., 2013). Interestingly, this function of α-catenin appears to occur at the level of β-catenin/TCF-occupied promoters and in the context of a histone H3 lysine K4 demethylase transcriptional repressor complex containing APC:CtBP:CoREST:LSD1 proteins. A second recent study confirmed earlier suggestions that endogenously expressed α-catenin can localize to the nucleus (El-Bahrawy, Talbot, Poulsom, & Alison, 2002; Giannini et al., 2000), but only under conditions where β-catenin is also nuclear, demonstrating that α-catenin nuclear function is largely β-catenin dependent (Daugherty et al., 2014). Like the study by Choi and colleagues, α-catenin can be found in a complex with both β-catenin and TCF on Wnt-responsive promoters, suggesting that α-catenin can limit β-catenin-mediated transcription at gene promoters, but mechanistically the studies diverge, probably because the latter study used APC mutant colon carcinoma cell lines. Indeed, Daugherty et al. found that the C-terminal actin-binding region of α-catenin is important for its capacity to inhibit β-catenin/TCF transcription, raising the possibility that α-catenin's ability to impact actin organization might be related to its function as a transcriptional inhibitor. Indeed, cell nuclei contain a substantial amounts of G-actin and β-actin incorporates into all three RNA polymerase complexes and some chromatin-remodeling complexes required for transcription (Hofmann et al., 2004; Hu, Wu, & Hernandez, 2004; Kukalev, Nord, Palmberg, Bergman, & Percipalle, 2005; Philimonenko et al., 2004) suggesting that nuclear proteins with a capacity to bind actin may affect gene expression. Supporting this idea, a nuclear-targeted form of α-catenin could induce the formation of nuclear F-actin filaments, while cells lacking α-catenin showed greater nuclear actin mobility using a GFP-tagged, nuclear-localized actin. In addition, formation of nuclear actin filaments correlated with reduced RNA synthesis and a more compact chromatin organization, suggesting cross talk with components that drive gene repression (such as the CtBP:CoREST:LSD1 described above). While the studies by Choi et al. and Daugherty et al.

present very different models of transcription inhibition by α-catenin, the models are not contradictory, and thus each set of observations likely inform different aspects of the transcription inhibitory mechanism. Moreover, evidence that the p120^{ctn} family of proteins can interact with and negatively regulate the REST/CoREST repressor complex (Lee et al., 2014) suggests that nuclear-localized catenins may broadly affect gene expression and cell differentiation through regulation of this complex (Fig. 4).

Other proteins that associate with cadherins, such as the LIM-family members Ajuba and Zyxin, further exhibit cytoskeletal as well as nuclear localizations and functions (reviewed in Kadrmas & Beckerle, 2004; Smith, Hoffman, & Beckerle, 2014). The physical interactions of Ajuba include that with α-catenin, occurring at cadherin junctions (Marie et al., 2003). In the nucleus, Ajuba inhibits growth and promotes endodermal differentiation in P19 embryonal cells (Kanungo, Pratt, Marie, & Longmore, 2000), and it also modulates certain growth-hormone receptors, as well as an unrelated transcription factor (Montoya-Durango et al., 2008). Further, acting upstream of nuclear entry, Ajuba negatively regulates Hippo signaling through functional interactions with kinases that determine YAP's nuclear transit (Das Thakur et al., 2010), while in Xenopus, Ajuba promotes meiotic maturation through ERK activation (Goyal et al., 1999). Thus, although additional insights are needed of how the functions of LIM-proteins at cadherin contacts are tied to those in the nucleus, a number of these proteins appear to be candidates in transducing cadherin-to-nuclear signals.

8. CADHERIN NUCLEAR SIGNALING VIA RTKs

In a number of systems, cadherins have been implicated in cross talk with receptor tyrosine kinases (RTKs), in some cases, through direct association with these receptors. Early interest in this area arose from findings that the time-course of contact-dependent inhibition of cell growth, mediated in part by cadherins, coincided with lessened RTK activity (Takahashi & Suzuki, 1996). There is also ample evidence that the cadherin/catenin complex is modified by RTKs (and their kinase effectors) (Daniel & Reynolds, 1997; Hoschuetzky, Aberle, & Kemler, 1994), which appears to change the affinity of the cadherin/β-catenin interface and thereby has consequences for both the adhesive and nuclear signaling function of β-catenin (reviewed in Daugherty & Gottardi, 2007). Both examples of cadherin nuclear signaling will be discussed below.

8.1. Cadherin-mediated inhibition of diverse RTKs

Early studies showed that the EGFR could colocalize with E-cadherin at apically localized adherens junctions (Chen, Solomon, Kui, & Soll, 2002), coassociate in immunoprecipitation assays (e.g., Hoschuetzky et al., 1994), and genetically interact in flies (Dumstrei, Wang, Shy, Tepass, & Hartenstein, 2002), suggesting a close relationship between these two proteins. Indeed, E-cadherin appears to limit EGFR activity at multiple levels. The cadherin ectodomain has been implicated in sterically preventing EGF-ligand binding (Qian, Karpova, Sheppard, McNally, & Lowy, 2004), while the cytodomain is required to suppress EGFR signaling downstream of its phosphoactivation (Perrais, Chen, Perez-Moreno, & Gumbiner, 2007). Other evidence suggests that physical interactions between the cadherin cytoplasmic domain, catenins, and the ERM protein, Merlin, are important for EGFR inhibition. Merlin associates with E-cadherin via α-catenin (Curto, Cole, Lallemand, Liu, & McClatchey, 2007; Gladden, Hebert, Schneeberger, & McClatchey, 2010) and acts as a link between cadherin-mediated contact maturation, the Par3 cell polarity complex, and the suppression of EGFR-mediated proliferation signals. Part of Merlin's activity is to restrict EGFR to a membrane compartment incapable of passing on a signal (e.g., by preventing endocytosis). Merlin is thought to do this by binding the PDZ-domain containing protein NHE-RF1 (Curto et al., 2007; Gladden et al., 2010), which is in complex with EGFR. Since NHE-RF1 binds analogously to other RTKs, additional RTKs may also be subject to modulation by cadherins. While E-cadherin can suppress signaling from RTKs other than EGFR, including IGF-1R, c-Met, and ErB2-4 (Qian et al., 2004; Vermeer et al., 2003), Merlin is selective for the EGFR (Curto et al., 2007), suggesting that the cadherin/catenin complex uses molecules functionally analogous to Merlin to limit signaling from other types of RTKs (Fig. 5).

Interestingly, cadherin homotypic adhesion between cells (i.e., in *trans*), often associated with cadherin clustering *in cis*, is not necessarily required for cadherins to influence RTKs (Cavallaro & Dejana, 2011). VE-cadherin's cytodomain can interact with VEGFR2, preventing its endocytosis and facilitating its dephosphorylation via the junctional phosphatase, DEP1 (Grazia Lampugnani et al., 2003). This reduces MAPK signaling, leading to growth inhibition (Lampugnani, Orsenigo, Gagliani, Tacchetti, & Dejana, 2006). VE-cadherin's association with VEGFR has also been tied to cell survival via the inhibition of apoptosis, as targeted truncation of VE-cadherin's β-catenin-binding region led to mouse embryo lethality,

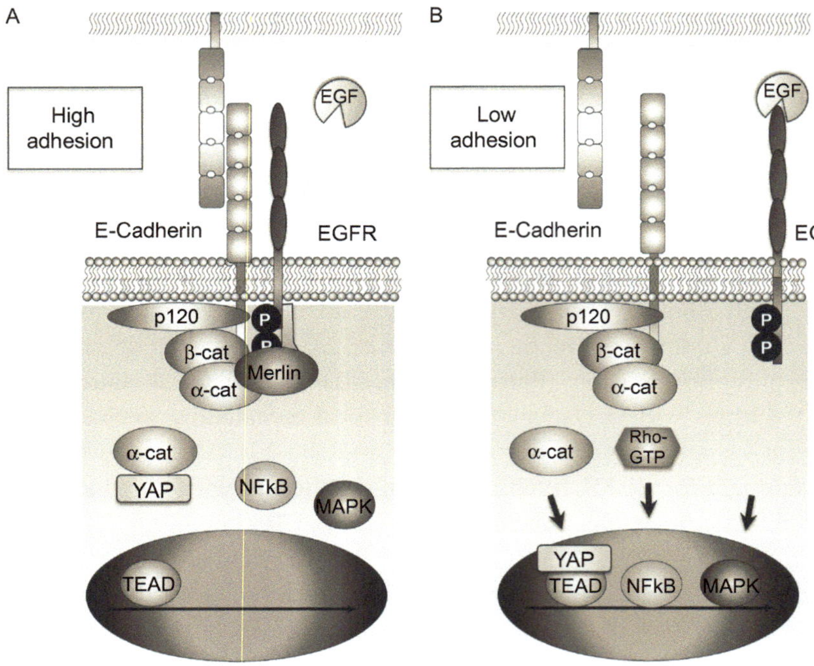

Figure 5 Cadherin signaling via RTKs, NFκB, and Hippo pathways. E-Cadherin in densely packed epithelial monolayers can inhibit access of EGF to the EGFR as well as downstream signaling from the EGFR via Merlin (A), as contrasted with less mature contacts characterized by lowered extents of *trans*-E-cadherin interactions, where the indicated nuclear signaling trajectories are enhanced (B). Cadherin engagement can also limit the nuclear accumulation of YAP and NFκB through α-catenin and Rho-dependent mechanisms, respectively.

in part, due to loss of responsiveness to survival signals from the VEGF-A ligand, leading to increased endothelial apoptosis (Carmeliet et al., 1999; Gavard, 2013). Thus, the catenin-binding domains of both E-cadherin and VE-cadherin are important for the regulation of their cognate RTKs. VE-cadherin has also been implicated in promoting or responding to signals from additional membrane receptors, including FGFR, TGFβ-R2, and the angiopoietin1 receptor, Tie2 (Fukuhara et al., 2008; Hayashi et al., 2013; Murakami et al., 2008; Rudini et al., 2008; Saharinen et al., 2008; Winderlich et al., 2009).

Another example where cadherins are involved in adhesion-independent signaling (observed even in single cells) is that of VE-cadherin bridging the mechanosensor PECAM to VEGFR2 (Shay-Salit et al., 2002; Tzima et al., 2005). Upon flow (shear-stress), PECAM activates Src-family kinases, while may be recruited from a preexisting cadherin-associated pool. Thus, mesenchymal cell types or shapes (e.g., even epithelial cells that are subconfluent) may be more sensitive to Wnt signals because cadherins are being more actively endocytosed, which appears important for β-catenin signaling. An attractive feature of this model is that it couples the dynamic state of cadherin cell–cell adhesive contacts with nuclear signaling functions that arise from those contacts, in this case taking place via β-catenin (Fig. 2B).

Remarkably, this "cortical release" model for β-catenin signaling activation may be evolutionarily conserved with a mechanism that controls cell division orientation in *Caenorhabditis elegans* (Kim et al., 2013). In worms, different forms of β-catenin-like proteins exist, with WRM1 being responsible for transducing noncanonical Wnt signals, as well as contributing to the orientation of cell division. While WRM1 does not bind cadherins or α-catenin to participate in cell adhesion (an activity carried out by HMP2), it is intriguing that WRM1's nuclear activity is regulated in part by its release from the cell cortex in response to Wnt and CDK1 signals. Thus, in the context of the Howard et al. study discussed above, perhaps β-catenin release from cadherins shares features with the cortical release of *C. elegans* WRM1.

6. THE OTHER CATENINS

While β-catenin may be the best-known example of a dual-function adhesion-nuclear signaling protein, it is important to recognize that other catenins appear to follow the same paradigm. For example, plakoglobin (γ-catenin), which is within the β-catenin subfamily and is typically associated with desmosomal cadherins, can also interact with E-cadherin under conditions where β-catenin is limiting (Huelsken et al., 2000). Like β-catenin, plakoglobin contains an N-terminal phosphodestruction box, is stabilized by Wnts (more modestly than β-catenin), and interacts with TCF family DNA-binding proteins to modulate gene expression (Karnovsky & Klymkowsky, 1995; Klymkowsky, Williams, Barish, Varmus, & Vourgourakis, 1999; Kolligs, Hu, Dang, & Fearon, 1999; Simcha et al., 1998; Williams, Barish, Klymkowsky, & Varmus, 2000; Zhurinsky, Shtutman, & Ben-Ze'ev, 2000). This occurs in a manner that is less well understood than for β-catenin, as plakoglobin appears to bind a distinct region on TCF that has context-dependent positive and negative effects on gene expression (Miravet et al., 2002; Solanas et al., 2004).

VE-cadherin assists PECAM in passing signals to VEGFR2, thereby enhancing PI3K activity and the NFκB pathway. The VE-cadherin enhancement of PI3K activity is also associated with cross talk between adherens and tight junctions through nuclear signaling by a FoxO family transcription factor (Taddei et al., 2008). In this study, VE-cadherin's enhancement of PI3K and Akt activity led to phosphorylation of FoxO1 and the release of its repressive activity at the promoter of claudin1, a component of tight junctions. Once cells become fully confluent, VE-cadherin clustering activates PIP3K and Akt signaling more strongly in releasing FoxO1 from DNA. A VE-cadherin mutant unable to bind β-catenin or interact with PIP3K was unable to produce such effects. In addition, VE-cadherin limits the signaling pool of β-catenin, as noted earlier. Once sequestered, β-catenin can no longer form a repressive complex together with FoxO1 and TCF4 (Taddei et al., 2008).

Like VE-cadherin's ability to facilitate VEGFR signaling discussed above, N-cadherin can stimulate FGFR signaling, even in single cells (Suyama, Shapiro, Guttman, & Hazan, 2002; reviewed in Radice, 2013). In this latter context, N-cadherin associates in *cis* with the extracellular domain of FGFR (within the same cell) (Utton, Eickholt, Howell, Wallis, & Doherty, 2001), preventing its ligand-mediated internalization and downregulation (Suyama et al., 2002). Lastly, E-cadherin-mediated cell–cell contact formation can lead to PI3K and Akt, activation, promoting cell survival in response to extracellular stimuli (McLachlan & Yap, 2007; Pece, Chiariello, Murga, & Gutkind, 1999), as well as EGFR-dependent, ligand-independent increases in MAPK activation (Pece & Gutkind, 2000). Taken together, the major varieties of classical cadherins (E-, N- and VE-types) play both facilitator and inhibitor roles toward their cognate RTKs (similar to E-cadherin's contribution to Wnt signaling). While there appears to be no current unified mechanism for these regulatory roles, interactions mediated by both extracellular and intracellular catenin-binding domains are clearly important for RTK regulation by cadherins (Fig. 5).

8.2. β-Catenin as a key target of RTKs and other membrane-activated kinases

Given that β-catenin is a dual-function adhesion/transcriptional coactivator protein, it has long been speculated that modifications that reduce the affinity of β-catenin for the cadherin and α-catenin could lead to enhanced nuclear signaling activity of β-catenin. We will largely discuss the more recent evidence favoring this model, since this area has been generally addressed in reviews elsewhere (Cavallaro & Dejana, 2011; Daugherty & Gottardi, 2007; Harris & Tepass, 2010; Hartsock & Nelson, 2008; Heuberger & Birchmeier, 2010; Meng & Takeichi, 2009; Nelson & Nusse, 2004; Niessen, Leckband, & Yap, 2011).

The best evidence for the aforementioned model involves a tyrosine phosphorylation in β-catenin at residue 654, which leads to reduced E-cadherin binding both *in vitro* (Roura et al., 1999) and *in vivo* (van Veelen et al., 2011), and enhanced nuclear signaling activity (Jean et al., 2009; Kim et al., 2009; Piedra et al., 2001; van Veelen et al., 2011; Xi et al., 2013). Activation of a number of kinases is associated with phosphorylation of β-catenin at Y654, including EGFR (Jean et al., 2009), HGF (David et al., 2008), TGFβ (Kim et al., 2009), and Src downstream of a variety of signals (Condello, Cao, & Matei, 2013; Lee et al., 2013; Sumiyoshi, Takahashi, Obata, Sugimoto, & Kohara, 2011; Xi et al., 2013). In most of these cases, pY654 β-catenin shows reduced association with cadherin contacts as well as enhanced nuclear accumulation and target-gene activation. Often missing from the discussion of these data, however, is the extent to which reduced binding of β-catenin to a cadherin *alone* is sufficient for β-catenin signaling, given that cadherin-free β-catenin is thought to be rapidly degraded by the destruction complex. We speculate, therefore, that these kinases may in parallel generate "Wnt-like" signals that inhibit GSK3 and ultimately inhibit the N-terminal phosphorylation and degradation of β-catenin, resulting in a signaling form of β-catenin that is refractory to sequestration by cadherins.

Other sites in β-catenin have also been implicated in reduced adhesive activity and enhanced nuclear signaling. Tyrosine phosphorylation of β-catenin residue 142 reduces its association with α-catenin (Piedra et al., 2003). This would lead to a form of β-catenin that may be better able to signal because of its reduced association with α-catenin, which can inhibit β-catenin signaling (discussed above) and/or enhanced association with transcriptional coactivator protein Bcl9 (Brembeck et al., 2004; Kramps et al., 2002), although the role of Y142 may be less important in (Hoffmans & Basler, 2007) or indirect toward Bcl9 (Sampietro et al., 2006). Lastly, tyrosine 489 in β-catenin has also been associated with reduced cadherin binding and nuclear signaling activity downstream of a Slit-Robo abl tyrosine kinase mechanism (Rhee, Buchan, Zukerberg, Lilien, & Balsamo, 2007).

While the above examples focus on the impact of phosphorylation for just a few sites in β-catenin, we have many more historic examples that are discussed above where signaling is coassociated with changes in the cadherin–catenin complex along with elevated β-catenin signaling, but where precise mechanisms are lacking, such as following the activation of the FGFR (Pai et al., 2008), EGFR (Galbiati et al., 2000; Lu, Ghosh, Wang, & Hunter, 2003), IGF-2 (Morali et al., 2001), and IGFR-type1 (Playford, Bicknell, Bodmer, & Macaulay, 2000). Conversely, inhibition of certain RTKs, such as ErbB2, coincides with canonical Wnt signal suppression and greater cell–cell adhesive integrity (Bonvini et al., 2001). Not surprisingly, phosphatases appear to counter the effects of tyrosine kinase activation, resulting in enhanced integrity of the cadherin–catenin complex and cell–cell adhesion, and in some cases reduced β-catenin signaling (Balsamo, Arregui, Leung, & Lilien, 1998; Hellberg, Burden-Gulley, Pietz, & Brady-Kalnay, 2002; Nawroth et al., 2002; Novellino et al., 2008; Xu et al., 2004; Yan et al., 2006). Whether the above effects are largely mediated through modification of β-catenin at Y142, 489, or 654 or other presently uncharacterized sites is not known. Moreover, $p120^{ctn}$ is also targeted by many of these same kinases (reviewed in Alema & Salvatore, 2007; Daniel & Reynolds, 1997) and phosphatases, with contributions of individual phosphorylation sites to adhesive and signaling functions just emerging. For example, PKCα activation is associated with $p120^{ctn}$'s phosphorylation at S879, resulting in its reduced association with VE-cadherin (Vandenbroucke St Amant et al., 2012), and conceivably increased $p120^{ctn}$ signaling. Likewise, as a consequence of ligand activation of the canonical Wnt pathway, CK1ε has been found to phosphorylate $p120^{ctn}$ at S268 and S269, resulting in its dissociation and enhanced nuclear activity in relieving Kaiso-mediated gene repression (Dann et al., 2014; Del Valle-Perez, Casagolda, et al., 2011). Other phosphorylation events are known to modulate $p120^{ctn}$'s nuclear signaling (Hong et al., 2010, 2012), but are not (yet) linked with reduced binding to cadherins.

9. CADHERIN NUCLEAR SIGNALING BY SMALL GTPases AND NFκB

As alluded to above, the canonical Wnt pathway is defined as being mediated via β-catenin. However, there are multiple distinct signaling trajectories not involving β-catenin that may be activated following Wnt-ligand binding to Frizzled/Lrp5/6. To what extent these trajectories are activated in parallel or independently of β-catenin is a matter of context and uncertainty, with cross talk (much of it antagonistic) occurring between them. Relative to that of β-catenin, these trajectories have been more difficult to study, as we possess fewer available endogenous readouts, synthetic reporters, or protein-based assays. The most prominent "noncanonical" pathways involve activation of small GTPases such as RhoA, Rac1, and Cdc42, often monitored via pull-down of their active GTP-bound forms, or indirectly, by scoring for phenotypic effects upon the actin or microtubule cytoskeletons, cell motility, or invasion (reviewed in Gomez-Orte, Saenz-Narciso, Moreno, & Cabello, 2013). Trajectories based upon intracellular calcium release also reside within the collection of noncanonical Wnt pathways, and there are indications of the involvement of trimeric G-protein pathways as well (Kilander et al., 2014; Luchtenborg et al., 2014).

In vivo, one of the examples of cadherin involvement with small GTPases comes from *Xenopus*, where the cytoplasmic region of cadherin11 activates Rho-GTPases at membrane regions via its interaction with the Rho guanine nucleotide-exchange factor (GEF) Trio (Kashef et al., 2009; van Rijssel & van Buul, 2012). In neural crest migration, the cytodomain of cadherin11 positively regulates RhoA, Rac and Cdc42 and is required for cell migrations into regions such as the pharyngeal pouches for formation of craniofacial cartilage (Kashef et al., 2009). As opposed to being solely enriched only along cell–cell borders in cranial neural crest explants, cadherin11 is also present at cell–extracellular matrix interfaces and cell protrusions, such as filopodia and lamellipodia, whose formation cadherin11 promotes.

VE-cadherin and VE-cadherin *trans*-interactions have likewise been linked to recruiting the Rho GEF Tiam1 to activate Rac in endothelial cells (Birukova et al., 2011; Lampugnani et al., 2002). A variety of studies have linked VE-cadherin to the modulation of Rac, Rho, and/or Cdc42 (Gavard & Gutkind, 2006; Mehta & Malik, 2006; van Nieuw Amerongen, van Delft, Vermeer, Collard, & van Hinsbergh, 2000; Wojciak-Stothard, Potempa, Eichholtz, & Ridley, 2001; Wojciak-Stothard & Ridley, 2003). Cadherins in some cases interact indirectly with small GTPases via catenins such as p120^{ctn}, which recruits the p190-RhoGEF to cadherins to enhance junctional integrity (Wildenberg et al., 2006). Through such effects, cadherins modulate small GTPases in manners that are relevant to cytoskeletal organization and force generation. Cadherin-dependent forces, for example, can include the generation of cortical tension that assists in driving many morphogenic events (reviewed in Nishimura & Takeichi, 2009; Niessen et al., 2011; Watanabe, Sato, & Kaibuchi, 2009).

In *Xenopus*, p120^{ctn} and δ-catenin regulate the small GTPases Rac and Rho during gastrulation (Ciesiolka et al., 2004; Fang et al., 2004; Gu et al., 2009). In mice, multiple studies have likewise indicated that this functional link is important for development, as exemplified in forming neurons (Elia, Yamamoto, Zang, & Reichardt, 2006). Together, small-GTPase modulation appears to be a signaling feature of cadherins and the p120^{ctn} subfamily (Anastasiadis, 2007; Hatzfeld, 2005; Kourtidis et al., 2013; Pieters, van Hengel, & van Roy, 2012; Ratheesh et al., 2013; Watanabe et al., 2009). While p120^{ctn}'s cytoskeletal effects can be pronounced, the consequent small-GTPase impact upon nuclear events has yet to be determined. As noted earlier, such effects are likely, even if occurring indirectly via changes in cell tension and shape. The plakophilin-catenin subfamily may also prove to engage in such nuclear signaling via small-GTPase effects. Although we do not discuss the plakophilins in depth, they are structurally similar to the other armadillo-domain-containing catenins and were first found in association with desmosomal cadherins (desmogleins and desmocollins) (Heid et al., 1994; reviewed in Bass-Zubek et al., 2009).

p120^{ctn} further engages in modulation of NFκB, a pathway having clear links to the nucleus. It is not currently known to what extent such regulation occurs only after p120^{ctn} departs from the cadherin complex, perhaps in response to phosphorylation events, versus when p120^{ctn} is still bound or conceivably oscillating on and off the cadherin (or newly synthesized). However, as for β-catenin, the signaling roles of p120^{ctn} are usually thought to be higher when p120^{ctn} is not cadherin bound. The knockout of p120^{ctn} in mouse skin and certain other tissues leads to elevated NFκB activity (Davis & Reynolds, 2006; Kobielak & Fuchs, 2006; Perez-Moreno et al., 2006). It is believed that this effect, and another one upon mitotic chromosome segregation, comes about via p120^{ctn}'s initial modulation of Rho, producing a downstream impact upon Rho-kinase and ultimately NFκB (Fig. 5).

In addition to cadherin modulation of NFκB signals via small GTPases such as Rho or Rac, more direct effects may be at play. In mammalian cultured cells, the exogenous expression of E-cadherin lowers β-catenin's and NFκB's association with gene targets such as fibronectin and Lef1 (Chen, Khan, Cukiernik, & Chakrabarti, 2003). Here, NFκB p65 appears to associate with the E-cadherin complex. In K-ras overexpressing cells, this association is lost, and such cells exhibit disrupted cell–cell junctions. Upon E-cadherin depletion, there is reduced NFκB association with cell–cell junctions as would be expected, and this is paralleled with an increase in

NFκB transcriptional activity (under both basal or TNFα conditions). The concept is that the E-cadherin complex is not only relevant to signaling conferred via β-catenin but also that via NFκB, with each acting on certain shared gene promoters (e.g., fibronectin and Lef1). This relationship is proposed to tether E-cadherin levels to cell fate choices, such as in EMT (Solanas et al., 2008). A positive feedback loop may be involved, since β-catenin and NFκB also promote the expression of Slug, ZEB1, and ZEB2, which are repressors of E-cadherin transcription (Chua et al., 2007; Conacci-Sorrell et al., 2003).

We have briefly mentioned these less traditionally discussed catenin trajectories because they are apt to provide additional opportunities for the cadherin–catenin complex to engage in nuclear signaling. In the context of *Xenopus* development, a number of labs have examined the functional relationship of C-cadherin with the PAPC-cadherin (Chen & Gumbiner, 2006; Chen, Koh, Yoder, & Gumbiner, 2009) and of these cadherins with the Frizzled receptor (Kraft, Berger, Wallkamm, Steinbeisser, & Wedlich, 2012). Upon the binding of the Wnt11 ligand to the Frizzled7 receptor, these cadherins are reported to bind Frizzled7 directly and to partition into distinct complexes (Medina, Swain, Kuerner, & Steinbeisser, 2004). Adhesion normally conferred by C-cadherin is lowered via a block to C-cadherin's lateral dimerization. As PAPC's expression is restricted to regions undergoing morphogenic movements in *Xenopus*, such as during gastrulation (Medina et al., 2004; Winklbauer, Medina, Swain, & Steinbeisser, 2001), PAPC lowers C-cadherin's adhesive function to enable needed cellular rearrangements. Further, PAPC's cytoplasmic domain interacts with Sprouty and ANR5 (Chung, Yamamoto, & Ueno, 2007; Wang et al., 2008) permitting PAPC to participate in PCP signaling via the modulation of small GTPases.

Other pathways are activated from cadherins that involve small GTPases. One relates to production of the IL6-cytokine, which binds cytokine receptors (e.g., gp130) on surrounding cells to initiate JAK-STAT signaling (Geletu et al., 2013). Cadherin ligation under high-density cell conditions, possibly mimicking cell-packing densities *in vivo*, affects Rho-family GTPase activity. The activation of Rac, for example, can lead to NFκB activation and thereby expression of the IL6-ligand. JAK-STAT signals often promote expression of cell survival gene products (e.g., Bcl-xL, survivin, Mcl1, myc, VEGF), and lower expression of p53. Such nuclear effects appear to be cadherin-dependent, since even a purified E-cadherin ectodomain fragment elicits such responses (Arulanandam et al., 2009).

Further, this relationship may be quite general among cadherins, as it has been observed for C- and M-cadherin (Charrasse et al., 2007; Noren, Niessen, Gumbiner, & Burridge, 2001), as well as N-cadherin and cadherin11 (Geletu et al., 2013). Under the conditions examined, IL6 production following cadherin-mediated homophilic ligation did not activate ERK1/2 (Arulanandam et al., 2009; Geletu et al., 2013; Raptis et al., 2000). Thus, in the context of cadherin-mediated high cell densities, the STAT pathway appears to a major factor in conferring cell survival.

10. CADHERIN NUCLEAR SIGNALING VIA PROTEOLYSIS

A number of studies show production of cadherin proteolytic fragments promotes cadherin-to-nuclear signaling, and in some cancer contexts, the nuclear entry and staining of the cadherin cytodomain has already been used in diagnosis (reviewed in Chetty & Serra, 2008; David & Rajasekaran, 2012). There are reports of extracellular proteases such as ADAMs and MMPs being involved, as well as intramembrane or intracellular proteases whose activities can be coupled to the initial occurrence of extracellular cleavage events, for example, cleavages subsequently occurring via presenilins or caspases (Damsky, Richa, Solter, Knudsen, & Buck, 1983; De Wever et al., 2007; Dusek et al., 2006; Lochter et al., 1997; Marambaud et al., 2002; Maretzky, Reiss, et al., 2005; Maretzky, Schulte, et al., 2005; McCusker, Cousin, Neuner, & Alfandari, 2009; Noe et al., 2001; Steinhusen et al., 2001; Symowicz et al., 2007; Vallorosi et al., 2000; reviewed in David & Rajasekaran, 2012; McCusker & Alfandari, 2009; Niessen et al., 2011). We will focus upon the intracellular cadherin fragments, but some of the extracellular cleavage products also appear to have biological activity in both development and pathology, for example, in relation to RTK activation (e.g., ErbB/HER2:HER3) and the consequent upregulation of MMP activity or inhibition of apoptosis (Fedor-Chaiken, Hein, Stewart, Brackenbury, & Kinch, 2003; Najy, Day, & Day, 2008; Inge et al., 2011; Grabowska, Sandhu, & Day, 2012; Nawrocki-Raby et al., 2003; Zuo et al., 2011). Further, *in vivo* exogenous expression of the ectodomain fragment of C-cadherin, for example, disrupts early gastrulation movements in *Xenopus*, an effect that may take place via altered aPKC and Rac activity in the recipient cells expressing endogenous intact C-cadherin (Seifert, Ibrahim, Stodtmeister, Winklbauer, & Niessen, 2009). In MDCK cells, aPKC associates with the cytoplasmic tail of E-cadherin, an interaction partially dependent on β-catenin. Incubation with E-cadherin's ectodomain increased the phosphorylation of aPKC and also its cytoplasmic association with native E-cadherin. aPKC and Rac are both involved in PCP pathways directing cell movements in development, so this work has suggested one means of cadherin outside-in signaling.

Intramembrane cleavage of cadherins has been evaluated for the release of active fragments to the cytoplasm or nucleus. One focus has been upon the presenilin protease, part of a membrane-associated complex that when misregulated promotes Alzheimer's disease. In the context of cadherins, while reducing their adhesive function (Marambaud et al., 2002), presenilin generates a cytodomain fragment of N-cadherin with nuclear signaling activity (Marambaud et al., 2003). In particular, the fragment enters the nucleus to bind CBP (Creb-binding protein), leading to CBP's degradation. This in turn lowers the ability of CBP's binding partner, CREB (cAMP response element-binding protein), to transactivate gene targets. Thus, in cases where presenilin is inactive, the CBP/CREB complex is more active.

As additionally resolved in mice deficient for presenilin, β-catenin is more active (Kang et al., 2002). Conversely, presenilin expression reduces β-catenin activity. This effect may occur indirectly through promoting β-catenin's phosphorylation via CK1 and GSK3β, and thereby reduced β-catenin stability (Kang et al., 1999, 2002; Soriano et al., 2001; Zhang et al., 1998). It is uncertain to what extent presenilin acts upon β-catenin that is associated with cadherin versus membrane localized via the destruction complex, as described (Maher et al., 2009).

Other proteases involved in the initiation of signals from the cadherin–catenin complex include the disintegrins and metalloproteases known as ADAMs. For example, ADAM10 colocalizes with N-cadherin in early mouse embryos, and they coassociate in cell line and *in vitro* contexts (Reiss et al., 2005). The extracellular cleavage of N-cadherin by ADAM10 first results in the generation of an extracellular fragment that retains some biological activities (Bixby & Zhang, 1990; Damsky et al., 1983; Paradies & Grunwald, 1993; Utton et al., 2001). The remaining membrane-bound fragment is then subject to an additional intramembrane cleavage step(s) by other proteases (Marambaud et al., 2003), such as gamma-secretase. In the case of N-cadherin, this results in lessened association with β-catenin and enhanced β-catenin cytoplasmic levels and nuclear activities (Fortini, 2002; Reiss et al., 2005; Shoval, Ludwig, & Kalcheim, 2007). The *in vivo* relevance was suggested in ADAM10-deficient mice embryos, where the loss of the N-cadherin C-terminal fragment dramatically reduced expression of an endogenous β-catenin reporter, cyclinD1. N-Cadherin proteolysis is initiated with the first extracellular cleavage step, which is itself responsive to the activation of PKC and calcium-mediated pathways. Thus, calcium influx during tissue remodeling or repair may act in part via shedding of the N-cadherin ectodomain in combination with cadherin cytodomain effects upon gene transcription. Such a scenario is likely to play out with additional cadherins as reported for E-cadherin in cancer cell-line contexts (Ito et al., 1999; Marambaud et al., 2002; Maretzky, Reiss, et al., 2005). Interestingly, p120^{ctn} appears to enhance the nuclear entry of an E-cadherin cytodomain fragment, with the fragment enhancing p120^{ctn}-facilitated relief of Kaiso-mediated target-gene repression and suppressing apoptosis (Ferber et al., 2008). Likewise, a fragment of E-cadherin arising as a consequence of presenilin 1 activity is reported to enhance β-catenin nuclear activity (Uemura et al., 2006). In the context of neurons, activation of the glutamate activated *N*-methyl-D-aspartate (NMDA) receptor, which contributes to long-term neuronal plasticity (as reviewed in Carroll & Zukin, 2002), leads to calcium influx and calpain activation, and the processing of both N-cadherin and β-catenin. Possibly initiated at the cadherin–catenin complex, the effects produced include enhanced β-catenin stability and signaling (Abe & Takeichi, 2007; reviewed in Arikkath & Reichardt, 2008).

In the developmental context of neural crest delamination in quail, bone morphogenic protein (BMP) signals result in N-cadherin fragmentation via ADAM10, leading to enhanced β-catenin expression and signaling, and thereby to enhanced cyclinD1 expression. Whereas such an N-cadherin cytodomain fragment might be expected to sequester β-catenin, another possibility in agreement with the observed downstream effects is that the fragment assists in protecting β-catenin from degradation until it reaches the nucleus to signal. Such a protection scenario is consistent with another study using the cytodomain E-cadherin (Simcha et al., 2001). In any case, interference with N-cadherin fragmentation compromised neural crest delamination. Prior to BMP signals, both the extracellular and intracellular domains of N-cadherin appear to be needed to maintain neural crest cells in a premigratory state (Shoval et al., 2007). Then, in initiating delamination and migration, BMP signals may act upon N-cadherin to both limit adhesive activity and coordinately enhance N-cadherin signaling output.

Desmosomal cadherins are essential in tissues that experience mechanical stress. In the intestine, cells differentiate as they progress along the crypt–villus axis. In the context of cadherin proteolytic fragments, the desmoglein2–cadherin was found to release a cytoplasmic fragment, likely resulting from a caspase, at the outset of apoptosis (Nava et al., 2007). Desmoglein's ectodomain is additionally shed via ADAM10 and 17 cleavage (Bech-Serra et al., 2006). Further reports point to desmosomal cadherins being fragmented, such as desmoglein1 (Dusek et al., 2006; Weiske et al., 2001), in addition to classic cadherins (Rios-Doria & Day, 2005; Rios-Doria et al., 2003). In the case of desmoglein2, its cytoplasmic fragment promotes apoptosis, perhaps as proposed, in concert with the physiological extrusion of cells from the intestinal villus or during inflammation (Fig. 6).

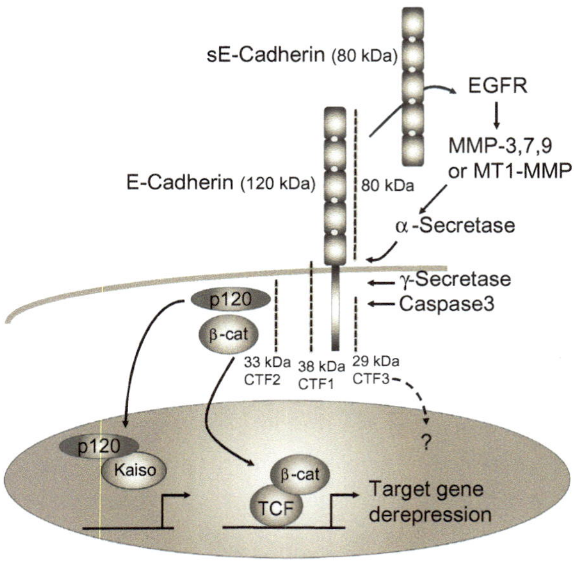

Figure 6 Cadherin signaling via proteolysis. Cadherins are cleaved at specific sites by proteases to generate fragments that are capable of transducing signals, either to the extracellular or to the intracellular space. The soluble extracellular E-cadherin fragment (sE-cadherin) can associate with intact E-cadherin present on other cells to alter cadherin-dependent cell properties inclusive of intracellular signaling (not shown). It also interacts with EGFR family members to activate MAPK signaling or metalloproteases, with the latter directly or indirectly resulting in the production of defined E-cadherin fragments. The intracellular C-terminal fragment number 2 (CTF2) is not membrane associated, so has been proposed to assist in the protection or translocation of p120^{ctn} or β-catenin to the nucleus. Here, p120^{ctn} displaces and thereby derepresses Kaiso-mediated repression of its gene targets, whereas β-catenin derepresses TCF/LEF-mediated gene repression via the recruitment of transcriptional coactivators. In the context of such selective proteolysis of E-cadherin, each catenin is thereby assisted in activating its respective gene targets.

11. PROTOCADHERIN SIGNALING TO THE NUCLEUS

The impact of the protocadherins on nuclear signaling is not as well studied as for the classical cadherins, even though the protocadherins constitute the largest group in the cadherin superfamily (reviewed in Kahr, Vandepoele, & van Roy, 2013; Weiner & Jontes, 2013). They typically have six to seven or more extracellular cadherin repeats, a single transmembrane pass, and interestingly, a cytoplasmic domain completely distinct from those of classical cadherins. Given that the cytodomains of protocadherins a and b are well conserved, their cytoplasmic domains likely execute conserved functions. They are not generally thought to interact with catenin proteins, although β-catenin has been reported to associate with protocadherin-PC (reviewed in Chen, Vacherot, et al., 2002; Vanhalst, Kools, Staes, van Roy, & Redies, 2005). Protocadherins are widely dispersed, with strong representation in neural tissues and those undergoing morphogenic movements (Chen & Gumbiner, 2006; Chen, Koh, et al., 2009; Nakao, Platek, Hirano, & Takeichi, 2008; Vanhalst et al., 2005; reviewed in Kahr et al., 2013). Some but not all have adhesive activity, possibly modulated by *cis*-interactions occurring between different protocadherin members to provide combinatorial arrangements that assist in specifying cell–cell associations and identity (e.g., of neurons; reviewed in Chen & Maniatis, 2013). Protocadherins further interact with classic cadherins (Biswas, Emond, & Jontes, 2010; Chen, Koh, et al., 2009; Chen, Vacherot, et al., 2002; Emond, Biswas, Blevins, & Jontes, 2011; Emond, Biswas, & Jontes, 2009; Yasuda et al., 2007), with which they have been found in *cis*- or *trans*-associations. Via mechanisms noted above involving classic cadherins, nuclear signaling might thus occur for protocadherins indirectly as a consequence of these relationships.

Protocadherins also associate with Frizzled7 (Berndt et al., 2011; Kraft et al., 2012; Medina et al., 2004), implicated in noncanonical Wnt signaling, as well as the phosphatase PP1α (Chen, Vacherot, et al., 2002), which has been examined in contexts of learning and memory (Mulkey, Endo, Shenolikar, & Malenka, 1994). Neurobiologists are further interested in protocadherins given some of their interactions with proteins such as Dab1 (Homayouni, Rice, & Curran, 2001), which acts in the Reelin signaling pathway to influence neuronal positioning (reviewed in Bar, Lambert de Rouvroit, & Goffinet, 2000), or with the protocadherin arcadlin, which modulates N-cadherin cell surface levels and thereby dendritic spine densities (Yamagata et al., 1999; Yasuda et al., 2007).

Other associations include those with the tyrosine kinase Ret (Schalm, Ballif, Buchanan, Phillips, & Maniatis, 2010) or with CK2β, Sprouty, ANR5, or FLRT3 (Chen, Koh, et al., 2009; Chung et al., 2007; Kietzmann, Wang, Weber, & Steinbeisser, 2012; Wang et al., 2008; reviewed in Chen & Maniatis, 2013). Protocadherin interactions assist in keeping Ret stable and responsive to its GNDF ligand (Schalm et al., 2010), where they have been found to inhibit signaling from Pyk2 (proline-rich tyrosine kinase) and Fak (Chen, Lu, et al., 2009; Garrett, Schreiner, Lobas, & Weiner, 2012), kinases having roles in small-GTPase modulation (RhoA, Rac1), as well as signaling to the nucleus. Protocadherin PAPC's interaction with Frizzled7 modulates noncanonical Wnt signaling (i.e., PCP signaling) as mediated via JNK, RhoA, and Rac, during tissue separation and elongation (i.e., convergence–extension movements) in *Xenopus* (Medina et al., 2004; Unterseher et al., 2004). While RhoA and Rac are often evaluated from the perspective of cytoskeletal control, they and JNK have the potential to pass signals directly to the nucleus. Likewise, interaction of the PAPC-protocadherin with CK2β facilitates the removal of β-catenin, making it more difficult for Wnt signals to inhibit β-catenin's destruction complex. This alters the transcriptional activity of canonical Wnt gene targets and thereby gastrulation in *Xenopus* (Kietzmann et al., 2012).

Protocadherin nuclear signaling through the MAPK pathway has also been suggested with the involvement of p38 MAPK and MAPKKK (even as such studies have also focused also on MAPKKK in the context of neuronal junctions) (Yasuda et al., 2007). Protocadherin7 (NF-protocadherin) has likewise been implicated in nuclear signaling via interactions with chromatin as well as the transcriptional regulator TAF1/Set (Heggem & Bradley, 2003; Murakami, Hijikata, Matsukawa, Ishikawa, & Yorifuji, 2006). Interestingly, however, the TAF1 interaction appears to occur in the cytoplasm, with work in *Xenopus* including examinations of cell–cell adhesion outcomes and proper neural fold and tube formation (Rashid, Newell, Shama, & Bradley, 2006). Finally, there are indications of protocadherins being cleaved by γ-secretase as described above for classic cadherins, to generate fragments that may transduce information to the nucleus (Bonn, Seeburg, & Schwarz, 2007; Haas, Frank, Veron, & Kemler, 2005; Hambsch, Grinevich, Seeburg, & Schwarz, 2005). Such cleavage may be predominant during development, with less cleavage found to occur as neurons differentiate (Buchanan, Schalm, & Maniatis, 2010).

12. ATYPICAL CADHERIN NUCLEAR SIGNALING

"Atypical" cadherins refer to a number of cadherin superfamily members possessing primary and folding structures that distinguish them from classical cadherins. Those most reported are in the invertebrate literature, where they were first characterized and include members such as Fat, Daschous, and Flamingo (vertebrate Celsr). In common with classical cadherins, Fat and Daschous span the plasma membrane once, but as is typical to this cadherin subfamily have considerably larger extracellular domains with a greater number of cadherin repeats, as well as cytodomains that differ from each other and classical cadherins. For some of the atypical cadherins, such as Flamingo, evidence suggests that their roles center on signaling, and perhaps less so on adhesion (Shimada, Usui, Yanagawa, Takeichi, & Uemura, 2001; Usui et al., 1999; Zhou, Goffinet, & Tissir, 2008; reviewed in Wu & Mlodzik, 2009). In all cases, the extracellular domains of atypical cadherins are highly variable (Oda, Tagawa, & Akiyama-Oda, 2005; reviewed in Harris & Tepass, 2010; Oda & Takeichi, 2011), suggesting the potential to engage in a wide variety of interactions.

As noted, much of the work upon these cadherins was initially conducted in *Drosophila*, with both similarities and significant departures later found in vertebrates. While direct signaling to the nucleus may occur from atypical cadherins, there appears to be no uniform mode of signaling. Rather, considerable findings suggest their contributions to the modulation of cytoskeletal architecture and functions in epithelial tissue patterning (e.g., from proximal-to-distal), and directing the morphogenesis or generation of epithelial, neuroepithelial, and neuronal forms (e.g., proper formation of the neural tube or guidance of dendrites or axons). For example, interaction of Fat with Daschous promotes the activation of Frizzled and pathways thought to be largely distinct (despite cross talk) from β-catenin stabilization (Yang, Axelrod, & Simon, 2002). While small GTPases and other mediators of the observed cytoskeletal effects are likely to modulate downstream gene activity, this has not yet been the emphasis of recent studies.

In *Drosophila*, however, clear genetic evidence indicates that the atypical Fat cadherin controls tissue size through the nuclear-directed Hippo pathway (Bennett & Harvey, 2006; Cho et al., 2006; Hamaratoglu et al., 2006; Silva, Tsatskis, Gardano, Tapon, & McNeill, 2006; Willecke et al., 2006). While most components of this pathway are conserved in vertebrates, one recent report suggests that the Fat–Hippo relationship is not conserved

(Bossuyt et al., 2014). Instead, it appears that differences arose in the Hippo pathway's upstream modulators between arthropods and vertebrates. Nonetheless, in invertebrates at least, there is clear precedence for nuclear signaling being dependent upon atypical cadherins.

Fat's putative connection to nuclear signaling was earlier made via the ERM-family protein Expanded, which together with Merlin modulates the EGFR and Hippo pathways (Bennett & Harvey, 2006; Curto et al., 2007; Gladden et al., 2010; Hamaratoglu et al., 2006; Silva et al., 2006; Willecke et al., 2006). In this nuclear-directed context, the definitive identity of upstream *Drosophila* ligands in relation to Fat remains in question. However, another *Drosophila* atypical cadherin, Daschous, is thought to functionally interact with Fat in certain contexts involving cell–cell interactions (Konsavage & Yochum, 2013; Rogulja, Rauskolb, & Irvine, 2008; reviewed in Matakatsu & Blair, 2006).

13. CADHERIN NUCLEAR SIGNALING VIA THE HIPPO PATHWAY

Evidence that cadherins functionally interact with the Hippo pathway has come from a variety of sources. In the context of a mammalian cancer cell-line deficient for E-cadherin, one of the key effectors of this pathway, Yap, is predominantly nuclear (Kim, Koh, Chen, & Gumbiner, 2011). When E-cadherin expression is restored, Yap moves to the cytoplasm in a cell density-dependent manner, and Yap target genes are no longer activated. The participation of the Hippo pathway in such cadherin and cell density-dependent effects upon cell growth was further made evident using beads coated with the E-cadherin ectodomain. Such beads normally have the capacity to inhibit cell growth, but not in the case where the Hippo pathway was inactivated through the removal of one of its components, or through the downstream expression of Yap (Kim et al., 2011). How cadherin mediates its effects upon Yap is not clear, but as noted below it may be via signals that involve α-catenin or other entities.

Even within a contiguous group of cells, the relationship of the cadherin complex to the Hippo pathway can vary according to cell density. For example, at the borders of a colony of cells grown in culture, cadherin-mediated cell–cell contacts and cell densities are often less mature than toward the center. This is reflected not only in the state of the contacts but also in the disposition of Hippo pathway components. Cells at the edge display a greater proportion of nuclear Yap and often display higher proliferative and migratory dispositions. Activation of the Hippo pathway, as seen in the phosphorylation of Yap by the Lats kinase as well as other measures, appears sensitive to the junctional status of cells. As noted, homophilic ligation of cadherins between cells (in *trans*) activates the Hippo kinase-mediated pathway to prevent Yap's nuclear entry (Gumbiner & Kim, 2014). Despite the uncertainty regarding how this occurs, the association of β-catenin and α-catenin with cadherin appears to be needed for cadherin sensing to be integrated with the Hippo pathway. Other junction-associated proteins are also likely to be involved in some of these contexts, including, for example, Merlin, Ajuba, and Kibra (Baumgartner, Poernbacher, Buser, Hafen, & Stocker, 2010; Das Thakur et al., 2010; Genevet, Wehr, Brain, Thompson, & Tapon, 2010; Gladden et al., 2010; Huang, Wu, Barrera, Matthews, & Pan, 2005; Morrison et al., 2001).

α-Catenin has been proposed to limit Yap activity in a manner that bypasses some of the upstream Hippo pathway components. For example, by interacting with Yap indirectly via 14-3-3σ, α-catenin prevents the PP2A phosphatase from dephosphorylating Yap, so that Yap is kept out of the nucleus (Schlegelmilch et al., 2011). In this context, α-catenin might inhibit Yap nuclear function in ways that are independent of α-catenin being associated with cadherin (Fig. 5). In addition, N-terminally phosphorylated β-catenin may also cross-regulate and inhibit Hippo pathway effectors, Yap and Taz (Azzolin et al., 2012; Rosenbluh et al., 2012). Lastly, the cytoskeleton and cell shape have also been studied with regard to their impact upon Hippo signal transduction (Dupont et al., 2011; Sansores-Garcia et al., 2011). Since cadherins and cell–cell junctions are integral to larger cytoskeletal structures and cell-shape considerations, such mechanics reflect yet another manner that signals originating at cadherin-based structures may be moved to the nucleus, a process termed mechanotransduction. Whatever the mechanism, upon the lessening of cadherin-mediated contacts, there is often a corresponding lessening of upstream Hippo signaling, such that downstream Yap enters the nucleus to bind a variety of transcription factors (Basu, Totty, Irwin, Sudol, & Downward, 2003; Chen, Loh, & Song, 2010; Komuro, Nagai, Navin, & Sudol, 2003; Varelas et al., 2010; Yagi, Chen, Shigesada, Murakami, & Ito, 1999), with a strong proportion modulating growth.

14. STEM CELL MAINTENANCE VIA CADHERIN NUCLEAR SIGNALING

Despite early established relationships between E-cadherin and β-catenin signaling (McEwen et al., 2012), and β-catenin signaling and stem cell maintenance (reviewed in Miki, Yasuda, & Kahn, 2011), evidence clearly shows that maintaining mESCs in the pluripotent state requires E-cadherin, where its loss drives entry into what is referred to as a "primed" state, termed mEpiESCs (reviewed Li et al., 2012; Pieters & van Roy, 2014). Human ESCs likewise exhibit the requirement of E-cadherin for self-renewal, pluripotency, and long-term survival (Li, Wang, et al., 2010). Remarkably, the key progenitor-pathway downstream of E-cadherin involves LIF-STAT signaling. LIF is a secreted cytokine that promotes stemness in mESCs, and interestingly, the ecto- or transmembrane domain of E-cadherin forms a complex with the transmembrane LIF receptor and its coreceptor GP130 (del Valle et al., 2013). This stabilizes the receptor complex at the protein level to promote JAK/Stat signaling (Hirai, Karian, & Kikyo, 2011). In E-cadherin-deficient mESCs, a loss in LIF responsiveness is observed (del Valle et al., 2013; Hawkins, Mohamet, Ritson, Merry, & Ward, 2012), while downstream STAT activation in such cells is able to maintain the stemness state (Matsuda et al., 1999) (Fig. 7).

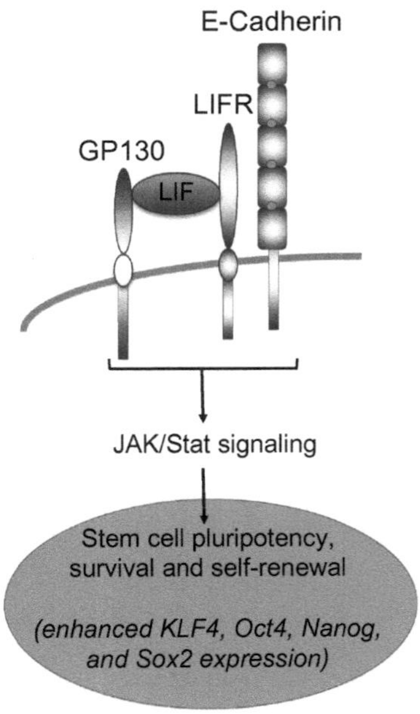

Figure 7 Cadherin signaling promotes JAK/STAT signaling for stem cell maintenance. Mouse and human ESCs require E-cadherin for pluripotency, self-renewal, and long-term survival. The transmembrane and ectodomain of E-cadherin interact with the LIF receptor (LIFR) and GP130 coreceptor. This association stabilizes the complex and thereby promotes JAK/Stat signaling and consequent stemness properties to the stem cells.

In mEpiESCs, N-cadherin in combination with E-cadherin is present at cell–cell contacts (Bao et al., 2009). N-Cadherin assists in maintaining the primed state by enabling activin signaling (Soncin et al., 2009). N-Cadherin has further been found to assist in maintaining the neural progenitor pool, through the modulation of delamination and differentiation, as described, for one case, above (reviewed Paulson et al., 2014; Rousso et al., 2012; Seki, Namba, Mochizuki, & Onodera, 2007; Yagita et al., 2009; Zhang et al., 2010).

Cadherins also help to organize stem cells. In the *Drosophila* ovary, for example, DE-cadherin allows for germline stem cells to make contact with cap cells forming the niche; here, the stemness-promoting Wnt ligand is expressed (Gonzalez-Reyes, 2003; Song & Xie, 2003; Song, Zhu, Doan, & Xie, 2002). Maintaining a stem cell niche in bone marrow is likewise dependent on stem cell attachment, but here the interaction occurs with osteoblast cells via N-cadherin (Zhang et al., 2003). Finally, cadherins play vital roles in stem cell differentiation (del Valle et al., 2013; Larue et al., 1996; Mohamet, Lea, & Ward, 2010; Moore, Radice, Dominis, & Kemler, 1999).

In stem cell contexts, cadherins might signal to the nucleus via catenins. However, thus far, the role of β-catenin has been discussed more in the context of the canonical Wnt pathway. How β-catenin operates in a signaling or other capacity is also under active discussion by stemness researchers (Anton, Kestler, & Kuhl, 2007; Lyashenko et al., 2011; Soncin et al., 2009). In the nucleus, events may involve β-catenin associating with transcription factors other than TCF/LEF. Some work suggests that β-catenin's adhesive function is more pertinent to its role in stemness (del Valle et al., 2013; Lyashenko et al., 2011), even as other interpretations have been offered (Wray et al., 2011; reviewed in Pieters & van Roy, 2014).

Lastly, recent work in mESCs suggests that p120^{ctn} might also require consideration. Roughly in line with known decreases in E-cadherin protein occurring in mESCs leaving their naive state, the signaling pool of p120^{ctn} rises. This was seen to derepress (activate) gene targets of the transcriptional repressor REST, to which p120^{ctn} binds, assisting in moving the mESCs toward a more differentiated state. While more work is required, E-cadherin depletion also resulted in elevation of these same gene targets, with p120^{ctn}'s specific involvement suggested by a partial rescue of the observed effects upon the codepletion of p120^{ctn} (but not β-catenin) (Lee et al., 2014).

15. DESMOSOMAL CADHERIN NUCLEAR SIGNALING

Desmoglein1–cadherin has been implicated in adhesion-independent signaling contributing to morphogenesis and differentiation of keratinocytes (Getsios et al., 2009). Here, via its cytoplasmic tail (when associated with the plasma membrane), Desmoglein1 suppressed EGFR activation and signaling, in particular the Erk1/2 trajectory, to facilitate differentiation (Getsios et al., 2009). Interestingly, the ectodomain of Desmoglein1 is not required for this activity, nor are Desmoglein1's associations with Desmocollin1–cadherin nor γ-catenin (plakoglobin). Desmoglein1 was concluded to act within an early instructive pathway directing keratinocytes through terminal differentiation, preventing Erk1/2 signaling in the suprabasal layers. Conceivably, this occurs via the direct interaction of Desmoglein1 with the EGFR, as they colocalize. Interestingly, Desmoglein2 expression in the suprabasal skin layer acts in a contrary manner to Desmoglein1, where Desmoglein2 ectopic presence appears to be functionally coupled to enhanced cell survival via increased Erk1/2 and NFκB activity (Brennan et al., 2007). In the case of another type of desmosomal cadherin, Desmocollin3, its ectopic expression was tied to increased β-catenin levels and transcriptional activity (Hardman et al., 2005). Other work has likewise linked desmosomal cadherins to proliferation, differentiation, and morphogenesis, but not always with the same outcomes (Chidgey et al., 2001; Elias et al., 2001; Eshkind et al., 2002; Merritt et al., 2002). Given that these desmosomal cadherins are normally expressed in the basal versus suprabasal regions of skin, each may provide a respective homeostatic balance for the activities of the other (Fig. 8).

In particular, the ectopic *in vivo* expression of Desmocollin3a and 3b in suprabasal keratinocytes of transgenic mice increases β-catenin stability and signaling by an unknown mechanism (Hardman et al., 2005). Again while the responsible biochemical links are uncertain, the expression of desmoglein2 in the same region likewise increased proliferation and cell survival, but apparently through effects upon multiple signaling pathways including EGFR and NFκB (Brennan et al., 2007).

There are additional indications of desmosomal cadherins signal to the nucleus in development and disease. One interpretation for ARVCs in humans followed the use of a mouse model in which Desmoplakin, a protein linking desmosomal cadherins to the intermediate cytoskeleton, was ablated. γ-Catenin (plakoglobin) was released from the desmosomal adhesion

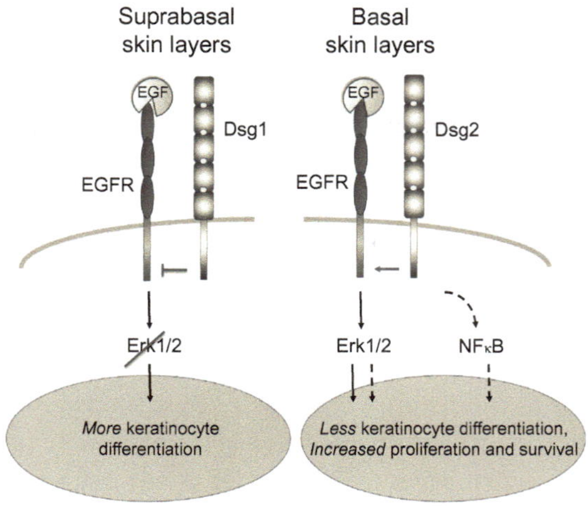

Figure 8 Desmosomal cadherin signaling in skin. In the suprabasal skin layers, the cytoplasmic domain of Desmoglein1 inhibits EGFR signaling via the Erk1/2 signaling trajectory (Getsios et al., 2009). This might occur via a direct interaction of the Desmoglein1 cytodomain with EGFR. Ectopic Desmoglein2 expression (involucrin promoter) in the suprabasal keratinocyte layer increases proliferation and cell survival, apparently via effects upon EGFR and NFκB signaling (Brennan et al., 2007). Misexpression of desmocollin3a and 3b in subrabasal keratinocytes leads to proliferation and differentiation defects, with increased β-catenin signaling normally seen only in basal keratinocytes (Hardman et al., 2005) (not shown).

complex and found to inhibit the expression of β-catenin gene targets, and thereby the canonical Wnt pathway. The loss of such essential Wnt signals is known to reduce the formation of muscle in favor of fatty tissue, fitting well with known features of the disease in humans (Ben-Ze'ev, Shtutman, & Zhurinsky, 2000; Garcia-Gras et al., 2006).

Work in intestinal cancer epithelial cells has likewise shown a functional interaction between Desmoglein2 and EGFR. The loss of Desmoglein2 showed reduced EGFR activation, and thereby reduced Erk and Src activity. Cell proliferation was concomitantly lessened, perhaps as a result of compensatory expression of Desmocollin2 followed by its association and inhibition of EGFR. Thus, Desmoglein2 and Desmocollin2 appear to have opposing roles in this cancer cell context, with Desmoglein2 promoting and Desmocollin2 inhibiting tumorigenesis (Kamekura et al., 2014).

Finally, it is interesting to consider that desmosomal cadherins also interact with catenin proteins, specifically plakophilin1–3. Additionally, a catenin found at both desmosomal and adherens junctions is γ-catenin (plakoglobin), with its signaling properties having been briefly mentioned earlier in this review. The plakophilin-catenins may possess roughly analogous roles to those of catenins present at the adherens junctions. That is, in addition to being at desmosomal junctions where they assist in linking Desmocollin- or Desmoglein-cadherins to the cytoskeleton (intermediate filaments), plakophilin1–3 are additionally present in the cytoplasm or nucleus of some cells (Mertens, Kuhn, & Franke, 1996). Plakophilin3 has been found in association with RNA-binding proteins such as G3BP, FXR1, and PABPC1 in cytoplasmic stress granules that arise and assist cells in response to adverse metabolic stimuli (Hofmann et al., 2006). Recently, plakophilin3 was found to bind the Ets-family transcription factor ETV1, with the functional interaction of these two proteins being supported in studies using *Xenopus* embryos (Munoz et al., 2014). Likewise, plakophilin2 interacts with a subunit of the RNA polymerase III complex (RPC155) (Mertens et al., 1996), possibly engaging in the modulation of ribosomal or tRNA transcription. Plakophilin2's entry to the nucleus is negatively regulated via direct phosphorylation by the kinase C-TAK, which produces a binding site for 14-3-3 to retain it within the cytoplasm (Muller, Ritt, Copeland, & Morrison, 2003). Additionally, plakophilin2 has been reported in association with β-catenin and to modestly enhance β-catenin's nuclear activity (Chen, Bonne, Hatzfeld, van Roy, & Green, 2002). These nuclear interactions of plakophilin2 or plakophilin3 leave open the possibility of cross talk between desmosomal cadherins and the nucleus, similar in concept to that discussed earlier for β-catenin and $p120^{ctn}$.

16. CONCLUSIONS

The cadherin family, which includes classic, atypical, and protocadherin forms, comprises a diverse protein family. Most engage in cell–cell adhesion, but a long-standing question has been to what extent the cadherins also engage in signaling. This review has focused in particular upon the involvement of cadherins in signaling to the nucleus. It occurs via established pathways, such as those involving RTK or Wnt components, but we have also touched upon additional pathways such as Hippo, that while developmentally central are still emerging. We have concerned ourselves little with signals directed primarily to the cytoplasm or cytoskeleton, despite their importance, including signals that largely rely upon mechanical considerations (such as involving "tensegrity"). In these latter cases, we have tried to direct the reader to other papers or reviews.

Our review of evidence from many investigators indicates that cadherins are integrated into the larger fabric of cell and developmental decisions and functions through an array of direct or indirect protein–protein associations. A central concept for consideration is that organismal fitness benefited from mechanisms that reported cell–cell adhesive states to the nucleus—contributing to one of many forms of outside-in signaling. Indeed, such reporting may have been selected for over the course of evolution. For example, signals involving or originating from cadherins can lead to the activation or repression of gene targets that affect proliferation and differentiation, and in other cases motility or cell death. As cadherins are expressed and function in most tissues, and at most developmental time points, as well as in tissue regeneration and homeostasis, the phenotypes that result from their loss or gain of function are many. Along with integrins, that are best known to interact with the extracellular matrix (but also some neighboring-cell receptors), or with other classes of cell–cell adhesion proteins, cadherins comprise one class of the multiple transmembrane systems capable of integrating the cell's outside world with that which lies within.

Since cadherins have adhesive as well as the noted signaling or additional functions, it can be challenging at the experimental level to discern if observed gain- or loss-of-function phenotypes are attributable mainly to altered adhesion-motility states or altered nuclear signaling (etc.). Indeed, since these functions engage each other in many cases, they are interdependent.

To recap, we have focused in this review on a few modalities of cadherin nuclear signal transduction. One involves the modulation by cadherins of other cell surface receptors, such as RTKs, or those that bind to Wnt ligands, such as Frizzled and Lrp5/6. Generally in this case, a series of indirect events transpire (such as promotion or inhibition of the MAPK cascade) that have an impact upon gene activity. A second type of cadherin-based mechanism involves events more proximal to the nucleus, such as direct interaction of cadherin cytodomains with proteins having the ability to enter the nucleus to alter gene activity. Examples here include the catenins and also DNA-binding transcription factors such NFκB. In some cases, we have seen that cadherin proteolytic fragments themselves exhibit the capacity to enter the nucleus to promote or inhibit gene programs.

Overall, our understanding of the larger picture of cadherin nuclear signaling is still quite modest. While many examples can be pointed out as summarized in this review, we lack an understanding at the systems level of the interrelationships that occur at the plasma membrane, cytoplasm, and nucleus. That is, while much information is known, we remain more in the position of imagining than comprehending how things fit together on a wider scale. This is of course a problem faced currently in all biological fields, and we thus look forward to the application of emerging or still-to-arise wet-bench technologies, as well as the development and wise application of more robust information processing and model testing.

ACKNOWLEDGMENTS

We regret not being able to discuss many studies relevant to this topic. P. D. M and C. J. G are supported by funds from the National Institutes of Health GM107079 and GM076561, respectively.

REFERENCES

Abe, K., & Takeichi, M. (2007). NMDA-receptor activation induces calpain-mediated beta-catenin cleavages for triggering gene expression. *Neuron*, *53*, 387–397.

Abreu-Blanco, M. T., Verboon, J. M., Liu, R., Watts, J. J., & Parkhurst, S. M. (2012). Drosophila embryos close epithelial wounds using a combination of cellular protrusions and an actomyosin purse string. *Journal of Cell Science*, *125*, 5984–5997.

Aktary, Z., Kulak, S., Mackey, J., Jahroudi, N., & Pasdar, M. (2013). Plakoglobin interacts with the transcription factor p53 and regulates the expression of 14-3-3sigma. *Journal of Cell Science*, *126*, 3031–3042.

Aktary, Z., & Pasdar, M. (2012). Plakoglobin: Role in tumorigenesis and metastasis. *International Journal of Cell Biology*, *2012*, 189521.

Aktary, Z., & Pasdar, M. (2013). Plakoglobin represses SATB1 expression and decreases in vitro proliferation, migration and invasion. *PLoS One*, *8*, e78388.

Alema, S., & Salvatore, A. M. (2007). p120 catenin and phosphorylation: Mechanisms and traits of an unresolved issue. *Biochimica et Biophysica Acta*, *1773*, 47–58.

Anastasiadis, P. Z. (2007). p120-ctn: A nexus for contextual signaling via Rho GTPases. *Biochimica et Biophysica Acta*, *1773*, 34–46.

Andrade, M. A., Petosa, C., O'Donoghue, S. I., Muller, C. W., & Bork, P. (2001). Comparison of ARM and HEAT protein repeats. *Journal of Molecular Biology*, *309*, 1–18.

Anton, R., Kestler, H. A., & Kuhl, M. (2007). Beta-catenin signaling contributes to stemness and regulates early differentiation in murine embryonic stem cells. *FEBS Letters*, *581*, 5247–5254.

Arikkath, J., & Reichardt, L. F. (2008). Cadherins and catenins at synapses: Roles in synaptogenesis and synaptic plasticity. *Trends in Neurosciences*, *31*, 487–494.

Arnsdorf, E. J., Tummala, P., & Jacobs, C. R. (2009). Non-canonical Wnt signaling and N-cadherin related beta-catenin signaling play a role in mechanically induced osteogenic cell fate. *PLoS One*, *4*, e5388.

Arulanandam, R., Vultur, A., Cao, J., Carefoot, E., Elliott, B. E., Truesdell, P. F., et al. (2009). Cadherin-cadherin engagement promotes cell survival via Rac1/Cdc42 and signal transducer and activator of transcription-3. *Molecular Cancer Research: MCR*, *7*, 1310–1327.

Azzolin, L., Zanconato, F., Bresolin, S., Forcato, M., Basso, G., Bicciato, S., et al. (2012). Role of TAZ as mediator of Wnt signaling. *Cell*, *151*, 1443–1456.

Balsamo, J., Arregui, C., Leung, T., & Lilien, J. (1998). The nonreceptor protein tyrosine phosphatase PTP1B binds to the cytoplasmic domain of N-cadherin and regulates the cadherin-actin linkage. *The Journal of Cell Biology*, *143*, 523–532.

Bao, S., Tang, F., Li, X., Hayashi, K., Gillich, A., Lao, K., et al. (2009). Epigenetic reversion of post-implantation epiblast to pluripotent embryonic stem cells. *Nature, 461*, 1292–1295.

Bar, I., Lambert de Rouvroit, C., & Goffinet, A. M. (2000). The evolution of cortical development. An hypothesis based on the role of the Reelin signaling pathway. *Trends in Neurosciences, 23*, 633–638.

Bass-Zubek, A. E., Godsel, L. M., Delmar, M., & Green, K. J. (2009). Plakophilins: Multifunctional scaffolds for adhesion and signaling. *Current Opinion in Cell Biology, 21*, 708–716.

Basu, S., Totty, N. F., Irwin, M. S., Sudol, M., & Downward, J. (2003). Akt phosphorylates the Yes-associated protein, YAP, to induce interaction with 14-3-3 and attenuation of p73-mediated apoptosis. *Molecular Cell, 11*, 11–23.

Baumgartner, R., Poernbacher, I., Buser, N., Hafen, E., & Stocker, H. (2010). The WW domain protein Kibra acts upstream of Hippo in Drosophila. *Developmental Cell, 18*, 309–316.

Bech-Serra, J. J., Santiago-Josefat, B., Esselens, C., Saftig, P., Baselga, J., Arribas, J., et al. (2006). Proteomic identification of desmoglein 2 and activated leukocyte cell adhesion molecule as substrates of ADAM17 and ADAM10 by difference gel electrophoresis. *Molecular and Cellular Biology, 26*, 5086–5095.

Benjamin, J. M., Kwiatkowski, A. V., Yang, C., Korobova, F., Pokutta, S., Svitkina, T., et al. (2010). AlphaE-catenin regulates actin dynamics independently of cadherin-mediated cell-cell adhesion. *The Journal of Cell Biology, 189*, 339–352.

Bennett, F. C., & Harvey, K. F. (2006). Fat cadherin modulates organ size in Drosophila via the Salvador/Warts/Hippo signaling pathway. *Current Biology, 16*, 2101–2110.

Ben-Ze'ev, A., Shtutman, M., & Zhurinsky, J. (2000). The integration of cell adhesion with gene expression: The role of beta-catenin. *Experimental Cell Research, 261*, 75–82.

Berger-Muller, S., & Suzuki, T. (2011). Seven-pass transmembrane cadherins: Roles and emerging mechanisms in axonal and dendritic patterning. *Molecular Neurobiology, 44*, 313–320.

Berndt, J. D., Aoyagi, A., Yang, P., Anastas, J. N., Tang, L., & Moon, R. T. (2011). Mindbomb 1, an E3 ubiquitin ligase, forms a complex with RYK to activate Wnt/beta-catenin signaling. *The Journal of Cell Biology, 194*, 737–750.

Bhatt, T., Rizvi, A., Batta, S. P., Kataria, S., & Jamora, C. (2013). Signaling and mechanical roles of E-cadherin. *Cell Communication & Adhesion, 20*, 189–199.

Bilic, J., Huang, Y. L., Davidson, G., Zimmermann, T., Cruciat, C. M., Bienz, M., et al. (2007). Wnt induces LRP6 signalosomes and promotes dishevelled-dependent LRP6 phosphorylation. *Science, 316*, 1619–1622.

Birukova, A. A., Zebda, N., Fu, P., Poroyko, V., Cokic, I., & Birukov, K. G. (2011). Association between adherens junctions and tight junctions via Rap1 promotes barrier protective effects of oxidized phospholipids. *Journal of Cellular Physiology, 226*, 2052–2062.

Biswas, S., Emond, M. R., & Jontes, J. D. (2010). Protocadherin-19 and N-cadherin interact to control cell movements during anterior neurulation. *The Journal of Cell Biology, 191*, 1029–1041.

Bixby, J. L., & Zhang, R. (1990). Purified N-cadherin is a potent substrate for the rapid induction of neurite outgrowth. *The Journal of Cell Biology, 110*, 1253–1260.

Blattler, A., Yao, L., Wang, Y., Ye, Z., Jin, V. X., & Farnham, P. J. (2013). ZBTB33 binds unmethylated regions of the genome associated with actively expressed genes. *Epigenetics Chromatin.*

Blitzer, J. T., & Nusse, R. (2006). A critical role for endocytosis in Wnt signaling. *BMC Cell Biology, 7*, 28.

Bonn, S., Seeburg, P. H., & Schwarz, M. K. (2007). Combinatorial expression of alpha- and gamma-protocadherins alters their presenilin-dependent processing. *Molecular and Cellular Biology, 27*, 4121–4132.

Bonvini, P., An, W. G., Rosolen, A., Nguyen, P., Trepel, J., Garcia de Herreros, A., et al. (2001). Geldanamycin abrogates ErbB2 association with proteasome-resistant beta-catenin in melanoma cells, increases beta-catenin-E-cadherin association, and decreases beta-catenin-sensitive transcription. *Cancer Research*, *61*, 1671–1677.

Bossuyt, W., Chen, C. L., Chen, Q., Sudol, M., McNeill, H., Pan, D., et al. (2014). An evolutionary shift in the regulation of the Hippo pathway between mice and flies. *Oncogene*, *33*, 1218–1228.

Brembeck, F. H., Schwarz-Romond, T., Bakkers, J., Wilhelm, S., Hammerschmidt, M., & Birchmeier, W. (2004). Essential role of BCL9-2 in the switch between beta-catenin's adhesive and transcriptional functions. *Genes & Development*, *18*, 2225–2230.

Brennan, D., Hu, Y., Joubeh, S., Choi, Y. W., Whitaker-Menezes, D., O'Brien, T., et al. (2007). Suprabasal Dsg2 expression in transgenic mouse skin confers a hyperproliferative and apoptosis-resistant phenotype to keratinocytes. *Journal of Cell Science*, *120*, 758–771.

Brieher, W. M., & Yap, A. S. (2013). Cadherin junctions and their cytoskeleton(s). *Current Opinion in Cell Biology*, *25*, 39–46.

Buchanan, S. M., Schalm, S. S., & Maniatis, T. (2010). Proteolytic processing of protocadherin proteins requires endocytosis. *Proceedings of the National Academy of Sciences of the United States of America*, *107*, 17774–17779.

Caca, K., Kolligs, F. T., Ji, X., Hayes, M., Qian, J., Yahanda, A., et al. (1999). Beta- and gamma-catenin mutations, but not E-cadherin inactivation, underlie T-cell factor/lymphoid enhancer factor transcriptional deregulation in gastric and pancreatic cancer. *Cell Growth & Differentiation*, *10*, 369–376.

Carmeliet, P., Lampugnani, M. G., Moons, L., Breviario, F., Compernolle, V., Bono, F., et al. (1999). Targeted deficiency or cytosolic truncation of the VE-cadherin gene in mice impairs VEGF-mediated endothelial survival and angiogenesis. *Cell*, *98*, 147–157.

Carroll, R. C., & Zukin, R. S. (2002). NMDA-receptor trafficking and targeting: Implications for synaptic transmission and plasticity. *Trends in Neurosciences*, *25*, 571–577.

Casagolda, D., Del Valle-Perez, B., Valls, G., Lugilde, E., Vinyoles, M., Casado-Vela, J., et al. (2010). A p120-catenin-CK1epsilon complex regulates Wnt signaling. *Journal of Cell Science*, *123*, 2621–2631.

Cavallaro, U., & Christofori, G. (2004). Cell adhesion and signalling by cadherins and Ig-CAMs in cancer. *Nature Reviews. Cancer*, *4*, 118–132.

Cavallaro, U., & Dejana, E. (2011). Adhesion molecule signalling: Not always a sticky business. *Nature Reviews. Molecular Cell Biology*, *12*, 189–197.

Cavey, M., & Lecuit, T. (2009). Molecular bases of cell-cell junctions stability and dynamics. *Cold Spring Harbor Perspectives in Biology*, *1*, a002998.

Charpentier, E., Lavker, R. M., Acquista, E., & Cowin, P. (2000). Plakoglobin suppresses epithelial proliferation and hair growth in vivo. *The Journal of Cell Biology*, *149*, 503–520.

Charrasse, S., Comunale, F., Fortier, M., Portales-Casamar, E., Debant, A., & Gauthier-Rouviere, C. (2007). M-cadherin activates Rac1 GTPase through the Rho-GEF trio during myoblast fusion. *Molecular Biology of the Cell*, *18*, 1734–1743.

Chen, X., Bonne, S., Hatzfeld, M., van Roy, F., & Green, K. J. (2002). Protein binding and functional characterization of plakophilin 2. Evidence for its diverse roles in desmosomes and beta-catenin signaling. *The Journal of Biological Chemistry*, *277*, 10512–10522.

Chen, X., & Gumbiner, B. M. (2006). Paraxial protocadherin mediates cell sorting and tissue morphogenesis by regulating C-cadherin adhesion activity. *The Journal of Cell Biology*, *174*, 301–313.

Chen, S., Khan, Z. A., Cukiernik, M., & Chakrabarti, S. (2003). Differential activation of NF-kappa B and AP-1 in increased fibronectin synthesis in target organs of diabetic complications. *American Journal of Physiology. Endocrinology and Metabolism*, *284*, E1089–E1097.

Chen, X., Koh, E., Yoder, M., & Gumbiner, B. M. (2009). A protocadherin-cadherin-FLRT3 complex controls cell adhesion and morphogenesis. *PLoS One*, *4*, e8411.

Chen, L., Loh, P. G., & Song, H. (2010). Structural and functional insights into the TEAD-YAP complex in the Hippo signaling pathway. *Protein & Cell*, *1*, 1073–1083.

Chen, J., Lu, Y., Meng, S., Han, M. H., Lin, C., & Wang, X. (2009). Alpha- and gamma-protocadherins negatively regulate PYK2. *The Journal of Biological Chemistry*, *284*, 2880–2890.

Chen, W. V., & Maniatis, T. (2013). Clustered protocadherins. *Development*, *140*, 3297–3302.

Chen, M. C., Solomon, T. E., Kui, R., & Soll, A. H. (2002). Apical EGF receptors regulate epithelial barrier to gastric acid: Endogenous TGF-alpha is an essential facilitator. *American Journal of Physiology. Gastrointestinal and Liver Physiology*, *283*, G1098–G1106.

Chen, M. W., Vacherot, F., De La Taille, A., Gil-Diez-De-Medina, S., Shen, R., Friedman, R. A., et al. (2002). The emergence of protocadherin-PC expression during the acquisition of apoptosis-resistance by prostate cancer cells. *Oncogene*, *21*, 7861–7871.

Chetty, R., & Serra, S. (2008). Nuclear E-cadherin immunoexpression: From biology to potential applications in diagnostic pathology. *Advances in Anatomic Pathology*, *15*, 234–240.

Chiasson, C. M., Wittich, K. B., Vincent, P. A., Faundez, V., & Kowalczyk, A. P. (2009). p120-catenin inhibits VE-cadherin internalization through a Rho-independent mechanism. *Molecular Biology of the Cell*, *20*, 1970–1980.

Chidgey, M., Brakebusch, C., Gustafsson, E., Cruchley, A., Hail, C., Kirk, S., et al. (2001). Mice lacking desmocollin 1 show epidermal fragility accompanied by barrier defects and abnormal differentiation. *The Journal of Cell Biology*, *155*, 821–832.

Cho, E., Feng, Y., Rauskolb, C., Maitra, S., Fehon, R., & Irvine, K. D. (2006). Delineation of a Fat tumor suppressor pathway. *Nature Genetics*, *38*, 1142–1150.

Choi, S. H., Estaras, C., Moresco, J. J., Yates, J. R., 3rd., & Jones, K. A. (2013). alpha-Catenin interacts with APC to regulate beta-catenin proteolysis and transcriptional repression of Wnt target genes. *Genes & Development*, *27*, 2473–2488.

Choi, H. J., Huber, A. H., & Weis, W. I. (2006). Thermodynamics of beta-catenin-ligand interactions: The roles of the N- and C-terminal tails in modulating binding affinity. *The Journal of Biological Chemistry*, *281*, 1027–1038.

Chu, M. L., Ahn, V. E., Choi, H. J., Daniels, D. L., Nusse, R., & Weis, W. I. (2013). Structural studies of Wnts and identification of an LRP6 binding site. *Structure*, *21*, 1235–1242.

Chua, H. L., Bhat-Nakshatri, P., Clare, S. E., Morimiya, A., Badve, S., & Nakshatri, H. (2007). NF-kappaB represses E-cadherin expression and enhances epithelial to mesenchymal transition of mammary epithelial cells: Potential involvement of ZEB-1 and ZEB-2. *Oncogene*, *26*, 711–724.

Chung, H. A., Yamamoto, T. S., & Ueno, N. (2007). ANR5, an FGF target gene product, regulates gastrulation in Xenopus. *Current Biology*, *17*, 932–939.

Ciesiolka, M., Delvaeye, M., Van Imschoot, G., Verschuere, V., McCrea, P., van Roy, F., et al. (2004). p120 catenin is required for morphogenetic movements involved in the formation of the eyes and the craniofacial skeleton in Xenopus. *Journal of Cell Science*, *117*, 4325–4339.

Ciruna, B., & Rossant, J. (2001). FGF signaling regulates mesoderm cell fate specification and morphogenetic movement at the primitive streak. *Developmental Cell*, *1*, 37–49.

Clevers, H., Loh, K. M., & Nusse, R. (2014). Stem cell signaling. An integral program for tissue renewal and regeneration: Wnt signaling and stem cell control. *Science*, *346*, 1248012.

Conacci-Sorrell, M., Simcha, I., Ben-Yedidia, T., Blechman, J., Savagner, P., & Ben-Ze'ev, A. (2003). Autoregulation of E-cadherin expression by cadherin-cadherin interactions: The roles of beta-catenin signaling, Slug, and MAPK. *The Journal of Cell Biology*, *163*, 847–857.

Condello, S., Cao, L., & Matei, D. (2013). Tissue transglutaminase regulates beta-catenin signaling through a c-Src-dependent mechanism. *FASEB Journal: Official Publication of the Federation of American Societies for Experimental Biology*, *27*, 3100–3112.

Cox, R. T., Kirkpatrick, C., & Peifer, M. (1996). Armadillo is required for adherens junction assembly, cell polarity, and morphogenesis during Drosophila embryogenesis. *The Journal of Cell Biology*, *134*, 133–148.

Cselenyi, C. S., Jernigan, K. K., Tahinci, E., Thorne, C. A., Lee, L. A., & Lee, E. (2008). LRP6 transduces a canonical Wnt signal independently of Axin degradation by inhibiting GSK3's phosphorylation of beta-catenin. *Proceedings of the National Academy of Sciences of the United States of America*, *105*, 8032–8037.

Curto, M., Cole, B. K., Lallemand, D., Liu, C. H., & McClatchey, A. I. (2007). Contact-dependent inhibition of EGFR signaling by Nf2/Merlin. *The Journal of Cell Biology*, *177*, 893–903.

Damsky, C. H., Richa, J., Solter, D., Knudsen, K., & Buck, C. A. (1983). Identification and purification of a cell surface glycoprotein mediating intercellular adhesion in embryonic and adult tissue. *Cell*, *34*, 455–466.

Daniel, J. M. (2007). Dancing in and out of the nucleus: p120(ctn) and the transcription factor Kaiso. *Biochimica et Biophysica Acta*, *1773*, 59–68.

Daniel, J. M., & Reynolds, A. B. (1997). Tyrosine phosphorylation and cadherin/catenin function. *Bioessays*, *19*, 883–891.

Daniel, J. M., & Reynolds, A. B. (1999). The catenin p120(ctn) interacts with Kaiso, a novel BTB/POZ domain zinc finger transcription factor. *Molecular and Cellular Biology*, *19*, 3614–3623.

Dann, S. G., Golas, J., Miranda, M., Shi, C., Wu, J., Jin, G., et al. (2014). p120 catenin is a key effector of a Ras-PKCvarepsilon oncogenic signaling axis. *Oncogene*, *33*, 1385–1394.

Dartsch, N., Schulte, D., Hagerling, R., Kiefer, F., & Vestweber, D. (2014). Fusing VE-cadherin to alpha-catenin impairs fetal liver hematopoiesis and lymph but not blood vessel formation. *Molecular and Cellular Biology*, *34*, 1634–1648.

Das Thakur, M., Feng, Y., Jagannathan, R., Seppa, M. J., Skeath, J. B., & Longmore, G. D. (2010). Ajuba LIM proteins are negative regulators of the Hippo signaling pathway. *Current Biology*, *20*, 657–662.

Daugherty, R. L., & Gottardi, C. J. (2007). Phospho-regulation of beta-catenin adhesion and signaling functions. *Physiology (Bethesda)*, *22*, 303–309.

Daugherty, R. L., Serebryannyy, L., Yemelyanov, A., Flozak, A. S., Yu, H. J., Kosak, S. T., et al. (2014). alpha-Catenin is an inhibitor of transcription. *Proceedings of the National Academy of Sciences of the United States of America*, *111*, 5260–5265.

David, J. M., & Rajasekaran, A. K. (2012). Dishonorable discharge: The oncogenic roles of cleaved E-cadherin fragments. *Cancer Research*, *72*, 2917–2923.

David, M. D., Yeramian, A., Dunach, M., Llovera, M., Canti, C., de Herreros, A. G., et al. (2008). Signalling by neurotrophins and hepatocyte growth factor regulates axon morphogenesis by differential beta-catenin phosphorylation. *Journal of Cell Science*, *121*, 2718–2730.

Davidson, G., Wu, W., Shen, J., Bilic, J., Fenger, U., Stannek, P., et al. (2005). Casein kinase 1 gamma couples Wnt receptor activation to cytoplasmic signal transduction. *Nature*, *438*, 867–872.

Davis, M. A., Ireton, R. C., & Reynolds, A. B. (2003). A core function for p120-catenin in cadherin turnover. *The Journal of Cell Biology*, *163*, 525–534.

Davis, M. A., & Reynolds, A. B. (2006). Blocked acinar development, E-cadherin reduction, and intraepithelial neoplasia upon ablation of p120-catenin in the mouse salivary gland. *Developmental Cell, 10*, 21–31.

De Wever, O., Derycke, L., Hendrix, A., De Meerleer, G., Godeau, F., Depypere, H., et al. (2007). Soluble cadherins as cancer biomarkers. *Clinical & Experimental Metastasis, 24*, 685–697.

del Valle, I., Rudloff, S., Carles, A., Li, Y., Liszewska, E., Vogt, R., et al. (2013). E-Cadherin is required for the proper activation of the Lifr/Gp130 signaling pathway in mouse embryonic stem cells. *Development, 140*, 1684–1692.

Del Valle-Perez, B., Arques, O., Vinyoles, M., de Herreros, A. G., & Dunach, M. (2011). Coordinated action of CK1 isoforms in canonical Wnt signaling. *Molecular and Cellular Biology, 31*, 2877–2888.

Del Valle-Perez, B., Casagolda, D., Lugilde, E., Valls, G., Codina, M., Dave, N., et al. (2011). Wnt controls the transcriptional activity of Kaiso through CK1epsilon-dependent phosphorylation of p120-catenin. *Journal of Cell Science, 124*, 2298–2309.

Dorsky, R. I., Sheldahl, L. C., & Moon, R. T. (2002). A transgenic Lef1/beta-catenin-dependent reporter is expressed in spatially restricted domains throughout zebrafish development. *Developmental Biology, 241*, 229–237.

Drees, F., Pokutta, S., Yamada, S., Nelson, W. J., & Weis, W. I. (2005). Alpha-catenin is a molecular switch that binds E-cadherin-beta-catenin and regulates actin-filament assembly. *Cell, 123*, 903–915.

DuFort, C. C., Paszek, M. J., & Weaver, V. M. (2011). Balancing forces: Architectural control of mechanotransduction. *Nature Reviews. Molecular Cell Biology, 12*, 308–319.

Dupont, S., Morsut, L., Aragona, M., Enzo, E., Giulitti, S., Cordenonsi, M., et al. (2011). Role of YAP/TAZ in mechanotransduction. *Nature, 474*, 179–183.

Dumstrei, K., Wang, F., Shy, D., Tepass, U., & Hartenstein, V. (2002). Interaction between EGFR signaling and DE-cadherin during nervous system morphogenesis. *Development, 129*(17), 3983–3994.

Dusek, R. L., Getsios, S., Chen, F., Park, J. K., Amargo, E. V., Cryns, V. L., et al. (2006). The differentiation-dependent desmosomal cadherin desmoglein 1 is a novel caspase-3 target that regulates apoptosis in keratinocytes. *The Journal of Biological Chemistry, 281*, 3614–3624.

Dusek, R. L., Godsel, L. M., Chen, F., Strohecker, A. M., Getsios, S., Harmon, R., et al. (2007). Plakoglobin deficiency protects keratinocytes from apoptosis. *The Journal of Investigative Dermatology, 127*, 792–801.

El-Bahrawy, M., Talbot, I., Poulsom, R., & Alison, M. (2002). Variable nuclear localization of alpha-catenin in colorectal carcinoma. *Laboratory Investigation, 82*, 1167–1174.

Elia, L. P., Yamamoto, M., Zang, K., & Reichardt, L. F. (2006). p120 catenin regulates dendritic spine and synapse development through Rho-family GTPases and cadherins. *Neuron, 51*, 43–56.

Elias, P. M., Matsuyoshi, N., Wu, H., Lin, C., Wang, Z. H., Brown, B. E., et al. (2001). Desmoglein isoform distribution affects stratum corneum structure and function. *The Journal of Cell Biology, 153*, 243–249.

Emond, M. R., Biswas, S., Blevins, C. J., & Jontes, J. D. (2011). A complex of Protocadherin-19 and N-cadherin mediates a novel mechanism of cell adhesion. *The Journal of Cell Biology, 195*, 1115–1121.

Emond, M. R., Biswas, S., & Jontes, J. D. (2009). Protocadherin-19 is essential for early steps in brain morphogenesis. *Developmental Biology, 334*, 72–83.

Eshkind, L., Tian, Q., Schmidt, A., Franke, W. W., Windoffer, R., & Leube, R. E. (2002). Loss of desmoglein 2 suggests essential functions for early embryonic development and proliferation of embryonal stem cells. *European Journal of Cell Biology, 81*, 592–598.

Fagotto, F. (2013). Looking beyond the Wnt pathway for the deep nature of beta-catenin. *EMBO Reports*, *14*, 422–433.
Fagotto, F., Funayama, N., Gluck, U., & Gumbiner, B. M. (1996). Binding to cadherins antagonizes the signaling activity of beta-catenin during axis formation in Xenopus. *The Journal of Cell Biology*, *132*, 1105–1114.
Fagotto, F., Gluck, U., & Gumbiner, B. M. (1998). Nuclear localization signal-independent and importin/karyopherin-independent nuclear import of beta-catenin. *Current Biology*, *8*, 181–190.
Fagotto, F., & Gumbiner, B. M. (1996). Cell contact-dependent signaling. *Developmental Biology*, *180*, 445–454.
Fang, D., Hawke, D., Zheng, Y., Xia, Y., Meisenhelder, J., Nika, H., et al. (2007). Phosphorylation of beta-catenin by AKT promotes beta-catenin transcriptional activity. *The Journal of Biological Chemistry*, *282*, 11221–11229.
Fang, X., Ji, H., Kim, S. W., Park, J. I., Vaught, T. G., Anastasiadis, P. Z., et al. (2004). Vertebrate development requires ARVCF and p120 catenins and their interplay with RhoA and Rac. *The Journal of Cell Biology*, *165*, 87–98.
Fang, X., Yu, S. X., Lu, Y., Bast, R. C., Jr., Woodgett, J. R., & Mills, G. B. (2000). Phosphorylation and inactivation of glycogen synthase kinase 3 by protein kinase A. *Proceedings of the National Academy of Sciences of the United States of America*, *97*, 11960–11965.
Farahani, E., Patra, H. K., Jangamreddy, J. R., Rashedi, I., Kawalec, M., Rao Pariti, R. K., et al. (2014). Cell adhesion molecules and their relation to (cancer) cell stemness. *Carcinogenesis*, *35*, 747–759.
Fedor-Chaiken, M., Hein, P. W., Stewart, J. C., Brackenbury, R., & Kinch, M. S. (2003). E-Cadherin binding modulates EGF receptor activation. *Cell Communication & Adhesion*, *10*, 105–118.
Ferber, E. C., Kajita, M., Wadlow, A., Tobiansky, L., Niessen, C., Ariga, H., et al. (2008). A role for the cleaved cytoplasmic domain of E-cadherin in the nucleus. *The Journal of Biological Chemistry*, *283*, 12691–12700.
Ferreira, A. C., Suriano, G., Mendes, N., Gomes, B., Wen, X., Carneiro, F., et al. (2011). E-Cadherin impairment increases cell survival through Notch-dependent upregulation of Bcl-2. *Human Molecular Genetics*, *21*, 334–343.
Ferreri, D. M., Minnear, F. L., Yin, T., Kowalczyk, A. P., & Vincent, P. A. (2008). N-cadherin levels in endothelial cells are regulated by monolayer maturity and p120 availability. *Cell Communication & Adhesion*, *15*, 333–349.
Formstone, C. J. (2010). 7TM-cadherins: Developmental roles and future challenges. *Advances in Experimental Medicine and Biology*, *706*, 14–36.
Fortini, M. E. (2002). Gamma-secretase-mediated proteolysis in cell-surface-receptor signalling. *Nature Reviews. Molecular Cell Biology*, *3*, 673–684.
Fukuhara, S., Sako, K., Minami, T., Noda, K., Kim, H. Z., Kodama, T., et al. (2008). Differential function of Tie2 at cell-cell contacts and cell-substratum contacts regulated by angiopoietin-1. *Nature Cell Biology*, *10*, 513–526.
Gagliardi, M., Hernandez, A., McGough, I. J., & Vincent, J. P. (2014). Inhibitors of endocytosis prevent Wnt/Wingless signalling by reducing the level of basal beta-Catenin/Armadillo. *Journal of Cell Science*, *127*, 4918–4926.
Galbiati, F., Volonte, D., Brown, A. M., Weinstein, D. E., Ben-Ze'ev, A., Pestell, R. G., et al. (2000). Caveolin-1 expression inhibits Wnt/beta-catenin/Lef-1 signaling by recruiting beta-catenin to caveolae membrane domains. *The Journal of Biological Chemistry*, *275*, 23368–23377.
Garcia-Gras, E., Lombardi, R., Giocondo, M. J., Willerson, J. T., Schneider, M. D., Khoury, D. S., et al. (2006). Suppression of canonical Wnt/beta-catenin signaling by nuclear plakoglobin recapitulates phenotype of arrhythmogenic right ventricular cardiomyopathy. *The Journal of Clinical Investigation*, *116*, 2012–2021.

Garrett, A. M., Schreiner, D., Lobas, M. A., & Weiner, J. A. (2012). Gamma-protocadherins control cortical dendrite arborization by regulating the activity of a FAK/PKC/MARCKS signaling pathway. *Neuron*, *74*, 269–276.

Gavard, J. (2013). Endothelial permeability and VE-cadherin: A wacky comradeship. *Cell Adhesion & Migration*, 7, 455–461.

Gavard, J., & Gutkind, J. S. (2006). VEGF controls endothelial-cell permeability by promoting the beta-arrestin-dependent endocytosis of VE-cadherin. *Nature Cell Biology*, *8*, 1223–1234.

Geis, K., Aberle, H., Kuhl, M., Kemler, R., & Wedlich, D. (1998). Expression of the Armadillo family member p120cas1B in Xenopus embryos affects head differentiation but not axis formation. *Development Genes and Evolution*, *207*, 471–481.

Geletu, M., Guy, S., Arulanandam, R., Feracci, H., & Raptis, L. (2013). Engaged for survival: From cadherin ligation to STAT3 activation. *JAK-STAT*, *2*, e27363.

Genevet, A., Wehr, M. C., Brain, R., Thompson, B. J., & Tapon, N. (2010). Kibra is a regulator of the Salvador/Warts/Hippo signaling network. *Developmental Cell*, *18*, 300–308.

Getsios, S., Simpson, C. L., Kojima, S., Harmon, R., Sheu, L. J., Dusek, R. L., et al. (2009). Desmoglein 1-dependent suppression of EGFR signaling promotes epidermal differentiation and morphogenesis. *The Journal of Cell Biology*, *185*, 1243–1258.

Giannini, A. L., Vivanco, M., & Kypta, R. M. (2000). Alpha-catenin inhibits beta-catenin signaling by preventing formation of a beta-catenin*T-cell factor*DNA complex. *The Journal of Biological Chemistry*, *275*, 21883–21888.

Gladden, A. B., Hebert, A. M., Schneeberger, E. E., & McClatchey, A. I. (2010). The NF2 tumor suppressor, Merlin, regulates epidermal development through the establishment of a junctional polarity complex. *Developmental Cell*, *19*, 727–739.

Goh, L. K., & Sorkin, A. (2013). Endocytosis of receptor tyrosine kinases. *Cold Spring Harbor Perspectives in Biology*, *5*, a017459.

Gomez, G. A., McLachlan, R. W., & Yap, A. S. (2011). Productive tension: Force-sensing and homeostasis of cell-cell junctions. *Trends in Cell Biology*, *21*, 499–505.

Gomez-Orte, E., Saenz-Narciso, B., Moreno, S., & Cabello, J. (2013). Multiple functions of the noncanonical Wnt pathway. *Trends in Genetics: TIG*, *29*, 545–553.

Gonzalez-Reyes, A. (2003). Stem cells, niches and cadherins: A view from Drosophila. *Journal of Cell Science*, *116*, 949–954.

Goyal, R. K., Lin, P., Kanungo, J., Payne, A. S., Muslin, A. J., & Longmore, G. D. (1999). Ajuba, a novel LIM protein, interacts with Grb2, augments mitogen-activated protein kinase activity in fibroblasts, and promotes meiotic maturation of Xenopus oocytes in a Grb2- and Ras-dependent manner. *Molecular and Cellular Biology*, *19*, 4379–4389.

Grabowska, M. M., Sandhu, B., & Day, M. L. (2012). EGF promotes the shedding of soluble E-cadherin in an ADAM10-dependent manner in prostate epithelial cells. *Cellular Signalling*, *24*, 532–538.

Graham, T. A., Weaver, C., Mao, F., Kimelman, D., & Xu, W. (2000). Crystal structure of a beta-catenin/Tcf complex. *Cell*, *103*, 885–896.

Grazia Lampugnani, M., Zanetti, A., Corada, M., Takahashi, T., Balconi, G., Breviario, F., et al. (2003). Contact inhibition of VEGF-induced proliferation requires vascular endothelial cadherin, beta-catenin, and the phosphatase DEP-1/CD148. *The Journal of Cell Biology*, *161*, 793–804.

Gu, D., Sater, A. K., Ji, H., Cho, K., Clark, M., Stratton, S. A., et al. (2009). Xenopus delta-catenin is essential in early embryogenesis and is functionally linked to cadherins and small GTPases. *Journal of Cell Science*, *122*, 4049–4061.

Gu, D., Tonthat, N. K., Lee, M., Ji, H., Bhat, K. P., Hollingsworth, F., et al. (2011). Caspase-3 cleavage links delta-catenin to the novel nuclear protein ZIFCAT. *The Journal of Biological Chemistry*, *286*, 23178–23188.

Gumbiner, B. M., & Kim, N. G. (2014). The Hippo-YAP signaling pathway and contact inhibition of growth. *Journal of Cell Science*, *127*, 709–717.

Ha, N. C., Tonozuka, T., Stamos, J. L., Choi, H. J., & Weis, W. I. (2004). Mechanism of phosphorylation-dependent binding of APC to beta-catenin and its role in beta-catenin degradation. *Molecular Cell, 15*, 511–521.

Haas, I. G., Frank, M., Veron, N., & Kemler, R. (2005). Presenilin-dependent processing and nuclear function of gamma-protocadherins. *The Journal of Biological Chemistry, 280*, 9313–9319.

Haglund, K., & Dikic, I. (2012). The role of ubiquitylation in receptor endocytosis and endosomal sorting. *Journal of Cell Science, 125*, 265–275.

Hamaratoglu, F., Willecke, M., Kango-Singh, M., Nolo, R., Hyun, E., Tao, C., et al. (2006). The tumour-suppressor genes NF2/Merlin and Expanded act through Hippo signalling to regulate cell proliferation and apoptosis. *Nature Cell Biology, 8*, 27–36.

Hambsch, B., Grinevich, V., Seeburg, P. H., & Schwarz, M. K. (2005). {gamma}-Protocadherins, presenilin-mediated release of C-terminal fragment promotes locus expression. *The Journal of Biological Chemistry, 280*, 15888–15897.

Hansen, S. D., Kwiatkowski, A. V., Ouyang, C. Y., Liu, H., Pokutta, S., Watkins, S. C., et al. (2013). alphaE-catenin actin-binding domain alters actin filament conformation and regulates binding of nucleation and disassembly factors. *Molecular Biology of the Cell, 24*, 3710–3720.

Hardman, M. J., Liu, K., Avilion, A. A., Merritt, A., Brennan, K., Garrod, D. R., et al. (2005). Desmosomal cadherin misexpression alters beta-catenin stability and epidermal differentiation. *Molecular and Cellular Biology, 25*, 969–978.

Harris, T. J., & Tepass, U. (2010). Adherens junctions: From molecules to morphogenesis. *Nature Reviews. Molecular Cell Biology, 11*, 502–514.

Hartsock, A., & Nelson, W. J. (2008). Adherens and tight junctions: Structure, function and connections to the actin cytoskeleton. *Biochimica et Biophysica Acta, 1778*, 660–669.

Hatanaka, K., Simons, M., & Murakami, M. (2011). Phosphorylation of VE-cadherin controls endothelial phenotypes via p120-catenin coupling and Rac1 activation. *American Journal of Physiology. Heart and Circulatory Physiology, 300*, H162–H172.

Hatzfeld, M. (2005). The p120 family of cell adhesion molecules. *European Journal of Cell Biology, 84*, 205–214.

Hatzfeld, M. (2007). Plakophilins: Multifunctional proteins or just regulators of desmosomal adhesion? *Biochimica et Biophysica Acta, 1773*, 69–77.

Hawkins, K., Mohamet, L., Ritson, S., Merry, C. L., & Ward, C. M. (2012). E-Cadherin and, in its absence, N-cadherin promotes Nanog expression in mouse embryonic stem cells via STAT3 phosphorylation. *Stem Cells, 30*, 1842–1851.

Hay, E., Laplantine, E., Geoffroy, V., Frain, M., Kohler, T., Muller, R., et al. (2009). N-cadherin interacts with axin and LRP5 to negatively regulate Wnt/beta-catenin signaling, osteoblast function, and bone formation. *Molecular and Cellular Biology, 29*, 953–964.

Hayashi, M., Majumdar, A., Li, X., Adler, J., Sun, Z., Vertuani, S., et al. (2013). VE-PTP regulates VEGFR2 activity in stalk cells to establish endothelial cell polarity and lumen formation. *Nature Communications, 4*, 1672.

Heasman, J., Crawford, A., Goldstone, K., Garner-Hamrick, P., Gumbiner, B., McCrea, P., et al. (1994). Overexpression of cadherins and underexpression of beta-catenin inhibit dorsal mesoderm induction in early Xenopus embryos. *Cell, 79*, 791–803.

Heggem, M. A., & Bradley, R. S. (2003). The cytoplasmic domain of Xenopus NF-protocadherin interacts with TAF1/set. *Developmental Cell, 4*, 419–429.

Heid, H. W., Schmidt, A., Zimbelmann, R., Schafer, S., Winter-Simanowski, S., Stumpp, S., et al. (1994). Cell type-specific desmosomal plaque proteins of the plakoglobin family: Plakophilin 1 (band 6 protein). *Differentiation; Research in Biological Diversity, 58*, 113–131.

Heinen, C. D. (2010). Genotype to phenotype: Analyzing the effects of inherited mutations in colorectal cancer families. *Mutation Research, 693*, 32–45.
Hellberg, C. B., Burden-Gulley, S. M., Pietz, G. E., & Brady-Kalnay, S. M. (2002). Expression of the receptor protein-tyrosine phosphatase, PTPmu, restores E-cadherin-dependent adhesion in human prostate carcinoma cells. *The Journal of Biological Chemistry, 277*, 11165–11173.
Hendriksen, J., Jansen, M., Brown, C. M., van der Velde, H., van Ham, M., Galjart, N., et al. (2008). Plasma membrane recruitment of dephosphorylated beta-catenin upon activation of the Wnt pathway. *Journal of Cell Science, 121*, 1793–1802.
Herzig, M., Savarese, F., Novatchkova, M., Semb, H., & Christofori, G. (2007). Tumor progression induced by the loss of E-cadherin independent of beta-catenin/Tcf-mediated Wnt signaling. *Oncogene, 26*, 2290–2298.
Heuberger, J., & Birchmeier, W. (2010). Interplay of cadherin-mediated cell adhesion and canonical Wnt signaling. *Cold Spring Harbor Perspectives in Biology, 2*, a002915.
Hirai, H., Karian, P., & Kikyo, N. (2011). Regulation of embryonic stem cell self-renewal and pluripotency by leukaemia inhibitory factor. *The Biochemical Journal, 438*, 11–23.
Hoffmans, R., & Basler, K. (2007). BCL9-2 binds Arm/beta-catenin in a Tyr142-independent manner and requires Pygopus for its function in Wg/Wnt signaling. *Mechanisms of Development, 124*, 59–67.
Hofmann, I., Casella, M., Schnolzer, M., Schlechter, T., Spring, H., & Franke, W. W. (2006). Identification of the junctional plaque protein plakophilin 3 in cytoplasmic particles containing RNA-binding proteins and the recruitment of plakophilins 1 and 3 to stress granules. *Molecular Biology of the Cell, 17*, 1388–1398.
Hofmann, W. A., Stojiljkovic, L., Fuchsova, B., Vargas, G. M., Mavrommatis, E., Philimonenko, V., et al. (2004). Actin is part of pre-initiation complexes and is necessary for transcription by RNA polymerase II. *Nature Cell Biology, 6*, 1094–1101.
Homayouni, R., Rice, D. S., & Curran, T. (2001). Disabled-1 interacts with a novel developmentally regulated protocadherin. *Biochemical and Biophysical Research Communications, 289*, 539–547.
Hong, J. Y., Park, J. I., Cho, K., Gu, D., Ji, H., Artandi, S. E., et al. (2010). Shared molecular mechanisms regulate multiple catenin proteins: Canonical Wnt signals and components modulate p120-catenin isoform-1 and additional p120 subfamily members. *Journal of Cell Science, 123*, 4351–4365.
Hong, J. Y., Park, J. I., Lee, M., Munoz, W. A., Miller, R. K., Ji, H., et al. (2012). Down's-syndrome-related kinase Dyrk1A modulates the p120-catenin-Kaiso trajectory of the Wnt signaling pathway. *Journal of Cell Science, 125*, 561–569.
Hoschuetzky, H., Aberle, H., & Kemler, R. (1994). Beta-catenin mediates the interaction of the cadherin-catenin complex with epidermal growth factor receptor. *The Journal of Cell Biology, 127*, 1375–1380.
Hosking, C. R., Ulloa, F., Hogan, C., Ferber, E. C., Figueroa, A., Gevaert, K., et al. (2007). The transcriptional repressor Glis2 is a novel binding partner for p120 catenin. *Molecular Biology of the Cell, 18*, 1918–1927.
Hosono, S., Gross, I., English, M. A., Hajra, K. M., Fearon, E. R., & Licht, J. D. (2000). E-cadherin is a WT1 target gene. *The Journal of Biological Chemistry, 275*, 10943–10953.
Howard, S., Deroo, T., Fujita, Y., & Itasaki, N. (2011). A positive role of cadherin in Wnt/beta-catenin signalling during epithelial-mesenchymal transition. *PLoS One, 6*, e23899.
Hu, P., Berkowitz, P., O'Keefe, E. J., & Rubenstein, D. S. (2003). Keratinocyte adherens junctions initiate nuclear signaling by translocation of plakoglobin from the membrane to the nucleus. *The Journal of Investigative Dermatology, 121*, 242–251.

Hu, P., O'Keefe, E. J., & Rubenstein, D. S. (2001). Tyrosine phosphorylation of human keratinocyte beta-catenin and plakoglobin reversibly regulates their binding to E-cadherin and alpha-catenin. *The Journal of Investigative Dermatology*, *117*, 1059–1067.

Hu, P., Wu, S., & Hernandez, N. (2004). A role for beta-actin in RNA polymerase III transcription. *Genes & Development*, *18*, 3010–3015.

Huang, J., Wu, S., Barrera, J., Matthews, K., & Pan, D. (2005). The Hippo signaling pathway coordinately regulates cell proliferation and apoptosis by inactivating Yorkie, the Drosophila Homolog of YAP. *Cell*, *122*, 421–434.

Huber, O., Korn, R., McLaughlin, J., Ohsugi, M., Herrmann, B. G., & Kemler, R. (1996). Nuclear localization of beta-catenin by interaction with transcription factor LEF-1. *Mechanisms of Development*, *59*, 3–10.

Huber, A. H., & Weis, W. I. (2001). The structure of the beta-catenin/E-cadherin complex and the molecular basis of diverse ligand recognition by beta-catenin. *Cell*, *105*, 391–402.

Huelsken, J., Vogel, R., Brinkmann, V., Erdmann, B., Birchmeier, C., & Birchmeier, W. (2000). Requirement for beta-catenin in anterior-posterior axis formation in mice. *The Journal of Cell Biology*, *148*, 567–578.

Hulpiau, P., & van Roy, F. (2009). Molecular evolution of the cadherin superfamily. *The International Journal of Biochemistry & Cell Biology*, *41*, 349–369.

Hulpiau, P., & van Roy, F. (2011). New insights into the evolution of metazoan cadherins. *Molecular Biology and Evolution*, *28*, 647–657.

Huveneers, S., & de Rooij, J. (2013). Mechanosensitive systems at the cadherin-F-actin interface. *Journal of Cell Science*, *126*, 403–413.

Hwang, S. G., Yu, S. S., Ryu, J. H., Jeon, H. B., Yoo, Y. J., Eom, S. H., et al. (2005). Regulation of beta-catenin signaling and maintenance of chondrocyte differentiation by ubiquitin-independent proteasomal degradation of alpha-catenin. *The Journal of Biological Chemistry*, *280*, 12758–12765.

Inge, L. J., Barwe, S. P., D'Ambrosio, J., Gopal, J., Lu, K., Ryazantsev, S., et al. (2011). Soluble E-cadherin promotes cell survival by activating epidermal growth factor receptor. *Experimental Cell Research*, *317*, 838–848.

Ito, K., Okamoto, I., Araki, N., Kawano, Y., Nakao, M., Fujiyama, S., et al. (1999). Calcium influx triggers the sequential proteolysis of extracellular and cytoplasmic domains of E-cadherin, leading to loss of beta-catenin from cell-cell contacts. *Oncogene*, *18*, 7080–7090.

Iyer, S., Ferreri, D. M., DeCocco, N. C., Minnear, F. L., & Vincent, P. A. (2004). VE-cadherin-p120 interaction is required for maintenance of endothelial barrier function. *American Journal of Physiology. Lung Cellular and Molecular Physiology*, *286*, L1143–L1153.

Jamora, C., DasGupta, R., Kocieniewski, P., & Fuchs, E. (2003). Links between signal transduction, transcription and adhesion in epithelial bud development. *Nature*, *422*, 317–322.

Jean, C., Blanc, A., Prade-Houdellier, N., Ysebaert, L., Hernandez-Pigeon, H., Al Saati, T., et al. (2009). Epidermal growth factor receptor/beta-catenin/T-cell factor 4/matrix metalloproteinase 1: A new pathway for regulating keratinocyte invasiveness after UVA irradiation. *Cancer Research*, *69*, 3291–3299.

Kadrmas, J. L., & Beckerle, M. C. (2004). The LIM domain: From the cytoskeleton to the nucleus. *Nature Reviews. Molecular Cell Biology*, *5*, 920–931.

Kahr, I., Vandepoele, K., & van Roy, F. (2013). Delta-protocadherins in health and disease. *Progress in Molecular Biology and Translational Science*, *116*, 169–192.

Kamekura, R., Kolegraff, K. N., Nava, P., Hilgarth, R. S., Feng, M., Parkos, C. A., et al. (2014). Loss of the desmosomal cadherin desmoglein-2 suppresses colon cancer cell proliferation through EGFR signaling. *Oncogene*, *33*, 4531–4536.

Kang, D. E., Soriano, S., Frosch, M. P., Collins, T., Naruse, S., Sisodia, S. S., et al. (1999). Presenilin 1 facilitates the constitutive turnover of beta-catenin: Differential activity of Alzheimer's disease-linked PS1 mutants in the beta-catenin-signaling pathway. *The Journal of Neuroscience*, *19*, 4229–4237.

Kang, D. E., Soriano, S., Xia, X., Eberhart, C. G., De Strooper, B., Zheng, H., et al. (2002). Presenilin couples the paired phosphorylation of beta-catenin independent of axin: Implications for beta-catenin activation in tumorigenesis. *Cell*, *110*, 751–762.

Kanungo, J., Pratt, S. J., Marie, H., & Longmore, G. D. (2000). Ajuba, a cytosolic LIM protein, shuttles into the nucleus and affects embryonal cell proliferation and fate decisions. *Molecular Biology of the Cell*, *11*, 3299–3313.

Karnovsky, A., & Klymkowsky, M. W. (1995). Anterior axis duplication in Xenopus induced by the over-expression of the cadherin-binding protein plakoglobin. *Proceedings of the National Academy of Sciences of the United States of America*, *92*, 4522–4526.

Kashef, J., Kohler, A., Kuriyama, S., Alfandari, D., Mayor, R., & Wedlich, D. (2009). Cadherin-11 regulates protrusive activity in Xenopus cranial neural crest cells upstream of Trio and the small GTPases. *Genes & Development*, *23*, 1393–1398.

Kietzmann, A., Wang, Y., Weber, D., & Steinbeisser, H. (2012). Xenopus paraxial protocadherin inhibits Wnt/beta-catenin signalling via casein kinase 2beta. *EMBO Reports*, *13*, 129–134.

Kilander, M. B., Petersen, J., Andressen, K. W., Ganji, R. S., Levy, F. O., Schuster, J., et al. (2014). Disheveled regulates precoupling of heterotrimeric G proteins to Frizzled 6. *FASEB Journal: Official Publication of the Federation of American Societies for Experimental Biology*, *28*, 2293–2305.

Kim, S., Ishidate, T., Sharma, R., Soto, M. C., Conte, D., Jr., Mello, C. C., et al. (2013). Wnt and CDK-1 regulate cortical release of WRM-1/beta-catenin to control cell division orientation in early Caenorhabditis elegans embryos. *Proceedings of the National Academy of Sciences of the United States of America*, *110*, E918–E927.

Kim, N. G., Koh, E., Chen, X., & Gumbiner, B. M. (2011). E-Cadherin mediates contact inhibition of proliferation through Hippo signaling-pathway components. *Proceedings of the National Academy of Sciences of the United States of America*, *108*, 11930–11935.

Kim, Y., Kugler, M. C., Wei, Y., Kim, K. K., Li, X., Brumwell, A. N., et al. (2009). Integrin alpha3beta1-dependent beta-catenin phosphorylation links epithelial Smad signaling to cell contacts. *The Journal of Cell Biology*, *184*, 309–322.

Kim, S. W., Park, J. I., Spring, C. M., Sater, A. K., Ji, H., Otchere, A. A., et al. (2004). Non-canonical Wnt signals are modulated by the Kaiso transcriptional repressor and p120-catenin. *Nature Cell Biology*, *6*, 1212–1220.

Klymkowsky, M. W., Williams, B. O., Barish, G. D., Varmus, H. E., & Vourgourakis, Y. E. (1999). Membrane-anchored plakoglobins have multiple mechanisms of action in Wnt signaling. *Molecular Biology of the Cell*, *10*, 3151–3169.

Kobielak, A., & Fuchs, E. (2006). Links between alpha-catenin, NF-kappaB, and squamous cell carcinoma in skin. *Proceedings of the National Academy of Sciences of the United States of America*, *103*, 2322–2327.

Kolligs, F. T., Hu, G., Dang, C. V., & Fearon, E. R. (1999). Neoplastic transformation of RK3E by mutant beta-catenin requires deregulation of Tcf/Lef transcription but not activation of c-myc expression. *Molecular and Cellular Biology*, *19*, 5696–5706.

Komuro, A., Nagai, M., Navin, N. E., & Sudol, M. (2003). WW domain-containing protein YAP associates with ErbB-4 and acts as a co-transcriptional activator for the carboxyl-terminal fragment of ErbB-4 that translocates to the nucleus. *The Journal of Biological Chemistry*, *278*, 33334–33341.

Konsavage, W. M., Jr., & Yochum, G. S. (2013). Intersection of Hippo/YAP and Wnt/beta-catenin signaling pathways. *Acta Biochimica et Biophysica Sinica*, *45*, 71–79.

Kourtidis, A., Ngok, S. P., & Anastasiadis, P. Z. (2013). p120 catenin: An essential regulator of cadherin stability, adhesion-induced signaling, and cancer progression. *Progress in Molecular Biology and Translational Science*, *116*, 409–432.

Kraft, B., Berger, C. D., Wallkamm, V., Steinbeisser, H., & Wedlich, D. (2012). Wnt-11 and Fz7 reduce cell adhesion in convergent extension by sequestration of PAPC and C-cadherin. *The Journal of Cell Biology*, *198*, 695–709.

Kramps, T., Peter, O., Brunner, E., Nellen, D., Froesch, B., Chatterjee, S., et al. (2002). Wnt/wingless signaling requires BCL9/legless-mediated recruitment of pygopus to the nuclear beta-catenin-TCF complex. *Cell*, *109*, 47–60.

Kukalev, A., Nord, Y., Palmberg, C., Bergman, T., & Percipalle, P. (2005). Actin and hnRNP U cooperate for productive transcription by RNA polymerase II. *Nature Structural & Molecular Biology*, *12*, 238–244.

Lam, L., Aktary, Z., Bishay, M., Werkman, C., Kuo, C. Y., Heacock, M., et al. (2012). Regulation of subcellular distribution and oncogenic potential of nucleophosmin by plakoglobin. *Oncogenesis*, *1*, e4.

Lampugnani, M. G., Orsenigo, F., Gagliani, M. C., Tacchetti, C., & Dejana, E. (2006). Vascular endothelial cadherin controls VEGFR-2 internalization and signaling from intracellular compartments. *The Journal of Cell Biology*, *174*, 593–604.

Lampugnani, M. G., Zanetti, A., Breviario, F., Balconi, G., Orsenigo, F., Corada, M., et al. (2002). VE-cadherin regulates endothelial actin activating Rac and increasing membrane association of Tiam. *Molecular Biology of the Cell*, *13*, 1175–1189.

Larue, L., Antos, C., Butz, S., Huber, O., Delmas, V., Dominis, M., et al. (1996). A role for cadherins in tissue formation. *Development*, *122*, 3185–3194.

Lee, M., Ji, H., Furuta, Y., Park, J. I., & McCrea, P. D. (2014). p120-catenin regulates REST and CoREST, and modulates mouse embryonic stem cell differentiation. *Journal of Cell Science*, *127*, 4037–4051.

Lee, D. J., Kang, D. H., Choi, M., Choi, Y. J., Lee, J. Y., Park, J. H., et al. (2013). Peroxiredoxin-2 represses melanoma metastasis by increasing E-cadherin/beta-catenin complexes in adherens junctions. *Cancer Research*, *73*, 4744–4757.

Li, L., Bennett, S. A., & Wang, L. (2012). Role of E-cadherin and other cell adhesion molecules in survival and differentiation of human pluripotent stem cells. *Cell Adhesion & Migration*, *6*, 59–70.

Li, J., Swope, D., Raess, N., Cheng, L., Muller, E. J., & Radice, G. L. (2011). Cardiac tissue-restricted deletion of plakoglobin results in progressive cardiomyopathy and activation of {beta}-catenin signaling. *Molecular and Cellular Biology*, *31*, 1134–1144.

Li, L., Wang, S., Jezierski, A., Moalim-Nour, L., Mohib, K., Parks, R. J., et al. (2010). A unique interplay between Rap1 and E-cadherin in the endocytic pathway regulates self-renewal of human embryonic stem cells. *Stem Cells*, *28*, 247–257.

Li, D., Zhou, J., Wang, L., Shin, M. E., Su, P., Lei, X., et al. (2010). Integrated biochemical and mechanical signals regulate multifaceted human embryonic stem cell functions. *The Journal of Cell Biology*, *191*, 631–644.

Lien, W. H., & Fuchs, E. (2014). Wnt some lose some: Transcriptional governance of stem cells by Wnt/beta-catenin signaling. *Genes & Development*, *28*, 1517–1532.

Lien, W. H., Klezovitch, O., Fernandez, T. E., Delrow, J., & Vasioukhin, V. (2006). alphaE-catenin controls cerebral cortical size by regulating the hedgehog signaling pathway. *Science*, *311*, 1609–1612.

Lochter, A., Galosy, S., Muschler, J., Freedman, N., Werb, Z., & Bissell, M. J. (1997). Matrix metalloproteinase stromelysin-1 triggers a cascade of molecular alterations that leads to stable epithelial-to-mesenchymal conversion and a premalignant phenotype in mammary epithelial cells. *The Journal of Cell Biology*, *139*, 1861–1872.

Lombardi, R., da Graca Cabreira-Hansen, M., Bell, A., Fromm, R. R., Willerson, J. T., & Marian, A. J. (2011). Nuclear plakoglobin is essential for differentiation of cardiac progenitor cells to adipocytes in arrhythmogenic right ventricular cardiomyopathy. *Circulation Research, 109*, 1342–1353.

Lu, Z., Ghosh, S., Wang, Z., & Hunter, T. (2003). Downregulation of caveolin-1 function by EGF leads to the loss of E-cadherin, increased transcriptional activity of beta-catenin, and enhanced tumor cell invasion. *Cancer Cell, 4*, 499–515.

Lu, M., Marsters, S., Ye, X., Luis, E., Gonzalez, L., & Ashkenazi, A. (2014). E-Cadherin couples death receptors to the cytoskeleton to regulate apoptosis. *Molecular Cell, 54*, 987–998.

Luchtenborg, A. M., Solis, G. P., Egger-Adam, D., Koval, A., Lin, C., Blanchard, M. G., et al. (2014). Heterotrimeric Go protein links Wnt-Frizzled signaling with ankyrins to regulate the neuronal microtubule cytoskeleton. *Development, 141*, 3399–3409.

Lyashenko, N., Winter, M., Migliorini, D., Biechele, T., Moon, R. T., & Hartmann, C. (2011). Differential requirement for the dual functions of beta-catenin in embryonic stem cell self-renewal and germ layer formation. *Nature Cell Biology, 13*, 753–761.

MacDonald, B. T., Tamai, K., & He, X. (2009). Wnt/beta-catenin signaling: Components, mechanisms, and diseases. *Developmental Cell, 17*, 9–26.

Maher, M. T., Flozak, A. S., Stocker, A., Chenn, A., & Gottardi, C. J. (2009). Activity of the beta-catenin phosphodestruction complex at cell-cell contacts is enhanced by cadherin-based adhesion. *The Journal of Cell Biology, 186*, 219–228.

Maiden, S. L., & Hardin, J. (2011). The secret life of alpha-catenin: Moonlighting in morphogenesis. *The Journal of Cell Biology, 195*, 543–552.

Maitre, J. L., & Heisenberg, C. P. (2013). Three functions of cadherins in cell adhesion. *Current Biology, 23*, R626–R633.

Major, M. B., Camp, N. D., Berndt, J. D., Yi, X., Goldenberg, S. J., Hubbert, C., et al. (2007). Wilms tumor suppressor WTX negatively regulates WNT/beta-catenin signaling. *Science, 316*, 1043–1046.

Marambaud, P., Shioi, J., Serban, G., Georgakopoulos, A., Sarner, S., Nagy, V., et al. (2002). A presenilin-1/gamma-secretase cleavage releases the E-cadherin intracellular domain and regulates disassembly of adherens junctions. *The EMBO Journal, 21*, 1948–1956.

Marambaud, P., Wen, P. H., Dutt, A., Shioi, J., Takashima, A., Siman, R., et al. (2003). A CBP binding transcriptional repressor produced by the PS1/epsilon-cleavage of N-cadherin is inhibited by PS1 FAD mutations. *Cell, 114*, 635–645.

Maretzky, T., Reiss, K., Ludwig, A., Buchholz, J., Scholz, F., Proksch, E., et al. (2005). ADAM10 mediates E-cadherin shedding and regulates epithelial cell-cell adhesion, migration, and beta-catenin translocation. *Proceedings of the National Academy of Sciences of the United States of America, 102*, 9182–9187.

Maretzky, T., Schulte, M., Ludwig, A., Rose-John, S., Blobel, C., Hartmann, D., et al. (2005). L1 is sequentially processed by two differently activated metalloproteases and presenilin/gamma-secretase and regulates neural cell adhesion, cell migration, and neurite outgrowth. *Molecular and Cellular Biology, 25*, 9040–9053.

Marie, H., Pratt, S. J., Betson, M., Epple, H., Kittler, J. T., Meek, L., et al. (2003). The LIM protein Ajuba is recruited to cadherin-dependent cell junctions through an association with alpha-catenin. *The Journal of Biological Chemistry, 278*, 1220–1228.

Matakatsu, H., & Blair, S. S. (2006). Separating the adhesive and signaling functions of the Fat and Dachsous protocadherins. *Development, 133*, 2315–2324.

Matsuda, T., Nakamura, T., Nakao, K., Arai, T., Katsuki, M., Heike, T., et al. (1999). STAT3 activation is sufficient to maintain an undifferentiated state of mouse embryonic stem cells. *The EMBO Journal, 18*, 4261–4269.

McCrea, P. D., & Park, J. I. (2007). Developmental functions of the P120-catenin sub-family. *Biochimica et Biophysica Acta, 1773*, 17–33.

McCusker, C. D., & Alfandari, D. (2009). Life after proteolysis: Exploring the signaling capabilities of classical cadherin cleavage fragments. *Communicative & Integrative Biology, 2,* 155–157.

McCusker, C., Cousin, H., Neuner, R., & Alfandari, D. (2009). Extracellular cleavage of cadherin-11 by ADAM metalloproteases is essential for Xenopus cranial neural crest cell migration. *Molecular Biology of the Cell, 20,* 78–89.

McEwen, A. E., Escobar, D. E., & Gottardi, C. J. (2012). Signaling from the adherens junction. *Sub-Cellular Biochemistry, 60,* 171–196.

McLachlan, R. W., & Yap, A. S. (2007). Not so simple: The complexity of phosphotyrosine signaling at cadherin adhesive contacts. *Journal of Molecular Medicine, 85,* 545–554.

Medina, A., Swain, R. K., Kuerner, K. M., & Steinbeisser, H. (2004). Xenopus paraxial protocadherin has signaling functions and is involved in tissue separation. *The EMBO Journal, 23,* 3249–3258.

Mege, R. M., Gavard, J., & Lambert, M. (2006). Regulation of cell-cell junctions by the cytoskeleton. *Current Opinion in Cell Biology, 18,* 541–548.

Mehta, D., & Malik, A. B. (2006). Signaling mechanisms regulating endothelial permeability. *Physiological Reviews, 86,* 279–367.

Meng, W., & Takeichi, M. (2009). Adherens junction: Molecular architecture and regulation. *Cold Spring Harbor Perspectives in Biology, 1,* a002899.

Merdek, K. D., Nguyen, N. T., & Toksoz, D. (2004). Distinct activities of the alpha-catenin family, alpha-catulin and alpha-catenin, on beta-catenin-mediated signaling. *Molecular and Cellular Biology, 24,* 2410–2422.

Merritt, A. J., Berika, M. Y., Zhai, W., Kirk, S. E., Ji, B., Hardman, M. J., et al. (2002). Suprabasal desmoglein 3 expression in the epidermis of transgenic mice results in hyperproliferation and abnormal differentiation. *Molecular and Cellular Biology, 22,* 5846–5858.

Mertens, C., Kuhn, C., & Franke, W. W. (1996). Plakophilins 2a and 2b: Constitutive proteins of dual location in the karyoplasm and the desmosomal plaque. *The Journal of Cell Biology, 135,* 1009–1025.

Mi, K., Dolan, P. J., & Johnson, G. V. (2006). The low density lipoprotein receptor-related protein 6 interacts with glycogen synthase kinase 3 and attenuates activity. *The Journal of Biological Chemistry, 281,* 4787–4794.

Miki, T., Yasuda, S. Y., & Kahn, M. (2011). Wnt/beta-catenin signaling in embryonic stem cell self-renewal and somatic cell reprogramming. *Stem Cell Reviews, 7,* 836–846.

Miravet, S., Piedra, J., Castano, J., Raurell, I., Franci, C., Dunach, M., et al. (2003). Tyrosine phosphorylation of plakoglobin causes contrary effects on its association with desmosomes and adherens junction components and modulates beta-catenin-mediated transcription. *Molecular and Cellular Biology, 23,* 7391–7402.

Miravet, S., Piedra, J., Miro, F., Itarte, E., Garcia de Herreros, A., & Dunach, M. (2002). The transcriptional factor Tcf-4 contains different binding sites for beta-catenin and plakoglobin. *The Journal of Biological Chemistry, 277,* 1884–1891.

Mohamet, L., Lea, M. L., & Ward, C. M. (2010). Abrogation of E-cadherin-mediated cellular aggregation allows proliferation of pluripotent mouse embryonic stem cells in shake flask bioreactors. *PLoS One, 5,* e12921.

Montell, D. J., Rorth, P., & Spradling, A. C. (1992). slow border cells, a locus required for a developmentally regulated cell migration during oogenesis, encodes Drosophila C/EBP. *Cell, 71,* 51–62.

Montoya-Durango, D. E., Velu, C. S., Kazanjian, A., Rojas, M. E., Jay, C. M., Longmore, G. D., et al. (2008). Ajuba functions as a histone deacetylase-dependent co-repressor for autoregulation of the growth factor-independent-1 transcription factor. *The Journal of Biological Chemistry, 283,* 32056–32065.

Moore, R., Radice, G. L., Dominis, M., & Kemler, R. (1999). The generation and in vivo differentiation of murine embryonal stem cells genetically null for either N-cadherin or N- and P-cadherin. *The International Journal of Developmental Biology*, *43*, 831–834.

Morali, O. G., Delmas, V., Moore, R., Jeanney, C., Thiery, J. P., & Larue, L. (2001). IGF-II induces rapid beta-catenin relocation to the nucleus during epithelium to mesenchyme transition. *Oncogene*, *20*, 4942–4950.

Morrison, H., Sherman, L. S., Legg, J., Banine, F., Isacke, C., Haipek, C. A., et al. (2001). The NF2 tumor suppressor gene product, merlin, mediates contact inhibition of growth through interactions with CD44. *Genes & Development*, *15*, 968–980.

Mulkey, R. M., Endo, S., Shenolikar, S., & Malenka, R. C. (1994). Involvement of a calcineurin/inhibitor-1 phosphatase cascade in hippocampal long-term depression. *Nature*, *369*, 486–488.

Muller, J., Ritt, D. A., Copeland, T. D., & Morrison, D. K. (2003). Functional analysis of C-TAK1 substrate binding and identification of PKP2 as a new C-TAK1 substrate. *The EMBO Journal*, *22*, 4431–4442.

Munoz, W. A., Lee, M., Miller, R. K., Ahmed, Z., Ji, H., Link, T. M., et al. (2014). Plakophilin-3 catenin associates with the ETV1/ER81 transcription factor to positively modulate gene activity. *PLoS One*, *9*, e86784.

Murakami, T., Hijikata, T., Matsukawa, M., Ishikawa, H., & Yorifuji, H. (2006). Zebrafish protocadherin 10 is involved in paraxial mesoderm development and somitogenesis. *Developmental Dynamics*, *235*, 506–514.

Murakami, M., Nguyen, L. T., Zhuang, Z. W., Moodie, K. L., Carmeliet, P., Stan, R. V., et al. (2008). The FGF system has a key role in regulating vascular integrity. *The Journal of Clinical Investigation*, *118*, 3355–3366.

Najy, A. J., Day, K. C., & Day, M. L. (2008). The ectodomain shedding of E-cadherin by ADAM15 supports ErbB receptor activation. *The Journal of Biological Chemistry*, *283*, 18393–18401.

Nakao, S., Platek, A., Hirano, S., & Takeichi, M. (2008). Contact-dependent promotion of cell migration by the OL-protocadherin-Nap1 interaction. *The Journal of Cell Biology*, *182*, 395–410.

Nava, P., Laukoetter, M. G., Hopkins, A. M., Laur, O., Gerner-Smidt, K., Green, K. J., et al. (2007). Desmoglein-2: A novel regulator of apoptosis in the intestinal epithelium. *Molecular Biology of the Cell*, *18*, 4565–4578.

Nawrocki-Raby, B., Gilles, C., Polette, M., Bruyneel, E., Laronze, J. Y., Bonnet, N., et al. (2003). Upregulation of MMPs by soluble E-cadherin in human lung tumor cells. *International Journal of Cancer*, *105*, 790–795.

Nawroth, R., Poell, G., Ranft, A., Kloep, S., Samulowitz, U., Fachinger, G., et al. (2002). VE-PTP and VE-cadherin ectodomains interact to facilitate regulation of phosphorylation and cell contacts. *The EMBO Journal*, *21*, 4885–4895.

Nelson, W. J., & Nusse, R. (2004). Convergence of Wnt, beta-catenin, and cadherin pathways. *Science*, *303*, 1483–1487.

Niessen, C. M., Leckband, D., & Yap, A. S. (2011). Tissue organization by cadherin adhesion molecules: Dynamic molecular and cellular mechanisms of morphogenetic regulation. *Physiological Reviews*, *91*, 691–731.

Niewiadomska, P., Godt, D., & Tepass, U. (1999). DE-Cadherin is required for intercellular motility during Drosophila oogenesis. *The Journal of Cell Biology*, *144*, 533–547.

Nishimura, T., & Takeichi, M. (2009). Remodeling of the adherens junctions during morphogenesis. *Current Topics in Developmental Biology*, *89*, 33–54.

Noe, V., Fingleton, B., Jacobs, K., Crawford, H. C., Vermeulen, S., Steelant, W., et al. (2001). Release of an invasion promoter E-cadherin fragment by matrilysin and stromelysin-1. *Journal of Cell Science*, *114*, 111–118.

Noren, N. K., Niessen, C. M., Gumbiner, B. M., & Burridge, K. (2001). Cadherin engagement regulates Rho family GTPases. *The Journal of Biological Chemistry*, *276*, 33305–33308.

Novellino, L., De Filippo, A., Deho, P., Perrone, F., Pilotti, S., Parmiani, G., et al. (2008). PTPRK negatively regulates transcriptional activity of wild type and mutated oncogenic beta-catenin and affects membrane distribution of beta-catenin/E-cadherin complexes in cancer cells. *Cellular Signalling*, *20*, 872–883.

Oda, H., Tagawa, K., & Akiyama-Oda, Y. (2005). Diversification of epithelial adherens junctions with independent reductive changes in cadherin form: Identification of potential molecular synapomorphies among bilaterians. *Evolution & Development*, *7*, 376–389.

Oda, H., & Takeichi, M. (2011). Evolution: Structural and functional diversity of cadherin at the adherens junction. *The Journal of Cell Biology*, *193*, 1137–1146.

Ohira, T., Gemmill, R. M., Ferguson, K., Kusy, S., Roche, J., Brambilla, E., et al. (2003). WNT7a induces E-cadherin in lung cancer cells. *Proceedings of the National Academy of Sciences of the United States of America*, *100*, 10429–10434.

Orsulic, S., Huber, O., Aberle, H., Arnold, S., & Kemler, R. (1999). E-Cadherin binding prevents beta-catenin nuclear localization and beta-catenin/LEF-1-mediated transactivation. *Journal of Cell Science*, *112*(Pt 8), 1237–1245.

Pai, R., Dunlap, D., Qing, J., Mohtashemi, I., Hotzel, K., & French, D. M. (2008). Inhibition of fibroblast growth factor 19 reduces tumor growth by modulating beta-catenin signaling. *Cancer Research*, *68*, 5086–5095.

Paradies, N. E., & Grunwald, G. B. (1993). Purification and characterization of NCAD90, a soluble endogenous form of N-cadherin, which is generated by proteolysis during retinal development and retains adhesive and neurite-promoting function. *Journal of Neuroscience Research*, *36*, 33–45.

Park, J. I., Kim, S. W., Lyons, J. P., Ji, H., Nguyen, T. T., Cho, K., et al. (2005). Kaiso/p120-catenin and TCF/beta-catenin complexes coordinately regulate canonical Wnt gene targets. *Developmental Cell*, *8*, 843–854.

Parnaud, G., Lavallard, V., Bedat, B., Matthey-Doret, D., Morel, P., Berney, T., et al. (2014). Cadherin engagement improves insulin secretion of single human beta-cells. *Diabetes*, Oct 2. pii: DB_140257. [Epub ahead of print]..

Paulson, A. F., Fang, X., Ji, H., Reynolds, A. B., & McCrea, P. D. (1999). Misexpression of the catenin p120(ctn)1A perturbs Xenopus gastrulation but does not elicit Wnt-directed axis specification. *Developmental Biology*, *207*, 350–363.

Paulson, A. F., Prasad, M. S., Thuringer, A. H., & Manzerra, P. (2014). Regulation of cadherin expression in nervous system development. *Cell Adhesion & Migration*, *8*, 19–28.

Pece, S., Chiariello, M., Murga, C., & Gutkind, J. S. (1999). Activation of the protein kinase Akt/PKB by the formation of E-cadherin-mediated cell-cell junctions. Evidence for the association of phosphatidylinositol 3-kinase with the E-cadherin adhesion complex. *The Journal of Biological Chemistry*, *274*, 19347–19351.

Pece, S., & Gutkind, J. S. (2000). Signaling from E-cadherins to the MAPK pathway by the recruitment and activation of epidermal growth factor receptors upon cell-cell contact formation. *The Journal of Biological Chemistry*, *275*, 41227–41233.

Perez-Moreno, M., Davis, M. A., Wong, E., Pasolli, H. A., Reynolds, A. B., & Fuchs, E. (2006). p120-catenin mediates inflammatory responses in the skin. *Cell*, *124*, 631–644.

Perrais, M., Chen, X., Perez-Moreno, M., & Gumbiner, B. M. (2007). E-Cadherin homophilic ligation inhibits cell growth and epidermal growth factor receptor signaling independently of other cell interactions. *Molecular Biology of the Cell*, *18*, 2013–2025.

Philimonenko, V. V., Zhao, J., Iben, S., Dingova, H., Kysela, K., Kahle, M., et al. (2004). Nuclear actin and myosin I are required for RNA polymerase I transcription. *Nature Cell Biology*, *6*, 1165–1172.

Philippova, M., Joshi, M. B., Kyriakakis, E., Pfaff, D., Erne, P., & Resink, T. J. (2009). A guide and guard: The many faces of T-cadherin. *Cellular Signalling*, *21*, 1035–1044.
Piao, S., Lee, S. H., Kim, H., Yum, S., Stamos, J. L., Xu, Y., et al. (2008). Direct inhibition of GSK3beta by the phosphorylated cytoplasmic domain of LRP6 in Wnt/beta-catenin signaling. *PLoS One*, *3*, e4046.
Piao, H. L., Yuan, Y., Wang, M., Sun, Y., Liang, H., & Ma, L. (2014). Alpha-catenin acts as a tumour suppressor in E-cadherin-negative basal-like breast cancer by inhibiting NF-kappaB signalling. *Nature Cell Biology*, *16*, 245–254.
Piedra, J., Martinez, D., Castano, J., Miravet, S., Dunach, M., & de Herreros, A. G. (2001). Regulation of beta-catenin structure and activity by tyrosine phosphorylation. *The Journal of Biological Chemistry*, *276*, 20436–20443.
Piedra, J., Miravet, S., Castano, J., Palmer, H. G., Heisterkamp, N., Garcia de Herreros, A., et al. (2003). p120 Catenin-associated Fer and Fyn tyrosine kinases regulate beta-catenin Tyr-142 phosphorylation and beta-catenin-alpha-catenin Interaction. *Molecular and Cellular Biology*, *23*, 2287–2297.
Pieters, T., van Hengel, J., & van Roy, F. (2012). Functions of p120ctn in development and disease. *Frontiers in Bioscience*, *17*, 760–783.
Pieters, T., & van Roy, F. (2014). Role of cell-cell adhesion complexes in embryonic stem cell biology. *Journal of Cell Science*, *127*, 2603–2613.
Pieters, T., van Roy, F., & van Hengel, J. (2012). Functions of p120ctn isoforms in cell-cell adhesion and intracellular signaling. *Frontiers in Bioscience*, *17*, 1669–1694.
Playford, M. P., Bicknell, D., Bodmer, W. F., & Macaulay, V. M. (2000). Insulin-like growth factor 1 regulates the location, stability, and transcriptional activity of beta-catenin. *Proceedings of the National Academy of Sciences of the United States of America*, *97*, 12103–12108.
Polakis, P. (2000). Wnt signaling and cancer. *Genes & Development*, *14*, 1837–1851.
Qian, X., Karpova, T., Sheppard, A. M., McNally, J., & Lowy, D. R. (2004). E-cadherin-mediated adhesion inhibits ligand-dependent activation of diverse receptor tyrosine kinases. *The EMBO Journal*, *23*, 1739–1748.
Radice, G. L. (2013). N-cadherin-mediated adhesion and signaling from development to disease: Lessons from mice. *Progress in Molecular Biology and Translational Science*, *116*, 263–289.
Raptis, L., Brownell, H. L., Vultur, A. M., Ross, G. M., Tremblay, E., & Elliott, B. E. (2000). Specific inhibition of growth factor-stimulated extracellular signal-regulated kinase 1 and 2 activation in intact cells by electroporation of a growth factor receptor-binding protein 2-Src homology 2 binding peptide. *Cell Growth & Differentiation*, *11*, 293–303.
Rashid, D., Newell, K., Shama, L., & Bradley, R. (2006). A requirement for NF-protocadherin and TAF1/Set in cell adhesion and neural tube formation. *Developmental Biology*, *291*, 170–181.
Ratheesh, A., Priya, R., & Yap, A. S. (2013). Coordinating Rho and Rac: The regulation of Rho GTPase signaling and cadherin junctions. *Progress in Molecular Biology and Translational Science*, *116*, 49–68.
Reiss, K., Maretzky, T., Ludwig, A., Tousseyn, T., de Strooper, B., Hartmann, D., et al. (2005). ADAM10 cleavage of N-cadherin and regulation of cell-cell adhesion and beta-catenin nuclear signalling. *The EMBO Journal*, *24*, 742–752.
Reynolds, A. B., & Carnahan, R. H. (2004). Regulation of cadherin stability and turnover by p120ctn: Implications in disease and cancer. *Seminars in Cell & Developmental Biology*, *15*, 657–663.
Rhee, J., Buchan, T., Zukerberg, L., Lilien, J., & Balsamo, J. (2007). Cables links Robo-bound Abl kinase to N-cadherin-bound beta-catenin to mediate Slit-induced modulation of adhesion and transcription. *Nature Cell Biology*, *9*, 883–892.

Rhee, J., Ryu, J. H., Kim, J. H., Chun, C. H., & Chun, J. S. (2012). alpha-Catenin inhibits beta-catenin-T-cell factor/lymphoid enhancing factor transcriptional activity and collagen type II expression in articular chondrocytes through formation of Gli3R.alpha-catenin.beta-catenin ternary complex. *The Journal of Biological Chemistry, 287*, 11751–11760.

Rios-Doria, J., & Day, M. L. (2005). Truncated E-cadherin potentiates cell death in prostate epithelial cells. *The Prostate, 63*, 259–268.

Rios-Doria, J., Day, K. C., Kuefer, R., Rashid, M. G., Chinnaiyan, A. M., Rubin, M. A., et al. (2003). The role of calpain in the proteolytic cleavage of E-cadherin in prostate and mammary epithelial cells. *The Journal of Biological Chemistry, 278*, 1372–1379.

Rogulja, D., Rauskolb, C., & Irvine, K. D. (2008). Morphogen control of wing growth through the Fat signaling pathway. *Developmental Cell, 15*, 309–321.

Rosenbluh, J., Nijhawan, D., Cox, A. G., Li, X., Neal, J. T., Schafer, E. J., et al. (2012). beta-Catenin-driven cancers require a YAP1 transcriptional complex for survival and tumorigenesis. *Cell, 151*, 1457–1473.

Roura, S., Miravet, S., Piedra, J., Garcia de Herreros, A., & Dunach, M. (1999). Regulation of E-cadherin/catenin association by tyrosine phosphorylation. *The Journal of Biological Chemistry, 274*, 36734–36740.

Rousso, D. L., Pearson, C. A., Gaber, Z. B., Miquelajauregui, A., Li, S., Portera-Cailliau, C., et al. (2012). Foxp-mediated suppression of N-cadherin regulates neuroepithelial character and progenitor maintenance in the CNS. *Neuron, 74*, 314–330.

Rudini, N., Felici, A., Giampietro, C., Lampugnani, M., Corada, M., Swirsding, K., et al. (2008). VE-cadherin is a critical endothelial regulator of TGF-beta signalling. *The EMBO Journal, 27*, 993–1004.

Ruzov, A., Hackett, J. A., Prokhortchouk, A., Reddington, J. P., Madej, M. J., Dunican, D. S., et al. (2009). The interaction of xKaiso with xTcf3: A revised model for integration of epigenetic and Wnt signalling pathways. *Development, 136*, 723–727.

Ruzov, A., Savitskaya, E., Hackett, J. A., Reddington, J. P., Prokhortchouk, A., Madej, M. J., et al. (2009). The non-methylated DNA-binding function of Kaiso is not required in early Xenopus laevis development. *Development, 136*, 729–738.

Saharinen, P., Eklund, L., Miettinen, J., Wirkkala, R., Anisimov, A., Winderlich, M., et al. (2008). Angiopoietins assemble distinct Tie2 signalling complexes in endothelial cell-cell and cell-matrix contacts. *Nature Cell Biology, 10*, 527–537.

Sampietro, J., Dahlberg, C. L., Cho, U. S., Hinds, T. R., Kimelman, D., & Xu, W. (2006). Crystal structure of a beta-catenin/BCL9/Tcf4 complex. *Molecular Cell, 24*, 293–300.

Sanson, B., White, P., & Vincent, J. P. (1996). Uncoupling cadherin-based adhesion from wingless signalling in Drosophila. *Nature, 383*, 627–630.

Sansores-Garcia, L., Bossuyt, W., Wada, K., Yonemura, S., Tao, C., Sasaki, H., et al. (2011). Modulating F-actin organization induces organ growth by affecting the Hippo pathway. *The EMBO Journal, 30*, 2325–2335.

Schackmann, R. C., Tenhagen, M., van de Ven, R. A., & Derksen, P. W. (2013). p120-catenin in cancer—Mechanisms, models and opportunities for intervention. *Journal of Cell Science, 126*, 3515–3525.

Schalm, S. S., Ballif, B. A., Buchanan, S. M., Phillips, G. R., & Maniatis, T. (2010). Phosphorylation of protocadherin proteins by the receptor tyrosine kinase Ret. *Proceedings of the National Academy of Sciences of the United States of America, 107*, 13894–13899.

Schepis, A., Sepich, D., & Nelson, W. J. (2012). Alpha-E-catenin regulates cell-cell adhesion and membrane blebbing during zebrafish epiboly. *Development, 139*, 537–546.

Schlegelmilch, K., Mohseni, M., Kirak, O., Pruszak, J., Rodriguez, J. R., Zhou, D., et al. (2011). Yap1 acts downstream of alpha-catenin to control epidermal proliferation. *Cell, 144*, 782–795.

Schmidt, A., & Jager, S. (2005). Plakophilins—Hard work in the desmosome, recreation in the nucleus? *European Journal of Cell Biology, 84*, 189–204.

Schneider, S., Herrenknecht, K., Butz, S., Kemler, R., & Hausen, P. (1993). Catenins in Xenopus embryogenesis and their relation to the cadherin-mediated cell-cell adhesion system. *Development, 118*, 629–640.

Sehgal, R. N., Gumbiner, B. M., & Reichardt, L. F. (1997). Antagonism of cell adhesion by an alpha-catenin mutant, and of the Wnt-signaling pathway by alpha-catenin in Xenopus embryos. *The Journal of Cell Biology, 139*, 1033–1046.

Seifert, K., Ibrahim, H., Stodtmeister, T., Winklbauer, R., & Niessen, C. M. (2009). An adhesion-independent, aPKC-dependent function for cadherins in morphogenetic movements. *Journal of Cell Science, 122*, 2514–2523.

Seki, T., Namba, T., Mochizuki, H., & Onodera, M. (2007). Clustering, migration, and neurite formation of neural precursor cells in the adult rat hippocampus. *The Journal of Comparative Neurology, 502*, 275–290.

Serio, R. N. (2014). Wnt of the two horizons: Putting stem cell self-renewal and cell fate determination into context. *Stem Cells and Development, 23*, 1975–1990.

Shay-Salit, A., Shushy, M., Wolfovitz, E., Yahav, H., Breviario, F., Dejana, E., et al. (2002). VEGF receptor 2 and the adherens junction as a mechanical transducer in vascular endothelial cells. *Proceedings of the National Academy of Sciences of the United States of America, 99*, 9462–9467.

Shimada, Y., Usui, T., Yanagawa, S., Takeichi, M., & Uemura, T. (2001). Asymmetric colocalization of Flamingo, a seven-pass transmembrane cadherin, and Dishevelled in planar cell polarization. *Current Biology, 11*, 859–863.

Shimamura, K., & Takeichi, M. (1992). Local and transient expression of E-cadherin involved in mouse embryonic brain morphogenesis. *Development, 116*, 1011–1019.

Shoval, I., Ludwig, A., & Kalcheim, C. (2007). Antagonistic roles of full-length N-cadherin and its soluble BMP cleavage product in neural crest delamination. *Development, 134*, 491–501.

Silva, E., Tsatskis, Y., Gardano, L., Tapon, N., & McNeill, H. (2006). The tumor-suppressor gene fat controls tissue growth upstream of expanded in the hippo signaling pathway. *Current Biology, 16*, 2081–2089.

Silvis, M. R., Kreger, B. T., Lien, W. H., Klezovitch, O., Rudakova, G. M., Camargo, F. D., et al. (2011). alpha-Catenin is a tumor suppressor that controls cell accumulation by regulating the localization and activity of the transcriptional coactivator Yap1. *Science Signaling, 4*, ra33.

Simcha, I., Kirkpatrick, C., Sadot, E., Shtutman, M., Polevoy, G., Geiger, B., et al. (2001). Cadherin sequences that inhibit beta-catenin signaling: A study in yeast and mammalian cells. *Molecular Biology of the Cell, 12*, 1177–1188.

Simcha, I., Shtutman, M., Salomon, D., Zhurinsky, J., Sadot, E., Geiger, B., et al. (1998). Differential nuclear translocation and transactivation potential of beta-catenin and plakoglobin. *The Journal of Cell Biology, 141*, 1433–1448.

Smith, M. A., Hoffman, L. M., & Beckerle, M. C. (2014). LIM proteins in actin cytoskeleton mechanoresponse. *Trends in Cell Biology, 24*, 575–583.

Smutny, M., & Yap, A. S. (2010). Neighborly relations: Cadherins and mechanotransduction. *The Journal of Cell Biology, 189*, 1075–1077.

Solanas, G., Miravet, S., Casagolda, D., Castano, J., Raurell, I., Corrionero, A., et al. (2004). beta-Catenin and plakoglobin N- and C-tails determine ligand specificity. *The Journal of Biological Chemistry, 279*, 49849–49856.

Solanas, G., Porta-de-la-Riva, M., Agusti, C., Casagolda, D., Sanchez-Aguilera, F., Larriba, M. J., et al. (2008). E-Cadherin controls beta-catenin and NF-kappaB transcriptional activity in mesenchymal gene expression. *Journal of Cell Science*, *121*, 2224–2234.

Soncin, F., Mohamet, L., Eckardt, D., Ritson, S., Eastham, A. M., Bobola, N., et al. (2009). Abrogation of E-cadherin-mediated cell-cell contact in mouse embryonic stem cells results in reversible LIF-independent self-renewal. *Stem Cells*, *27*, 2069–2080.

Song, X., & Xie, T. (2003). Wingless signaling regulates the maintenance of ovarian somatic stem cells in Drosophila. *Development*, *130*, 3259–3268.

Song, X., Zhu, C. H., Doan, C., & Xie, T. (2002). Germline stem cells anchored by adherens junctions in the Drosophila ovary niches. *Science*, *296*, 1855–1857.

Soriano, S., Kang, D. E., Fu, M., Pestell, R., Chevallier, N., Zheng, H., et al. (2001). Presenilin 1 negatively regulates beta-catenin/T cell factor/lymphoid enhancer factor-1 signaling independently of beta-amyloid precursor protein and notch processing. *The Journal of Cell Biology*, *152*, 785–794.

Spring, C. M., Kelly, K. F., O'Kelly, I., Graham, M., Crawford, H. C., & Daniel, J. M. (2005). The catenin p120ctn inhibits Kaiso-mediated transcriptional repression of the beta-catenin/TCF target gene matrilysin. *Experimental Cell Research*, *305*, 253–265.

Stamos, J. L., & Weis, W. I. (2013). The beta-catenin destruction complex. *Cold Spring Harbor Perspectives in Biology*, *5*, a007898.

Steinhusen, U., Weiske, J., Badock, V., Tauber, R., Bommert, K., & Huber, O. (2001). Cleavage and shedding of E-cadherin after induction of apoptosis. *The Journal of Biological Chemistry*, *276*, 4972–4980.

Stepniak, E., Radice, G. L., & Vasioukhin, V. (2009). Adhesive and signaling functions of cadherins and catenins in vertebrate development. *Cold Spring Harbor Perspectives in Biology*, *1*, a002949.

Su, Y., Fu, C., Ishikawa, S., Stella, A., Kojima, M., Shitoh, K., et al. (2008). APC is essential for targeting phosphorylated beta-catenin to the SCFbeta-TrCP ubiquitin ligase. *Molecular Cell*, *32*, 652–661.

Suh, E. K., & Gumbiner, B. M. (2003). Translocation of beta-catenin into the nucleus independent of interactions with FG-rich nucleoporins. *Experimental Cell Research*, *290*, 447–456.

Sumiyoshi, E., Takahashi, S., Obata, H., Sugimoto, A., & Kohara, Y. (2011). The beta-catenin HMP-2 functions downstream of Src in parallel with the Wnt pathway in early embryogenesis of C. elegans. *Developmental Biology*, *355*, 302–312.

Sun, Z., Parrish, A. R., Hill, M. A., & Meininger, G. A. (2014). N-cadherin, a vascular smooth muscle cell-cell adhesion molecule: Function and signaling for vasomotor control. *Microcirculation*, *21*, 208–218.

Suyama, K., Shapiro, I., Guttman, M., & Hazan, R. B. (2002). A signaling pathway leading to metastasis is controlled by N-cadherin and the FGF receptor. *Cancer Cell*, *2*, 301–314.

Swope, D., Li, J., & Radice, G. L. (2013). Beyond cell adhesion: The role of armadillo proteins in the heart. *Cellular Signalling*, *25*, 93–100.

Symowicz, J., Adley, B. P., Gleason, K. J., Johnson, J. J., Ghosh, S., Fishman, D. A., et al. (2007). Engagement of collagen-binding integrins promotes matrix metalloproteinase-9-dependent E-cadherin ectodomain shedding in ovarian carcinoma cells. *Cancer Research*, *67*, 2030–2039.

Taddei, A., Giampietro, C., Conti, A., Orsenigo, F., Breviario, F., Pirazzoli, V., et al. (2008). Endothelial adherens junctions control tight junctions by VE-cadherin-mediated upregulation of claudin-5. *Nature Cell Biology*, *10*, 923–934.

Taelman, V. F., Dobrowolski, R., Plouhinec, J. L., Fuentealba, L. C., Vorwald, P. P., Gumper, I., et al. (2010). Wnt signaling requires sequestration of glycogen synthase kinase 3 inside multivesicular endosomes. *Cell*, *143*, 1136–1148.

Takahashi, N., Ishihara, S., Takada, S., Tsukita, S., & Nagafuchi, A. (2000). Posttranscriptional regulation of alpha-catenin expression is required for Wnt signaling in L cells. *Biochemical and Biophysical Research Communications, 277*, 691–698.

Takahashi, K., & Suzuki, K. (1996). Density-dependent inhibition of growth involves prevention of EGF receptor activation by E-cadherin-mediated cell-cell adhesion. *Experimental Cell Research, 226*, 214–222.

Tanneberger, K., Pfister, A. S., Brauburger, K., Schneikert, J., Hadjihannas, M. V., Kriz, V., et al. (2011). Amer1/WTX couples Wnt-induced formation of PtdIns(4,5)P2 to LRP6 phosphorylation. *The EMBO Journal, 30*, 1433–1443.

Tian, Q., Feetham, M. C., Tao, W. A., He, X. C., Li, L., Aebersold, R., et al. (2004). Proteomic analysis identifies that 14-3-3zeta interacts with beta-catenin and facilitates its activation by Akt. *Proceedings of the National Academy of Sciences of the United States of America, 101*, 15370–15375.

Tran, N. L., Adams, D. G., Vaillancourt, R. R., & Heimark, R. L. (2002). Signal transduction from N-cadherin increases Bcl-2. Regulation of the phosphatidylinositol 3-kinase/Akt pathway by homophilic adhesion and actin cytoskeletal organization. *The Journal of Biological Chemistry, 277*, 32905–32914.

Twiss, F., & de Rooij, J. (2013). Cadherin mechanotransduction in tissue remodeling. *Cellular and Molecular Life Sciences, 70*, 4101–4116.

Tzima, E., Irani-Tehrani, M., Kiosses, W. B., Dejana, E., Schultz, D. A., Engelhardt, B., et al. (2005). A mechanosensory complex that mediates the endothelial cell response to fluid shear stress. *Nature, 437*, 426–431.

Uemura, K., Kihara, T., Kuzuya, A., Okawa, K., Nishimoto, T., Bito, H., et al. (2006). Activity-dependent regulation of beta-catenin via epsilon-cleavage of N-cadherin. *Biochemical and Biophysical Research Communications, 345*, 951–958.

Unterseher, F., Hefele, J. A., Giehl, K., De Robertis, E. M., Wedlich, D., & Schambony, A. (2004). Paraxial protocadherin coordinates cell polarity during convergent extension via Rho A and JNK. *The EMBO Journal, 23*, 3259–3269.

Usui, T., Shima, Y., Shimada, Y., Hirano, S., Burgess, R. W., Schwarz, T. L., et al. (1999). Flamingo, a seven-pass transmembrane cadherin, regulates planar cell polarity under the control of Frizzled. *Cell, 98*, 585–595.

Utton, M. A., Eickholt, B., Howell, F. V., Wallis, J., & Doherty, P. (2001). Soluble N-cadherin stimulates fibroblast growth factor receptor dependent neurite outgrowth and N-cadherin and the fibroblast growth factor receptor co-cluster in cells. *Journal of Neurochemistry, 76*, 1421–1430.

Vallorosi, C. J., Day, K. C., Zhao, X., Rashid, M. G., Rubin, M. A., Johnson, K. R., et al. (2000). Truncation of the beta-catenin binding domain of E-cadherin precedes epithelial apoptosis during prostate and mammary involution. *The Journal of Biological Chemistry, 275*, 3328–3334.

Valls, G., Codina, M., Miller, R. K., Del Valle-Perez, B., Vinyoles, M., Caelles, C., et al. (2012). Upon Wnt stimulation, Rac1 activation requires Rac1 and Vav2 binding to p120-catenin. *Journal of Cell Science, 125*, 5288–5301.

van de Wetering, M., Barker, N., Harkes, I. C., van der Heyden, M., Dijk, N. J., Hollestelle, A., et al. (2001). Mutant E-cadherin breast cancer cells do not display constitutive Wnt signaling. *Cancer Research, 61*, 278–284.

van Nieuw Amerongen, G. P., van Delft, S., Vermeer, M. A., Collard, J. G., & van Hinsbergh, V. W. (2000). Activation of RhoA by thrombin in endothelial hyperpermeability: Role of Rho kinase and protein tyrosine kinases. *Circulation Research, 87*, 335–340.

van Rijssel, J., & van Buul, J. D. (2012). The many faces of the guanine-nucleotide exchange factor trio. *Cell Adhesion & Migration, 6*, 482–487.

van Roy, F. (2014). Beyond E-cadherin: Roles of other cadherin superfamily members in cancer. *Nature Reviews*, *14*, 121–134.

van Veelen, W., Le, N. H., Helvensteijn, W., Blonden, L., Theeuwes, M., Bakker, E. R., et al. (2011). Beta-catenin tyrosine 654 phosphorylation increases Wnt signalling and intestinal tumorigenesis. *Gut*, *60*, 1204–1212.

Vandenbroucke St Amant, E., Tauseef, M., Vogel, S. M., Gao, X. P., Mehta, D., Komarova, Y. A., et al. (2012). PKCalpha activation of p120-catenin serine 879 phospho-switch disassembles VE-cadherin junctions and disrupts vascular integrity. *Circulation Research*, *111*, 739–749.

Vanhalst, K., Kools, P., Staes, K., van Roy, F., & Redies, C. (2005). delta-Protocadherins: A gene family expressed differentially in the mouse brain. *Cellular and Molecular Life Sciences*, *62*, 1247–1259.

Varelas, X., Miller, B. W., Sopko, R., Song, S., Gregorieff, A., Fellouse, F. A., et al. (2010). The Hippo pathway regulates Wnt/beta-catenin signaling. *Developmental Cell*, *18*, 579–591.

Vasioukhin, V., Bauer, C., Degenstein, L., Wise, B., & Fuchs, E. (2001). Hyperproliferation and defects in epithelial polarity upon conditional ablation of alpha-catenin in skin. *Cell*, *104*, 605–617.

Vermeer, P. D., Einwalter, L. A., Moninger, T. O., Rokhlina, T., Kern, J. A., Zabner, J., et al. (2003). Segregation of receptor and ligand regulates activation of epithelial growth factor receptor. *Nature*, *422*, 322–326.

Vinyoles, M., Del Valle-Perez, B., Curto, J., Vinas-Castells, R., Alba-Castellon, L., Garcia de Herreros, A., et al. (2014). Multivesicular GSK3 sequestration upon Wnt signaling is controlled by p120-catenin/cadherin interaction with LRP5/6. *Molecular Cell*, *53*, 444–457.

Wang, Y., Janicki, P., Koster, I., Berger, C. D., Wenzl, C., Grosshans, J., et al. (2008). Xenopus paraxial protocadherin regulates morphogenesis by antagonizing Sprouty. *Genes & Development*, *22*, 878–883.

Watanabe, T., Sato, K., & Kaibuchi, K. (2009). Cadherin-mediated intercellular adhesion and signaling cascades involving small GTPases. *Cold Spring Harbor Perspectives in Biology*, *1*, a003020.

Weiner, J. A., & Jontes, J. D. (2013). Protocadherins, not prototypical: A complex tale of their interactions, expression, and functions. *Frontiers in Molecular Neuroscience*, *6*, 4.

Weiske, J., Schoneberg, T., Schroder, W., Hatzfeld, M., Tauber, R., & Huber, O. (2001). The fate of desmosomal proteins in apoptotic cells. *The Journal of Biological Chemistry*, *276*, 41175–41181.

Wildenberg, G. A., Dohn, M. R., Carnahan, R. H., Davis, M. A., Lobdell, N. A., Settleman, J., et al. (2006). p120-catenin and p190RhoGAP regulate cell-cell adhesion by coordinating antagonism between Rac and Rho. *Cell*, *127*, 1027–1039.

Willecke, M., Hamaratoglu, F., Kango-Singh, M., Udan, R., Chen, C. L., Tao, C., et al. (2006). The fat cadherin acts through the hippo tumor-suppressor pathway to regulate tissue size. *Current Biology*, *16*, 2090–2100.

Willert, J., Epping, M., Pollack, J. R., Brown, P. O., & Nusse, R. (2002). A transcriptional response to Wnt protein in human embryonic carcinoma cells. *BMC Developmental Biology*, *2*, 8.

Williams, B. O., Barish, G. D., Klymkowsky, M. W., & Varmus, H. E. (2000). A comparative evaluation of beta-catenin and plakoglobin signaling activity. *Oncogene*, *19*, 5720–5728.

Williamson, L., Raess, N. A., Caldelari, R., Zakher, A., de Bruin, A., Posthaus, H., et al. (2006). Pemphigus vulgaris identifies plakoglobin as key suppressor of c-Myc in the skin. *The EMBO Journal*, *25*, 3298–3309.

Winderlich, M., Keller, L., Cagna, G., Broermann, A., Kamenyeva, O., Kiefer, F., et al. (2009). VE-PTP controls blood vessel development by balancing Tie-2 activity. *The Journal of Cell Biology, 185*, 657–671.

Winklbauer, R., Medina, A., Swain, R. K., & Steinbeisser, H. (2001). Frizzled-7 signalling controls tissue separation during Xenopus gastrulation. *Nature, 413*, 856–860.

Wojciak-Stothard, B., Potempa, S., Eichholtz, T., & Ridley, A. J. (2001). Rho and Rac but not Cdc42 regulate endothelial cell permeability. *Journal of Cell Science, 114*, 1343–1355.

Wojciak-Stothard, B., & Ridley, A. J. (2003). Shear stress-induced endothelial cell polarization is mediated by Rho and Rac but not Cdc42 or PI 3-kinases. *The Journal of Cell Biology, 161*, 429–439.

Wong, M. H., Rubinfeld, B., & Gordon, J. I. (1998). Effects of forced expression of an NH2-terminal truncated beta-Catenin on mouse intestinal epithelial homeostasis. *The Journal of Cell Biology, 141*, 765–777.

Woodhead, G. J., Mutch, C. A., Olson, E. C., & Chenn, A. (2006). Cell-autonomous beta-catenin signaling regulates cortical precursor proliferation. *The Journal of Neuroscience, 26*, 12620–12630.

Wray, J., Kalkan, T., Gomez-Lopez, S., Eckardt, D., Cook, A., Kemler, R., et al. (2011). Inhibition of glycogen synthase kinase-3 alleviates Tcf3 repression of the pluripotency network and increases embryonic stem cell resistance to differentiation. *Nature Cell Biology, 13*, 838–845.

Wu, G., Huang, H., Garcia Abreu, J., & He, X. (2009). Inhibition of GSK3 phosphorylation of beta-catenin via phosphorylated PPPSPXS motifs of Wnt coreceptor LRP6. *PLoS One, 4*, e4926.

Wu, J., & Mlodzik, M. (2009). A quest for the mechanism regulating global planar cell polarity of tissues. *Trends in Cell Biology, 19*, 295–305.

Wu, X., Tu, X., Joeng, K. S., Hilton, M. J., Williams, D. A., & Long, F. (2008). Rac1 activation controls nuclear localization of beta-catenin during canonical Wnt signaling. *Cell, 133*, 340–353.

Xi, Y., Wei, Y., Sennino, B., Ulsamer, A., Kwan, I., Brumwell, A. N., et al. (2013). Identification of pY654-beta-catenin as a critical co-factor in hypoxia-inducible factor-1alpha signaling and tumor responses to hypoxia. *Oncogene, 32*, 5048–5057.

Xiao, K., Garner, J., Buckley, K. M., Vincent, P. A., Chiasson, C. M., Dejana, E., et al. (2005). p120-catenin regulates clathrin-dependent endocytosis of VE-cadherin. *Molecular Biology of the Cell, 16*, 5141–5151.

Xu, G., Craig, A. W., Greer, P., Miller, M., Anastasiadis, P. Z., Lilien, J., et al. (2004). Continuous association of cadherin with beta-catenin requires the non-receptor tyrosine-kinase Fer. *Journal of Cell Science, 117*, 3207–3219.

Yagi, R., Chen, L. F., Shigesada, K., Murakami, Y., & Ito, Y. (1999). A WW domain-containing yes-associated protein (YAP) is a novel transcriptional co-activator. *The EMBO Journal, 18*, 2551–2562.

Yagita, Y., Sakurai, T., Tanaka, H., Kitagawa, K., Colman, D. R., & Shan, W. (2009). N-cadherin mediates interaction between precursor cells in the subventricular zone and regulates further differentiation. *Journal of Neuroscience Research, 87*, 3331–3342.

Yamagata, K., Andreasson, K. I., Sugiura, H., Maru, E., Dominique, M., Irie, Y., et al. (1999). Arcadlin is a neural activity-regulated cadherin involved in long term potentiation. *The Journal of Biological Chemistry, 274*, 19473–19479.

Yan, H. X., Yang, W., Zhang, R., Chen, L., Tang, L., Zhai, B., et al. (2006). Protein-tyrosine phosphatase PCP-2 inhibits beta-catenin signaling and increases E-cadherin-dependent cell adhesion. *The Journal of Biological Chemistry, 281*, 15423–15433.

Yanagawa, S., Lee, J. S., Haruna, T., Oda, H., Uemura, T., Takeichi, M., et al. (1997). Accumulation of Armadillo induced by Wingless, Dishevelled, and dominant-negative

Zeste-White 3 leads to elevated DE-cadherin in Drosophila clone 8 wing disc cells. *The Journal of Biological Chemistry*, *272*, 25243–25251.

Yang, C. H., Axelrod, J. D., & Simon, M. A. (2002). Regulation of Frizzled by fat-like cadherins during planar polarity signaling in the Drosophila compound eye. *Cell*, *108*, 675–688.

Yasuda, S., Tanaka, H., Sugiura, H., Okamura, K., Sakaguchi, T., Tran, U., et al. (2007). Activity-induced protocadherin arcadlin regulates dendritic spine number by triggering N-cadherin endocytosis via TAO2beta and p38 MAP kinases. *Neuron*, *56*, 456–471.

Zantek, N. D., Azimi, M., Fedor-Chaiken, M., Wang, B., Brackenbury, R., & Kinch, M. S. (1999). E-Cadherin regulates the function of the EphA2 receptor tyrosine kinase. *Cell Growth & Differentiation*, *10*, 629–638.

Zeng, X., Tamai, K., Doble, B., Li, S., Huang, H., Habas, R., et al. (2005). A dual-kinase mechanism for Wnt co-receptor phosphorylation and activation. *Nature*, *438*, 873–877.

Zhang, Z., Hartmann, H., Do, V. M., Abramowski, D., Sturchler-Pierrat, C., Staufenbiel, M., et al. (1998). Destabilization of beta-catenin by mutations in presenilin-1 potentiates neuronal apoptosis. *Nature*, *395*, 698–702.

Zhang, J., Niu, C., Ye, L., Huang, H., He, X., Tong, W. G., et al. (2003). Identification of the haematopoietic stem cell niche and control of the niche size. *Nature*, *425*, 836–841.

Zhang, J., Shemezis, J. R., McQuinn, E. R., Wang, J., Sverdlov, M., & Chenn, A. (2013). AKT activation by N-cadherin regulates beta-catenin signaling and neuronal differentiation during cortical development. *Neural Development*, *8*, 7.

Zhang, J., Woodhead, G. J., Swaminathan, S. K., Noles, S. R., McQuinn, E. R., Pisarek, A. J., et al. (2010). Cortical neural precursors inhibit their own differentiation via N-cadherin maintenance of beta-catenin signaling. *Developmental Cell*, *18*, 472–479.

Zhou, L., Goffinet, A. M., & Tissir, F. (2008). Role of the cadherin Celsr3 in the connectivity of the cerebral cortex. *Médecine Sciences: M/S*, *24*, 1025–1027.

Zhurinsky, J., Shtutman, M., & Ben-Ze'ev, A. (2000). Differential mechanisms of LEF/TCF family-dependent transcriptional activation by beta-catenin and plakoglobin. *Molecular and Cellular Biology*, *20*, 4238–4252.

Zuo, J. H., Zhu, W., Li, M. Y., Li, X. H., Yi, H., Zeng, G. Q., et al. (2011). Activation of EGFR promotes squamous carcinoma SCC10A cell migration and invasion via inducing EMT-like phenotype change and MMP-9-mediated degradation of E-cadherin. *Journal of Cellular Biochemistry*, *112*, 2508–2517.

CHAPTER SIX

Nectins and Nectin-Like Molecules in Development and Disease

Kenji Mandai[*,†], Yoshiyuki Rikitake[†,‡,§], Masahiro Mori[†,¶,||], Yoshimi Takai[*,†,1]

[*]Division of Pathogenetic Signaling, Kobe University Graduate School of Medicine, Kobe, Japan
[†]CREST, Japan Science and Technology Agency, Kobe, Japan
[‡]Division of Signal Transduction, Department of Biochemistry and Molecular Biology, Kobe University Graduate School of Medicine, Kobe, Japan
[§]Division of Cardiovascular Medicine, Department of Internal Medicine, Kobe University Graduate School of Medicine, Kobe, Japan
[¶]Division of Neurophysiology, Department of Physiology and Cell Biology, Kobe University Graduate School of Medicine, Kobe, Japan
[||]Faculty of Health Sciences, Kobe University Graduate School of Health Sciences, Kobe, Japan
[1]Corresponding author: e-mail address: ytakai@med.kobe-u.ac.jp

Contents

Current Topics in Developmental Biology, Volume 112
ISSN 0070-2153
http://dx.doi.org/10.1016/bs.ctdb.2014.11.019

Abstract

Nectins and nectin-like molecules (Necls)/Cadms are Ca^{2+}-independent immunoglobulin superfamily cell adhesion molecules, expressed in most cell types. Nectins mediate not only homotypic but also heterotypic cell-cell adhesion, in contrast to classic cadherins which participate only in homophilic adhesion. Nectins and Necls function in organogenesis of the eye, inner ear, tooth, and cerebral cortex and in a variety of developmental processes including spermatogenesis, axon guidance, synapse formation, and myelination. They are also involved in various diseases, such as viral infection, hereditary ectodermal dysplasia, Alzheimer's disease, autism spectrum disorder, and cancer. Thus, nectins and Necls are crucial for both physiology and pathology. This review summarizes recent advances in research on these cell adhesion molecules in development and pathogenesis.

1. INTRODUCTION

Cell–cell adhesion is a fundamental process for development of multicellular organisms. Perturbation of this process leads to the disruption of homeostasis and dysontogenesis. Perturbation of cell–cell adhesion can also be a cause of disease. In multicellular organisms, cell–cell adhesion occurs in a homotypic or heterotypic manner; e.g., the homotypic cell–cell adhesion is formed between two neighboring epithelial cells, and the heterotypic cell–cell adhesion is formed between differentiating germ cells and Sertoli cells in the testis. In mammalian epithelial cells, cell–cell adhesion is mediated by specialized junctions, notably of tight junctions (TJ), adherens junctions (AJ), and desmosomes. Specific cell adhesion molecules (CAMs) mediate cell–cell adhesion at each of the adhesion machineries. The members of the nectin family, originally identified as α-herpes virus receptors, are CAMs localized at AJ. Nectins are Ca^{2+}-independent immunoglobulin superfamily CAMs (Reymond et al., 2001; Satoh-Horikawa et al., 2000; Takahashi et al., 1999) and contribute to a variety of cell–cell adhesion, acting cooperatively with or independently of cadherins (Takai, Ikeda, Ogita, & Rikitake, 2008; Takai, Miyoshi, Ikeda, & Ogita, 2008). Later, nectin-like molecules (Necls), which have domain structures similar to those of nectins, were defined as a gene family related to nectins. Necl-5 and -2 were originally reported as a human poliovirus receptor (pvr; Mendelsohn, Wimmer, & Racaniello, 1989) and a tumor suppressor in human non-small-cell lung cancer (Kuramochi et al., 2001), respectively. Necls have many nomenclatures, and Necl-1 to -4 were recently renamed Cadm-3, -1, -2, and -4, respectively (Table 1), but Necls are used in this chapter. The members of nectins and Necls interact in *trans-homo* and *trans-hetero* with each other and in *cis* with growth factor receptors and integrins. Moreover, nectins and Necls are associated with human diseases, such as hereditary ectodermal dysplasia, Alzheimer's disease, autism spectrum disorder, and cancer. In the following pages, we introduce the physical and functional properties of nectins and Necls and then summarize recent advances in understanding the role of these molecules in development and disease.

2. PHYSICAL PROPERTIES

2.1. Molecular properties

The nectin family and Necl family members are type-I single-pass transmembrane proteins. The nectin family is composed of four members, nectin-1 to -4, while the Necl family is composed of five members, Necl-1 to -5. All the members share the same domain organization: an extracellular region with three Ig-like domains, a single transmembrane region, and a cytoplasmic tail region (Fig. 1). Most of the nectin family and Necl family members have multiple splicing variants. The Ensembl genome database (Flicek et al., 2014) shows that mouse nectin-1 to -4 have one, two, eight, and three protein-coding splicing variants, respectively, and that mouse Necl-1 to -5 have two, six, five, one, and one protein-coding splicing variants, respectively.

2.2. Binding properties of the extracellular regions

CAMs form *trans*-dimers to *trans*-interact with each other, thereby mediating cell-cell adhesion. All cadherins and nectins form *trans-homo*-dimers through their extracellular regions to *trans-homo*-interact with each other (i.e., interactions between the same members on opposing cells; Fig. 2). However, cadherins do not form *trans-hetero*-dimers through their extracellular regions to heterophilically *trans*-interact among the members (i.e., interactions between the different members on opposing cells). In contrast, nectins form *trans-hetero*-dimers among the members in various combinations. This *trans-hetero*-dimer formation is a molecular basis for the unique functions of nectins

Table 1 Nectin and Necl family members

Member	Symbol	Alias	Related developmental process	Related disease
Nectin-1	*pvrl1*	PRR1, HveC, and CD111	Inner ear development, cerebral cortex development, axon guidance, synapse formation, and tooth development	Viral infection, ectodermal dysplasia, and stress-related mental disorders[a]
Nectin-2	*pvrl2*	PRR2, HveB, Mph, and CD112	Spermatogenesis	Viral infection, Alzheimer's disease, and cancer
Nectin-3	*pvrl3*	PRR3 and CD113	Spermatogenesis, ocular development, inner ear development, cerebral cortex development, axon guidance, synapse formation, and tooth development	Stress-related mental disorders[a]
Nectin-4	*pvrl4*	PRR4		Viral infection, ectodermal dysplasia, and cancer
Necl-1	*Cadm3*	Tsll1 and SynCAM 3	Myelination	Alzheimer's disease[a] and cancer[a]
Necl-2	*Cadm1*	Igsf4, RA175, SgIGSF, Tslc1, SynCAM 1, and SynCAM	Spermatogenesis, axon guidance, and synapse formation	Autism spectrum disorder and cancer
Necl-3	*Cadm2*	SynCAM 2	Axon guidance	Autism spectrum disorder and attention-deficit hyperactivity disorder
Necl-4	*Cadm4*	Tsll2 and SynCAM 4	Myelination	Cancer[a]
Necl-5	*pvr*	CD155 and Tage4		Viral infection and cancer

[a]The relationship with these diseases is only shown in mouse models.

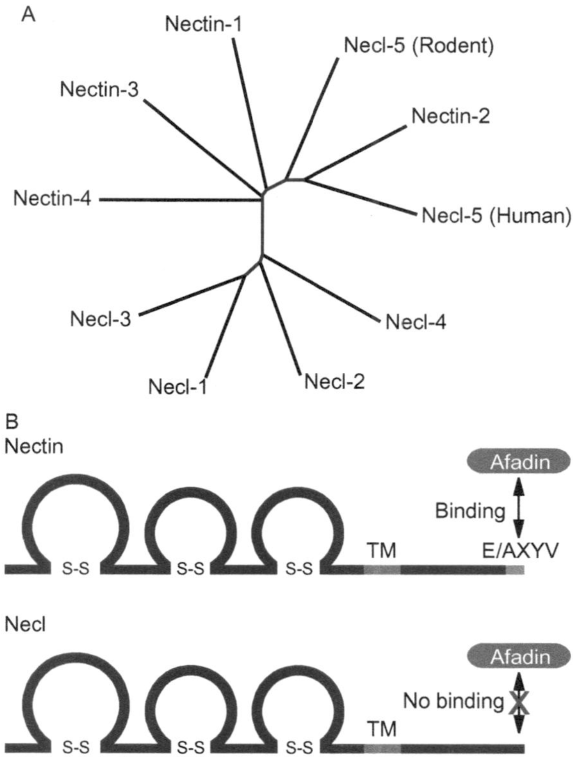

Figure 1 Nectin and Necl protein family. (A) A molecular phylogenetic tree of rodent nectins and Necls. The branch lengths are proportional to the number of amino acid changes. Necl-5 is more similar to nectins than to other Necl members. (B) Molecular structures of nectins and Necls. Nectins and Necls share three Ig-like extracellular domains, a transmembrane region (TM), and a cytoplasmic domain. The nectin C-terminus contains interaction motifs (E/AxYV, where x represents any amino acid) that allow interaction with afadin. Necls do not interact with afadin.

(Mizoguchi et al., 2002; Okabe et al., 2004; Ozaki-Kuroda et al., 2002; Togashi et al., 2011). Among various combinations, the heterophilic *trans*-interaction between nectin-1 and -3 is stronger than those of any other combinations. Nectins homophilically interact in *cis* on the same cell surface, and this *cis*-dimer formation has been shown to be essential for the *trans*-interactions of nectins (Momose et al., 2002). However, recent structural analysis of nectins precludes the head-to-head *trans*-interaction of *cis*-dimers. Instead, it is indicated that the *trans*-interaction of nectins is mediated by a *trans*-dimer through a conserved interface among nectins (Harrison et al., 2012; Liu et al., 2012; Samanta et al., 2012). Currently, the reason for the contradiction is not yet understood.

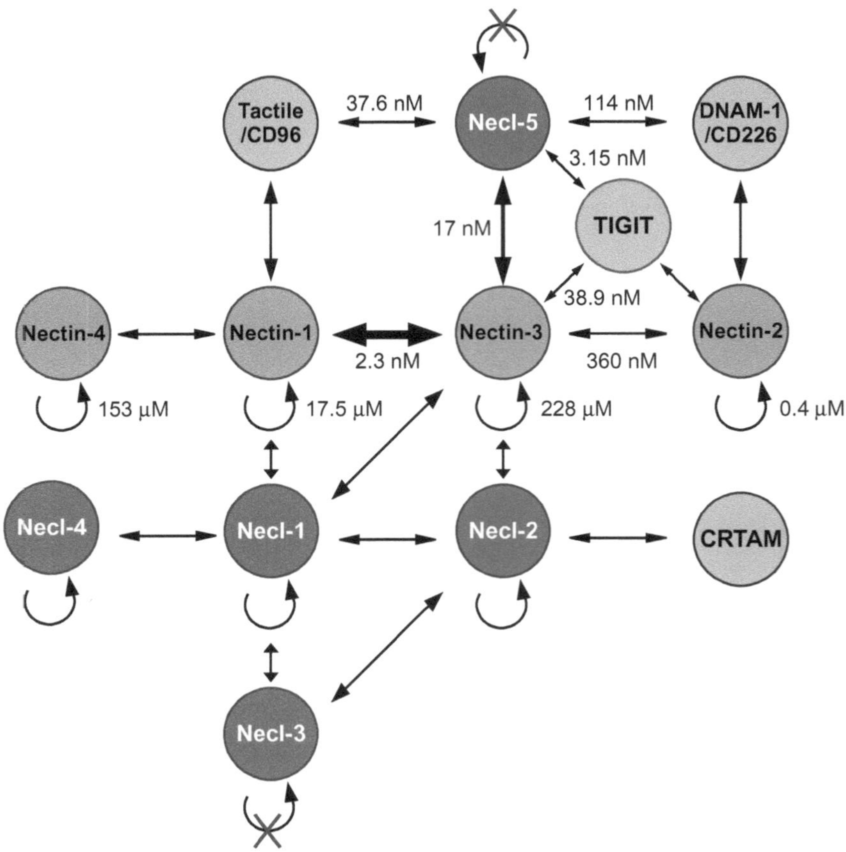

Figure 2 *trans*-Interaction of nectins, Necls, and other Ig-like molecules. Homophilic (looped arrows) and heterophilic (double-headed arrows) *trans*-interactions are shown. Heterophilic *trans*-interaction between nectin-1 and -3 is the strongest followed by that between nectin-3 and Necl-5 and that between nectin-2 and -3. Crossing bars indicate absence of *trans-homo*-interaction. Tactile, T-cell-activated increased late expression; DNAM-1, DNAX accessory molecule 1; TIGIT, T cell immunoreceptor with Ig and ITIM domains; CRTAM, class I-restricted T-cell-associated molecule. Values besides arrows are K_d for homophilic and heterophilic interactions.

Necl-1, -2, -4, and -5 form *cis-homo*-dimers on the same cell surface (Ikeda et al., 2003; Kakunaga et al., 2005; Shingai et al., 2003; Williams et al., 2006). However, it is currently not known whether *cis-homo*-dimer formation is necessary for Necls to *trans*-interact. Necl-1, -2, and -4 form *trans-homo*-dimers and homophilically *trans*-interact, but Necl-3 and -5 do not (Fig. 2; Galibert et al., 2005; Ikeda et al., 2003; Kakunaga et al., 2005; Maurel et al., 2007; Niederkofler, Baeriswyl, Ott, & Stoeckli, 2010; Spiegel et al., 2007; Williams et al., 2006). All Necls form *trans-hetero*-dimers to heterophilically *trans*-interact among the members and also with the members of the nectin family in various combinations (Galibert et al., 2005; Kakunaga et al., 2005; Maurel et al., 2007; Niederkofler et al., 2010; Pellissier, Gerber, Bauer, Ballivet, & Ossipow, 2007; Spiegel et al., 2007).

Nectins and Necls as well as cadherins are expressed in tissues and organs, but nectins and Necls are furthermore expressed in lymphocytes, such as natural killer (NK) cells and $CD8^+$ T cells, although cadherins are not expressed in these cells. They heterophilically interact in *trans* through their extracellular regions with other immunoglobulin superfamily CAMs composed of one to three Ig-like domains, such as T-cell-activated increased late expression (Tactile)/CD96, class I-restricted T-cell-associated molecule (CRTAM), DNAX accessory molecule 1 (DNAM-1)/CD226, and T-cell immunoreceptor with Ig and ITIM domains (TIGIT; Fig. 2) (Boles, Barchet, Diacovo, Cella, & Colonna, 2005; Bottino et al., 2003; Fuchs, Cella, Giurisato, Shaw, & Colonna, 2004; Stanietsky et al., 2009). Nectins and Necls regulate the functions of NK cells and $CD8^+$ T cells through these interactions.

In addition, nectins interact with growth factor receptors; e.g., nectin-3 interacts in *cis* with the platelet-derived growth factor (PDGF) receptor (Kanzaki et al., 2008); nectin-1 interacts in *cis* with the fibroblast growth factor (FGF) receptor (Bojesen et al., 2012). Nectin-3 is involved in the cell survival signal downstream of the PDGF receptor, and nectin-1 enhances the FGF receptor-induced neurite elongation (Bojesen et al., 2012). Furthermore, nectin-1 and -3, but not nectin-2, physically interact in *cis* with integrin $\alpha_v\beta_3$ at cell–cell adhesion sites, and this interaction is implicated in a cross talk between cell–matrix and cell–cell junctions (Sakamoto et al., 2006; Sakamoto, Ogita, Komura, & Takai, 2008). Necls also interact in *cis* with growth factor receptors and integrins (Mizutani et al., 2011). Necl-2 interacts in *cis* with integrin $\alpha_6\beta_4$, and this interaction inhibits the disassembly of integrin $\alpha_6\beta_4$ from hemidesmosomes, eventually stabilizing them. Necl-2 also interacts in *cis* with ErbB3, and this interaction inhibits the ligand-induced, ErbB3-mediated activation of Rac and Akt, resulting in the inhibition of cell movement and survival (Kawano, Ikeda, Kishimoto, Ogita, & Takai, 2009; Mizutani et al., 2011). Moreover, Necl-5 interacts in *cis* with integrin $\alpha_v\beta_3$ and growth factor receptors, such as the PDGF receptor and the vascular endothelial growth factor (VEGF) receptor, and these interactions are implicated in directional cell movement and VEGF-induced angiogenesis, respectively (Amano et al., 2008; Kinugasa et al., 2012; Minami et al., 2007).

2.3. Binding properties of the intracellular regions

The intracellular region of nectins and Necls binds many proteins. Nectins bind afadin, an actin filament (F-actin)-binding protein, which connects nectins to the actin cytoskeleton (Mandai et al., 1997; Satoh-Horikawa et al., 2000; Takahashi et al., 1999). This binding is mediated between the four amino acids in the cytoplasmic tails of nectins and the PDZ domain of afadin (Mandai et al., 1997; Satoh-Horikawa et al., 2000; Takahashi et al., 1999) and is required for nectins to cluster at cell–cell adhesion sites (Miyahara et al., 2000) and for a cadherin–catenin complex to be recruited to nectin-based cell–cell adhesion sites (Tachibana et al., 2000). Nectins bind Par-3, a cell polarity protein, which forms a complex with the other cell polarity proteins, aPKC and Par-6, and regulates cell polarization (Ohno, 2001; Takekuni et al., 2003). This binding is mediated between the four amino acids in the cytoplasmic tail of nectins and the PDZ domain of Par-3. Other intracellular binding partners of nectins include PICK1 (protein interacting with PRKCA1) (Reymond, Garrido-Urbani, Borg, Dubreuil, & Lopez, 2005), MUPP1 (multiple PDZ domain protein), and PATJ (Pals1-associated TJ protein) (Adachi et al., 2009). These interactions are required for the regulation of TJ and AJ in epithelial cells. Nectins interact with membrane-palmitoylated protein 3 (MPP3) family members, and the interactions regulate trafficking and processing of nectins (Dudak, Kim, Cheong, Federoff, & Lim, 2011). Nectin-2 and -4 interact with zyxin, and this interaction localizes zyxin to cell–cell adhesion sites (Gregory Call et al., 2011). Nectin-1 and -3 interact with willin, and this interaction localizes willin to cell–cell adhesion sites (Ishiuchi & Takeichi, 2012).

Necls do not directly bind afadin but bind many other proteins. Necl-1 and -2 bind the membrane-associated guanylate kinase family members, Pals2, Dlg3/MPP3, and calcium/calmodulin-associated Ser/Thr kinase, while Necl-2 additionally binds the tumor suppressor gene product DAL1, a band 4.1 family member protein, which connects Necl-2 to the actin cytoskeleton (Fukuhara, Shimizu, Kawakatsu, Fukuhara, & Takai, 2003; Ikeda et al., 2003; Kakunaga et al., 2005; Shingai et al., 2003; Yageta et al., 2002). These interactions may be required to localize Necls to cell–cell adhesion sites. Moreover, the cytoplasmic regions of Necl-2 and -4 bind the protein tyrosine phosphatase PTPN13, which inhibits the heregulin-induced ErbB2/ErbB3 signaling (Kawano et al., 2009; Sugiyama et al., 2013).

3. FUNCTIONAL PROPERTIES

Cell proliferation, differentiation, movement, and adhesion are fundamental cellular processes in the ontogenesis of multicellular organisms. In developing tissues, newly born cells migrate to particular places through directional cell movement and/or random cell movement and then form cell–cell adhesions with neighboring cells to form tissues. Nectins and Necls play important roles in these processes.

3.1. Cell proliferation

Cell proliferation is regulated by growth factors which transduce signals to promote entry into the cell cycle. Necls are involved in the interface between the growth factor signaling and the cell cycle. Necl-5 enhances the PDGF-induced cell proliferation by shortening the period of the G (0)/G(1) phase of the cell cycle (Kakunaga et al., 2004). Necl-5 also enhances the PDGF-induced activation of the Ras–Raf–MEK–ERK pathway and consequently upregulates cyclins D2 and E and downregulates p27Kip1 (Hirota, Irie, Okamoto, Ikeda, & Takai, 2005). Moreover, Necl-5 regulates the VEGF-induced angiogenesis by controlling the interaction of VEGF receptor 2 with integrin $\alpha_v\beta_3$ and the VEGF receptor 2-mediated activation of downstream proangiogenic and survival signals, including Rap1, Akt, and endothelial nitric oxide synthase (Kinugasa et al., 2012). Furthermore, Necl-5 interacts with sprouty2, which prevents it from being tyrosine-phosphorylated by c-Src (Kajita, Ikeda, Tamaru, & Takai, 2007). Sprouty is a negative regulator of growth factor-induced signaling for cell proliferation (Christofori, 2003; Kim & Bar-Sagi, 2004). When sprouty is tyrosine-phosphorylated by c-Src in response to growth factors, it inhibits the growth factor-induced activation of Ras and its downstream Raf–MEK–ERK signaling (Kim & Bar-Sagi, 2004). When Necl-5 is downregulated on the cell surface by *trans*-interaction with nectin-3 at cell–cell adhesion sites, sprouty2 that is released from Necl-5 is phosphorylated by c-Src, and then inhibits the PDGF-induced activation of Ras. This inhibition further suppresses *de novo* synthesis of Necl-5. Thus, this system is at least one of the mechanisms underlying contact inhibition of cell proliferation.

3.2. Cell differentiation

Nectins and Necls regulate cell differentiation and subsequent tissue morphogenesis. For example, as discussed in Section 4.1, nectin-2 and -3 and Necl-2 are localized at the junction between spermatogenic cells and Sertoli cells and regulate differentiation of spermatogenic cells. Newborn nectin-1-deficient mouse pups show shiny and slightly reddish skin. The amount of loricrin, one of the differentiation markers and also a major component of cornified cell envelopes, is markedly reduced in the epidermis of the mutant mice. Thus, nectin-1 is required for the differentiation of epidermis (Wakamatsu et al., 2007). Necl-4 plays a role for Schwann cell differentiation and subsequent myelination *in vitro* (Maurel et al., 2007). In addition, mutations in human *nectin-1* and *-4* are responsible for ectodermal dysplasia syndrome and ectodermal dysplasia–syndactyly syndrome, respectively, as described in Section 6.2. They are autosomal recessive disorders of which symptoms include unusual faces, cleft lip/palate, dental anomalies, syndactyly, and, in some cases, intellectual disability. Thus, nectins also regulate tissue morphogenesis.

3.3. Cell movement

There are two types of cell movement, directional cell movement and random cell movement. An example of directional cell movement is the cell movement toward a source of a chemoattractant. When cells move, they collide with each other, and in most cases, these cells cease their movement in the direction of the collision. This process is called contact inhibition of cell movement. As for the mechanisms, growth factor receptors and integrins are involved in cell movement, and cadherins are involved in contact inhibition. In addition to these molecules, nectins and Necls play roles in these processes (Takai, Miyoshi, et al., 2008). Necl-5 interacts with both integrin $\alpha_v\beta_3$ and the PDGF receptor and stimulates directional cell movement by enhancing both integrin $\alpha_v\beta_3$- and PDGF receptor-induced signaling pathways, including activation of Rac, which regulates the formation of peripheral ruffles and focal complexes (Amano et al., 2008; Minami et al., 2007; Miyata et al., 2009). In addition, Necl-5 attracts growing microtubules to the plasma membrane of the leading edge of moving cells (Minami et al., 2010). Necl-5 is also involved in the contact inhibition of cell movement. When two moving cells collide with each other, Necl-5 on the surface of one cell heterophilically interacts in *trans* with nectin-3, which may be diffusely distributed along the surface of the other cell, initiating the formation of cell–cell junctions (Ikeda et al., 2003). The *trans*-interaction of Necl-5 with nectin-3 is transient; thus, Necl-5 is downregulated and endocytosed from the plasma membrane in a clathrin-dependent manner, leading to reduction of cell movement (Fujito et al., 2005). On the other hand, nectin-3 dissociated from Necl-5 is retained on the plasma membrane and subsequently interacts in *trans* with nectin-1, which most feasibly interacts in *trans* with nectin-3 (Ikeda et al., 2003). Then, cadherin is recruited to the nectin-based cell adhesion sites, eventually establishing AJ (Takai, Ikeda, et al., 2008; Fig. 3). Therefore, the cell–cell contact-induced *trans*-interaction of Necl-5 with nectin-3 and subsequent downregulation of Necl-5 are at least one of the mechanisms of the contact-mediated inhibition of cell movement (Fujito et al., 2005). After establishment of AJ, integrin $\alpha_v\beta_3$ is inactivated, but it continues to colocalize with nectins at AJ (Sakamoto et al., 2008). All members of the nectin

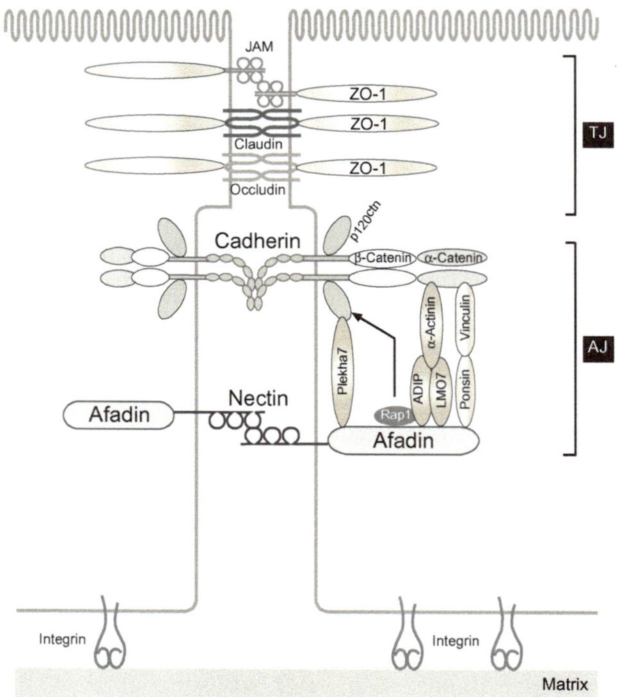

Figure 3 The nectin–afadin complex at cell–cell adhesion of epithelial cells. The nectin–afadin complex localizes at AJ and is connected with the cadherin–catenin complex through the molecular linkages including ponsin–vinculin, α-actinin–ADIP, and α-actinin–LMO7, which are associated with F-actin. Activated Rap1 induces the association of afadin with p120ctn, which increases the *trans*-interacting activity of E-cadherin, leading to stabilization of AJ. Afadin is required for the formation of AJ and TJ. AJ, adherens junction; TJ, tight junction.

family can potentially interact with PTPμ through their extracellular regions, and the *trans*-interactions of nectins enhance its phosphatase activity, leading to inactivation of integrin $\alpha_v\beta_3$ (Sakamoto et al., 2008). In this way, nectins function in the inactivation of integrin $\alpha_v\beta_3$ at AJ. In addition to the cell–cell contact-mediated downregulation of Necl-5, this nectin-induced inactivation of integrin $\alpha_v\beta_3$ provides an additional mechanism for contact inhibition of cell movement.

3.4. Cell adhesion

AJ is one of the adhesion machineries in many mammalian cell types, including epithelial cells, endothelial cells, and fibroblasts. The core molecular components of AJ are the cadherin–catenin and nectin–afadin complexes (Takai, Ikeda, et al., 2008; Takeichi, 1991; Fig. 3). Afadin and nectins are characterized by their strict localization at AJ in epithelial cells, whereas cadherins, α-catenin, β-catenin, and p120^{ctn} are widely distributed along the lateral plasma membrane in epithelial cells (Mandai et al., 1997; Takai, Ikeda, et al., 2008). The direct or indirect binding partners of afadin, such as vinculin, ponsin, and plekha7, are also strictly colocalized at AJ with afadin and nectins (Mandai et al., 1999; Meng, Mushika, Ichii, & Takeichi, 2008), while vinculin and ponsin are also localized at cell–matrix junction. Available evidence indicates that nectins initiate cell–cell adhesion, followed by the recruitment of the complex of E-cadherin and α- and β-catenins to the nectin-based cell–cell adhesion sites through afadin without *trans*-interaction by E-cadherin. Subsequently, recruited E-cadherin molecules *trans*-interact with each other to establish AJ (Honda et al., 2003; Tachibana et al., 2000; Takai, Ikeda, et al., 2008). In the experiments using the inhibitors of nectin-1-based cell–cell adhesions (recombinant proteins comprising either the extracellular region of nectin-3 or glycoprotein D, an envelope protein of herpes simplex virus [HSV] type 1), they prevent the formation of E-cadherin-based cell–cell adhesion in epithelial cells and fibroblasts (Honda et al., 2003). Moreover, afadin knockout and knockdown inhibit the formation of the cadherin-based AJ and TJ as well as cell polarity (Ikeda et al., 1999; Ooshio et al., 2010). On the other hand, in α-catenin knockdown cells, the deficiency of α-catenin inhibits the formation of cadherin-based AJ, but not that of nectin-based AJ (Pokutta, Drees, Takai, Nelson, & Weis, 2002; Tachibana et al., 2000). Thus, the nectin–afadin complex is essential for the initiation and formation of cell–cell adhesion. The association between nectin and cadherin molecules is physically mediated by afadin, α-catenin, and their binding proteins, such as ponsin, LMO7, ADIP, vinculin, α-actinin, and ZO-1 (Takai, Ikeda, et al., 2008). The nectin-induced intracellular signals and the recruitment of the cadherin–catenin system to the nectin-based cell–cell adhesion sites further enhance the recruitment of the nectin–afadin complex to the cell–cell contact sites. This recruitment then enhances the nectin–nectin interaction, which in turn enhances the nectin–afadin interaction. Thus, the recruitment of afadin to the cell–cell contact sites facilitates the formation of AJ in a positive feedback manner (Kurita, Ogita, & Takai, 2011). In this way, the signaling induced by nectins and cadherins facilitates the clustering of the CAMs and their cell–cell adhesion activity. In addition to the formation of AJ and TJ, nectin-1 and -3 regulate the formation of desmosomes (Barron et al., 2008; Yoshida, Miyoshi, Takai, & Thesleff, 2010); however, the role of nectins in the formation of gap junctions remains unknown.

4. DEVELOPMENT

In developing tissues, cells proliferate, differentiate, move, and form cell–cell adhesions between homotypic cells as well as heterotypic cells. As discussed above, cadherins have been characterized as CAMs mediating homotypic cell–cell adhesions, because they interact in *trans* almost exclusively homophilically between the same members (Takeichi, 1991). However, the molecular mechanisms mediating heterotypic cell–cell adhesions had remained elusive for a long time. Nectins and Necls interact in *trans* both homophilically and heterophilically between the same and different members with some exceptions, suggesting that nectins and Necls are good candidate molecules mediating heterotypic cell–cell adhesions. Recently, evidence has been accumulated, showing that nectins and Necls play important roles in tissue development by mediating heterotypic cell–cell adhesions *in vivo*. In this section, our current knowledge of roles for nectins and Necls in various development processes is summarized.

4.1. Spermatogenesis

Spermatogenesis is the process in which spermatozoa are produced from male germ cells. Spermatogenic cells attach to Sertoli cells, which provide structural and metabolic support to the developing spermatogenic cells during differentiation. Nectin-2 and -3 and Necl-2 are required for the interaction between spermatogenic cells and Sertoli cells. Sertoli cells and spermatids express nectin-2 and -3, respectively, both of which are connected to the actin cytoskeleton (Fig. 4A). Afadin is expressed in Sertoli cells;

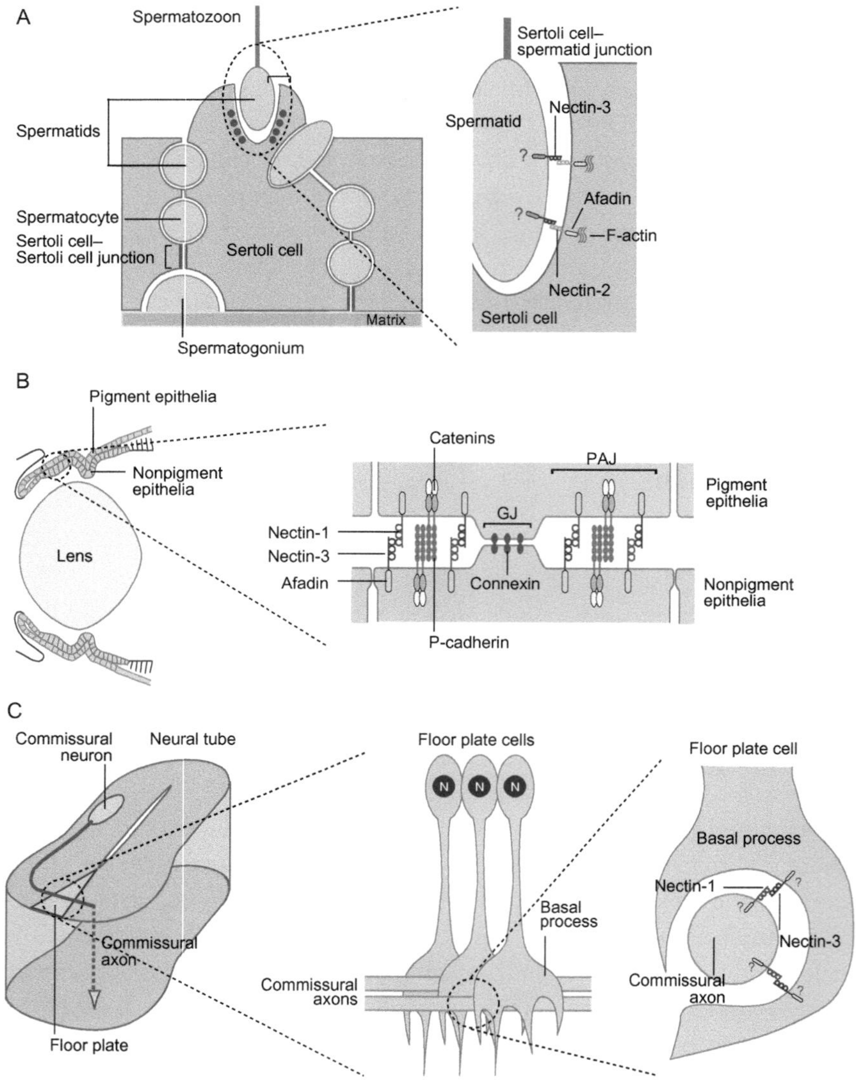

Figure 4 Various types of heterotypic cell–cell adhesion mediated by nectins. (A) Localization of nectin-2 and -3 at Sertoli cell–spermatid junctions in the testis. (B) Localization of nectin-1 and -3 at the contact site between the pigment cell and non-pigment cell layers of the ciliary epithelium in the eye. GJ, gap junction; puncta adherentia junction. (C) Localization of nectin-1 and -3 at the contact site between commissural axons and floor plate cells in the neural tube. F-actin is not shown in (B) and (C). (See the color plate.)

thus, afadin connects nectin-2 to the actin cytoskeleton. However, it is unknown whether afadin is present in the spermatids. Disruption of a *nectin-2* gene results in male infertility (Bouchard et al., 2000). The nectin-2-deficient spermatozoa are morphologically aberrant with defects in nuclear and cytoskeletal morphology and mitochondrial localization (Mueller, Rosenquist, Takai, Bronson, & Wimmer, 2003). Remarkably, they fail to penetrate oocytes. Thus, the nectin–afadin complex plays essential roles in coupling cell–cell adhesion and the cortical actin scaffold at Sertoli–spermatid junctions that is critical for spermatid development and in subsequent sperm morphogenesis (Ozaki-Kuroda et al., 2002). As expected, nectin-3-deficient males also exhibit infertility. The mutant mice have defects in the later steps of sperm morphogenesis including distorted nuclei and abnormal distribution of mitochondria, as well as defects in the localization of nectin-2 at the Sertoli–spermatid junctions, indicating that the heterophilic *trans*-interaction between nectin-2 and -3 is essential for the formation and maintenance of Sertoli–spermatid junctions (Inagaki et al., 2006). Additionally, Necl-2/SgIGSF is exclusively expressed in spermatogenic cells and binds to Sertoli cells *in vitro* (Wakayama et al., 2003). Necl-2/Tslc1-deficient male mice are also infertile, and Necl-2 is required during an early stage of spermatogenesis (van der Weyden et al., 2006).

4.2. Eye development

The eye is derived from the neural tube. The optic vesicle derived from the neural tube develops into an optic cup, which then develops into the retina, ciliary marginal zone, the ciliary epithelium, the pigmented epithelium, and the iris. Nectin-1 and -3 are localized at the apex–apex junctions between the pigment and nonpigment cell layers of the ciliary epithelia. Both the nectin-1-deficient mice and the nectin-3-deficient mice show microphthalmia and separation of the apex–apex adhesion between the pigment and nonpigment epithelia of the ciliary body (Inagaki et al., 2005). The heterophilic *trans*-interaction between nectin-1 and -3 mediates the apex–apex adhesion between the pigment and nonpigment cell layers of the ciliary epithelia, and this nectin-mediated adhesion is essential for the morphogenesis of the ciliary body (Fig. 4B). In addition, transgenic mice expressing a soluble form of the first Ig-like domain of nectin-1, but not that of the entire ectodomain, show microphthalmia and lack of vitreous body (Yoshida et al., 2008). Furthermore, there is a report of a patient with a balanced translocation, 46,XY,t(1;3)(q31.3;q13.13), whose nectin-3 expression is significantly reduced. The patient has severe bilateral congenital cataracts, abnormalities in the central nervous system, and mild developmental delay (Lachke et al., 2012). Thus, *nectin-3* is a critical gene for human ocular development.

4.3. Inner ear development

The organ of Corti of the mammalian inner ear contains sensory hair cells and supporting cells in the auditory sensory epithelia. These cells are arranged to form a checkerboard-like cellular pattern. However, cellular and molecular mechanisms that produce this characteristic arrangement of cells had remained unknown for a long time. In the mouse organ of Corti, hair cells express nectin-1 and supporting cells express nectin-3; in addition, both cells express nectin-2 (Fig. 5). The *trans*-interaction between nectin-1 and -3 mediates the heterotypic adhesion between these two types of cells and establishes the checkerboard-like cellular pattern; this pattern is lost due to aberrant attachment between sensory hair cells in both nectin-1-deficient

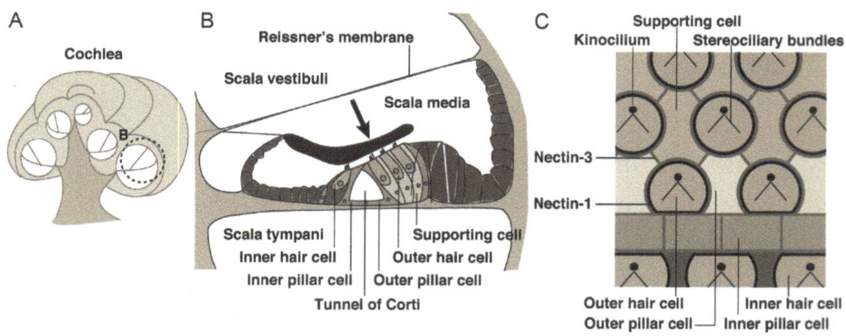

Figure 5 Formation of the checkerboard-like cellular pattern in the organ of Corti. (A) Schematic illustration of the cochlea. The area encircled by dashed dots is enlarged in (B). (B) Cross-sectional illustration of the organ of Corti. The cochlea includes three chambers, and the organ of Corti is located in the scala media. A single row of inner hair cells is located on the medial side of the epithelium, whereas three rows of outer hair cells are located more laterally. The inner and outer hair cell regions are separated by the tunnel of Corti, which is surrounded by single rows of inner and outer pillar cells. An arrow indicates the direction of view shown in (C). (C) Luminal surface illustration of the auditory epithelium in the organ of Corti. Outer hair cells, inner hair cells, and various types of supporting cells are arranged in a checkerboard-like cellular pattern in the auditory epithelium. Nectin-1 and -3 are expressed exclusively by hair cells and supporting cells, respectively. Therefore, the *trans*-interaction between nectin-1 and -3 occurs at the boundaries between these cells. (See the color plate.)

mice and nectin-3-deficient mice (Togashi et al., 2011). Moreover, in the nectin-3-deficient mice, abnormal positioning of the kinocilium and misorientation and dysmorphology of the hair bundles are observed in aberrantly attached sensory hair cells (Fukuda et al., 2014). Thus, the *trans*-interaction between nectin-1 and -3 is critical not only for the checkerboard-like cellular pattern formation but also for positioning of the kinocilium and the morphology and orientation of stereociliary bundle in hair cells.

4.4. Tooth development

Teeth develop from the tooth germ, which is an aggregation of cells derived from the first branchial arch and the neural crest. The tooth germ is composed of the enamel organ, the dental papilla, and the dental follicle. Moreover, the enamel organ is composed of the outer enamel epithelium, inner enamel epithelium, stellate reticulum, and stratum intermedium and gives rise to ameloblasts, which produce enamel and become a part of the reduced enamel epithelium. In developing teeth, nectin-1 and -3 are strongly expressed at the interface between the maturation-stage ameloblasts and the underlying cells of the stratum intermedium (Barron et al., 2008; Yoshida et al., 2010). Nectin-1-deficient mice exhibit defective amelogenesis of their incisor teeth, which are prone to wear and breakage. This defect appears to result from loss of adhesive contact between mature ameloblasts and the underlying stratum intermedium. At this interface in wild-type mice, numerous, large desmosomes are present; however, in the mutant mice, the desmosomes are smaller and less numerous. Thus, nectin-1 regulates desmosome assembly and is required for normal enamel mineralization (Barron et al., 2008). Nectin-1 could function cooperatively with nectin-3 for the desmosome assembly at the interface between ameloblasts and the stratum intermedium. At postnatal day 10, the upper molars of *nectin-1;nectin-3* compound mutants exhibit conical crown shape and retarded enamel maturation. Moreover, the desmosomal junctions between ameloblasts and stratum intermedium are significantly reduced in the compound mutants. However, these phenotypes are not observed in nectin-1-deficient mice. Thus, the heterophilic interaction between nectin-1 and -3 recruits desmosomal junctions, which are required for proper enamel formation (Yoshida et al., 2010). Although the reason why the contradiction occurs between these studies (Barron et al., 2008; Yoshida et al., 2010) is not yet understood, it can be concluded that nectins are required for desmosome assembly and for enamel mineralization.

4.5. Cerebral cortex development

In mouse, from embryonic days (E) 11.5 to E17.5, the dorsolateral neuroepithelium in the rostral neural tube expands. Progenitor cells reside in the layers adjacent to the ventricle and produce the projection neurons of the neocortex. These postmitotic neurons tangentially and radially migrate in the developing neocortex. Nectin-3 is expressed in callosal projection neurons in the layer II/III of the mouse cerebral cortex (Molyneaux et al., 2009). Moreover, Cajal–Retzius cells express nectin-1, whereas radially migrating neurons express nectin-3. By the experiments using knockdown and dominant negative forms of nectins, it is shown that the nectin-1- and -3-mediated *trans*-interaction between Cajal–Retzius cells and migrating neurons critically regulates radial migration of the neurons in concert with reelin and N-cadherin (Gil-Sanz et al., 2013).

4.6. Axon guidance

Axon guidance is mediated by a variety of extracellular guidance cues that include secreted factors, such as netrins and semaphorins, and CAMs, such as cadherins and ephrins. In the developing spinal cord, commissural axons grow toward the ventral midline, cross the floor plate, and then abruptly change their trajectory from the circumferential to the longitudinal axis (Fig. 4C). This axon guidance is mediated by the contacts between commissural axons and the basal processes of floor plate cells. Cadherins do not localize at the contact sites, while the members of nectins and Necls localize there and serve as CAMs. The commissural axons express nectin-1, while the floor plate cells express nectin-3. In the cultured rat hindbrain, the inhibitors of nectin-1-based cell–cell adhesions cause abnormal fasciculation of the commissural axons and impairment of the contacts and induce failure of longitudinal turns of the commissural axons at the contralateral sides. Thus, the *trans*-interaction between nectin-1 and -3 plays an important role in the guidance of the commissural axons (Okabe et al., 2004). In addition to these nectins, Necls are also involved in the guidance of commissural axons. The floor plate cells express Necl-3, while the commissural axons express Necl-2. Knockdown of Necl-3 in one half of the spinal cord induces path-finding error of the commissural axons at the midline in the developing chick spinal cord *in vivo*. Mainly, axons fail to turn rostrally after crossing the floor plate.

The *trans*-interaction between Necl-2 and -3 mediates the turning response in postcrossing commissural axons (Niederkofler et al., 2010). The weak *trans*-interactions between the members of nectins and Necls, instead of the strong adhesion mediated by cadherins, might be advantageous for commissural axons to elongate continuously, while the axons are attached to the floor plate cells.

4.7. Synapse formation

Synapses are asymmetric homotypic cell–cell adhesions. The synapse consists of at least two types of intercellular junctions: synaptic junction and puncta adherentia junction (PAJ). Synaptic junction is accompanied by an active zone and a postsynaptic density, whereas PAJ devoids of these structures. PAJ is rather similar to AJ of epithelial cells and is regarded as a mechanical adhesion site between a presynaptic terminal and a postsynapse. At the mossy fiber–CA3 synapses, synapses formed between mossy fiber terminals and dendrites of CA3 pyramidal neurons, both the synaptic junction and PAJ are highly developed and are remodeled in an activity-dependent manner (Amaral & Dent, 1981). N-cadherin and αN- and β-catenins are localized symmetrically at both the presynaptic and postsynaptic sides of PAJ, whereas nectin-1 and -3 are localized asymmetrically at the presynaptic and postsynaptic sides of PAJ, respectively (Mizoguchi et al., 2002). In both the nectin-1-deficient mice and the nectin-3-deficient mice, the numbers of PAJ at the mossy fiber–CA3 synapses are reduced. In addition, the abnormal mossy fiber trajectory is observed. These results indicate that nectins are involved in the formation of PAJ and the proper mossy fiber trajectory (Honda et al., 2006). Necl-2 also regulates synapse formation *in vivo*. Necl-2/SynCAM 1 is localized to excitatory, but not inhibitory, neurons and mossy fiber–CA3 terminals in the hippocampus. Overexpression of Necl-2/SynCAM 1 in transgenic mice increases the number of excitatory synapses, while loss of Necl-2/SynCAM 1 results in fewer excitatory synapses. Moreover, Necl-2/SynCAM 1 also functions at mature synapses to alter their plasticity by regulating long-term depression (Robbins et al., 2010). Furthermore, the retina in Necl-2/SynCAM 1-deficient mice shows fewer fully assembled, triadic rod ribbon synapses. In addition, rod synapse ribbons are shortened in the mutant mice, and protein levels of Ribeye, a major structural component of ribbons, are reduced in Necl-2/SynCAM 1-deficient retina. Thus, Necl-2/SynCAM 1 regulates the structural and functional integrity of ribbon synapses in the rod visual pathway in the retina

(Ribic, Liu, Crair, & Biederer, 2014). Additionally, Necl-2 interacts with ErbB4 and regulates the development of excitatory synapses via the regulation of ErbB4 activity in GABAergic inhibitory neurons (Yamada et al., 2013).

4.8. Myelination

Myelin is an insulating sheath that wraps the axon of a neuron. Schwann cells and oligodendrocytes supply the myelin for the peripheral nervous system and the central nervous system, respectively. Necl-1 is expressed at the contact sites among axons, their terminals and glial cell processes that cooperatively form axon bundles, synapses, and myelinated axons, and it is likely to serve as a CAM at the nonjunctional cell–cell contact sites of the nervous tissues (Kakunaga et al., 2005). Necl-1 plays an important role in the initiation of axonal myelination process in the central nervous system, as myelination is delayed in the optic nerve and spinal cord in the Necl-1-deficient mice (Park et al., 2008). Necl-3 also acts as a heterotypic CAM, which is localized at the interface between the axon shaft and the myelin sheath and functionally and selectively interacts with oligodendrocytes (Pellissier et al., 2007). Necl-4 in Schwann cells serves as the glial-binding partner for axonal Necl-1, and the interaction between these two CAMs mediates axon–glial contacts during myelination in peripheral nerves. Moreover, Necl-4 plays a role for Schwann cell differentiation and subsequent myelination *in vitro* (Maurel et al., 2007; Spiegel et al., 2007). However, it is reported that Necl-4 is dispensable for axonal myelination in the developing nervous system (Zhu et al., 2013). This discrepancy could be caused by different assay systems used to draw their conclusions. In addition, Necl-4/Cadm4-deficient mice display focal hypermyelination, abnormal axon–glial contact, and redistribution of ion channels along the axon. Necl-4 regulates growth of the myelin unit and organization of the underlying axonal membrane (Golan et al., 2013). Thus, both in the central and peripheral nervous systems, Necls are involved in myelination by mediating heterotypic cell-cell adhesions.

5. DISEASES

Several members of nectins and Necls had been identified as viral receptors, before finding their physiological functions as CAMs. Until now, accumulating evidence shows the relationship of nectins and Necls with human diseases including hereditary ectodermal dysplasia, Alzheimer's disease, autism spectrum disorder, and cancer.

5.1. Viral infection

Viral infection initiates by adhering of virus to its specific receptor protein expressed on cells. Thus, some viruses display distinct tissue tropism. For example, herpes virus and poliovirus show a tropism for neurons. *Nectin-1* and *-2* are isolated as the pvr-related genes and named *PRR1* and *PRR2*, respectively (Eberle, Dubreuil, Mattei, Devilard, & Lopez, 1995; Lopez et al., 1995). However, neither PRR1 nor PRR2 has thus far been shown to serve as a poliovirus receptor. PRR1 and PRR2 were later shown to serve as receptors for α-herpes viruses, facilitating their entry and intercellular spreading, and renamed HveC and HveB, respectively (Geraghty, Krummenacher, Cohen, Eisenberg, & Spear, 1998; Spear, Eisenberg, & Cohen, 2000). Human nectin-1 allows entry of all α-herpes viruses so far tested, including HSV types 1 and 2, pseudorabies virus, and bovine herpes virus type 1 (Geraghty et al., 1998). In HSV disease, the intercellular spreading of the virus significantly contributes to the pathogenesis. The interaction of nectin-1 with afadin increases the efficiency of intercellular spreading, but not the entry, of HSV-1, whereas the E-cadherin–catenin system increases the efficiency of both the entry and intercellular spreading of HSV-1 (Sakisaka et al., 2001). Although herpes virus shows a tropism for neurons, nectin-1 is not essential for infection of the nervous system (Taylor et al., 2007). Nectin-4 is recently identified as the epithelial cell receptor for the measles virus (Muhlebach et al., 2011; Noyce et al., 2011). Moreover, nectin-4 is identified as the epithelial receptor for canine distemper virus and peste des petits ruminants virus (Birch et al., 2013; Noyce, Delpeut, & Richardson, 2013; Pratakpiriya et al., 2012). Canine distemper virus and peste des petits ruminants virus along with measles virus are members of the *Morbillivirus* genus. Nectin-4 interacts with the viral attachment protein with high affinity through its membrane-distal domain. As for Necls, human Necl-5 was originally isolated as a receptor for poliovirus and was named pvr (Koike et al., 1990; Mendelsohn et al., 1989). Poliovirus is the causative agent of an acute disease of the central nervous system, poliomyelitis. Poliovirus is a neurotrophic virus that produces severe lesions selectively in the central nervous system, particularly in the motor neurons. Recently, it is speculated that there are Necl-5/pvr-dependent and -independent poliovirus trafficking pathways toward the central nervous system (Ohka, Nihei, Yamazaki, & Nomoto, 2012). Thus, nectins and Necls are not only CAMs but also viral receptors and play a critical role in the pathogenesis of neurotrophic viral infections. As for nectin-3 and Necl-1 to -4, they have not been identified as viral receptors.

5.2. Ectodermal dysplasia

Mutations in the *nectin-1* gene have been shown to be responsible for an autosomal recessive cleft lip/palate–ectodermal dysplasia syndrome, which has been also called Zlotogora–Ogur syndrome and Margarita Island ectodermal dysplasia (Sozen et al., 2001; Suzuki et al., 2000). It is clinically characterized by cleft lip/palate, hidrotic ectodermal dysplasia, developmental defects of the hands, and, in some cases, intellectual disability (Suzuki, Bustos, & Spritz, 1998). Mutations are located at Trp185 in the second Ig-like domain and at Gly323 in the third Ig-like domain in the *nectin-1* gene, resulting in a nonsense mutation or a frameshift. However, the nectin-1-deficient mice do not develop cleft lip/palate (Barron et al., 2008; Yoshida et al., 2010, 2012). Thus, the exact role for nectin-1 in the formation of lip and palate still remains elusive. On the other hand, in the Tgf-β3-deficient mouse, cleft palate is observed, and the fusion of palatal shelves is rescued by adding Tgf-β3 exogenously into cultures of the mutant palate (Taya, O'Kane, & Ferguson, 1999). Nectin-1 is expressed in the medial edge epithelium; however, this expression pattern is lost in the Tgf-β3-deficient mice, suggesting that nectin-1 is involved in the initial adhesion of the opposing medial edge epithelium (Martinez-Sanz et al., 2008). Thus, nectin-1 is not essential but required for the fusion of palatal shelves. Mutations in human *nectin-4* also cause an ectodermal dysplasia–syndactyly syndrome-1 that is characterized by the combination of hair and tooth abnormalities, alopecia, and cutaneous syndactyly (Brancati et al., 2010). Mutations are located at Thr185Met in the second Ig-like domain and at Arg284Gln and Pro304 in the third Ig-like domain in the *nectin-4* gene. The Arg284Gln mutation results in abnormal splicing, and the Pro304 mutation results in a frameshift. Other mutations, Pro212Arg and Val242Met in the second Ig-like domain, are also reported (Fortugno et al., 2014; Jelani, Chishti, & Ahmad, 2011).

5.3. Alzheimer's disease

Alzheimer's disease is a common form of disease with dementia. Recent three genome-wide association studies have shown significant association of single nucleotide polymorphisms (SNPs) in the *nectin-2/pvrl2* locus with Alzheimer's disease (Harold et al., 2009; Logue et al., 2011; Takei et al., 2009): (1) SNPs in a genomic region encompassing *pvrl2*, *tomm40*, *apoe*, and *apoc1* associate with late onset Alzheimer's disease around the *apoe* (apolipoprotein E) locus in Japanese (Takei et al., 2009); (2) an SNP in the 3′-UTR region of *nectin-2* is one of the 13 genome-wide significant SNPs at the *apoe* locus, which are associated with Alzheimer's disease (Harold et al., 2009); and (3) one of the SNPs in *nectin-2* in the region of the *apoe* gene is significantly associated with Alzheimer's disease after adjusting for the confounding effects of *apoe* genotype in African Americans (Logue et al., 2011). Moreover, nectin-3 is depleted from the stratum lacunosum moleculare in the CA1 area of the hippocampus, in transgenic mice that express wild-type or mutant human protein Tau, either chronically or subacutely (Maurin et al., 2013). Although it has not been shown that the nectin-3 expression is decreased in Alzheimer brain, the decreased expression of nectin-3 in the stratum lacunosum moleculare is an early marker of impaired transport, and eventual synaptic problems, caused by tauopathy. As for the relationship of Necls with Alzheimer's disease, Necl-1 expression is significantly upregulated in pyroglutamate-modified amyloid β-expressing transgenic mice (TBA42 mice; Yang, Wittnam, Zubarev, & Bayer, 2013). These results suggest that nectins and Necl-1 are implicated in pathology of Alzheimer's disease.

5.4. Neurodevelopmental disorder

Necls have been reported to play roles in autism spectrum disorder (ASD) and attention-deficit hyperactivity disorder (ADHD). Mutations in a *Necl-2/Cadm1* gene are found in male Caucasian ASD patients and their family members (Zhiling et al., 2008). These mutations are located at His246Asn and Tyr251Ser in the third Ig-like domain and result in more susceptibility to the cleavage and/or degradation of the Necl-2/Cadm1 protein. Currently, the functions of the third Ig-like domain of Necl-2 are not known. The loss-of-function analyses using Necl-2/Cadm1-deficient mice demonstrate that Necl-2/Cadm1 plays roles in regulating emotional behaviors, social behaviors, and motor performances (Takayanagi et al., 2010). Thus, Necl-2/Cadm1 is implicated in psychiatric diseases such as autism. In addition, disruption of Necl-2/SynCAM 1-dependent astroglial function results in behavioral abnormalities similar to those described in animal model of ADHD (Sandau, Alderman, Corfas, Ojeda, & Raber, 2012). In addition to *Necl-2*,

Necl-3/Cadm2 is also identified as an ASD candidate gene through genome-wide analysis (Casey et al., 2012). Furthermore, *Necl-3/Cadm2* is shown to associate with hyperactivity and hyperactivity/impulsivity by a genome-wide association study for ADHD (Albayrak et al., 2013). However, pathophysiological roles of Necls in these disorders have not been found.

5.5. Stress-related mental disorders

Stress-related mental disorders are induced by acute or prolonged anxiety. To respond to mental stresses, the brain reacts through the autonomic nervous and endocrine systems. Recently, nectin-1 and -3 are shown to be involved in this process, using rodent models. The expression of nectin-3 in the CA3 area of the hippocampus is regulated by chronic stress in a corticotropin-releasing hormone receptor 1-dependent fashion and is associated with spatial memory and dendritic complexity (Wang et al., 2011). In addition, hippocampal nectin-3 is necessary for the effects of stress on memory and structural plasticity (Wang et al., 2013). Thus, the "corticotropin-releasing hormone"–"corticotropin-releasing hormone receptor 1" system interacts with the nectin–afadin complex and cooperatively mediates the stress-induced effects on memory and structural plasticity. In addition, nectin-1 recruitment to the perisynapse in the ventral hippocampus plays an important role in the formation of contextual fear memories (Fantin, van der Kooij, Grosse, Krummenacher, & Sandi, 2013). Thus, nectin-1 and -3 and afadin in the hippocampus can be a new drug target linked to memory.

5.6. Cancer

Recently, evidence has accumulated, showing that nectins and Necls contribute to tumorigenesis. Nectin-2 is overexpressed in clinical breast and ovarian cancer tissues (Oshima et al., 2013). Nectin-4 is also overexpressed in breast carcinoma (Fabre-Lafay et al., 2005; Pavlova et al., 2013), in non-small-cell lung cancers (Takano et al., 2009), and in ovarian cancers (Derycke et al., 2010). *Necl-2/Tslc1* serves as a tumor suppressor in human non-small-cell lung cancer (Kuramochi et al., 2001). Rodent Necl-5/pvr is also shown to be overexpressed in rat and mouse colon carcinoma (Chadeneau, LeMoullac, & Denis, 1994; Chadeneau et al., 1996). Necl-5 is expressed at low levels in many adult tissues, but it becomes overexpressed in many carcinomas (Chadeneau et al., 1994, 1996; Ikeda et al., 2004), including human colorectal carcinoma and malignant glioma (Masson et al., 2001). Necl-5 overexpression correlates with the malignant phenotype in cutaneous melanoma (Bevelacqua et al., 2012) and is crucial for invasiveness of lung cancer cells (Tane et al., 2013). Overexpression of Necl-5 in cancer cells is responsible at least partly for the enhanced motility and proliferation of them (Ikeda et al., 2003, 2004). Interestingly, tumorigenicity is more efficiently suppressed in colon cancer cells overexpressing both Necl-1 and -4 than the cells expressing either of them alone *in vitro* (Raveh, Gavert, Spiegel, & Ben-Ze'ev, 2009). This suppression of tumorigenicity is associated with enhanced rates of apoptosis of the cancer cells. Nectin-2 and Necl-2 and -5 are involved in cancer immunity by modulating signaling in certain T cells and NK cells, as shown below (Boles et al., 2005; Fuchs et al., 2004; Pende et al., 2005; Stanietsky et al., 2009). NK cells recognize Necl-5/pvr through an additional receptor, Tactile/CD96. Tactile/CD96 promotes NK cell adhesion to target cells expressing Necl-5/pvr and stimulates cytotoxicity of activated NK cells. As Necl-5/pvr is highly expressed in certain tumors, this receptor system may be critical for NK cell recognition of tumors through Tactile/CD96 expressed in NK cells (Fuchs et al., 2004). Moreover, the interaction between DNAM-1 expressed in NK cells and Necl-5/pvr or nectin-2 expressed in target cells enhances the NK-mediated lysis of tumor cells (Pende et al., 2005). Furthermore, NK cells and $CD8^+$ T cells recognize Necl-2 through a receptor known as CRTAM, which is expressed only on activated NK cells and T cells. The interaction between CRTAM and Necl-2 promotes cytotoxicity of NK cells and interferon-γ secretion of $CD8^+$ T cells *in vitro* as well as NK cell-mediated rejection of tumors expressing Necl-2 *in vivo*. This is one of the mechanisms of tumor suppression mediated by Necl-2/tslc1 (Boles et al., 2005). TIGIT is expressed in all human NK cells and binds Necl-5/pvr and nectin-2/pvrl2, which are expressed in target cells. Through these interactions, TIGIT inhibits NK cytotoxicity (Stanietsky et al., 2009).

6. CONCLUSIONS AND PERSPECTIVES

In this chapter, we have described the roles of nectins and Necls in development and disease. To understand the mechanisms of their roles in development, the molecular property that most of the nectins and Necls both homophilically and heterophilically *trans*-interact among the members is crucially important. This property is responsible for heterotypic interaction of cells, such as the interaction between Sertoli cells and spermatid, and the interaction of cells in the auditory sensory epithelia that form the checkerboard-like cellular pattern. The property of nectins that they are connected to cadherin–catenin cell adhesion system through afadin is also important (Takai, Ikeda, et al., 2008). Through this molecular linkage, nectins regulate formation of cell-cell adhesion cooperatively with cadherin during development. Nectins and Necls also regulate cell migration and proliferation. In addition to development, nectins and Necls are implicated in pathophysiology of many diseases. Members of nectins and Necls have been reported as viral receptors. Mutations in the nectin genes can be the causes of hereditary ectodermal dysplasia. SNPs in the *nectin-2* gene locus associate with Alzheimer's disease. Moreover, the members of nectins and Necls are involved in development of ASD and many types of cancers. Furthermore, nectin-2 and Necl-2 and -5 are involved in cancer immunity by modulating signaling in certain T cells and NK cells. As introduced in this chapter, the physiological and pathological roles of nectins and Necls will be continuously revealed in future. To accelerate the research furthermore, various mouse models including conditional mutant and knock-in mouse lines will be powerful tools. In addition, various unbiased high-throughput analyses and human genetic studies to identify disease-linked variants of genes will be of great help. These analyses will promote our understanding of *in vivo* functions of nectins and Necls and contribute to translational researches on these molecules.

The remaining fundamental questions concerning the biological functions of nectins and Necls include their specific roles in cell division of progenitor cells, cell differentiation, establishment of planar cell polarity, synaptic plasticity, neural circuit formation, formation of heterotypic cell-cell adhesions, and modulation of ligand–receptor signaling. Moreover, molecular mechanisms of pathophysiological roles for nectins and Necls in the diseases described above are also intriguing.

ACKNOWLEDGMENTS

We thank our colleagues and collaborators for their great contributions and excellent achievements. This work was supported by Grants-in-Aid for Scientific Research from the Japan Society for the Promotion of Science, Grants-in-Aid for Scientific Research on Innovative Areas from the Ministry of Education, Culture, Sports, Science and Technology (MEXT), Japan, and the grant from Japan Foundation for Applied Enzymology.

REFERENCES

Adachi, M., Hamazaki, Y., Kobayashi, Y., Itoh, M., Tsukita, S., & Furuse, M. (2009). Similar and distinct properties of MUPP1 and Patj, two homologous PDZ domain-containing tight-junction proteins. *Molecular and Cellular Biology*, *29*(9), 2372–2389.

Albayrak, O., Putter, C., Volckmar, A. L., Cichon, S., Hoffmann, P., Nothen, M. M., et al. (2013). Common obesity risk alleles in childhood attention-deficit/hyperactivity disorder. *American Journal of Medical Genetics Part B: Neuropsychiatric Genetics*, *162B*(4), 295–305.

Amano, H., Ikeda, W., Kawano, S., Kajita, M., Tamaru, Y., Inoue, N., et al. (2008). Interaction and localization of Necl-5 and PDGF receptor beta at the leading edges of moving NIH3T3 cells: Implications for directional cell movement. *Genes to Cells*, *13*(3), 269–284.

Amaral, D. G., & Dent, J. A. (1981). Development of the mossy fibers of the dentate gyrus: I. A light and electron microscopic study of the mossy fibers and their expansions. *The Journal of Comparative Neurology*, *195*(1), 51–86.

Barron, M. J., Brookes, S. J., Draper, C. E., Garrod, D., Kirkham, J., Shore, R. C., et al. (2008). The cell adhesion molecule nectin-1 is critical for normal enamel formation in mice. *Human Molecular Genetics*, *17*(22), 3509–3520.

Bevelacqua, V., Bevelacqua, Y., Candido, S., Skarmoutsou, E., Amoroso, A., Guarneri, C., et al. (2012). Nectin like-5 overexpression correlates with the malignant phenotype in cutaneous melanoma. *Oncotarget*, *3*(8), 882–892.

Birch, J., Juleff, N., Heaton, M. P., Kalbfleisch, T., Kijas, J., & Bailey, D. (2013). Characterization of ovine Nectin-4, a novel peste des petits ruminants virus receptor. *Journal of Virology*, *87*(8), 4756–4761.

Bojesen, K. B., Clausen, O., Rohde, K., Christensen, C., Zhang, L., Li, S., et al. (2012). Nectin-1 binds and signals through the fibroblast growth factor receptor. *The Journal of Biological Chemistry*, *287*(44), 37420–37433.

Boles, K. S., Barchet, W., Diacovo, T., Cella, M., & Colonna, M. (2005). The tumor suppressor TSLC1/NECL-2 triggers NK-cell and $CD8^+$ T-cell responses through the cell-surface receptor CRTAM. *Blood*, *106*(3), 779–786.

Bottino, C., Castriconi, R., Pende, D., Rivera, P., Nanni, M., Carnemolla, B., et al. (2003). Identification of PVR (CD155) and Nectin-2 (CD112) as cell surface ligands for the human DNAM-1 (CD226) activating molecule. *The Journal of Experimental Medicine*, *198*(4), 557–567.

Bouchard, M. J., Dong, Y., McDermott, B. M., Jr., Lam, D. H., Brown, K. R., Shelanski, M., et al. (2000). Defects in nuclear and cytoskeletal morphology and mitochondrial localization in spermatozoa of mice lacking nectin-2, a component of cell–cell adherens junctions. *Molecular and Cellular Biology*, *20*(8), 2865–2873.

Brancati, F., Fortugno, P., Bottillo, I., Lopez, M., Josselin, E., Boudghene-Stambouli, O., et al. (2010). Mutations in PVRL4, encoding cell adhesion molecule nectin-4, cause ectodermal dysplasia–syndactyly syndrome. *American Journal of Human Genetics*, *87*(2), 265–273.

Casey, J. P., Magalhaes, T., Conroy, J. M., Regan, R., Shah, N., Anney, R., et al. (2012). A novel approach of homozygous haplotype sharing identifies candidate genes in autism spectrum disorder. *Human Genetics*, *131*(4), 565–579.

Chadeneau, C., LeMoullac, B., & Denis, M. G. (1994). A novel member of the immunoglobulin gene superfamily expressed in rat carcinoma cell lines. *The Journal of Biological Chemistry*, *269*(22), 15601–15605.

Chadeneau, C., LeMoullac, B., LeCabellec, M., Mattei, M., Meflah, K., & Denis, M. G. (1996). Isolation and chromosomal location of mE4, a novel murine gene of the immunoglobulin superfamily. *Mammalian Genome*, 7(8), 636–637.

Christofori, G. (2003). Split personalities: The agonistic antagonist Sprouty. *Nature Cell Biology*, *5*(5), 377–379.

Derycke, M. S., Pambuccian, S. E., Gilks, C. B., Kalloger, S. E., Ghidouche, A., Lopez, M., et al. (2010). Nectin 4 overexpression in ovarian cancer tissues and serum: Potential role as a serum biomarker. *American Journal of Clinical Pathology*, *134*(5), 835–845.

Dudak, A., Kim, J., Cheong, B., Federoff, H. J., & Lim, S. T. (2011). Membrane palmitoylated proteins regulate trafficking and processing of nectins. *European Journal of Cell Biology*, *90*(5), 365–375.

Eberle, F., Dubreuil, P., Mattei, M. G., Devilard, E., & Lopez, M. (1995). The human PRR2 gene, related to the human poliovirus receptor gene (PVR), is the true homolog of the murine MPH gene. *Gene*, *159*(2), 267–272.

Fabre-Lafay, S., Garrido-Urbani, S., Reymond, N., Goncalves, A., Dubreuil, P., & Lopez, M. (2005). Nectin-4, a new serological breast cancer marker, is a substrate for tumor necrosis factor-alpha-converting enzyme (TACE)/ADAM-17. *The Journal of Biological Chemistry*, *280*(20), 19543–19550.

Fantin, M., van der Kooij, M. A., Grosse, J., Krummenacher, C., & Sandi, C. (2013). A key role for nectin-1 in the ventral hippocampus in contextual fear memory. *PLoS One*, *8*(2), e56897.

Flicek, P., Amode, M. R., Barrell, D., Beal, K., Billis, K., Brent, S., et al. (2014). Ensembl 2014. *Nucleic Acids Research*, *42*(Database issue), D749–D755.

Fortugno, P., Josselin, E., Tsiakas, K., Agolini, E., Cestra, G., Teson, M., et al. (2014). Nectin-4 mutations causing ectodermal dysplasia with syndactyly perturb the rac1 pathway and the kinetics of adherens junction formation. *The Journal of Investigative Dermatology*, *134*(8), 2146–2153.

Fuchs, A., Cella, M., Giurisato, E., Shaw, A. S., & Colonna, M. (2004). Cutting edge: CD96 (tactile) promotes NK cell–target cell adhesion by interacting with the poliovirus receptor (CD155). *Journal of Immunology*, *172*(7), 3994–3998.

Fujito, T., Ikeda, W., Kakunaga, S., Minami, Y., Kajita, M., Sakamoto, Y., et al. (2005). Inhibition of cell movement and proliferation by cell–cell contact-induced interaction of Necl-5 with nectin-3. *The Journal of Cell Biology*, *171*(1), 165–173.

Fukuda, T., Kominami, K., Wang, S., Togashi, H., Hirata, K., Mizoguchi, A., et al. (2014). Aberrant cochlear hair cell attachments caused by Nectin-3 deficiency result in hair bundle abnormalities. *Development*, *141*(2), 399–409.

Fukuhara, A., Shimizu, K., Kawakatsu, T., Fukuhara, T., & Takai, Y. (2003). Involvement of nectin-activated Cdc42 small G protein in organization of adherens and tight junctions in Madin–Darby canine kidney cells. *The Journal of Biological Chemistry*, *278*(51), 51885–51893.

Galibert, L., Diemer, G. S., Liu, Z., Johnson, R. S., Smith, J. L., Walzer, T., et al. (2005). Nectin-like protein 2 defines a subset of T-cell zone dendritic cells and is a ligand for class-I-restricted T-cell-associated molecule. *The Journal of Biological Chemistry*, *280*(23), 21955–21964.

Geraghty, R. J., Krummenacher, C., Cohen, G. H., Eisenberg, R. J., & Spear, P. G. (1998). Entry of alphaherpesviruses mediated by poliovirus receptor-related protein 1 and poliovirus receptor. *Science*, *280*(5369), 1618–1620.

Gil-Sanz, C., Franco, S. J., Martinez-Garay, I., Espinosa, A., Harkins-Perry, S., & Muller, U. (2013). Cajal–Retzius cells instruct neuronal migration by coincidence signaling between secreted and contact-dependent guidance cues. *Neuron*, *79*(3), 461–477.

Golan, N., Kartvelishvily, E., Spiegel, I., Salomon, D., Sabanay, H., Rechav, K., et al. (2013). Genetic deletion of Cadm4 results in myelin abnormalities resembling Charcot–Marie–Tooth neuropathy. *The Journal of Neuroscience*, *33*(27), 10950–10961.

Gregory Call, S., Brereton, D., Bullard, J. T., Chung, J. Y., Meacham, K. L., Morrell, D. J., et al. (2011). A zyxin–nectin interaction facilitates zyxin localization to cell–cell adhesions. *Biochemical and Biophysical Research Communications*, *415*(3), 485–489.

Harold, D., Abraham, R., Hollingworth, P., Sims, R., Gerrish, A., Hamshere, M. L., et al. (2009). Genome-wide association study identifies variants at CLU and PICALM associated with Alzheimer's disease. *Nature Genetics*, *41*(10), 1088–1093.

Harrison, O. J., Vendome, J., Brasch, J., Jin, X., Hong, S., Katsamba, P. S., et al. (2012). Nectin ectodomain structures reveal a canonical adhesive interface. *Nature Structural & Molecular Biology*, *19*(9), 906–915.

Hirota, T., Irie, K., Okamoto, R., Ikeda, W., & Takai, Y. (2005). Transcriptional activation of the mouse Necl-5/Tage4/PVR/CD155 gene by fibroblast growth factor or oncogenic Ras through the Raf–MEK–ERK–AP-1 pathway. *Oncogene*, *24*(13), 2229–2235.

Honda, T., Sakisaka, T., Yamada, T., Kumazawa, N., Hoshino, T., Kajita, M., et al. (2006). Involvement of nectins in the formation of puncta adherentia junctions and the mossy fiber trajectory in the mouse hippocampus. *Molecular and Cellular Neurosciences*, *31*(2), 315–325.

Honda, T., Shimizu, K., Kawakatsu, T., Yasumi, M., Shingai, T., Fukuhara, A., et al. (2003). Antagonistic and agonistic effects of an extracellular fragment of nectin on formation of E-cadherin-based cell–cell adhesion. *Genes to Cells*, *8*(1), 51–63.

Ikeda, W., Kakunaga, S., Itoh, S., Shingai, T., Takekuni, K., Satoh, K., et al. (2003). Tage4/Nectin-like molecule-5 heterophilically trans-interacts with cell adhesion molecule Nectin-3 and enhances cell migration. *The Journal of Biological Chemistry*, *278*(30), 28167–28172.

Ikeda, W., Kakunaga, S., Takekuni, K., Shingai, T., Satoh, K., Morimoto, K., et al. (2004). Nectin-like molecule-5/Tage4 enhances cell migration in an integrin-dependent, Nectin-3-independent manner. *The Journal of Biological Chemistry*, *279*(17), 18015–18025.

Ikeda, W., Nakanishi, H., Miyoshi, J., Mandai, K., Ishizaki, H., Tanaka, M., et al. (1999). Afadin: A key molecule essential for structural organization of cell–cell junctions of polarized epithelia during embryogenesis. *The Journal of Cell Biology*, *146*(5), 1117–1132.

Inagaki, M., Irie, K., Ishizaki, H., Tanaka-Okamoto, M., Miyoshi, J., & Takai, Y. (2006). Role of cell adhesion molecule nectin-3 in spermatid development. *Genes to Cells*, *11*(9), 1125–1132.

Inagaki, M., Irie, K., Ishizaki, H., Tanaka-Okamoto, M., Morimoto, K., Inoue, E., et al. (2005). Roles of cell-adhesion molecules nectin 1 and nectin 3 in ciliary body development. *Development*, *132*(7), 1525–1537.

Ishiuchi, T., & Takeichi, M. (2012). Nectins localize willin to cell–cell junctions. *Genes to Cells*, *17*(5), 387–397.

Jelani, M., Chishti, M. S., & Ahmad, W. (2011). Mutation in PVRL4 gene encoding nectin-4 underlies ectodermal-dysplasia–syndactyly syndrome (EDSS1). *Journal of Human Genetics*, *56*(5), 352–357.

Kajita, M., Ikeda, W., Tamaru, Y., & Takai, Y. (2007). Regulation of platelet-derived growth factor-induced Ras signaling by poliovirus receptor Necl-5 and negative growth regulator Sprouty2. *Genes to Cells*, *12*(3), 345–357.

Kakunaga, S., Ikeda, W., Itoh, S., Deguchi-Tawarada, M., Ohtsuka, T., Mizoguchi, A., et al. (2005). Nectin-like molecule-1/TSLL1/SynCAM3: A neural tissue-specific immunoglobulin-like cell–cell adhesion molecule localizing at non-junctional contact sites of presynaptic nerve terminals, axons and glia cell processes. *Journal of Cell Science*, *118*(Pt 6), 1267–1277.

Kakunaga, S., Ikeda, W., Shingai, T., Fujito, T., Yamada, A., Minami, Y., et al. (2004). Enhancement of serum- and platelet-derived growth factor-induced cell proliferation by Necl-5/Tage4/poliovirus receptor/CD155 through the Ras–Raf–MEK–ERK signaling. *The Journal of Biological Chemistry*, *279*(35), 36419–36425.

Kanzaki, N., Ogita, H., Komura, H., Ozaki, M., Sakamoto, Y., Majima, T., et al. (2008). Involvement of the nectin–afadin complex in PDGF-induced cell survival. *Journal of Cell Science*, *121*(Pt 12), 2008–2017.

Kawano, S., Ikeda, W., Kishimoto, M., Ogita, H., & Takai, Y. (2009). Silencing of ErbB3/ErbB2 signaling by immunoglobulin-like Necl-2. *The Journal of Biological Chemistry*, *284*(35), 23793–23805.

Kim, H. J., & Bar-Sagi, D. (2004). Modulation of signalling by Sprouty: A developing story. *Nature Reviews. Molecular Cell Biology*, *5*(6), 441–450.
Kinugasa, M., Amano, H., Satomi-Kobayashi, S., Nakayama, K., Miyata, M., Kubo, Y., et al. (2012). Necl-5/poliovirus receptor interacts with VEGFR2 and regulates VEGF-induced angiogenesis. *Circulation Research*, *110*(5), 716–726.
Koike, S., Horie, H., Ise, I., Okitsu, A., Yoshida, M., Iizuka, N., et al. (1990). The poliovirus receptor protein is produced both as membrane-bound and secreted forms. *The EMBO Journal*, *9*(10), 3217–3224.
Kuramochi, M., Fukuhara, H., Nobukuni, T., Kanbe, T., Maruyama, T., Ghosh, H. P., et al. (2001). TSLC1 is a tumor-suppressor gene in human non-small-cell lung cancer. *Nature Genetics*, *27*(4), 427–430.
Kurita, S., Ogita, H., & Takai, Y. (2011). Cooperative role of nectin–nectin and nectin–afadin interactions in formation of nectin-based cell–cell adhesion. *The Journal of Biological Chemistry*, *286*(42), 36297–36303.
Lachke, S. A., Higgins, A. W., Inagaki, M., Saadi, I., Xi, Q., Long, M., et al. (2012). The cell adhesion gene PVRL3 is associated with congenital ocular defects. *Human Genetics*, *131*(2), 235–250.
Liu, J., Qian, X., Chen, Z., Xu, X., Gao, F., Zhang, S., et al. (2012). Crystal structure of cell adhesion molecule nectin-2/CD112 and its binding to immune receptor DNAM-1/CD226. *Journal of Immunology*, *188*(11), 5511–5520.
Logue, M. W., Schu, M., Vardarajan, B. N., Buros, J., Green, R. C., Go, R. C., et al. (2011). A comprehensive genetic association study of Alzheimer disease in African Americans. *Archives of Neurology*, *68*(12), 1569–1579.
Lopez, M., Eberle, F., Mattei, M. G., Gabert, J., Birg, F., Bardin, F., et al. (1995). Complementary DNA characterization and chromosomal localization of a human gene related to the poliovirus receptor-encoding gene. *Gene*, *155*(2), 261–265.
Mandai, K., Nakanishi, H., Satoh, A., Obaishi, H., Wada, M., Nishioka, H., et al. (1997). Afadin: A novel actin filament-binding protein with one PDZ domain localized at cadherin-based cell-to-cell adherens junction. *The Journal of Cell Biology*, *139*(2), 517–528.
Mandai, K., Nakanishi, H., Satoh, A., Takahashi, K., Satoh, K., Nishioka, H., et al. (1999). Ponsin/SH3P12: An l-afadin- and vinculin-binding protein localized at cell–cell and cell–matrix adherens junctions. *The Journal of Cell Biology*, *144*(5), 1001–1017.
Martinez-Sanz, E., Del Rio, A., Barrio, C., Murillo, J., Maldonado, E., Garcillan, B., et al. (2008). Alteration of medial-edge epithelium cell adhesion in two Tgf-beta3 null mouse strains. *Differentiation*, *76*(4), 417–430.
Masson, D., Jarry, A., Baury, B., Blanchardie, P., Laboisse, C., Lustenberger, P., et al. (2001). Overexpression of the CD155 gene in human colorectal carcinoma. *Gut*, *49*(2), 236–240.
Maurel, P., Einheber, S., Galinska, J., Thaker, P., Lam, I., Rubin, M. B., et al. (2007). Nectin-like proteins mediate axon Schwann cell interactions along the internode and are essential for myelination. *The Journal of Cell Biology*, *178*(5), 861–874.
Maurin, H., Seymour, C. M., Lechat, B., Borghgraef, P., Devijver, H., Jaworski, T., et al. (2013). Tauopathy differentially affects cell adhesion molecules in mouse brain: Early down-regulation of nectin-3 in stratum lacunosum moleculare. *PLoS One*, *8*(5), e63589.
Mendelsohn, C. L., Wimmer, E., & Racaniello, V. R. (1989). Cellular receptor for poliovirus: Molecular cloning, nucleotide sequence, and expression of a new member of the immunoglobulin superfamily. *Cell*, *56*(5), 855–865.
Meng, W., Mushika, Y., Ichii, T., & Takeichi, M. (2008). Anchorage of microtubule minus ends to adherens junctions regulates epithelial cell–cell contacts. *Cell*, *135*(5), 948–959.
Minami, Y., Ikeda, W., Kajita, M., Fujito, T., Amano, H., Tamaru, Y., et al. (2007). Necl-5/poliovirus receptor interacts in cis with integrin alphaVbeta3 and regulates its clustering and focal complex formation. *The Journal of Biological Chemistry*, *282*(25), 18481–18496.

Minami, A., Mizutani, K., Waseda, M., Kajita, M., Miyata, M., Ikeda, W., et al. (2010). Necl-5/PVR enhances PDGF-induced attraction of growing microtubules to the plasma membrane of the leading edge of moving NIH3T3 cells. *Genes to Cells*, *15*(11), 1123–1135.

Miyahara, M., Nakanishi, H., Takahashi, K., Satoh-Horikawa, K., Tachibana, K., & Takai, Y. (2000). Interaction of nectin with afadin is necessary for its clustering at cell–cell contact sites but not for its cis dimerization or trans interaction. *The Journal of Biological Chemistry*, *275*(1), 613–618.

Miyata, M., Rikitake, Y., Takahashi, M., Nagamatsu, Y., Yamauchi, Y., Ogita, H., et al. (2009). Regulation by afadin of cyclical activation and inactivation of Rap1, Rac1, and RhoA small G proteins at leading edges of moving NIH3T3 cells. *The Journal of Biological Chemistry*, *284*(36), 24595–24609.

Mizoguchi, A., Nakanishi, H., Kimura, K., Matsubara, K., Ozaki-Kuroda, K., Katata, T., et al. (2002). Nectin: An adhesion molecule involved in formation of synapses. *The Journal of Cell Biology*, *156*(3), 555–565.

Mizutani, K., Kawano, S., Minami, A., Waseda, M., Ikeda, W., & Takai, Y. (2011). Interaction of nectin-like molecule 2 with integrin alpha6beta4 and inhibition of disassembly of integrin alpha6beta4 from hemidesmosomes. *The Journal of Biological Chemistry*, *286*(42), 36667–36676.

Molyneaux, B. J., Arlotta, P., Fame, R. M., MacDonald, J. L., MacQuarrie, K. L., & Macklis, J. D. (2009). Novel subtype-specific genes identify distinct subpopulations of callosal projection neurons. *The Journal of Neuroscience*, *29*(39), 12343–12354.

Momose, Y., Honda, T., Inagaki, M., Shimizu, K., Irie, K., Nakanishi, H., et al. (2002). Role of the second immunoglobulin-like loop of nectin in cell–cell adhesion. *Biochemical and Biophysical Research Communications*, *293*(1), 45–49.

Mueller, S., Rosenquist, T. A., Takai, Y., Bronson, R. A., & Wimmer, E. (2003). Loss of nectin-2 at Sertoli–spermatid junctions leads to male infertility and correlates with severe spermatozoan head and midpiece malformation, impaired binding to the zona pellucida, and oocyte penetration. *Biology of Reproduction*, *69*(4), 1330–1340.

Muhlebach, M. D., Mateo, M., Sinn, P. L., Prufer, S., Uhlig, K. M., Leonard, V. H., et al. (2011). Adherens junction protein nectin-4 is the epithelial receptor for measles virus. *Nature*, *480*(7378), 530–533.

Niederkofler, V., Baeriswyl, T., Ott, R., & Stoeckli, E. T. (2010). Nectin-like molecules/SynCAMs are required for post-crossing commissural axon guidance. *Development*, *137*(3), 427–435.

Noyce, R. S., Bondre, D. G., Ha, M. N., Lin, L. T., Sisson, G., Tsao, M. S., et al. (2011). Tumor cell marker PVRL4 (nectin 4) is an epithelial cell receptor for measles virus. *PLoS Pathogens*, 7(8), e1002240.

Noyce, R. S., Delpeut, S., & Richardson, C. D. (2013). Dog nectin-4 is an epithelial cell receptor for canine distemper virus that facilitates virus entry and syncytia formation. *Virology*, *436*(1), 210–220.

Ohka, S., Nihei, C., Yamazaki, M., & Nomoto, A. (2012). Poliovirus trafficking toward central nervous system via human poliovirus receptor-dependent and -independent pathway. *Frontiers in Microbiology*, *3*, 147.

Ohno, S. (2001). Intercellular junctions and cellular polarity: The PAR–aPKC complex, a conserved core cassette playing fundamental roles in cell polarity. *Current Opinion in Cell Biology*, *13*(5), 641–648.

Okabe, N., Shimizu, K., Ozaki-Kuroda, K., Nakanishi, H., Morimoto, K., Takeuchi, M., et al. (2004). Contacts between the commissural axons and the floor plate cells are mediated by nectins. *Developmental Biology*, *273*(2), 244–256.

Ooshio, T., Kobayashi, R., Ikeda, W., Miyata, M., Fukumoto, Y., Matsuzawa, N., et al. (2010). Involvement of the interaction of afadin with ZO-1 in the formation of tight junctions in Madin–Darby canine kidney cells. *The Journal of Biological Chemistry*, *285*(7), 5003–5012.

Oshima, T., Sato, S., Kato, J., Ito, Y., Watanabe, T., Tsuji, I., et al. (2013). Nectin-2 is a potential target for antibody therapy of breast and ovarian cancers. *Molecular Cancer*, *12*, 60.

Ozaki-Kuroda, K., Nakanishi, H., Ohta, H., Tanaka, H., Kurihara, H., Mueller, S., et al. (2002). Nectin couples cell–cell adhesion and the actin scaffold at heterotypic testicular junctions. *Current Biology*, *12*(13), 1145–1150.

Park, J., Liu, B., Chen, T., Li, H., Hu, X., Gao, J., et al. (2008). Disruption of Nectin-like 1 cell adhesion molecule leads to delayed axonal myelination in the CNS. *The Journal of Neuroscience*, *28*(48), 12815–12819.

Pavlova, N. N., Pallasch, C., Elia, A. E., Braun, C. J., Westbrook, T. F., Hemann, M., et al. (2013). A role for PVRL4-driven cell–cell interactions in tumorigenesis. *Elife*, *2*, e00358.

Pellissier, F., Gerber, A., Bauer, C., Ballivet, M., & Ossipow, V. (2007). The adhesion molecule Necl-3/SynCAM-2 localizes to myelinated axons, binds to oligodendrocytes and promotes cell adhesion. *BMC Neuroscience*, *8*, 90.

Pende, D., Bottino, C., Castriconi, R., Cantoni, C., Marcenaro, S., Rivera, P., et al. (2005). PVR (CD155) and Nectin-2 (CD112) as ligands of the human DNAM-1 (CD226) activating receptor: Involvement in tumor cell lysis. *Molecular Immunology*, *42*(4), 463–469.

Pokutta, S., Drees, F., Takai, Y., Nelson, W. J., & Weis, W. I. (2002). Biochemical and structural definition of the l-afadin- and actin-binding sites of alpha-catenin. *The Journal of Biological Chemistry*, *277*(21), 18868–18874.

Pratakpiriya, W., Seki, F., Otsuki, N., Sakai, K., Fukuhara, H., Katamoto, H., et al. (2012). Nectin4 is an epithelial cell receptor for canine distemper virus and involved in neurovirulence. *Journal of Virology*, *86*(18), 10207–10210.

Raveh, S., Gavert, N., Spiegel, I., & Ben-Ze'ev, A. (2009). The cell adhesion nectin-like molecules (Necl) 1 and 4 suppress the growth and tumorigenic ability of colon cancer cells. *Journal of Cellular Biochemistry*, *108*(1), 326–336.

Reymond, N., Fabre, S., Lecocq, E., Adelaide, J., Dubreuil, P., & Lopez, M. (2001). Nectin4/PRR4, a new afadin-associated member of the nectin family that trans-interacts with nectin1/PRR1 through V domain interaction. *The Journal of Biological Chemistry*, *276*(46), 43205–43215.

Reymond, N., Garrido-Urbani, S., Borg, J. P., Dubreuil, P., & Lopez, M. (2005). PICK-1: A scaffold protein that interacts with Nectins and JAMs at cell junctions. *FEBS Letters*, *579*(10), 2243–2249.

Ribic, A., Liu, X., Crair, M. C., & Biederer, T. (2014). Structural organization and function of mouse photoreceptor ribbon synapses involve the immunoglobulin protein synaptic cell adhesion molecule 1. *The Journal of Comparative Neurology*, *522*(4), 900–920.

Robbins, E. M., Krupp, A. J., Perez de Arce, K., Ghosh, A. K., Fogel, A. I., Boucard, A., et al. (2010). SynCAM 1 adhesion dynamically regulates synapse number and impacts plasticity and learning. *Neuron*, *68*(5), 894–906.

Sakamoto, Y., Ogita, H., Hirota, T., Kawakatsu, T., Fukuyama, T., Yasumi, M., et al. (2006). Interaction of integrin alpha(v)beta3 with nectin. Implication in cross-talk between cell–matrix and cell–cell junctions. *The Journal of Biological Chemistry*, *281*(28), 19631–19644.

Sakamoto, Y., Ogita, H., Komura, H., & Takai, Y. (2008). Involvement of nectin in inactivation of integrin alpha(v)beta(3) after the establishment of cell–cell adhesion. *The Journal of Biological Chemistry*, *283*(1), 496–505.

Sakisaka, T., Taniguchi, T., Nakanishi, H., Takahashi, K., Miyahara, M., Ikeda, W., et al. (2001). Requirement of interaction of nectin-1alpha/HveC with afadin for efficient cell–cell spread of herpes simplex virus type 1. *Journal of Virology*, *75*(10), 4734–4743.

Samanta, D., Ramagopal, U. A., Rubinstein, R., Vigdorovich, V., Nathenson, S. G., & Almo, S. C. (2012). Structure of Nectin-2 reveals determinants of homophilic and heterophilic interactions that control cell–cell adhesion. *Proceedings of the National Academy of Sciences of the United States of America, 109*(37), 14836–14840.
Sandau, U. S., Alderman, Z., Corfas, G., Ojeda, S. R., & Raber, J. (2012). Astrocyte-specific disruption of SynCAM1 signaling results in ADHD-like behavioral manifestations. *PLoS One*, 7(4), e36424.
Satoh-Horikawa, K., Nakanishi, H., Takahashi, K., Miyahara, M., Nishimura, M., Tachibana, K., et al. (2000). Nectin-3, a new member of immunoglobulin-like cell adhesion molecules that shows homophilic and heterophilic cell–cell adhesion activities. *The Journal of Biological Chemistry, 275*(14), 10291–10299.
Shingai, T., Ikeda, W., Kakunaga, S., Morimoto, K., Takekuni, K., Itoh, S., et al. (2003). Implications of nectin-like molecule-2/IGSF4/RA175/SgIGSF/TSLC1/SynCAM1 in cell–cell adhesion and transmembrane protein localization in epithelial cells. *The Journal of Biological Chemistry, 278*(37), 35421–35427.
Sozen, M. A., Suzuki, K., Tolarova, M. M., Bustos, T., Fernandez Iglesias, J. E., & Spritz, R. A. (2001). Mutation of PVRL1 is associated with sporadic, non-syndromic cleft lip/palate in northern Venezuela. *Nature Genetics, 29*(2), 141–142.
Spear, P. G., Eisenberg, R. J., & Cohen, G. H. (2000). Three classes of cell surface receptors for alphaherpesvirus entry. *Virology, 275*(1), 1–8.
Spiegel, I., Adamsky, K., Eshed, Y., Milo, R., Sabanay, H., Sarig-Nadir, O., et al. (2007). A central role for Necl4 (SynCAM4) in Schwann cell–axon interaction and myelination. *Nature Neuroscience, 10*(7), 861–869.
Stanietsky, N., Simic, H., Arapovic, J., Toporik, A., Levy, O., Novik, A., et al. (2009). The interaction of TIGIT with PVR and PVRL2 inhibits human NK cell cytotoxicity. *Proceedings of the National Academy of Sciences of the United States of America, 106*(42), 17858–17863.
Sugiyama, H., Mizutani, K., Kurita, S., Okimoto, N., Shimono, Y., & Takai, Y. (2013). Interaction of Necl-4/CADM4 with ErbB3 and integrin alpha6 beta4 and inhibition of ErbB2/ErbB3 signaling and hemidesmosome disassembly. *Genes to Cells, 18*(6), 519–528.
Suzuki, K., Bustos, T., & Spritz, R. A. (1998). Linkage disequilibrium mapping of the gene for Margarita Island ectodermal dysplasia (ED4) to 11q23. *American Journal of Human Genetics, 63*(4), 1102–1107.
Suzuki, K., Hu, D., Bustos, T., Zlotogora, J., Richieri-Costa, A., Helms, J. A., et al. (2000). Mutations of PVRL1, encoding a cell–cell adhesion molecule/herpesvirus receptor, in cleft lip/palate–ectodermal dysplasia. *Nature Genetics, 25*(4), 427–430.
Tachibana, K., Nakanishi, H., Mandai, K., Ozaki, K., Ikeda, W., Yamamoto, Y., et al. (2000). Two cell adhesion molecules, nectin and cadherin, interact through their cytoplasmic domain-associated proteins. *The Journal of Cell Biology, 150*(5), 1161–1176.
Takahashi, K., Nakanishi, H., Miyahara, M., Mandai, K., Satoh, K., Satoh, A., et al. (1999). Nectin/PRR: An immunoglobulin-like cell adhesion molecule recruited to cadherin-based adherens junctions through interaction with Afadin, a PDZ domain-containing protein. *The Journal of Cell Biology, 145*(3), 539–549.
Takai, Y., Ikeda, W., Ogita, H., & Rikitake, Y. (2008). The immunoglobulin-like cell adhesion molecule nectin and its associated protein afadin. *Annual Review of Cell and Developmental Biology, 24*, 309–342.
Takai, Y., Miyoshi, J., Ikeda, W., & Ogita, H. (2008). Nectins and nectin-like molecules: Roles in contact inhibition of cell movement and proliferation. *Nature Reviews. Molecular Cell Biology, 9*(8), 603–615.

Takano, A., Ishikawa, N., Nishino, R., Masuda, K., Yasui, W., Inai, K., et al. (2009). Identification of nectin-4 oncoprotein as a diagnostic and therapeutic target for lung cancer. *Cancer Research*, *69*(16), 6694–6703.

Takayanagi, Y., Fujita, E., Yu, Z., Yamagata, T., Momoi, M. Y., Momoi, T., et al. (2010). Impairment of social and emotional behaviors in Cadm1-knockout mice. *Biochemical and Biophysical Research Communications*, *396*(3), 703–708.

Takei, N., Miyashita, A., Tsukie, T., Arai, H., Asada, T., Imagawa, M., et al. (2009). Genetic association study on in and around the APOE in late-onset Alzheimer disease in Japanese. *Genomics*, *93*(5), 441–448.

Takeichi, M. (1991). Cadherin cell adhesion receptors as a morphogenetic regulator. *Science*, *251*(5000), 1451–1455.

Takekuni, K., Ikeda, W., Fujito, T., Morimoto, K., Takeuchi, M., Monden, M., et al. (2003). Direct binding of cell polarity protein PAR-3 to cell–cell adhesion molecule nectin at neuroepithelial cells of developing mouse. *The Journal of Biological Chemistry*, *278*(8), 5497–5500.

Tane, S., Maniwa, Y., Hokka, D., Tauchi, S., Nishio, W., Okita, Y., et al. (2013). The role of Necl-5 in the invasive activity of lung adenocarcinoma. *Experimental and Molecular Pathology*, *94*(2), 330–335.

Taya, Y., O'Kane, S., & Ferguson, M. W. (1999). Pathogenesis of cleft palate in TGF-beta3 knockout mice. *Development*, *126*(17), 3869–3879.

Taylor, J. M., Lin, E., Susmarski, N., Yoon, M., Zago, A., Ware, C. F., et al. (2007). Alternative entry receptors for herpes simplex virus and their roles in disease. *Cell Host & Microbe*, *2*(1), 19–28.

Togashi, H., Kominami, K., Waseda, M., Komura, H., Miyoshi, J., Takeichi, M., et al. (2011). Nectins establish a checkerboard-like cellular pattern in the auditory epithelium. *Science*, *333*(6046), 1144–1147.

van der Weyden, L., Arends, M. J., Chausiaux, O. E., Ellis, P. J., Lange, U. C., Surani, M. A., et al. (2006). Loss of TSLC1 causes male infertility due to a defect at the spermatid stage of spermatogenesis. *Molecular and Cellular Biology*, *26*(9), 3595–3609.

Wakamatsu, K., Ogita, H., Okabe, N., Irie, K., Tanaka-Okamoto, M., Ishizaki, H., et al. (2007). Up-regulation of loricrin expression by cell adhesion molecule nectin-1 through Rap1–ERK signaling in keratinocytes. *The Journal of Biological Chemistry*, *282*(25), 18173–18181.

Wakayama, T., Koami, H., Ariga, H., Kobayashi, D., Sai, Y., Tsuji, A., et al. (2003). Expression and functional characterization of the adhesion molecule spermatogenic immunoglobulin superfamily in the mouse testis. *Biology of Reproduction*, *68*(5), 1755–1763.

Wang, X. D., Chen, Y., Wolf, M., Wagner, K. V., Liebl, C., Scharf, S. H., et al. (2011). Forebrain CRHR1 deficiency attenuates chronic stress-induced cognitive deficits and dendritic remodeling. *Neurobiology of Disease*, *42*(3), 300–310.

Wang, X. D., Su, Y. A., Wagner, K. V., Avrabos, C., Scharf, S. H., Hartmann, J., et al. (2013). Nectin-3 links CRHR1 signaling to stress-induced memory deficits and spine loss. *Nature Neuroscience*, *16*(6), 706–713.

Williams, Y. N., Masuda, M., Sakurai-Yageta, M., Maruyama, T., Shibuya, M., & Murakami, Y. (2006). Cell adhesion and prostate tumor-suppressor activity of TSLL2/IGSF4C, an immunoglobulin superfamily molecule homologous to TSLC1/IGSF4. *Oncogene*, *25*(10), 1446–1453.

Yageta, M., Kuramochi, M., Masuda, M., Fukami, T., Fukuhara, H., Maruyama, T., et al. (2002). Direct association of TSLC1 and DAL-1, two distinct tumor suppressor proteins in lung cancer. *Cancer Research*, *62*(18), 5129–5133.

Yamada, A., Inoue, E., Deguchi-Tawarada, M., Matsui, C., Togawa, A., Nakatani, T., et al. (2013). Necl-2/CADM1 interacts with ErbB4 and regulates its activity in GABAergic neurons. *Molecular and Cellular Neurosciences*, *56*, 234–243.

Yang, H., Wittnam, J. L., Zubarev, R. A., & Bayer, T. A. (2013). Shotgun brain proteomics reveals early molecular signature in presymptomatic mouse model of Alzheimer's disease. *Journal of Alzheimer's Disease*, *37*(2), 297–308.

Yoshida, T., Miyoshi, J., Takai, Y., & Thesleff, I. (2010). Cooperation of nectin-1 and nectin-3 is required for normal ameloblast function and crown shape development in mouse teeth. *Developmental Dynamics*, *239*(10), 2558–2569.

Yoshida, M., Shimono, Y., Togashi, H., Matsuzaki, K., Miyoshi, J., Mizoguchi, A., et al. (2012). Periderm cells covering palatal shelves have tight junctions and their desquamation reduces the polarity of palatal shelf epithelial cells in palatogenesis. *Genes to Cells*, *17*(6), 455–472.

Yoshida, K., Tomioka, Y., Kase, S., Morimatsu, M., Shinya, K., Ohno, S., et al. (2008). Microphthalmia and lack of vitreous body in transgenic mice expressing the first immunoglobulin-like domain of nectin-1. *Graefe's Archive for Clinical and Experimental Ophthalmology*, *246*(4), 543–549.

Zhiling, Y., Fujita, E., Tanabe, Y., Yamagata, T., Momoi, T., & Momoi, M. Y. (2008). Mutations in the gene encoding CADM1 are associated with autism spectrum disorder. *Biochemical and Biophysical Research Communications*, *377*(3), 926–929.

Zhu, Y., Li, H., Li, K., Zhao, X., An, T., Hu, X., et al. (2013). Necl-4/SynCAM-4 is expressed in myelinating oligodendrocytes but not required for axonal myelination. *PLoS One*, *8*(5), e64264.

CHAPTER SEVEN

Anchors and Signals: The Diverse Roles of Integrins in Development

Aidan P. Maartens, Nicholas H. Brown[1]
Department of Physiology, Development and Neuroscience, The Gurdon Institute, University of Cambridge, Cambridge, United Kingdom
[1]Corresponding author: e-mail address: nb117@cam.ac.uk

Contents

Abstract

Integrins mediate cell adhesion by providing a link between the actin cytoskeleton and the extracellular matrix. As well as acting to anchor cells, integrin adhesions provide sensory input via mechanotransduction and synergism with signaling pathways, and provide the cell with the conditions necessary for differentiation in a permissive manner. In this review, we explore how integrins contribute to development, and what this tells us about how they work. From a signaling perspective, the influence of integrins on cell viability and fate is muted in a developmental context as compared to cell culture. Integrin phenotypes tend to arise from a failure of normally specified cells to create tissues properly, due to defective adhesion. The diversity of integrin functions in development shows how cell adhesion is continuously adjusted, both within and between animals, to fit developmental purpose.

Current Topics in Developmental Biology, Volume 112
ISSN 0070-2153
http://dx.doi.org/10.1016/bs.ctdb.2014.11.020

1. INTRODUCTION

1.1. Developmental biology: Signaling and mechanics

The core problems of developmental biology—cell fate determination, tissue patterning, morphogenesis—remained opaque in terms of underlying mechanism until the revolution in molecular biology allowed a move beyond a description of the event, and uncovered the gene expression patterns and protein interactions that drive development. The discipline became a "science of arrows" (Gilbert, 1998), with a strong focus on how signaling ligands from one cell could elicit fate changes in another, working through conserved signal transduction pathways that culminated in altered gene expression. Dynamic intercellular signaling provided a mechanism for how patterns could emerge from initially uniform tissues, and how one tissue could induce the fate of another (Freeman & Gurdon, 2002).

The "executive" or "decision-making" side of development is only half the story, however; the initiation of a gene expression program marks the intention, not the act. Of equal importance is the question of how cell fate is transformed into cell form (Larsen & Mclaughlin, 1987; Wieschaus, 1996), and thus an understanding of development also has to include how changes in cell mechanics contribute to the building of bodies. A suite of cell behaviors drives development: individual and collective cell migration; proliferation and apoptosis; neighbor exchanges and shape changes. These behaviors commonly involve regulation of the cytoskeleton and the membrane (Lecuit, Lenne, & Munro, 2011), and are co-ordinated to drive higher order morphogenesis as tissues are sculpted into functional forms.

We thus come to a conception of development as orchestrated by chemical signaling pathways, and executed by "realizators" (Garcia-Bellido, 1975) that act to change cell and tissue form. This model, distinguishing signaling from mechanics, can be augmented when we consider how the mechanical environment of the cell feeds back to influence various aspects of cell behavior (Bershadsky, Balaban, & Geiger, 2003; Mammoto, Mammoto, & Ingber, 2013). The ability to sense changes in force is clearly crucial for any cell to read its environment. The intracellular response to these changes is mediated by a process termed mechanotransduction, which translates changes in force into chemical signals. Mechanotransduction takes many forms, from stretch activated ion channels in the membrane to changes in protein conformation under tension (DuFort, Paszek, & Weaver, 2011).

The mechanical components of a cell's environment can influence cell choices in a similar manner to the chemical components.

Cell adhesion lies at the nexus of signaling and mechanics, as adhesions are both sites of structural connection between the cell and the outside world, and sites of environmental sensing and signaling. The critical importance of cell adhesion to metazoans is self-evident; cells need to stick together to form tissues. Development involves both dynamic changes in adhesion, as seen in transitions between epithelia and mesenchyme (Ciruna & Rossant, 2001), and the formation of stable adhesions that provide tissue integrity, exemplified by *Drosophila* muscle development (Brown et al., 2002; Zervas, Gregory, & Brown, 2001). Adhesion takes two primary forms in animals; direct adhesion between cells mediated by cadherins (Halbleib & Nelson, 2006) and adhesion between the cell and the extracellular matrix (ECM), which is mediated by integrins and the focus of this review.

1.2. The identification of integrins and their exploration in cell culture and whole animal models

Work in the 1960s to 1980s revealed that cells migrating in dishes interact with the substrate at specific sites of the plasma membrane that are also sites of actin attachment on the cytoplasmic side (Abercrombie, 1980). The first proteins identified at these sites were cytoplasmic, for example, vinculin (Burridge & Feramisco, 1980; Geiger, 1979) and the src kinase (Rohrschneider, 1980). In the following years, the transmembrane receptors that bound the ECM, integrins, were identified in multiple cell types and across many species (Hynes, 1987), and subsequently found to mediate adhesion in diverse developmental and pathological contexts in addition to cell migration. The basic architecture of the cell–matrix adhesion was thus revealed; ECM ligand—transmembrane receptor—cytoplasmic protein complex—actin. This architecture, which is also found in cadherin adhesions (although as homophilic adhesion molecules the extracellular ligand is the same transmembrane receptor on the adjacent cell), provides a continuous physical link between the external and internal worlds of the cell, allowing forces generated within the cell to be transmitted to the outside world, and vice versa (Fig. 1).

Integrins function as heterodimers of α and β subunits (for a more detailed account of the structure and biochemistry of integrins, see Campbell & Humphries, 2011). The β subunits can bind to more than one α subunit, and some α subunits can bind to more than one β subunit

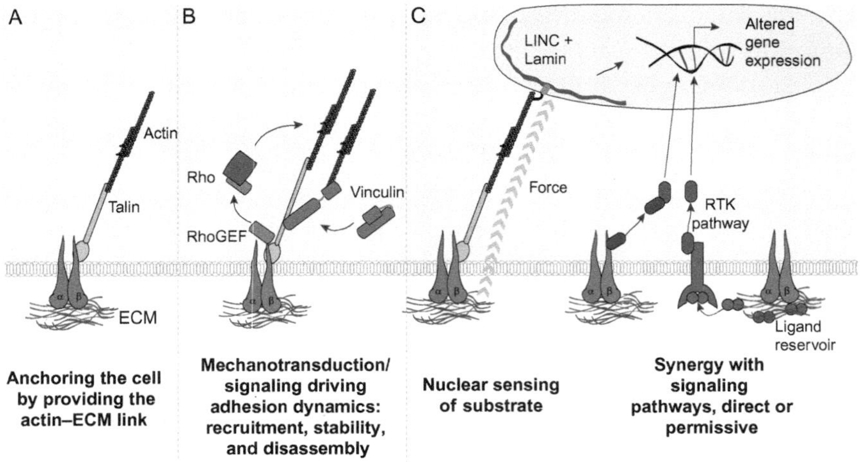

Figure 1 Adhesion architecture, mechanics, and signaling. (A) Basic architecture of an integrin adhesion. (B) Signaling interactions that feed back into adhesion dynamics. RhoGEF recruitment promotes RhoGTPase activation, and a key target of Rho activity is the actin cytoskeleton. Force on the adhesion can also stretch talin, leading to recruitment of vinculin and providing additional links to actin. (C) Integrin influence on gene expression. Actin can connect to the LINC/lamin complex, allowing force transduced through integrins to translate into nuclear deformation and changes in gene expression. Integrins may also recruit proteins that influence signaling pathways, or play a permissive role by maintaining proximity between the cell and an ECM rich in ligands. Abbreviations: ECM, extracellular matrix; LINC, linker of nucleoskeleton and cytoskeleton complex; RTK, receptor tyrosine kinase. (See the color plate.)

(Fig. 2). Different heterodimers provide functional diversity by interacting with distinct ligands, and this can be developmentally regulated by differential expression of the subunits (Meighan & Schwarzbauer, 2008). For example, the *Drosophila* αPS1βPS heterodimer binds to laminin, while the αPS2βPS heterodimer binds to molecules containing the RGD peptide, such as thrombospondin and tiggrin (Fogerty et al., 1994; Gotwals, Fessler, Wehrli, & Hynes, 1994; Subramanian, Wayburn, Bunch, & Volk, 2007). Transcriptional switching between αPS1 and αPS2 is observed at a specific point in egg chamber development, and this switch leads to the recruitment of the adapter protein tensin (Delon & Brown, 2009). The number of subunits per species increases with organismal complexity; worms have two α and one β subunit, flies have five α and two β subunits, while mammals have 18 α and 8 β subunits, which can assemble into 24 heterodimer pairs (Fig. 2; Bouvard et al., 2001; Brown, 2000; Humphries et al., 2006).

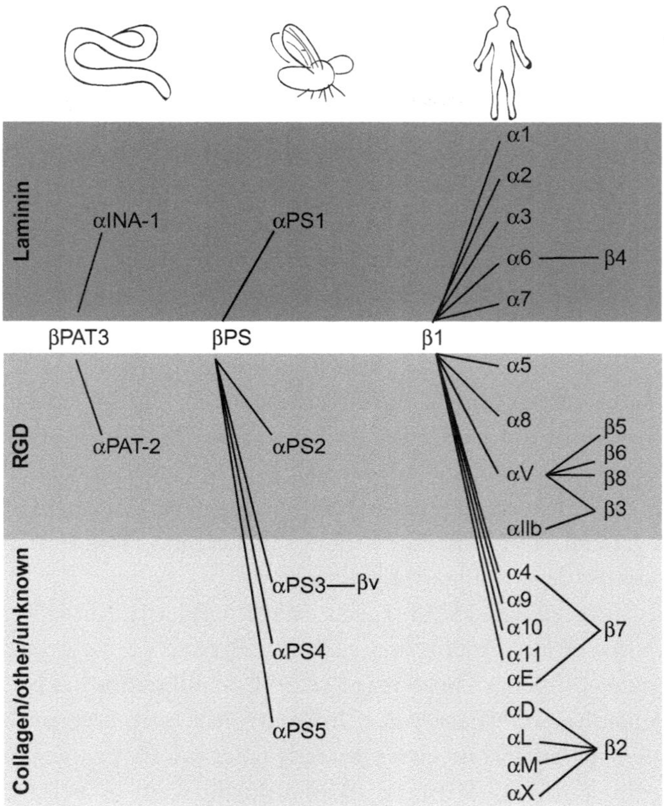

Figure 2 Integrin heterodimer types and ligand binding in worm, fly, and man. Heterodimer pairs are indicated by black lines connecting α and β subunits. Primary ligand binding is split into laminin-binding (green (dark gray in the print version)), RGD-binding (blue (gray in the print version)), and binding to collagen, other ligands such as other cell adhesion molecules, and currently unknown factors (cream (light gray in the print version)). Main β subunits can make multiple heterodimers, while α subunits are less promiscuous. *Adapted from Humphries, Byron, and Humphries (2006) and Bouvard et al. (2001).*

Integrins bind a variety of proteins with both their extracellular and intracellular domains, and can also bind "in *cis*" to other transmembrane proteins such as growth factor receptors. The extracellular ligands are predominantly ECM proteins such as laminin, fibronectin, and collagen. Originally considered mere intercellular filler, the ECM, a highly crosslinked meshwork of insoluble proteins, is now known to contribute to multiple aspects of cell and tissue biology during development (Brown, 2011; Rozario &

DeSimone, 2010). Integrins not only bind ECM proteins but also help assemble them into functional, organized matrices. For example, the assembly of a fibronectin matrix from component dimers depends on integrin binding both in cell culture (Akiyama, Yamada, Chen, & Yamada, 1989) and in embryos (Koshida et al., 2005; Marsden & DeSimone, 2001). An ordered ECM can act as a tract for cell migration, define tissue boundaries, and provide physical instructive cues to the cell (Brown, 2011; Rozario & DeSimone, 2010). The demonstration that the ECM can move in concert with moving cells during avian primitive streak formation (Zamir, Rongish, & Little, 2008) and in *Xenopus* extracts (Davidson, Dzamba, Keller, & Desimone, 2008), and that the basement membrane can slide to facilitate tissue interactions in worms (Ihara et al., 2011), emphasizes the dynamic nature of the ECM; we cannot always think of it as a stationary reference point for cell behaviors. As well as functions mediated through integrins, the ECM can interact with numerous other membrane receptors, provide a general, supportive function for tissues, and provide a reservoir of signaling molecules to influence cell fate.

On the intracellular side, integrins do not bind actin directly, but indirectly via some of the proteins they recruit with their short cytoplasmic tails. The dynamic complex of proteins at cell–ECM adhesions has been termed the integrin adhesome (Zaidel-Bar, Itzkovitz, Ma'ayan, Iyengar, & Geiger, 2007). A crucial adhesome component is talin, which can simultaneously bind the cytoplasmic tail of integrin β subunits and actin, as well as mediating recruitment of a number of other proteins to the adhesion (Critchley, 2009). Depleting talin in fibroblasts leads to the failure of normal adhesion formation *in vitro* (Zhang et al., 2008), and talin loss causes phenotypes that are identical to loss of βPS-containing integrins in *Drosophila* (Brown et al., 2002) and β1-containing integrins in mouse skeletal muscle (Conti, Monkley, Wood, Critchley, & Muller, 2009), suggesting that all integrin adhesive functions require talin.

Aside from proteins with actin binding or regulating activity, the intracellular adhesome also contains proteins that function as scaffolds, promoting interactions between other proteins (for example, paxillin (Deakin & Turner, 2008)), and proteins that enzymatically modify proteins to change their activity, such as kinases (for example, focal adhesion kinase (Mitra, Hanson, & Schlaepfer, 2005)). The precise attribution of function to any single protein in the adhesome can be confounded by the degree of interconnectivity (Zaidel-Bar et al., 2007). In general, integrins only provide a functional output when they are bound to proteins on both sides of the plasma membrane (though there may be exceptions to this rule; for example, unliganded integrins can play a role in promoting cell death (Stupack, Puente, Boutsaboualoy, Storgard, & Cheresh, 2001)). We can thus think of the "functional unit" as an integrin bound to other proteins in both intracellular and extracellular environments.

The ease with which large integrin adhesions can be observed and perturbed in cell culture led to such studies dominating integrin research. The role of integrins and their many associated proteins in coordinating the movement of migrating cells—commonly fibroblasts migrating on two-dimensional substrates such as gels or artificial matrices—has become one of the most intensively studied processes in cell biology (Huttenlocher & Horwitz, 2011). Other modes of cell migration have also been revealed in different cell types, utilizing distinct cytoskeletal, membrane and adhesion mechanisms, and indeed requirements for integrins; the 2D fibroblast model is, clearly, not universally applicable to cell migration (Friedl, Sahai, Weiss, & Yamada, 2012; Huttenlocher & Horwitz, 2011). Furthermore, many migration events *in vivo* are collective, and through complex three-dimensional environments, such that the physical requirements and constraints might differ considerably (Friedl & Gilmour, 2009). Finally, the way that integrins work in cell migration may differ to its multitude of other roles in tissue construction, where the adhesion turnover will be a lot slower than that seen in quickly migrating cells (Bokel & Brown, 2002). For example, muscle development in *Drosophila* involves the progressive stabilization of adhesions, as seen by reductions in protein mobility, in response to increases in force from muscle contraction (Pines et al., 2012; Yuan, Fairchild, Perkins, & Tanentzapf, 2010). Thus, although cells migrating in culture have taught us a lot about how integrins work, the lessons need not all be universal.

Alongside work on cells in culture, studies that addressed the role of integrins during development also mushroomed, with analysis of genetic knockouts or knockdowns in a number of model organisms (Brower, 2003; Wickstrom, Radovanac, & Fassler, 2011), and, in certain models, tissue-specific knockouts. As we will discover, integrin adhesion regulates a whole host of developmental processes, with both similarities and distinctions between the ways integrins work in different developmental contexts within and between model organisms, and indeed between studies using cells in culture versus those in intact animals (Bokel & Brown, 2002). Developmental defects arising from integrin loss include defects in morphogenesis (tissues are not formed properly), structural defects (tissue integrity is lost), and in some cases, alterations to basic cellular properties (cells do not divide, survive, or differentiate properly).

1.3. Integrins: Signaling and mechanics

The discovery that integrin adhesions contained both actin-binding proteins and proteins like kinases led to the current model that integrins perform two key functions; first, provide a mechanical link between the ECM and cytoskeleton, anchoring the cell and transducing force; and second, influence cell behavior by regulating cell signaling (Fig. 1). The signaling side of integrin function has many component parts, depending on how restrictive one wishes to be with the definition of a signal. A strict definition may confine this to effects on signal transduction pathways that terminate in the nucleus. This can happen by a number of means, from "*cis*" interactions with membrane receptors enhancing their activation, to the recruitment and activation of adhesome components that themselves influence a given pathway (Streuli & Akhtar, 2009). A broader definition of signaling might involve all of the protein interactions that do not play a direct role in linking to the cytoskeleton, for example, recruitment cascades. The complexity is that many of these interactions may feed back onto the mechanical link. For example, adhesion signaling crucially impinges on RhoGTPases, multifunctional proteins that play a role multiple cell behaviors (Huveneers & Danen, 2009). Among their primary targets are regulators of the actin cytoskeleton, and hence they are important in regulating the mechanical link provided by the adhesion; signaling feeds into mechanics, and vice versa. Signaling can also control local features of adhesion within the cell, such as adhesion polarity and localized disassembly. In a sense, any signaling aspect of integrin function is necessarily downstream of the anchorage; you cannot have a signaling platform without the ECM–actin link.

As determined *in vitro*, both integrin and cadherin adhesions are sites of mechanotransduction (le Duc et al., 2010; Ross et al., 2013); the structures that bear force in the cell play a key role in sensing that force (Schwartz, 2010). One of the key responses of adhesions to increases in tension is reinforcement, which can be driven by changes in protein recruitment and cytoskeleton dynamics, and allows the adhesion to withstand force across it. While integrins themselves may behave as mechanosensors—force influences the strength of interaction between $\alpha5\beta1$ and fibronectin, for example (Friedland, Lee, & Boettiger, 2009)—certain adhesome proteins certainly do. Force across talin leads to the exposure of cryptic vinculin-binding sites and subsequent binding of vinculin (del Rio et al., 2009), and vinculin itself regulates protein composition in response to changes in tension across the adhesion via a conformational change (Carisey et al., 2013).

Aside from shoring up the adhesion in response to stress, integrin-mediated mechanotransduction may influence cell fate choices. Integrins allow stem cells to sense the stiffness of the matrix and make decisions on when and how to differentiate accordingly (Engler, Sen, Sweeney, & Discher, 2006; Trappmann et al., 2012); the mechanical properties of the cell's environment can thus play a similarly inductive role as chemical signals. In the context of development, mechanical forces have been shown to both augment cell signaling to reinforce cell fate decisions, and even in some cases play a primary instructive role (Miller & Davidson, 2013). Here, integrins could be crucial in mediating mechanical induction, but examples from development are limited.

Thus, during development, phenotypes following integrin removal could result from putative mechanical or signaling functions of integrins, or both. Do we see integrins primarily as anchors, or as "interactive information interfaces" (Geiger & Yamada, 2011) that allow the cell to sense and influence its environment? In this review, we take examples of integrin function from the development of various model organisms—mice, worms, flies, zebrafish, chickens, frogs—and consider how they give us a better understanding of how integrins work. Given the depth of the literature we do not hope to be comprehensive, rather to identify threads that link functions in different organisms, and give a picture of the diversity of processes in which integrins are implicated. We start by asking whether integrins regulate basic cellular properties of viability and fate, and then move on to some of the morphogenetic processes influenced by integrin function.

2. INTEGRINS IN CHOICES OF LIFE AND DEATH: ANIMAL LETHAL, NOT ALWAYS CELL LETHAL

Metazoan development requires integrins; mutations in some or all of the α or β subunits are lethal in flies (reviewed in Brown, 2000), worms (Williams & Waterston, 1994), mice (reviewed in Bouvard et al., 2001), and zebrafish (Julich, Geisler, Holley, & Tübingen 2000 Screen Consortium, 2005), while nongenetic reductions in integrin function perturb development in frogs (Marsden & DeSimone, 2001), chickens (Rallis,

Pinchin, & Ish-Horowicz, 2010; Tucker, 2004), and sea urchins (Marsden & Burke, 1998). Mutations in the core components of the ECM and the adhesome are also lethal or deleterious in various contexts.

The fact that integrins are required for the development of multiple species is consistent with the high degree of conservation of integrin and adhesome genes across metazoa (Hynes & Zhao, 2000). High conservation emphasizes the utility of model organisms for adhesion research. It also raises the possibility that integrin adhesion was a key invention that facilitated metazoan multicellularity. However, the expansion of genome sequencing quickly challenged this idea; integrins are found in the genomes of the closest eukaryotic relatives of the metazoa (Sebe-Pedros et al., 2013; Suga et al., 2013), suggesting that the role of integrins in tissue construction was co-opted from a previous role(s) in nonmetazoan eukaryotes. Indeed, many adhesome components predate the integrins, as seen in the slime mold *Dictyostelium discoideum*, which has talin and paxillin but no integrin (Bukharova et al., 2005; Kreitmeier, Gerisch, Heizer, & Mullertaubenberger, 1995). While many of the closest eukaryotic relatives of the metazoa live unicellular lives, others have been shown to transition to simple forms of multicellularity, either by cell division as in the choanoflagellate *Salpingoeca rosetta* (Fairclough, Dayel, & King, 2010), or aggregation as in the filasterian *Capsaspora owczarzaki* (Sebe-Pedros et al., 2013). Thus, the ancestral integrin may have played a role in the transition from unicellular to colonial lifestyles (Richter & King, 2013). Alternatively, some other function among its myriad roles in metazoan development may represent the ancestral role, for example, cell migration (Huttenlocher & Horwitz, 2011), phagocytosis (Dupuy & Caron, 2008), or ciliogenesis (Antoniades, Stylianou, & Skourides, 2014). Defining this function may help our understanding of integrins in metazoan development.

The lethality of integrin mutants raises the question of whether integrin adhesion is required for cell viability in general. In early cell culture studies, epithelial cell detachment from the matrix was shown to promote a form of programmed cell death termed anoikis (Frisch & Francis, 1994). Anoikis provides a mechanism to prevent detached cells from ectopically reintegrating into other tissue sites, and hence its role in cancer metastasis is a keen area of study (Chiarugi & Giannoni, 2008). That most cells cannot survive when in suspension is testament to the importance of anchorage to a substrate-to-cell function (Discher, Janmey, & Wang, 2005). In a distinct process, termed integrin-mediated death, unliganded integrins promote cell death via caspase recruitment (Stupack et al., 2001). Both cases implicate integrins in providing a survival signal to cells, and the loss of suitable ECM binding leads to the loss of this signal, initiating apoptosis. However, the role of anoikis or integrin-mediated death during development is unclear, as is their relationship to other ways integrins can influence cell viability, for example, by crosstalk with growth factors. As discussed below, what is clear is that in some developmental contexts loss of integrins increases cell death, while in others it does not.

Broad loss of integrin function can be achieved by knocking out the β subunits, as a single β typically forms multiple heterodimers, while single α subunits typically form one. Mice with a null mutation in the gene encoding β1 fail to implant and arrest very early in development (Fassler & Meyer, 1995), complicating efforts to investigate its role in cell viability (although embryoid bodies and new *in vitro* culturing techniques can help in this regard (Bedzhov & Zernicka-Goetz, 2014; Liu et al., 2009)). Researchers can thus turn to α subunit knockdowns, or reduction of β1 activity in a tissue-specific manner. For example, perturbation of α4β1 function leads to increased apoptosis of neurons in the developing chick retina (Leu et al., 2004) and neural crest (Testaz & Duband, 2001). Loss of α*5* in mice led to apoptosis specifically in the neural crest cells, even though the mesoderm showed the most drastic phenotypes (Goh, Yang, & Hynes, 1997); in this case, different cell types show different apoptosis sensitivity following integrin loss, in a manner not necessarily correlating with phenotypic severity. Knockdown of zebrafish α5 caused apoptosis in both the hyoid arches of the head skeleton (Crump, Swartz, & Kimmel, 2004), and the otic (but not trigmental) placode (Bhat & Riley, 2011). In the otic placode, providing increased FGF8 rescued cell viability, but not defects in cell migration (Bhat & Riley, 2011); in this case, normal adhesion is not a prerequisite for viability. While these studies are complicated by issues of subunit redundancy, they show how when apoptosis is increased, it is not total or universal. This supports the notion that integrins act as a "rheostat," modulating or enhancing the effects of dedicated survival pathways without providing a definitive or necessary signal (Streuli & Akhtar, 2009).

Cells isolated from the murine mammary gland require contact with a basement membrane for viability, and function-blocking integrin antibodies suppress the "viability" signal provided by matrix contact (Boudreau, Sympson, Werb, & Bissell, 1995; Pullan et al., 1996), consistent with an anoikis mechanism. However, when β1 integrin was genetically removed from the mouse mammary epithelium *in vivo* (Naylor et al., 2005), or in three-dimensional primary cultures (Akhtar & Streuli, 2013), no increase in apoptosis was seen. Thus, the viability of epithelial cells cultured on two-dimensional substrates is more sensitive to loss of integrins. A similar scenario is seen in *Drosophila*, where loss of epithelial-ECM contact on the whole does not lead to cell death. Large mutant clones lacking the βPS subunit can be generated in the wing and follicle cells, showing that cell viability is not dramatically affected (Brabant, Fristrom, Bunch, & Brower, 1996; Brower & Jaffe, 1989; Devenport & Brown, 2004). In these cases (and in the case of the murine mammary epithelium), integrin null cells appear to remain in contact with one another, and thus the situation is not representative of the culture dish, where nonadherent cells float off into the media and die. Anoikis is counteracted by other systems of adhesion that maintain cell-to-cell contacts and a minimal level of epithelial integrity in the absence of integrins.

In other developmental contexts, loss of integrins does not affect cell survival. In worms and flies, while loss of the β subunit leads to various defects in development and embryonic lethality, global upregulation of apoptosis has not been reported. In mice lacking both α3 and α6 integrin subunits, various limb defects arose, including the failure to separate the digits coincident with a reduction in the apoptosis that normally separates digits (De Arcangelis, Mark, Kreidberg, Sorokin, & Georges-Labouesse, 1999). Knocking out both α5 and αV integrins in zebrafish reduced proliferation but did not induce apoptosis (Dray et al., 2013). Finally, loss of β1 integrin in the neural crest precursors of the enteric nervous system led to a failure of migration and gut colonization, but no increase in apoptosis (Breau et al., 2006). In these cases, to quote Breau et al., "the complex environment *in vivo* probably counteracts the apoptotic effect due to the loss of β1 integrins *in vitro*." The stripped down *in vitro* environment thus sharpens cellular sensitivity to integrin loss.

Recently, a noncanonical link between integrins and apoptosis has been reported from worms. Programmed cell death is a widespread and key component of development, allowing the sculpting of tissues as well as getting rid of unwanted tissues once they have fulfilled their purpose (Fuchs & Steller, 2011). Worms lack professional phagocytes, and dying cells are engulfed by neighboring cells in the tissue (Reddien & Horvitz, 2004). Loss of either of the two α integrin subunits (αINA-1 or αPAT-2) resulted in an increase in the number of cell corpses seen early in development (Hsieh, Hsu, Jiang, & Wu, 2012; Hsu & Wu, 2010). This does not reflect a requirement for integrins in cell viability, however, rather a role for integrins in the engulfment of apoptotic cells during development; integrins were enriched in the membrane protrusions that engulf the dying cell, and in their absence, engulfment was substantially delayed (Hsieh et al., 2012; Hsu & Wu, 2010). The extracellular domain of either α subunit was shown to cluster on apoptotic cells when overexpressed, but what the integrins are recognizing on the target cell has yet to be determined. The involvement of integrins in phagocytosis in a number of other contexts (Dupuy & Caron, 2008) raises the possibility that it might reflect an ancestral function, and one that may have premetazoan origins. Ancient integrins would thus be cast as cellular sensors, rather than cell anchors. Additionally, this study suggests integrins may contribute to corpse clearance, rather than cell viability, in other contexts in which apoptosis has been reported.

Thus, in many developmental contexts, a dedicated signal from integrins to cell survival has not been found.

3. INTEGRINS AND CELL FATE CHOICES

3.1. Mechanotransduction of matrix properties by stem cells

Metazoan development involves the segregation of cells into distinct cell fates, and the subsequent morphological differentiation of the cells to construct different tissues. Although, the role of integrins in the latter process is clear, for example, by adhering different cell types together via an intervening matrix, how integrins influence cell fate choices is less certain. Instructive induction is canonically mediated by extracellular ligands from a variety of peptide families, for example, WNT, Hedgehog, Notch, and TGFβ. The idea that matrix properties, transduced by integrins, play an instructive role is gaining traction, particularly in the stem cell field (Watt & Huck, 2013). Mesenchymal stem cells can be coaxed down distinct differentiation pathways (neurogenic, myogenic, osteogenic) solely by altering the stiffness of the substrate on which they are cultured (Engler et al., 2006). Substrate stiffness is sensed by integrins binding to the ECM and pulling on it, and the resultant mechanical feedback plays a key role in the fate choices taken by the stem cell (Trappmann et al., 2012).

Within the cell, mechanotransduction pathways can run to the nucleus via effects on signaling cascades such as ERK/MAPK (Trappmann et al., 2012), or more directly via transmitting tension to the nuclear lamin proteins, leading to alterations in nuclear stiffness and architecture and consequently changes to gene expression (Swift & Discher, 2014; Swift et al., 2013).

The latter case is intriguing because a cytoskeletal link between the adhesion and the nucleus can elicit a much quicker change than chemical signaling cascades (Mammoto et al., 2013); this more "direct" mechanotransduction may help the cell respond quickly to environmental alterations. The cell also need not be passive with respect to its environment; mesenchymal stem cells can alter their ECM environment via expression of matrix metalloproteinases, and the altered environment then feeds back into integrin activation and increased cellular tension via RhoGTPases (Tang et al., 2013). This then influences the nuclear translocation of the YAP/TAZ transcriptional regulators, resulting in altered gene expression to promote osteogenic differentiation (Tang et al., 2013). In this study, the loss of matrix remodeling *in vivo* leads to a loss of osteogenesis as stem cell differentiation is biased toward adipogenesis and chondrogenesis. These cases provide tantalizing examples of how integrins sense the mechanics of the environment and transmit this information to the nucleus to influence cell fate. The importance of such mechanisms during development has yet to be extensively investigated.

3.2. Stem cells *in vivo*: Niche integrity via anchorage

Integrins are also implicated in multiple stages of stem cell biology in developing organisms. Stem cells reside in niches, a catch-all term to describe the environment that maintains the stem cell in its undifferentiated state; daughter cells that leave the niche progress to differentiate. Niche architecture relies on adhesion between the stem cell and either the ECM directly or supporting cells that themselves are attached to the ECM (Morrison & Spradling, 2008). Perturbation of adhesion leads to defective niche architecture and hence defective stem cell biology (Chen, Lewallen, & Xie, 2013; Marthiens, Kazanis, Moss, Long, & Ffrench-Constant, 2010).

Work in *Drosophila* has uncovered roles for integrins in stem cell biology in three contexts: the testes, the ovary, and the intestine. In the testes, two populations of stem cells, germline, and somatic, contact a stromal niche comprised of hub cells, and after division stem identity is retained by the daughter that contacts the hub (Ellis & Tanentzapf, 2010). The hub itself is anchored to the anterior tip of the testes by integrin binding to surrounding ECM and, in the absence of integrins, the hub is mispositioned (Tanentzapf, Devenport, Godt, & Brown, 2007). In the absence of the niche, there is an increase in stem cell differentiation eventually leading to a depletion of the stem cell pool. A similar, albeit weaker, failure of tip cell adhesion is seen following loss of Lasp (Lee, Zhou, Kim,

Kalbfleisch, & Schock, 2008), an actin-binding protein and a component of focal adhesions in cell culture (Lin et al., 2004), which thus may be part of the machinery that, within the cell, links ECM-bound integrins to actin. Niche architecture is established and maintained by integrin anchorage of supporting cells.

In the fly ovary, follicle stem cells give rise to the follicular epithelium, a tissue that surrounds and shapes the developing oocyte (Bilder & Haigo, 2012). In this case, the stem cells are anchored directly to the ECM, and in the absence of integrins, lose contact with it and do not self renew (O'Reilly, Lee, & Simon, 2008). Stem cells in this niche are influenced by a number of signaling pathways (Vied, Reilein, Field, & Kalderon, 2012), and thus, in the absence of integrins, this cocktail of signals is not received properly. In both examples from the fly gonads, integrins ensure stem cells are in the right place to receive self-renewal signals; cell fate *per se* is not directly regulated by integrins.

In the *Drosophila* intestine, stem cells contact the basement membrane directly and continuously replenish the epithelium in which they reside. Knockdown of integrins αPS1, αPS2 or βPS with RNAi in the adult intestine led to an expansion in the stem cell pool (Goulas, Conder, & Knoblich, 2012). By observing the orientation of the mitotic spindle and the localization of the Par complex, which is asymmetrically inherited to promote differentiation, the authors propose that integrins ensure that only one of the daughters remains anchored to the basement membrane, while the other inherits differentiation fate (Goulas et al., 2012). In another *Drosophila* study however, knockdown of βPS with a separate RNAi, or complete genetic ablation of αPS1 or βPS (but not αPS2) in MARCM clones, led to a failure of stem cell proliferation, rather than an expansion of the stem pool (Lin et al., 2013). These results suggest that stem cell identity requires anchorage via integrins, even though differentiation steps are unhindered; in the absence of anchorage, differentiation is favored over self-renewal. A recent study has proposed one way that anchoring of the intestinal stem cell to the matrix could influence stem cell behavior, by ensuring localized contact with self-renewal signals. BMP ligands provided by other cells are enriched at the basal surface of the epithelium, and when the basement membrane component Viking (a *Drosophila* Type IV Collagen) was reduced genetically, this enrichment was lost and the stem cell population was expanded (Tian & Jiang, 2014). The authors propose that collagen provides a sink that traps BMP ligands and (a situation also observed in the early embryo (Wang, Harris, Bayston, & Ashe, 2008)), restricting the self-renewal signal to the basal surface. In this "permissive" scenario, integrin anchorage ensures that stem cells receive the right signals. Further work is required to clarify precisely how integrins work in stem cells of the fly gut.

Integrins also contribute to niche architecture in vertebrates. Neurons in the mammalian neocortex derive from neural stem cells that predominantly show a characteristically bipolar morphology; apical processes contact the ventricular surface and basal processes contact the plial basement membrane (Marthiens et al., 2010). The ventricular zone, where the nucleus resides, provides the niche, giving the signals necessary for stem cell maintenance. Both apical and basal process adhesions appear to be mediated by integrins. Mutant mice lacking the integrin α6 subunit displayed a separation between basal processes and the basement membrane (Haubst, Georges-Labouesse, De Arcangelis, Mayer, & Gotz, 2006), although fate and proliferation was unaffected. To target the apical side, β1-blocking antibodies were injected into the ventricle, and led to detachment of the apical process from the ventricle, as well as an increase in proliferation away from the ventricular surface, altered spindle orientation, "bent" basal processes, and cortical layering defects (Loulier et al., 2009). While disentangling the many phenotypes has its caveats, a parsimonious interpretation is that niche architecture relies on apical adhesion to the ventricle, and perturbation of this adhesion results in a series of defects caused by the stem cells not being kept in the niche (Loulier et al., 2009).

A similar stem cell detachment takes place in the subventricular zone of the adult brain. Here, stem cells and their daughter transient-amplifying cells adhere to a laminin-rich matrix that surrounds the vasculature (Shen et al., 2008; Tavazoie et al., 2008), and loss of this adhesion, via function-blocking antibodies injected into the ventricle, led to the detachment of the cells from the vasculature, and a resultant change in proliferation behavior (Shen et al., 2008). In these examples, integrins provide a critical permissive function for stem cells and the balance between self-renewal and differentiation.

3.3. Induction *in vivo*: Beyond the niche

Outside of the stem cell niche, while most developmental work supports a primarily permissive role for integrins in induction, there are certain examples where integrins may act more directly, including oligodendrocyte development where at multiple stages integrin crosstalk with growth factor receptors provides cues and signals for correct differentiation (Baron, Colognato, & Ffrench-Constant, 2005). Additionally, in the mammary gland, laminin–integrin interactions and a downstream pathway comprising ILK and Rac1 are crucial for epithelial cells to produce milk (Streuli, 2009).

Recent work from zebrafish provides an intriguing example of how ECM–integrin interactions could influence induction. Early zebrafish development involves the induction of mesoderm by endoderm, and coculturing experiments suggested that this induction was at least in part mediated by fibronectin. The expression of fibronectin was upregulated in endoderm cells, and knockdown led to reduced induction both in the coculturing experiments and *in vivo* as seen by assaying cardiac gene expression (Cheng et al., 2013). The fibronectin–integrin interaction is proposed to influence mesodermal induction by modulating Wnt signaling, and because a fibronectin matrix exists between endoderm and mesoderm in many species, the authors propose that it may reflect an evolutionary conserved mechanism (Cheng et al., 2013).

An evolutionarily conserved role for mechanical force in mesoderm induction has also recently been reported. Mesoderm is induced in a specific area of tissue that is being compressed by the onset of epiboly in gastrulation, and this can be tracked by observing the nuclear translocation of β catenin (Brunet et al., 2013). Gains and losses in force led to stepwise changes in β catenin nuclear translocation, and this scenario was mimicked in *Drosophila* tissues (Brunet et al., 2013; Farge, 2003). Whether the mechanical forces provided by epiboly also act upon the fibronectin–integrin interaction reported by Cheng et al. is an intriguing possibility, but it is worth remembering that integrins are not the only means by which the cell can sense force.

A recent study on heart development in the sea squirt *Ciona intestinalis* points toward a role of increased adhesion in aiding induction. Precardiac founder cells divide asymmetrically to produce smaller, founder cells that follow the cardiac lineage and larger cells that become tail muscle (Davidson, 2007). The smaller founder cells invade the epidermis just prior to mitosis, and talin is found enriched at this ventral membrane (Norton, Cooley, Islam, Cota, & Davidson, 2013). While excess adhesion provided by active Rap GTPase expanded cardiac induction, loss of adhesion using a dominant negative Rap reduced it, and this effect could be rescued by providing excess integrins (Norton et al., 2013). How might increased adhesion influence induction? The authors propose that adhesion provides a regional enhancement of inductive signals provided by FGF. In this scenario therefore, just as we saw for the stem cell niche, integrins enhance inductive signals provided by "classical" signaling pathways.

In multiple other developmental contexts, loss of integrins does not alter cell fate. In the absence of integrins in flies and worms, muscle specification and patterning is normal, but functional attachments to adjacent muscles or the epidermis are not formed (Brown, 2000). Integrin–fibronectin interactions regulate somitogenesis and trunk elongation in zebrafish, and axis extension and neurulation in *Xenopus*, without changing the cell fate patterning of the tissues (Dray et al., 2013; Julich et al., 2005; Koshida et al., 2005; Lallier & Desimone, 2000; Marsden & DeSimone, 2003). Finally, cells mutant for β1 contribute to most tissues in chimeric mice (Fassler & Meyer, 1995), showing how they are competent to respond to inductive signals. These examples show how, in many contexts, integrins are firmly in the "realizator" camp of developmental factors, acting downstream of fate determining mechanisms to construct and maintain tissues; any feedback they give to the cell does not alter its fate. Properly specified cells do not, or rather cannot, behave properly. The rest of the review focuses on how this occurs.

4. INTEGRINS IN SINGLE AND COLLECTIVE CELL MIGRATION: BEYOND THE FIBROBLAST MODEL

Migration, whether by single cells, small groups of cells, or tissues *en masse*, is a crucial component of development, as cells are often not born where they need to end up and organs require the coming together of tissues of different origins. The critical importance of integrins to cell migration on 2D substrates has been firmly established (Huttenlocher & Horwitz, 2011), but the particular features of this migration event are not universally shared. Even before the molecular delineation of adhesions, it was noted that the large focal adhesions linked to actin cables seen in fibroblasts on 2D are not as discernable in other contexts, or other migrating cell types (Abercrombie, 1980). When fibroblasts are embedded in three-dimensional matrices, cell morphology during movement also changes, and more obviously resembles migration *in vivo* (Bard & Hay, 1975). The fibroblasts end up looking more like fibroblasts, and focal adhesions are less discernable (though not absent (Kubow & Horwitz, 2011)).

Recent work has also swelled the repertoire of cell migration strategies, from leukocytes that can rapidly migrate toward and enter the lymph nodes normally in the absence of integrins (Lammermann et al., 2008), to tumor cells that migrate via polarized water inflow/outflow in confined environments (Stroka et al., 2014), to fibroblasts that, in 3D matrices, pull the nucleus in the direction of travel, increasing cytoplasmic pressure and driving the formation of protrusions known as lobopodia (Petrie, Koo, & Yamada, 2014). Additionally, collective cell migration, where cells have to stay in contact while they move, provides a different set of challenges compared to single-cell migration (Friedl & Gilmour, 2009). In keeping with the diversity of cell migration strategies, integrins are utilized in distinct ways depending on developmental context.

4.1. Integrins powering and modulating cell movement, cell autonomously, and nonautonomously

In numerous developmental contexts, loss or reduction in integrin function perturbs migration (Fig. 3). Primordial germ cell (PGC) migration is a crucial migration event across animals, as germ cells navigate in a complex path through tissues of the developing organism to their destination in the gonads (Richardson & Lehmann, 2010). In chimeric mice, while PGCs lacking β1 were specified correctly, they failed to colonize the gonad from their birthplace in the hindgut, demonstrating a failure of migration (Anderson et al., 1999). Dorsal frontrunner cells in the developing zebrafish heart fail to form clusters and fail to migrate properly when αVβ1 is knocked down (Ablooglu, Tkachenko, Kang, & Shattil, 2010). During *Drosophila* oogenesis, the follicular epithelium surrounds the developing oocyte, which undergoes an elongation phase to produce an ellipsoid egg (Bilder & Haigo, 2012). Live imaging has revealed the follicle cells migrate in a polarized manner, using the surrounding basement membrane as a substrate, and driving the rotation of the entire egg chamber during its elongation phase (Haigo & Bilder, 2011). Integrins are required both for the migration—mutant clones lacking βPS stayed stationary—and coincident polarization of the basement membrane (Haigo & Bilder, 2011). A key feature of this migration appears to be polarized integrin activity at the leading edge; cells mutant for the *misshapen* kinase showed elevated integrin expression and collagen IV accumulation at the lagging edge, and a resultant failure to migrate (Lewellyn, Cetera, & Horne-Badovinac, 2013). Thus, too much integrin activity in the wrong place, as well as too little, hinders migration in the follicle cells.

In other cases, integrins coordinate, rather than just power, migration. The placodes of the vertebrate head give rise to sensory structures and are formed by a combination of cell migration and rearrangements. When the zebrafish α5 subunit was knocked down by a morpholino, placode development in the posterior head region was defective (Bhat & Riley, 2011). Live tracking revealed that individual cells normally migrated from the adjacent ectoderm into the placode in a directed manner, while α5

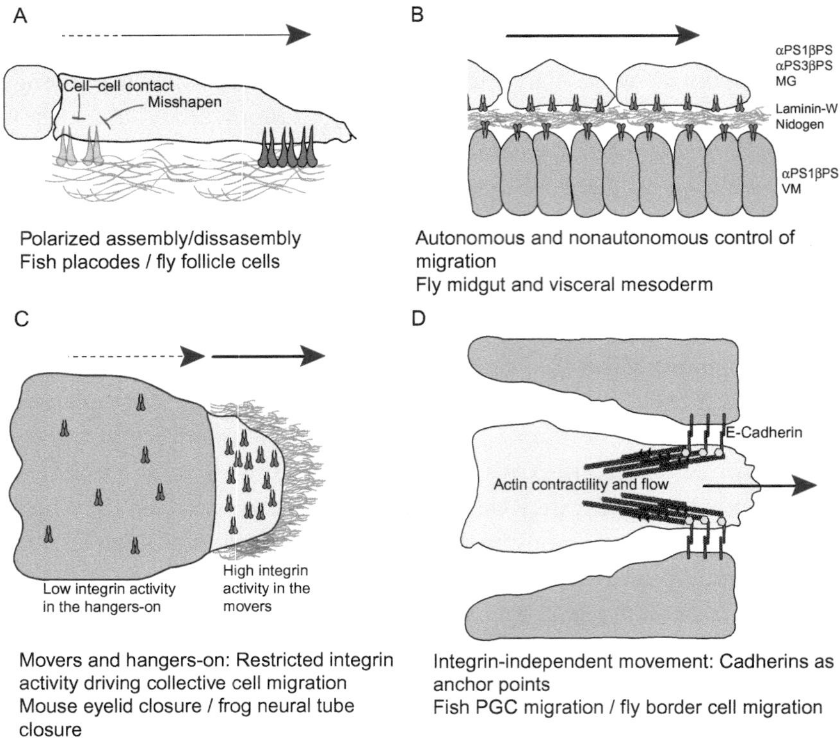

Figure 3 Integrins in cell migration. (A) Cell-autonomous control of migration via polarized assembly/disassembly along the axis of movement. Disassembly can be controled by cell–cell contact, or activation of inhibitory kinases. (B) During fly embryogenesis, integrins control migration cell autonomously in the MG, while helping to construct a track for migration in the VM. Different heterodimers help mediate these roles. (C) During collective cell migration, restricted integrin activity in the tissue can lead to wholesale migration. (D) Integrin-independent movement can occur via cadherins that provide traction points for movement. Abbreviations: MG, midgut; VM, visceral mesoderm; PGC, primordial germ cell.

morphant cells followed more erratic and meandering pathways, even though the distance traveled (hence the capacity to migrate) was not reduced (Bhat & Riley, 2011). The effect was cell autonomous, as shown by transplantation studies. As cells still failed to migrate in a wild-type ECM context, this rules out a scenario whereby integrin mutant cells cannot migrate due to a failure of ECM reorganization. Failure to organize the ECM is seen in many other integrin mutant contexts (see Dray et al., 2013; Marsden & DeSimone, 2001; Urbano, Dominguez-Gimenez, Estrada, & Martin-Bermudo, 2011).

Cell-autonomous requirements for integrins in migration may reflect polarized adhesion assembly and disassembly along the axis of movement. Placodes are formed near to cells of the neural crest, a migratory population, and placode migration is stimulated by neural crest arrival (Theveneau et al., 2013). Cultured populations of cells revealed a chase and run mechanism, whereby neural crest cells are attracted to the placodes, which themselves are repulsed by contact with the crest. While transient contact between the cell populations is mediated by cadherin contacts, focal adhesions are reduced at the site of contact and enriched in the "leading edge" of the placode, presumably driving migration in the direction opposite to the neural crest (Theveneau et al., 2013). How contact regulates adhesion disassembly is unclear, but is a fascinating example of the crosstalk between cell–cell and cell–ECM adhesions, which is critical in many developmental contexts (Weber, Bjerke, & DeSimone, 2011). In the absence of polarized adhesion, migration may be undirected.

Cell-autonomous requirements for integrins in migration can supplemented by nonautonomous roles in building the tracks for migration. *Drosophila* hemocytes, macrophages that display stereotypic migration patterns, migrated slower when integrins were lost using cell-specific RNAi or genetic loss of function with the MARCM system (Comber et al., 2013; Moreira, Jacinto, & Prag, 2013). In the embryo, whole animal βPS mutants reveal an additional requirement in the ventral nerve cord upon which the hemocytes migrate (Comber et al., 2013), showing how integrins also contribute to migration by sculpting the environment in which the cells move.

A similar interplay between migrating cells and their substrates can be seen during collective cell migration in the fly embryo. Primordial midgut cells migrate over the visceral mesoderm, and αPS2 is required in the visceral mesoderm to provide a substrate, while αPS1 and αPS3 are required in the migrating cells (Martin-Bermudo, Alvarez-Garcia, & Brown, 1999). Other migratory cell populations also use the visceral mesoderm track, namely the tracheal cells (Boube, Martin-Bermudo, Brown, & Casanova, 2001), and the caudal visceral mesoderm that contributes to the muscles surrounding the gut (Urbano et al., 2011). In the visceral mesoderm, αPS2 is required to organize a coherent ECM containing laminin W and nidogen (Urbano et al., 2011), which provides the road for migration. This is a neat example of how different integrin heterodimers can mediate cell-autonomous (αPS1/αPS3 in the midgut, trachea, and caudal visceral mesoderm) and non-autonomous (αPS2 in the visceral mesoderm) roles in migration.

Integrins may also contribute to building cell migration roads in the vertebrate brain. The adult brain harbors proliferative zones from which newly born neurons migrate; for example, neurons born in the subventricular zone migrate as chains of cells within tubes of glia, along the rostral migratory stream to the olfactory bulb. When the β1 integrin subunit gene was specifically deleted from neuronal progenitors, migration was hindered, with more retained in the rostral stream and some showing abnormal invasion into other regions of the brain (Belvindrah, Hankel, Walker, Patton, & Muller, 2007). The formation of cell chains and glial tubes is also disrupted (Belvindrah et al., 2007), and because both the "convoy" of cells and the "road" on which they are migrating are disrupted, it is hard to test the site of requirement in this context. A case for integrins primarily controlling the architecture of the "road" has been made for granule neurons in the cerebellum. Cell-type specific gene targeting using different Cre lines revealed defects in granule cell precursor migration when β1 was depleted from glia, but not the granule cells themselves (Frick et al., 2012).

Integrins can function nonautonomously during collective cell migration by promoting migration in a subset of cells within the tissue. *Xenopus* neural tube closure involves movement of both the dorsal neural ectoderm and the multilayered nonneural epidermis that covers the rest of the embryo. In this epidermis, the deepest cells, those in contact with the basement membrane, moved fastest and "ahead" of the more superficial ones (Morita et al., 2012). β1 depletion led to an overall slowdown in migration, while E-Cadherin knockdown led to an increased lag between the deep and superficial cells (Morita et al., 2012). This leads to a model whereby deep cells, interacting with fibronectin via integrins, tow along the superficial cells to which they are adhered via Cadherins.

Towing is also observed during embryonic eyelid closure in mice, although this works by a different mechanism. Eyelid closure involves the migration of the epidermis and its underlying dermis over the cornea. Live imaging experiments revealed that cells at the front of the eyelid adopted a mesenchymal morphology, elongating parallel to the axis of the front and intercalating in this direction (Heller, Kumar, Grill, & Fuchs, 2014). This intercalation provides a force that tows the adjacent, passively moving epidermis to complete eyelid closure. Front cells express high levels of α5β1 integrin and fibronectin, and when these were depleted, cell movement was reduced and eyelid closure failed (Heller et al., 2014). Just how a localized integrin–fibronectin interaction is utilized to promote intercalation and provide the force that drags the rest of the epithelium along is an intriguing question. As with the frog neural tube, it is an example of how localized integrin activity can drive a tissue movement that includes passive hangers-on. We can make an analogy with a single migrating cell, in which restricted integrin activity polarizes movement.

4.2. Integrin-independent cell movement *in vivo*

One of the surprising aspects about development is that in many contexts, integrins play no apparent role in cell migration. Thus, muscle migration to target attachment sites does not require integrins in *Drosophila* (Wright, 1960) or *C. elegans* (Hresko, Williams, & Waterston, 1994). In the worm, laminin is required (Viveiros, Hutter, & Moerman, 2011), suggesting another laminin receptor regulates this migration event. PGC migration in flies does not require integrins (Devenport & Brown, 2004). A key event in this migration is the crossing of the endoderm, and, rather than actively invading an intact epithelium, it appears that the germ cells take advantage of gaps between the cells as the endoderm undergoes an epithelial to mesenchymal transition (Seifert & Lehmann, 2012). The ability of the PGC cells to squeeze through tissues in an integrin-independent manner shows similarities to leukocytes (Lammermann et al., 2008).

Where integrins are not involved, direct cell–cell adhesion can suffice to power migration, particularly where the migration pathway is made up of other cells. During *Drosophila* oogenesis, border cells migrate as a cluster through the nurse cells reach the oocyte, and this is driven by dynamic cadherin adhesion, both between the border cells and the surrounding nurse cells, and within the border cluster itself (Cai et al., 2014); however, there is no requirement for integrins (Devenport & Brown, 2004). PGC migration in fish is integrin-independent, and the cells move via a blebbing mechanism, coupling actin flow to cadherin adhesions formed with other cells in the environment (Kardash et al., 2010). In these cases, cadherins do what integrins have been demonstrated to do in culture; couple actin contractility and flow to the ligands outside of the cell.

Thus, the roles of integrin in cell migration during development are more diverse than simply powering movement. Integrins are also required to build roads for migration, to coordinate groups of migrating cells, and in some cases, are not required at all. Recent work from cell culture shows how the cell tailors its migration strategy to environmental context (Petrie, Gavara, Chadwick, & Yamada, 2012), and this might explain the varied requirements for integrins revealed by genetic studies.

5. BUILDING TISSUES AND ORGANS: SNAGS AND ANCHORS

Binding between the extracellular domains of integrins and ECM ligands can have profound consequences for both the cell and its environment. During development, integrins can either act to snag dispersing ECM components and enrich them in one region of the tissue, or act as anchors to halt migration in moving cells (Fig. 4).

Loss of integrins can lead to a loss of normal accumulation and patterning of the ECM. In frog gastrulation, expression of dominant negative β1 integrin suppressed fibronectin matrix assembly on the blastocoel roof (Marsden & DeSimone, 2001). In zebrafish development, trunk elongation is perturbed and fibronectin reduced and misaligned when α5 and αV are coincidently knocked down (Dray et al., 2013), and integrin expression in myocardial precursors, downstream of the *snail* transcription factor, is required for the formation of a fibronectin layer in the developing heart (Qiao et al., 2014).

A similar process occurs in zebrafish somitogenesis. Genetic screens for mutations that cause defects in somites recovered mutations in the genes encoding the integrin α5 subunit and fibronectin (Julich et al., 2005; Koshida et al., 2005). The absence of α5 caused the fusion of anterior somites, without altering somite patterning, and loss of fibronectin accumulation at somite borders (Julich et al., 2005; Koshida et al., 2005). Integrin localization to the somite boundary was shown to precede fibrillogenesis (Julich, Mould, Koper, & Holley, 2009). Thus, a localized integrin snag is required to trap fibronectin and promote fibrillogenesis. The specific localization of fibrillogenesis to somite boundaries (rather than all cell–cell boundaries) appears to arise as a combined result of trans-inactivation, where integrins on adjacent cells reciprocally inhibit clustering by a currently unclear mechanism, and specifically localized activation by Eph/Ephrin signaling that overcomes trans-inactivation at the boundaries (Julich et al., 2009). In this case, localized integrin activity, in an "inside–out" manner, promotes ECM accumulation, and this ECM is required to hold the tissue together. Integrin–fibronectin interaction also influences somitogenesis in frogs (Kragtorp & Miller, 2007), mice (Yang et al., 1999), and chickens (Rallis et al., 2010), although in chickens a distinct mechanism, of integrins acting through Wnt signaling, has been proposed.

Localized integrin upregulation can also act to stop movement of an ECM. Basement membranes provide integrity and boundaries to tissues,

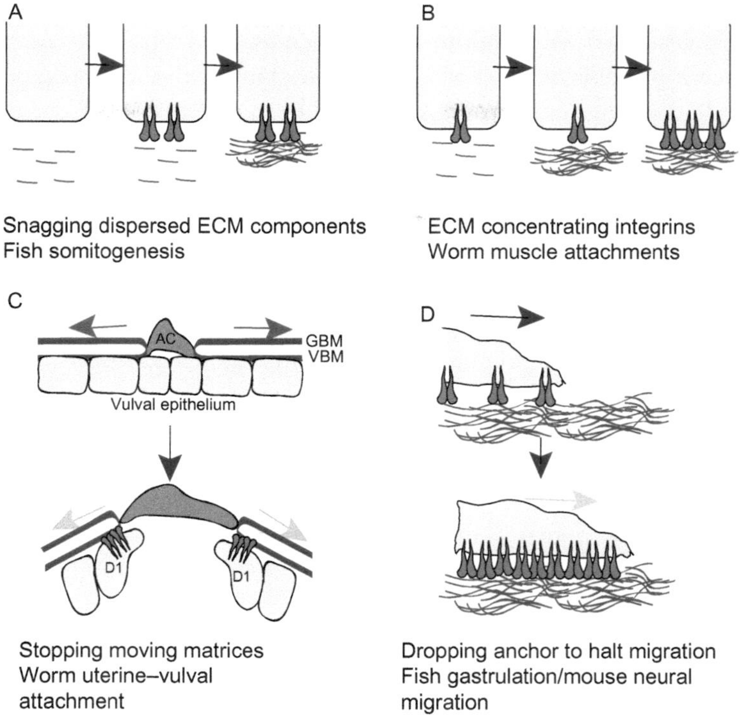

Figure 4 Integrins as snags and anchors for ECM. (A) Local upregulation of integrins can help to snag and concentrate dispersed ECM components. (B) Localized enrichment of ECM components can concentrate integrins. (C) Upregulation of integrins can stop moving matrices. During anchor cell invasion in the worm, integrins are upregulated to stop the sliding of a bilayered basement membrane between the gonadal and uterine epithelia. Drawing after Matus et al. (2014). (D) Increased integrin activity can stop movement, by providing too much adhesion. Abbreviations: AC, anchor cell; GBM, gonadal basement membrane; VBM, ventral basement membrane.

but in certain developmental events, and indeed in many pathological contexts, cells need to break through it to get to the other side (Kelley, Lohmer, Hagedorn, & Sherwood, 2014). During gonad development in the worm, the uterine and vulval tissues are initially separated by a bilayered basement membrane. A specialized cell within the gonad, the anchor cell, invades this membrane to establish vulval–uterine contact, and this leads to the formation of a gap in the membrane. The αINA1βPAT3 integrin controls this developmental process in two stages. During the invasion step, it functions in the anchor cell to mediate formation of the invasive membrane, via plasma membrane targeting of various components (Hagedorn et al., 2009).

Anchor cell invasion promotes a novel behavior in the basement membrane, as visualized with laminin and collagen tagged with photoconvertible fluorescent proteins; it slides across the underlying vulval cells (Ihara et al., 2011). This sliding is in part driven by regulated patterns of cell divisions in the underlying vulval tissue (Matus et al., 2014), and always stops at the same place, such that the hole made by the invasion event is limited in size. The vulval cells express αINA1βPAT3, and its depletion caused an over-expansion of the basement membrane hole, indicating that integrins play a role in arresting basement membrane sliding (Ihara et al., 2011). Intriguingly, the enrichment of vulval cell integrins to the edge of the membrane gap required prior enrichment of laminin. Thus, increased ECM at the edge of the sliding basement membrane localizes integrin, which in turn stabilizes the sliding membrane (Matus et al., 2014). This is a good example of the back-and-forth interactions occurring between integrins and ECM components. The role of ECM in concentrating integrins has long been appreciated from cell culture studies, and has also been documented during development, for example, in worm muscles, where integrin recruitment to attachment sites requires the extracellular ligand perlecan (Hresko et al., 1994).

Migrating cells can also upregulate integrins to stop their migration. In cell culture, while disassembly of integrins at the rear of the cell is required for cell movement (Huttenlocher & Horwitz, 2011), vinculin-stimulated increases in integrin activity (as assayed by conformation-specific antibodies) slows down migration (Carisey et al., 2013). Similarly, a failure to downregulate integrins in the cell rear stops follicle cell migration (Lewellyn et al., 2013). In the reverse scenario, migrating neurons in the mouse brain "overshoot" their target when integrins are downregulated (Georges-Labouesse, Mark, Messaddeq, & Gansmuller, 1998). Similar overshooting is observed during zebrafish gastrulation, where endodermal precursors migrate anteriorly in a manner dependent on chemokine receptors. When integrin–ECM interactions were perturbed with RGD peptides, the endoderm overshot its target, displacing to the anterior (Nair & Schilling, 2008). These examples show how localized stimulation of adhesion can be used to halt the progress of migration.

6. INTEGRINS IN EPITHELIA: CELL POLARITY AND DIVISION ORIENTATION

Our final examples from development come from the role of integrins in epithelial cell biology. On their basal surface, epithelial cells contact the

ECM via integrins, while the apical surface can either contact other cells, in the case of a multilayered epithelium, contact an apical ECM, as is the case in the arthropod epidermis, or face the lumen, in the case of tube-forming epithelia. The basal contact mediated by integrins can play a fundamental role in telling the cell which way is down, and thus establishing apicobasal polarity (Lee & Streuli, 2014).

Recent work from the mammary gland identifies a pathway between integrins, polarity, and the formation of the lumen. Loss of integrin from the mammary epithelium both in the animal and in explant cultures causes a failure in lumen formation (Akhtar & Streuli, 2013). Cultured wild-type ascini gradually establish apicobasal polarity, with the apical marker and polarity regulator aPKC being progressively lost from the outside and enriched on the luminal side of the ascini. In the absence of β1 integrin, aPKC stayed on the outside of the cells, and the lumen was not formed (Akhtar & Streuli, 2013). Integrins and the effector protein ILK were required to establish the polarization of microtubules along the cell's apical–basal axis, and perturbing microtubules prevented apical proteins from being removed from the basal compartment and apically transcytosed. Integrins were relocated to the top surface and apical markers lost when basement membranes were ectopically overlain on the apical side (consistent with the redistribution of the apical markers requiring integrin). The inversion of epithelial polarity by activating integrins on the apical side suggests that integrin activation on one surface leads to the definition of apical on the surface opposite it, via a transcytosis mechanism. This is borne out by work studying *C. elegans* pharynx formation; laminin accumulates on the outside of the developing pharyngeal cyst, colocalizing with integrins, and is required for the polarized localization of the apical marker Par6 to the lumenal side (Rasmussen, Reddy, & Priess, 2012). Whether a similar ILK-microtubule axis is in play in this system has yet to be determined.

Once polarity has been established, the orientation of cell divisions can play a key role in tissue geometry. Regulating which cells divide and where the daughters end up can influence the topology of the epithelium and drive morphogenesis (Gillies & Cabernard, 2011). For example, spindle orientation with respect to the plane of the epithelium can determine whether the daughter cell is maintained within the basal layer of the epithelium or released from it (parallel vs. perpendicular to the plane of the epithelium). Multiple studies have reported defects in spindle orientation in this dimension following the loss of integrins. Mammalian lungs are formed from extensively branched monolayered epithelia. When integrins are deleted from the epithelium, branching is lost as the epithelium becomes multilayered (Chen & Krasnow, 2012). While in wild-type lungs, most cells orient with their mitotic spindles parallel to the lumen, integrin loss led to an increase in perpendicular divisions without altering overall cell division rates (Chen & Krasnow, 2012).

Deviations to normal spindle orientations are also seen when integrin function is perturbed in other systems, in the neocortical niche that houses neural stem cells, perturbation of β1 integrin from the apical surface led to a reduction in divisions that occur horizontally with respect to the ventricle (Loulier et al., 2009); in *Xenopus* gastrulation, deep cells of the blastocoel roof require integrin-mediated interaction with a fibronectin matrix to ensure cell divisions are parallel to the roof (Marsden & DeSimone, 2001); in the murine skin, β1 mutant cells showed an unpolarized spindle orientation and random segregation of differentiation-determining Par factors (Lechler & Fuchs, 2005); and in the mammary epithelium, depletion of β1 from the basal layer randomized the orientation of cell division (Taddei et al., 2008). The underlying cell biological mechanism may differ in each case.

As with most aspects of integrins in development, there are always counter examples to challenge the universality of a given phenomenon. In *Drosophila*, the follicular epithelium that surrounds the developing oocyte, cells tend to orient their spindles parallel to the epithelium (Bergstralh, Lovegrove, & St Johnston, 2013; Fernandez-Minan, Martin-Bermudo, & Gonzalez-Reyes, 2007). As in the mouse lung, loss of integrins in the epithelium leads to multilayering, but in integrin null cells that are still part of the monolayer, the cell division plane is still roughly parallel to the epithelium (Bergstralh et al., 2013). This suggests that multilayering is not a consequence of altered spindle orientation, but some other facet of integrin function in maintaining epithelial integrity, in contrast to the mouse, where cell division orientation changes precede multilayering (Chen & Krasnow, 2012). As we saw in our discussion of anoikis, in the absence of integrins other mechanisms may provide cues to overcome any issues with the axis of division.

7. OUTLOOK

Animal development provides a seemingly inexhaustible supply of events in which integrins play an important role. As we have seen, integrins establish the ECM–actin link to stick cells together, provide the means of migration, and sculpt tissues. In this sense, the role of integrins in many contexts can be explained without the need to invoke much signaling; the anchorage itself is the primary functional output of the protein, and the most obvious integrin phenotypes tend to be adhesion phenotypes, not cell fate phenotypes. The signaling side of integrin function impinges importantly on adhesion dynamics, involving the response of adhesion to changes in force or the regulation of adhesions with respect to things like the direction of movement. It may also influence basic cellular features like polarity.

Where integrins do help augment cell fate choices, the next question is how this works. Within the context of development, it is difficult to distinguish a direct signal transduction pathway mediated by integrins from a more permissive role of providing the necessary environment for the realization of the fate choice. A permissive role could be achieved by a variety of mechanisms, including maintaining the proximity between the cells releasing an inductive signal and the receiving cells, as has been demonstrated in stem cell niches, or facilitating the reception of the signal by providing a scaffold for components of diverse signaling networks. The notion that integrins may be a crucial component linking changes in the ECM to changes in nuclear architecture and hence gene expression has yet to be thoroughly explored in a developmental context.

ACKNOWLEDGMENTS

The authors wish to apologize to colleagues whose work has been omitted due to space constraints. The Brown lab acknowledges the funding it has received from the Wellcome Trust, the BBSRC, and the Human Frontier Science Program.

REFERENCES

Abercrombie, M. (1980). Croonian lecture, 1978—Crawling movement of metazoan cells. *Proceedings of the Royal Society Series B Biological Sciences*, *207*(1167), 129–147. http://dx.doi.org/10.1098/Rspb.1980.0017.

Ablooglu, A. J., Tkachenko, E., Kang, J. A., & Shattil, S. J. (2010). Integrin alpha V is necessary for gastrulation movements that regulate vertebrate body asymmetry. *Development*, *137*(20), 3449–3458. http://dx.doi.org/10.1242/Dev.045310.

Akhtar, N., & Streuli, C. H. (2013). An integrin-ILK-microtubule network orients cell polarity and lumen formation in glandular epithelium. *Nature Cell Biology*, *15*(1), 17–27. http://dx.doi.org/10.1038/Ncb2646.

Akiyama, S. K., Yamada, S. S., Chen, W. T., & Yamada, K. M. (1989). Analysis of fibronectin receptor function with monoclonal-antibodies—Roles in cell-adhesion, migration, matrix assembly, and cytoskeletal organization. *Journal of Cell Biology*, *109*(2), 863–875. http://dx.doi.org/10.1083/Jcb.109.2.863.

Anderson, R., Fassler, R., Georges-Labouesse, E., Hynes, R. O., Bader, B. L., Kreidberg, J. A., et al. (1999). Mouse primordial germ cells lacking beta 1 integrins enter the germline but fail to migrate normally to the gonads. *Development*, *126*(8), 1655–1664.

Antoniades, I., Stylianou, P., & Skourides, P. A. (2014). Making the connection: Ciliary adhesion complexes anchor basal bodies to the actin cytoskeleton. *Developmental Cell, 28*(1), 70–80. http://dx.doi.org/10.1016/J.Devcel.2013.12.003.

Bard, J. B. L., & Hay, E. D. (1975). Behavior of fibroblasts from developing avian cornea—Morphology and movement insitu and invitro. *Journal of Cell Biology, 67*(2), 400–418. http://dx.doi.org/10.1083/Jcb.67.2.400.

Baron, W., Colognato, H., & Ffrench-Constant, C. (2005). Integrin-growth factor interactions as regulators of oligodendroglial development and function. *Glia, 51*(1), 80. http://dx.doi.org/10.1002/Glia.20238, (vol. 49, p. 467, 2005).

Bedzhov, I., & Zernicka-Goetz, M. (2014). Self-organizing properties of mouse pluripotent cells initiate morphogenesis upon implantation. *Cell, 156*(5), 1032–1044. http://dx.doi.org/10.1016/J.Cell.2014.01.023.

Belvindrah, R., Hankel, S., Walker, J., Patton, B. L., & Muller, U. (2007). beta 1 integrins control the formation of cell chains in the adult rostral migratory stream. *Journal of Neuroscience, 27*(10), 2704–2717. http://dx.doi.org/10.1523/Jneurosci.2991-06.2007.

Bergstralh, D. T., Lovegrove, H. E., & St Johnston, D. (2013). Discs large links spindle orientation to apical-basal polarity in drosophila epithelia. *Current Biology, 23*(17), 1707–1712. http://dx.doi.org/10.1016/J.Cub.2013.07.017.

Bershadsky, A. D., Balaban, N. Q., & Geiger, B. (2003). Adhesion-dependent cell mechanosensitivity. *Annual Review of Cell and Developmental Biology, 19*, 677–695. http://dx.doi.org/10.1146/Annurev.Cellbio.19.111301.153011.

Bhat, N., & Riley, B. B. (2011). Integrin-alpha 5 coordinates assembly of posterior cranial placodes in zebrafish and enhances Fgf-dependent regulation of Otic/epibranchial cells. *Plos One, 6*(12), e27778. http://dx.doi.org/10.1371/journal.pone.0027778.

Bilder, D., & Haigo, S. L. (2012). Expanding the morphogenetic repertoire: Perspectives from the Drosophila egg. *Developmental Cell, 22*(1), 12–23. http://dx.doi.org/10.1016/J.Devcel.2011.12.003.

Bokel, C., & Brown, N. H. (2002). Integrins in development: Moving on, responding to, and sticking to the extracellular matrix. *Developmental Cell, 3*(3), 311–321. http://dx.doi.org/10.1016/S1534-5807(02)00265-4.

Boube, M., Martin-Bermudo, M. D., Brown, N. H., & Casanova, J. (2001). Specific tracheal migration is mediated by complementary expression of cell surface proteins. *Genes & Development, 15*(12), 1554–1562. http://dx.doi.org/10.1101/Gad.195501.

Boudreau, N., Sympson, C. J., Werb, Z., & Bissell, M. J. (1995). Suppression of ice and apoptosis in mammary epithelial-cells by extracellular-matrix. *Science, 267*(5199), 891–893. http://dx.doi.org/10.1126/Science.7531366.

Bouvard, D., Brakebusch, C., Gustafsson, E., Aszodi, A., Bengtsson, T., Berna, A., et al. (2001). Functional consequences of integrin gene mutations in mice. *Circulation Research, 89*(3), 211–223. http://dx.doi.org/10.1161/Hh1501.094874.

Brabant, M. C., Fristrom, D., Bunch, T. A., & Brower, D. L. (1996). Distinct spatial and temporal functions for PS integrins during Drosophila wing morphogenesis. *Development, 122*(10), 3307–3317.

Breau, M. A., Pietri, T., Eder, O., Blanche, M., Brakebusch, C., Fassler, R., et al. (2006). Lack of beta 1 integrins in enteric neural crest cells leads to a Hirschsprung-like phenotype. *Development, 133*(9), 1725–1734. http://dx.doi.org/10.1242/Dev.02346.

Brower, D. L. (2003). Platelets with wings: The maturation of Drosophila integrin biology. *Current Opinion in Cell Biology, 15*(5), 607–613. http://dx.doi.org/10.1016/S0955-0674(03)00102-9.

Brower, D. L., & Jaffe, S. M. (1989). Requirement for integrins during Drosophila wing development. *Nature, 342*(6247), 285–287. http://dx.doi.org/10.1038/342285a0.

Brown, N. H. (2000). Cell-cell adhesion via the ECM: Integrin genetics in fly and worm. *Matrix Biology, 19*(3), 191–201. http://dx.doi.org/10.1016/S0945-053x(00)00064-0.

Brown, N. H. (2011). Extracellular matrix in development: Insights from mechanisms conserved between invertebrates and vertebrates. *Cold Spring Harbor Perspectives in Biology*, *3*, a005082. http://dx.doi.org/10.1101/cshperspect.a005082.
Brown, N. H., Gregory, S. L., Rickoll, W. L., Fessler, L. I., Prout, M., White, R. A. H., et al. (2002). Talin is essential for integrin function in Drosophila. *Developmental Cell*, *3*(4), 569–579. http://dx.doi.org/10.1016/S1534-5807(02)00290-3.
Brunet, T., Bouclet, A., Ahmadi, P., Mitrossilis, D., Driquez, B., Brunet, A. C., et al. (2013). Evolutionary conservation of early mesoderm specification by mechanotransduction in Bilateria. *Nature Communications*, *4*, 2821. http://dx.doi.org/10.1038/Ncomms3821.
Bukharova, T., Weijer, G., Bosgraaf, L., Dormann, D., van Haastert, P. J., & Weijer, C. J. (2005). Paxillin is required for cell-substrate adhesion, cell sorting and slug migration during Dictyostelium development. *Journal of Cell Science*, *118*(20), 4913 (vol. 118, p. 4295, 2005).
Burridge, K., & Feramisco, J. R. (1980). Micro-injection and localization of a 130 k-protein in living fibroblasts—Relationship to actin and fibronectin. *Cell*, *19*(3), 587–595. http://dx.doi.org/10.1016/S0092-8674(80)80035-3.
Cai, D. F., Chen, S. C., Prasad, M., He, L., Wang, X. B., Choesmel-Cadamuro, V., et al. (2014). Mechanical feedback through E-cadherin promotes direction sensing during collective cell migration. *Cell*, *157*(5), 1146–1159. http://dx.doi.org/10.1016/J.Cell.2014.03.045.
Campbell, I. D., & Humphries, M. J. (2011). Integrin structure, activation, and interactions. *Cold Spring Harbor Perspectives in Biology*, *3*, a004994. http://dx.doi.org/10.1101/cshperspect.a004994.
Carisey, A., Tsang, R., Greiner, A. M., Nijenhuis, N., Heath, N., Nazgiewicz, A., et al. (2013). Vinculin regulates the recruitment and release of core focal adhesion proteins in a force-dependent manner. *Current Biology*, *23*(4), 271–281. http://dx.doi.org/10.1016/J.Cub.2013.01.009.
Chen, J. C., & Krasnow, M. A. (2012). Integrin beta 1 suppresses multilayering of a simple epithelium. *Plos One*, 7(12), e52886. http://dx.doi.org/10.1371/journal.pone.0052886.
Chen, S. Y., Lewallen, M., & Xie, T. (2013). Adhesion in the stem cell niche: Biological roles and regulation. *Development*, *140*(2), 255–265. http://dx.doi.org/10.1242/Dev.083139.
Cheng, P., Andersen, P., Hassel, D., Kaynak, B. L., Limphong, P., Juergensen, L., et al. (2013). Fibronectin mediates mesendodermal cell fate decisions. *Development*, *140*(12), 2587–2596. http://dx.doi.org/10.1242/Dev.089052.
Chiarugi, P., & Giannoni, E. (2008). Anoikis: A necessary death program for anchorage-dependent cells. *Biochemical Pharmacology*, *76*(11), 1352–1364. http://dx.doi.org/10.1016/J.Bcp.2008.07.023.
Ciruna, B., & Rossant, J. (2001). FGF signaling regulates mesoderm cell fate specification and morphogenetic movement at the primitive streak. *Developmental Cell*, *1*(1), 37–49. http://dx.doi.org/10.1016/S1534-5807(01)00017-X.
Comber, K., Huelsmann, S., Evans, I., Sanchez-Sanchez, B. J., Chalmers, A., Reuter, R., et al. (2013). A dual role for the beta PS integrin myospheroid in mediating Drosophila embryonic macrophage migration. *Journal of Cell Science*, *126*(15), 3475–3484. http://dx.doi.org/10.1242/Jcs.129700.
Conti, F. J., Monkley, S. J., Wood, M. R., Critchley, D. R., & Muller, U. (2009). Talin 1 and 2 are required for myoblast fusion, sarcomere assembly and the maintenance of myotendinous junctions. *Development*, *136*(21), 3597–3606. http://dx.doi.org/10.1242/Dev.035857.
Critchley, D. R. (2009). Biochemical and structural properties of the integrin-associated cytoskeletal protein talin. *Annual Review of Biophysics*, *38*, 235–254. http://dx.doi.org/10.1146/Annurev.Biophys.050708.133744.

Crump, J. G., Swartz, M. E., & Kimmel, C. B. (2004). An integrin-dependent role of pouch endoderm in hyoid cartilage development. *Plos Biology*, *2*(9), 1432–1445. http://dx.doi.org/10.1371/journal.pbio.0020244.

Davidson, B. (2007). Ciona intestinalis as a model for cardiac development. *Seminars in Cell & Developmental Biology*, *18*(1), 16–26. http://dx.doi.org/10.1016/J.Semcdb.2006.12.007.

Davidson, L. A., Dzamba, B. D., Keller, R., & Desimone, D. W. (2008). Live imaging of cell protrusive activity, and extracellular matrix assembly and remodeling during morphogenesis in the frog, Xenopus laevis. *Developmental Dynamics*, *237*(10), 2684–2692. http://dx.doi.org/10.1002/Dvdy.21600.

De Arcangelis, A., Mark, M., Kreidberg, J., Sorokin, L., & Georges-Labouesse, E. (1999). Synergistic activities of alpha 3 and alpha 6 integrins are required during apical ectodermal ridge formation and organogenesis in the mouse. *Development*, *126*(17), 3957–3968.

Deakin, N. O., & Turner, C. E. (2008). Paxillin comes of age. *Journal of Cell Science*, *121*(15), 2435–2444. http://dx.doi.org/10.1242/Jcs.018044.

del Rio, A., Perez-Jimenez, R., Liu, R. C., Roca-Cusachs, P., Fernandez, J. M., & Sheetz, M. P. (2009). Stretching single talin rod molecules activates vinculin binding. *Science*, *323*(5914), 638–641. http://dx.doi.org/10.1126/Science.1162912.

Delon, I., & Brown, N. H. (2009). The integrin adhesion complex changes its composition and function during morphogenesis of an epithelium. *Journal of Cell Science*, *122*(23), 4363–4374. http://dx.doi.org/10.1242/Jcs.055996.

Devenport, D., & Brown, N. H. (2004). Morphogenesis in the absence of integrins: Mutation of both Drosophila beta subunits prevents midgut migration. *Development*, *131*(21), 5405–5415. http://dx.doi.org/10.1242/Dev.01427.

Discher, D. E., Janmey, P., & Wang, Y. L. (2005). Tissue cells feel and respond to the stiffness of their substrate. *Science*, *310*(5751), 1139–1143. http://dx.doi.org/10.1126/Science.1116995.

Dray, N., Lawton, A., Nandi, A., Julich, D., Emonet, T., & Holley, S. A. (2013). Cell-fibronectin interactions propel vertebrate trunk elongation via tissue mechanics. *Current Biology*, *23*(14), 1335–1341. http://dx.doi.org/10.1016/J.Cub.2013.05.052.

DuFort, C. C., Paszek, M. J., & Weaver, V. M. (2011). Balancing forces: Architectural control of mechanotransduction. *Nature Reviews Molecular Cell Biology*, *12*(5), 308–319. http://dx.doi.org/10.1038/Nrm3112.

Dupuy, A. G., & Caron, E. (2008). Integrin-dependent phagocytosis—Spreading from microadhesion to new concepts. *Journal of Cell Science*, *121*(11), 1773–1783. http://dx.doi.org/10.1242/Jcs.018036.

Ellis, S. J., & Tanentzapf, G. (2010). Integrin-mediated adhesion and stem-cell-niche interactions. *Cell and Tissue Research*, *339*(1), 121–130. http://dx.doi.org/10.1007/S00441-009-0828-4.

Engler, A. J., Sen, S., Sweeney, H. L., & Discher, D. E. (2006). Matrix elasticity directs stem cell lineage specification. *Cell*, *126*(4), 677–689. http://dx.doi.org/10.1016/J.Cell.2006.06.044.

Fairclough, S. R., Dayel, M. J., & King, N. (2010). Multicellular development in a choanoflagellate. *Current Biology*, *20*(20), R875–R876. http://dx.doi.org/10.1016/J.Cub.2010.09.014.

Farge, E. (2003). Mechanical induction of twist in the Drosophila foregut/stomodeal primordium. *Current Biology*, *13*(16), 1365–1377. http://dx.doi.org/10.1016/S0960-9822(03)00576-1.

Fassler, R., & Meyer, M. (1995). Consequences of lack of beta-1 integrin gene-expression in mice. *Genes & Development*, *9*(15), 1896–1908. http://dx.doi.org/10.1101/Gad.9.15.1896.

Fernandez-Minan, A., Martin-Bermudo, M. D., & Gonzalez-Reyes, A. (2007). Integrin signaling regulates spindle orientation in Drosophila to preserve the follicular-epithelium monolayer. *Current Biology*, *17*(8), 683–688. http://dx.doi.org/10.1016/J.Cub.2007.02.052.

Fogerty, F. J., Fessler, L. I., Bunch, T. A., Yaron, Y., Parker, C. G., Nelson, R. E., et al. (1994). Tiggrin, a novel Drosophila extracellular-matrix protein that functions as a ligand for Drosophila alpha(Ps2)beta(Ps) integrins. *Development*, *120*(7), 1747–1758.

Freeman, M., & Gurdon, J. B. (2002). Regulatory principles of developmental signaling. *Annual Review of Cell and Developmental Biology*, *18*, 515–539. http://dx.doi.org/10.1146/Annurev.Cellbio.18.012502.083458.

Frick, A., Grammel, D., Schmidt, F., Poschl, J., Priller, M., Pagella, P., et al. (2012). Proper cerebellar development requires expression of beta 1-integrin in Bergmann glia, but not in granule neurons. *Glia*, *60*(5), 820–832. http://dx.doi.org/10.1002/Glia.22314.

Friedl, P., & Gilmour, D. (2009). Collective cell migration in morphogenesis, regeneration and cancer. *Nature Reviews Molecular Cell Biology*, *10*(7), 445–457. http://dx.doi.org/10.1038/Nrm2720.

Friedl, P., Sahai, E., Weiss, S., & Yamada, K. M. (2012). New dimensions in cell migration. *Nature Reviews Molecular Cell Biology*, *13*(11), 743–747. http://dx.doi.org/10.1038/Nrm3459.

Friedland, J. C., Lee, M. H., & Boettiger, D. (2009). Mechanically activated integrin switch controls alpha(5)beta(1) function. *Science*, *323*(5914), 642–644. http://dx.doi.org/10.1126/Science.1168441.

Frisch, S. M., & Francis, H. (1994). Disruption of epithelial cell-matrix interactions induces apoptosis. *Journal of Cell Biology*, *124*(4), 619–626. http://dx.doi.org/10.1083/Jcb.124.4.619.

Fuchs, Y., & Steller, H. (2011). Programmed cell death in animal development and disease. *Cell*, *147*(4), 742–758. http://dx.doi.org/10.1016/J.Cell.2011.10.033.

Garcia-Bellido, A. (1975). Genetic control of wing disc development in Drosophila. *Ciba Foundation Symposium*, *29*, 161–182.

Geiger, B. (1979). 130 k protein from chicken gizzard—Its localization at the termini of microfilament bundles in cultured chicken-cells. *Cell*, *18*(1), 193–205. http://dx.doi.org/10.1016/0092-8674(79)90368-4.

Geiger, B., & Yamada, K. M. (2011). Molecular architecture and function of matrix adhesions. *Cold Spring Harbor Perspectives in Biology*, *3*, a005033. http://dx.doi.org/10.1101/cshperspect.a005033.

Georges-Labouesse, E., Mark, M., Messaddeq, N., & Gansmuller, A. (1998). Essential role of alpha 6 integrins in cortical and retinal lamination. *Current Biology*, *8*(17), 983–986. http://dx.doi.org/10.1016/S0960-9822(98)70402-6.

Gilbert, S. F. (1998). Conceptual breakthroughs in developmental biology. *Journal of Biosciences*, *23*(3), 169–176. http://dx.doi.org/10.1007/Bf02720017.

Gillies, T. E., & Cabernard, C. (2011). Cell division orientation in animals. *Current Biology*, *21*(15), R599–R609. http://dx.doi.org/10.1016/J.Cub.2011.06.055.

Goh, K. L., Yang, J. T., & Hynes, R. O. (1997). Mesodermal defects and cranial neural crest apoptosis in alpha 5 integrin-null embryos. *Development*, *124*(21), 4309–4319.

Gotwals, P. J., Fessler, L. I., Wehrli, M., & Hynes, R. O. (1994). Drosophila Ps1 integrin is a laminin receptor and differs in ligand specificity from Ps2. *Proceedings of the National Academy of Sciences of the United States of America*, *91*(24), 11447–11451. http://dx.doi.org/10.1073/Pnas.91.24.11447.

Goulas, S., Conder, R., & Knoblich, J. A. (2012). The par complex and integrins direct asymmetric cell division in adult intestinal stem cells. *Cell Stem Cell*, *11*(4), 529–540. http://dx.doi.org/10.1016/J.Stem.2012.06.017.

Hagedorn, E. J., Yashiro, H., Ziel, J. W., Ihara, S., Wang, Z., & Sherwood, D. R. (2009). Integrin acts upstream of netrin signaling to regulate formation of the anchor cell's invasive membrane in C. elegans. *Developmental Cell, 17*(2), 187–198. http://dx.doi.org/10.1016/J.Devcel.2009.06.006.

Haigo, S. L., & Bilder, D. (2011). Global tissue revolutions in a morphogenetic movement controlling elongation. *Science, 331*(6020), 1071–1074. http://dx.doi.org/10.1126/Science.1199424.

Halbleib, J. M., & Nelson, W. J. (2006). Cadherins in development: Cell adhesion, sorting, and tissue morphogenesis. *Genes & Development, 20*(23), 3199–3214. http://dx.doi.org/10.1101/Gad.1486806.

Haubst, N., Georges-Labouesse, E., De Arcangelis, A., Mayer, U., & Gotz, M. (2006). Basement membrane attachment is dispensable for radial glial cell fate and for proliferation, but affects positioning of neuronal subtypes. *Development, 133*(16), 3245–3254. http://dx.doi.org/10.1242/Dev.02486.

Heller, E., Kumar, K. V., Grill, S. W., & Fuchs, E. (2014). Forces generated by cell intercalation tow epidermal sheets in mammalian tissue morphogenesis. *Developmental Cell, 28*(6), 617–632. http://dx.doi.org/10.1016/J.Devcel.2014.02.011.

Hresko, M. C., Williams, B. D., & Waterston, R. H. (1994). Assembly of body-wall muscle and muscle-cell attachment structures in Caenorhabditis elegans. *Journal of Cell Biology, 124*(4), 491–506. http://dx.doi.org/10.1083/Jcb.124.4.491.

Hsieh, H. H., Hsu, T. Y., Jiang, H. S., & Wu, Y. C. (2012). Integrin alpha PAT-2/CDC-42 signaling is required for muscle-mediated clearance of apoptotic cells in Caenorhabditis elegans. *Plos Genetics, 8*(5), e1002663. http://dx.doi.org/10.1371/Journal.Pgen.1002663.

Hsu, T. Y., & Wu, Y. C. (2010). Engulfment of apoptotic cells in C. elegans is mediated by integrin alpha/SRC signaling. *Current Biology, 20*(6), 477–486. http://dx.doi.org/10.1016/J.Cub.2010.01.062.

Humphries, J. D., Byron, A., & Humphries, M. J. (2006). Integrin ligands at a glance. *Journal of Cell Science, 119*(19), 3901–3903. http://dx.doi.org/10.1242/Jcs.03098.

Huttenlocher, A., & Horwitz, A. R. (2011). Integrins in cell migration. *Cold Spring Harbor Perspectives in Biology, 3*, a005074. http://dx.doi.org/10.1101/cshperspect.a005074.

Huveneers, S., & Danen, E. H. J. (2009). Adhesion signaling—Crosstalk between integrins, Src and Rho. *Journal of Cell Science, 122*(8), 1059–1069. http://dx.doi.org/10.1242/Jcs.039446.

Hynes, R. O. (1987). Integrins—A family of cell-surface receptors. *Cell, 48*(4), 549–554. http://dx.doi.org/10.1016/0092-8674(87)90233-9.

Hynes, R. O., & Zhao, Q. (2000). The evolution of cell adhesion. *Journal of Cell Biology, 150*(2), F89–F95. http://dx.doi.org/10.1083/Jcb.150.2.F89.

Ihara, S., Hagedorn, E. J., Morrissey, M. A., Chi, Q. Y., Motegi, F., Kramer, J. M., et al. (2011). Basement membrane sliding and targeted adhesion remodels tissue boundaries during uterine-vulval attachment in Caenorhabditis elegans. *Nature Cell Biology, 13*(6), 641–651. http://dx.doi.org/10.1038/Ncb2233.

Julich, D., Geisler, R., Holley, S. A., & Tübingen 2000 Screen Consortium (2005). Integrin alpha 5 and delta/notch signaling have complementary spatiotemporal requirements during zebrafish somitogenesis. *Developmental Cell, 8*(4), 575–586. http://dx.doi.org/10.1016/J.Devcel.2005.01.016.

Julich, D., Mould, A. P., Koper, E., & Holley, S. A. (2009). Control of extracellular matrix assembly along tissue boundaries via Integrin and Eph/Ephrin signaling. *Development, 136*(17), 2913–2921. http://dx.doi.org/10.1242/Dev.038935.

Kardash, E., Reichman-Fried, M., Maitre, J. L., Boldajipour, B., Papusheva, E., Messerschmidt, E. M., et al. (2010). A role for Rho GTPases and cell-cell adhesion in single-cell motility *in vivo*. *Nature Cell Biology, 12*(1), 47–53. http://dx.doi.org/10.1038/Ncb2003.

Kelley, L. C., Lohmer, L. L., Hagedorn, E. J., & Sherwood, D. R. (2014). Traversing the basement membrane *in vivo*: A diversity of strategies. *Journal of Cell Biology, 204*(3), 291–302. http://dx.doi.org/10.1083/Jcb.201311112.
Koshida, S., Kishimoto, Y. K., Ustumi, H., Shimizu, T., Furutani-Seiki, M., Kondoh, H., et al. (2005). Integrin alpha 5-dependent fibronectin accumulation for maintenance of somite boundaries in zebrafish embryos. *Developmental Cell, 8*(4), 587–598. http://dx.doi.org/10.1016/J.Devcel.2005.03.006.
Kragtorp, K. A., & Miller, J. R. (2007). Integrin alpha 5 is required for somite rotation and boundary formation in Xenopus. *Developmental Dynamics, 236*(9), 2713–2720. http://dx.doi.org/10.1002/Dvdy.21280.
Kreitmeier, M., Gerisch, G., Heizer, C., & Mullertaubenberger, A. (1995). A talin homolog of dictyostelium rapidly assembles at the leading-edge of cells in response to chemoattractant. *Journal of Cell Biology, 129*(1), 179–188. http://dx.doi.org/10.1083/Jcb.129.1.179.
Kubow, K. E., & Horwitz, A. R. (2011). Reducing background fluorescence reveals adhesions in 3D matrices. *Nature Cell Biology, 13*(1), 3–5.
Lallier, T. E., & Desimone, D. W. (2000). Separation of neural induction and neurulation in Xenopus. *Developmental Biology, 225*(1), 135–150. http://dx.doi.org/10.1006/Dbio.2000.9833.
Lammermann, T., Bader, B. L., Monkley, S. J., Worbs, T., Wedlich-Soldner, R., Hirsch, K., et al. (2008). Rapid leukocyte migration by integrin-independent flowing and squeezing. *Nature, 453*(7191), 51–55. http://dx.doi.org/10.1038/Nature06887.
Larsen, E., & Mclaughlin, H. M. G. (1987). The morphogenetic alphabet—lessons for simple-minded genes. *Bioessays*, 7(3), 130–132. http://dx.doi.org/10.1002/Bies.950070309.
le Duc, Q., Shi, Q. M., Blonk, I., Sonnenberg, A., Wang, N., Leckband, D., et al. (2010). Vinculin potentiates E-cadherin mechanosensing and is recruited to actin-anchored sites within adherens junctions in a myosin II-dependent manner. *Journal of Cell Biology, 191*(4), 891. http://dx.doi.org/10.1083/Jcb.20100114920101020c (vol. 189, p. 1107, 2010).
Lechler, T., & Fuchs, E. (2005). Asymmetric cell divisions promote stratification and differentiation of mammalian skin. *Nature, 437*(7056), 275–280. http://dx.doi.org/10.1038/Nature03922.
Lecuit, T., Lenne, P. F., & Munro, E. (2011). Force generation, transmission, and integration during cell and tissue morphogenesis. *Annual Review of Cell and Developmental Biology, 27*(27), 157–184. http://dx.doi.org/10.1146/Annurev-Cellbio-100109-104027.
Lee, J. L., & Streuli, C. H. (2014). Integrins and epithelial cell polarity. *Journal of Cell Science, 127*(15), 3217–3225. http://dx.doi.org/10.1242/Jcs.146142.
Lee, S., Zhou, L. L., Kim, J., Kalbfleisch, S., & Schock, F. (2008). Lasp anchors the Drosophila male stem cell niche and mediates spermatid individualization. *Mechanisms of Development, 125*(9–10), 768–776. http://dx.doi.org/10.1016/J.Mod.2008.06.012.
Leu, S. T., Jacques, S. A. L., Wingerd, K. L., Hikita, S. T., Tolhurst, E. C., Pring, J. L., et al. (2004). Integrin alpha-4 beta 1 function is required for cell survival in developing retina. *Developmental Biology, 276*(2), 416–430. http://dx.doi.org/10.1016/J.Ydbio.2004.09.003.
Lewellyn, L., Cetera, M., & Horne-Badovinac, S. (2013). Misshapen decreases integrin levels to promote epithelial motility and planar polarity in Drosophila. *Journal of Cell Biology, 200*(6), 721–729. http://dx.doi.org/10.1083/Jcb.201209129.
Lin, Y. H., Park, Z. Y., Lin, D., Brahmbhatt, A. A., Rio, M. C., Yates, J. R., et al. (2004). Regulation of cell migration and survival by focal adhesion targeting of Lasp-1. *Journal of Cell Biology, 165*(3), 421–432. http://dx.doi.org/10.1083/Jcb.200311045.
Lin, G. N., Zhang, X., Ren, J., Pang, Z. M., Wang, C. H., Xu, N., et al. (2013). Integrin signaling is required for maintenance and proliferation of intestinal stem cells in Drosophila. *Developmental Biology, 377*(1), 177–187. http://dx.doi.org/10.1016/J.Ydbio.2013.01.032.

Liu, J., He, X. W., Corbett, S. A., Lowry, S. F., Graham, A. M., Fassler, R., et al. (2009). Integrins are required for the differentiation of visceral endoderm. *Journal of Cell Science*, *122*(2), 233–242. http://dx.doi.org/10.1242/Jcs.037663.

Loulier, K., Lathia, J. D., Marthiens, V., Relucio, J., Mughal, M. R., Tang, S. C., et al. (2009). beta 1 integrin maintains integrity of the embryonic neocortical stem cell niche. *Plos Biology*, 7(8), e1000176. http://dx.doi.org/10.1371/Journal.Pbio.1000176.

Mammoto, T., Mammoto, A., & Ingber, D. E. (2013). Mechanobiology and developmental control. *Annual Review of Cell and Developmental Biology*, *29*(29), 27–61. http://dx.doi.org/10.1146/Annurev-Cellbio-101512-122340.

Marsden, M., & Burke, R. D. (1998). The beta L integrin subunit is necessary for gastrulation in sea urchin embryos. *Developmental Biology*, *203*(1), 134–148. http://dx.doi.org/10.1006/Dbio.1998.9033.

Marsden, M., & DeSimone, D. W. (2001). Regulation of cell polarity, radial intercalation and epiboly in Xenopus: Novel roles for integrin and fibronectin. *Development*, *128*(18), 3635–3647.

Marsden, M., & DeSimone, D. W. (2003). Integrin-ECM interactions regulate cadherin-dependent cell adhesion and are required for convergent extension in Xenopus. *Current Biology*, *13*(14), 1182–1191. http://dx.doi.org/10.1016/S0960-9822(03)00433-0.

Marthiens, V., Kazanis, I., Moss, L., Long, K., & Ffrench-Constant, C. (2010). Adhesion molecules in the stem cell niche—More than just staying in shape? *Journal of Cell Science*, *123*(10), 1613–1622. http://dx.doi.org/10.1242/Jcs.054312.

Martin-Bermudo, M. D., Alvarez-Garcia, I., & Brown, N. H. (1999). Migration of the Drosophila primordial midgut cells requires coordination of diverse PS integrin functions. *Development*, *126*(22), 5161–5169.

Matus, D. Q., Chang, E., Makohon-Moore, S. C., Hagedorn, M. A., Chi, Q. Y., & Sherwood, D. R. (2014). Cell division and targeted cell cycle arrest opens and stabilizes basement membrane gaps. *Nature Communications*, *5*, 4184. http://dx.doi.org/10.1038/Ncomms5184.

Meighan, C. M., & Schwarzbauer, J. E. (2008). Temporal and spatial regulation of integrins during development. *Current Opinion in Cell Biology*, *20*(5), 520–524. http://dx.doi.org/10.1016/J.Ceb.2008.05.010.

Miller, C. J., & Davidson, L. A. (2013). The interplay between cell signalling and mechanics in developmental processes. *Nature Reviews Genetics*, *14*(10), 733–744. http://dx.doi.org/10.1038/Nrg3513.

Mitra, S. K., Hanson, D. A., & Schlaepfer, D. D. (2005). Focal adhesion kinase: In command and control of cell motility. *Nature Reviews Molecular Cell Biology*, *6*(1), 56–68. http://dx.doi.org/10.1038/Nrm1549.

Moreira, C. G. A., Jacinto, A., & Prag, S. (2013). Drosophila integrin adhesion complexes are essential for hemocyte migration *in vivo*. *Biology Open*, *2*, 795–801. http://dx.doi.org/10.1242/bio.20134564.

Morita, H., Kajiura-Kobayashi, H., Takagi, C., Yamamoto, T. S., Nonaka, S., & Ueno, N. (2012). Cell movements of the deep layer of non-neural ectoderm underlie complete neural tube closure in Xenopus. *Development*, *139*(8), 1417–1426. http://dx.doi.org/10.1242/Dev.073239.

Morrison, S. J., & Spradling, A. C. (2008). Stem cells and niches: Mechanisms that promote stem cell maintenance throughout life. *Cell*, *132*(4), 598–611. http://dx.doi.org/10.1016/J.Cell.2008.01.038.

Nair, S., & Schilling, T. F. (2008). Chemokine signaling controls endodermal migration during zebrafish gastrulation. *Science*, *322*(5898), 89–92. http://dx.doi.org/10.1126/Science.1160038.

Naylor, M. J., Li, N., Cheung, J., Lowe, E. T., Lambert, E., Marlow, R., et al. (2005). Ablation of beta 1 integrin in mammary epithelium reveals a key role for integrin in glandular

morphogenesis and differentiation. *Journal of Cell Biology, 171*(4), 717–728. http://dx.doi.org/10.1083/Jcb.200503144.
Norton, J., Cooley, J., Islam, A. F. M. T., Cota, C. D., & Davidson, B. (2013). Matrix adhesion polarizes heart progenitor induction in the invertebrate chordate Ciona intestinalis. *Development, 140*(6), 1301–1311. http://dx.doi.org/10.1242/Dev.085548.
O'Reilly, A. M., Lee, H. H., & Simon, M. A. (2008). Integrins control the positioning and proliferation of follicle stem cells in the Drosophila ovary. *Journal of Cell Biology, 182*(4), 801–815. http://dx.doi.org/10.1083/Jcb.200710141.
Petrie, R. J., Gavara, N., Chadwick, R. S., & Yamada, K. M. (2012). Nonpolarized signaling reveals two distinct modes of 3D cell migration. *Journal of Cell Biology, 197*(3), 439–455. http://dx.doi.org/10.1083/Jcb.201201124.
Petrie, R. J., Koo, H., & Yamada, K. M. (2014). Generation of compartmentalized pressure by a nuclear piston governs cell motility in a 3D matrix. *Science, 345*(6200), 1062–1065. http://dx.doi.org/10.1126/science.1256965.
Pines, M., Das, R., Ellis, S. J., Morin, A., Czerniecki, S., Yuan, L., et al. (2012). Mechanical force regulates integrin turnover in Drosophila *in vivo*. *Nature Cell Biology, 14*(9), 935–943. http://dx.doi.org/10.1038/Ncb2555.
Pullan, S., Wilson, J., Metcalfe, A., Edwards, G. M., Goberdhan, N., Tilly, J., et al. (1996). Requirement of basement membrane for the suppression of programmed cell death in mammary epithelium. *Journal of Cell Science, 109*, 631–642.
Qiao, L. J., Gao, H. W., Zhang, T., Jing, L. L., Xiao, C., Xiao, Y., et al. (2014). Snail modulates the assembly of fibronectin via alpha 5 integrin for myocardial migration in zebrafish embryos. *Scientific Reports, 4*, 4470. http://dx.doi.org/10.1038/Srep04470.
Rallis, C., Pinchin, S. M., & Ish-Horowicz, D. (2010). Cell-autonomous integrin control of Wnt and Notch signalling during somitogenesis. *Development, 137*(21), 3591–3601. http://dx.doi.org/10.1242/Dev.050070.
Rasmussen, J. P., Reddy, S. S., & Priess, J. R. (2012). Laminin is required to orient epithelial polarity in the C. elegans pharynx. *Development, 139*(11), 2050–2060. http://dx.doi.org/10.1242/Dev.078360.
Reddien, P. W., & Horvitz, H. R. (2004). The engulfment process of programmed cell death in Caenorhabditis elegans. *Annual Review of Cell and Developmental Biology, 20*, 193–221. http://dx.doi.org/10.1146/Annurev.Cellbio.20.022003.114619.
Richardson, B. E., & Lehmann, R. (2010). Mechanisms guiding primordial germ cell migration: Strategies from different organisms. *Nature Reviews Molecular Cell Biology, 11*(1), 37–49. http://dx.doi.org/10.1038/Nrm2815.
Richter, D. J., & King, N. (2013). The genomic and cellular foundations of animal origins. *Annual Review of Genetics, 47*(47), 509–537. http://dx.doi.org/10.1146/Annurev-Genet-111212-133456.
Rohrschneider, L. R. (1980). Adhesion plaques of rous-sarcoma virus-transformed cells contain the Src gene-product. *Proceedings of the National Academy of Sciences of the United States of America-Biological Sciences*, 77(6), 3514–3518. http://dx.doi.org/10.1073/Pnas.77.6.3514.
Ross, T. D., Coon, B. G., Yun, S., Baeyens, N., Tanaka, K., Ouyang, M. X., et al. (2013). Integrins in mechanotransduction. *Current Opinion in Cell Biology, 25*(5), 613–618. http://dx.doi.org/10.1016/J.Ceb.2013.05.006.
Rozario, T., & DeSimone, D. W. (2010). The extracellular matrix in development and morphogenesis: A dynamic view. *Developmental Biology, 341*(1), 126–140. http://dx.doi.org/10.1016/J.Ydbio.2009.10.026.
Schwartz, M. A. (2010). Integrins and extracellular matrix in mechanotransduction. *Cold Spring Harbor Perspectives in Biology, 2*, a005066. http://dx.doi.org/10.1101/cshperspect.a005066.
Sebe-Pedros, A., Irimia, M., del Campo, J., Parra-Acero, H., Russ, C., Nusbaum, C., et al. (2013). Regulated aggregative multicellularity in a close unicellular relative of metazoa. *Elife, 2*, e02187. http://dx.doi.org/10.7554/eLife.01287.

Seifert, J. R. K., & Lehmann, R. (2012). Drosophila primordial germ cell migration requires epithelial remodeling of the endoderm. *Development, 139*(12), 2101–2106. http://dx.doi.org/10.1242/Dev.078949.

Shen, Q., Wang, Y., Kokovay, E., Lin, G., Chuang, S. M., Goderie, S. K., et al. (2008). Adult SVZ stem cells lie in a vascular niche: A quantitative analysis of niche cell-cell interactions. *Cell Stem Cell, 3*(3), 289–300. http://dx.doi.org/10.1016/J.Stem.2008.07.026.

Streuli, C. H. (2009). Integrins and cell-fate determination. *Journal of Cell Science, 122*(2), 171–177. http://dx.doi.org/10.1242/Jcs.018945.

Streuli, C. H., & Akhtar, N. (2009). Signal co-operation between integrins and other receptor systems. *Biochemical Journal, 418*, 491–506. http://dx.doi.org/10.1042/Bj20081948.

Stroka, K., Jiang, H., Chen, S.-H., Tong, Z., Wirtz, D., Sun, S., et al. (2014). Water permeation drives tumor cell migration in confined microenvironments. *Cell, 157*(3), 611–623. http://dx.doi.org/10.1016/j.cell.2014.02.052.

Stupack, D. G., Puente, X. S., Boutsaboualoy, S., Storgard, C. M., & Cheresh, D. A. (2001). Apoptosis of adherent cells by recruitment of caspase-8 to unligated integrins. *Journal of Cell Biology, 155*(3), 459–470. http://dx.doi.org/10.1083/Jcb.200106070.

Subramanian, A., Wayburn, B., Bunch, T., & Volk, T. (2007). Thrombospondin-mediated adhesion is essential for the formation of the myotendinous junction in Drosophila. *Development, 134*(7), 1269–1278. http://dx.doi.org/10.1242/Dev.000406.

Suga, H., Chen, Z. H., de Mendoza, A., Sebe-Pedros, A., Brown, M. W., Kramer, E., et al. (2013). The Capsaspora genome reveals a complex unicellular prehistory of animals. *Nature Communications, 4*, 2325. http://dx.doi.org/10.1038/Ncomms3325.

Swift, J., & Discher, D. E. (2014). The nuclear lamina is mechano-responsive to ECM elasticity in mature tissue. *Journal of Cell Science, 127*(14), 3005–3015. http://dx.doi.org/10.1242/Jcs.149203.

Swift, J., Ivanovska, I. L., Buxboim, A., Harada, T., Dingal, P. C. D. P., Pinter, J., et al. (2013). Nuclear lamin-A scales with tissue stiffness and enhances matrix-directed differentiation. *Science, 341*(6149), 1240104. http://dx.doi.org/10.1126/Science.1240104.

Taddei, I., Deugnier, M. A., Faraldo, M. M., Petit, V., Bouvard, D., Medina, D., et al. (2008). beta 1 integrin deletion from the basal compartment of the mammary epithelium affects stem cells. *Nature Cell Biology, 10*(6), 716–722. http://dx.doi.org/10.1038/Ncb1734.

Tanentzapf, G., Devenport, D., Godt, D., & Brown, N. H. (2007). Integrin-dependent anchoring of a stem-cell niche. *Nature Cell Biology, 9*(12), 1413–1418. http://dx.doi.org/10.1038/Ncb1660.

Tang, Y., Rowe, R. G., Botvinick, E. L., Kurup, A., Putnam, A. J., Seiki, M., et al. (2013). MT1-MMP-dependent control of skeletal stem cell commitment via a beta 1-integrin/YAP/TAZ signaling axis. *Developmental Cell, 25*(4), 402–416. http://dx.doi.org/10.1016/J.Devcel.2013.04.011.

Tavazoie, M., Van der Veken, L., Silva-Vargas, V., Louissaint, M., Colonna, L., Zaidi, B., et al. (2008). A specialized vascular niche for adult neural stem cells. *Cell Stem Cell, 3*(3), 279–288. http://dx.doi.org/10.1016/J.Stem.2008.07.025.

Testaz, S., & Duband, J. L. (2001). Central role of the alpha 4 beta 1 integrin in the coordination of avian truncal neural crest cell adhesion, migration, and survival. *Developmental Dynamics, 222*(2), 127–140. http://dx.doi.org/10.1002/Dvdy.1181.

Theveneau, E., Steventon, B., Scarpa, E., Garcia, S., Trepat, X., Streit, A., et al. (2013). Chase-and-run between adjacent cell populations promotes directional collective migration. *Nature Cell Biology, 15*(7), 763–772. http://dx.doi.org/10.1038/Ncb2772.

Tian, A. G., & Jiang, J. (2014). Intestinal epithelium-derived BMP controls stem cell self-renewal in Drosophila adult midgut. *Elife, 3*, e01857. http://dx.doi.org/10.7554/eLife.01857.

Trappmann, B., Gautrot, J. E., Connelly, J. T., Strange, D. G. T., Li, Y., Oyen, M. L., et al. (2012). Extracellular-matrix tethering regulates stem-cell fate. *Nature Materials, 11*(8), 742. http://dx.doi.org/10.1038/Nmat3387 (vol. 11, p. 642, 2012).
Tucker, R. P. (2004). Antisense knockdown of the Œ≤1 integrin subunit in the chicken embryo results in abnormal neural crest cell development. *The International Journal of Biochemistry & Cell Biology, 36*(6), 1135–1139.http://dx.doi.org/10.1016/j.biocel.2004.01.010.
Urbano, J. M., Dominguez-Gimenez, P., Estrada, B., & Martin-Bermudo, M. D. (2011). PS integrins and laminins: Key regulators of cell migration during Drosophila embryogenesis. *Plos One, 6*(9), e23893. http://dx.doi.org/10.1371/journal.pone.0023893.
Vied, C., Reilein, A., Field, N. S., & Kalderon, D. (2012). Regulation of stem cells by intersecting gradients of long-range niche signals. *Developmental Cell, 23*(4), 836–848. http://dx.doi.org/10.1016/J.Devcel.2012.09.010.
Viveiros, R., Hutter, H., & Moerman, D. G. (2011). Membrane extensions are associated with proper anterior migration of muscle cells during Caenorhabditis elegans embryogenesis. *Developmental Biology, 358*(1), 189–200. http://dx.doi.org/10.1016/J.Ydbio.2011.07.026.
Wang, X. M., Harris, R. E., Bayston, L. J., & Ashe, H. L. (2008). Type IV collagens regulate BMP signalling in Drosophila. *Nature, 455*(7209), 72–77. http://dx.doi.org/10.1038/Nature07214.
Watt, F. M., & Huck, W. T. S. (2013). Role of the extracellular matrix in regulating stem cell fate. *Nature Reviews Molecular Cell Biology, 14*(8), 467–473. http://dx.doi.org/10.1038/Nrm3620.
Weber, G. F., Bjerke, M. A., & DeSimone, D. W. (2011). Integrins and cadherins join forces to form adhesive networks. *Journal of Cell Science, 124*(8), 1183–1193. http://dx.doi.org/10.1242/Jcs.064618.
Wickstrom, S. A., Radovanac, K., & Fassler, R. (2011). Genetic analyses of integrin signaling. *Cold Spring Harbor Perspectives in Biology, 3*, a005116. http://dx.doi.org/10.1101/cshperspect.a005116.
Wieschaus, E. (1996). From molecular patterns to morphogenesis—The lessons from studies on the fruit fly Drosophila (nobel lecture). *Angewandte Chemie, International Edition in English, 35*(19), 2188–2194. http://dx.doi.org/10.1002/anie.199621881.
Williams, B. D., & Waterston, R. H. (1994). Genes critical for muscle development and function in Caenorhabditis-elegans identified through lethal mutations. *Journal of Cell Biology, 124*(4), 475–490. http://dx.doi.org/10.1083/Jcb.124.4.475.
Wright, T. R. F. (1960). The phenogenetics of the embryonic mutant, lethal myospheroid, in Drosophila melanogaster. *Journal of Experimental Zoology, 143*(1), 77–99. http://dx.doi.org/10.1002/Jez.1401430107.
Yang, J. T., Bader, B. L., Kreidberg, J. A., Ullmann-Cullere, M., Trevithick, J. E., & Hynes, R. O. (1999). Overlapping and independent functions of fibronectin receptor integrins in early mesodermal development. *Developmental Biology, 215*(2), 264–277. http://dx.doi.org/10.1006/Dbio.1999.9451.
Yuan, L., Fairchild, M. J., Perkins, A. D., & Tanentzapf, G. (2010). Analysis of integrin turnover in fly myotendinous junctions. *Journal of Cell Science, 123*(6), 939–946. http://dx.doi.org/10.1242/Jcs.063040.
Zaidel-Bar, R., Itzkovitz, S., Ma'ayan, A., Iyengar, R., & Geiger, B. (2007). Functional atlas of the integrin adhesome. *Nature Cell Biology, 9*(8), 858–868. http://dx.doi.org/10.1038/Ncb0807-858.
Zamir, E. A., Rongish, B. J., & Little, C. D. (2008). The ECM moves during primitive streak formation-computation of ECM versus cellular motion. *Plos Biology, 6*(10), 2163–2171. http://dx.doi.org/10.1371/journal.pbio.0060247.

Zervas, C. G., Gregory, S. L., & Brown, N. H. (2001). Drosophila integrin-linked kinase is required at sites of integrin adhesion to link the cytoskeleton to the plasma membrane. *Journal of Cell Biology*, *152*(5), 1007–1018. http://dx.doi.org/10.1083/Jcb.152.5.1007.

Zhang, X., Jiang, G., Cai, Y., Monkley, S. J., Critchley, D. R., & Sheetz, M. P. (2008). Talin depletion reveals independence of initial cell spreading from integrin activation and traction. *Nature Cell Biology*, *10*(9), 1062–1068. http://dx.doi.org/10.1038/Ncb1765.

CHAPTER EIGHT

Epithelial–Mesenchymal Transitions: From Cell Plasticity to Concept Elasticity

Pierre Savagner[*,1]
*IRCM, Institut de Recherche en Cancérologie de Montpellier, INSERM U896, Institut régional du cancer Université Montpellier1, Montpellier, France
[1]Corresponding author: e-mail address: pierre.savagner@icm.unicancer.fr

Contents

Abstract

Epithelial–mesenchymal transition (EMT) is a developmental cellular process occurring during early embryo development, including gastrulation and neural crest cell migration. It can be broken down in distinct functional steps: (1) loss of baso-apical polarization characterized by cytoskeleton, tight junctions, and hemidesmosomes remodeling; (2) individualization of cells, including a decrease in cell–cell adhesion forces, (3) emergence of motility, and (4) invasive properties, including passing through the sub-epithelial basement membrane. These phases occur in an uninterrupted process, without requiring mitosis, in an order and with a degree of completion dictated by the microenvironment. The whole process reflects the activation of specific transcription factor families, called EMT transcription factors. Several mechanisms can combine to induce EMT. Some are reversible, involving growth factors and cytokines and/or environmental signals including extracellular matrix and local physical conditions. Others are irreversible, such as genomic alterations during carcinoma progression, along a selective and irreversible clonal drift. In carcinomas, these signals can converge to initiate a metastable phenotype. In this state, similarly to activated keratinocytes during re-epithelialization, cells can initiate a cohort migration and engage into a transient and reversible EMT controlled by the local environment prior to efficient intravasation and metastasis. EMT transcription factors also participate in cancer progression by

Current Topics in Developmental Biology, Volume 112
ISSN 0070-2153
http://dx.doi.org/10.1016/bs.ctdb.2014.11.021

inducing apoptosis resistance and maintaining stem-like properties exposed in tumor recurrences. These properties, very important on a clinical point of view, are not intrinsically linked to EMT, but can share common pathways.

1. BACK TO THE ORIGINS: DEFINING EMT

The epithelial–mesenchymal transition (EMT) concept evolved from an encounter between *in vitro* studies of epithelial cells individualizing and migrating in response to a signal and *in vivo* observations of epithelial cells individualizing and migrating away from epithelial sheets during "classic" developmental stages. *In vitro* cell models included original observations on lens epithelial cells grown in collagen gels, that gave rise to the concept (Greenburg & Hay, 1982; Hay, 1968); bladder carcinoma cells activated by collagen or FGF1 (Valles et al., 1990); and MDCK cells activated by HGF/SF (Stoker, Gherardi, Perryman, & Gray, 1987; Weidner, Behrens, Vandekerckhove, & Birchmeier, 1990). EMT was found to be mostly reversible in these cell lines, cells reversing along a mesenchymal–epithelial transition (MET) when the specific EMT stimuli were removed. Classic developmental EMT stages comprised gastrulation, neural crest cell migration, and heart morphogenesis. The EMT concept evolved mostly based on these findings, by pulling together common morphological features: downregulation of cell–cell adhesion structures, cytoskeleton remodeling following a collapse of the basal–apical polarity, and emergence of cell motility. Early molecular markers were defined accordingly to these models: E-cadherin downregulation, *de novo* vimentin expression, cytokeratin downregulation, actin and polarity reorganization, and cell motility.

With the exponential growth of publications dedicated to EMT in the following years, a number of markers, pathways, and gene families have since been linked to EMT. An early finding was the requirement for specific transcriptional events. Several transcription factor families were rapidly incriminated in the process, including the *Snail, Twist*, and *Zeb* families. They were dubbed EMT mastergenes since they were originally described in this context. This may be a "EMT-centric" view since their main physiological functions remain to be established. However, the fact remains that at least one of them is expressed in every developmental EMT situations. EMT transcription factors (EMT-TF) have in common the capacity to recognize E-boxes in various targets genes, including *E-cadherin*, through distinct mechanisms and transcriptional complexes. However, they also get involved in other molecular pathways. In the last ten years, they were found to be involved in controlling very different functions including apoptosis and stemness, through distinct mechanisms (Brabletz, 2012; Creighton, Chang, & Rosen, 2010; Fang et al., 2011; Scheel & Weinberg, 2012). This has lead to a functional link between the EMT process and these processes. If this link can be observed in various *in vivo* situations and *in vitro* models, it must be emphasized that this only reflects shared pathways and should not be considered as a dogmatic concept. There is a risk of inflating the "EMT concept" to a potentially misleading globalizing and monolithic cellular process combining the regulation of EMT (in the original meaning), cell stemness, differentiation, motility and apoptotic death, controlled by a single set of genes.

One way to avoid oversimplifying the "EMT" concept is by generating a classification of EMT and EMT-like processes. Several typings have been proposed. A sensible one is based on embryo development, discriminating the successive and constitutive EMT–MET cycles during early embryogenesis and organogenesis (Thiery, Acloque, Huang, & Nieto, 2009), resulting in primary, secondary, tertiary, and quaternary EMT. It is tempting to speculate that each of these successive phases could reflect a distinct genomic or epigenetic control level, mediated by common transcriptional mediators destabilizing a primordial epithelial phenotype (Lim & Thiery, 2012). In fact, the hypothesis of a "default" epithelial phenotype was proposed some time ago (Frisch, 1997) to emphasize the idea that some epithelial gate-keepers mechanisms had to be breached to obtain cellular remodeling. As detailed further, this idea is now supported by an increasing number of observations and should have an interesting impact in the cancer field.

Therefore, we suggest that the use of the term "EMT" should be restricted using a few defining functional criteria to avoid confusion. We feel the EMT process should be defined by (1) a loss of baso-apical polarization linked to cytoskeleton, tight junctions, and hemidesmosome remodeling; (2) the individualization of cells, linked to a decrease in cell–cell adhesion forces; (3) the emergence of motility; and (4) the display of invasive properties, combined with the passage of cells through a subepithelial basement membrane. In addition, these phases should occur in an uninterrupted process, without requiring mitosis. Various pathways have been found to induce, transduce and/or maintain this EMT process. Most of them, thus far, have been found to involve at least one member of the EMT-TF family.

However, these genes are also involved in separate cell functions and their expression cannot be considered as an unequivocal EMT marker.

2. EMT IS A MORPHOGENIC DEVELOPMENTAL PROCESS, OR IS IT?

Cell phenotype modulation can be traced to the very early developmental stages in all metazoans. The first divisions from the zygote give rise to rounded cells showing no apparent integrated cell–cell adhesion structures; the *Zona pellucida* apparently suffices to maintain early blastomere proximity (De Vries et al., 2004). However, these 2/4/8 cells express adhesion molecules such as E-cadherin and Desmocollin 3 at the membrane level and Desmocollin 3 is required for preimplantation (Den, Cheng, Merched-Sauvage, & Koch, 2006). The first cell–cell adhesion structures involving cytoskeletal connection appear at the time of compaction, when the mouse embryo comprises eight blastomeres (Morali, Savagner, & Larue, 2005). Compaction can be considered as the time when an individual multicellular organism defines itself as a unit. Further cell divisions and EMT phases will not threaten this essential self-identity. Features of epithelial differentiation and polarization are mobilized at these founding stages defining the first epithelium, the trophectoderm, in various mammalian species (Reima, Lehtonen, Virtanen, & Flechon, 1993).

The first embryonic EMT takes place during gastrulation and has been extensively reviewed (Lim & Thiery, 2012; Nakaya & Sheng, 2008; Solnica-Krezel & Sepich, 2012). Very briefly, during this relatively fast event, lasting about 24 h in mammals, presumptive mesodermal and endodermal cell populations migrate inside the ectodermal sheet to evolve into the three primary layers. Interestingly, the level of cell cohesion between migrating prefated mesodermal cells is variable among species. Cells are compacted in frog with strong intercellular interactions, but more individualized in zebrafish and chicken. The prefated distinct cell populations express distinct migration modes and cell phenotypes during gastrulation. In frog, EMT takes place after the nascent mesodermal cells have migrated as a coherent mass through the blastopore in a process called involution (Solnica-Krezel & Sepich, 2012). In amniotes, EMT precedes this move and nascent mesodermal cells pass inside the primitive streak as individual cells in a process called ingression.

Therefore, the link between EMT and cell motility depends on context and cell individualization is not a prerequisite for motility. The challenge remains to decipher specific pathways mediating the cellular responses involved in gastrulation, from cell fating to motility as thoroughly reviewed in recent reviews (Heisenberg & Solnica-Krezel, 2008; Keller, 2005; Leptin, 2005; Solnica-Krezel, 2005).

An important feature is that the primary cell population is a polarized epithelium involving a basement membrane that has to be broken down in most species (Nakaya & Sheng, 2013). The timing for this degradation is specific. It actually occurs prior to cell polarity disruption during chick embryo gastrulation, a process that involves RhoA and microtubule dynamics (Nakaya & Sheng, 2008). Overall, the gastrulation process involves massive cell migration. This large scale shift can start prior to EMT stage, as seen with the so-called "Polonaise" cell movements in chicken, or can start later, in link with the EMT/cell individualization stage in mice (Nowotschin & Hadjantonakis, 2010). Therefore, EMT is only a part of the global process and appears to play a distinct role in different species.

Accordingly, cadherins are regulated during this process, with differences among species reflecting cell behavior (Hammerschmidt & Wedlich, 2008). In *Xenopus*, the classical cadherin C-cadherin is necessary for involution (Kim, Yamamoto, Bouwmeester, Agius, & Robertis, 1998). *C-cadherin* is then downregulated, through an interaction with paraxial protocadherin that is required for the convergent extension movements (Chen & Gumbiner, 2006; Marsden & DeSimone, 2003). *E-cadherin* is downregulated in amniotes, but upregulated in zebrafish, following internalization, and appears to be involved in supporting cell elongation and migration (Montero et al., 2005). In mouse, a shift is observed in the cadherin expression pattern during gastrulation, between E-cadherin and N-cadherin, another classic cadherin that is expressed by mesodermal cells and associated with invasion and motility (Hatta & Takeichi, 1986). N-cadherin can promote both adhesion and motility depending on the cellular context. Interestingly, cell–cell adhesion and motility properties are mediated by distinct sites of N-cadherin, showing that these properties can be regulated independently (Derycke & Bracke, 2004). The downregulation of *E-cadherin* in amniotes involves EMT-TFs, with specific differences: *Snail* appears to be essential for gastrulation in mice, whereas *Slug* appears to be necessary during chicken gastrulation. Considering the difference in cell population behavior observed in these two species during early gastrulation, it is tempting to hypothesize that these two genes could regulate distinct processes: Snail could target primarily cell–cell adhesion and Slug be more involved in cohesive cell group migration, like that described in other physiological situations. Accordingly, during gastrulation, mesodermal cells from *Snail*-deficient mice maintain a baso-apical polarity and express E-cadherin. However, these embryos do not survive past this stage (Carver, Jiang, Lan, Oram, & Gridley, 2001). Members of the *Zeb* family are expressed early in zebrafish and also contribute to gastrulation by modulating cell–cell adhesion, targeting *E-cadherin* and *EpCAM*, as well as *Snail* genes (Vannier, Mock, Brabletz, & Driever, 2013). Members of the *Twist* family are not expressed before gastrulation and therefore do not appear to participate to mesodermal induction (Barnes & Firulli, 2009).

The next classic example of EMT is the formation of the neural crest cells. This process has also been studied and reviewed extensively (Kerosuo & Bronner-Fraser, 2012; Lim & Thiery, 2012; Theveneau & Mayor, 2012) (see Chapter 'Embryonic Cell–Cell Adhesion: A Key Player in Collective Neural Crest Migration' by Elias H. Barriga and Roberto Mayor in this volume). Briefly, during and after neural tube closure, precursor cells delaminate from border regions of the neural fold or the newly formed neural tube. This migration proceeds along the anterior–posterior axis in coordination with the formation of somites and lasts for several days. Neural crest cells form streams of individual cells following specific migratory routes. These streams reach their final destination where cells differentiate into numerous derivatives including the peripheral nervous system (neurons and glial cells), endocrine cells, cranial cartilages and bones, tendons, smooth muscle cells, and melanocytes. Several populations of neural crest cells can be distinguished, including cranial and trunk neural crest cells. Interestingly, these populations differ in the signaling pathways controlling their delamination and migration modes. They also differ in their fate and participate in distinct tissues and organs. Transplantation experiments indicate that they are mostly committed from the predelamination stage but maintain plasticity (Le Douarin, Creuzet, Couly, & Dupin, 2004).

One important distinction with gastrulation is that the neuroepithelial cells in most species no longer express classic epithelial characteristics at the time of neural tube formation: predelaminating neural plate cells do not express cytokeratins and desmosomes in chicken and mammals (Bennett, 1987; Page, 1989). Nevertheless, these cells display a clear baso-apical polarization with adherens and tight junctions. The delamination of cranial neural crest cells is massive and involves a full individualization in chicken and mice. Cells engage in an EMT, detach from each other, lose their apico-basal polarity and start migrating to form three main streams. In *Xenopus*, cranial neural crest cells initiate a delamination and start migrating before undergoing an EMT, resulting in an intermediate phenotype (reviewed in Aybar & Mayor, 2002; Theveneau & Mayor, 2012). The basement membrane also plays a distinct role in the migration process. It constitutes an obstacle for cranial neural crest cells and must be degraded to allow migration (Monsonego-Ornan et al., 2012). At the trunk level, a local and temporary depletion of the membrane precedes and presumably contributes to the delamination of neural crest cells (Sternberg & Kimber, 1986). However, most neural crest cells do not appear to be crossing through basement membranes during migration. They actually use basement membranes, such as the ectoderm basement membrane, for migratory guidance (Duband & Thiery, 1982).

Cadherins also participate in the process: a switch from E- to N-cadherin is described during the neural tube formation (reviewed in Taneyhill and Schiffmacher, 2013). N-cadherin and cadherin-6 are expressed in the neural tube and are downregulated in the most dorsal region, to be replaced by cadherin-7 in chick or cadherin-11 in the frog, prior to neural crest cell emigration (Kerosuo & Bronner-Fraser, 2012; Nakagawa & Takeichi, 1998). The downregulation of *cadherin-6*, mediated by Slug in chick, induces cell loss of polarity. Slug is an early marker of neural crest cell induction and *Slug* antisense mRNA inhibits cranial neural crest cell migration in chicken (Nieto, Sargent, Wilkinson, & Cooke, 1994). In mouse, Slug and Snail are only expressed after neural crest cell delamination but play a role in further migration as attested by the strong craniofacial disorders in double mutants (Murray, Oram, & Gridley, 2007). Among other EMT-TFs only Zeb2/SIP1 appears to be required for delamination (Van de Putte et al., 2003). Other transcription factors also play important roles in targeting cell–cell adhesion, polarity, neural crest cell maintenance, competence, pluripotency, and survival, including Sox 9/10, FoxD3, Ets1. Clearly, the emigration process involves more than adherens junction downregulation (Lim & Thiery, 2012). Similarly to gastrulation, the order of events allowing neural crest cell depolarization, basement membrane degradation, delamination and migration vary with species, reinforcing the idea that these stages, integral components of the EMT process, are regulated by distinct pathways.

This relative independence of EMT modules that we discussed during gastrulation and neural crest cell emergence is also found during other EMT phases. For example, during sclerotome formation, the ventral aspect of the somite engages in an EMT which also involves cell dissociation and individual migration. Unlike the situation during gastrulation in chick, the presclerotome cells downregulate adherens junctions before the somite basement membrane is degraded (Duband et al., 1987).

In another classic example, during heart morphogenesis, several phases of EMT and MET succeed each other to generate the endothelial lining, with cells delaminating basally to give rise to the endocardial cushion and its derivatives. The basement membrane is hardly significant during this transition (Kitten, Markwald, & Bolender, 1987), even though similar EMT pathways are engaged, involving specific members of the TGFbeta family, each conveys specific cellular responses. At the transcriptional level, Snail and Zeb families are causally involved in endothelial cell–cell separation, and mesenchymal formation (reviewed in Person, Klewer, & Runyan, 2005).

During organogenesis, elements of the EMT process are found in most morphogenetic events. Cells can organize in cohesive sheets, or as individualized cells prior to differentiation. EMT pathways can be involved in these processes but may result in distinct cellular responses, not equivalent to an EMT in terms of cell fating and global behavior. For example, during mammary gland morphogenesis, an invasive structure called terminal end bud (TEB) leads the tubule progression through the mammary fat pad. The TEB is composed of a superficial monolayer of cohesive "cap cells" expressing specific cytokeratins and P-cadherin, a cadherin associated with motility (Ribeiro et al., 2013). Cap cells overlie a poorly organized mass of inner cells called "body cells." Body cells are partially polarized, actively rearranging, and moving around. They can locally align their apical domains to establish local polarity and transient microlumens (Ewald et al., 2012; Godde, Galea, Elsum, & Humbert, 2010). The formation of the main tubule lumen results from coordinated apoptotic activity at the rear aspect of the TEB as it moves forward. Tubule progression is supported by a strong proliferative activity in the outer layers, mostly the cap cells. An uninterrupted basal membrane can be detected all around the tubule (Williams & Daniel, 1983). This membrane is thinner, but uninterrupted at the front aspect of the TEB. Cell migration as a whole proceeds from proliferation but does not involve an individualization of the cells. In fact, hampering with cell–cell adhesion by targeting E- or P-cadherin disrupts the normal tubule growth process (Daniel, Strickland, & Friedmann, 1995). Epithelial differentiation occurs in the outer layers at the rear aspect of the progressing TEB and gives rise to two epithelial layers making for the emerging tubule wall: the basal-myoepithelial and luminal epithelial cells expressing distinct cytokeratins and possessing distinct physiological functions. Myoepithelial cells directly face the basement membrane and display hemidesmosomes not seen in cap cells or body cells (Ormerod & Rudland, 1986). Tubule fragments bear the ability to resume branching when transplanted *in vivo*, demonstrating stemness potential. Several subpopulations of stem-like cells have been characterized, and appear to be mostly located in the basal layer, among the myoepithelial cells (Fu, Lindeman, & Visvader, 2014). Overall, the process does not involve EMT processing, but EMT-TF Slug is causally involved, contributing to cell differentiation, stemness, apoptosis, and proliferation, i.e., the main cellular responses involved in this morphogenic process (Fu et al., 2014; Guo et al., 2012; Nassour et al., 2012). This constitutes an example of EMT-independent involvement of a classic EMT transcription factor during morphogenesis.

In other cell types such as blood cells during hematopoiesis, Slug and Zeb2 play constitutive roles in preserving subpopulations of cells from apoptosis, without any EMT module involvement (Sun, Shao, Bai, Wang, & Wu, 2010; Wu et al., 2005). The closely related gene *Snail* is expressed in inflammatory macrophages, apparently playing a role as a migration mediator, but obviously not mediating an EMT process (Hotz, Visekruna, Buhr, & Hotz, 2010). Snail also protects mammary epithelial cells from T-cell mediated lysis (Akalay et al., 2013). Similarly, Snail, Twist, and Zeb factors are involved in non-EMT functions, during chondrogenesis (Seki et al., 2002; Barnes & Firulli, 2009), hematopoiesis (Dong et al., 2014; Goossens et al., 2011) and more generally lineage specification (Cakouros, Raices, Gronthos, & Glackin, 2010). Modulation of cell–cell adhesion during these processes can be an integral and transient part of the process, but EMT stages do not appear to be involved.

In conclusion, analyzing developmental EMT stages supports the view that we presented above, that the EMT process can be broken down in several functional modules (Levayer & Lecuit, 2008), independently regulated and directly affecting cell phenotype: Loss of baso-apical polarity, down-regulation of cell–cell adhesion forces, motility, and invasiveness. These modules involve tissue-specific pathways dictating an order of progression that is case-specific. Organogenesis also requires cell plasticity and can mobilize some of these modules, resulting in an intermediate phenotype. We suggested earlier the term of metastable phenotype to describe cells expressing the Modules 1 and 3, engaging in a cohesive migration after significant cytoskeleton remodeling and the loss of baso-apical polarity (Fig. 1). This appellation includes the notion that this phenotype is maintained by an unstable, transient, and evolving molecular mechanism. An example is provided by activated keratinocytes during cutaneous re-epithelialization. These cells go through a transient and active phenotype modulation for 24–72 h, involving cytoskeleton modulation (cytokeratin and integrin

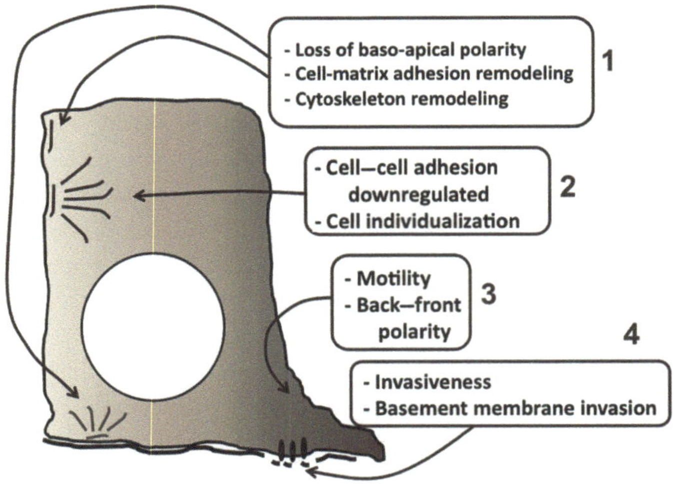

Figure 1 An archetypic EMT process can be analyzed as four distinct functional modules: Loss of baso-apical polarity, cell individualization, motility, and invasiveness.

switch), hemidesmosome dissolution, and cohesive migration with maintenance of desmosomes and adherens junctions (reviewed in Arnoux, Come, Kusewitt, & Savagner, 2005; Leopold, Vincent, & Wang, 2012). A different metastable phenotype is expressed by the mammary TEB inner cells, migrating, and rearranging transiently before differentiating into luminal cells on the side of the wall. These cells express adherens junctions, but no direct contact to stroma. Their migration pattern, similarly to the keratinocytes, involve mostly cell–cell contact. Migration is powered by proliferation, reflecting mostly cap cell activity, leading an invasive tissue invasion, but maintaining an intact basement membrane (Fig. 2).

3. CONTROLLING EMT OR BEING CONTROLLED BY EMT

EMT is induced *in vivo* by multiple signals that have been extensively reviewed in recent articles (De Craene & Berx, 2013; Nieto & Cano, 2012; Zheng & Kang, 2014). Briefly, most peptide growth factors (e.g. FGF, EGF, HGF, TGFβ), cytokines, differentiation factors (Wnt, Notch, SHH, NFκB pathways), and hormones can induce EMT, as well as extracellular matrix components (collagen), and physical microenvironment (hypoxia, oxydative and metabolic stress, UV light). Activation signals are typically transduced to the nucleus, through various pathways, to generate a

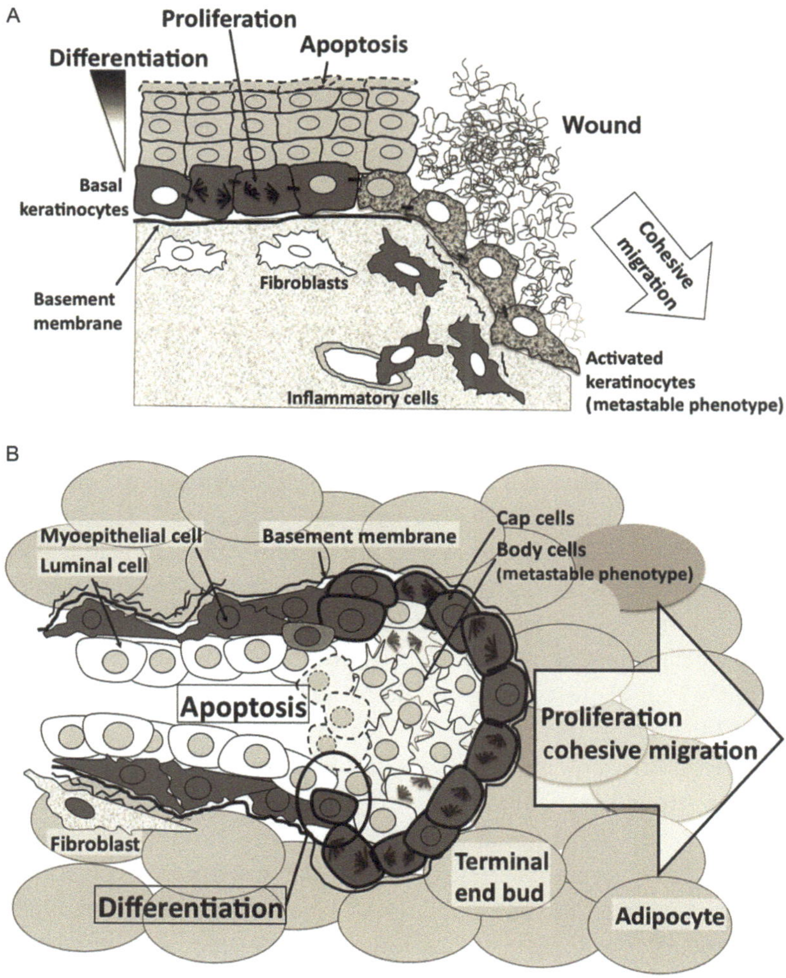

Figure 2 Physiological morphogenetic processes can share some functional modules with EMT without involving cell individualization. These processes produce cells with a metastable phenotype involving transient new properties including motility and phenotype plasticity. (A) During cutaneous wound healing re-epithelialization, activation of basal keratinocytes induces a cohort migration characterizing a metastable phenotype. (B) During mammary tubulogenesis, the terminal end bud is a transient structure composed of cap and body cells expressing cell–cell adhesion structures. Body cells are poorly polarized, motile, and progressively reach a differentiation stage as they reverse from a metastable phenotype. (See the color plate.)

transcriptional activation involving a growing lists of transcriptional modulator families including Snail, Twist, Zeb, but also FOX, SOX, E47, KLF8, Brachyury, HMG2a, Six1, Zeppo, Goosecoid, Gata3, Pit-1 (compiled in De Craene & Berx, 2013; Lamouille, Xu, & Derynck, 2014; Lim &

Thiery, 2012; Nieto & Cano, 2012; Savagner, 2001; Zheng & Kang, 2014). Depending on the context, these factors can repress and/or activate target genes, resulting in cellular responses leading to EMT. A recurrent question has been to identify specific effectors of EMT. It is always tempting to elaborate linear mechanistic models to explain a biological process, including an activating signal, a transducer, and an effector. But this can be deceptive, as is particularly true for the EMT pathways, which have been found to be redundant and circular. As explored more recently, the EMT-TF control each other and can be controlled by their target genes (including E-cadherin) (Casas et al., 2011; Dhasarathy, Phadke, Mav, Shah, & Wade, 2011). Reciprocally, activating factors such as Bmp/Wnt or NF-κB pathways can be themselves targeted by EMT mastergenes in a typically homeostatic mode (Shi, Severson, Yang, Wedlich, & Klymkowsky, 2011; Zhang & Klymkowsky, 2009).

Therefore, pathways can only be explored within a specific cellular and environmental context. The nature and configuration of the extracellular substrate (2D vs. 3D) dictates specific mechanical tensions at the level of the membrane, inducing signaling, cytoskeleton remodeling, and organization. For example, collagen 3D matrix can induce EMT in cells expressing loose cell–cell connections. It requires additional signaling events in order to induce EMT in more cohesive epithelial cells (Katz et al., 2011; Shamir et al., 2014; Tucker, Boyer, Valles, & Thiery, 1991). Overall, effector molecules are mostly involved in the four modules we defined earlier: (1) Molecular structures maintaining baso-apical cell polarization, including cell-matrix and cytoskeleton organization and composition (Crumbs, PAR, and Scribble complexes); (2) molecular structures involved in cell motility and matrix adhesion (dynamic linkage between integrins and actin microfilaments); (3) Cell–cell adhesion structures (adherens junctions and desmosomes) providing the cell–cell adhesion forces; and (4) molecular complexes involved in invasiveness (including MMPs). Clearly, none of these pathways is specific for EMT and they cross-react through mechanical and molecular signaling. For example, actin organization controls and mediate the dynamics of the cytoskeleton, engaging cell–cell junctional as well as motility structures (Michael & Yap, 2013). It is therefore very important when studying EMT pathways, to use an appropriate model allowing the survey of all these modules. For example, conclusions from experiments using classic plastic 2D substrates must be taken with caution since the system imposes a rigid 2D substrate and is not adjusted to evaluate invasiveness.

EMT-specific regulation has been scrutinized at all levels, first focusing on EMT-TF. These factors are transcriptionally activated by most of the inducers mentioned above, including growth/differentiation factors, matrix components and the physical microenvironment. They are themselves targeted by other transcription factors, such as Elf5, KLF4, Sox3 in a tissue-specific mode (Zheng & Kang, 2014). More recently, they were found to respond to metabolic (Dong et al., 2013) and redox conditions (Giannoni, Parri, & Chiarugi, 2012). In addition, EMT-TF are regulated at the posttranslational level by ubiquitination, involving GSK3β in several pathways and complexes (Voutsadakis, 2012; Wu, Li, et al., 2012). Chaperone proteins can also protect EMT-TF and therefore promote EMT. For example, LOXL2/3 binds and stabilizes Snail (Peinado et al., 2005). Noncoding RNAs have been found to target sets of genes associated with the epithelial or mesenchymal phenotype, providing potentially a more systemic control of phenotype (Wright, Richer, & Goodall, 2010; Zhang & Ma, 2012). Some miRNA, such as miR200 and miR34, directly target EMT-TF and play an epithelial safe-keeping role. Epithelial phenotype is also regulated by alternate splicing. Several genes involved in EMT such as *p120 catenin, FGFR2*, and *CD44* are regulated by alternate splicing, under control from RNA binding proteins ESRP1, ESRP2. These genes appear to control an epithelial splicing program that is now being explored systematically (Shapiro et al., 2011). Interestingly, *ESRP1/2* are themselves transcriptional targets of Snail and Zeb factors (Biamonti, Bonomi, Gallo, & Ghigna, 2012; Brown et al., 2011; De Craene & Berx, 2013; Warzecha & Carstens, 2012) providing yet again a retrocontrol mode. The next level of EMT regulation is that of chromatin conformation, regulated by several factors such as HDAC proteins, found to form complexes with EMT-TF and modify chromatin configuration to allow transcriptional regulation and repression of E-boxes, main target of EMT-TF (Wu, Tsai, Wu, Teng, & Wu, 2012). Ezh2, an histone methyltransferase controlled by EMT-inducer Sox4 also contributes to EMT (Tiwari et al., 2013). Conversely, histone acetyltransferase and coactivator CBP was found to preserve epithelial phenotype (Abell et al., 2011; Wu, Tsai, et al., 2012).

Interestingly, each of these three control levels: miRNA, alternate splicing, and chromatin conformation have been suggested to control epithelial phenotype from a global perspective, beyond tissue-specific considerations. It will be intriguing to see which one or which collaborative effort actually plays this role during development and at which stage.

4. REVISITING EMT IN CANCER

The EMT process has been suggested to play a role in cancer since the early formulation of the concept (Boyer, Tucker, Valles, Franke, & Thiery, 1989; Weidner et al., 1990). It is presumed to play a critical role during metastasis initiation by transiently empowering a subpopulation of responsive cells to migrate individually and intravasate (Vanharanta & Massague, 2013) thereby becoming circulating stem cells. To a confusing extent, it is also evoked to explain the partially dedifferentiated phenotype of most carcinoma cells and their invasive behavior locally. This dedifferentiated phenotype has been recognized by pathologists for a long time as a part of inherent tumor heterogeneity. EMT has been scrutinized and evaluated phenotypically in all types of carcinoma to establish semi-quantitative classifications (Bates & Mercurio, 2005; Klymkowsky & Savagner, 2009; Roxanis, 2013). In breast cancer for example, cells express a wide range of dedifferentiation, and can be classified into (a) cytokeratin-positive/E-cad-positive cohesive cells organized in pseudotubules in invasive ductal carcinoma, displaying intercellular junctional complexes (desmosomes, adherens junction), and apical polarity; (b) cytokeratin-positive/E-cad-negative isolated cells, pathognomonic in infiltrating lobular carcinomas where they display Indian file migration; and (c) cytokeratin-negative/E-cad-negative carcinosarcoma cells expressing a fully mesenchymal phenotype (Klymkowsky & Savagner, 2009). These phenotypes epitomize progressive but stabilized stages toward the full EMT phenotype. However, in the absence of direct proof, the actual dedifferentiation process and the occurrence of EMT stages in vivo are subject to lively debate (Chui, 2013; Klymkowsky & Savagner, 2009; Ledford, 2011; Roxanis, 2013; Tarin, Thompson, & Newgreen, 2005).

One initial approach to address this issue is to look for the expression pattern of EMT-TF in carcinoma. However, during the growth and progression of breast and colorectal tumors, immunolocalization studies show overall a very versatile expression of EMT-TF, found in tumor as well as stroma cells (Becker et al., 2007; Come et al., 2006; Geradts et al., 2011; Soini et al., 2011). This is where the controversy starts: it is difficult to define the origin of these stroma cells and there is no unequivocal evidence from an histological perspective that they derive from cytokeratin-positive carcinoma cells. Experimental studies using reporter genes coupled to specific epithelial/mesenchymal promoters have shown that this progression can occur in mice models (Trimboli et al., 2008), but the clinical data are more difficult to interpret. Genomic analysis has been used to show that micro-dissected samples from stromal and epithelial areas in tumor biopsies may share a common origin in a significant but limited number of cases, but these studies need to be carried out on a much larger scale with defined tumor subtypes to be conclusive in a clinical perspective (Moinfar et al., 2000; Trimboli et al., 2008).

Also, when considering the partially dedifferentiated phenotype expressed by most of these carcinoma cells, the controversy is fueled by a sometimes ambiguous vocabulary about the process involved in this switch of phenotype. Indeed, there are at least two very distinct mechanisms that allow the emergence of the partial to advanced EMT-like phenotype encountered in tumors (Fig. 3):

(1) The EMT "*sensu stricto*." This is a rapid (24–72 h) and often reversible transition characterized *in vitro* in numerous cell models. It is independent, and sometimes antagonistic to, proliferation. This process is intrinsically difficult to demonstrate *in vivo* because the implied loss of cytokeratin impedes the validation of an epithelial origin. It is

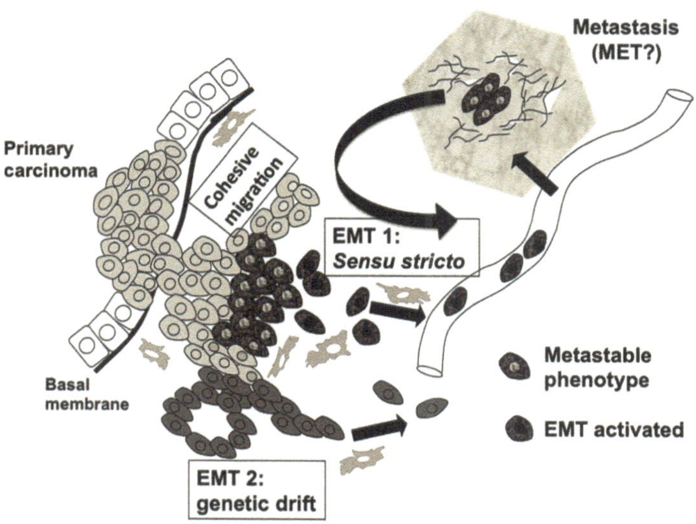

Figure 3 During carcinoma progression, a genetic drift engender transformed cells with new properties. Among them, the acquisition of a metastable phenotype confers to the cells a plasticity necessary to the metastatic process. (See the color plate.)

partially exemplified *in vivo* by re-epithelializing keratinocytes during wound healing (Arnoux et al., 2005). This transient metastable phenotype only represents a partial EMT since activated keratinocytes express cytokeratin, desmosomes, and migrate in cohesive cohorts. Intravital microscopy has shown that transient phases of full individualization can occur in carcinomas following local signaling (Giampieri et al., 2009).

(2) Alternatively, the dedifferentiated phenotype of most tumor cells appears to result from a progressive clonal progression through successive generations of cells displaying genomic alterations. It is recognized that the heterogeneity found in tumors reflects the progressive emergence of clones. These clones are subjected to a drastic selection process reflecting the restrictive growth conditions of tumors (oxygen and nutrient scarcity, hyperconfluency, immune and stromal reactions). This selection for surviving and proliferating cells reflects irreversible mutations. A loss of cytokeratins can follow this process in some clones. However, metastatic foci generally show a more differentiated phenotype, suggesting that the process mediating their emergence is not absolute, is potentially reversible and reflects more a metastable phenotype, as defined above, than a fully mesenchymal phenotype (Brabletz, 2012). Indeed circulating cancer cells appear isolated but also form small clusters, implying cell–cell adhesion forces, thought to be precursors for metastatic foci (Aceto et al., 2014).

Tumors are highly heterogeneous: the above two processes can combine, resulting in all levels of phenotype encountered in carcinomas. In the absence of direct evidence in clinical samples, we propose to use the term "EMT-like" phenotype to describe partially dedifferentiated tumor cell phenotypes described by pathologists, and avoid potentially misleading speculation on the process(es) responsible for the emergence of this phenotype.

Following the wave of EMT-related publications, several groups have analyzed and compared transcriptional signatures for EMT, based on *in vitro* models and *in vivo* clinical samples. The goal was to obtain a prognostic indicator for cancer, by assessing the extent of cells expressing an EMT-like phenotype. This method evaluated the global EMT extent in the cell population that was sampled. It does not take into account local subpopulations that would go through, transiently or not, an EMT and are diluted in the rest of the tumor. EMT signatures were found to predict response to treatment in nonsmall cell lung carcinoma (NSCLC) (Byers et al., 2013), as well as EGFR inhibitor resistance pattern and metastatic outcome (Bryant et al., 2012). In colorectal samples, a strong correlation between EMT and a molecular subtype was described (Loboda et al., 2011). Conversely, a lung metastasis signature was found to link the Wnt pathway, stemness, and EMT in basal-like breast cancer (DiMeo et al., 2009). More recently, a global EMT scoring method was established from ovarian, breast, bladder, lung, colorectal, and gastric cancers. This methodology was found to distinguish EMT from stemness, and made clear that previous connections found between EMT-TF and drug resistance (Dave, Mittal, Tan, & Chang, 2012; Kurrey et al., 2009), strongly depend on the tumor molecular type (Tan et al., 2014).

5. ARE CANCER CELLS REACTIVATING AN EMBRYONIC PROCESS OR BARELY SURVIVING?

During initial tumor growth, very early events can drive phenotypic drift. Some EMT-TF are already expressed in the host tissue, playing a specific and non-EMT related role, such as Slug in breast epithelial cells. But in most cases, expression occurs *de novo*, reflecting new tumor-related signaling. The initial transforming event is believed to be associated with oncogene activation and/or tumor suppressor gene suppression. Most oncogenes such as c-Myc can induce EMT-TF expression (Cho, Cho, Lee, & Kang, 2010). Tumor suppressor genes *p53, BRCA1*, and *RB* also regulate EMT-TF, by distinct mechanisms (Jiang, Jones, et al., 2011). For example, BRCA1 regulates Slug protein lifetime. *BRCA1* mutation, predominant in the basal tumors, results in Slug overexpression by tumor cells (Proia et al., 2011). Typically, *Snail* genes themselves also repress *BRCA* at the transcriptional level by combining with chromatin demethylase LSD1 (Wu, Tsai, et al., 2012). The regulation may also involve miRNAs. For example, P53 activates miRNA-200, repressing EMT genes. Inactivation of *p53* causes a deficiency in miRNA-200 resulting in EMT inducing genes (Chang et al., 2011).

As the tumor grows, a local microenvironment materializes with a stromal reaction involving activated fibroblast cells and immune cells recruited locally. Both cell types express growth factors and cytokines known to activate EMT transcription factor expression directly (Fuxe & Karlsson, 2012;

Gao, Vahdat, Wong, Chang, & Mittal, 2012). Later on, the tumor microenvironment may drive local hypoxia and nutrient deficiency in tumor cells, inducing HIF1 expression, and therefore directly activating EMT-TF (Jiang, Tang, & Liang, 2011). Adjustments in energy metabolism favor glycolysis over aerobic respiration (Warburg effect), also inducing EMT (Lin et al., 2012). Finally, stromal extracellular matrix molecules surrounding invasive tumor cells directly affects phenotype, through integrin signalization. Substrate stiffness also triggers YAP/TAZ activation, another way to activate EMT-TF (Dupont et al., 2011). In summary, the specific conditions linked to carcinomas emergence and progression include many signaling pathways prone to activate EMT-TF, and therefore controlling the tumor phenotype and progression.

As discussed above, EMT-TF are also involved in cellular responses that do not belong to the EMT process, such as apoptosis control (Frisch, Schaller, & Cieply, 2013; Tiwari, Gheldof, Tatari, & Christofori, 2012), stemness, and differentiation control (Mani et al., 2008; Morel et al., 2008; Scheel & Weinberg, 2012). This is also demonstrated during carcinoma recurrence (Moody et al., 2005). However, the most global molecular EMT signature published so far could discern EMT from stemness profiles in tumors, suggesting distinct functional patterns (Tan et al., 2014). Similarly, if some apoptotic pathways can be prevented by EMT-TFs, this is not a general feature of apoptosis. It will be very important to decipher which cell responses are actually controlled by EMT-TF in carcinomas and which ones are the most meaningful. The antiapoptotic effect could ultimately prove to be more clinically relevant for metastasis and recurrences than the EMT profile.

6. EMT WITH OR WITHOUT CADHERINS: A CANCER METASTABLE PHENOTYPE

In most carcinomas, the majority of the cells maintain some level of cohesiveness and express some epithelial markers such as cytokeratins. The early targets of dedifferentiation are the structures involved in basal–apical polarity, including tight junctions and hemidesmosomes, defined as Module 1 in the EMT process. The resultant phenotype, that we call a metastable phenotype, can lead to motile behavior (Module 3) using so-called "mesenchymal" or blebbing migration mode depending on the microenvironment (Friedl & Wolf, 2010). Migration can be individual or collective, depending of the balance between cell–matrix and cell–cell adhesion forces.

E-cadherin has been described to be an early and reliable target for EMT-TFs (Batlle et al., 2000; Cano et al., 2000). It is considered generally as a tumor suppressor gene and is inactivated or delocalized in numerous tumor types (Berx et al., 1995; Birchmeier, 1995; Cheng et al., 2001; Christofori & Semb, 1999; Mareel, Bracke, & Van Roy, 1994; Semb & Christofori, 1998). However, the link with invasiveness and EMT is not unequivocal and appear to be very tumor type-specific (Come et al., 2006; Hollestelle et al., 2013; Rakha et al., 2013). In fact, cohort migration is now recognized as a frequent invasion mode for carcinoma. It is mediated by cadherins and contribute significantly to invasion, specially in the lymphatic vessels (reviewed in Thiery, 2009). This migration can be accompanied by invasiveness (Module 4), (Nabeshima et al., 2000).

Completion of individualization (Module 2) may occur during the migratory process, following local and transient activation, and completing an EMT process leading to isolated circulating cancer cells and metastasis (Giampieri et al., 2009). Loss of E-cadherin contributes clearly to this step. However, individualization may occur in presence of E-cadherin, without impeding subsequent migration, even when this individualization is induced by the overexpression of an EMT transcriptional factor such as Slug (Savagner, Yamada, & Thiery, 1997) or Twist (Shamir et al., 2014). A switch in the expression pattern of distinct EMT-TFs can also lead to E-cadherin downregulation and promote invasiveness in melanoma cells (Caramel et al., 2013). Conversely, metastatic foci typically express cytokeratin and E-cadherin, suggesting that the process selecting metastatic cells is also able to turn down the EMT signaling that contributed to intra/extravasation, as shown for Twist in a mouse model (Tsai, Donaher, Murphy, Chau, & Yang, 2012). This step has been described as a MET, even though it has not been analyzed beyond the characterization of the end-product: the metastatic foci. To reconcile this step with the concept of EMT, as induced by oncogenic transformation and clonal selection, there is a need to hypothesize that invasive cells, beyond stemness properties and apoptosis resistance, maintain enough plasticity to engage back in cell–cell adhesion and baso-apical polarity. This ability is preserved in re-epithelializing keratinocytes expressing a metastable phenotype and repolarizing progressively to rebuild an epithelium. We suggest it could be carried by metastatic cells, with an involvement of EMT-TFs achieving the proper metastable phenotype.

In addition to E-cadherin, other classical cadherins including N- and P-cadherins are expressed in carcinoma cells as a feature of their normal differentiation status or as a transformation-linked feature, reflecting specific activation and typically a coincidental E-cadherin downregulation. The switch does not have to be total and the balance between E-cadherin/N-cadherin or E-cadherin/P-cadherin may actually control cell migratory abilities (Nieman, Prudoff, Johnson, & Wheelock, 1999; Ribeiro et al., 2013). P-cadherin overexpression can induce migratory properties, apparently along a pathway mobilizing p120-catenin, then Src and small-GTPases triggering actin filaments dynamics (Ribeiro et al., 2013).

7. CONCLUSION

In conclusion, based on classic developmental examples and on multilevel cancer observations, EMT cannot be analyzed as a single linear program. It should be considered as a combination of several cellular response modules taking place in an order and with a degree of completion controlled by distinct pathways. Behind these modules, several transcription factors (EMT-TF) can provide a unifying signal, but each of these factors also carry specific roles, resulting in different outcomes. In carcinoma, EMT phenotypes can be triggered by various inherent processes, transient and reversible or not reversible, linked sometimes very loosely to the EMT process when they reflect genomic events. Some of these pathways can induce only one or more functional EMT modules, resulting in a partial EMT and a metastable phenotype reflected by cohort migration and plasticity. In addition, EMT-TF also control several responses very relevant to cancer progression, including apoptosis control and stemness properties. This functional repertoire goes beyond EMT but make them integrative key players to understand cell dynamics and fate during carcinoma progression.

REFERENCES

Abell, A. N., Jordan, N. V., Huang, W., Prat, A., Midland, A. A., Johnson, N. L., et al. (2011). MAP3K4/CBP-regulated H2B acetylation controls epithelial-mesenchymal transition in trophoblast stem cells. *Cell Stem Cell*, *8*(5), 525–537.

Aceto, N., Bardia, A., Miyamoto, D. T., Donaldson, M. C., Wittner, B. S., Spencer, J. A., et al. (2014). Circulating tumor cell clusters are oligoclonal precursors of breast cancer metastasis. *Cell*, *158*(5), 1110–1122.

Akalay, I., Janji, B., Hasmim, M., Noman, M. Z., Andre, F., De Cremoux, P., et al. (2013). Epithelial-to-mesenchymal transition and autophagy induction in breast carcinoma promote escape from T-cell-mediated lysis. *Cancer Research*, *73*(8), 2418–2427.

Arnoux, V., Come, C., Kusewitt, D., & Savagner, P. (2005). Cutaneous wound healing: A partial and reversible EMT. In P. Savagner (Ed.), *Rise and fall of epithelial phenotype: Concepts of epithelial-mesenchymal transition* (pp. 111–134). Austin, TX: Landes Biosciences.
Aybar, M. J., & Mayor, R. (2002). Early induction of neural crest cells: Lessons learned from frog, fish and chick. *Current Opinion in Genetics & Development*, *12*(4), 452–458.
Barnes, R. M., & Firulli, A. B. (2009). A twist of insight—The role of Twist-family bHLH factors in development. *International Journal of Developmental Biology*, *53*(7), 909–924.
Bates, R. C., & Mercurio, A. M. (2005). The epithelial-mesenchymal transition (EMT) and colorectal cancer progression. *Cancer Biology & Therapy*, *4*(4), 365–370.
Batlle, E., Sancho, E., Franci, C., Dominguez, D., Monfar, M., Baulida, J., et al. (2000). The transcription factor snail is a repressor of E-cadherin gene expression in epithelial tumour cells. *Nature Cell Biology*, *2*(2), 84–89.
Becker, K. F., Rosivatz, E., Blechschmidt, K., Kremmer, E., Sarbia, M., & Hofler, H. (2007). Analysis of the E-cadherin repressor Snail in primary human cancers. *Cells, Tissues, Organs*, *185*(1–3), 204–212.
Bennett, G. S. (1987). Changes in intermediate filament composition during neurogenesis. *Current Topics in Developmental Biology*, *21*, 151–183.
Berx, G., Cleton-Jansen, A. M., Nollet, F., de Leeuw, W. J., van de Vijver, M., Cornelisse, C., et al. (1995). E-cadherin is a tumour/invasion suppressor gene mutated in human lobular breast cancers. *EMBO Journal*, *14*(24), 6107–6115.
Biamonti, G., Bonomi, S., Gallo, S., & Ghigna, C. (2012). Making alternative splicing decisions during epithelial-to-mesenchymal transition (EMT). *Cellular and Molecular Life Sciences*, *69*(15), 2515–2526.
Birchmeier, W. (1995). E-cadherin as a tumor (invasion) suppressor gene. *Bioessays*, *17*(2), 97–99.
Boyer, B., Tucker, G. C., Valles, A. M., Franke, W. W., & Thiery, J. P. (1989). Rearrangements of desmosomal and cytoskeletal proteins during the transition from epithelial to fibroblastoid organization in cultured rat bladder carcinoma cells. *Journal of Cell Biology*, *109*(4 Pt. 1), 1495–1509.
Brabletz, T. (2012). To differentiate or not—Routes towards metastasis. *Nature Reviews Cancer*, *12*(6), 425–436.
Brown, R. L., Reinke, L. M., Damerow, M. S., Perez, D., Chodosh, L. A., Yang, J., et al. (2011). CD44 splice isoform switching in human and mouse epithelium is essential for epithelial-mesenchymal transition and breast cancer progression. *Journal of Clinical Investigation*, *121*(3), 1064–1074.
Bryant, J. L., Britson, J., Balko, J. M., Willian, M., Timmons, R., Frolov, A., et al. (2012). A microRNA gene expression signature predicts response to erlotinib in epithelial cancer cell lines and targets EMT. *British Journal of Cancer*, *106*(1), 148–156.
Byers, L. A., Diao, L., Wang, J., Saintigny, P., Girard, L., Peyton, M., et al. (2013). An epithelial-mesenchymal transition gene signature predicts resistance to EGFR and PI3K inhibitors and identifies Axl as a therapeutic target for overcoming EGFR inhibitor resistance. *Clinical Cancer Research*, *19*(1), 279–290.
Cakouros, D., Raices, R. M., Gronthos, S., & Glackin, C. A. (2010). Twist-ing cell fate: Mechanistic insights into the role of twist in lineage specification/differentiation and tumorigenesis. *Journal of Cellular Biochemistry*, *110*(6), 1288–1298.
Cano, A., Perez-Moreno, M. A., Rodrigo, I., Locascio, A., Blanco, M. J., del Barrio, M. G., et al. (2000). The transcription factor snail controls epithelial-mesenchymal transitions by repressing E-cadherin expression. *Nature Cell Biology*, *2*(2), 76–83.
Caramel, J., Papadogeorgakis, E., Hill, L., Browne, G. J., Richard, G., Wierinckx, A., et al. (2013). A switch in the expression of embryonic EMT-inducers drives the development of malignant melanoma. *Cancer Cell*, *24*(4), 466–480.

Carver, E. A., Jiang, R., Lan, Y., Oram, K. F., & Gridley, T. (2001). The mouse snail gene encodes a key regulator of the epithelial-mesenchymal transition. *Molecular and Cellular Biology*, *21*(23), 8184–8188.

Casas, E., Kim, J., Bendesky, A., Ohno-Machado, L., Wolfe, C. J., & Yang, J. (2011). Snail2 is an essential mediator of Twist1-induced epithelial mesenchymal transition and metastasis. *Cancer Research*, *71*(1), 245–254.

Chang, C. J., Chao, C. H., Xia, W., Yang, J. Y., Xiong, Y., Li, C. W., et al. (2011). p53 regulates epithelial-mesenchymal transition and stem cell properties through modulating miRNAs. *Nature Cell Biology*, *13*(3), 317–323.

Chen, X., & Gumbiner, B. M. (2006). Paraxial protocadherin mediates cell sorting and tissue morphogenesis by regulating C-cadherin adhesion activity. *Journal of Cell Biology*, *174*(2), 301–313.

Cheng, C. W., Wu, P. E., Yu, J. C., Huang, C. S., Yue, C. T., Wu, C. W., et al. (2001). Mechanisms of inactivation of E-cadherin in breast carcinoma: Modification of the two-hit hypothesis of tumor suppressor gene. *Oncogene*, *20*(29), 3814–3823.

Cho, K. B., Cho, M. K., Lee, W. Y., & Kang, K. W. (2010). Overexpression of c-myc induces epithelial mesenchymal transition in mammary epithelial cells. *Cancer Letters*, *293*(2), 230–239.

Christofori, G., & Semb, H. (1999). The role of the cell-adhesion molecule E-cadherin as a tumour-suppressor gene. *Trends in Biochemical Sciences*, *24*(2), 73–76.

Chui, M. H. (2013). Insights into cancer metastasis from a clinicopathologic perspective: Epithelial-Mesenchymal Transition is not a necessary step. *International Journal of Cancer*, *132*(7), 1487–1495.

Come, C., Magnino, F., Bibeau, F., De Santa Barbara, P., Becker, K. F., Theillet, C., et al. (2006). Snail and slug play distinct roles during breast carcinoma progression. *Clinical Cancer Research*, *12*(18), 5395–5402.

Creighton, C. J., Chang, J. C., & Rosen, J. M. (2010). Epithelial-mesenchymal transition (EMT) in tumor-initiating cells and its clinical implications in breast cancer. *Journal of Mammary Gland Biology and Neoplasia*, *15*(2), 253–260.

Daniel, C. W., Strickland, P., & Friedmann, Y. (1995). Expression and functional role of E- and P-cadherins in mouse mammary ductal morphogenesis and growth. *Developmental Biology*, *169*(2), 511–519.

Dave, B., Mittal, V., Tan, N. M., & Chang, J. C. (2012). Epithelial-mesenchymal transition, cancer stem cells and treatment resistance. *Breast Cancer Research*, *14*(1), 202.

De Craene, B., & Berx, G. (2013). Regulatory networks defining EMT during cancer initiation and progression. *Nature Reviews Cancer*, *13*(2), 97–110.

De Vries, W. N., Evsikov, A. V., Haac, B. E., Fancher, K. S., Holbrook, A. E., Kemler, R., et al. (2004). Maternal beta-catenin and E-cadherin in mouse development. *Development*, *131*(18), 4435–4445.

Den, Z., Cheng, X., Merched-Sauvage, M., & Koch, P. J. (2006). Desmocollin 3 is required for pre-implantation development of the mouse embryo. *Journal of Cell Science*, *119*(Pt. 3), 482–489.

Derycke, L. D., & Bracke, M. E. (2004). N-cadherin in the spotlight of cell-cell adhesion, differentiation, embryogenesis, invasion and signalling. *International Journal of Developmental Biology*, *48*(5–6), 463–476.

Dhasarathy, A., Phadke, D., Mav, D., Shah, R. R., & Wade, P. A. (2011). The transcription factors Snail and Slug activate the transforming growth factor-beta signaling pathway in breast cancer. *PLoS One*, *6*(10), e26514.

DiMeo, T. A., Anderson, K., Phadke, P., Fan, C., Perou, C. M., Naber, S., et al. (2009). A novel lung metastasis signature links Wnt signaling with cancer cell self-renewal and epithelial-mesenchymal transition in basal-like breast cancer. *Cancer Research*, *69*(13), 5364–5373.

Dong, C. Y., Liu, X. Y., Wang, N., Wang, L. N., Yang, B. X., Ren, Q., et al. (2014). Twist-1, a novel regulator of hematopoietic stem cell self-renewal and myeloid lineage development. *Stem Cells*, *32*(12), 3173–3182.

Dong, C., Yuan, T., Wu, Y., Wang, Y., Fan, T. W., Miriyala, S., et al. (2013). Loss of FBP1 by Snail-mediated repression provides metabolic advantages in basal-like breast cancer. *Cancer Cell*, *23*(3), 316–331.

Duband, J. L., Dufour, S., Hatta, K., Takeichi, M., Edelman, G. M., & Thiery, J. P. (1987). Adhesion molecules during somitogenesis in the avian embryo. *Journal of Cell Biology*, *104*(5), 1361–1374.

Duband, J. L., & Thiery, J. P. (1982). Distribution of fibronectin in the early phase of avian cephalic neural crest cell migration. *Developmental Biology*, *93*(2), 308–323.

Dupont, S., Morsut, L., Aragona, M., Enzo, E., Giulitti, S., Cordenonsi, M., et al. (2011). Role of YAP/TAZ in mechanotransduction. *Nature*, *474*(7350), 179–183.

Ewald, A. J., Huebner, R. J., Palsdottir, H., Lee, J. K., Perez, M. J., Jorgens, D. M., et al. (2012). Mammary collective cell migration involves transient loss of epithelial features and individual cell migration within the epithelium. *Journal of Cell Science*, *125*(Pt. 11), 2638–2654.

Fang, X., Cai, Y., Liu, J., Wang, Z., Wu, Q., Zhang, Z., et al. (2011). Twist2 contributes to breast cancer progression by promoting an epithelial-mesenchymal transition and cancer stem-like cell self-renewal. *Oncogene*, *30*(47), 4707–4720.

Friedl, P., & Wolf, K. (2010). Plasticity of cell migration: A multiscale tuning model. *Journal of Cell Biology*, *188*(1), 11–19.

Frisch, S. M. (1997). The epithelial cell default-phenotype hypothesis and its implications for cancer. *Bioessays*, *19*, 705–709.

Frisch, S. M., Schaller, M., & Cieply, B. (2013). Mechanisms that link the oncogenic epithelial-mesenchymal transition to suppression of anoikis. *Journal of Cell Science*, *126*(Pt. 1), 21–29.

Fu, N., Lindeman, G. J., & Visvader, J. E. (2014). The mammary stem cell hierarchy. *Current Topics in Developmental Biology*, *107*, 133–160.

Fuxe, J., & Karlsson, M. C. (2012). TGF-beta-induced epithelial-mesenchymal transition: A link between cancer and inflammation. *Seminars in Cancer Biology*, *22*(5–6), 455–461.

Gao, D., Vahdat, L. T., Wong, S., Chang, J. C., & Mittal, V. (2012). Microenvironmental regulation of epithelial-mesenchymal transitions in cancer. *Cancer Research*, *72*(19), 4883–4889.

Geradts, J., de Herreros, A. G., Su, Z., Burchette, J., Broadwater, G., & Bachelder, R. E. (2011). Nuclear Snail1 and nuclear ZEB1 protein expression in invasive and intraductal human breast carcinomas. *Human Pathology*, *42*(8), 1125–1131.

Giampieri, S., Manning, C., Hooper, S., Jones, L., Hill, C. S., & Sahai, E. (2009). Localized and reversible TGFbeta signalling switches breast cancer cells from cohesive to single cell motility. *Nature Cell Biology*, *11*(11), 1287–1296.

Giannoni, E., Parri, M., & Chiarugi, P. (2012). EMT and oxidative stress: A bidirectional interplay affecting tumor malignancy. *Antioxidants & Redox Signaling*, *16*(11), 1248–1263.

Godde, N. J., Galea, R. C., Elsum, I. A., & Humbert, P. O. (2010). Cell polarity in motion: Redefining mammary tissue organization through EMT and cell polarity transitions. *Journal of Mammary Gland Biology and Neoplasia*, *15*(2), 149–168.

Goossens, S., Janzen, V., Bartunkova, S., Yokomizo, T., Drogat, B., Crisan, M., et al. (2011). The EMT regulator Zeb2/Sip1 is essential for murine embryonic hematopoietic stem/progenitor cell differentiation and mobilization. *Blood*, *117*(21), 5620–5630.

Greenburg, G., & Hay, E. D. (1982). Epithelia suspended in collagen gels can lose polarity and express characteristics of migrating mesenchymal cells. *Journal of Cell Biology*, *95*(1), 333–339.

Guo, W., Keckesova, Z., Donaher, J. L., Shibue, T., Tischler, V., Reinhardt, F., et al. (2012). Slug and Sox9 cooperatively determine the mammary stem cell state. *Cell, 148*(5), 1015–1028.

Hammerschmidt, M., & Wedlich, D. (2008). Regulated adhesion as a driving force of gastrulation movements. *Development, 135*(22), 3625–3641.

Hatta, K., & Takeichi, M. (1986). Expression of N-cadherin adhesion molecules associated with early morphogenetic events in chick development. *Nature, 320*(6061), 447–449.

Hay, E. D. (1968). Organization and fine structure of epithelium and mesenchyme in the developing chick embryo. In R. Fleischmajer & R. E. Billingham (Eds.), *Epithelial-mesenchymal interactions; 18th hahnemann symposium*. Baltimore: Williams and Wilkins.

Heisenberg, C. P., & Solnica-Krezel, L. (2008). Back and forth between cell fate specification and movement during vertebrate gastrulation. *Current Opinion in Genetics & Development, 18*(4), 311–316.

Hollestelle, A., Peeters, J. K., Smid, M., Timmermans, M., Verhoog, L. C., Westenend, P. J., et al. (2013). Loss of E-cadherin is not a necessity for epithelial to mesenchymal transition in human breast cancer. *Breast Cancer Research and Treatment, 138*(1), 47–57.

Hotz, B., Visekruna, A., Buhr, H. J., & Hotz, H. G. (2010). Beyond epithelial to mesenchymal transition: A novel role for the transcription factor Snail in inflammation and wound healing. *Journal of Gastrointestinal Surgery, 14*(2), 388–397.

Jiang, Z., Jones, R., Liu, J. C., Deng, T., Robinson, T., Chung, P. E., et al. (2011). RB1 and p53 at the crossroad of EMT and triple-negative breast cancer. *Cell Cycle, 10*(10), 1563–1570.

Jiang, J., Tang, Y. L., & Liang, X. H. (2011). EMT: A new vision of hypoxia promoting cancer progression. *Cancer Biology & Therapy, 11*(8), 714–723.

Katz, E., Dubois-Marshall, S., Sims, A. H., Gautier, P., Caldwell, H., Meehan, R. R., et al. (2011). An in vitro model that recapitulates the epithelial to mesenchymal transition (EMT) in human breast cancer. *PLoS One, 6*(2), e17083.

Keller, R. (2005). Cell migration during gastrulation. *Current Opinion in Cell Biology, 17*(5), 533–541.

Kerosuo, L., & Bronner-Fraser, M. (2012). What is bad in cancer is good in the embryo: Importance of EMT in neural crest development. *Seminars in Cell & Developmental Biology, 23*(3), 320–332.

Kim, S. H., Yamamoto, A., Bouwmeester, T., Agius, E., & Robertis, E. M. (1998). The role of paraxial protocadherin in selective adhesion and cell movements of the mesoderm during Xenopus gastrulation. *Development, 125*(23), 4681–4690.

Kitten, G. T., Markwald, R. R., & Bolender, D. L. (1987). Distribution of basement membrane antigens in cryopreserved early embryonic hearts. *Anatomical Record, 217*(4), 379–390.

Klymkowsky, M. W., & Savagner, P. (2009). Epithelial-mesenchymal transition: A cancer researcher's conceptual friend and foe. *American Journal of Pathology, 174*(5), 1588–1593.

Kurrey, N. K., Jalgaonkar, S. P., Joglekar, A. V., Ghanate, A. D., Chaskar, P. D., Doiphode, R. Y., et al. (2009). Snail and slug mediate radioresistance and chemoresistance by antagonizing p53-mediated apoptosis and acquiring a stem-like phenotype in ovarian cancer cells. *Stem Cells, 27*(9), 2059–2068.

Lamouille, S., Xu, J., & Derynck, R. (2014). Molecular mechanisms of epithelial-mesenchymal transition. *Nature Reviews Molecular Cell Biology, 15*(3), 178–196.

Le Douarin, N. M., Creuzet, S., Couly, G., & Dupin, E. (2004). Neural crest cell plasticity and its limits. *Development, 131*, 4637–4650.

Ledford, H. (2011). Cancer theory faces doubts. *Nature, 472*(7343), 273.

Leopold, P. L., Vincent, J., & Wang, H. (2012). A comparison of epithelial-to-mesenchymal transition and re-epithelialization. *Seminars in Cancer Biology, 22*(5–6), 471–483.

Leptin, M. (2005). Gastrulation movements: The logic and the nuts and bolts. *Developmental Cell, 8*(3), 305–320.
Levayer, R., & Lecuit, T. (2008). Breaking down EMT. *Nature Cell Biology, 10*(7), 757–759.
Lim, J., & Thiery, J. P. (2012). Epithelial-mesenchymal transitions: Insights from development. *Development, 139*(19), 3471–3486.
Lin, C. C., Cheng, T. L., Tsai, W. H., Tsai, H. J., Hu, K. H., Chang, H. C., et al. (2012). Loss of the respiratory enzyme citrate synthase directly links the Warburg effect to tumor malignancy. *Science Reports, 2*, 785.
Loboda, A., Nebozhyn, M. V., Watters, J. W., Buser, C. A., Shaw, P. M., Huang, P. S., et al. (2011). EMT is the dominant program in human colon cancer. *BMC Medical Genomics, 4*, 9.
Mani, S. A., Guo, W., Liao, M. J., Eaton, E. N., Ayyanan, A., Zhou, A. Y., et al. (2008). The epithelial-mesenchymal transition generates cells with properties of stem cells. *Cell, 133*(4), 704–715.
Mareel, M., Bracke, M., & Van Roy, F. (1994). Invasion promoter versus invasion suppressor molecules: The paradigm of E-cadherin. *Molecular Biology Reports, 19*(1), 45–67.
Marsden, M., & DeSimone, D. W. (2003). Integrin-ECM interactions regulate cadherin-dependent cell adhesion and are required for convergent extension in Xenopus. *Current Biology, 13*(14), 1182–1191.
Michael, M., & Yap, A. S. (2013). The regulation and functional impact of actin assembly at cadherin cell–cell adhesions. *Seminars in Cell & Developmental Biology, 24*(4), 298–307.
Moinfar, F., Man, Y. G., Arnould, L., Bratthauer, G. L., Ratschek, M., & Tavassoli, F. A. (2000). Concurrent and independent genetic alterations in the stromal and epithelial cells of mammary carcinoma: Implications for tumorigenesis. *Cancer Research, 60*(9), 2562–2566.
Monsonego-Ornan, E., Kosonovsky, J., Bar, A., Roth, L., Fraggi-Rankis, V., Simsa, S., et al. (2012). Matrix metalloproteinase 9/gelatinase B is required for neural crest cell migration. *Developmental Biology, 364*(2), 162–177.
Montero, J. A., Carvalho, L., Wilsch-Brauninger, M., Kilian, B., Mustafa, C., & Heisenberg, C. P. (2005). Shield formation at the onset of zebrafish gastrulation. *Development, 132*(6), 1187–1198.
Moody, S. E., Perez, D., Pan, T. C., Sarkisian, C. J., Portocarrero, C. P., Sterner, C. J., et al. (2005). The transcriptional repressor Snail promotes mammary tumor recurrence. *Cancer Cell, 8*(3), 197–209.
Morali, O. G., Savagner, P., & Larue, L. (2005). Epithelium mesenchyme transitions are crucial morphogenetic events occurring during early development. In P. Savagner (Ed.), *Rise and fall of epithelial phenotype: Concepts of epithelial-mesenchymal transition*. Georgetown, TX; New York, NY: Landes Bioscience; Kluwer Academic.
Morel, A. P., Lievre, M., Thomas, C., Hinkal, G., Ansieau, S., & Puisieux, A. (2008). Generation of breast cancer stem cells through epithelial-mesenchymal transition. *PLoS One, 3*(8), e2888.
Murray, S. A., Oram, K. F., & Gridley, T. (2007). Multiple functions of Snail family genes during palate development in mice. *Development, 134*(9), 1789–1797.
Nabeshima, K., Inoue, T., Shimao, Y., Okada, Y., Itoh, Y., Seiki, M., et al. (2000). Front-cell-specific expression of membrane-type 1 matrix metalloproteinase and gelatinase A during cohort migration of colon carcinoma cells induced by hepatocyte growth factor/scatter factor. *Cancer Research, 60*(13), 3364–3369.
Nakagawa, S., & Takeichi, M. (1998). Neural crest emigration from the neural tube depends on regulated cadherin expression. *Development, 125*(15), 2963–2971.
Nakaya, Y., & Sheng, G. (2008). Epithelial to mesenchymal transition during gastrulation: An embryological view. *Development, Growth & Differentiation, 50*(9), 755–766.

Nakaya, Y., & Sheng, G. (2013). EMT in developmental morphogenesis. *Cancer Letters*, *341*(1), 9–15.

Nassour, M., Idoux-Gillet, Y., Selmi, A., Come, C., Faraldo, M. L., Deugnier, M. A., et al. (2012). Slug controls stem/progenitor cell growth dynamics during mammary gland morphogenesis. *PLoS One*, 7(12), e53498.

Nieman, M. T., Prudoff, R. S., Johnson, K. R., & Wheelock, M. J. (1999). N-cadherin promotes motility in human breast cancer cells regardless of their E-cadherin expression. *Journal of Cell Biology*, *147*(3), 631–644.

Nieto, M. A., & Cano, A. (2012). The epithelial-mesenchymal transition under control: Global programs to regulate epithelial plasticity. *Seminars in Cancer Biology*, *22*(5–6), 361–368.

Nieto, M. A., Sargent, M. G., Wilkinson, D. G., & Cooke, J. (1994). Control of cell behavior during vertebrate development by Slug, a zinc finger gene. *Science*, *264*(5160), 835–839.

Nowotschin, S., & Hadjantonakis, A. K. (2010). Cellular dynamics in the early mouse embryo: From axis formation to gastrulation. *Current Opinion in Genetics & Development*, *20*(4), 420–427.

Ormerod, E. J., & Rudland, P. S. (1986). Regeneration of mammary glands in vivo from isolated mammary ducts. *Journal of Embryology and Experimental Morphology*, *96*, 229–243.

Page, M. (1989). Changing patterns of cytokeratins and vimentin in the early chick embryo. *Development*, *105*(1), 97–107.

Peinado, H., Iglesias-de, Del Carmen, la Cruz, M., Olmeda, D., Csiszar, K., Fong, K. S., et al. (2005). A molecular role for lysyl oxidase-like 2 enzyme in snail regulation and tumor progression. *EMBO Journal*, *24*(19), 3446–3458.

Person, A. D., Klewer, S. E., & Runyan, R. B. (2005). Cell biology of cardiac cushion development. *International Review of Cytology*, *243*, 287–335.

Proia, T. A., Keller, P. J., Gupta, P. B., Klebba, I., Jones, A. D., Sedic, M., et al. (2011). Genetic predisposition directs breast cancer phenotype by dictating progenitor cell fate. *Cell Stem Cell*, *8*(2), 149–163.

Rakha, E. A., Teoh, T. K., Lee, A. H., Nolan, C. C., Ellis, I. O., & Green, A. R. (2013). Further evidence that E-cadherin is not a tumour suppressor gene in invasive ductal carcinoma of the breast: An immunohistochemical study. *Histopathology*, *62*(5), 695–701.

Reima, I., Lehtonen, E., Virtanen, I., & Flechon, J. E. (1993). The cytoskeleton and associated proteins during cleavage, compaction and blastocyst differentiation in the pig. *Differentiation*, *54*(1), 35–45.

Ribeiro, A. S., Sousa, B., Carreto, L., Mendes, N., Nobre, A. R., Ricardo, S., et al. (2013). P-cadherin functional role is dependent on E-cadherin cellular context: A proof of concept using the breast cancer model. *Journal of Pathology*, *229*(5), 705–718.

Roxanis, I. (2013). Occurrence and significance of epithelial-mesenchymal transition in breast cancer. *Journal of Clinical Pathology*, *66*(6), 517–521.

Savagner, P. (2001). Leaving the neighborhood: Molecular mechanisms involved during epithelial-mesenchymal transition. *Bioessays*, *23*(10), 912–923.

Savagner, P., Yamada, K. M., & Thiery, J. P. (1997). The zinc-finger protein slug causes desmosome dissociation, an initial and necessary step for growth factor-induced epithelial-mesenchymal transition. *Journal of Cell Biology*, *137*(6), 1403–1419.

Scheel, C., & Weinberg, R. A. (2012). Cancer stem cells and epithelial-mesenchymal transition: Concepts and molecular links. *Seminars in Cancer Biology*, *22*(5–6), 396–403.

Seki, K., Fujimori, T., Savagner, P., Hata, A., Aikawa, T., Ogata, N., et al. (2002). Mouse Snail-related transcription repressors regulate chondrocyte extracellular matrix, type II collagen and aggrecan. *Journal of Biological Chemistry*, *278*, 41862–41870.

Semb, H., & Christofori, G. (1998). The tumor-suppressor function of E-cadherin. *American Journal of Human Genetics*, *63*(6), 1588–1593.

Shamir, E. R., Pappalardo, E., Jorgens, D. M., Coutinho, K., Tsai, W. T., Aziz, K., et al. (2014). Twist1-induced dissemination preserves epithelial identity and requires E-cadherin. *Journal of Cell Biology, 204*(5), 839–856.
Shapiro, I. M., Cheng, A. W., Flytzanis, N. C., Balsamo, M., Condeelis, J. S., Oktay, M. H., et al. (2011). An EMT-driven alternative splicing program occurs in human breast cancer and modulates cellular phenotype. *PLoS Genetics*, 7(8), e1002218.
Shi, J., Severson, C., Yang, J., Wedlich, D., & Klymkowsky, M. W. (2011). Snail2 controls mesodermal BMP/Wnt induction of neural crest. *Development, 138*(15), 3135–3145.
Soini, Y., Tuhkanen, H., Sironen, R., Virtanen, I., Kataja, V., Auvinen, P., et al. (2011). Transcription factors zeb1, twist and snai1 in breast carcinoma. *BMC Cancer, 11*, 73.
Solnica-Krezel, L. (2005). Conserved patterns of cell movements during vertebrate gastrulation. *Current Biology, 15*(6), R213–R228.
Solnica-Krezel, L., & Sepich, D. S. (2012). Gastrulation: Making and shaping germ layers. *Annual Review of Cell and Developmental Biology, 28*, 687–717.
Sternberg, J., & Kimber, S. J. (1986). The relationship between emerging neural crest cells and basement membranes in the trunk of the mouse embryo: A TEM and immunocytochemical study. *Journal of Embryology and Experimental Morphology, 98*, 251–268.
Stoker, M., Gherardi, E., Perryman, M., & Gray, J. (1987). Scatter factor is a fibroblast-derived modulator of epithelial cell mobility. *Nature, 327*(6119), 239–242.
Sun, Y., Shao, L., Bai, H., Wang, Z. Z., & Wu, W. S. (2010). Slug deficiency enhances self-renewal of hematopoietic stem cells during hematopoietic regeneration. *Blood, 115*(9), 1709–1717.
Tan, T. Z., Miow, Q. H., Miki, Y., Noda, T., Mori, S., Huang, R. Y., et al. (2014). Epithelial-mesenchymal transition spectrum quantification and its efficacy in deciphering survival and drug responses of cancer patients. *EMBO Molecular Medicine*.
Taneyhill, L. A., & Schiffmacher, A. T. (2013). Cadherin dynamics during neural crest cell ontogeny. *Progress in Molecular Biology and Translational Science, 116*, 291–315.
Tarin, D., Thompson, E. W., & Newgreen, D. F. (2005). The fallacy of epithelial mesenchymal transition in neoplasia. *Cancer Research, 65*(14), 5996–6000, discussion 6000–5991.
Theveneau, E., & Mayor, R. (2012). Neural crest delamination and migration: from epithelium-to-mesenchyme transition to collective cell migration. *Developmental Biology, 366*(1), 34–54.
Thiery, J. P. (2009). Metastasis: Alone or together? *Current Biology, 19*(24), R1121–R1123.
Thiery, J. P., Acloque, H., Huang, R. Y., & Nieto, M. A. (2009). Epithelial-mesenchymal transitions in development and disease. *Cell, 139*(5), 871–890.
Tiwari, N., Gheldof, A., Tatari, M., & Christofori, G. (2012). EMT as the ultimate survival mechanism of cancer cells. *Seminars in Cancer Biology, 22*(3), 194–207.
Tiwari, N., Tiwari, V. K., Waldmeier, L., Balwierz, P. J., Arnold, P., Pachkov, M., et al. (2013). Sox4 is a master regulator of epithelial-mesenchymal transition by controlling Ezh2 expression and epigenetic reprogramming. *Cancer Cell, 23*(6), 768–783.
Trimboli, A. J., Fukino, K., de Bruin, A., Wei, G., Shen, L., Tanner, S. M., et al. (2008). Direct evidence for epithelial-mesenchymal transitions in breast cancer. *Cancer Research, 68*(3), 937–945.
Tsai, J. H., Donaher, J. L., Murphy, D. A., Chau, S., & Yang, J. (2012). Spatiotemporal regulation of epithelial-mesenchymal transition is essential for squamous cell carcinoma metastasis. *Cancer Cell, 22*(6), 725–736.
Tucker, G. C., Boyer, B., Valles, A. M., & Thiery, J. P. (1991). Combined effects of extracellular matrix and growth factors on NBT-II rat bladder carcinoma cell dispersion. *Journal of Cell Science, 100*(Pt. 2), 371–380.
Valles, A. M., Boyer, B., Badet, J., Tucker, G. C., Barritault, D., & Thiery, J. P. (1990). Acidic fibroblast growth factor is a modulator of epithelial plasticity in a rat bladder

carcinoma cell line. *Proceedings of the National Academy of Sciences of the United States of America*, *87*(3), 1124–1128.
Van de Putte, T., Maruhashi, M., Francis, A., Nelles, L., Kondoh, H., Huylebroeck, D., et al. (2003). Mice lacking ZFHX1B, the gene that codes for Smad-interacting protein-1, reveal a role for multiple neural crest cell defects in the etiology of Hirschsprung disease-mental retardation syndrome. *American Journal of Human Genetics*, *72*(2), 465–470.
Vanharanta, S., & Massague, J. (2013). Origins of metastatic traits. *Cancer Cell*, *24*(4), 410–421.
Vannier, C., Mock, K., Brabletz, T., & Driever, W. (2013). Zeb1 regulates E-cadherin and Epcam (epithelial cell adhesion molecule) expression to control cell behavior in early zebrafish development. *Journal of Biological Chemistry*, *288*(26), 18643–18659.
Voutsadakis, I. A. (2012). Ubiquitination and the Ubiquitin-Proteasome System as regulators of transcription and transcription factors in epithelial mesenchymal transition of cancer. *Tumour Biology*, *33*(4), 897–910.
Warzecha, C. C., & Carstens, R. P. (2012). Complex changes in alternative pre-mRNA splicing play a central role in the epithelial-to-mesenchymal transition (EMT). *Seminars in Cancer Biology*, *22*(5–6), 417–427.
Weidner, K. M., Behrens, J., Vandekerckhove, J., & Birchmeier, W. (1990). Scatter factor: Molecular characteristics and effect on the invasiveness of epithelial cells. *Journal of Cell Biology*, *111*(5 Pt. 1), 2097–2108.
Williams, J. M., & Daniel, C. W. (1983). Mammary ductal elongation: Differentiation of myoepithelium and basal lamina during branching morphogenesis. *Developmental Biology*, *97*(2), 274–290.
Wright, J. A., Richer, J. K., & Goodall, G. J. (2010). microRNAs and EMT in mammary cells and breast cancer. *Journal of Mammary Gland Biology and Neoplasia*, *15*(2), 213–223.
Wu, W. S., Heinrichs, S., Xu, D., Garrison, S. P., Zambetti, G. P., Adams, J. M., et al. (2005). Slug antagonizes p53-mediated apoptosis of hematopoietic progenitors by repressing puma. *Cell*, *123*(4), 641–653.
Wu, Z. Q., Li, X. Y., Hu, C. Y., Ford, M., Kleer, C. G., & Weiss, S. J. (2012). Canonical Wnt signaling regulates Slug activity and links epithelial-mesenchymal transition with epigenetic Breast Cancer 1, Early Onset (BRCA1) repression. *Proceedings of the National Academy of Sciences of the United States of America*, *109*(41), 16654–16659.
Wu, C. Y., Tsai, Y. P., Wu, M. Z., Teng, S. C., & Wu, K. J. (2012). Epigenetic reprogramming and post-transcriptional regulation during the epithelial-mesenchymal transition. *Trends in Genetics*, *28*(9), 454–463.
Zhang, C., & Klymkowsky, M. W. (2009). Unexpected functional redundancy between Twist and Slug (Snail2) and their feedback regulation of NF-kappaB via Nodal and Cerberus. *Developmental Biology*, *331*(2), 340–349.
Zhang, J., & Ma, L. (2012). MicroRNA control of epithelial-mesenchymal transition and metastasis. *Cancer Metastasis Reviews*, *31*(3–4), 653–662.
Zheng, H., & Kang, Y. (2014). Multilayer control of the EMT master regulators. *Oncogene*, *33*(14), 1755–1763.

CHAPTER NINE

Embryonic Cell–Cell Adhesion: A Key Player in Collective Neural Crest Migration

Elias H. Barriga, Roberto Mayor[1]
Cell and Developmental Biology Department, University College London, London, United Kingdom
[1]Corresponding author: e-mail address: r.mayor@ucl.ac.uk

Contents

Abstract

Cell migration is essential for morphogenesis, adult tissue remodeling, wound healing, and cancer cell migration. Cells can migrate as individuals or groups. When cells migrate in groups, cell–cell interactions are crucial in order to promote the coordinated behavior, essential for collective migration. Interestingly, recent evidence has shown that cell–cell interactions are also important for establishing and maintaining the directionality of these migratory events. We focus on neural crest cells, as they possess extraordinary migratory capabilities that allow them to migrate and colonize tissues all over the embryo. Neural crest cells undergo an epithelial-to-mesenchymal transition at the same time than perform directional collective migration. Cell–cell adhesion has been shown to be an important source of planar cell polarity and cell coordination during collective movement. We also review molecular mechanisms underlying cadherin turnover, showing how the modulation and dynamics of cell–cell adhesions are crucial in order to maintain tissue integrity and collective migration *in vivo*. We conclude that cell–cell

Current Topics in Developmental Biology, Volume 112
ISSN 0070-2153
http://dx.doi.org/10.1016/bs.ctdb.2014.11.023

adhesion during embryo development cannot be considered as simple passive resistance to force, but rather participates in signaling events that determine important cell behaviors required for cell migration.

1. INTRODUCTION

1.1. Neural crest cells

1.1.1 Neural crest formation

The neural crest is a multipotent and highly migratory embryonic cell population, whose development and migration are fundamental for proper formation of a plethora of tissues within vertebrate embryos. Neural crest cells are formed at the border of the neural plate, at the interface between the neural ectoderm and the epidermis (Erickson & Reedy, 1998; Milet & Monsoro-Burq, 2012; Steventon, Carmona-Fontaine, & Mayor, 2005). Neural crest formation occurs at the intersection of many signaling pathways, such as Wnt, retinoic acid, Notch, fibroblast growth factor (FGF), and bone morphogenetic protein (BMP). These signals trigger the expression of a gene regulatory network that defines the neural crest as a territory, and it is also required for subsequent steps of neural crest development (Marchant, Linker, Ruiz, Guerrero, & Mayor, 1998; Mayor, Guerrero, & Martinez, 1997; Mayor, Morgan, & Sargent, 1995; Prasad, Sauka-Spengler, & LaBonne, 2012; Sauka-Spengler & Bronner-Fraser, 2008; Steventon, Araya, Linker, Kuriyama, & Mayor, 2009; Stuhlmiller & Garcia-Castro, 2012). During neurulation, the neural crest cells start to actively migrate in order to invade distinct tissues and differentiate into diverse cell types, including pigment cells, cranial cartilage and bones, enteric nervous system, heart outflow tract, among other tissues (Bhatt, Diaz, & Trainor, 2013; Noden, 1983; Pavan & Raible, 2012; Sasselli, Pachnis, & Burns, 2012; Fig. 1).

1.1.2 Neural crest cell migration

In order to reach their target tissues, neural crest cells delaminate from the neural tube by undergoing epithelial-to-mesenchymal transition (EMT). EMT is a process by which epithelial cells lose their apicobasal polarity and cell–cell adhesion, and gain migratory and invasive capabilities (Kerouso & Bronner-Fraser, 2012; Lim & Thiery, 2012; Thiery & Sleeman, 2006). Neural crest cells perform an EMT by changing the profile of cadherin expression, where specific type-I cadherins are switched on or

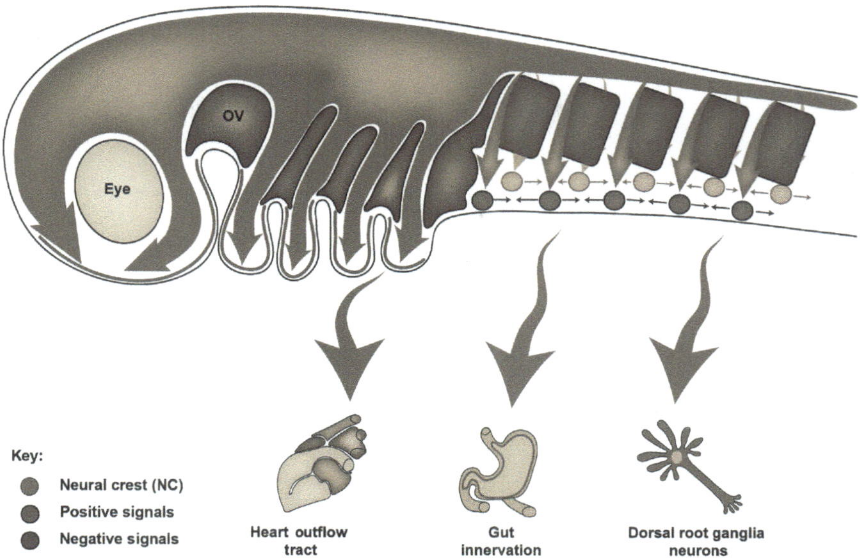

Figure 1 Neural crest cell directional migration. Illustration of a hypothetical vertebrate embryo. Neural crest cells are shown in purple, restrictive signals in red, and attractant cues are represented in green. Thick arrows represent migratory stream of cephalic neural crest cell, migrating around the eye and the otic vesicle (OV). Neural crest cells streams follow chemotactic signals (green) and avoid chemorepulsive cues (red). Trunk neural crest cells are shown as thinner streams that avoid restrictive signals secreted by the somite (red). Migratory neural crest cells reach their target tissues and differentiate to distinct cell types. Among other tissues, neural crest cells contribute to the heart outflow tract, gut innervations, and dorsal root ganglia neurons. (See the color plate.)

off, and type-II cadherins start being expressed (Mayor & Theveneau, 2013; Strobl-Mazzullaa & Bronner, 2012) (Fig. 2). In *Xenopus*, it has been reported that presumptive cephalic neural crest cells express mainly the type-I E-cadherin and N-cadherin as adhesion molecules. At the onset of their migration, E-cadherin expression is repressed and N-cadherin expression levels are slightly reduced, in a mechanism controlled by the transcription factor Twist (Barriga, Maxwell, Reyes, & Mayor, 2013; Theveneau et al., 2013). Similarly, the transcription factor Sip1 is required in two phases of chick neural crest EMT. Sip1 represses the expression of E-cadherin in the presumptive neural crest, thereby allowing cell detachment. During migration, Sip1 activates the expression of N-cadherin and is important for maintaining N-cadherin expression (Rogers, Saxena, & Bronner, 2013). Similarly, Snail2 is essential to allow chick cephalic neural crest cell

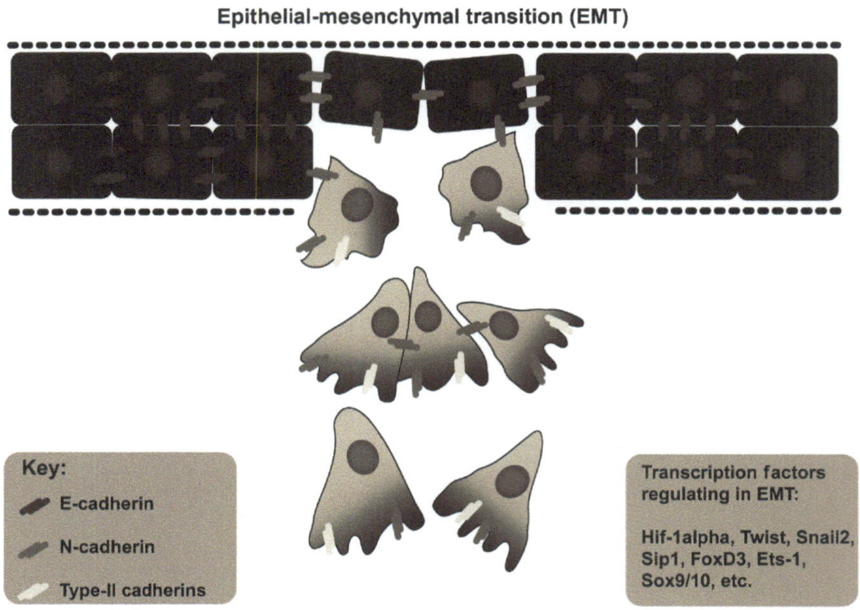

Figure 2 Cadherins switch during neural crest cells epithelial-to-mesenchymal transition (EMT). Events occurring during neural crest cell EMT. Presumptive neural crest cells form an epithelium-like group of cells along the lateral edge of the neural tube. Before migration, neural crest cells express E-cadherin (red parallel lines) and N-cadherin (green parallel lines) as cell–cell adhesion molecules. Prior to migration, neural crest cells undergo EMT. This process involves transcriptional repression of E-cadherin, to allow cells to separate from the neuroectoderm and ectoderm, via delamination. The expression levels of N-cadherin are slightly but not completely reduced, allowing cells to form transient contacts that mediate signaling during collective migration. Once migrating, neural crest cells start to express type-II cadherins such as Cadherin-11, -7, -19, and -20 (yellow parallel lines), this proteins are described to promote protrusive activity, and to be involved in modulating contact inhibition of locomotion (CIL). In the lower right corner are listed some of the master transcription factors that control neural crest cells EMT in different animal models. (See the color plate.)

EMT by transcriptionally repressing Cadherin-6B (Taneyhill, Coles, & Bronner, 2007). Likewise, Ets-1 controls neural crest cell EMT by working together with Snail2 in the repression of Cadherin-6B (Theveneau, Duband, & Altabef, 2007). Type-II cadherins such as Cadherin-11, Cadherin-7, Cadherin-19, and Cadherin-20 are expressed in the migratory neural crest, but their functions in these cells are just starting to be studied (Chalpe, Prasad, Henke, & Paulson, 2010; Faulkner-Jones et al., 1999; Moore & Larue, 2004; Nakagawa & Takeichi, 1995, 1998; Pla et al., 2001; Takahashi & Osumi, 2005).

Once delaminated, neural crest cells follow stereotypical pathways and form migratory streams (Alfandari, Cousin, & Marsden, 2010; Mayor & Theveneau, 2013; Minoux & Rijli, 2010; Fig. 1). These streams are generated by a combination of different signals that restrict neural crest cells from entering neighboring territories (chemorepulsion) (Fig. 1). Trunk neural crest cells express EphB3 receptors that response to ephrin-B1, which is secreted by caudal somitic cells. This expression profile results in a segmented pattern of migratory streams of neural crest cells that cover the rostral but not the caudal portion of somites (Krull, 1998; Yu & Moens, 2005a; Fig. 1). In the cephalic region, stream formation has been attributed to a similar interaction between ephrins and their Eph receptors as well as class 3 semaphorins and their receptors neuropilin and plexin (Gammill, Gonzalez, & Bronner-Fraser, 2007; Smith, Robinson, Patel, & Wilkinson, 1997; Yu & Moens, 2005b).

When migrating through these streams, neural crest cells display high directionality toward their target tissues. However, the mechanisms by which such great directionality is achieved are just starting to be understood. External cues were long thought as driving forces of directional migration. However, recent finding suggest that directional migration emerges as consequence of cell–cell interactions, instead of long-range external signals.

Neural crest cells perform collective cell migration, which can be defined as the coordinated migration of cells as tight clusters or loose groups, where cooperation between cells contributes to their overall directionality (Etienne-Manneville, 2014; Friedl, Sahai, Weiss, & Yamada, 2012; Revenu et al., 2014; Rørth, 2009; Theveneau & Mayor, 2012). In several migrating collectives, cadherin-based cell contacts were thought to simply control cell–cell attachment, but it is now clear that cadherin-based cell contacts are also key regulators of an exquisite molecular communication that allow cells to polarize and collectively migrate with high directionality (Carmona-Fontaine et al., 2008; Desai, Gao, Raghavan, Liu, & Chen, 2009; Dupin, Camand, & Etienne-Manneville, 2009; Friedl & Gilmour, 2009; Revenu et al., 2014; Theveneau et al., 2010). Thus, in this chapter, we provide an overview on planar cell polarity (PCP) (front–rear) and cell–cell adhesion dynamics, to then go on to discuss about the role of cadherin-based cell contacts as a source of cell adhesion and polarity during neural crest cell directional collective migration.

1.2. Front–rear planar cell polarity

One of the keystones of directional migration is the acquisition of front–rear polarity. During migration, this involves protrusion formation at the front and retraction at the rear of the cell. Based mainly on *in vitro* analysis of single cells migrating on a 2D surface, it has been established that this front–rear polarity is dependent on the localized polymerization of actin, which is usually controlled by the activity of the Rho family of small GTPases (Friedl & Wolf, 2009; Guilluy, Garcia-Mata, & Burridge, 2011; Ridley et al., 2003).

Rac1 GTPase activity regulates actin polymerization during lamellipodial extension by directly activating WAVE family of actin nucleators, and phosphoinositides that are required for the activity of actin-nucleating proteins, such as the N-WASP-Arp2/3 complex (Burridge & Wennerberg, 2004; Ridley et al., 1992; Rohatgi, Nollau, Ho, Kirschner, & Mayer, 2001). Cdc42 GTPase activity has been mainly observed at the very tip of the cell front, controlling actin polymerization during filopodium formation (Itoh et al., 2002). On the other hand, the activation of RhoA small GTPase at the rear of the cell promotes stress fiber and focal adhesions formation (Burridge & Wennerberg, 2004; Ridley & Hall, 1992; Fig. 3). Furthermore, when RhoA activates its effector Rho kinase (ROCK), it promotes actomyosin-based contractility, and protrusion collapse via activation of PTEN, which works as a phosphoinositide phosphatase (Chrzanowska-Wodnicka & Burridge, 1996; Li et al., 2005; Ridley, 2011). The mutual antagonism between RhoA and Rac1 is described to be the source of their polarizing functions. This is a complex interaction but in a simplified model, Rac1 inhibits RhoA at the front, by activating p190RhoGAP (Bustos, Forget, Settleman, & Hansen, 2008). Conversely, RhoA inhibition of Rac1 at the rear occurs via ROCK phosphorylation of FilGAP, which, in turn, stimulates its RacGAP activity and leads to Rac1 inactivation (Ohta, Hartwig, & Stossel, 2006). However, recent evidence suggests that the distribution of RhoA and Rac1 activity is more complex, as RhoA can also be found at the front of the cell (Machacek et al., 2009).

Cells can polarize by the following extrinsic and intrinsic cues (Drubin & Nelson, 1996). Single migratory cells can polarize and migrate toward chemotactic cues *in vivo*; one of the best examples of this extrinsically promoted polarity is the attraction of progenitor germ cells toward Sdf1 (Raz, 2005). Nevertheless, the homeostatic distribution of the cytoskeleton components and the distribution of small GTPase activities provide migratory cells with intrinsic self-organizing properties that determine front–rear polarity, independent of external cues (Burke et al., 2014; Mitchison, 1992; reviewed in Woodham & Machesky, 2014). A good example of this occurs during human neutrophil migration: quiescent neutrophils exposed to uniform concentrations of chemoattractant still acquire front–rear polarity. Furthermore, when a chemoattractant is placed behind neutrophils that are already polarized, these

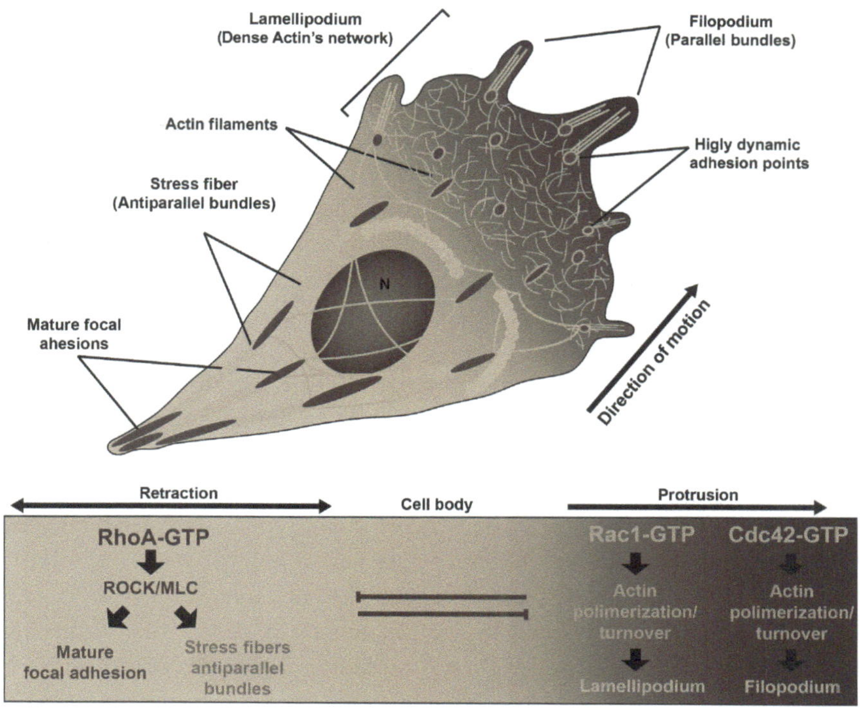

Figure 3 Small GTPases controlling cell polarity and cytoskeletal reorganization. A migratory cell shows polarized activities of Cdc42, Rac1, and RhoA small GTPases. At the rear of the polarized cell, RhoA activity (pail purple (light gray in the print version) in the gradient) promotes contractile activity by influencing the formation of stress fibers (yellow (white in the print version) stripes) and mature cell–substrate adhesions (red (dark gray in the print version) ellipses). Toward the front of the cell Rac1 activity controls lamellipodium formation by promoting actin assembly (yellow (white in the print version) lines) and the formation of nascent cell–substrate adhesion (red (dark gray in the print version) dots). At the very front of the polarized cell, Cdc42 is thought to control filopodium and nascent adhesion formation.

cells perform u-turns, rather than repolarize, in order to migrate toward the attractant (Xu et al., 2003; Zigmond, Levitsky, & Kreel, 1981). These observations suggest that, in these cells, planar polarity is established intrinsically—the chemotactic signal merely provides guidance.

During collective migration cells also migrate by following external cues, however, cells within the group are actively communicating and coordinating their migration. These communications are described as an emergent feature of cadherin-based cell contact and confer polarity and directionality to the migratory group (Friedl, Hegerfeldt, & Tusch, 2004; Rørth, 2012; Winklbauer, Selchow, Nagel, & Angres, 1992).

1.3. Cell–cell adhesion during migration

Cell–cell adhesion is a key component of directional collective migration, providing cell–cell attachment and front–rear polarity to the migrating collectives. As such, cell–cell adhesion appears to be essential for embryos to develop, wounds to heal, adult tissues to remodel, and cancer cells to invade. While cell–cell adhesions can include tight and adherens junctions (AJs) (Farquhar & Palade, 1963; Meng & Takeichi, 2009; Tsukita, Furuse, & Itoh, 2001), in this review, we will focus on the analysis of cadherin-based cell–cell AJs as a source of directionality and cell polarity during neural crest cell collective migration.

AJs are defined as a type of cell–cell adhesion structure in which the homophilic interactions between the extracellular domains of cadherin molecules connect the plasma membranes of neighboring cells. These types of cell–cell adhesions where initially described in epithelial cells, but they are also present in a wide range of nonepithelial cell types, such as neural crest cells, zebrafish blastoderm, and many metastatic cancer cells (Arboleda-Estudillo et al., 2010; Cavey & Lecuit, 2009; Ilina & Friedl, 2012; McKeown, Wallace, & Anderson, 2013). Since the degree of cell spacing is crucial for tissue function, AJs can have different morphologies and conformation depending on the cell type where they are found (Cavey & Lecuit, 2009; Meng & Takeichi, 2009). AJs are produced in three main steps. (i) Initiation: this step is defined as an opportunistic event that results from the exploratory activity of a cell, using large protrusions to survey the environment and scan for other cells nearby. Upon cell–cell contact, cadherin molecules start to accumulate at the contact points; these cadherins initiate a calcium-dependent homophilic interaction that anchors the cells together (Cavey & Lecuit, 2009; Kametani & Takeichi, 2007). During the very first moments of contact, a transient peak of Rac1 and Arp2/3 activities is observed; Rac1 activity allows the cells to keep exploring, looking for new contact points. This peak is rapidly inhibited by the activation of RhoA at the contact site (Perez, Tamada, Sheetz, & James Nelson, 2008; Yamada & Nelson, 2007). This localized RhoA activity also mediates the disassembly of actin filaments from the lamellipodium, replacing them with actin bundles with slow turnover dynamics (Baum & Georgiou, 2011; Vasioukhin, Bauer, Yin, & Fuchs, 2000; Yonemura, 2011). (ii) Lateral expansion: it is thought that AJs expand through the cell–cell contact in two steps. First, laterally repeated rounds of activation and downregulation of Rac1/RhoA small GTPases activities (described above) occur outward from the initial point of cadherin contact. This leads to a spotted pattern of actin polymerization along the contact.

In a second step, myosin-II-mediated contraction at the edge of the contact exerts pulling forces that facilitate the mechanical expansion of the cell–cell contacts (Cavey & Lecuit, 2009). (iii) AJs stabilization: this step is characterized by a downregulation of actin bundle turnover at the contact point, which helps to stabilize the interaction of the junctional complexes (cadherin/catenin) to the cytoskeleton and to maintain cell–cell adhesion (Pilot, Philippe, Lemmers, & Lecuit, 2006; Zandy, Playford, & Pendergast, 2007).

Since Rac1 and RhoA activities are essential for the formation and expansion of the AJs (Braga, Machesky, Hall, & Hotchin, 1997), the activity of these small GTPases must be carefully balanced in order to avoid premature contractility that might destabilize nascent contacts, or excessive contractility that could disassemble mature AJs (Jou & Nelson, 1998). While it is important to stabilize AJs in some contexts, other situations may require a higher rate of junctional turnover. For example, during morphogenetic rearrangements such as wound healing, AJs must be remodeled. Consequently, the cadherin–actin interaction in these situations must be readily disassembled. Thus, the balance of AJs dynamics between stabilization and disassembly must be carefully controlled, and it appears that early signaling events mediated by cadherin-based cell contacts play an important role in this process. In consequence, the early signaling induced by cadherin-based cell contacts, and the balance and dynamic between its players, will determine if the nascent AJs will be stabilized or dissembled.

2. CELL–CELL ADHESION DURING NEURAL CREST CELL DIRECTIONAL COLLECTIVE MIGRATION

2.1. Epithelial-to-mesenchymal transition

EMT has often been thought of as a linear process in which epithelial cells completely lose their stable junctions and apicobasal polarity in order to become highly migratory mesenchymal cells. Current knowledge, however, has shown that EMT is nonlinear and reversible, and as such it involves several steps until its completion (Ahlstrom & Erickson, 2009; Duband, 2010; Thiery, Acloque, Huang, & Nieto, 2009; Thiery & Sleeman, 2006). During EMT cells can take on a wide range of behaviors, from a purely epithelial arrangement with strong cell–cell adhesions, to a very flexible group of mesenchymal cells with transient contacts, to solitary mesenchymal cells in which cell–cell adhesions and apicobasal polarity have been completely lost (Acloque, Adams, Fishwick, Bronner-Fraser, & Nieto, 2009; Ahlstrom & Erickson, 2009; Baum, Settleman, & Quinlan, 2008; Duband, 2010; Theveneau & Mayor, 2012).

During neural crest cell delamination, cells perform a partial EMT and migrate as a fluid-like group of mesenchymal cells in which cell–cell contacts are retained and based mainly on N-cadherin (Kuriyama et al., 2014; Theveneau et al., 2010; Fig. 2). Even though these interactions are transient, they are sufficient to allow for the emergence of cooperative and collective behaviors and have been demonstrated to be essential for collective migration (Monier-Gavelle & Duband, 1997; Piloto & Schilling, 2010).

2.2. Contact inhibition of locomotion

Neural crest cells have been shown to perform contact inhibition of locomotion (CIL), a cellular behavior where cell–cell contacts play a crucial role in providing planar polarity and directionality to the migratory group (Theveneau & Mayor, 2012). CIL is the process by which a cell stops moving after contacting another cell, and it is described as occurring in two phases: collapse of cell protrusion upon contact that transiently arrests cell migration and subsequent repolarization in the opposite direction, with colliding cells eventually moving away from each other (Abercrombie & Dunn, 1975; Abercrombie & Heaysman, 1953, 1954; Carmona-Fontaine et al., 2008; Heckman, 2009; Fig. 4B). In a migratory group of cells, CIL prevents protrusion formation between neighbors and polarizes the group so that the protrusive activity occurs in those cells at the leading edges (Shih & Yamada, 2012).

A role for cell–cell adhesion in regulating directional movements of neural crest cells, and CIL as a driving force of this process, was first proposed in the 1980s (Davis & Trinkaus, 1981; Erickson, 1985; Rovassio, Delouvee, Yamada, Timpl, & Thiery, 1983). Subsequent analysis of protrusive activity after collisions of chick trunk neural crest cells further strengthened the notion that neural crest cells undergo CIL (Teddy & Kulesa, 2004). More recently, the requirement of N-cadherin-dependent cell contacts and the molecular components underlying CIL were demonstrated *in vivo* for *Xenopus* and zebrafish embryos (Carmona-Fontaine & Mayor, 2008; Theveneau et al., 2010; Fig. 4B). When two neural crest cells collide, N-cadherin locally activates the noncanonical Wnt/PCP pathway. This leads to RhoA activation which, in turn, represses Rac1 activity at the contact point. The result is that cell protrusions are established on the opposite side of the cell, prompting the colliding cells to move away from each other (Carmona-Fontaine et al., 2008; Theveneau et al., 2010; Theveneau & Mayor, 2011).

This phenomenon is not unique to the neural crest. Indeed, cadherin-dependent cell contacts provide cell polarity in other types of mesenchymal

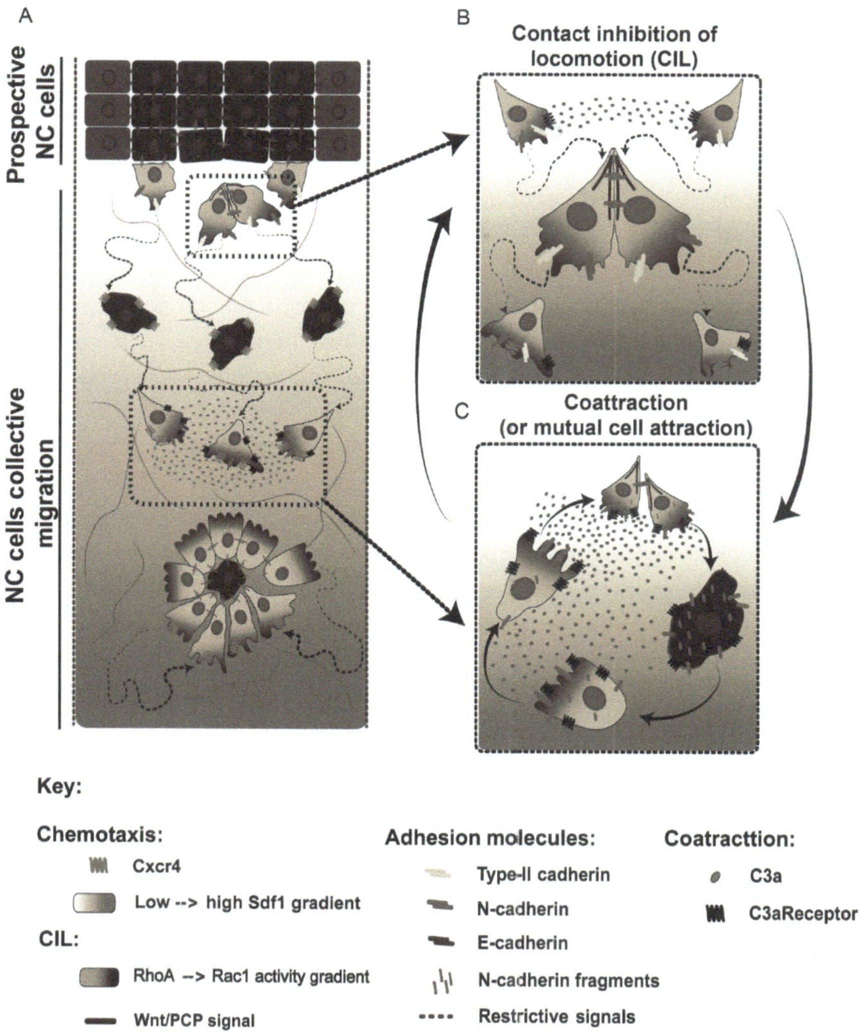

Figure 4 Integration of chemotaxis, contact inhibition of locomotion (CIL), and coattraction along neural crest cells collective migration. (A) Events occurring within a stream of migratory neural crest cells. Restrictive signals are shown as parallel dotted red lines; chemotactic cues (Sdf1) are represented as a green gradient (with high concentration of Sdf1 in green). After prospective neural crest cells undergo EMT, cells experience CIL, repolarize, and disperse. If Sdf1 levels are low neural crest cells disperse and become polarized single cells. Sdf1 can guide polarized cells and, by coattraction, these cells come together to form transient cell–cell contacts. Upon contact, cells undergo CIL, repolarize, and move away from each other again. Repetitive cycle of CIL and coattraction are proposed to maintain cells migrating in a collective and directional manner. (B) Contact inhibition of locomotion (CIL). CIL is a cellular behavior by which upon contact, two migratory cells stop their migration and repolarize in opposite directions.

(Continued)

cells. For instance, astrocytes perform directional collective migration toward wound areas in response to inflammatory signals. In these cells, N-cadherin-dependent cell contacts promote tissue stability and planar polarity by recruiting p120-catenin to the contact area, perhaps regulating small GTPases activity in a similar mechanism as the one described in neural crest and epithelial cells (Camand, Peglion, Osmani, Sanson, & Etienne-Manneville, 2011; Dupin et al., 2009; Peglion, Llense, & Etienne-Manneville, 2014; Zegers et al., 2003). In epithelial cells, cadherin-dependent cell contacts have been showed to maintain tissue organization and control cell polarity during sheet migration. MDCK epithelial cells express high levels of E-cadherin that sequesters p120 at cell–cell contacts, promoting RhoA activation, and allowing tissue organization by avoiding internal cells to overlap. Rac1, in contrast, is activated at the leading edge to promote protrusive activity and migration (Grosheva, Shtutman, Elbaum, & Bershadsky, 2000; Jou & Nelson, 1998; Zegers et al., 2003; Pollack, Runyan, & Mostov, 1998). Single migratory cells also exhibit CIL, as described in *Drosophila* macrophages (hemocytes). After their formation, these cells become evenly distributed along the embryo by following well-defined routes. CIL appears to be responsible for polarizing these cells and coordinates this highly ordered distribution (Davis et al., 2012; Stramer, Dunn, Davis, & Mayor, 2013; Stramer et al., 2010).

The cytoskeleton also plays an important role in CIL, but our knowledge about how cell–cell adhesion molecules and small GTPases control cytoskeleton dynamics is just starting to be expanded (Gao & Chen, 2010; Harris & Tepass, 2010). Cadherins and Dishevelled (Dsh, a downstream effector of Wnt signaling) regulate microtubules dynamics; this regulation has been shown to be important for CIL in *Drosophila* hemocytes and chick fibroblasts (Kadir, Astin, Tahtamouni, Martin, & Nobes, 2011; Stramer et al., 2010).

Figure 4—Cont'd At the molecular level, two neural crest cells collide and upon contact they form very transient N-cadherin-based cell–cell adhesions (green lines). N-cadherin generates local Wnt/PCP signals (blue), which in turn activate RhoA and repress Rac1 at the contact point to repolarize the cells in opposite directions. Type-II cadherins (yellow lines) have also been shown to be involved in CIL. (C) Mutual cell attraction or coattraction. During coattraction, neural crest cells mutually attract each other by expressing and secreting the complement fragment C3a (light blue dots) and the cognate receptor C3aR (black barrels). After experiencing CIL, a neural crest cell moves away from the group. Soon after, the lost cell repolarizes and migrates back toward the group following a C3a gradient. When the cell is docked back to the group, it establishes transient N-cadherin-dependent cell contacts again and the cycle is repeated. (See the color plate.)

It was recently shown that Par3 modulates CIL in *Xenopus* and zebrafish neural crest cells. Upon contact, N-cadherin is localized to the contact point and recruits Par3, which in turn inhibit Trio (a Rac1-GEF), leading to microtubule catastrophe and protrusion collapse (Moore et al., 2013).

Type-II cadherins such as Cadherin-11, Cadherin-7, Cadherin-19, and Cadherin-20 are also expressed during neural crest cell migration, but their function during collective migration is just starting to be understood (Taneyhill, 2008; Fig. 2). *Xenopus* Cadherin-11 (Xcad-11) was first shown to regulate protrusive activity in the cephalic neural crest by modulating Rho GTPase activity via an interaction with GEF-Trio (Kashef et al., 2009). A recent report shows that Xcad-11 is involved in CIL, where it modulates the repulsive response of colliding neural crest cells by mediating cell–cell adhesion, but the mechanism underlying its adhesive function in CIL remains to be elucidated (Becker, Mayor, & Kashef, 2013). S180 cells transfected with chick Cadherin-7 form cell–cell interactions with low adhesiveness by forming less stable complexes with β-catenin; in the same cells Cadherin-7 was demonstrated to promote cell–cell adhesion, ROCK activation, and actomyosin contractility, in an integrin-dependent manner (Chu et al., 2006; Dufour, Beauvais-Jouneau, Delouvée, & Thiery, 1999; Martinez-Rico, Pincet, Thiery, & Dufour, 2010).

2.3. Chemotaxis

Chemotaxis toward Sdf1 facilitates directional migration during *Xenopus* and chick neural crest cell collective migration (Rezzoug, Seelan, Bhattacherjee, Greene, & Pisano, 2011; Theveneau et al., 2010), and N-cadherin-dependent cell contacts are essential for this chemotaxis to occur (Theveneau et al., 2010). Single cells barely respond when exposed to a chemoattractant source. But responsiveness increases with cell density, and larger groups of cells can chemotax more efficiently than single cells. When blocking CIL, by inhibiting N-cadherin or Wnt/PCP, chemotaxis is drastically reduced. Consistently, Sdf1 is able to increase Rac1 activity at the leading edge, but only when cells are already polarized by CIL. Thus, cells polarized by CIL respond more efficiently to Sdf1 than those in which CIL is blocked (Theveneau et al., 2010). These observations suggest that cadherin-based cell interactions allow the emergence of a collective migratory behavior, where cell density allows the collective to perform tasks that single cells would not achieve by themselves. Thus, local interaction between cells allows the collective to self-organize and preestablish a polarity that will be then oriented by an external cue (Sdf1), whereas single cells will "feel" surrounded by external cues and would not be able to detect a gradient. A similar example is observed during zebrafish lateral line migration, where Sdf1 facilitates directional migration of an already self-organized group of cells (Haas & Gilmour, 2006; Rørth, 2009).

2.4. Coattraction

Given that CIL repolarizes cells after contact, one might predict that it would drive cell dispersion and cause a reduction in cell density. How, then, can neural crest cells still migrate collectively? *In vivo*, cells are restricted to specific migratory paths by repulsive signals secreted from their neighbor tissues, contributing to the maintenance of a compact cell cluster (Theveneau & Mayor, 2012). Despite this, in absence of any restrictive signal *in vitro*, neural crest cells still migrate as a group (Barriga et al., 2013; Theveneau et al., 2010). Thus, some other force must enable neural crest cells to stay together. In *Xenopus laevis*, it was described that neural crest cells can mutually attract in a process called coattraction. During coattraction, neural crest cells express the complement fragment C3a and the cognate receptor C3aR (Carmona-Fontaine et al., 2011). C3a/C3aR form a well-characterized chemotactic axis in the immune system (Ricklin, Hajishengallis, Yang, & Lambris, 2010). Since the neural crest cells secrete C3a, its concentration is high in areas with a high density of neural crest cells. Thus, if a C3aR positive cell escapes from the group after CIL, it will be rapidly attracted back to the group by following the local gradient of C3a (Fig. 4C). When the cells are docked back to the group, N-cadherin-dependent cell contacts promote a new round of CIL (Carmona-Fontaine et al., 2011; Fig. 4C). In this way, a balance between CIL and coattraction is assumed to maintain neural crest cells collective migration (Fig. 4A–C). Taken together, these observations reveal that cell–cell adhesions do not work just by mechanically coupling cells. Instead, contact-mediated cell interaction works together with a variety of cellular processes and behaviors, in order to maintain cells migrating in a collective and directional manner.

2.5. Cell–cell adhesion molecule turnover

Collective directional migration of mesenchymal-like cells such as mesodermal cells, astrocytes, and neural crest cells relies on the formation of N-cadherin-based cell contacts (Theveneau & Mayor, 2012). Consequently, modulation of N-cadherin expression levels is crucial in order to allow these cells to properly migrate and reach their targets. Cadherin expression can be controlled at the transcriptional or posttranslational level. During EMT, cadherin transcripts are highly regulated, but once migrating, it appears that the regulation occurs more at the protein level via endocytosis or proteolysis, depending on the cell type (Morita & Heisenberg, 2013; Schiffmacher, Padmanabhan, Jhingory, & Taneyhill, 2014). In chick embryos, for example, ingressing mesoderm cells perform collective migration and express N-cadherin (Chuai, Hughes, & Weijer, 2012; Inuzuka, Redies, & Takeichi, 1991). It has been shown that in these cells, PDGF can modulate N-cadherin protein levels with no effect on its mRNA. This occurs in an Akt-dependent manner, although the details of the mechanism remain unknown (Chuai et al., 2012; Yang, Chrisman, & Wijer, 2008).

In astrocytes, the mechanism underlying N-cadherin protein modulation was recently unveiled. Astrocytes migrate as a mesenchymal group of cells by using N-cadherin-based transient cell contacts. In order to maintain planar polarity and efficient migration, the group performs continuous treadmilling of their cell adhesions at the lateral side of the leading cells (Dupin et al., 2009). In this example, N-cadherin is internalized by endocytosis at the rear of these cells and then recycled to the front, where it accumulates to be reincorporated to new AJs. The polarized recycling of N-cadherin is controlled by GSK3 phosphorylation of p120-catenin at the rear of the cell; this phosphorylation event promotes the dissociation of p120-catenin from the AJs and subsequent N-cadherin endocytosis (Peglion et al., 2014).

Zebrafish blastoderm cells exhibit strong adhesion based on E-cadherin expression. During gastrulation, this adhesion has to be modulated in order to allow blastodermal cells to acquire a dynamic behavior, similar to EMT (Borghi, Lowndes, Maruthamuthu, Gardel, & Nelson, 2010; Song et al., 2013; Ulrich et al., 2005). In this case, it was found that the transcription factor Pou5f1/Oct4 activates the expression of EGF, which in turns promotes E-cadherin endocytosis, possibly via p120-catenin (Song et al., 2013).

The dynamics of N-cadherin-based cell contacts is also regulated in neural crest cells. In *Xenopus* cephalic neural crest cells, lysophosphatidic acid (LPA) and its receptor (LPAR2) stimulate N-cadherin endocytosis (Kuriyama et al., 2014). Cadherins levels can also be modulated by proteolysis: recent findings show that Cadherin-6b can be also processed by ADAMs metalloproteases in chick neural crest cells (Schiffmacher et al., 2014).

Together, these finding show that modulating cadherin levels is crucial in order to ensure the emergence of collective behaviors that promote tissue integrity, fluidity, and directionality during collective migration.

3. CONCLUDING REMARKS

Here, we reviewed the involvement of cell–cell adhesion in different processes and behaviors occurring along collective directional migration. Cell–cell adhesion is frequently understood as a process by which cells remain together, however, here we depicted emergent functions arising from cell–cell interactions, which allow migrating collectives to perform tasks that single cells cannot. An important aspect of this function that has been widely documented for cadherins is the ability of these molecules to activate specific local signals. For migrating cells, the modulation of small GTPases by cadherin engagement plays a fundamental role in controlling migratory properties, such as front–rear polarity, actin polarization, and focal adhesion turnover. In the light of the functions discussed here, cell–cell adhesion during embryo development cannot be considered as simply a passive resistance to force, but rather a signaling event that is vital for proper collective cell behavior.

ACKNOWLEDGMENTS

We thank Dr J. Leslie for comments on the manuscript. The work performed at R. M. lab mentioned in this review was supported by grants from MRC (MR/J000655/1) and BBSRC to R. M. E. B. was supported by QuanTissue Exchange Grant 4505.

REFERENCES

Abercrombie, M., & Dunn, G. A. (1975). Adhesions of fibroblasts to substratum during contact inhibition observed by interference reflection microscopy. *Experimental Cell Research*, *92*, 57–62.

Abercrombie, M., & Heaysman, J. (1953). Observations on the social behaviour of cells in tissue culture. I. Speed of movement of chick heart fibroblasts in relation to their mutual contacts. *Experimental Cell Research*, *5*, 111–113.

Abercrombie, M., & Heaysman, J. (1954). Observations of the social behavior of cells in tissue culture. II "Monolayering" of fibroblasts. *Experimental Cell Research*, *6*, 293–306.

Acloque, H., Adams, M., Fishwick, K., Bronner-Fraser, M., & Nieto, A. (2009). Epithelial-mesenchymal transitions: The importance of changing cell state in development and disease. *Journal of Clinical Investigation*, *119*, 1438–1449.

Ahlstrom, J. D., & Erickson, C. A. (2009). The neural crest epithelial-mesenchymal transition in 4D: A 'tail' of multiple non-obligatory cellular mechanisms. *Development*, *136*, 1801–1812.

Alfandari, D., Cousin, H., & Marsden, M. (2010). Mechanism of *Xenopus* cranial neural crest cell migration. *Cell Adhesion & Migration*, *4*, 553–560.

Arboleda-Estudillo, Y., Krieg, M., Stühmer, J., Licata, N., Muller, D., & Heisenberg, C. P. (2010). Movement directionality in collective migration of germ layer progenitors. *Current Biology*, *20*, 161–169.

Barriga, E. H., Maxwell, P. H., Reyes, A. E., & Mayor, R. (2013). The hypoxia factor Hif-1α controls neural crest chemotaxis and epithelial to mesenchymal transition. *Journal of Cell Biology*, *201*, 759–776.

Baum, B., & Georgiou, M. (2011). Dynamics of adherens junctions in epithelial establishment, maintenance, and remodeling. *Journal of Cell Biology*, *192*, 907–917.

Baum, B., Settleman, J., & Quinlan, M. P. (2008). Transitions between epithelial and mesenchymal states in development and disease. *Seminars in Cell & Developmental Biology*, *19*, 294–308.

Becker, S., Mayor, R., & Kashef, J. (2013). Cadherin-11 mediates contact inhibition of locomotion during *Xenopus* neural crest cell migration. *PLoS One*, *8*, e85717.

Bhatt, S., Diaz, R., & Trainor, P. A. (2013). Signals and switches in mammalian neural crest cell differentiation. *Cold Spring Harbor Perspectives in Biology*, *5*, a008326.

Borghi, N., Lowndes, M., Maruthamuthu, V., Gardel, M. L., & Nelson, W. J. (2010). Regulation of cell motile behaviour by crosstalk between cadherin- and integrin-mediated adhesions. *Proceedings of the National Academy of Sciences of the United States of America*, *107*, 13324–13329.

Braga, V., Machesky, L. M., Hall, A., & Hotchin, N. A. (1997). The small GTPases Rho and Rac are required for the establishment of cadherin-dependent cell–cell contacts. *Journal of Cell Biology*, *137*, 1421–1431.

Burke, T., Christensen, J., Barone, E., Suarez, C., Sirotkin, V., & Kova, D. R. (2014). Homeostatic actin cytoskeleton networks are regulated by assembly factor competition for monomers. *Current Biology*, *24*, 579–585.

Burridge, K., & Wennerberg, K. (2004). Rho and Rac take center stage. *Cell*, *116*, 167–179.

Bustos, R. I., Forget, M. A., Settleman, J. E., & Hansen, S. H. (2008). Coordination of Rho and Rac GTPase function *via* p190B RhoGAP. *Current Biology*, *18*, 1606–1611.

Camand, E., Peglion, F., Osmani, N., Sanson, M., & Etienne-Manneville, S. (2011). N-cadherin expression level modulates integrin-mediated polarity and strongly impacts on the speed and directionality of glial cell migration. *Journal of Cell Science*, *125*, 844–857.

Carmona-Fontaine, C., Matthews, H. K., Kuriyama, S., Moreno, M., Dunn, G. A., Parsons, M., et al. (2008). Contact inhibition of locomotion *in vivo* controls neural crest directional migration. *Nature*, *456*, 957–961.

Carmona-Fontaine, C., & Mayor, R. (2008). Directional cell migration *in vivo*: Wnt at the crest. *Cell Adhesion & Migration*, *2*, 240–242.

Carmona-Fontaine, C., Theveneau, E., Tzekou, A., Tada, M., Woods, M., Page, K. M., et al. (2011). Complement fragment C3a controls mutual cell attraction during collective cell migration. *Developmental Cell*, *21*, 1026–1037.

Cavey, M., & Lecuit, T. (2009). Molecular bases of cell–cell junctions stability and dynamics. *Cold Spring Harbor Perspectives in Biology*, *1*, a002998.

Chalpe, A. B., Prasad, M., Henke, A. J., & Paulson, A. F. (2010). Regulation of cadherin expression in the chicken neural crest by the Wnt/b-catenin signaling pathway. *Cell Adhesion & Migration*, *4*, 431–438.

Chrzanowska-Wodnicka, M., & Burridge, K. (1996). Rho-stimulated contractility drives the formation of stress fibers and focal adhesions. *Journal of Cell Biology*, *133*, 1403–1415.

Chu, Y. S., Eder, O., Thomas, W., Simcha, I., Pincet, F., Ze'ev, A. B., et al. (2006). Prototypical type I E-cadherin and type II Cadherin-7 mediate very distinct adhesiveness through their extracellular domains. *Journal of Biological Chemistry*, *281*, 2901–2910.

Chuai, M., Hughes, D., & Weijer, C. (2012). Collective epithelial and mesenchymal cell migration during gastrulation. *Current Genomics*, *13*, 267–277.
Davis, J. R., Huang, C. Y., Zanet, J., Harrison, S., Rosten, E., Cox, S., et al. (2012). Emergence of embryonic pattern through contact inhibition of locomotion. *Development*, *139*, 4555–4560.
Davis, E. M., & Trinkaus, J. P. (1981). Significance of cell-to-cell contacts for the directional movement of neural crest cells within a hydrated collagen lattice. *Journal of Embryology and Experimental Morphology*, *63*, 29–51.
Desai, R. A., Gao, L., Raghavan, S., Liu, W. F., & Chen, C. S. (2009). Cell polarity triggered by cell–cell adhesion via E-cadherin. *Journal of Cell Science*, *122*, 905–911.
Drubin, D. G., & Nelson, W. J. (1996). Origins of cell polarity. *Cell*, *84*, 335–344.
Duband, J. L. (2010). Diversity in the molecular and cellular strategies of epithelium-to-mesenchyme transitions: Insights from the neural crest. *Cell Adhesion & Migration*, *4*, 458–482.
Dufour, S., Beauvais-Jouneau, A., Delouvée, A., & Thiery, J. P. (1999). Differential function of N-cadherin and Cadherin-7 in the control of embryonic cell motility. *Journal of Cell Biology*, *146*, 501–516.
Dupin, I., Camand, E., & Etienne-Manneville, S. (2009). Classical cadherins control nucleus and centrosome position and cell polarity. *Journal of Cell Biology*, *185*, 779–786.
Erickson, C. A. (1985). Control of neural crest cell dispersion in the trunk of the avian embryo. *Developmental Biology*, *111*, 138–157.
Erickson, C. A., & Reedy, M. V. (1998). Neural crest development: The interplay between morphogenesis and cell differentiation. *Current Topics in Developmental Biology*, *40*, 177–209.
Etienne-Manneville, S. (2014). Neighborly relations during collective migration. *Current Opinion in Cell Biology*, *30*, 51–59.
Farquhar, M. G., & Palade, G. E. (1963). Junctional complexes in various epithelia. *Journal of Cell Biology*, *17*, 375–412.
Faulkner-Jones, B. E., Godinho, L. N., Reese, B. E., Pasquini, G. F., Ruefli, A., & Tan, S. (1999). Cloning and expression of mouse Cadherin-7, a type-II cadherin isolated from the developing eye. *Molecular and Cellular Neurosciences*, *14*, 1–16.
Friedl, P., & Gilmour, D. (2009). Collective cell migration in morphogenesis, regeneration and cancer. *Nature Reviews. Molecular Cell Biology*, *10*, 445–457.
Friedl, P., Hegerfeldt, Y., & Tusch, M. (2004). Collective cell migration in morphogenesis and cancer. *The International Journal of Developmental Biology*, *48*, 441–449.
Friedl, P., Sahai, E., Weiss, S., & Yamada, K. M. (2012). New dimensions in cell migraton. *Nature Reviews. Molecular Cell Biology*, *13*, 743–747.
Friedl, P., & Wolf, K. (2009). Plasticity of cell migration: A multiscale tuning model. *Journal of Cell Biology*, *188*, 11–19.
Gammill, L. S., Gonzalez, C., & Bronner-Fraser, M. (2007). Neuropilin 2/semaphorin 3F signaling is essential for cranial neural crest migration and trigeminal ganglion condensation. *Developmental Neurobiology*, *67*, 47–56.
Gao, C., & Chen, Y. G. (2010). Dishevelled: The hub of Wnt signaling. *Cellular Signalling*, *22*, 717–727.
Grosheva, I., Shtutman, M., Elbaum, M., & Bershadsky, A. (2000). p120-catenin affects cell motility *via* modulation of activity of Rho-family GTPases: A link between cell–cell contact formation and regulation of cell locomotion. *Journal of Cell Science*, *114*, 695–707.
Guilluy, C., Garcia-Mata, R., & Burridge, K. (2011). Rho protein crosstalk: Another social network? *Trends in Cell Biology*, *21*, 718–726.
Haas, P., & Gilmour, D. (2006). Chemokine signaling mediates self-organizing tissue migration in the zebrafish lateral line. *Developmental Cell*, *10*, 673–680.
Harris, T. J., & Tepass, U. (2010). Adherens junctions: From molecules to morphogenesis. *Nature Reviews. Molecular Cell Biology*, *11*, 502–514.

Heckman, C. A. (2009). Contact inhibition revisited. *Journal of Cellular Physiology*, *220*, 574–575.

Ilina, O., & Friedl, P. (2012). Mechanisms of collective cell migration at a glance. *Journal of Cell Science*, *122*, 3203–3208.

Inuzuka, H., Redies, C., & Takeichi, M. (1991). Differential expression of R- and N-cadherin in neural and mesodermal tissues during early chicken development. *Development*, *113*, 959–967.

Itoh, R. E., Kurokawa, K., Ohba, Y., Yoshizaki, H., Mochizuki, N., & Matsuda, M. (2002). Activation of Rac and Cdc42 video imaged by fluorescent resonance energy transfer-based single-molecule probes in the membrane of living cells. *Molecular and Cellular Biology*, *22*, 6582–6591.

Jou, T., & Nelson, W. J. (1998). Effects of regulated expression of mutant RhoA and Rac1 small GTPases on the development of epithelial (MDCK) cell polarity. *Journal of Cell Biology*, *142*, 85–100.

Kadir, S., Astin, J. W., Tahtamouni, L., Martin, P., & Nobes, C. D. (2011). Microtubule remodelling is required for the front-rear polarity switch during contact inhibition of locomotion. *Journal of Cell Science*, *124*, 2642–2653.

Kametani, Y., & Takeichi, M. (2007). Basal-to-apical cadherin flow at cell junctions. *Nature Cell Biology*, *9*, 92–98.

Kashef, J., Köhler, A., Kuriyama, S., Alfandari, D., Mayor, R., & Wedlish, D. (2009). Cadherin-11 regulates protrusive activity in Xenopus cranial neural crest cells upstream of Trio and the small GTPases. *Genes & Development*, *23*, 1393–1398.

Kerouso, L., & Bronner-Fraser, M. (2012). What is bad in cancer is good in the embryo: Importance of EMT in neural crest development. *Seminars in Cell & Developmental Biology*, *23*, 320–332.

Krull, C. (1998). Inhibitory interactions in the patterning of trunk neural crest migration. *Annals of the New York Academy of Sciences*, *23*, 13–22.

Kuriyama, S., Theveneau, E., Benedetto, A., Parsons, M., Tanaka, M., Charras, G., et al. (2014). *In vivo* collective cell migration requires an LPAR2-dependent increase in tissue fluidity. *Journal of Cell Biology*, *206*, 113–127.

Li, Z., Dong, X., Wang, Z., Liu, W., Deng, N., Ding, Y., et al. (2005). Regulation of PTEN by Rho small GTPases. *Nature Cell Biology*, *7*, 399–404.

Lim, J., & Thiery, J. P. (2012). Epithelial-mesenchymal transitions: Insights from development. *Development*, *139*, 3471–3486.

Machacek, M., Hodgson, L., Welch, C., Elliott, H., Pertz, O., Nalbant, P., et al. (2009). Coordination of Rho GTPase activities during cell protrusion. *Nature*, *461*, 99–103.

Marchant, L., Linker, C., Ruiz, P., Guerrero, N., & Mayor, R. (1998). The inductive properties of mesoderm suggest that the neural crest cells are specified by a BMP gradient. *Developmental Biology*, *198*, 319–329.

Martinez-Rico, C., Pincet, F., Thiery, J. P., & Dufour, S. (2010). Integrins stimulate E-cadherin-mediated intercellular adhesion by regulating Src-kinase activation and actomyosin contractility. *Journal of Cell Sciences*, *123*, 712–722.

Mayor, R., Guerrero, N., & Martinez, C. (1997). Role of FGF and noggin in neural crest induction. *Developmental Biology*, *189*, 1–12.

Mayor, R., Morgan, R., & Sargent, M. G. (1995). Induction of the prospective neural crest of *Xenopus*. *Development*, *121*, 767–777.

Mayor, R., & Theveneau, E. (2013). The neural crest. *Development*, *140*, 2247–2251.

McKeown, S. J., Wallace, A. S., & Anderson, R. B. (2013). Expression and function of cell adhesion molecules during neural crest migration. *Developmental Biology*, *373*, 244–257.

Meng, W., & Takeichi, M. (2009). Adherens junction: Molecular architecture and regulation. *Cold Spring Harbor Perspectives in Biology*, *1*, a002899.

Milet, C., & Monsoro-Burq, A. H. (2012). Neural crest induction at the neural plate border in vertebrales. *Developmental Biology*, *366*, 22–33.
Minoux, M., & Rijli, F. M. (2010). Molecular mechanisms of cranial neural crest cell migration and patterning in craniofacial development. *Development*, *137*, 2605–2621.
Mitchison, T. J. (1992). Self-organization of polymer-motor systems in the cytoskeleton. *Philosophical Transactions of the Royal Society of London. Series B, Biological Sciences*, *336*, 99–106.
Monier-Gavelle, F., & Duband, J. L. (1997). Cross talk between adhesion molecules: Control of N-cadherin activity by intracellular signals elicited by beta1 and beta3 integrins in migrating neural crest cells. *Journal of Cell Biology*, *137*, 1663–1681.
Moore, R., & Larue, L. (2004). Cell surface molecules and truncal neural crest ontogeny: A perspective. *Birth Defects Research. Part C, Embryo Today: Reviews*, *72*, 140–150.
Moore, R., Theveneau, E., Pozzi, S., Alexandre, P., Richardson, J., Merks, A., et al. (2013). Par3 controls neural crest migration by promoting microtubule catastrophe during contact inhibition of locomotion. *Development*, *140*, 4763–4775.
Morita, H., & Heisenberg, C. P. (2013). Holding on and letting go: Cadherin turnover in cell intercalation. *Developmental Cell*, *24*, 567–569.
Nakagawa, S., & Takeichi, M. (1995). Neural crest cell–cell adhesion controlled by sequential and subpopulation-specific expression of novel cadherins. *Development*, *121*, 1321–1332.
Nakagawa, S., & Takeichi, M. (1998). Neural crest emigration from the neural tube depends on regulated cadherin expression. *Development*, *125*, 2963–2971.
Noden, D. M. (1983). The role of the neural crest in patterning of avian cranial skeletal, connective, and muscle tissues. *Developmental Biology*, *96*, 144–165.
Ohta, Y., Hartwig, J. H., & Stossel, T. P. (2006). FilGAP, a Rho- and ROCK-regulated GAP for Rac binds filamin A to control actin remodelling. *Nature Cell Biology*, *8*, 803–814.
Pavan, W. J., & Raible, D. W. (2012). Specification of neural crest into sensory neuron and melanocyte lineages. *Developmental Biology*, *366*, 55–63.
Peglion, F., Llense, F., & Etienne-Manneville, S. (2014). Adherens junction treadmilling during collective migration. *Nature Cell Biology*, *16*, 639–651.
Perez, T. D., Tamada, M., Sheetz, M. P., & James Nelson, W. J. (2008). Immediate-early signaling induced by E-cadherin engagement and adhesion. *Journal of Biological Chemistry*, *283*, 5014–5022.
Pilot, F., Philippe, J. M., Lemmers, C., & Lecuit, T. (2006). Spatial control of actin organization at adherens junctions by the synaptotagmin-like protein Btsz. *Nature*, *442*, 580–584.
Piloto, S., & Schilling, T. F. (2010). Ovo1 links Wnt signaling with N-cadherin localization during neural crest migration. *Development*, *137*, 1981–1990.
Pla, P., Moore, R., Morali, O. G., Grille, S., Martinozzi, S., Delmas, V., et al. (2001). Cadherins in neural crest cell development and transformation. *Journal of Cellular Physiology*, *189*, 121–132.
Pollack, A. L., Runyan, R. B., & Mostov, K. E. (1998). Morphogenetic mechanisms of epithelial tubulogenesis: MDCK cell polarity is transiently rearranged without loss of cell–cell contact during scatter factor/hepatocyte growth factor-induced tubulogenesis. *Developmental Biology*, *204*, 64–79.
Prasad, M. S., Sauka-Spengler, T., & LaBonne, C. (2012). Induction of the neural crest state: Control of stem cell attributes by gene regulatory, post-transcriptional and epigenetic interactions. *Developmental Biology*, *366*, 10–21.
Raz, E. (2005). Germ cells: Sex and repression in mice. *Current Biology*, *15*, 600–603.

Revenu, C., Streichan, S., Donà, E., Lecaudey, V., Lars, H., & Gilmour, D. (2014). Quantitative cell polarity imaging defines leader-to-follower transitions during collective migration and the key role of microtubule-dependent adherens junction formation. *Development*, *141*, 1282–1291.

Rezzoug, F., Seelan, R., Bhattacherjee, V., Greene, G., & Pisano, M. (2011). Chemokine-mediated migration of mesencephalic neural crest cells. *Cytokine*, *56*, 760–768.

Ricklin, D., Hajishengallis, G., Yang, K., & Lambris, J. D. (2010). Complement: A key system for immune surveillance and homeostasis. *Nature Immunology*, *11*, 785–797.

Ridley, A. J. (2011). Life at the leading edge. *Cell*, *145*, 1012–1022.

Ridley, A. J., & Hall, A. (1992). The small GTP-binding protein rho regulates the assembly of focal adhesions and actin stress fibers in response to growth factors. *Cell*, *70*, 389–399.

Ridley, A. J., Schwartz, M. A., Burridge, K., Firtel, R. A., Ginsberg, M. H., Borisy, G., et al. (2003). Cell migration: Integrating signals from front to back. *Science*, *302*, 1704–1709.

Rogers, C. D., Saxena, A., & Bronner, M. (2013). Sip1 mediates an E-cadherin-to-N-cadherin switch during cranial neural crest EMT. *Journal of Cell Biology*, *203*, 835–847.

Rohatgi, R., Nollau, P., Ho, H. Y., Kirschner, M. W., & Mayer, B. J. (2001). Nck and phosphatidylinositol 4,5-bisphosphate synergistically activate actin polymerization through the N-WASP-Arp2/3 pathway. *Journal of Biological Chemistry*, *276*, 26448–26452.

Rørth, P. (2009). Collective cell migration. *Annual Review of Cell and Developmental Biology*, *25*, 407–429.

Rørth, P. (2012). Fellow travellers: Emergent properties of collective cell migration. *EMBO Reports*, *13*, 984–991.

Rovassio, R. A., Delouvee, A., Yamada, K. M., Timpl, R., & Thiery, J. P. (1983). Neural crest cell migration: Requirements for exogenous fibronectin and high cell density. *Journal of Cell Biology*, *96*, 462–473.

Sasselli, V., Pachnis, V., & Burns, A. J. (2012). The enteric nervous system. *Developmental Biology*, *366*, 64–73.

Sauka-Spengler, T., & Bronner-Fraser, M. (2008). A gene regulatory network orchestrates neural crest formation. *Nature Reviews. Molecular Cell Biology*, *9*, 557–568.

Schiffmacher, A. T., Padmanabhan, R., Jhingory, S., & Taneyhill, L. A. (2014). Cadherin-6B is proteolytically processed during epithelial-to-mesenchymal transitions of the cranial neural crest. *Molecular Biology of the Cell*, *25*, 41–54.

Shih, W., & Yamada, S. (2012). N-cadherin-mediated cell–cell adhesion promotes cell migration in a three-dimensional matrix. *Cell Adhesion & Migration*, *6*, 513–517.

Smith, A., Robinson, V., Patel, K., & Wilkinson, D. G. (1997). The EphA4 and EphB1 receptor tyrosine kinases and ephrin-B2 ligand regulate targeted migration of branchial neural crest cells. *Current Biology*, *7*, 561–570.

Song, S., Eckerle, S., Onichtchouk, D., Marrs, J. A., Nitschke, R., & Driever, W. (2013). Pou5f1-dependent EGF expression controls E-cad endocytosis, cell adhesion, and zebrafish epiboly movements. *Developmental Cell*, *24*, 486–501.

Steventon, B., Araya, C., Linker, C., Kuriyama, S., & Mayor, R. (2009). Differential requirements of BMP and Wnt signalling during gastrulation and neurulation define two steps in neural crest induction. *Development*, *136*, 771–779.

Steventon, B., Carmona-Fontaine, C., & Mayor, R. (2005). Genetic network during neural crest induction: From cell specification to cell survival. *Seminars in Cell & Developmental Biology*, *16*, 647–654.

Stramer, B. M., Dunn, G. A., Davis, J. R., & Mayor, R. (2013). Rediscovering contact inhibition in the embryo. *Journal of Microscopy*, *251*, 206–211.

Stramer, B., Moreira, S., Millard, T., Evans, I., Huang, C. Y., Sabet, O., et al. (2010). Clasp-mediated microtubule bundling regulates persistent motility and contact repulsion in Drosophila macrophages *in vivo*. *Journal of Cell Biology*, *189*, 681–689.

Strobl-Mazzullaa, P. H., & Bronner, M. (2012). Epithelial to mesenchymal transition: New and old insights from the classical neural crest model. *Seminars in Cancer Biology*, *22*, 411–416.

Stuhlmiller, T. J., & Garcia-Castro, M. I. (2012). Current perspectives of the signaling pathways directing neural crest induction. *Cellular and Molecular Life Sciences*, *69*, 3715–3737.

Takahashi, M., & Osumi, N. (2005). Identification of a novel type II classical cadherin: Rat cadherin19 is expressed in the cranial ganglia and Schwann cell precursors during development. *Developmental Dynamics*, *232*, 200–208.

Taneyhill, L. A. (2008). To adhere or not to adhere: The role of cadherins in neural crest development. *Cell Adhesion & Migration*, *2*, 223–230.

Taneyhill, L. A., Coles, E. G., & Bronner, M. (2007). Snail2 directly represses cadherin6B during epithelial-to-mesenchymal transitions of the neural crest. *Development*, *134*, 1481–1490.

Teddy, J. M., & Kulesa, P. M. (2004). *In vivo* evidence for short- and long-range cell communication in cranial neural crest cells. *Development*, *131*, 6141–6151.

Theveneau, E., Duband, J. L., & Altabef, M. (2007). Ets-1 confers cranial features on neural crest delamination. *PLoS One*, *2*, e1142.

Theveneau, E., Marchant, L., Kuriyama, S., Gull, M., Moepps, B., Parsons, M., et al. (2010). Collective chemotaxis requires contact-dependent cell polarity. *Developmental Cell*, *19*, 39–53.

Theveneau, E., & Mayor, R. (2011). Collective cell migration of the cephalic neural crest: The art of integrating information. *Genesis*, *49*, 164–176.

Theveneau, E., & Mayor, R. (2012). Neural crest delamination and migration: From epithelium-to-mesenchyme transition to collective cell migration. *Developmental Biology*, *366*, 34–54.

Theveneau, E., Steventon, B., Scarpa, E., Garcia, S., Trepat, X., Streit, A., et al. (2013). Chase-and-run between adjacent cell populations promotes directional collective migration. *Nature Cell Biology*, *15*, 763–772.

Thiery, J. P., Acloque, H., Huang, R. Y., & Nieto, M. A. (2009). Epithelial-mesenchymal transitions in development and disease. *Cell*, *139*, 871–890.

Thiery, J. P., & Sleeman, J. P. (2006). Complex networks orchestrate epithelial–mesenchymal transitions. *Nature Reviews. Molecular Cell Biology*, *7*, 131–142.

Tsukita, S., Furuse, M., & Itoh, M. (2001). Multifunctional strands in tight junctions. *Nature Reviews. Molecular Cell Biology*, *2*, 285–293.

Ulrich, F., Krieg, M., Schotz, E. M., Link, V., Castanon, I., Schnabel, V., et al. (2005). Wnt11 functions in gastrulation by controlling cell cohesion through Rab5c and E-cadherin. *Developmental Cell*, *9*, 555–564.

Vasioukhin, V., Bauer, C., Yin, M., & Fuchs, E. (2000). Directed actin polymerization is the driving force for epithelial cell–cell adhesion. *Cell*, *100*, 209–219.

Winklbauer, R., Selchow, A., Nagel, M., & Angres, B. (1992). Cell interaction and its role in mesoderm cell migration during Xenopus gastrulation. *Developmental Dynamics*, *195*, 290–302.

Woodham, E. F., & Machesky, L. M. (2014). Polarised cell migration: Intrinsic and extrinsic drivers. *Current Opinion in Cell Biology*, *30*, 25–32.

Xu, J., Wang, F., Van Keymeulen, A., Herzmark, P., Straight, A., Kelly, K., et al. (2003). Divergent signals and cytoskeletal assemblies regulate self-organizing polarity in neutrophils. *Cell*, *114*, 201–214.

Yamada, S., & Nelson, W. J. (2007). Localized zones of Rho and Rac activities drive initiation and expansion of epithelial cell–cell adhesion. *Journal of Cell Biology*, *178*, 517–527.

Yang, X., Chrisman, H., & Wijer, C. (2008). PDGF signalling controls the migration of mesoderm cells during chick gastrulation by regulating N-cadherin expression. *Development*, *135*, 3521–3530.

Yonemura, S. (2011). Cadherin–actin interactions at adherens junctions. *Current Opinion in Cell Biology*, *23*, 515–522.

Yu, H., & Moens, C. (2005a). Semaphorin signaling guides cranial neural crest cell migration in zebrafish. *Developmental Biology*, *208*, 373–385.

Yu, H., & Moens, C. B. (2005b). Semaphorin signaling guides cranial neural crest cell migration in zebrafish. *Developmental Biology*, *280*, 373–385.

Zandy, N., Playford, M., & Pendergast, N. M. (2007). Abl tyrosine kinases regulate cell–cell adhesion through Rho GTPases. *Proceedings of the National Academy of Sciences of the United States of America*, *104*, 17686–17691.

Zegers, M., Forget, M., Chernoff, J., Mostov, K., Beest, M., & Hansen, S. (2003). Pak1 and PIX regulate contact inhibition during epithelial wound healing. *EMBO Journal*, *22*, 4155–4165.

Zigmond, S. H., Levitsky, H. I., & Kreel, B. J. (1981). Cell polarity: An examination of its behavioral expression and its consequences for polymorphonuclear leukocyte chemotaxis. *Journal of Cell Biology*, *89*, 585–592.

CHAPTER TEN

VE-cadherin in Vascular Development: A Coordinator of Cell Signaling and Tissue Morphogenesis

Anne Karine Lagendijk, Benjamin M. Hogan[1]
Institute for Molecular Bioscience, The University of Queensland, Brisbane, Queensland, Australia
[1]Corresponding author: e-mail address: b.hogan@imb.uq.edu.au

Contents

Abstract

Blood and lymphatic vessels make up the vascular system of vertebrates and are lined by specialized endothelial cells. The connections between endothelial cells are formed by adhesion molecules and are essential to maintain cell–cell adhesion, cell–cell communication, and the integrity of our vascular tubes. One key adhesion molecule is the adherens junctional protein vascular endothelial cadherin (VE-cadherin). In addition to its role in endothelial adhesion, it is emerging that this protein is actively involved in modulating key cellular signaling cascades within endothelial cells and can control the behavior of endothelial cells during development and morphogenesis. We describe key studies that highlight VE-cadherin as a regulatory hub in endothelial cell signaling during angiogenesis, vessel morphogenesis, and vascular development.

Current Topics in Developmental Biology, Volume 112
ISSN 0070-2153
http://dx.doi.org/10.1016/bs.ctdb.2014.11.024

1. VASCULAR CELL ADHESION

Our vascular tree consists of several specialized major vessel types: arteries, veins, and lymphatic vessels. These vessels are lined with morphologically and functionally distinct endothelial cells. The endothelial cells function as mediators of fluid homeostasis, hormone distribution, immune cell trafficking, and nutrient distribution. Within our plumbed and integrated vasculature, the cellular adhesion of endothelial cells is crucial for all of these major functions.

In blood vessels, robust adhesion is necessary to retain the circulating blood within the vessel lumen while the endothelial wall is under pressure. However, the balance between adhesion and permeability within blood vessel junctions is highly adaptable since the junction needs to maintain the potential to weaken, for example, when transmigration of leukocytes occurs (Vestweber, 2012). In contrast to the blood vasculature, the lymphatic system does not transport blood cells but instead transports excessive fluid from peripheral tissues back into the circulatory system. Adhesion of lymphatic endothelial cells is weaker than in blood vessels and junctions are constructed in a stereotypical "button-like" configuration to ensure trafficking of not only fluid but also lipids and immune cells between the extracellular space and the vessel lumen (Baluk et al., 2007). These examples highlight both the functional importance of adhesive junctions and their heterogeneity in the mature vasculature.

Similar to epithelial cells, adhesion in the endothelium is established by interactions between transmembrane proteins present on the surface of neighboring cells. There are three types of junctions present in endothelial cells, namely tight junctions, adherens junctions, and gap junctions (ECs lack desmosomes) that can be distinguished spatially based on distinctive membrane localization but also by the expression of unique adhesion protein complexes. The principal adhesion protein establishing cell-to-cell contacts in adherens junctions of endothelial cells is vascular endothelial cadherin (VE-cadherin). VE-cadherin was first identified in 1991 amongst seven other cadherin clones as "clone 5" (cadherin-5), which was found in endothelial cells (Suzuki, Sano, & Tanihara, 1991). Given the selective expression of cadherin-5 at the junctions of endothelial cells, the cadherin-5 was renamed into VE-cadherin in 1995 (Breviario et al., 1995). The role of adherens junctional complexes and in particular the protein VE-cadherin will be the focus of this chapter.

2. VE-CADHERIN AS A PHYSICAL ADAPTOR

Like other members of the cadherin superfamily, the VE-cadherin protein consists of a large extracellular domain (ectodomain) with five cadherin repeats (EC1–EC5), a single-pass transmembrane domain and a cytoplasmic tail (Lampugnani et al., 1992; Tanihara et al., 1994; Fig. 1). Adhesion is established by homophilic interactions between the extracellular regions of dimers of VE-cadherin on adjoining cells. The extracellular domain is stabilized contextually by binding of calcium ions to the cadherin repeats (Hatta, Okada, & Takeichi, 1985; Takeichi, 1977; Vestweber & Kemler, 1984). By using antibodies against EC1 and EC4 it has been shown that both domains are required in angiogenesis contexts and EC1 also regulates endothelial integrity (Corada et al., 2001, 2002). While it is clear that these domains are essential for cell–cell adhesion, it is not understood how this occurs at a structural level. Intercellular signals at cell-to-cell contacts can be transferred by VE-cadherin to the inside of the cell via transmembrane and cytoplasmic binding partners (Vincent, Xiao, Buckley, & Kowalczyk, 2004). These bridging partners interconnect the actomyosin cytoskeleton to junctional complexes at the membrane, ensuring a structurally integrated cell within a cohesive multicellular tissue.

Cytoplasmic cadherin binding partners were initially identified in the 1980s and 1990s to bind to the epithelial E-cadherin and included the catenins α-catenin, β-catenin, and γ-catenin (Ozawa, Baribault, & Kemler, 1989; Ozawa & Kemler, 1992). Interference with cadherin–catenin linkage

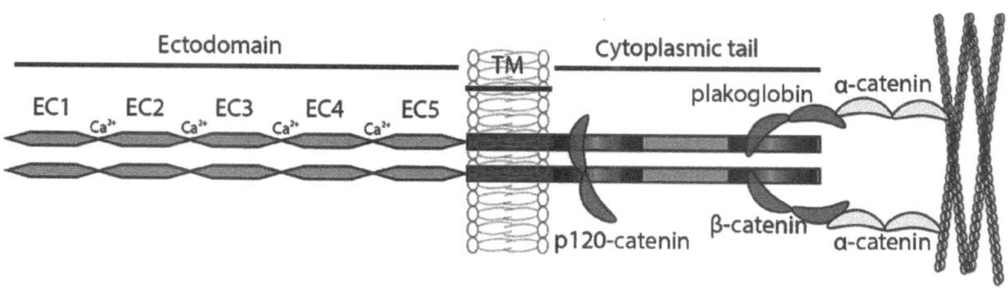

Figure 1 *Structural representation of a VE-cadherin homodimer at the adherens junction.* The VE-cadherin protein consists of an extracellular domain (ectodomain) with five cadherin repeats (EC1–EC5) and Ca^{2+} binding sites, a transmembrane domain (TM) and a cytoplasmic tail. Major binding partners of VE-cadherin, namely p120-catenin, β-catenin, and plakoglobin all bind to the cytoplasmic tail of VE-cadherin. β-catenin and plakoglobin also bind to α-catenin, which connects to the actomyosin machinery of the cell.

revealed the necessity of this interaction to facilitate cell-to-cell adhesion (Nagafuchi & Takeichi, 1988; Navarro et al., 1995; Ozawa, Ringwald, & Kemler, 1990). In endothelial cells, the cytoplasmic tail of VE-cadherin has been shown to predominantly interact with catenins, namely p120-catenin, β-catenin, and plakoglobin (previously γ-catenin) (reviewed in Giannotta, Trani, & Dejana, 2013; Fig. 1). β-catenin and plakoglobin form the structural bridges between VE-cadherin and the actin cytoskeleton (Fig. 1) while p120-catenin was initially identified as an Src kinase substrate (Reynolds, Herbert, Cleveland, Berg, & Gaut, 1992) and is shown to be involved in cadherin clustering (Yap, Brieher, & Gumbiner, 1997) and actin organization to enhance adhesion (Thoreson et al., 2000). Indirectly, β-catenin can also link VE-cadherin to α-catenin and the release of α-catenin from the VE-cadherin complex at the junctions can induce actin bundling in the cytoplasm (Noda et al., 2010; Rimm, Koslov, Kebriaei, Cianci, & Morrow, 1995; Fig. 1).

Depending on the state of the endothelium, interactions between VE-cadherin and these intracellular binding partners are regulated in a manner that will result in an appropriate cellular response. In newly formed vessels the endothelial cells are in a more active state, cell-to-cell contacts are less stable, which leads to cellular responses that can enhance processes like migration and proliferation. Whereas in more mature endothelial cells cell-to-cell junctions are more stable which is necessary to maintain vascular structure, thereby inhibiting processes like cellular expansion and inducing interaction with perivascular cells (reviewed in Dejana & Vestweber, 2013; Gaengel, Genove, Armulik, & Betsholtz, 2009).

3. MODULATION OF VE-CADHERIN BY KINASES, PHOSPHATASES, AND SMALL GTPases

Since the discovery of catenin binding, additional cytoplasmic cadherin interacting proteins have been identified biochemically and a hierarchy of direct and indirect linkages has been established (reviewed in Vincent et al., 2004). While the focus of this chapter is the active role of VE-cadherin in developmental and morphogenesis processes, and the intracellular VE-cadherin interactions have been expertly reviewed in detail elsewhere (Dejana & Vestweber, 2013; Giannotta et al., 2013; Vincent et al., 2004), it is important to briefly describe the complexity of some of these interactions and introduce their potential to influence endothelial signaling pathways described later in the chapter.

The VE-cadherin cytoplasmic tail is regulated by a range of phosphatases and kinases, which modify protein function, signaling, and junctional permeability (reviewed in Dejana & Vestweber, 2013; Kuppers, Vockel, Nottebaum, & Vestweber, 2014). Key tyrosine residues in the VE-cadherin tail have been identified, mainly in cultured cells, including Y645, Y658, Y685, Y731, and Y733 (reviewed in Kuppers et al., 2014). The most prominent kinase involved is Src: a major regulator of cell–matrix and cell–cell interactions. Src is phosphorylated in response to diverse signals including integrin signaling (Huveneers & Danen, 2009; Mitra & Schlaepfer, 2006), interstitial pressure (Planas-Paz et al., 2012), and shear stress (Davis, Cai, Drummond, & Harrison, 2001). Src phosphorylation at the junctions generally results in VE-cadherin dissociation from the junctions and thus increased permeability (Fig. 2). Orsenigo et al. for the first time revealed the relevance of Y658 and Y685 phosphorylation *in vivo* by using antibodies labeling phosphorylated VE-cadherin proteins at these specific residues (Orsenigo et al., 2012). Remarkably, venous endothelial cells are phosphorylated at these sites while arterial cells are not (Orsenigo et al.,

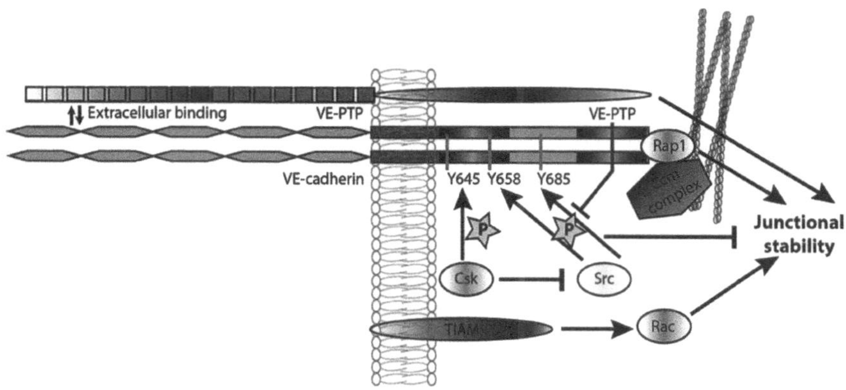

Figure 2 *A selection of functional interactions of VE-cadherin with phosphatases, kinases, and small GTPases.* VE-cadherin interacts with the endothelial-specific phosphatase VE-PTP via the ectodomain. This interaction stabilizes VE-cadherin at the junctions via inhibition of phosphorylation of the VE-cadherin cytoplasmic tail. Src kinase has been shown *in vivo* to phosphorylate Y658 and Y685, which results in reduced stability of VE-cadherin at the junction. Csk has been shown to inhibit this activity of Src via phosphorylation of Y645. The adherens junction-associated TIAM can enhance junctional stability of VE-cadherin via activation of the small GTPase Rac. Another small GTPase Rap1, in association with the CCM protein complex, forms a connection between the VE-cadherin cytoplasmic tail and the actin cytoskeleton thereby promoting junctional stability of VE-cadherin.

2012). When culturing these cells, this difference was lost, highlighting that VE-cadherin phosphorylation is dependent on factors present *in vivo* and contextually regulated. This process is under the influence of shear stress and requires Src in a regulatory event that increases permeability (Fig. 2; Orsenigo et al., 2012).

Dephosphorylation of VE-cadherin by protein–tyrosine–phosphatases (PTPs) on the other hand mostly enhances junctional strength (reviewed in Dejana & Vestweber, 2013; Kuppers et al., 2014). Well studied in the context of leukocyte extravasation is the vascular restricted vascular endothelial PTPs (VE-PTP) (Baumer et al., 2006; Fachinger, Deutsch, & Risau, 1999; Fukuhara et al., 2008; Saharinen et al., 2008). VE-PTP binds to the ectodomain of VE-cadherin (Fig. 2; Nawroth et al., 2002) and functions counteractive to vascular endothelial growth factor (VEGF) by dephosphorylating tyrosine residues in the cytoplasmic tail of VE-cadherin normally phosphorylated upon VEGF activity via Src (Fig. 2; Nawroth et al., 2002). This interaction stabilizes VE-cadherin complexes at the junctions while dissociation of VE-PTP from VE-cadherin increases vascular permeability (Fig. 2; Nottebaum et al., 2008). Loss of VE-PTP during mouse embryonic development causes embryonic lethality from E9.5 to E10 (Baumer et al., 2006; Fachinger et al., 1999), stages reminiscent of loss of VE-cadherin itself (detailed further below) (Carmeliet et al., 1999; Gory-Faure et al., 1999). To investigate the relevance of VE-cadherin–VE-PTP interaction for diapedesis Broermann and colleagues generated genetic knock-in mice expressing modified VE-cadherin and VE-PTP proteins (Broermann et al., 2011). Both proteins contained an additional C-terminal domain susceptible to compound-induced stabilization of their interaction (Broermann et al., 2011). This study demonstrated that the VE-cadherin–VE-PTP complex stabilizes endothelial junctions. Increased permeability, normally observed during LPS-induced inflammation and upon VEGF stimulation, requires dissociation of this complex (Broermann et al., 2011). As a consequence, stabilization of VE-cadherin–VE-PTP in knock-in models inhibits leukocyte extravasation (Broermann et al., 2011).

In addition to these prominent interactions, it has also been shown that the binding of kinase Csk to Y645 can lead to the inactivation of Src activity (Fig. 2) and that of the focal adhesion kinase (FAK) at adherens junctions can influence β-catenin dissociation from VE-cadherin in angiogenesis (Baumeister et al., 2005; Duan, Imamoto, & Fong, 2004). Of great interest in the context of intracellular signaling in development and morphogenesis, VE-cadherin is capable of triggering Akt–PI3K signaling (Taddei et al.,

2008), although how prevalent this is in developmental or morphogenesis contexts remains unclear. Taken together, these observations indicate that the phosphorylation of VE-cadherin and associated proteins represents a dynamic regulatory landscape with great potential to regulate important cellular and tissue level outcomes.

A number of small GTPases are crucial for regulating endothelial cell morphology, migration, and activity. In the context of the VE-cadherin intracellular domains, several important regulators are associated. Rap1 interacts with VE-cadherin and controls the interaction of the junctions with the actin cytoskeleton (Fig. 2; Glading, Han, Stockton, & Ginsberg, 2007; Sakurai et al., 2006). Rap1 prominently associates with the cerebral cavernous malformations protein 1 (CCM1) and underpins vascular integrity in the pathophysiology of the inherited vascular disorder cerebral cavernous malformations (Glading & Ginsberg, 2010; Glading et al., 2007). This complex acts in part by the modulation of Raf1 and Rho kinase (Glading & Ginsberg, 2010; Glading et al., 2007). Another important GTPase interaction is with the RacGEF TIAM. TIAM is associated with adherens junctions and is involved in the activation of Rac, which stabilizes the junctions (Fig. 2; Lampugnani et al., 2002). Finally, it is worth noting that p120-catenin, when free of the adherens junctional complex, can act as an inhibitor of Rho (Anastasiadis et al., 2000; Reynolds & Carnahan, 2004). These observations together highlight the complexity of the small GTPase regulatory network and that VE-cadherin acts as a hub in this network.

In epithelia, PAR3 and PAR6 polarity proteins and atypical protein kinase C (aPKC) control cell polarity and junctional maturation. The functional relevance of the interaction between VE-cadherin and the PAR/aPKC polarity complex is far from understood. However, it is established that the cytoplasmic domain of VE-cadherin interacts directly with both PAR3 and PAR6 (Iden et al., 2006). These proteins are important effectors of CCM1 function during lumen formation and polarization of cultured endothelial cells (Lampugnani et al., 2010). In addition, the presence of aPKC is necessary *in vitro* as an effector of Rac1 and Cdc42 (Iden et al., 2006; Koh, Mahan, & Davis, 2008). These examples and those mentioned above provide a deeper appreciation of the environment in which the intracellular domain of VE-cadherin functions. This list is far from exhaustive and further interaction networks have been summarized elsewhere (Dejana & Vestweber, 2013; Giannotta et al., 2013; Kuppers et al., 2014). This chapter will now delve into the organismal level requirements for VE-cadherin in development and morphogenesis and how this crucial protein interacts with major vascular signaling pathways.

4. VE-CADHERIN IN VESSEL DEVELOPMENT AND MORPHOGENESIS

There is strong evidence that VE-cadherin plays active roles in the formation of new vessel networks. Several animal models have been used to explore the physiological and pathophysiological roles of VE-cadherin. Two independent studies described the analysis of a loss of VE-cadherin function mouse model by inducing deletion of early exons in the *VE-cadherin* gene (Carmeliet et al., 1999; Gory-Faure et al., 1999). An additional loss-of-function model was described in which VE-cadherin was truncated in the cytoplasmic tail resulting in a failure to bind to β-catenin (Carmeliet et al., 1999). In the developing mouse embryo, a primary vascular plexus gives rise to the later vascular network. Upon initial arterial–venous differentiation in the primary plexus, the arterial and venous cells undergo a series of molecular and morphological changes yielding a vascular network with specialized types of vessels (Corada, Morini, & Dejana, 2014; Harvey & Oliver, 2004; Swift & Weinstein, 2009). *VE-cadherin* heterozygous mice were indistinguishable from wild-type littermates and survived to adulthood where no effects on fertility were observed (Carmeliet et al., 1999; Gory-Faure et al., 1999). However, both the loss of VE-cadherin models or the introduction of the truncated VE-cadherin protein resulted in the same phenotypes (Carmeliet et al., 1999; Gory-Faure et al., 1999). Homozygous mutant embryos died at midgestation with severe cardiovascular defects. At E10.5, homozygous mutant embryos were shown to be light in color and small in size with additional morphological defects like reduced somite number, enlargement of the pericardial sac, and reduced cardiac contractility (Carmeliet et al., 1999; Gory-Faure et al., 1999).

Detailed analysis of E8.0 mutant embryos found them to be indistinguishable from heterozygous and wild-type siblings, showing that VE-cadherin is not required for angioblasts to differentiate into endothelial cells or for initial vasculogenesis to occur (Carmeliet et al., 1999; Gory-Faure et al., 1999). Shortly after, at E8.5, mutant vessels were severely affected, showing a failure to lumenize or an expanded vessel diameter. Connections between endothelial and endocardial cells were lost and detachment of cells was observed (Carmeliet et al., 1999; Gory-Faure et al., 1999). Later, during stages of sprouting angiogenesis, detachment of endothelial cells increased further and sprouting angiogenesis was severely impaired in mutants giving rise to a reduced number of irregular dilated sprouts compared with wild-type littermates (Carmeliet et al., 1999; Gory-Faure et al., 1999). By

E9.5, regression and collapse of mutant vessels result in a complete cardiovascular failure leading to necrosis in the embryo proper (Carmeliet et al., 1999; Gory-Faure et al., 1999).

In the zebrafish, partial knockdown of VE-cadherin using morpholinos did not affect vasculogenesis of the major head and trunk vessels (Montero-Balaguer et al., 2009). However, sprouting angiogenesis was impaired in VE-cadherin morphants resulting in a reduction in the number of sprouts (Montero-Balaguer et al., 2009). Interestingly, in partial knockdown scenarios, the sprouts that did emerge were prone to hemorrhaging (Montero-Balaguer et al., 2009). This suggests that the weakened adherens junctions were not capable of maintaining vessel integrity when under the pressure of blood flow. In addition, non-lumenized vessels were described that failed to anastomose to the perfused vasculature and this failure to anastomose led to vessel regression (Montero-Balaguer et al., 2009). These observations suggested that there may be multiple functions for VE-cadherin in developing vessels. In line with this observation but perhaps more telling, a recently described zebrafish VE-cadherin mutant has highlighted an interesting link between junctional adhesion and anastomosis of blood vessels in the head (Lenard et al., 2013). The authors show that in wild-type embryos, anastomosis is initiated by the formation of a single intercellular junction that forms when filopodia of two opposing tip cells make first contact. As a result of this contact, remaining filopodial extensions from the touching tip cells collapse, indicating that anastomosis is initiated by establishing a junctional contact (Lenard et al., 2013). Ultimately, the cells differentiate to achieve a less active state and form stable junctions (Lenard et al., 2013). However, when VE-cadherin function is lost, continuous unstable filopodial protrusions extend between the tip cells suggesting that VE-cadherin is essential to trigger intracellular signals that identify cell-to-cell recognition and anastomosis of normal lumenized vessels. This intriguing observation may indicate an important function in contact-dependent adhesion.

5. VE-CADHERIN AND VASCULAR SIGNALING DURING ANGIOGENESIS AND MORPHOGENESIS

5.1 VEGF signaling, FGF signaling, and angiogenic VE-cadherin

The process of angiogenesis is defined as the outgrowth of new vessels from pre-existing vasculature. The main growth factors that instruct proangiogenic behavior of endothelial cells are members of the vascular endothelial growth factor (VEGF) family. Six VEGF factors have been identified to date, namely VEGFA, VEGFB, VEGFC, VEGFD, VEGFE, and placental growth factor (PlGF). The VEGF ligands can interact with three different tyrosine kinase receptors, VEGF receptors 1 to 3 (also referred to as Flt1, Flk1 and Flt4). Many different combinations of VEGF ligand–receptor interaction have been described (reviewed in Lohela, Bry, Tammela, & Alitalo, 2009). In the context of angiogenesis, the interaction between VEGFA and VEGFR2 has been studied in the most detail and can be considered highly potent to drive sprouting angiogenesis. The importance of VEGFA as a ligand to drive angiogenesis is underlined by the fact that heterozygous loss of *Vegfa* function in mice results in embryonic lethality (Carmeliet et al., 1996; Ferrara et al., 1996). The clinical importance of VEGFA in vascular pathologies and therapeutics has been extensively documented (Ferrara, 2000; Ferrara & Alitalo, 1999).

In 1977, before the discovery of the genetic signals that drive angiogenesis, Ausprunk and Folkman were the first to describe the process of angiogenic sprouting in the context of tumor growth (Ausprunk & Folkman, 1977). Interestingly, as one of their observations by electron microscopy they define a weakening of the cell-to-cell contacts in endothelial cells of the angiogenic sprout. In line with these observations, the ligand that we know as VEGFA was initially identified as vascular permeability factor (VPF; Senger et al., 1983; Senger, Perruzzi, Feder, & Dvorak, 1986). Since then it has been widely appreciated that junctional integrity is reduced in VEGF-activated endothelial cells, mainly by a reduced presence of VE-cadherin in adherens junctions. Mechanistically, work in cultured cells has shown that upon activation by VEGFA, VEGFR2 dimers are formed which result in Src-mediated phosphorylation of Vav2 (a guanine nucleotide-exchange factor) which causes activation of the small GTPase Rac (Fig. 3). Activated Rac induces p21-activated kinase (PAK)-mediated phosphorylation of a conserved serine residue (S665) in the cytoplasmic tail of VE-cadherin leading to recruitment of β-arrestin2 (Fig. 3; Gavard & Gutkind, 2006). This signaling pathway activates arrestin-dependent endocytosis of VE-cadherin and thereby reduces cell–cell adhesion (Fig. 3). This reduction in adhesion is thought to permit normal sprouting angiogenesis (Gavard & Gutkind, 2006). In addition, it has been shown *in vitro* that VE-cadherin can directly influence the level of internalization of actively signaling VEGFR2 (Lampugnani, Orsenigo, Gagliani, Tacchetti, & Dejana, 2006). Here, the mechanism of modulation of VEGFR2 signaling involves the activity of the p120-catenin and β-catenin-associated phosphatase DEP-1 (Fig. 3; Lampugnani et al., 2006) but it is unclear how pervasive this mechanism is *in vivo*. Hence, while VE-cadherin downregulation is

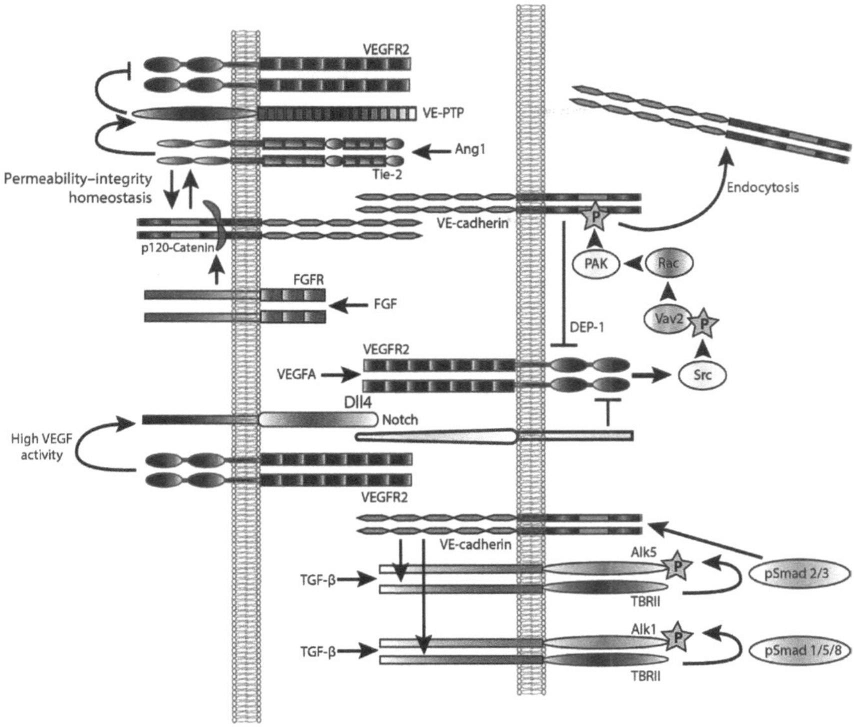

Figure 3 *Schematic representation of relationships between major vascular signaling pathways and VE-cadherin.* VEGFR2 activation by VEGFA results in an activation of Src kinase, which subsequently phosphorylates Vav2, ultimately activating PAK (via Rac). PAK phosphorylation of the VE-cadherin cytoplasmic tail triggers endocytosis of the protein and thus enhances junctional permeability. Reciprocally VE-cadherin can dampen VEGFR2 activity via activation of the phosphatase DEP-1. Another mechanism to inhibit VEGF signaling activity is via active Dll4–Notch signaling which dampens the VEGF-induced effect at VE-cadherin. Ang–Tie2 signaling can also inhibit VEGF-induced signaling via binding to VE-PTP. Levels of active Tie2 and VE-cadherin at the junctions were shown to maintain a balance in permeability and integrity of the junctions. FGF signals enhance VE-cadherin stability by promoting VE-cadherin presence at the junctions and also by enhancing the interaction between VE-cadherin and p120-catenin. Finally, TGF-β ligands in endothelial cells activate TBRII receptors that form active receptor complexes with both Alk1 and Alk5 (by phosphorylation) resulting in active pSmad2/3 and pSmad 1/5/8 transcriptional complexes. pSmad 2/3 activation has been shown to be required for VE-cadherin junctional stability and VE-cadherin in return enhances the formation of both TBRII–TBRI complexes upon TGF-β activation. (See the color plate.)

permissive in angiogenesis, it can also modulate angiogenic outcomes in response to VEGFA by directly controlling intracellular signaling.

Another potent family of growth factors and receptors that induce proliferation, migration, and differentiation in cultured endothelial cells is the fibroblast growth factor (FGF) family. A large number of FGF's are encoded in mammalian genomes and they act via four tyrosine kinase FGF receptors (FGFR1–4). In vascular development *in vivo*, a deep understanding of FGF function has been complicated by ligand and receptor redundancy (reviewed in Gerwins, Skoldenberg, & Claesson-Welsh, 2000). However, FGF's have been shown to influence coronary artery formation (Lavine et al., 2006), tumor angiogenesis (reviewed in Brooks, Kilgour, & Smith, 2012; Katoh & Nakagama, 2014) and in conditional endothelial FGFR1/2 double knockout mice a crucial requirement has been recently established in neoangiogenesis in wound repair (Oladipupo et al., 2014). Of particular interest, Murakami et al. utilized soluble ligand traps for FGF ligands as well as dominant negative forms of FGF receptors *in vivo* and in cultured endothelial cells (Murakami et al., 2008). They found that FGF signaling is necessary to maintain VE-cadherin and the maintenance of VE-cadherin–p120-catenin binding at the junctions (Fig. 3; Murakami et al., 2008; Oladipupo et al., 2014). It has also been established that VE-cadherin can be co-immunoprecipitated with FGFR1 and DEP-1 and that VE-cadherin can exert a negative influence on FGF signaling (Giampietro et al., 2012). These studies highlight that VE-cadherin can act as a modulator if both FGF and VEGF signaling in endothelial cells and argue for its role and the importance of the adherens junctions as a signaling hub in vascular development. It remains to be determined the contexts in which such modulation contributes to morphogenesis in developmental angiogenesis.

5.2 Notch signaling and VE-cadherin in angiogenesis

The interplay between pro-angiogenic VEGF signaling and inhibitory Notch signaling acts as a deterministic mechanism to resolve which endothelial cells undergo sprouting in developing vessels. A gradient of VEGF ligand activates the endothelial cells within the parent vessel. The cell that receives the highest level of VEGF induction becomes motile, polarized, and migratory. To limit the number of sprouting cells, high VEGF signaling activates the expression of the Notch ligand, delta-like-4 (Dll4) (Fig. 3), which binds to Notch receptors expressed on neighboring cells (Fig. 3). Activation of Notch signaling in the adjacent cells reduces their responsiveness to VEGF (Fig. 3). As a result, VEGF-induced signaling is high in the initial activated cell and this cell is a tip cell: the leading cell in a sprout. The cells that follow the tip cell are called stalk cells. In stalk cells, levels of VEGF signaling are lower, due to activation of Dll4–Notch signaling.

This tip–stalk cell model was first introduced in 2003 (Gerhardt et al., 2003) and it has since been refined to an appreciation that the tip and stalk cell states are transient. Tip and stalk cells swap position and cell behaviors dependent on a continuously changing balance between proangiogenic and anti-angiogenic factors (Arima et al., 2011; Jakobsson et al., 2010).

Of particular pertinence is the recent work from Bentley et al. that demonstrates a crucial role for the adherens junctions and VE-cadherin in modulating the Notch–VEGF signaling interplay (Bentley et al., 2014). First, the authors performed *in silico* predictions to simulate angiogenesis based on current models from embryoid body sprouting. It was predicted that differential adhesion by VE-cadherin at the junction in response to cellular levels of VEGFR2 signaling and cell-autonomous formation of contractile protrusions are required in angiogenic cell behavior. Data showed that activation of endothelial cells by VEGF correlated with a serrated appearance of the junctions (Bentley et al., 2014). Reducing Notch signaling *in vitro* and *in vivo* demonstrated that Notch is crucial to maintain junctional heterogeneity (Bentley et al., 2014). This inhibition of Notch activity resulted in the increased abundance of active junctions. Over activation of the pathway with a constitutively active form of the Notch1 intracellular domain (NICD) resulted in less active junctions (Bentley et al., 2014). Similarly, reduction of VEGFA signaling reduced the presence of activated junctions. Ultimately, the authors showed that junctional dynamics are modulated by Notch–VEGF signaling (Bentley et al., 2014). FRAP experiments *in vitro* found a decreased turnover of junctional VE-cadherin–GFP proteins of cells experiencing high Notch activation. These results suggest that the reduction seen in cell mixing when levels of VEGF or Notch are affected, probably occurs due to a failure of differential junctional activity of VE-cadherin, which normally generates heterogeneity in cell behavior (Bentley et al., 2014). Importantly, in establishing a direct mechanistic link with VE-cadherin activity and morphogenesis, Src-mediated phosphorylation of VE-cadherin (Fig. 3) was demonstrated to be required to drive this cell mixing (Bentley et al., 2014). T-cell specific adaptor protein (TSAd) is required for recruiting Src to phosphorylated VEGFR2 receptors (Matsumoto et al., 2005; Sun et al., 2012), and inducing VE-cadherin endocytosis. In TSAd deficient mice there was a shift toward less active endothelial cell junctions consistent with reduced downstream responsiveness to VEGFA (Bentley et al., 2014).

This work brought to light that VEGF and Notch are not simple on/off switches but that subtle differences in pathway activities are what decides cell behavior. This requires the participation of VE-cadherin at the junctions, which permits the cells to change state and allows for a continuous spectrum of cell behavior within angiogenic sprouts. This heterogeneity is essential for angiogenesis and again VE-cadherin acts as a hub for the decisive signaling events.

5.3 Angiopoietin–Tie signaling and junctional regulation

Like VEGFR signaling, the Angiopoietin–Tie signaling pathway is largely selective in function to the endothelium and is essential in directing the development of the cardiovascular system during embryogenesis (reviewed in Fagiani & Christofori, 2013; Thomas & Augustin, 2009). Not surprisingly perhaps, this pathway has also been shown to act in part via modulating the cellular activity of VE-cadherin in studies described below.

Currently, four Angiopoietin ligands have been identified; Angiopoietin-1 (Ang-1), Angiopoietin-2 (Ang-2), Angiopoietin-3 (Ang-3), and Angiopoietin-4 (Ang-4) with Ang-1 and -2 being of most relevance here. These Angiopoietin ligands directly interact with and activate Tie-2 tyrosine kinase receptors. In addition, via binding to Tie-2 positioned in Tie-1/Tie-2 heterodimers, the Angiopoietins can also activate Tie-1 signaling. Both Tie-1 and Tie-2 receptors are expressed on endothelial cells and a subclass of macrophages (reviewed in Fagiani & Christofori, 2013; Thomas & Augustin, 2009). Ang-1 is required for vascular morphogenesis during embryogenesis with Ang-1 deficient mice exhibiting a disturbed vascular network, dilated vessels that lack basement membrane and perivascular connections from E11.5. Ang-2 mutants display a failure in the postnatal remodeling and maturation of the lymphatic vasculature (Dellinger et al., 2008; Gale et al., 2002). In addition, adult Ang-2 deficient mice have compromised vascular functions in the context of vascular diseases (Fiedler et al., 2006). Tie-1 and Tie-2 knockout studies in mice have revealed that neither receptor is essential for early vasculogenesis but both are required for vascular network angiogenesis and elaboration during morphogenesis. Tie-1 null embryos display milder phenotypes than Tie-2 null embryos. In Tie-1 mutant pups angiogenesis still occurs but the integrity of the newly formed vessels is perturbed. When Tie-2 function is lost, a malformed vascular network develops with phenotypes highly reminiscent of the loss of Ang-1 function (Puri, Rossant, Alitalo, Bernstein, & Partanen, 1995; Sato et al., 1995). Tie-1/Tie-2 double mutants embryos resemble the Tie-2 knockout animals. Both Tie-1 and Tie-2, therefore, function in endothelial proliferation and maintenance but the requirements are temporally distinct.

The activation of Tie2 signaling by Ang1 in endothelial cells *in vitro* leads to the translocation of the signaling complex to the adherens junctions when cell–cell adhesion is present (Saharinen et al., 2008). In the absence of cell–cell adhesion through adherens junctions, the Ang1–Tie2 signaling complex localizes to cell–matrix attachment sites and elicits distinct intracellular signaling outcomes. In the context of active Ang1–Tie2 signaling at adherens junctions, Tie2 binds to VE-PTP (Fig. 3) and active signaling influences cellular permeability. Importantly in this context, this Tie2-dependent function of VE-PTP was also shown to modulate the phosphorylation state of VEGFR2 in a study that demonstrated the importance of the junctions in establishing the relationship between Tie2–VE-PTP signaling and VEGFR2 signaling outputs in endothelial cells (Hayashi et al., 2013). VEGFR2 was found to be dephosphorylated at the junctions in a Tie2-dependent manner that is mediated by VE-PTP (Fig. 3). Taken together, these studies suggest that Tie2 signaling complexes likely influence junctional activity through VE-PTP interactions and that junctional VE-cadherin can probably influence Tie2-mediated as well as VEGFR2-mediated intracellular signaling.

In vivo, there is additional evidence that Ang–Tie signaling directly influences adherens junctions. In zebrafish, Tie2 is not required for the formation and function of the blood or lymphatic vasculature system under normal conditions (Gjini et al., 2011). However, exposure of *tie2* mutant vasculature to atorvastatin, a drug that induces hemorrhaging in wild-type littermates, showed that loss of Tie2 enhances vessel integrity. Mechanistically, VE-cadherin was shown to be required for the enhanced vascular integrity in *tie2* mutants using a genetic epistasis approach (Gjini et al., 2011). These results suggest that to maintain pro-permeability and pro-integrity signals a Tie2–VE-cadherin interaction is required (Fig. 3).

In the context of lymphatic vascular adherens junctions, the link is more direct. In the recent study by Zheng et al., Ang2 inhibition was shown to inhibit the maturation of lymphatic vascular junctions (Zheng et al., 2014). Normally, lymphatic endothelial cells in immature vessels have continuous adherens junctions between cells but these junctions become progressively discontinuous as the vessels mature. Ultimately the junctions formed between lymphatic endothelial cells in functional vessels are distinctly "button-like" in nature (Baluk et al., 2007). With the loss of Ang2 activation, these junctions remain continuous, failing to take on the typical "button-like" morphology. It was found that Ang2 is needed for normal phosphorylation of Y685 on VE-cadherin (Zheng et al., 2014), which controls permeability (Orsenigo et al., 2012). This mechanism appears to be active in multiple contexts, being observed in developing initial lymphatics and mature collecting lymphatic vessels using both Ang2 blocking antibodies and knockout mice (Zheng et al., 2014). Taken together with the above findings, these studies suggest the possibility of an integral and reciprocal modulatory role of junctional VE-cadherin and Tie-receptor-mediated endothelial cell signaling.

5.4 TGF-β–Smad signaling and VE-cadherin

The transforming growth factor-β (TGF-β) family of secreted polypeptides includes a large group of over 40 proteins of which the signaling driven by TGF-β, bone morphogenetic proteins (BMPs), and activin growth factors has been most extensively studied (reviewed in Feng & Derynck, 2005). TGF-β ligands bind to the constitutively active TGF-β type II receptor (TBRII). This interaction results in the recruitment and phosphorylation of the TGF-β type I receptor (TBRI/activin receptor-like kinase 5—ALK5). Activation of the TBRI/ALK5 receptors results in recruitment and phosphorylation of receptor associated-Smads (R-Smads) 2 and 3. Phosphorylation leads to dissociation of the R-Smads, which then form a complex with Smad4 within the cell. This trimeric complex acts in the nucleus with cell specific cofactors to regulate transcription (Attisano & Wrana, 2002; Derynck & Zhang, 2003; Derynck, Zhang, & Feng, 1998; Itoh, Itoh, Goumans, & Ten Dijke, 2000). In the vasculature, TGF-β ligands can exert a range of effects depending on the context of the responding endothelial cell (Pepper, 1997). Interestingly, specialized TGF-β signaling is at play in the endothelium through an additional TGF-β type I receptor, ALK1 (Goumans et al., 2002; Oh et al., 2000), which is predominantly expressed in endothelial cells in the context of angiogenesis (Goumans, Lebrin, & Valdimarsdottir, 2003; Roelen, van Rooijen, & Mummery, 1997). As a result, when endothelial cells are stimulated with TGF-β, not only R-Smads 2/3 but also R-Smads 1/5/8, become activated via ALK5 and ALK1, respectively (Fig. 3). Phenotypes of knockout mouse models for both the ligand TGF-β1 (Dickson et al., 1995; Kulkarni et al., 1993; Shull et al., 1992) and the type II and I receptors TBRII (Oshima, Oshima, & Taketo, 1996), Alk1 (Oh et al., 2000) and Alk5 (Goumans, Valdimarsdottir, et al., 2003) revealed that all are required in processes that include angiogenesis, vessel maintenance, and pericyte recruitment (reviewed in Goumans & Mummery, 2000). For the type I receptors, a hierarchy was observed in Alk5 mutant mice where Alk5 is required for TGF-β signaling via Alk1 (Goumans, Valdimarsdottir, et al., 2003; Goumans et al., 2002; Larsson et al., 2001).

VE-cadherin at the junctions can modulate the outcome of TGF-β signaling in both cultured endothelial cells and in cancer cell lines (Labelle et al., 2008; Rudini et al., 2008). VE-cadherin mutant cultured cells were shown to be less able to respond to TGF-β stimulation. Phosphorylation levels of both R-Smads 2/3 (via Alk5) and R-Smads 1/5/8 (via Alk1) were reduced when VE-cadherin was not present. VE-cadherin binds directly to TBRII and Alk5 based on co-immunoprecipitation studies. These VE-cadherin-TGF-β receptor interactions were shown to be increased upon TGF-β stimulation (Rudini et al., 2008). Further experiments have indicated that VE-cadherin interaction promotes the occurrence of active TBRII–TBRI complexes, increasing the amount of phosphorylated R-Smads in the nucleus of endothelial cells upon TGF-β stimulation (Rudini et al., 2008). Given that numerous studies have found that the TGF-β pathway is indispensable for the normal embryonic development of the vasculature, the biochemical capabilities of VE-cadherin to control this signaling pathway are highly suggestive of *in vivo* relevance. Indeed, one recent study highlighted that upon endothelial-specific knockout of R-Smads 2 and 3, VE-cadherin localization to the endothelial junctions is impaired (Fig. 3; Itoh et al., 2012). In these mutant mice that specifically lack vascular TGF-β–Alk5–Smad2/3 signaling, the vascular tree is properly patterned but the vessels fail to mature, observed by a reduced recruitment of pericytes and unstable endothelial junctions (Itoh et al., 2012). These defects lead to leakage of blood from the vasculature, which ultimately results in embryonic death from E10.5 (Itoh et al., 2012). This certainly suggests the importance of TGF-β upstream of VE-cadherin. However, the specific *in vivo* contexts in which VE-cadherin/TGF-β cross talk is active and the relative importance of this mechanism during angiogenesis or developmental processes remains to be fully understood.

5.5 VE-cadherin and the potential for regulation of other signaling events

The binding to or modulation of VE-cadherin by a number of important intracellular signaling proteins (as summarized above) suggests significant potential for signaling interactions in numerous contexts. Two particularly intriguing possibilities, while not understood with the mechanistic detail such as those above, are explored briefly here.

Mechanotransduction is the interpretation of mechanical stimuli by a cell and their conversion into meaningful signaling outcomes.

Mechanotransduction can occur at the level of cadherin family proteins by well-defined mechanisms in epithelial cells (reviewed in Gomez, McLachlan, & Yap, 2011; Leckband, le Duc, Wang, & de Rooij, 2011; Smutny & Yap, 2010). In endothelial cells, the intracellular kinase Src phosphorylates the VE-cadherin intracellular tail at a series of functional tyrosine residues (reviewed in Kuppers et al., 2014). Src can be recruited by ligand-mediated signaling (e.g., by VEGF signaling) but also by mechanical stimuli originating at the cell–cell interface or integrin-mediated mechanical interactions at the cell–matrix interface (reviewed in Huveneers & Danen, 2009). As such, through at least this mechanism, cellular signaling at the adherens junctions can mediate processes of mechanotransduction. Recent data emerging from studies of the endothelial response to shear stress have revealed that VE-cadherin itself is under significant tension and together with PECAM-1 and VEGFR2 it forms a mechanosensory complex that can activate intracellular signals like Src kinase itself (Conway et al., 2013; Tzima et al., 2005). Within the mechanosensory complex VE-cadherin-mediated tension is not responsive to flow-mediated shear stress but tension across PECAM-1 is highly responsive (Conway et al., 2013). Therefore, VE-cadherin is thought to function as a structural component of the mechanosensory complex, while PECAM-1 and VEGFR2 transduce intracellular signals (Conway et al., 2013; Tzima et al., 2005). Whether VE-cadherin has the potential to modify signaling in response to changes in junctional tension might be context dependent. Given that most current data correlating force through VE-cadherin with transduction events derives from the study of cultured endothelial cells, it remains unclear the cellular contexts *in vivo* and also the mechanical stimuli *in vivo* which may be important. Nevertheless, the possibility that force responsive mechanotransduction through VE-cadherin may contribute to morphogenetic events is intriguing and plausible based on mechanisms characterized *in vitro*.

Other pathways that can be modulated by junctional localization and have prominent roles in vascular development and disease are the canonical and noncanonical Wnt signaling pathways. Both β-catenin and p120-catenin are components of the Wnt signaling pathway that localize to either the nucleus or the adherens junctions and bind intracellularly to VE-cadherin (reviewed in Nelson & Nusse, 2004; Perez-Moreno & Fuchs, 2006; Valenta, Hausmann, & Basler, 2012). It is unclear to what degree junctional VE-cadherin might influence the amount of available β-catenin or p120-catenin for function in the nucleus and if such a mechanism is utilized. However, several recent studies have shown roles for Wnt signaling in developmental angiogenesis and so the concept remains intriguing (Corada et al., 2010; Gore et al., 2011; Korn et al., 2014). More directly implicated, the effectors of the noncanonical Wnt signaling (planar cell polarity, PCP) pathway are found to bind in a complex with E-cadherin, polarity effectors, and aPKC in fibroblasts and mediate the cellular interrelationship between polarity and adhesion (Castelli et al., 2013; Georgiou, Marinari, Burden, & Baum, 2008; Yan et al., 2006). There is clearly an integral relationship between these mechanisms and the function of the noncanonical Wnt pathway. In endothelial cells, we and others have recently found that Pkd1 mutants display an early loss of both lymphatic endothelial cell polarity and junctional maturation (Coxam et al., 2014; Outeda et al., 2014). Pkd1 is a large transmembrane protein that is found in a complex with E-cadherin and polarity complexes (aPKC, PAR3) and influences PCP signaling outcomes in some cell types. It remains to be specifically determined if Pkd1 may act at this level in endothelial cells but if so this would suggest a role for junctional VE-cadherin in noncanonical Wnt signaling, reiterating the mechanism at play in nephric tubule formation (Castelli et al., 2013). More directly, but in a discretely different endothelial context, during lymphatic valve formation, PCP signaling controls the adherens junctional rearrangements that are essential for valve formation—a distinct lymphatic endothelial morphogenesis event. Specifically, well-characterized PCP signaling components Vangl2 and the homolog of Flamingo (Celsr1) were found to be essential for cellular reorientation and rearrangements during lymphatic valve formation (Tatin et al., 2013). Celsr1 acts cell autonomously and localizes with VE-cadherin at the adherens junctions in order to negatively regulate VE-cadherin-mediated junctional stabilization (Tatin et al., 2013). This study clearly demonstrated a direct regulatory relationship between VE-cadherin at the endothelial junctions and noncanonical Wnt signaling (Tatin et al., 2013). While this is the case in a distinct cellular context, future work is now needed to explore the extent and influence of this mechanism more generally.

6. CONCLUSIONS

In this chapter we highlight key *in vivo* observations in vascular development of mechanisms by which VE-cadherin acts in important signaling pathways. This influential role of VE-cadherin can control cell and tissue morphogenesis during development and most likely in a variety of vascular disease settings.

VE-cadherin abundance and function at the junctions are altered by mechanisms of phosphorylation, regulated endocytosis, and transcriptional regulation. VE-cadherin can bind both extracellular and intracellular protein binding partners thereby forming functional complexes that hold the potential to convey structural changes to the junctions or to activate intracellular signals. In addition to several well-described examples, we suggest other cascades for which the adherens junctions are likely to act as modulatory hubs, including mechanotransduction events, noncanonical Wnt, and cell polarity pathways. The question of how changes in tension across VE-cadherin contribute to various aspects of vessel development and function remains to be further investigated and the relevance of physical cues needs to be deciphered in relation to our growing understanding of growth factor driven signaling.

Given the important functions for VE-cadherin-mediated signaling modifications during development, vessel function, and in crucial disease processes such as inflammation, it is clear that the study of this key protein and its associated signaling partners will continue to intrigue and expand in breadth into the future. Ultimately, a deeper understanding of VE-cadherin has significant potential to open up new therapeutic avenues that target signaling rather than just the physical adhesive properties of the vascular adherens junctions.

ACKNOWLEDGEMENTS

A. K. L. is supported by NHMRC APP1010489 and a University of Queensland Postdoctoral Fellowship. B. M. H. is supported by an ARC Future Fellowship FT100100165.

REFERENCES

Anastasiadis, P. Z., Moon, S. Y., Thoreson, M. A., Mariner, D. J., Crawford, H. C., Zheng, Y., et al. (2000). Inhibition of RhoA by p120 catenin. *Nature Cell Biology*, *2*(9), 637–644. http://dx.doi.org/10.1038/35023588.

Arima, S., Nishiyama, K., Ko, T., Arima, Y., Hakozaki, Y., Sugihara, K., et al. (2011). Angiogenic morphogenesis driven by dynamic and heterogeneous collective endothelial cell movement. *Development*, *138*(21), 4763–4776. http://dx.doi.org/10.1242/dev.068023.

Attisano, L., & Wrana, J. L. (2002). Signal transduction by the TGF-beta superfamily. *Science*, *296*(5573), 1646–1647. http://dx.doi.org/10.1126/science.1071809.

Ausprunk, D. H., & Folkman, J. (1977). Migration and proliferation of endothelial cells in preformed and newly formed blood vessels during tumor angiogenesis. *Microvascular Research*, *14*(1), 53–65.

Baluk, P., Fuxe, J., Hashizume, H., Romano, T., Lashnits, E., Butz, S., et al. (2007). Functionally specialized junctions between endothelial cells of lymphatic vessels. *Journal of Experimental Medicine*, *204*(10), 2349–2362. http://dx.doi.org/10.1084/jem.20062596.

Baumeister, U., Funke, R., Ebnet, K., Vorschmitt, H., Koch, S., & Vestweber, D. (2005). Association of Csk to VE-cadherin and inhibition of cell proliferation. *EMBO Journal, 24*(9), 1686–1695. http://dx.doi.org/10.1038/sj.emboj.7600647.

Baumer, S., Keller, L., Holtmann, A., Funke, R., August, B., Gamp, A., et al. (2006). Vascular endothelial cell-specific phosphotyrosine phosphatase (VE-PTP) activity is required for blood vessel development. *Blood, 107*(12), 4754–4762. http://dx.doi.org/10.1182/blood-2006-01-0141.

Bentley, K., Franco, C. A., Philippides, A., Blanco, R., Dierkes, M., Gebala, V., et al. (2014). The role of differential VE-cadherin dynamics in cell rearrangement during angiogenesis. *Nature Cell Biology, 16*(4), 309–321. http://dx.doi.org/10.1038/ncb2926.

Breviario, F., Caveda, L., Corada, M., Martin-Padura, I., Navarro, P., Golay, J., et al. (1995). Functional properties of human vascular endothelial cadherin (7B4/cadherin-5), an endothelium-specific cadherin. *Arteriosclerosis, Thrombosis, and Vascular Biology, 15*(8), 1229–1239.

Broermann, A., Winderlich, M., Block, H., Frye, M., Rossaint, J., Zarbock, A., et al. (2011). Dissociation of VE-PTP from VE-cadherin is required for leukocyte extravasation and for VEGF-induced vascular permeability in vivo. *Journal of Experimental Medicine, 208*(12), 2393–2401. http://dx.doi.org/10.1084/jem.20110525.

Brooks, A. N., Kilgour, E., & Smith, P. D. (2012). Molecular pathways: Fibroblast growth factor signaling: A new therapeutic opportunity in cancer. *Clinical Cancer Research, 18*(7), 1855–1862. http://dx.doi.org/10.1158/1078-0432.CCR-11-0699.

Carmeliet, P., Ferreira, V., Breier, G., Pollefeyt, S., Kieckens, L., Gertsenstein, M., et al. (1996). Abnormal blood vessel development and lethality in embryos lacking a single VEGF allele. *Nature, 380*(6573), 435–439. http://dx.doi.org/10.1038/380435a0.

Carmeliet, P., Lampugnani, M. G., Moons, L., Breviario, F., Compernolle, V., Bono, F., et al. (1999). Targeted deficiency or cytosolic truncation of the VE-cadherin gene in mice impairs VEGF-mediated endothelial survival and angiogenesis. *Cell, 98*(2), 147–157.

Castelli, M., Boca, M., Chiaravalli, M., Ramalingam, H., Rowe, I., Distefano, G., et al. (2013). Polycystin-1 binds Par3/aPKC and controls convergent extension during renal tubular morphogenesis. *Nature Communications, 4*, 2658. http://dx.doi.org/10.1038/ncomms3658.

Conway, D. E., Breckenridge, M. T., Hinde, E., Gratton, E., Chen, C. S., & Schwartz, M. A. (2013). Fluid shear stress on endothelial cells modulates mechanical tension across VE-cadherin and PECAM-1. *Current Biology, 23*(11), 1024–1030. http://dx.doi.org/10.1016/j.cub.2013.04.049.

Corada, M., Liao, F., Lindgren, M., Lampugnani, M. G., Breviario, F., Frank, R., et al. (2001). Monoclonal antibodies directed to different regions of vascular endothelial cadherin extracellular domain affect adhesion and clustering of the protein and modulate endothelial permeability. *Blood, 97*(6), 1679–1684.

Corada, M., Morini, M. F., & Dejana, E. (2014). Signaling pathways in the specification of arteries and veins. *Arteriosclerosis, Thrombosis, and Vascular Biology*. http://dx.doi.org/10.1161/ATVBAHA.114.303218.

Corada, M., Nyqvist, D., Orsenigo, F., Caprini, A., Giampietro, C., Taketo, M. M., et al. (2010). The Wnt/beta-catenin pathway modulates vascular remodeling and specification by upregulating Dll4/Notch signaling. *Developmental Cell, 18*(6), 938–949. http://dx.doi.org/10.1016/j.devcel.2010.05.006.

Corada, M., Zanetta, L., Orsenigo, F., Breviario, F., Lampugnani, M. G., Bernasconi, S., et al. (2002). A monoclonal antibody to vascular endothelial-cadherin inhibits tumor angiogenesis without side effects on endothelial permeability. *Blood, 100*(3), 905–911.

Coxam, B., Sabine, A., Bower, N. I., Smith, K. A., Pichol-Thievend, C., Skoczylas, R., et al. (2014). Pkd1 regulates lymphatic vascular morphogenesis during development. *Cell Reports*, 7(3), 623–633. http://dx.doi.org/10.1016/j.celrep.2014.03.063.

Davis, M. E., Cai, H., Drummond, G. R., & Harrison, D. G. (2001). Shear stress regulates endothelial nitric oxide synthase expression through c-Src by divergent signaling pathways. *Circulation Research*, *89*(11), 1073–1080.

Dejana, E., & Vestweber, D. (2013). The role of VE-cadherin in vascular morphogenesis and permeability control. *Progress in Molecular Biology and Translational Science*, *116*, 119–144. http://dx.doi.org/10.1016/B978-0-12-394311-8.00006-6.

Dellinger, M., Hunter, R., Bernas, M., Gale, N., Yancopoulos, G., Erickson, R., et al. (2008). Defective remodeling and maturation of the lymphatic vasculature in Angiopoietin-2 deficient mice. *Developmental Biology*, *319*(2), 309–320. http://dx.doi.org/10.1016/j.ydbio.2008.04.024.

Derynck, R., & Zhang, Y. E. (2003). Smad-dependent and Smad-independent pathways in TGF-beta family signalling. *Nature*, *425*(6958), 577–584. http://dx.doi.org/10.1038/nature02006.

Derynck, R., Zhang, Y., & Feng, X. H. (1998). Smads: Transcriptional activators of TGF-beta responses. *Cell*, *95*(6), 737–740.

Dickson, M. C., Martin, J. S., Cousins, F. M., Kulkarni, A. B., Karlsson, S., & Akhurst, R. J. (1995). Defective haematopoiesis and vasculogenesis in transforming growth factor-beta 1 knock out mice. *Development*, *121*(6), 1845–1854.

Duan, L. J., Imamoto, A., & Fong, G. H. (2004). Dual roles of the C-terminal Src kinase (Csk) during developmental vascularization. *Blood*, *103*(4), 1370–1372. http://dx.doi.org/10.1182/blood-2003-05-1701.

Fachinger, G., Deutsch, U., & Risau, W. (1999). Functional interaction of vascular endothelial-protein-tyrosine phosphatase with the angiopoietin receptor Tie-2. *Oncogene*, *18*(43), 5948–5953. http://dx.doi.org/10.1038/sj.onc.1202992.

Fagiani, E., & Christofori, G. (2013). Angiopoietins in angiogenesis. *Cancer Letters*, *328*(1), 18–26. http://dx.doi.org/10.1016/j.canlet.2012.08.018.

Feng, X. H., & Derynck, R. (2005). Specificity and versatility in tgf-beta signaling through Smads. *Annual Review of Cell and Developmental Biology*, *21*, 659–693. http://dx.doi.org/10.1146/annurev.cellbio.21.022404.142018.

Ferrara, N. (2000). VEGF: An update on biological and therapeutic aspects. *Current Opinion in Biotechnology*, *11*(6), 617–624.

Ferrara, N., & Alitalo, K. (1999). Clinical applications of angiogenic growth factors and their inhibitors. *Nature Medicine*, *5*(12), 1359–1364. http://dx.doi.org/10.1038/70928.

Ferrara, N., Carver-Moore, K., Chen, H., Dowd, M., Lu, L., O'Shea, K. S., et al. (1996). Heterozygous embryonic lethality induced by targeted inactivation of the VEGF gene. *Nature*, *380*(6573), 439–442. http://dx.doi.org/10.1038/380439a0.

Fiedler, U., Reiss, Y., Scharpfenecker, M., Grunow, V., Koidl, S., Thurston, G., et al. (2006). Angiopoietin-2 sensitizes endothelial cells to TNF-alpha and has a crucial role in the induction of inflammation. *Nature Medicine*, *12*(2), 235–239. http://dx.doi.org/10.1038/nm1351.

Fukuhara, S., Sako, K., Minami, T., Noda, K., Kim, H. Z., Kodama, T., et al. (2008). Differential function of Tie2 at cell-cell contacts and cell-substratum contacts regulated by angiopoietin-1. *Nature Cell Biology*, *10*(5), 513–526. http://dx.doi.org/10.1038/ncb1714.

Gaengel, K., Genove, G., Armulik, A., & Betsholtz, C. (2009). Endothelial-mural cell signaling in vascular development and angiogenesis. *Arteriosclerosis, Thrombosis, and Vascular Biology*, *29*(5), 630–638. http://dx.doi.org/10.1161/ATVBAHA.107.161521.

Gale, N. W., Thurston, G., Hackett, S. F., Renard, R., Wang, Q., McClain, J., et al. (2002). Angiopoietin-2 is required for postnatal angiogenesis and lymphatic patterning, and only the latter role is rescued by Angiopoietin-1. *Developmental Cell*, *3*(3), 411–423.

Gavard, J., & Gutkind, J. S. (2006). VEGF controls endothelial-cell permeability by promoting the beta-arrestin-dependent endocytosis of VE-cadherin. *Nature Cell Biology*, *8*(11), 1223–1234. http://dx.doi.org/10.1038/ncb1486.

Georgiou, M., Marinari, E., Burden, J., & Baum, B. (2008). Cdc42, Par6, and aPKC regulate Arp2/3-mediated endocytosis to control local adherens junction stability. *Current Biology, 18*(21), 1631–1638. http://dx.doi.org/10.1016/j.cub.2008.09.029.

Gerhardt, H., Golding, M., Fruttiger, M., Ruhrberg, C., Lundkvist, A., Abramsson, A., et al. (2003). VEGF guides angiogenic sprouting utilizing endothelial tip cell filopodia. *Journal of Cell Biology, 161*(6), 1163–1177. http://dx.doi.org/10.1083/jcb.200302047.

Gerwins, P., Skoldenberg, E., & Claesson-Welsh, L. (2000). Function of fibroblast growth factors and vascular endothelial growth factors and their receptors in angiogenesis. *Critical Reviews in Oncology/Hematology, 34*(3), 185–194.

Giampietro, C., Taddei, A., Corada, M., Sarra-Ferraris, G. M., Alcalay, M., Cavallaro, U., et al. (2012). Overlapping and divergent signaling pathways of N-cadherin and VE-cadherin in endothelial cells. *Blood, 119*(9), 2159–2170. http://dx.doi.org/10.1182/blood-2011-09-381012.

Giannotta, M., Trani, M., & Dejana, E. (2013). VE-cadherin and endothelial adherens junctions: Active guardians of vascular integrity. *Developmental Cell, 26*(5), 441–454. http://dx.doi.org/10.1016/j.devcel.2013.08.020.

Gjini, E., Hekking, L. H., Kuchler, A., Saharinen, P., Wienholds, E., Post, J. A., et al. (2011). Zebrafish Tie-2 shares a redundant role with Tie-1 in heart development and regulates vessel integrity. *Disease Models & Mechanisms, 4*(1), 57–66. http://dx.doi.org/10.1242/dmm.005033.

Glading, A. J., & Ginsberg, M. H. (2010). Rap1 and its effector KRIT1/CCM1 regulate beta-catenin signaling. *Disease Models & Mechanisms, 3*(1–2), 73–83. http://dx.doi.org/10.1242/dmm.003293.

Glading, A., Han, J., Stockton, R. A., & Ginsberg, M. H. (2007). KRIT-1/CCM1 is a Rap1 effector that regulates endothelial cell cell junctions. *Journal of Cell Biology, 179*(2), 247–254. http://dx.doi.org/10.1083/jcb.200705175.

Gomez, G. A., McLachlan, R. W., & Yap, A. S. (2011). Productive tension: Force-sensing and homeostasis of cell-cell junctions. *Trends in Cell Biology, 21*(9), 499–505. http://dx.doi.org/10.1016/j.tcb.2011.05.006.

Gore, A. V., Swift, M. R., Cha, Y. R., Lo, B., McKinney, M. C., Li, W., et al. (2011). Rspo1/Wnt signaling promotes angiogenesis via Vegfc/Vegfr3. *Development, 138*(22), 4875–4886. http://dx.doi.org/10.1242/dev.068460.

Gory-Faure, S., Prandini, M. H., Pointu, H., Roullot, V., Pignot-Paintrand, I., Vernet, M., et al. (1999). Role of vascular endothelial-cadherin in vascular morphogenesis. *Development, 126*(10), 2093–2102.

Goumans, M. J., Lebrin, F., & Valdimarsdottir, G. (2003). Controlling the angiogenic switch: A balance between two distinct TGF-b receptor signaling pathways. *Trends in Cardiovascular Medicine, 13*(7), 301–307.

Goumans, M. J., & Mummery, C. (2000). Functional analysis of the TGFbeta receptor/Smad pathway through gene ablation in mice. *International Journal of Developmental Biology, 44*(3), 253–265.

Goumans, M. J., Valdimarsdottir, G., Itoh, S., Lebrin, F., Larsson, J., Mummery, C., et al. (2003). Activin receptor-like kinase (ALK)1 is an antagonistic mediator of lateral TGFbeta/ALK5 signaling. *Molecular Cell, 12*(4), 817–828.

Goumans, M. J., Valdimarsdottir, G., Itoh, S., Rosendahl, A., Sideras, P., & ten Dijke, P. (2002). Balancing the activation state of the endothelium via two distinct TGF-beta type I receptors. *EMBO Journal, 21*(7), 1743–1753. http://dx.doi.org/10.1093/emboj/21.7.1743.

Harvey, N. L., & Oliver, G. (2004). Choose your fate: Artery, vein or lymphatic vessel? *Current Opinion in Genetics & Development, 14*(5), 499–505. http://dx.doi.org/10.1016/j.gde.2004.07.005.

Hatta, K., Okada, T. S., & Takeichi, M. (1985). A monoclonal antibody disrupting calcium-dependent cell-cell adhesion of brain tissues: Possible role of its target antigen in animal pattern formation. *Proceedings of the National Academy of Sciences of the United States of America*, *82*(9), 2789–2793.

Hayashi, M., Majumdar, A., Li, X., Adler, J., Sun, Z., Vertuani, S., et al. (2013). VE-PTP regulates VEGFR2 activity in stalk cells to establish endothelial cell polarity and lumen formation. *Nature Communications*, *4*, 1672. http://dx.doi.org/10.1038/ncomms2683.

Huveneers, S., & Danen, E. H. (2009). Adhesion signaling—Crosstalk between integrins, Src and Rho. *Journal of Cell Science*, *122*(Pt 8), 1059–1069. http://dx.doi.org/10.1242/jcs.039446.

Iden, S., Rehder, D., August, B., Suzuki, A., Wolburg-Buchholz, K., Wolburg, H., et al. (2006). A distinct PAR complex associates physically with VE-cadherin in vertebrate endothelial cells. *EMBO Reports*, 7(12), 1239–1246. http://dx.doi.org/10.1038/sj.embor.7400819.

Itoh, F., Itoh, S., Adachi, T., Ichikawa, K., Matsumura, Y., Takagi, T., et al. (2012). Smad2/Smad3 in endothelium is indispensable for vascular stability via S1PR1 and N-cadherin expressions. *Blood*, *119*(22), 5320–5328. http://dx.doi.org/10.1182/blood-2011-12-395772.

Itoh, S., Itoh, F., Goumans, M. J., & Ten Dijke, P. (2000). Signaling of transforming growth factor-beta family members through Smad proteins. *European Journal of Biochemistry*, *267*(24), 6954–6967.

Jakobsson, L., Franco, C. A., Bentley, K., Collins, R. T., Ponsioen, B., Aspalter, I. M., et al. (2010). Endothelial cells dynamically compete for the tip cell position during angiogenic sprouting. *Nature Cell Biology*, *12*(10), 943–953. http://dx.doi.org/10.1038/ncb2103.

Katoh, M., & Nakagama, H. (2014). FGF receptors: Cancer biology and therapeutics. *Medicinal Research Reviews*, *34*(2), 280–300. http://dx.doi.org/10.1002/med.21288.

Koh, W., Mahan, R. D., & Davis, G. E. (2008). Cdc42- and Rac1-mediated endothelial lumen formation requires Pak2, Pak4 and Par3, and PKC-dependent signaling. *Journal of Cell Science*, *121*(Pt 7), 989–1001. http://dx.doi.org/10.1242/jcs.020693.

Korn, C., Scholz, B., Hu, J., Srivastava, K., Wojtarowicz, J., Arnsperger, T., et al. (2014). Endothelial cell-derived non-canonical Wnt ligands control vascular pruning in angiogenesis. *Development*, *141*(8), 1757–1766. http://dx.doi.org/10.1242/dev.104422.

Kulkarni, A. B., Huh, C. G., Becker, D., Geiser, A., Lyght, M., Flanders, K. C., et al. (1993). Transforming growth factor beta 1 null mutation in mice causes excessive inflammatory response and early death. *Proceedings of the National Academy of Sciences of the United States of America*, *90*(2), 770–774.

Kuppers, V., Vockel, M., Nottebaum, A. F., & Vestweber, D. (2014). Phosphatases and kinases as regulators of the endothelial barrier function. *Cell and Tissue Research*, *355*(3), 577–586. http://dx.doi.org/10.1007/s00441-014-1812-1.

Labelle, M., Schnittler, H. J., Aust, D. E., Friedrich, K., Baretton, G., Vestweber, D., et al. (2008). Vascular endothelial cadherin promotes breast cancer progression via transforming growth factor beta signaling. *Cancer Research*, *68*(5), 1388–1397. http://dx.doi.org/10.1158/0008-5472.CAN-07-2706.

Lampugnani, M. G., Orsenigo, F., Gagliani, M. C., Tacchetti, C., & Dejana, E. (2006). Vascular endothelial cadherin controls VEGFR-2 internalization and signaling from intracellular compartments. *Journal of Cell Biology*, *174*(4), 593–604. http://dx.doi.org/10.1083/jcb.200602080.

Lampugnani, M. G., Orsenigo, F., Rudini, N., Maddaluno, L., Boulday, G., Chapon, F., et al. (2010). CCM1 regulates vascular-lumen organization by inducing endothelial polarity. *Journal of Cell Science*, *123*(Pt 7), 1073–1080. http://dx.doi.org/10.1242/jcs.059329.

Lampugnani, M. G., Resnati, M., Raiteri, M., Pigott, R., Pisacane, A., Houen, G., et al. (1992). A novel endothelial-specific membrane protein is a marker of cell-cell contacts. *Journal of Cell Biology*, *118*(6), 1511–1522.

Lampugnani, M. G., Zanetti, A., Breviario, F., Balconi, G., Orsenigo, F., Corada, M., et al. (2002). VE-cadherin regulates endothelial actin activating Rac and increasing membrane association of Tiam. *Molecular Biology of the Cell*, *13*(4), 1175–1189. http://dx.doi.org/10.1091/mbc.01-07-0368.

Larsson, J., Goumans, M. J., Sjostrand, L. J., van Rooijen, M. A., Ward, D., Leveen, P., et al. (2001). Abnormal angiogenesis but intact hematopoietic potential in TGF-beta type I receptor-deficient mice. *EMBO Journal*, *20*(7), 1663–1673. http://dx.doi.org/10.1093/emboj/20.7.1663.

Lavine, K. J., White, A. C., Park, C., Smith, C. S., Choi, K., Long, F., et al. (2006). Fibroblast growth factor signals regulate a wave of Hedgehog activation that is essential for coronary vascular development. *Genes & Development*, *20*(12), 1651–1666. http://dx.doi.org/10.1101/gad.1411406.

Leckband, D. E., le Duc, Q., Wang, N., & de Rooij, J. (2011). Mechanotransduction at cadherin-mediated adhesions. *Current Opinion in Cell Biology*, *23*(5), 523–530. http://dx.doi.org/10.1016/j.ceb.2011.08.003.

Lenard, A., Ellertsdottir, E., Herwig, L., Krudewig, A., Sauteur, L., Belting, H. G., et al. (2013). In vivo analysis reveals a highly stereotypic morphogenetic pathway of vascular anastomosis. *Developmental Cell*, *25*(5), 492–506. http://dx.doi.org/10.1016/j.devcel.2013.05.010.

Lohela, M., Bry, M., Tammela, T., & Alitalo, K. (2009). VEGFs and receptors involved in angiogenesis versus lymphangiogenesis. *Current Opinion in Cell Biology*, *21*(2), 154–165. http://dx.doi.org/10.1016/j.ceb.2008.12.012.

Matsumoto, T., Bohman, S., Dixelius, J., Berge, T., Dimberg, A., Magnusson, P., et al. (2005). VEGF receptor-2 Y951 signaling and a role for the adapter molecule TSAd in tumor angiogenesis. *EMBO Journal*, *24*(13), 2342–2353. http://dx.doi.org/10.1038/sj.emboj.7600709.

Mitra, S. K., & Schlaepfer, D. D. (2006). Integrin-regulated FAK-Src signaling in normal and cancer cells. *Current Opinion in Cell Biology*, *18*(5), 516–523. http://dx.doi.org/10.1016/j.ceb.2006.08.011.

Montero-Balaguer, M., Swirsding, K., Orsenigo, F., Cotelli, F., Mione, M., & Dejana, E. (2009). Stable vascular connections and remodeling require full expression of VE-cadherin in zebrafish embryos. *PLoS One*, *4*(6), e5772. http://dx.doi.org/10.1371/journal.pone.0005772.

Murakami, M., Nguyen, L. T., Zhuang, Z. W., Moodie, K. L., Carmeliet, P., Stan, R. V., et al. (2008). The FGF system has a key role in regulating vascular integrity. *Journal of Clinical Investigation*, *118*(10), 3355–3366. http://dx.doi.org/10.1172/JCI35298.

Nagafuchi, A., & Takeichi, M. (1988). Cell binding function of E-cadherin is regulated by the cytoplasmic domain. *EMBO Journal*, 7(12), 3679–3684.

Navarro, P., Caveda, L., Breviario, F., Mandoteanu, I., Lampugnani, M. G., & Dejana, E. (1995). Catenin-dependent and -independent functions of vascular endothelial cadherin. *Journal of Biological Chemistry*, *270*(52), 30965–30972.

Nawroth, R., Poell, G., Ranft, A., Kloep, S., Samulowitz, U., Fachinger, G., et al. (2002). VE-PTP and VE-cadherin ectodomains interact to facilitate regulation of phosphorylation and cell contacts. *EMBO Journal*, *21*(18), 4885–4895.

Nelson, W. J., & Nusse, R. (2004). Convergence of Wnt, beta-catenin, and cadherin pathways. *Science*, *303*(5663), 1483–1487. http://dx.doi.org/10.1126/science.1094291.

Noda, K., Zhang, J., Fukuhara, S., Kunimoto, S., Yoshimura, M., & Mochizuki, N. (2010). Vascular endothelial-cadherin stabilizes at cell-cell junctions by anchoring to circumferential actin bundles through alpha- and beta-catenins in cyclic AMP-Epac-Rap1 signal-activated endothelial cells. *Molecular Biology of the Cell*, *21*(4), 584–596. http://dx.doi.org/10.1091/mbc.E09-07-0580.

Nottebaum, A. F., Cagna, G., Winderlich, M., Gamp, A. C., Linnepe, R., Polaschegg, C., et al. (2008). VE-PTP maintains the endothelial barrier via plakoglobin and becomes dissociated from VE-cadherin by leukocytes and by VEGF. *Journal of Experimental Medicine*, *205*(12), 2929–2945. http://dx.doi.org/10.1084/jem.20080406.

Oh, S. P., Seki, T., Goss, K. A., Imamura, T., Yi, Y., Donahoe, P. K., et al. (2000). Activin receptor-like kinase 1 modulates transforming growth factor-beta 1 signaling in the regulation of angiogenesis. *Proceedings of the National Academy of Sciences of the United States of America*, *97*(6), 2626–2631.

Oladipupo, S. S., Smith, C., Santeford, A., Park, C., Sene, A., Wiley, L. A., et al. (2014). Endothelial cell FGF signaling is required for injury response but not for vascular homeostasis. *Proceedings of the National Academy of Sciences of the United States of America*, *111*(37), 13379–13384. http://dx.doi.org/10.1073/pnas.1324235111.

Orsenigo, F., Giampietro, C., Ferrari, A., Corada, M., Galaup, A., Sigismund, S., et al. (2012). Phosphorylation of VE-cadherin is modulated by haemodynamic forces and contributes to the regulation of vascular permeability in vivo. *Nature Communications*, *3*, 1208. http://dx.doi.org/10.1038/ncomms2199.

Oshima, M., Oshima, H., & Taketo, M. M. (1996). TGF-beta receptor type II deficiency results in defects of yolk sac hematopoiesis and vasculogenesis. *Developmental Biology*, *179*(1), 297–302. http://dx.doi.org/10.1006/dbio.1996.0259.

Outeda, P., Huso, D. L., Fisher, S. A., Halushka, M. K., Kim, H., Qian, F., et al. (2014). Polycystin signaling is required for directed endothelial cell migration and lymphatic development. *Cell Reports*, 7(3), 634–644. http://dx.doi.org/10.1016/j.celrep.2014.03.064.

Ozawa, M., Baribault, H., & Kemler, R. (1989). The cytoplasmic domain of the cell adhesion molecule uvomorulin associates with three independent proteins structurally related in different species. *EMBO Journal*, *8*(6), 1711–1717.

Ozawa, M., & Kemler, R. (1992). Molecular organization of the uvomorulin-catenin complex. *Journal of Cell Biology*, *116*(4), 989–996.

Ozawa, M., Ringwald, M., & Kemler, R. (1990). Uvomorulin-catenin complex formation is regulated by a specific domain in the cytoplasmic region of the cell adhesion molecule. *Proceedings of the National Academy of Sciences of the United States of America*, *87*(11), 4246–4250.

Pepper, M. S. (1997). Transforming growth factor-beta: Vasculogenesis, angiogenesis, and vessel wall integrity. *Cytokine & Growth Factor Reviews*, *8*(1), 21–43.

Perez-Moreno, M., & Fuchs, E. (2006). Catenins: Keeping cells from getting their signals crossed. *Developmental Cell*, *11*(5), 601–612. http://dx.doi.org/10.1016/j.devcel.2006.10.010.

Planas-Paz, L., Strilic, B., Goedecke, A., Breier, G., Fassler, R., & Lammert, E. (2012). Mechanoinduction of lymph vessel expansion. *EMBO Journal*, *31*(4), 788–804. http://dx.doi.org/10.1038/emboj.2011.456.

Puri, M. C., Rossant, J., Alitalo, K., Bernstein, A., & Partanen, J. (1995). The receptor tyrosine kinase TIE is required for integrity and survival of vascular endothelial cells. *EMBO Journal*, *14*(23), 5884–5891.

Reynolds, A. B., & Carnahan, R. H. (2004). Regulation of cadherin stability and turnover by p120ctn: Implications in disease and cancer. *Seminars in Cell & Developmental Biology*, *15*(6), 657–663. http://dx.doi.org/10.1016/j.semcdb.2004.09.003.

Reynolds, A. B., Herbert, L., Cleveland, J. L., Berg, S. T., & Gaut, J. R. (1992). p120, a novel substrate of protein tyrosine kinase receptors and of p60v-src, is related to cadherin-binding factors beta-catenin, plakoglobin and armadillo. *Oncogene*, 7(12), 2439–2445.

Rimm, D. L., Koslov, E. R., Kebriaei, P., Cianci, C. D., & Morrow, J. S. (1995). Alpha 1(E)-catenin is an actin-binding and -bundling protein mediating the attachment of F-actin to the membrane adhesion complex. *Proceedings of the National Academy of Sciences of the United States of America*, *92*(19), 8813–8817.

Roelen, B. A., van Rooijen, M. A., & Mummery, C. L. (1997). Expression of ALK-1, a type 1 serine/threonine kinase receptor, coincides with sites of vasculogenesis and angiogenesis in early mouse development. *Developmental Dynamics*, *209*(4), 418–430. http://dx.doi.org/10.1002/(SICI)1097-0177(199708)209:4<418::AID-AJA9>3.0.CO;2-L.

Rudini, N., Felici, A., Giampietro, C., Lampugnani, M., Corada, M., Swirsding, K., et al. (2008). VE-cadherin is a critical endothelial regulator of TGF-beta signalling. *EMBO Journal*, *27*(7), 993–1004. http://dx.doi.org/10.1038/emboj.2008.46.

Saharinen, P., Eklund, L., Miettinen, J., Wirkkala, R., Anisimov, A., Winderlich, M., et al. (2008). Angiopoietins assemble distinct Tie2 signalling complexes in endothelial cell-cell and cell-matrix contacts. *Nature Cell Biology*, *10*(5), 527–537. http://dx.doi.org/10.1038/ncb1715.

Sakurai, A., Fukuhara, S., Yamagishi, A., Sako, K., Kamioka, Y., Masuda, M., et al. (2006). MAGI-1 is required for Rap1 activation upon cell-cell contact and for enhancement of vascular endothelial cadherin-mediated cell adhesion. *Molecular Biology of the Cell*, *17*(2), 966–976. http://dx.doi.org/10.1091/mbc.E05-07-0647.

Sato, T. N., Tozawa, Y., Deutsch, U., Wolburg-Buchholz, K., Fujiwara, Y., Gendron-Maguire, M., et al. (1995). Distinct roles of the receptor tyrosine kinases Tie-1 and Tie-2 in blood vessel formation. *Nature*, *376*(6535), 70–74. http://dx.doi.org/10.1038/376070a0.

Senger, D. R., Galli, S. J., Dvorak, A. M., Perruzzi, C. A., Harvey, V. S., & Dvorak, H. F. (1983). Tumor cells secrete a vascular permeability factor that promotes accumulation of ascites fluid. *Science*, *219*(4587), 983–985.

Senger, D. R., Perruzzi, C. A., Feder, J., & Dvorak, H. F. (1986). A highly conserved vascular permeability factor secreted by a variety of human and rodent tumor cell lines. *Cancer Research*, *46*(11), 5629–5632.

Shull, M. M., Ormsby, I., Kier, A. B., Pawlowski, S., Diebold, R. J., Yin, M., et al. (1992). Targeted disruption of the mouse transforming growth factor-beta 1 gene results in multifocal inflammatory disease. *Nature*, *359*(6397), 693–699. http://dx.doi.org/10.1038/359693a0.

Smutny, M., & Yap, A. S. (2010). Neighborly relations: Cadherins and mechanotransduction. *Journal of Cell Biology*, *189*(7), 1075–1077. http://dx.doi.org/10.1083/jcb.201005151.

Sun, Z., Li, X., Massena, S., Kutschera, S., Padhan, N., Gualandi, L., et al. (2012). VEGFR2 induces c-Src signaling and vascular permeability in vivo via the adaptor protein TSAd. *Journal of Experimental Medicine*, *209*(7), 1363–1377. http://dx.doi.org/10.1084/jem.20111343.

Suzuki, S., Sano, K., & Tanihara, H. (1991). Diversity of the cadherin family: Evidence for eight new cadherins in nervous tissue. *Cell Regulation*, *2*(4), 261–270.

Swift, M. R., & Weinstein, B. M. (2009). Arterial-venous specification during development. *Circulation Research*, *104*(5), 576–588. http://dx.doi.org/10.1161/CIRCRESAHA.108.188805.

Taddei, A., Giampietro, C., Conti, A., Orsenigo, F., Breviario, F., Pirazzoli, V., et al. (2008). Endothelial adherens junctions control tight junctions by VE-cadherin-mediated upregulation of claudin-5. *Nature Cell Biology*, *10*(8), 923–934. http://dx.doi.org/10.1038/ncb1752.

Takeichi, M. (1977). Functional correlation between cell adhesive properties and some cell surface proteins. *Journal of Cell Biology*, *75*(2 Pt 1), 464–474.

Tanihara, H., Kido, M., Obata, S., Heimark, R. L., Davidson, M., St John, T., et al. (1994). Characterization of cadherin-4 and cadherin-5 reveals new aspects of cadherins. *Journal of Cell Science*, *107*(Pt 6), 1697–1704.

Tatin, F., Taddei, A., Weston, A., Fuchs, E., Devenport, D., Tissir, F., et al. (2013). Planar cell polarity protein Celsr1 regulates endothelial adherens junctions and directed cell rearrangements during valve morphogenesis. *Developmental Cell*, *26*(1), 31–44. http://dx.doi.org/10.1016/j.devcel.2013.05.015.

Thomas, M., & Augustin, H. G. (2009). The role of the Angiopoietins in vascular morphogenesis. *Angiogenesis*, *12*(2), 125–137. http://dx.doi.org/10.1007/s10456-009-9147-3.

Thoreson, M. A., Anastasiadis, P. Z., Daniel, J. M., Ireton, R. C., Wheelock, M. J., Johnson, K. R., et al. (2000). Selective uncoupling of p120(ctn) from E-cadherin disrupts strong adhesion. *Journal of Cell Biology*, *148*(1), 189–202.

Tzima, E., Irani-Tehrani, M., Kiosses, W. B., Dejana, E., Schultz, D. A., Engelhardt, B., et al. (2005). A mechanosensory complex that mediates the endothelial cell response to fluid shear stress. *Nature*, *437*(7057), 426–431. http://dx.doi.org/10.1038/nature03952.

Valenta, T., Hausmann, G., & Basler, K. (2012). The many faces and functions of beta-catenin. *EMBO Journal*, *31*(12), 2714–2736. http://dx.doi.org/10.1038/emboj.2012.150.

Vestweber, D. (2012). Novel insights into leukocyte extravasation. *Current Opinion in Hematology*, *19*(3), 212–217. http://dx.doi.org/10.1097/MOH.0b013e3283523e78.

Vestweber, D., & Kemler, R. (1984). Rabbit antiserum against a purified surface glycoprotein decompacts mouse preimplantation embryos and reacts with specific adult tissues. *Experimental Cell Research*, *152*(1), 169–178.

Vincent, P. A., Xiao, K., Buckley, K. M., & Kowalczyk, A. P. (2004). VE-cadherin: Adhesion at arm's length. *American Journal of Physiology. Cell Physiology*, *286*(5), C987–C997. http://dx.doi.org/10.1152/ajpcell.00522.2003.

Yan, H. X., Yang, W., Zhang, R., Chen, L., Tang, L., Zhai, B., et al. (2006). Protein-tyrosine phosphatase PCP-2 inhibits beta-catenin signaling and increases E-cadherin-dependent cell adhesion. *Journal of Biological Chemistry*, *281*(22), 15423–15433. http://dx.doi.org/10.1074/jbc.M602607200.

Yap, A. S., Brieher, W. M., & Gumbiner, B. M. (1997). Molecular and functional analysis of cadherin-based adherens junctions. *Annual Review of Cell and Developmental Biology*, *13*, 119–146. http://dx.doi.org/10.1146/annurev.cellbio.13.1.119.

Zheng, W., Nurmi, H., Appak, S., Sabine, A., Bovay, E., Korhonen, E. A., et al. (2014). Angiopoietin 2 regulates the transformation and integrity of lymphatic endothelial cell junctions. *Genes & Development*, *28*(14), 1592–1603. http://dx.doi.org/10.1101/gad.237677.114.

CHAPTER ELEVEN

Adhesion in Mammary Development: Novel Roles for E-Cadherin in Individual and Collective Cell Migration

Eliah R. Shamir*,†, Andrew J. Ewald*,†,1

*Department of Cell Biology, Center for Cell Dynamics, Johns Hopkins University School of Medicine, Baltimore, Maryland, USA

†Department of Oncology, Johns Hopkins University School of Medicine, Baltimore, Maryland, USA

[1]Corresponding author: e-mail address: aewald2@jhmi.edu

Contents

Abstract

Epithelial tissues are essential for barrier function, secretion, and regulation of fluid transport. Their function requires cell polarity and cell–cell adhesion, mediated through intercellular junctions. Conversely, disruption of adhesion and polarity is thought to drive cancer progression. The mammary gland is an important model for cell adhesion due to its postnatal hormonally regulated development; ducts undergo branching morphogenesis in response to steroid hormones during puberty. These hormonal signals induce a transition from simple to stratified architecture, initiated by asymmetric luminal

Current Topics in Developmental Biology, Volume 112
ISSN 0070-2153
http://dx.doi.org/10.1016/bs.ctdb.2014.12.001

cell divisions. Ductal elongation is accomplished by this multilayered, low-polarity epithelium, and polarity is reestablished as elongation ceases. The requirement for cell adhesion has been tested in 3D culture and *in vivo*, using gene deletion, knockdown, and misexpression in both developmental and homeostatic contexts. Attention has focused on E-cadherin, the major classical cadherin in luminal epithelial cells. Classic studies revealed a requirement for E-cadherin during lactation, and E-cadherin loss is widely posited to promote metastasis. However, recent findings demonstrated a broader requirement for E-cadherin during branching morphogenesis and homeostasis and also, surprisingly, in epithelial dissemination. These studies suggest that long-standing models of the role of adhesion in epithelial biology need to be revisited. Advances in inducible gene expression and knockdown, CRISPR/Cas9 technology, and fluorescent labeling of genetically modified cells offer the opportunity to test the roles of diverse adhesion systems and to develop a mechanistic understanding of how cell adhesion regulates development and cancer.

1. INTRODUCTION

Epithelium is one of the fundamental animal tissues and lines the cavities and surfaces of the body. Epithelial tissues were first defined anatomically based on their organization into layers of tightly connected cells with apicobasal polarity (Virchow, 1860). Despite this characteristic appearance, it has been challenging to develop a simple molecular definition that encompasses all epithelial cells. There is no single molecular marker expressed specifically in all epithelial cells, nor any master epithelial gene that is required for the genesis or function of all epithelial tissues (Davies & Garrod, 1997). However, despite differences in specific expression, adhesion proteins and their associated cell–cell junctions are a recognizable, shared feature of diverse epithelial organs (Nelson, 2003). Cell–cell adhesion is also required for the establishment and maintenance of distinct apical and basolateral membrane domains, which define epithelial cell polarity (Nelson, 2003). It is therefore reasonable to anticipate that adhesion proteins are important regulators of normal and pathologic epithelial biology.

Here, we review the role of cell–cell adhesion in mammary development and highlight the contributions of major adhesion molecules to specific developmental processes. We also briefly discuss the potential functions of adhesion proteins in breast cancer invasion and metastasis. We focus on E-cadherin, the major classical cadherin expressed in luminal epithelial cells, which is classified as a tumor and invasion suppressor in multiple epithelial cancers (van Roy & Berx, 2008). Recent data suggest novel roles for

E-cadherin that challenge our traditional notions of adhesion proteins as limiting cell motility and migration. Supporting this concept, most breast tumors and their metastases are E-cadherin positive (Kowalski, Rubin, & Kleer, 2003), and studies have conflicted on the correlation between E-cadherin status, histologic grade, metastasis, and prognosis (Gamallo et al., 1993; Hunt, Douglas-Jones, Jasani, Morgan, & Pignatelli, 1997; Lipponen, Saarelainen, Ji, Aaltomaa, & Syrjanen, 1994; Oka et al., 1993; Siitonen et al., 1996; Tan et al., 1999). Finally, we revisit the definition of an epithelial cell and speculate on the limits of "epithelial" cell behaviors.

2. CELL–CELL ADHESION IN MAMMARY DEVELOPMENT: THE MAJOR PLAYERS

2.1 From simple to stratified: Transitions in adhesion during morphogenesis

The mammary gland has long served as a valuable model system for studying cell adhesion in epithelial morphogenesis and tumor biology (Boussadia, Kutsch, Hierholzer, Delmas, & Kemler, 2002; Daniel, Strickland, & Friedmann, 1995; Gjorevski & Nelson, 2011; Knudsen & Wheelock, 2005; McNally & Martin, 2011; Nanba, Nakanishi, & Hieda, 2001; Runswick, O'Hare, Jones, Streuli, & Garrod, 2001). Mature mammary ducts exhibit simple epithelial architecture, with a bilayer of inner luminal and outer myoepithelial cells, each expressing distinct adhesion proteins (Daniel et al., 1995; Fig. 1). Luminal epithelial cells connect to each other through tight junctions, desmosomes, gap junctions, and adherens junctions (Pitelka, Hamamoto, Duafala, & Nemanic, 1973). Luminal and myoepithelial cells link through both desmosomes and gap junctions, and myoepithelial cells bind to the basement membrane through hemidesmosomes (Pitelka et al., 1973). However, these simple ducts arise from a multilayered, relatively unpolarized embryonic mammary placode (Hogg, Harrison, & Tickle, 1983; Nanba et al., 2001), and there are characteristic variations in the quantity and organization of intercellular junctions during periods of active morphogenesis (Ewald, Brenot, Duong, Chan, & Werb, 2008; Ewald et al., 2012; Pitelka et al., 1973).

From birth through puberty, mammary ducts are essentially quiescent and retain their polarized, bilayered organization (Hogg et al., 1983; Huebner, Lechler, & Ewald, 2014). The majority of branching morphogenesis occurs during puberty, regulated by steroid hormone and receptor

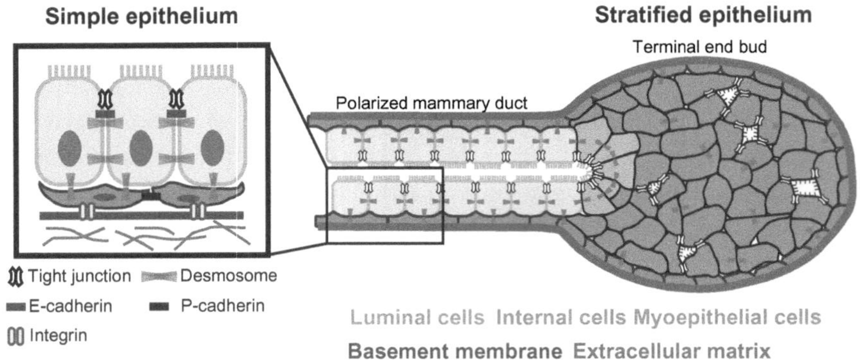

Figure 1 Normal transitions in adhesion during epithelial branching morphogenesis. Mammary epithelium initiates branching morphogenesis postnatally. Tube elongation is accomplished by a stratified terminal end bud, which contains many internal luminal cells that lack apicobasal polarity and display reduced numbers of intercellular junctions. The epithelium at the rear polarizes to a bilayered, simple ductal architecture consisting of an inner layer of luminal cells and a basal layer of myoepithelial cells. Epithelial cells in the ducts are connected by many cell–cell junctions. *Schematic adapted from original by Robert Huebner, with permission.* (See the color plate.)

tyrosine kinase signaling (Hennighausen & Robinson, 2005; Sternlicht, 2006; Sternlicht, Kouros-Mehr, Lu, & Werb, 2006). These signals induce a transition from simple to stratified architecture and lead to the formation of specialized terminal end buds (TEBs) at the tips of each duct (Hinck & Silberstein, 2005; Huebner et al., 2014; Silberstein & Daniel, 1982; Fig. 1). Recent studies revealed that this transition is initiated by asymmetric cell divisions within the luminal epithelial layer and that the products of these cell divisions lack apicobasal polarity and have few intercellular junctions (Ewald et al., 2012; Huebner et al., 2014). Mammary tubulogenesis is accomplished by these migratory, proliferative, unpolarized cells (Ewald et al., 2008, 2012; Huebner et al., 2014; reviewed in Huebner & Ewald, 2014).

2.2 Adherens junctions

A major contributor to epithelial integrity is the adherens junction, a supramolecular structure that mediates homophilic intercellular interactions (Takeichi, 2014). The most frequently studied molecular components of the adherens junction are the adhesion receptor E-cadherin and the associated cytosolic proteins α-catenin, β-catenin, and p120 catenin (Takeichi,

2014). However, the mature adherens junction involves additional transmembrane adhesion receptors and cytosolic proteins, such as nectin and afadin, and the core adhesive elements can exist on the membrane in both junctional and nonjunctional configurations (Niessen & Gottardi, 2008). Additionally, the connection of the adherens junction to the actin cytoskeleton is dependent on both cell contact and mechanical force (Buckley et al., 2014; Yamada & Nelson, 2007). In the mammary epithelial bilayer, cadherin isoforms are expressed in a cell type-specific manner: luminal cells express E-cadherin, and myoepithelial cells express P-cadherin (Daniel et al., 1995). Here, we discuss the roles of these two cadherins and the catenins in mammary epithelial development.

Early functional analyses of cell adhesion frequently relied on function perturbing antibodies. Classic work in kidney epithelial cells in culture revealed that antibody-based disruption of E-cadherin coordinately inhibits the formation of adherens junctions, tight junctions, and desmosomes (Gumbiner, Stevenson, & Grimaldi, 1988). These reagents can also be utilized *in vivo*, by implanting antibody-soaked beads in the mammary fat pad. This approach revealed that disruption of E-cadherin specifically affects the luminal epithelium, while disruption of P-cadherin specifically affects the myoepithelium (Daniel et al., 1995). Further supporting the importance of differential cadherin expression in mammary epithelial development, engineered combinations of human luminal and myoepithelial cells self-organize into bilayers in culture, and this organization is disrupted by addition of antibodies targeting E-cadherin or P-cadherin (Chanson et al., 2011). Below, we review the subsequent genetic analyses of mammary-specific deletion or misexpression of adhesion molecules and identify remaining gaps in our knowledge.

2.2.1 *Early requirement for* E-cadherin

E-cadherin is broadly expressed in epithelial cells from early embryonic stages through to mature organs. Accordingly, it was reasonable to expect that it was required for the formation of epithelial tissues. Consistent with that concept, *E-cadherin* null embryos initially compact due to residual maternal E-cadherin but fail to form a trophectoderm and blastocyst cavity and die at the time of implantation (Larue, Ohsugi, Hirchenhain, & Kemler, 1994; Riethmacher, Brinkmann, & Birchmeier, 1995). Electron microscopy analysis of *E-cadherin* null embryos has demonstrated that cell–cell contacts are structurally distorted and form irregular interdigitating membranes, yet they maintain desmosomes and tight junctions (Riethmacher et al.,

1995). Interestingly, these interdigitating membranes are morphologically similar to those observed connecting normal mammary epithelial cells during periods of active morphogenesis, suggesting that ductal elongation may involve partial disassembly of adherens junctions (Ewald et al., 2012). These studies established an essential role for *E-cadherin*, but its very early requirement prevented analysis at postnatal stages.

2.2.2 Strategies for genetic analysis in the postnatal mammary gland

To overcome this challenge, investigators have relied on cell type-specific expression of *Cre recombinase* to conditionally delete genes. In the mammary gland, most studies have relied on the mouse mammary tumor virus (MMTV) long terminal repeat (Wagner et al., 2001) and whey acidic protein (WAP) (Wagner et al., 1997) promoters. These tools have been very productive and have enabled the analysis of mammary-specific requirements for many genes (McNally & Martin, 2011). However, several challenges have emerged that limit the ability of either line to generate "perfect" mammary-specific gene deletions. The first is that both promoters exhibit a degree of mosaicism within the epithelial compartment, resulting in a varying mixture of wild-type and recombined cells at different stages. The second is the varying timing of Cre activity; depending on the founder line and strain background, the MMTV promoter becomes active beginning in embryogenesis, whereas the WAP promoter becomes active during the second half of pregnancy (Wagner et al., 2001, 1997). However, both promoters are most active during late pregnancy and lactation, which has meant that effects of gene ablation on pubertal branching morphogenesis have been less frequently characterized. Importantly, differences in the timing of gene deletion in similarly targeted cell populations can result in divergent phenotypes. For example, conditional loss of p53 and E-cadherin in alveolar progenitor cells (via the MMTV promoter) induces invasive lobular carcinoma (ILC) (Derksen et al., 2011, 2006); however, loss of p53 and E-cadherin in mature alveolar cells (via the WAP promoter) does not result in tumor formation (Kotb, Hierholzer, & Kemler, 2011). Finally, recent studies from multiple investigators reported significant lactational defects in mice expressing the *MMTV-Cre* transgene from the A founder line (Robinson & Hennighausen, 2011; Yuan, Wang, Pao, Anderson, & Gu, 2011). Even accounting for these limitations, existing promoter-Cre transgenic lines have been essential in enabling an analysis of the role of cell adhesion in mammary development.

2.2.3 *Postnatal analysis of* E-cadherin *function in the mammary gland*

An early application of this approach was expression of a truncated form of *E-cadherin* under the MMTV promoter to test the specific contribution of E-cadherin's cytoplasmic domain to mammary development (Delmas et al., 1999). In the virgin and pregnant gland, overexpression of the cytoplasmic domain induces precocious alveolar formation and differentiation but no histologic adhesion defects. In contrast, in the lactating gland, the cytoplasmic domain exerts a dominant-negative effect on cell–cell adhesion, cell polarity, and the integrity of the basement membrane (Delmas et al., 1999). Importantly, transgene activation is highest during lactation, and variation in protein levels of E-cadherin's cytoplasmic domain may account for the discrepancy in effects on cell adhesion and morphology at different stages of development.

Conditional gene deletion was next used to test the consequences of E-cadherin loss in the pregnant and lactating mammary gland (Fig. 2A and D; Boussadia et al., 2002). MMTV-Cre-mediated recombination in *E-cadherin*$^{fl/fl}$ mice induces *E-cadherin* deletion in differentiating alveolar epithelium, which impairs terminal differentiation during late pregnancy. The gland develops normally until about 16–18 days of pregnancy, after which there is significantly reduced milk protein production and massive apoptosis at parturition, similar to an involuting gland (Boussadia et al., 2002). E-cadherin thus plays an essential role in the survival and function of alveolar epithelial cells. However, E-cadherin loss did not result in mammary tumor formation (Boussadia et al., 2002). In two subsequent studies, conditional *E-cadherin* deletion was driven by *K14-Cre* expression, which has low, stochastic activity in the mammary epithelium, and by *WAP-Cre* expression, which has patchy activity in the virgin gland in addition to high activity in the lactating gland (Derksen et al., 2011, 2006). Neither mouse model yields observable developmental defects in virgin, pregnant, or parous mice or results in mammary tumors (Derksen et al., 2011, 2006). Importantly, no E-cadherin^{-} ducts were observed, which was inferred to result from rapid elimination of E-cadherin^{-} cells by apoptosis (Derksen et al., 2011, 2006).

Collectively, these genetic approaches demonstrated a requirement for E-cadherin during lactation but did not test the consequences of E-cadherin loss in the virgin gland. Due to low levels of Cre-mediated deletion during puberty and the lack of reliable markers for recombined cells, the fate of E-cadherin^{-} cells during branching morphogenesis remained unknown. To address this question, a recent study used fluorescently labeled genetic mosaic analysis, 3D organotypic culture, and

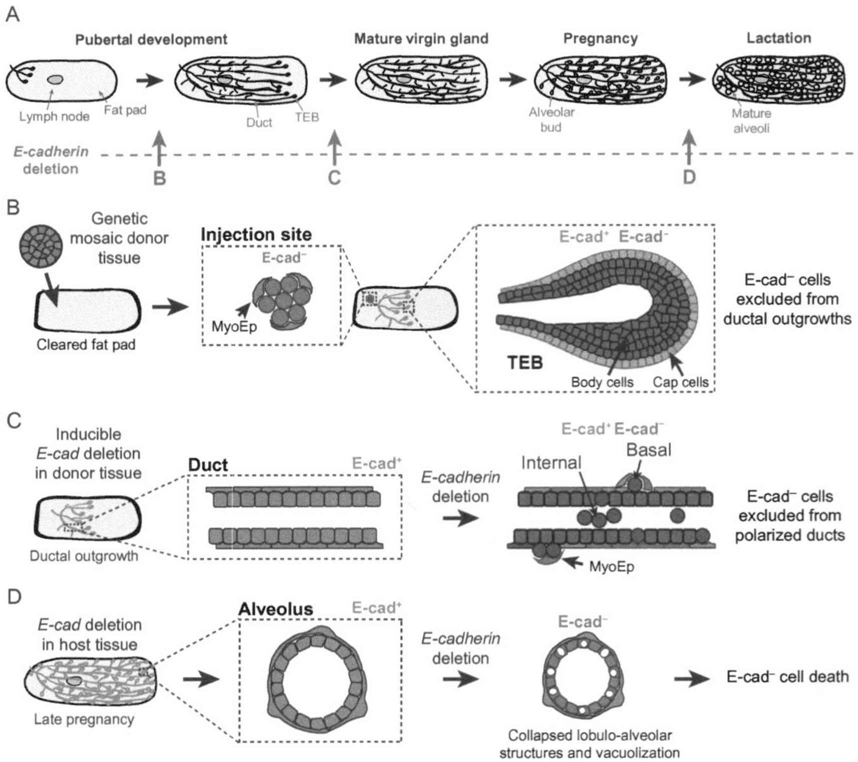

Figure 2 E-cadherin is required at multiple stages of mammary epithelial development. (A) Mammary pubertal branching morphogenesis initiates at approximately 3 weeks postnatal. A stratified terminal end bud (TEB) leads the elongation front, and a full ductal network forms over 7 weeks. During pregnancy, lactogenic hormones induce the ductal epithelium to form specialized structures called alveoli that mature to secrete milk during lactation. Different genetic approaches have been used to test the consequences of E-cadherin loss during distinct stages of development. (B) E-cadherin is required for branching morphogenesis (Shamir et al., 2014). E-cadherin$^-$ cells (green) within transplanted genetic mosaic epithelium are excluded from a ductal network elaborated exclusively by E-cadherin$^+$ cells. (C) E-cadherin is required for maintenance of epithelial architecture (Shamir et al., 2014). E-cadherin$^-$ cells (green) are extruded apically and basally from mature E-cadherin$^+$ epithelial ducts. (D) E-cadherin is required for terminal differentiation and cell survival of alveoli in the lactating gland (Boussadia et al., 2002). E-cadherin loss induces lobuloalveolar collapse and massive apoptosis. (See the color plate.)

orthotopic transplantation to test the consequences of E-cadherin loss on mammary branching and maintenance of epithelial architecture (Fig. 2A–C; Shamir et al., 2014). The Cre biosensor *mT/mG* was used to distinguish the fate of E-cadherin$^-$ cells in genetic mosaic epithelium. *E-cadherin* deletion in transplanted donor tissue results in failure of E-cadherin$^-$ cells to contribute to ductal outgrowths, which are elaborated by E-cadherin^{+} cells (Fig. 2B; Shamir et al., 2014). E-cadherin^{-} cells are viable for at least 6 weeks *in vivo* but are excluded both from elongating, multilayered buds and from polarized, bilayered ducts. *E-cadherin* deletion in polarized epithelial ducts results in the extrusion of E-cadherin^{-} cells apically, into the ductal lumen, and basally, onto the ductal surface (Fig. 2C; Shamir et al., 2014). These data reveal a requirement for E-cadherin in branching morphogenesis and in duct homeostasis *in vivo*.

Consistent with *in vivo* findings, in 3D culture, E-cadherin^{-} cells can locally invade past basement membrane (Fig. 3C); however, they rarely disseminate (Shamir et al., 2014). By ultrastructural analysis, cells in these invasive groups are still connected by desmosomes (Shamir et al., 2014). E-cadherin loss thus disrupts epithelial polarity and organization but is not sufficient to induce dissemination *in vivo* or *in vitro*. It remains an open question whether the observed cell–cell junctions functionally limit the dissemination of E-cadherin^{-} cells. Interestingly, basally positioned E-cadherin^{-} cells are consistently covered by myoepithelial cells, both in 3D culture and *in vivo*. This result raises the alternate possibility that myoepithelial cells resist the dissemination of E-cadherin^{-} luminal epithelial cells. This scenario is plausible as the presence of an intact myoepithelium is the feature that clinically distinguishes ductal carcinoma *in situ* from invasive breast cancer (Hu et al., 2008; Polyak, 2010; Polyak & Hu, 2005). Furthermore, even focal disruptions in the myoepithelium correlate with poor patient outcomes in breast cancer (Man et al., 2003), and myoepithelial cells have been proposed as cellular suppressors of tumor cell invasion (Polyak & Hu, 2005; Sternlicht & Barsky, 1997). Future studies are needed to test this concept through myoepithelium-specific genetic perturbation.

2.2.4 P-cadherin

P-cadherin is a classical cadherin specifically expressed in myoepithelial cells (Daniel et al., 1995). The exception is during late pregnancy and lactation, when luminal epithelial cells secrete high levels of a soluble fragment of P-cadherin in human milk (Soler, Russo, Russo, & Knudsen, 2002). P-cadherin-deficient mice are viable, and the females are both fertile and can nurse their young (Radice et al., 1997). This result was somewhat surprising given the normally high levels of expression of *P-cadherin* in the placenta and in the myoepithelium (Radice et al., 1997). However, P-cadherin loss induces precocious differentiation in the postpubescent virgin mammary gland, with appearance of alveolar-like buds that express milk proteins (Radice et al., 1997). Over time, P-cadherin^{-} females develop focal

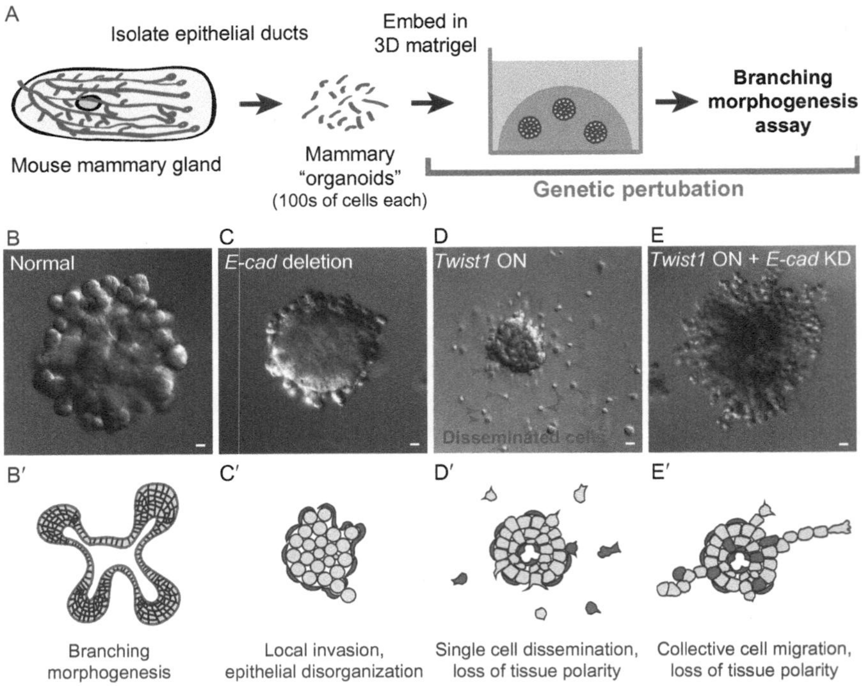

Figure 3 E-cadherin is required for Twist1-induced single cell dissemination. 3D organotypic culture of primary mouse mammary epithelium was used to isolate the effects of single gene changes associated with the epithelial–mesenchymal transition. (A) In the assay, fragments of mammary ducts, called "organoids," are explanted into a basement membrane-rich extracellular matrix. Genetic perturbations, including Cre-lox-based gene deletion, shRNA knockdown, and tet-inducible gene expression, are performed *in vitro*. (B) Normal organoids branch in response to growth factor. (C) *E-cadherin* deletion blocks branching and induces epithelial disorganization. (D) *Twist1* expression blocks branching and induces single cell dissemination. (E) *E-cadherin* knockdown inhibits Twist1-induced single cell dissemination. Bars, 20 μm. *Images ©Shamir et al. (2014). Originally published in* The Journal of Cell Biology, *doi: http://dx.doi.org/10.1083/jcb.201306088.*

hyperplasia in the luminal epithelial compartment; however, P-cadherin⁻ mice do not develop carcinomas (Radice et al., 1997). P-cadherin therefore serves as an important regulator of proliferation and differentiation in the mammary gland. Importantly, these data demonstrate that loss of a myoepithelium-specific protein can induce a phenotype in the luminal epithelial cells, revealing a functional cross talk between epithelial lineages. It remains unclear whether the consequences of P-cadherin loss result directly from a defect in P-cadherin-mediated signaling or from a more general disruption in cell–cell adhesion and tissue organization.

2.2.5 Catenins

Inside the cell, cadherins are linked to the actin cytoskeleton by the cadherin–catenin core complex, which includes α-catenin, β-catenin, and p120-catenin (Takeichi, 2014). Genetic testing of the roles for catenins in the mammary gland supports the concept that the adherens junction is required for pubertal development and epithelial cell survival during lactation.

α-Catenin connects to the cadherin complex through β-catenin and also binds to F-actin. Loss of α-catenin causes embryonic lethality due to disruption of the trophoblast epithelium (Torres et al., 1997). As such, mammary-specific *α-catenin* gene deletion was induced in *α-catenin*$^{fl/fl}$ mice by expressing Cre under control of either the WAP or MMTV promoter (Nemade et al., 2004). Loss of α-catenin impaired functional differentiation and polarization in alveolar epithelial cells (Nemade et al., 2004). Reduced milk protein gene expression resulted in failure to thrive among offspring of these mice. Epithelial clusters at parturition lacked central lumens and lipid droplets and displayed increased cell death, similar to the mammary epithelium at involution (Nemade et al., 2004). This combination of impaired milk production and increased apoptosis closely parallels the phenotypic consequences of conditional *E-cadherin* deletion (Boussadia et al., 2002). Moreover, as with *E-cadherin*, deletion of *α-catenin* does not result in hyperproliferation or tumor formation.

Importantly, *α-catenin* deletion induced by MMTV-Cre has no detectable effect on branching morphogenesis (Nemade et al., 2004). Nevertheless, the known mosaic expression of the *MMTV-Cre* transgene (Wagner et al., 2001) suggests the possibility that cell competition with α-catenin^{+} cells may have obscured a defect. Fluorescent labeling of genetically modified E-cadherin^{-} cells recently enabled demonstration of a previously unrecognized requirement for E-cadherin in branching morphogenesis (Shamir et al., 2014). Analogous approaches will likely reveal additional roles for adhesion molecules such as α-catenin in developmental stages other than lactation, during which the MMTV and WAP promoters are most active.

β-Catenin's role in mammary development and cancer has largely been studied in the context of its signaling function as an essential element in the canonical Wnt pathway, not its structural function as a component of the adherens junction (Hatsell, Rowlands, Hiremath, & Cowin, 2003). For example, a stabilized, constitutively active form of β-catenin expressed in luminal epithelial cells (*MMTV::ΔN89-β-catenin* mice) induces precocious lobuloalveolar development and differentiation in male and female virgin mice, incomplete involution following cessation of lactation, and development of adenocarcinomas in all female mice early in life (Imbert, Eelkema, Jordan, Feiner, & Cowin, 2001). A constitutively active form of β-catenin expressed in basal myoepithelial cells (*K5::ΔN57-β-catenin* mice) induces precocious lobuloalveolar development in pregnancy, sustained luminal cell proliferation during lactation and accelerated involution, and development of basal-type hyperplasia and invasive carcinomas (Teuliere et al., 2005). Consistent with these findings, expression of a dominant-negative β-catenin (*β-eng*) that retains normal cell–cell adhesion properties but lacks the cytoplasmic signaling domain inhibits lobuloalveolar development and induces apoptosis (Tepera, McCrea, & Rosen, 2003). We are not aware of any studies that have specifically tested the consequences of *β-catenin* deletion on mammary epithelial development. Importantly, deletion of *E-cadherin* or *α-catenin* in normal mammary ducts *in vivo* has not been reported to result in nuclear localization of β-catenin or Wnt activation (Boussadia et al., 2002; Nemade et al., 2004). Disruption of the adherens junction in otherwise normal mammary epithelial cells therefore does not necessarily activate β-catenin signaling.

p120-catenin (p120) regulates cadherin levels at the cell surface by physically blocking cadherin endocytosis and subsequent degradation (Kowalczyk & Reynolds, 2004). The requirement for p120 in the mammary gland was tested through MMTV-Cre-mediated recombination in *p120*$^{fl/fl}$ mice. p120$^-$ cells are excluded from TEBs upon genetic mosaic ablation at puberty, followed by rapid elimination of p120$^-$ cells and a transient delay in ductal outgrowth that is accomplished by p120$^+$ cells (Kurley et al., 2012). Sorting of both luminal and myoepithelial p120$^-$ cells is observed in nascent TEBs, with shedding of K8$^+$p120$^-$ cells into the lumen and accumulation of SMA$^+$p120$^-$ cells in the subcapsular space. This observation suggests a functional requirement for catenin-based cell–cell junctions in myoepithelial cells. Finally, transplanted p120$^-$ cells fail to reconstitute the mammary gland. These findings directly parallel the consequences of conditional *E-cadherin* deletion and demonstrate the importance of labeling of the recombined cells: in genetic mosaics, p120$^-$ and E-cadherin$^-$ cells are excluded apically and basally from normal epithelium and fail to contribute to ductal outgrowths, whereas wild-type cells in the same epithelium successfully reconstitute the ductal network (Kurley et al., 2012; Shamir et al., 2014). p120 and E-cadherin thus have nonredundant roles in mammary cell–cell adhesion and in the collective migration required for normal development.

Moreover, as loss of p120 induces concurrent depletion of E-cadherin (Kurley et al., 2012), the absence of cadherin stability may serve as the ultimate driver of the phenotype.

2.3 Desmosomes

Desmosomes are observed at luminal–luminal, myoepithelial–myoepithelial, and luminal–myoepithelial cell–cell contacts, and their number shifts dynamically during morphogenesis (Pitelka et al., 1973). Resting or nonsecretory mammary ducts have an abundance of desmosomes (Pitelka et al., 1973). In contrast, during lactation and involution, alveoli lose nearly all desmosomes and adherens junctions and establish a new network of tight junctions (Pitelka et al., 1973). Interestingly, desmosomes appear to be the junctions that remain in multiple contexts of reduced cell–cell adhesion: (a) in the normal, stratified, elongating bud (Ewald et al., 2012) and (b) in chains of migratory, but collective, E-cadherin^{-} cells (Shamir et al., 2014). Nevertheless, the requirement for desmosomal proteins in mammary epithelial development has not been tested genetically.

Desmocollins (Dscs) and desmogleins (Dsgs), the desmosomal cadherins, each exist as three genetic isoforms that are differentially expressed in the mammary epithelial bilayer: *Dsc2* and *Dsg2* are ubiquitously expressed, *Dsc3* and *Dsg3* are expressed only in myoepithelial cells, and *Dsc1* and *Dsg1* are absent (Runswick et al., 2001). The function of desmosomal adhesion in cell positioning was tested *in vitro* by treating human mammary epithelial cells with blocking peptides against cell adhesion recognition (CAR) sites on the desmosomal cadherins (Runswick et al., 2001). Combined, but not individual, Dsc2 and Dsg2 CAR peptides inhibit luminal–luminal cell adhesion, and combined, but not individual, Dsc3 and Dsg3 peptides disrupt proper cell type-specific positioning, suggesting a role for specific isoforms in mediating polarized sorting of luminal and myoepithelial cells (Runswick et al., 2001). In contrast, a separate study concluded that downregulation of Dsc3 and Dsg3 facilitates branching morphogenesis in a primary mammary 3D culture model (Basham et al., 2013). Further elucidation of the precise role of desmosomal adhesion *in vivo* will likely require genetic ablation of various desmosomal proteins both in isolation and in combination. Major open questions include whether desmosome number is tightly regulated across distinct morphogenetic events and whether desmosomes play an essential compensatory function for adherens junctions.

3. CELL–CELL ADHESION IN BREAST CANCER

3.1 E-cadherin: An invasion suppressor?

E-cadherin is considered a suppressor of tumor invasion and metastasis across epithelial cancers (Beavon, 2000). E-cadherin function can be lost through mutation, gene deletion, or transcriptional repression (Berx & Van Roy, 2001). Classic experiments in human carcinoma cell lines demonstrated that those with an "epithelioid" morphology generally are E-cadherin^{+} and non-invasive into collagen gels, whereas those with a "fibroblastoid" morphology are E-cadherin^{-} and invasive (Frixen et al., 1991). Furthermore, forced *E-cadherin* expression was sufficient to inhibit the invasiveness of human breast carcinoma cells (Frixen et al., 1991; Vleminckx, Vakaet, Mareel, Fiers, & van Roy, 1991).

In the breast, E-cadherin loss is a characteristic feature of ILC (Berx et al., 1995; Moll, Mitze, Frixen, & Birchmeier, 1993). In the mouse, loss of E-cadherin synergizes with loss of p53 to accelerate tumor formation and metastasis (Derksen et al., 2006). At the tissue level, loss of E-cadherin and p53 is accompanied by anoikis resistance and increased angiogenesis, and the tumor itself has a remarkable histologic resemblance to human ILC (Derksen et al., 2006). This study revealed that E-cadherin can function as both a tumor and metastasis suppressor. Nevertheless, loss of E-cadherin is commonly observed in benign lobular carcinoma *in situ* as well as ILC, suggesting that E-cadherin loss is not sufficient for invasion (Vos et al., 1997). Interestingly, germline *E-cadherin* mutations are rarely observed in patients with invasive lobular breast cancer but are implicated in hereditary diffuse gastric cancer (Paredes et al., 2012).

Importantly, while E-cadherin loss is characteristic of ILC, lobular carcinomas represent only about 10% of breast tumors (Arpino, Bardou, Clark, & Elledge, 2004). Most primary breast tumors and their distant metastases are in fact E-cadherin positive (Kowalski et al., 2003). Specifically, E-cadherin is detectable in most ductal carcinomas, the most common histologic subtype of breast cancer (Moll et al., 1993). The dual presence of E-cadherin in the primary tumor and the metastatic sites can still be harmonized with a view of E-cadherin as an obligate invasion suppressor if there is a transient loss of E-cadherin as the cancer cells are traveling to distant organs (Polyak & Weinberg, 2009). Arguing against this concept is the observation that E-cadherin levels are normal or higher in tumor microemboli of inflammatory breast cancer, a highly aggressive and poor prognosis breast cancer subtype (Kleer, van Golen, Braun, & Merajver, 2001).

The natural next question is whether E-cadherin levels in the primary tumor correlate with lymph node involvement, metastasis, and clinical outcomes. Several studies have identified an association between reduced or absent E-cadherin staining and increased invasiveness, higher rates of metastasis, and poor clinical outcomes (Berx & van Roy, 2009; Hunt et al., 1997; Oka et al., 1993; Siitonen et al., 1996). Conversely, multiple studies have found an association between E-cadherin status and histologic type without detecting a significant, independent correlation between E-cadherin status and outcomes (Gamallo et al., 1993; Lipponen et al., 1994). Finally, some studies have noted an independent correlation between strong E-cadherin staining and poor patient survival (Tan et al., 1999). Collectively, these data reveal that *E-cadherin* expression usefully distinguishes histologic subtypes but that its relationship to clinical outcomes, such as metastasis or survival, is not yet sufficiently well understood. From an underlying biological mechanism perspective, it is possible that complex spatial and temporal dynamics in *E-cadherin* expression could mask real correlations with survival. Alternatively, it is possible that there are both E-cadherin^{+} and E-cadherin^{-} paths to metastasis.

3.2 Cadherin switching

One mechanism for regulating cadherin dynamics during epithelial invasion is cadherin switching, in which cells induce expression of a cadherin isoform not normally present in that cell type (Wheelock, Shintani, Maeda, Fukumoto, & Johnson, 2008). The most commonly observed example is gain of N-cadherin, with or without concurrent loss of E-cadherin (Wheelock et al., 2008). *N-cadherin* expression is normally restricted to neural tissue and mesenchymal cells. Its expression in human mammary tumor cell lines correlates with increased invasiveness and motility *in vitro* (Nieman, Prudoff, Johnson, & Wheelock, 1999) and increased metastasis in nude mice *in vivo* (Hazan, Phillips, Qiao, Norton, & Aaronson, 2000). In these tumor cell lines, N-cadherin-mediated induction of cell migration was found to be independent of E-cadherin status; forced *E-cadherin* expression in N-cadherin^{+} cells did not reduce motility (Nieman et al., 1999).

Based on these observations, the expectation was that ectopic *N-cadherin* expression in mammary epithelial cells would promote invasion *in vivo*, regardless of *E-cadherin* expression. Surprisingly, *N-cadherin* expression in the pregnant and lactating mammary epithelium, driven by the MMTV promoter, has no detectable phenotype (Knudsen, Sauer, Johnson, & Wheelock, 2005). Mid- and late-stage pregnant mice have normal mammary gland morphology and function, retain *E-cadherin* and *β-catenin* expression, and do not develop spontaneous tumors (Knudsen et al., 2005). A separate study tested whether N-cadherin could functionally replace E-cadherin in the mammary gland (Kotb et al., 2011). A knock-in of *N-cadherin* at the *E-cadherin* locus, combined with conditional deletion of the second *E-cadherin* allele during late pregnancy, induces massive apoptosis of alveolar cells and a lactation defect more severe than E-cadherin loss alone (Kotb et al., 2011). After multiple lactation cycles, mammary glands develop fibrocystic change that frequently progresses to tumor formation in combination with conditional deletion of one *p53* allele (Kotb et al., 2011). Collectively, these studies demonstrate that expression of *N-cadherin* is insufficient to drive invasion or tumor formation and that E-cadherin is specifically required for epithelial integrity and cannot be substituted by N-cadherin.

A second plausible basis for the cadherin switch is the gain of P-cadherin in luminal epithelial cells. P-cadherin is the dominant classical cadherin in myoepithelial cells, and its expression is low or undetectable in normal luminal epithelial cells (Chanson et al., 2011; Daniel et al., 1995). Human breast cancers typically originate in luminal cells but frequently express *P-cadherin*, and *P-cadherin* expression is associated with high histologic grade and poor clinical outcomes (Paredes et al., 2005; Soler, Knudsen, Salazar, Han, & Keshgegian, 1999). To test the effect of inappropriate P-cadherin in normal adult luminal cells, transgenic mice were generated with *P-cadherin* expression driven by the MMTV promoter. Mammary glands with ectopic *P-cadherin* expression maintain membrane-localized E-cadherin and β-catenin; have no apparent defects in branching morphogenesis, ductal architecture, lactation, or involution; and do not develop mammary tumors (Radice, Sauer, Kostetskii, Peralta Soler, & Knudsen, 2003).

Taken together, misexpression of either *N-cadherin* or *P-cadherin* in normal mammary epithelium, without loss of E-cadherin, is not sufficient to disrupt mammary gland function or induce tumor formation. It is worth investigating whether another cadherin, such as cadherin 6 or 11, would be capable of inducing invasion or dissemination in the mammary epithelium (Gheldof & Berx, 2013; Jia, Liu, Hansen, Ter Beest, & Zegers, 2011). For example, experimental induction of cadherin 8, which is normally expressed in neurons, is sufficient to induce cyst formation in kidney epithelium (Kher et al., 2011). Alternately, the capacity of inappropriate cadherin expression to promote invasion and metastasis may require a decrease in E-cadherin levels and/or other as yet unknown molecular events. The cadherin switching phenomena observed in culture remain an attractive conceptual framework for regulating cell motility, but they have proven difficult to replicate *in vivo* in the mammary epithelium.

3.3 Rethinking the epithelial-to-mesenchymal transition

A second, related model is that cells fundamentally change their identity and undergo an epithelial-to-mesenchymal transition (EMT) in order to detach and migrate away from a tissue as a single cell (Greenburg & Hay, 1982; Hay & Zuk, 1995; Thiery, 2002). This concept is grounded in observations of cellular and molecular dynamics during embryogenesis (Acloque, Adams, Fishwick, Bronner-Fraser, & Nieto, 2009; Thiery, Acloque, Huang, & Nieto, 2009), such as neural crest formation, and has been proposed as a mechanism for cancer metastasis (Polyak & Weinberg, 2009; Thiery et al., 2009; Yang & Weinberg, 2008). The EMT paradigm originated from classic experiments by Elizabeth Hay, who observed that mature epithelia explanted into collagen I gels lost polarity and migrated into the matrix as single cells with a strong morphological resemblance to embryonic fibroblasts, a mesenchymal cell type (Greenburg & Hay, 1982; Hay & Zuk, 1995). These morphological observations formed the basis for the concept of "loss of the epithelial phenotype," which expanded to include loss of epithelium-specific gene expression (Thiery, 2002). In turn, E-cadherin is often considered as the archetypal "caretaker" of the epithelial phenotype, and its loss has been used to invoke a broader change in cellular differentiation state (Peinado, Olmeda, & Cano, 2007).

A challenge to testing the EMT model in human cancer is that different investigators have used different criteria to define whether an EMT has occurred (see also the "Epithelial-Mesenchymal transitions: from cell plasticity to concept elasticity" by Pierre Savagner in this volume). While Hay proposed a fundamentally morphological definition, most current investigators define the molecular basis of EMT in terms of a transcriptional program consisting of downregulation of *E-cadherin*, catenins, and cytokeratins and upregulation of the mesenchymal markers *N-cadherin*, *vimentin*, and *fibronectin* (Peinado et al., 2007). Just as no single gene is specific to all epithelial tissues, no single marker definitively identifies an EMT in all circumstances. Nevertheless, EMT transcription factors such as Snail1, Slug, and Twist1 are collectively thought to operate through these molecular changes, and their expression has been correlated with invasion and metastasis in breast cancer patients and in experimental cancer models (Blanco et al., 2002; Martin, Goyal, Watkins, & Jiang, 2005; Mironchik et al., 2005; Peinado et al., 2007; Yang et al., 2004).

The varied use of morphological versus molecular criteria, or a combination of the two, has resulted in tension across studies as to when and whether an EMT has occurred. The concept of an epithelium is, at root, very morphological, and cell escape during metastasis would appear to require a dramatic break from this morphology. Nevertheless, the frequent absence of concurrent EMT-associated molecular changes in metastatic tumors has led some investigators to propose a "partial" EMT. For example, the commonly used MMTV-PyMT mouse model of breast cancer, which histologically resembles invasive ductal carcinoma and is highly metastatic to the lung, does not induce an EMT, as tumors are positive for E-cadherin and cytokeratins and negative for fibronectin (Guy, Cardiff, & Muller, 1992; Trimboli et al., 2008). Even a mouse model of ILC, which is E-cadherin negative, retains expression of the luminal epithelial cytokeratin K8 and does not upregulate vimentin (Derksen et al., 2006). In addition, human primary breast tumor cells positive for the EMT transcription factor Slug can express high levels of E-cadherin (Come et al., 2006). Such discrepancies indicate that "EMT" has different meanings depending on the investigator's chosen definition and does not precisely or universally explain how epithelial tumor cells disseminate. Furthermore, lineage analysis of mesenchymal gene expression within epithelial cells revealed that mouse models of breast cancer can present clear evidence of a molecular EMT during metastasis (e.g., WAP-Myc), whereas other models accomplish metastasis while maintaining epithelial gene expression (e.g., MMTV-Neu and MMTV-PyMT) (Trimboli et al., 2008). The concept of EMT in human breast cancer remains controversial, with some pathologists arguing that they do not see evidence to support a shift toward mesenchymal cell identity in cancer cells (Tarin, 2005). The most direct evidence for the molecular changes associated with EMT are in *Myc*-overexpressing experimental mouse models (Trimboli et al., 2008) and in the claudin-low subtype of human breast cancer (Hennessy et al., 2009; Taube et al., 2010). Furthermore, the molecular program of EMT may not correlate with a transition to single cell organization, as a recent study exhaustively characterized the invasive boundary of breast tumors with the stroma and observed frequent collective invasion and essentially no isolated single cancer cells (Bronsert et al., 2014).

To resolve the direct effects of EMT-associated molecular changes on cell- and tissue-level behavior, a recent study directly contrasted the effect of *E-cadherin* deletion with *Twist1* expression in an organotypic culture model of primary mouse mammary epithelium (Fig. 3A–E; Shamir et al., 2014). Loss of E-cadherin induced single-file invasion past basement membrane but only rare dissemination in 3D culture or *in vivo* (Fig. 3C). In contrast, Twist1 induced rapid and robust dissemination of single cells that were cytokeratin positive and migrated with dynamic, forward-oriented protrusions (Fig. 3D; Shamir et al., 2014). Contrary to expectation, Twist1 did not result in a loss of epithelial-specific gene expression or significantly regulate any canonical EMT genes, such as *E-cadherin*, *N-cadherin*, and *vimentin* (Shamir et al., 2014). Instead, Twist1 regulated a set of genes that collectively reprogram the composition of the extracellular space and the cell's interactions with the extracellular matrix (ECM) (Shamir et al., 2014). Surprisingly, E-cadherin was still localized to the membranes of cells at every stage of dissemination, and *E-cadherin* knockdown inhibited dissemination of Twist1$^+$ cells (Shamir et al., 2014). Rather, loss of E-cadherin promoted more cohesive migration of chains of Twist1$^+$, cytokeratin$^+$ cells (Fig. 3E). The exact levels of E-cadherin may thus critically regulate whether Twist1 induces single cell dissemination or collective epithelial invasion.

3.4 "Unexpected" roles for E-cadherin

Classic portrayals of E-cadherin as solely a tumor and/or invasion suppressor may be too narrow. In certain contexts, E-cadherin appears to promote cell migration, invasion, and even tumor progression (Rodriguez, Lewis-Tuffin, & Anastasiadis, 2012). In the developing zebrafish embryo, E-cadherin-mediated traction forces are required for single cell migration of chemokine-guided germ cells, and interfering with E-cadherin disrupts F-actin dynamics and organization at the cellular front and inhibits cell motility (Kardash et al., 2010). This study demonstrated that the precise levels of E-cadherin are important, as too much E-cadherin results in slower than normal migration, whereas loss of E-cadherin results in no migration (Kardash et al., 2010). Analogous regulation may have implications for the mechanisms of tumor cell motility *in vivo*. A recent study of developmental collective cell migration of the border cells in Drosophila revealed a novel role for E-cadherin as an integrator of mechanical signals and demonstrated its requirement for directional migration of cell clusters (Cai et al., 2014).

E-cadherin is also associated with aggressive cancer in multiple organs. In addition to being normal or increased in inflammatory breast cancer, E-cadherin is expressed in most ovarian carcinomas, which are derived from normally E-cadherin$^-$ ovarian surface epithelium (Auersperg et al., 1999; Kleer et al., 2001). Among a small subset of glioblastomas, E-cadherin is associated with decreased survival and correlates with increased invasiveness *in vivo* in orthotopic mouse xenografts (Lewis-Tuffin et al., 2010). The function of E-cadherin in such tumors and validation of its potential role in promoting invasion remains a major open area for future research (Rodriguez et al., 2012). It is also important to distinguish whether E-cadherin's role in cell migration is primarily through direct control of intercellular adhesion or instead through regulation of intracellular signaling networks (Chu, Boley, Moraes, Barsky, & Robertson, 2013).

4. CAN A MIGRATORY SINGLE CELL OR CELL CLUSTER BE "EPITHELIAL"?

As our conception of metastasis broadens to include the invasion of E-cadherin^{+} cancer cells, an important definitional question arises: can we consider the resulting migratory single cells and cell groups to be epithelial? The answer to this question really rests on whether we define epithelial identity at the tissue or cellular level. In a homeostatic context, individual cells are components of an epithelial tissue and could not accomplish any of its essential barrier, secretory, or absorptive functions in isolation. At this functional level, it seems clear that an isolated single cell cannot be a functional barrier or an absorptive epithelium.

Yet the answer is subtler when considering the defining morphologic and molecular properties of an epithelium. Tight cell–cell adhesion represents one such property and has traditionally been thought of as serving mostly a structural or barrier function in epithelial ducts. However, mammary epithelial cells maintain at least some intercellular adhesive junctions during active migratory processes (Ewald et al., 2012; Shamir et al., 2014). Data from genetic mouse models and patient samples now increasingly support a more nuanced portrait of the roles of cell–cell adhesion proteins in normal and pathologic cellular contexts. In particular, there is not a simple relationship between loss of E-cadherin and invasive or disseminative behavior in normal or tumor mammary epithelium (Rodriguez et al., 2012). However, it is clear that migratory single cells can maintain membrane-localized E-cadherin and β-catenin and a normal complement of mRNAs for the genes encoding core epithelial functions (Shamir et al., 2014). Consistent with this observation, *E-cadherin* is expressed in most invasive breast cancers and, remarkably, is required for efficient single cell dissemination of Twist1^{+} cells (Kowalski et al., 2003; Shamir et al., 2014).

Given these observations, should we consider disseminated cells or groups "epithelial"? The community will have to settle this issue, likely through active new research and careful consideration of definitions. Our answer, based on our data, is that epithelial cells clearly are capable of migration within their tissue during developmental contexts (Ewald et al., 2008, 2012); that epithelial cells can be induced to disseminate into the ECM by changes in the composition, organization, or adhesive density of the microenvironment (Beck, Singh, Rothenberg, Elisseeff, & Ewald, 2013; Nguyen-Ngoc et al., 2012; Nguyen-Ngoc & Ewald, 2013); that simple molecular interventions, such as *Twist1* expression, can induce dissemination of cells that maintain epithelial gene expression (Shamir et al., 2014); and that metastasis can occur in cells that retain many features of their epithelial origin (Cheung, Gabrielson, Werb, & Ewald, 2013; Nguyen-Ngoc et al., 2012). It may therefore be most useful to expand our concept of epithelial cells and tissues to include different characteristic states of epithelial function and behavior, including specifically a migratory or morphogenetically active epithelial state (Ewald et al., 2008). Considered from this perspective, it becomes important to determine the relative frequency with which tumors recapitulate epithelial migratory states versus mesenchymal fates or migratory states.

4.1 Dissemination: A transition in the substrate for migration

Epithelium is most often conceptualized in terms of quiescent ducts or sheets. However, building epithelial ductal networks involves tube elongation and branching, a dynamic process driven by the collective migration of adherent groups of cells (Andrew & Ewald, 2010). In the mammary gland, ultrastructural analysis and confocal imaging of stratification and bud elongation have revealed that cells within these groups lack apicobasal polarity, have reduced cell–cell junctions, are individually motile, and can appear elongated and even protrusive (Ewald et al., 2012). The presence of such features in normal cells during development suggests that motility rests within the spectrum of normal epithelial cell behavior. Epithelial dissemination may therefore not be an "invented" feature of metastatic cancer; rather, epithelial cells may be able to activate and modulate a conserved migratory program. In turn, dissemination might best be explained by a transition in the substrate for migration: in a normal migratory *epithelium*, the substrate is adjacent cells; in a disseminating migratory *epithelial cell*, the substrate is the surrounding ECM (Fig. 4A–B). Consistent with this concept, mammary epithelial cell dissemination induced

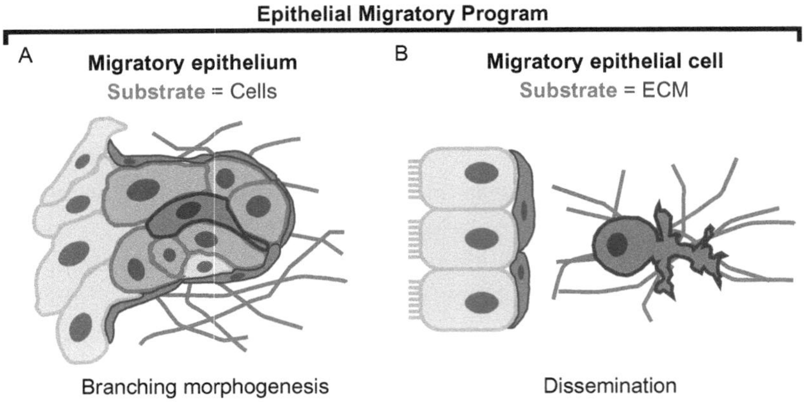

Figure 4 Is there are conserved epithelial migratory program? Dissemination might be explained by a switch in the substrate of migration from adjacent cells to ECM. (A) Epithelial cells collectively migrate as an elongating bud during mammary branching morphogenesis. Within the multilayer, an individual epithelial cell (green; dark gray shade in the print version) can appear elongated and protrusive. (B) During Twist1-induced dissemination, a single epithelial cell (green; dark gray shade in the print version) migrates through the ECM with amoeboid morphology and extensive protrusions at the leading front.

by Twist1 primarily involves transcriptional changes in genes that reside in the extracellular compartment. Thus, dissemination may primarily involve a change in the type of adhesion proteins on the cell surface and the composition of the immediate extracellular environment.

4.2 Balancing cell–cell and cell–matrix adhesion

Epithelial cells do not exist in isolation; rather, the epithelium is surrounded by connective tissue, which changes in organization and composition in cancer. Cell adhesion to the ECM via integrins has an established role in regulating mammary epithelial differentiation and morphogenesis (Muschler & Streuli, 2010). The composition and mechanical properties of the ECM also modulate cell behaviors in normal development and tumor progression (Ghajar & Bissell, 2008; Schedin & Keely, 2011). A collagen I-rich ECM, which models a more adhesive, tumorigenic microenvironment, induces protrusive invasion and dissemination even in normal mammary epithelium (Nguyen-Ngoc et al., 2012). Furthermore, the microenvironment can modulate phenotypic outcomes of genetic changes. For example, loss of P-cadherin induces hyperbranching of mammary epithelium cultured in a basement membrane-rich ECM but promotes excess, sustained myoepithelial dissemination when the same epithelium is cultured in collagen I (Nguyen-Ngoc et al., 2012).

We propose that it is useful to integrate these data within a free energy model for dissemination, such that cells leave the epithelial group whenever the energy of cell–matrix adhesion exceeds the energy of cell–cell adhesion. Our model allows for different categories of dissemination-inducing signals: broadly, cell-intrinsic cues and cell-extrinsic cues. The former encompasses transcriptional changes regulating the cell's interaction with the extracellular space, such as those induced by Twist1 (Shamir et al., 2014). The latter includes many examples of microenvironmental cues, such as a collagen I ECM or synthetic matrices containing adhesive peptides, which can promote invasion and dissemination in normal epithelial tissues (Beck et al., 2013; Nguyen-Ngoc et al., 2012). Notably, microenvironment-driven dissemination is the origin of the EMT concept described by Elizabeth Hay (Greenburg & Hay, 1982). Thus, we anticipate that there are multiple avenues to reach the same cellular outcome of dissemination.

4.3 Novel functions for classic molecules: Next steps in adhesion biology

To date, our knowledge of cell–cell adhesion in normal mammary gland development and homeostasis has been mostly confined to E-cadherin and P-cadherin, in part due to the technical and financial difficulties of developing new loss-of-function reagents. CRISPR/Cas9 technology now offers the opportunity to systematically and rapidly assay the function of many classes of adhesion molecules, including various components of adherens junctions and desmosomes that are dysregulated in breast cancer (Leary et al., 2008). Due to the temporal constraints of *in vivo* gene inactivation approaches with the MMTV and WAP promoters, there is much that remains unknown about the function of adhesion proteins during pubertal branching morphogenesis. Genetic perturbation in organotypic cultures, combined with *in vivo* validation with fluorescent reporters, has introduced a valuable tool to circumvent these limitations. Inducible shRNA or gene deletion techniques further enable the study of loss of cell–cell adhesion molecules in established ducts.

Looking forward, a major unanswered question is how do adhesion proteins regulate and promote migration? The precise levels of particular cell–cell junctions may play an unappreciated role in modulating epithelial cell behaviors. For example, mammary end buds contain a reduced number of desmosomes, but the requirement for desmosomes in bud elongation remains untested, and the presence of too many desmosomes may even inhibit branching (Basham et al., 2013). Moreover, in cancer, the idea that adhesion proteins principally act as invasion suppressors is incompatible with many experimental and clinical observations. Cellular context, including posttranslational modifications and protein turnover, may critically regulate junction dynamics and cell motility and can collaborate with the microenvironment to alter tissue-level phenotypes. We envision that future investigations in adhesion biology have the potential to uncover novel roles for many epithelial adhesion molecules in both development and disease.

ACKNOWLEDGMENTS

E. R. S. was supported by an Isaac and Lucille Hay Graduate Fellowship, endowed by Elizabeth "Betty" Hay. E. R. S. and A. J. E. were supported by a Research Scholar Grant (RSG-12-141-01-CSM) from the American Cancer Society, by funds from the NIH/NCI (P30 CA006973), and by a grant from the Breast Cancer Research Foundation. The authors thank Jamie Davies and David Garrod for stimulating discussions of epithelial identity and epithelial morphogenesis.

REFERENCES

Acloque, H., Adams, M. S., Fishwick, K., Bronner-Fraser, M., & Nieto, M. A. (2009). Epithelial-mesenchymal transitions: The importance of changing cell state in development and disease. *The Journal of Clinical Investigation*, *119*, 1438–1449.

Andrew, D. J., & Ewald, A. J. (2010). Morphogenesis of epithelial tubes: Insights into tube formation, elongation, and elaboration. *Developmental Biology*, *341*, 34–55.

Arpino, G., Bardou, V. J., Clark, G. M., & Elledge, R. M. (2004). Infiltrating lobular carcinoma of the breast: Tumor characteristics and clinical outcome. *Breast Cancer Research*, *6*, R149–R156.

Auersperg, N., Pan, J., Grove, B. D., Peterson, T., Fisher, J., Maines-Bandiera, S., et al. (1999). E-cadherin induces mesenchymal-to-epithelial transition in human ovarian surface epithelium. *Proceedings of the National Academy of Sciences of the United States of America*, *96*, 6249–6254.

Basham, K. J., Kieffer, C., Shelton, D. N., Leonard, C. J., Bhonde, V. R., Vankayalapati, H., et al. (2013). Chemical genetic screen reveals a role for desmosomal adhesion in mammary branching morphogenesis. *The Journal of Biological Chemistry*, *288*, 2261–2270.

Beavon, I. R. (2000). The E-cadherin-catenin complex in tumour metastasis: Structure, function and regulation. *European Journal of Cancer*, *36*, 1607–1620.

Beck, J. N., Singh, A., Rothenberg, A. R., Elisseeff, J. H., & Ewald, A. J. (2013). The independent roles of mechanical, structural and adhesion characteristics of 3D hydrogels on the regulation of cancer invasion and dissemination. *Biomaterials*, *34*, 9486–9495.

Berx, G., Cleton-Jansen, A. M., Nollet, F., de Leeuw, W. J., van de Vijver, M., Cornelisse, C., et al. (1995). E-cadherin is a tumour/invasion suppressor gene mutated in human lobular breast cancers. *The EMBO Journal*, *14*, 6107–6115.

Berx, G., & Van Roy, F. (2001). The E-cadherin/catenin complex: An important gatekeeper in breast cancer tumorigenesis and malignant progression. *Breast Cancer Research*, *3*, 289–293.

Berx, G., & van Roy, F. (2009). Involvement of members of the cadherin superfamily in cancer. *Cold Spring Harbor Perspectives in Biology*, *1*, a003129.

Blanco, M. J., Moreno-Bueno, G., Sarrio, D., Locascio, A., Cano, A., Palacios, J., et al. (2002). Correlation of snail expression with histological grade and lymph node status in breast carcinomas. *Oncogene*, *21*, 3241–3246.

Boussadia, O., Kutsch, S., Hierholzer, A., Delmas, V., & Kemler, R. (2002). E-cadherin is a survival factor for the lactating mouse mammary gland. *Mechanisms of Development*, *115*, 53–62.

Bronsert, P., Enderle-Ammour, K., Bader, M., Timme, S., Kuehs, M., Csanadi, A., et al. (2014). Cancer cell invasion and EMT marker expression: A three-dimensional study of the human cancer-host interface. *The Journal of Pathology*, *234*, 410–422.

Buckley, C. D., Tan, J., Anderson, K. L., Hanein, D., Volkmann, N., Weis, W. I., et al. (2014). Cell adhesion. The minimal cadherin-catenin complex binds to actin filaments under force. *Science*, *346*, 1254211.

Cai, D., Chen, S. C., Prasad, M., He, L., Wang, X., Choesmel-Cadamuro, V., et al. (2014). Mechanical feedback through E-cadherin promotes direction sensing during collective cell migration. *Cell*, *157*, 1146–1159.

Chanson, L., Brownfield, D., Garbe, J. C., Kuhn, I., Stampfer, M. R., Bissell, M. J., et al. (2011). Self-organization is a dynamic and lineage-intrinsic property of mammary epithelial cells. *Proceedings of the National Academy of Sciences of the United States of America*, *108*, 3264–3269.

Cheung, K. J., Gabrielson, E., Werb, Z., & Ewald, A. J. (2013). Collective invasion in breast cancer requires a conserved basal epithelial program. *Cell*, *155*, 1639–1651.

Chu, K., Boley, K. M., Moraes, R., Barsky, S. H., & Robertson, F. M. (2013). The paradox of E-cadherin: Role in response to hypoxia in the tumor microenvironment and regulation of energy metabolism. *Oncotarget*, *4*, 446–462.

Come, C., Magnino, F., Bibeau, F., De Santa Barbara, P., Becker, K.-F., Theillet, C., et al. (2006). Snail and slug play distinct roles during breast carcinoma progression. *Clinical Cancer Research*, *12*, 5395–5402.

Daniel, C. W., Strickland, P., & Friedmann, Y. (1995). Expression and functional role of E- and P-cadherins in mouse mammary ductal morphogenesis and growth. *Developmental Biology*, *169*, 511–519.

Davies, J. A., & Garrod, D. R. (1997). Molecular aspects of the epithelial phenotype. *Bioessays*, *19*, 699–704.

Delmas, V., Pla, P., Feracci, H., Thiery, J. P., Kemler, R., & Larue, L. (1999). Expression of the cytoplasmic domain of E-cadherin induces precocious mammary epithelial alveolar formation and affects cell polarity and cell-matrix integrity. *Developmental Biology*, *216*, 491–506.

Derksen, P. W. B., Braumuller, T. M., van der Burg, E., Hornsveld, M., Mesman, E., Wesseling, J., et al. (2011). Mammary-specific inactivation of E-cadherin and p53 impairs functional gland development and leads to pleomorphic invasive lobular carcinoma in mice. *Disease Models & Mechanisms*, *4*, 347–358.

Derksen, P. W. B., Liu, X. L., Saridin, F., van der Gulden, H., Zevenhoven, J., Evers, B., et al. (2006). Somatic inactivation of E-cadherin and p53 in mice leads to metastatic lobular mammary carcinoma through induction of anoikis resistance and angiogenesis. *Cancer Cell*, *10*, 437–449.

Ewald, A. J., Brenot, A., Duong, M., Chan, B. S., & Werb, Z. (2008). Collective epithelial migration and cell rearrangements drive mammary branching morphogenesis. *Developmental Cell*, *14*, 570–581.

Ewald, A. J., Huebner, R. J., Palsdottir, H., Lee, J. K., Perez, M. J., Jorgens, D. M., et al. (2012). Mammary collective cell migration involves transient loss of epithelial features and individual cell migration within the epithelium. *Journal of Cell Science*, *125*, 2638–2654.

Frixen, U. H., Behrens, J., Sachs, M., Eberle, G., Voss, B., Warda, A., et al. (1991). E-cadherin-mediated cell-cell adhesion prevents invasiveness of human carcinoma cells. *The Journal of Cell Biology*, *113*, 173–185.

Gamallo, C., Palacios, J., Suarez, A., Pizarro, A., Navarro, P., Quintanilla, M., et al. (1993). Correlation of E-cadherin expression with differentiation grade and histological type in breast carcinoma. *The American Journal of Pathology*, *142*, 987–993.

Ghajar, C. M., & Bissell, M. J. (2008). Extracellular matrix control of mammary gland morphogenesis and tumorigenesis: Insights from imaging. *Histochemistry and Cell Biology*, *130*, 1105–1118.

Gheldof, A., & Berx, G. (2013). Cadherins and epithelial-to-mesenchymal transition. *Progress in Molecular Biology and Translational Science*, *116*, 317–336.

Gjorevski, N., & Nelson, C. M. (2011). Integrated morphodynamic signalling of the mammary gland. *Nature Reviews. Molecular Cell Biology*, *12*, 581–593.

Greenburg, G., & Hay, E. D. (1982). Epithelia suspended in collagen gels can lose polarity and express characteristics of migrating mesenchymal cells. *The Journal of Cell Biology*, *95*, 333–339.

Gumbiner, B., Stevenson, B., & Grimaldi, A. (1988). The role of the cell adhesion molecule uvomorulin in the formation and maintenance of the epithelial junctional complex. *The Journal of Cell Biology*, *107*, 1575–1587.

Guy, C. T., Cardiff, R. D., & Muller, W. J. (1992). Induction of mammary tumors by expression of polyomavirus middle T oncogene: A transgenic mouse model for metastatic disease. *Molecular and Cellular Biology*, *12*, 954–961.

Hatsell, S., Rowlands, T., Hiremath, M., & Cowin, P. (2003). Beta-catenin and Tcfs in mammary development and cancer. *Journal of Mammary Gland Biology and Neoplasia*, *8*, 145–158.

Hay, E. D., & Zuk, A. (1995). Transformations between epithelium and mesenchyme: Normal, pathological, and experimentally induced. *American Journal of Kidney Diseases*, *26*, 678–690.

Hazan, R. B., Phillips, G. R., Qiao, R. F., Norton, L., & Aaronson, S. A. (2000). Exogenous expression of N-cadherin in breast cancer cells induces cell migration, invasion, and metastasis. *The Journal of Cell Biology*, *148*, 779–790.

Hennessy, B. T., Gonzalez-Angulo, A. M., Stemke-Hale, K., Gilcrease, M. Z., Krishnamurthy, S., Lee, J. S., et al. (2009). Characterization of a naturally occurring breast cancer subset enriched in epithelial-to-mesenchymal transition and stem cell characteristics. *Cancer Research*, *69*, 4116–4124.

Hennighausen, L., & Robinson, G. W. (2005). Information networks in the mammary gland. *Nature Reviews. Molecular Cell Biology*, *6*, 715–725.

Hinck, L., & Silberstein, G. B. (2005). Key stages in mammary gland development: The mammary end bud as a motile organ. *Breast Cancer Research*, *7*, 245–251.

Hogg, N. A., Harrison, C. J., & Tickle, C. (1983). Lumen formation in the developing mouse mammary gland. *Journal of Embryology and Experimental Morphology*, *73*, 39–57.

Hu, M., Yao, J., Carroll, D. K., Weremowicz, S., Chen, H., Carrasco, D., et al. (2008). Regulation of in situ to invasive breast carcinoma transition. *Cancer Cell*, *13*, 394–406.

Huebner, R. J., & Ewald, A. J. (2014). Cellular foundations of mammary tubulogenesis. *Seminars in Cell and Developmental Biology*, *31*, 124–131.

Huebner, R. J., Lechler, T., & Ewald, A. J. (2014). Developmental stratification of the mammary epithelium occurs through symmetry-breaking vertical divisions of apically positioned luminal cells. *Development*, *141*, 1085–1094.

Hunt, N. C., Douglas-Jones, A. G., Jasani, B., Morgan, J. M., & Pignatelli, M. (1997). Loss of E-cadherin expression associated with lymph node metastases in small breast carcinomas. *Virchows Archiv*, *430*, 285–289.

Imbert, A., Eelkema, R., Jordan, S., Feiner, H., & Cowin, P. (2001). Delta N89 beta-catenin induces precocious development, differentiation, and neoplasia in mammary gland. *The Journal of Cell Biology*, *153*, 555–568.

Jia, L., Liu, F., Hansen, S. H., Ter Beest, M. B., & Zegers, M. M. (2011). Distinct roles of cadherin-6 and E-cadherin in tubulogenesis and lumen formation. *Molecular Biology of the Cell*, *22*(12), 2031–2041.

Kardash, E., Reichman-Fried, M., Maitre, J. L., Boldajipour, B., Papusheva, E., Messerschmidt, E. M., et al. (2010). A role for Rho GTPases and cell-cell adhesion in single-cell motility in vivo. *Nature Cell Biology*, *12*, 47–53, sup pp. 1–11.

Kher, R., Sha, E. C., Escobar, M. R., Andreoli, E. M., Wang, P., Xu, W. M., et al. (2011). Ectopic expression of cadherin 8 is sufficient to cause cyst formation in a novel 3D collagen matrix renal tubule culture. *American Journal of Physiology. Cell Physiology*, *301*, C99–C105.

Kleer, C. G., van Golen, K. L., Braun, T., & Merajver, S. D. (2001). Persistent E-cadherin expression in inflammatory breast cancer. *Modern Pathology*, *14*, 458–464.

Knudsen, K. A., Sauer, C., Johnson, K. R., & Wheelock, M. J. (2005). Effect of N-cadherin misexpression by the mammary epithelium in mice. *Journal of Cellular Biochemistry*, *95*, 1093–1107.

Knudsen, K. A., & Wheelock, M. J. (2005). Cadherins and the mammary gland. *Journal of Cellular Biochemistry*, *95*, 488–496.

Kotb, A. M., Hierholzer, A., & Kemler, R. (2011). Replacement of E-cadherin by N-cadherin in the mammary gland leads to fibrocystic changes and tumor formation. *Breast Cancer Research*, *13*, R104.

Kowalczyk, A. P., & Reynolds, A. B. (2004). Protecting your tail: Regulation of cadherin degradation by p120-catenin. *Current Opinion in Cell Biology*, *16*, 522–527.

Kowalski, P. J., Rubin, M. A., & Kleer, C. G. (2003). E-cadherin expression in primary carcinomas of the breast and its distant metastases. *Breast Cancer Research*, *5*, R217–R222.

Kurley, S. J., Bierie, B., Carnahan, R. H., Lobdell, N. A., Davis, M. A., Hofmann, I., et al. (2012). p120-catenin is essential for terminal end bud function and mammary morphogenesis. *Development*, *139*, 1754–1764.

Larue, L., Ohsugi, M., Hirchenhain, J., & Kemler, R. (1994). E-cadherin null mutant embryos fail to form a trophectoderm epithelium. *Proceedings of the National academy of Sciences of the United States of America*, *91*, 8263–8267.

Leary, R. J., Lin, J. C., Cummins, J., Boca, S., Wood, L. D., Parsons, D. W., et al. (2008). Integrated analysis of homozygous deletions, focal amplifications, and sequence alterations in breast and colorectal cancers. *Proceedings of the National Academy of Sciences of the United States of America*, *105*, 16224–16229.

Lewis-Tuffin, L. J., Rodriguez, F., Giannini, C., Scheithauer, B., Necela, B. M., Sarkaria, J. N., et al. (2010). Misregulated E-cadherin expression associated with an aggressive brain tumor phenotype. *PLoS One*, *5*, e13665.

Lipponen, P., Saarelainen, E., Ji, H., Aaltomaa, S., & Syrjanen, K. (1994). Expression of E-cadherin (E-CD) as related to other prognostic factors and survival in breast cancer. *The Journal of Pathology*, *174*, 101–109.

Man, Y. G., Tai, L., Barner, R., Vang, R., Saenger, J. S., Shekitka, K. M., et al. (2003). Cell clusters overlying focally disrupted mammary myoepithelial cell layers and adjacent cells within the same duct display different immunohistochemical and genetic features: Implications for tumor progression and invasion. *Breast Cancer Research*, *5*, R231–R241.

Martin, T. A., Goyal, A., Watkins, G., & Jiang, W. G. (2005). Expression of the transcription factors snail, slug, and twist and their clinical significance in human breast cancer. *Annals of Surgical Oncology*, *12*, 488–496.

McNally, S., & Martin, F. (2011). Molecular regulators of pubertal mammary gland development. *Annals of Medicine*, *43*, 212–234.

Mironchik, Y., Winnard, P. T., Vesuna, F., Kato, Y., Wildes, F., Pathak, A. P., et al. (2005). Twist overexpression induces in vivo angiogenesis and correlates with chromosomal instability in breast cancer. *Cancer Research*, *65*, 10801–10809.

Moll, R., Mitze, M., Frixen, U. H., & Birchmeier, W. (1993). Differential loss of E-cadherin expression in infiltrating ductal and lobular breast carcinomas. *The American Journal of Pathology*, *143*, 1731–1742.

Muschler, J., & Streuli, C. H. (2010). Cell-matrix interactions in mammary gland development and breast cancer. *Cold Spring Harbor Perspectives in Biology*, *2*, a003202.

Nanba, D., Nakanishi, Y., & Hieda, Y. (2001). Changes in adhesive properties of epithelial cells during early morphogenesis of the mammary gland. *Development, Growth & Differentiation*, *43*, 535–544.

Nelson, W. J. (2003). Epithelial cell polarity from the outside looking in. *News in Physiological Sciences*, *18*, 143–146.

Nemade, R. V., Bierie, B., Nozawa, M., Bry, C., Smith, G. H., Vasioukhin, V., et al. (2004). Biogenesis and function of mouse mammary epithelium depends on the presence of functional alpha-catenin. *Mechanisms of Development*, *121*, 91–99.

Nguyen-Ngoc, K.-V., Cheung, K. J., Brenot, A., Shamir, E. R., Gray, R. S., Hines, W. C., et al. (2012). ECM microenvironment regulates collective migration and local dissemination in normal and malignant mammary epithelium. *Proceedings of the National Academy of Sciences of the United States of America*, *109*, E2595–E2604.

Nguyen-Ngoc, K. V., & Ewald, A. J. (2013). Mammary ductal elongation and myoepithelial migration are regulated by the composition of the extracellular matrix. *Journal of Microscopy*, *251*, 212–223.

Nieman, M. T., Prudoff, R. S., Johnson, K. R., & Wheelock, M. J. (1999). N-cadherin promotes motility in human breast cancer cells regardless of their E-cadherin expression. *The Journal of Cell Biology*, *147*, 631–644.

Niessen, C. M., & Gottardi, C. J. (2008). Molecular components of the adherens junction. *Biochimica et Biophysica Acta*, *1778*, 562–571.

Oka, H., Shiozaki, H., Kobayashi, K., Inoue, M., Tahara, H., Kobayashi, T., et al. (1993). Expression of E-cadherin cell adhesion molecules in human breast cancer tissues and its relationship to metastasis. *Cancer Research*, *53*, 1696–1701.

Paredes, J., Albergaria, A., Oliveira, J. T., Jerónimo, C., Milanezi, F., & Schmitt, F. C. (2005). P-cadherin overexpression is an indicator of clinical outcome in invasive breast carcinomas and is associated with CDH3 promoter hypomethylation. *Clinical Cancer Research*, *11*, 5869–5877.

Paredes, J., Figueiredo, J., Albergaria, A., Oliveira, P., Carvalho, J., Ribeiro, A. S., et al. (2012). Epithelial E- and P-cadherins: Role and clinical significance in cancer. *Biochimica et Biophysica Acta*, *1826*, 297–311.

Peinado, H., Olmeda, D., & Cano, A. (2007). Snail, Zeb and bHLH factors in tumour progression: An alliance against the epithelial phenotype? *Nature Reviews. Cancer*, *7*, 415–428.

Pitelka, D. R., Hamamoto, S. T., Duafala, J. G., & Nemanic, M. K. (1973). Cell contacts in the mouse mammary gland. I. Normal gland in postnatal development and the secretory cycle. *The Journal of Cell Biology*, *56*, 797–818.

Polyak, K. (2010). Molecular markers for the diagnosis and management of ductal carcinoma in situ. *Journal of the National Cancer Institute. Monographs*, *2010*, 210–213.

Polyak, K., & Hu, M. (2005). Do myoepithelial cells hold the key for breast tumor progression? *Journal of Mammary Gland Biology and Neoplasia*, *10*, 231–247.

Polyak, K., & Weinberg, R. A. (2009). Transitions between epithelial and mesenchymal states: Acquisition of malignant and stem cell traits. *Nature Reviews. Cancer, 9*, 265–273.
Radice, G. L., Ferreira-Cornwell, M. C., Robinson, S. D., Rayburn, H., Chodosh, L. A., Takeichi, M., et al. (1997). Precocious mammary gland development in P-cadherin-deficient mice. *The Journal of Cell Biology, 139*, 1025–1032.
Radice, G. L., Sauer, C. L., Kostetskii, I., Peralta Soler, A., & Knudsen, K. A. (2003). Inappropriate P-cadherin expression in the mouse mammary epithelium is compatible with normal mammary gland function. *Differentiation, 71*, 361–373.
Riethmacher, D., Brinkmann, V., & Birchmeier, C. (1995). A targeted mutation in the mouse E-cadherin gene results in defective preimplantation development. *Proceedings of the National Academy of Sciences of the United States of America, 92*, 855–859.
Robinson, G. W., & Hennighausen, L. (2011). MMTV-Cre transgenes can adversely affect lactation: Considerations for conditional gene deletion in mammary tissue. *Analytical Biochemistry, 412*, 92–95.
Rodriguez, F. J., Lewis-Tuffin, L. J., & Anastasiadis, P. Z. (2012). E-cadherin's dark side: Possible role in tumor progression. *Biochimica et Biophysica Acta, 1826*, 23–31.
Runswick, S. K., O'Hare, M. J., Jones, L., Streuli, C. H., & Garrod, D. R. (2001). Desmosomal adhesion regulates epithelial morphogenesis and cell positioning. *Nature Cell Biology, 3*, 823–830.
Schedin, P., & Keely, P. J. (2011). Mammary gland ECM remodeling, stiffness, and mechanosignaling in normal development and tumor progression. *Cold Spring Harbor Perspectives in Biology, 3*(1), a003228.
Shamir, E. R., Pappalardo, E., Jorgens, D. M., Coutinho, K., Tsai, W. T., Aziz, K., et al. (2014). Twist1-induced dissemination preserves epithelial identity and requires E-cadherin. *The Journal of Cell Biology, 204*, 839–856.
Siitonen, S. M., Kononen, J. T., Helin, H. J., Rantala, I. S., Holli, K. A., & Isola, J. J. (1996). Reduced E-cadherin expression is associated with invasiveness and unfavorable prognosis in breast cancer. *American Journal of Clinical Pathology, 105*, 394–402.
Silberstein, G. B., & Daniel, C. W. (1982). Glycosaminoglycans in the basal lamina and extracellular matrix of the developing mouse mammary duct. *Developmental Biology, 90*, 215–222.
Soler, A. P., Knudsen, K. A., Salazar, H., Han, A. C., & Keshgegian, A. A. (1999). P-cadherin expression in breast carcinoma indicates poor survival. *Cancer, 86*, 1263–1272.
Soler, A. P., Russo, J., Russo, I. H., & Knudsen, K. A. (2002). Soluble fragment of P-cadherin adhesion protein found in human milk. *Journal of Cellular Biochemistry, 85*, 180–184.
Sternlicht, M. D. (2006). Key stages in mammary gland development: The cues that regulate ductal branching morphogenesis. *Breast Cancer Research, 8*, 201.
Sternlicht, M. D., & Barsky, S. H. (1997). The myoepithelial defense: A host defense against cancer. *Medical Hypotheses, 48*, 37–46.
Sternlicht, M. D., Kouros-Mehr, H., Lu, P., & Werb, Z. (2006). Hormonal and local control of mammary branching morphogenesis. *Differentiation, 74*, 365–381.
Takeichi, M. (2014). Dynamic contacts: Rearranging adherens junctions to drive epithelial remodelling. *Nature Reviews. Molecular Cell Biology, 15*, 397–410.
Tan, D. S., Potts, H. W., Leong, A. C., Gillett, C. E., Skilton, D., Harris, W. H., et al. (1999). The biological and prognostic significance of cell polarity and E-cadherin in grade I infiltrating ductal carcinoma of the breast. *The Journal of Pathology, 189*, 20–27.
Tarin, D. (2005). The fallacy of epithelial mesenchymal transition in neoplasia. *Cancer Research, 65*, 5996–6001.
Taube, J. H., Herschkowitz, J. I., Komurov, K., Zhou, A. Y., Gupta, S., Yang, J., et al. (2010). Core epithelial-to-mesenchymal transition interactome gene-expression

signature is associated with claudin-low and metaplastic breast cancer subtypes. *Proceedings of the National Academy of Sciences of the United States of America, 107*, 15449–15454.
Tepera, S. B., McCrea, P. D., & Rosen, J. M. (2003). A beta-catenin survival signal is required for normal lobular development in the mammary gland. *Journal of Cell Science, 116*, 1137–1149.
Teuliere, J., Faraldo, M. M., Deugnier, M. A., Shtutman, M., Ben-Ze'ev, A., Thiery, J. P., et al. (2005). Targeted activation of beta-catenin signaling in basal mammary epithelial cells affects mammary development and leads to hyperplasia. *Development, 132*, 267–277.
Thiery, J. P. (2002). Epithelial–mesenchymal transitions in tumour progression. *Nature Reviews. Cancer, 2*, 442–454.
Thiery, J. P., Acloque, H., Huang, R. Y. J., & Nieto, M. A. (2009). Epithelial-mesenchymal transitions in development and disease. *Cell, 139*, 871–890.
Torres, M., Stoykova, A., Huber, O., Chowdhury, K., Bonaldo, P., Mansouri, A., et al. (1997). An alpha-E-catenin gene trap mutation defines its function in preimplantation development. *Proceedings of the National Academy of Sciences of the United States of America, 94*, 901–906.
Trimboli, A. J., Fukino, K., de Bruin, A., Wei, G., Shen, L., Tanner, S. M., et al. (2008). Direct evidence for epithelial-mesenchymal transitions in breast cancer. *Cancer Research, 68*, 937–945.
van Roy, F., & Berx, G. (2008). The cell-cell adhesion molecule E-cadherin. *Cellular and Molecular Life Sciences, 65*, 3756–3788.
Virchow, R. (1860). *Cellular pathology: As based upon physiological and pathological histology. Twenty lectures delivered in the Pathological institute of Berlin during the months of February, March and April, 1858* (F. Chance, Trans.). New York: R. M. De Witt.
Vleminckx, K., Vakaet, L., Jr., Mareel, M., Fiers, W., & van Roy, F. (1991). Genetic manipulation of E-cadherin expression by epithelial tumor cells reveals an invasion suppressor role. *Cell, 66*, 107–119.
Vos, C. B., Cleton-Jansen, A. M., Berx, G., de Leeuw, W. J., ter Haar, N. T., van Roy, F., et al. (1997). E-cadherin inactivation in lobular carcinoma in situ of the breast: An early event in tumorigenesis. *British Journal of Cancer, 76*, 1131–1133.
Wagner, K. U., McAllister, K., Ward, T., Davis, B., Wiseman, R., & Hennighausen, L. (2001). Spatial and temporal expression of the Cre gene under the control of the MMTV-LTR in different lines of transgenic mice. *Transgenic Research, 10*, 545–553.
Wagner, K. U., Wall, R. J., St-Onge, L., Gruss, P., Wynshaw-Boris, A., Garrett, L., et al. (1997). Cre-mediated gene deletion in the mammary gland. *Nucleic Acids Research, 25*, 4323–4330.
Wheelock, M. J., Shintani, Y., Maeda, M., Fukumoto, Y., & Johnson, K. R. (2008). Cadherin switching. *Journal of Cell Science, 121*, 727–735.
Yamada, S., & Nelson, W. J. (2007). Localized zones of Rho and Rac activities drive initiation and expansion of epithelial cell-cell adhesion. *The Journal of Cell Biology, 178*, 517–527.
Yang, J., Mani, S. A., Donaher, J. L., Ramaswamy, S., Itzykson, R. A., Come, C., et al. (2004). Twist, a master regulator of morphogenesis, plays an essential role in tumor metastasis. *Cell, 117*, 927–939.
Yang, J., & Weinberg, R. A. (2008). Epithelial-mesenchymal transition: At the crossroads of development and tumor metastasis. *Developmental Cell, 14*, 818–829.
Yuan, T. C., Wang, Y. P., Pao, L., Anderson, S. M., & Gu, H. H. (2011). Lactation defect in a widely used MMTV-Cre transgenic line of mice. *PLoS One, 6*, e19233.

CHAPTER TWELVE

Cell Adhesion in Epidermal Development and Barrier Formation

Kaelyn D. Sumigray[*,†], **Terry Lechler**[*,†,1]
[*]Department of Dermatology, Duke University Medical Center, Durham, North Carolina, USA
[†]Department of Cell Biology, Duke University Medical Center, Durham, North Carolina, USA
[1]Corresponding author: e-mail address: terry.lechler@duke.edu

Contents

Abstract

Cell–cell adhesions are necessary for structural integrity and barrier formation of the epidermis. Here, we discuss insights from genetic and cell biological studies into the roles of individual cell–cell junctions and their composite proteins in regulating epidermal development and function. In addition to individual adhesive functions, we will discuss emerging ideas on mechanosensation/transduction of junctions in the epidermis, noncanonical roles for adhesion proteins, and crosstalk/interdependencies between the junctional systems. These studies have revealed that cell adhesion proteins are connected to many aspects of tissue physiology including growth control, differentiation, and inflammation.

Current Topics in Developmental Biology, Volume 112
ISSN 0070-2153
http://dx.doi.org/10.1016/bs.ctdb.2014.11.027

1. INTRODUCTION

Cell–cell adhesion is essential for the formation and maintenance of multicellular tissues, particularly epithelia. Adhesion is provided by specialized cell–cell junctions, primarily adherens junctions (AJ), desmosomes, and tight junctions. In addition to forming physical connections between cells, these junctions organize and regulate cytoskeletal elements and modulate signaling pathways to regulate tissue development, structure, and physiology. Although cell adhesion has been extensively studied in cultured cells, genetic approaches have revealed the underlying physiologic functions of individual components and unexpected noncanonical roles of junctional proteins. Junctions do not function independently and it is becoming increasingly apparent that junctional crosstalk, interdependencies, and compensation are necessary for tissue robustness.

The mouse epidermis is a prime model to study the roles of cell–cell junctions in tissue architecture and physiology. The epidermis is a stratified epithelium that performs several essential protective functions. It is specialized to protect from water loss, dehydration, and toxin entry into the body. To accomplish this, the epidermis must form and maintain a tight barrier between the organism and its environment and withstand large amounts of mechanical stress on a daily basis. Essential to this barrier function is the proper establishment of cell–cell adhesion. Here, we discuss the roles for cell–cell adhesion in epidermal development and barrier function. Rather than a comprehensive review of the field, we have selected various new and under-discussed aspects of epidermal adhesion in addition to a basic description of roles of cell adhesion molecules in the epidermis elucidated by genetic studies.

2. DEVELOPMENT OF THE STRATIFIED EPIDERMIS

The mouse epidermis is derived from surface ectoderm positioned atop a basement membrane that commits itself to an epidermal cell fate around embryonic day 9.5 (e9.5). Expression of the transcription factor p63, a master regulator of epidermal commitment, is required for the conversion from keratin 8/18-positive ectoderm to keratin 5/14-positive epidermis (Byrne, Tainsky, & Fuchs, 1994; Mills et al., 1999; Pellegrini et al., 2001; Yang et al., 1999). This newly committed layer of cells becomes the basal layer of the epidermis. Around e13.5, the epidermis begins to stratify in an anterior to posterior wave across the body. As additional cell layers form, they begin a terminal differentiation program. Cells progress through the spinous and granular layers before finally dying to produce cornified envelopes. The cornified envelopes are composed of highly cross-linked lipids and proteins that seal the epidermis to create the outside–in barrier (Steven & Steinert, 1994).

Cell–cell adhesions—AJs, desmosomes, and tight junctions—have distinct localization patterns in the epidermis. Basal cells are polarized and have AJs and desmosomes along their lateral and apical membranes. This is distinct from simple epithelial cells, which have nonadhesive apical surfaces. In suprabasal cells, AJs and desmosomes are found on the entire cell surface. Desmosome density and composition markedly change as cells differentiate. In contrast, tight junctions only assemble in granular cells and are not found surrounding the entire cell, but exist in a planar polygonal network (Furuse et al., 2002; Morita et al., 1998; Schluter, Wepf, Moll, & Franke, 2004).

Both human mutations and genetic ablation studies in mice have demonstrated roles for all three adhesive structures in epidermal function. Each of the junctions has canonical roles, but additional functions are now beginning to be appreciated. Tight junctions provide the inside–out barrier essential to prevent water loss, desmosomes provide mechanical strength, and AJs coordinate many diverse aspects of epidermal physiology. It is also becoming increasingly clear that not only does each cell–cell junction have its own independent functions, but also that there is crosstalk and interplay between the various junctional complexes that is important to give the epidermis full barrier activity (Godsel et al., 2010; Lewis et al., 1997; Sumigray, Foote, & Lechler, 2012; Tunggal et al., 2005).

2.1. Periderm function—An antiadhesive?

While cell adhesion is essential for epidermal function, adhesions in the wrong context can be detrimental, and mechanisms have evolved to prevent inappropriate adhesions between epidermal sheets. Simple epithelia have a nonadhesive apical domain that is separated from adhesive lateral domains, thus preventing epithelial sheet fusion. However, suprabasal epidermal cells are able to contact other cells on all surfaces. This could create a problem during development if suprabasal surface cells on apposed epidermal sheets fused. To prevent this, the developing epidermis generates a superficial epithelial cell layer that covers it—the periderm. Periderm prevents pathological epithelial adhesions during embryogenesis. The periderm is derived

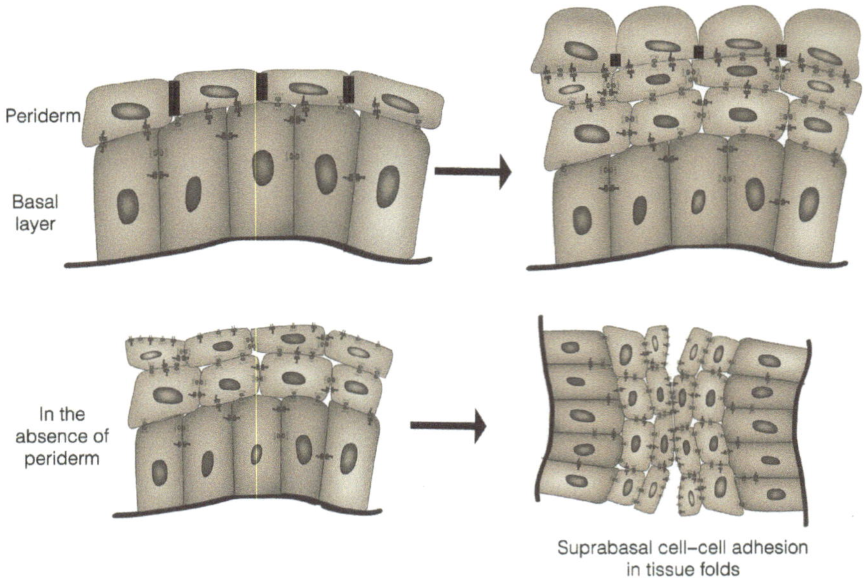

Figure 1 Periderm function in the embryo. The periderm is found on top of the embryonic epidermis where it forms tight junctions. In the absence of periderm, suprabasal epithelial cells are competent to interact with other cells, resulting in epidermal sheet fusions. (See the color plate.)

from the committed single-layered epidermis, but is molecularly distinct. It is first seen at around e11 in the mouse, and it is sloughed off when the cornified layer forms around e17 (M'Boneko & Merker, 1988). The periderm is essentially a simple epithelium with tight junctions that sits on top of the epidermis and prevents fusions. Classic cytological studies predicted a non-adhesive function for the periderm (Maconnachie, 1979; M'Boneko & Merker, 1988). This was recently confirmed by both genetic and toxin-induced loss of the periderm, which resulted in oral and digit epithelium fusions (Richardson et al., 2014), demonstrating that this layer is critical for proper development (see Fig. 1).

3. ADHERENS JUNCTIONS

AJs are cadherin-based junctions that link the F-actin cytoskeleton to the plasma membrane. Transmembrane cadherins, most prominently E-cadherin in the epidermis, form the physical linkages between cells through both lateral dimerization on the same cell and adhesion dimerization between E-cadherin molecules on adjacent cells (Brieher, Yap, & Gumbiner, 1996;

Ozawa, 2002; Tomschy, Fauser, Landwehr, & Engel, 1996). The E-cadherin cytoplasmic tail binds the Armadillo repeat-containing proteins β-catenin and p120-catenin. β-Catenin binds α-catenin, which links the F-actin cytoskeleton to the junction. Deletion of some individual AJ components in the epidermis results in cell adhesion defects, which will be described below. Interestingly, deletion of catenins also results in phenotypes that revealed unexpected roles for these junctional proteins in cell signaling regulation. The signaling roles for these proteins will be described later in the review.

3.1. Epidermal adhesion defects upon loss of AJ proteins

E-cadherin is the primary cadherin expressed in the interfollicular epidermis, though P-cadherin is also expressed, most prominently in hair follicles (Jamora, DasGupta, Kocieniewski, & Fuchs, 2003; Muller-Rover et al., 1999). Loss of E-cadherin in the epidermis has variable effects depending on mouse strain background and time of gene ablation in the epidermis. Phenotypes range from epidermal hyperproliferation (in aging mice) and defective hair follicles to embryonic lethality due to tight-junction defects (Tinkle, Lechler, Pasolli, & Fuchs, 2004; Tunggal et al., 2005; Young et al., 2003). P-cadherin is upregulated upon E-cadherin deletion in many epidermal cell types, suggesting that P-cadherin can compensate for many, but not all, of E-cadherin's functions. An exception is the hair follicle's inner root sheath cells, which do not upregulate P-cadherin, resulting in degeneration (Tinkle et al., 2004). The partial functional substitution of P-cadherin for E-cadherin was clearly demonstrated in mice with loss of both E- and P-cadherin (Tinkle, Pasolli, Stokes, & Fuchs, 2008). This resulted in neonatal lethality associated with barrier defects and some epidermal microblistering, demonstrating that expression of classical cadherins is crucial for epidermal development and integrity.

Loss of individual catenins results in distinct phenotypes. Both in cultured cells and intact epidermis, p120-catenin regulates cortical cadherin levels (Davis, Ireton, & Reynolds, 2003; Ireton et al., 2002; Thoreson et al., 2000; Xiao et al., 2003; Yap, Niessen, & Gumbiner, 1998). p120-Catenin stabilizes cadherins by binding to their cytoplasmic tails, where it blocks a conserved endocytic signal in cadherin (Nanes et al., 2012). Surprisingly, however, epidermal loss of p120-catenin does not result in overt adhesion or barrier defects, despite a significant decrease in cortical and total levels of AJ proteins (Perez-Moreno et al., 2006). Thus, p120-catenin is not required for proper epidermal development, and its loss reveals that the epidermis tolerates substantial decreases in AJ levels. However, the p120-catenin knockout mice do develop increased inflammatory responses in adulthood, as discussed later.

Initial studies on epidermal-specific β-catenin loss of function mice suggested it did not have an essential role in the epidermis. While viable, these mice have profound defects in hair follicle formation (Huelsken, Vogel, Erdmann, Cotsarelis, & Birchmeier, 2001). β-Catenin's role in hair follicle formation is now attributed to its role in the Wnt-signaling pathway, which will not be discussed here, but has been reviewed elsewhere (Lee & Tumbar, 2012; Millar, 2002). It has been hypothesized that the β-catenin paralog plakoglobin can functionally substitute for β-catenin loss at the AJ. In support of this, plakoglobin incorporates into AJs in heart muscle upon β-catenin ablation (Huelsken et al., 2001; Ray, Foote, & Lechler, 2013). More recent evidence suggests that β-catenin's function in Wnt signaling is important for interfollicular epidermis proliferation and that plakoglobin can rescue some, but not all, of β-catenin's AJ functions (discussed in more detail in Section 3.3; Choi et al., 2013; Lim et al., 2013; Ray et al., 2013).

α-Catenin does not directly bind to E-cadherin, but is incorporated into the AJ through interactions with β-catenin (Aberle et al., 1994). In the epidermis, loss of α-catenin results in severe phenotypes; regions of the skin become completely denuded, while others peel. Additionally, loss of α-catenin induces cell dissociation, massive hyperproliferation, and loss of epidermal polarity (Kobielak & Fuchs, 2006; Lim et al., 2013; Vasioukhin, Bauer, Degenstein, Wise, & Fuchs, 2001). While some of these defects are also seen in E/P-cadherin-null mice, the phenotype is quite distinct. The hyperproliferation, in particular, is unique to loss of α-catenin and has been the subject of much study, as discussed below.

Taken together, these studies suggest that there is not a consistent phenotype associated with loss of AJ components in the skin. While combined loss of E/P-cadherin resulted in cell–cell separations, this is not seen under conditions where most, but not all of the AJs are lost, and the phenotype depends on individual AJ component levels and localization. This raises an important point as to what assays are relevant for analyzing AJ function in the epidermis. Clearly, changes in protein levels are an easy readout, but large changes do not necessarily correlate with perturbed adhesion. Similarly, AJ turnover can be used to look at stability (Delva & Kowalczyk, 2009; Foote, Sumigray, & Lechler, 2013; Georgiou, Marinari, Burden, & Baum, 2008), but this suffers from the same problems. As discussed later, tight-junction defects appear to be a more sensitive readout of AJ function, but there can be many other causes of these defects. *In vitro* mechanical assays, which have not been used to examine mutant skin to date, are functional (Huen et al., 2002), and could reveal how significantly AJs contribute to epidermal mechanics. However, assays to examine mechanical robustness need to be developed and applied to mutant skin to more precisely define the contributions of AJs in the epidermis.

3.2. AJs link actin to the cell cortex

The canonical function of the AJ is to form a link between cells and to the underlying F-actin cytoskeleton. However, how local F-actin is assembled, organized and how it dynamically associates with AJs remains an area of active experimentation (Brieher & Yap, 2013).

In cultured cells, significant evidence suggests that the actin-nucleating Arp2/3 complex mediates F-actin assembly at the AJ (Tang & Brieher, 2012; Verma et al., 2012). However, epidermal-specific loss of an Arp2/3 complex component did not result in detectable AJ defects (Zhou et al., 2013). Cortical F-actin was still present, as were AJ proteins—even the turnover of AJs at the cortex remained unaltered. Previous work in cultured keratinocytes suggested that formin family members may be the relevant nucleators, though this has not yet been tested in intact tissue (Kobielak, Pasolli, & Fuchs, 2004). It is surprising that two very different nucleators (one branched, one straight) could accomplish the same function. Two possible explanations are that different types of AJs (for example, spot AJs vs. zonula adherens (ZA)) have different F-actin organization, and therefore, different nucleators. An alternative is that AJs can reorganize existing F-actin networks to meet their needs, and both Arp2/3 and formin networks can be utilized.

The simplest model for how AJs link actin to the cortex is directly through α-catenin, as it can bind both F-actin and β-catenin. However, α-catenin is unable to simultaneously bind both these proteins in solution (Drees, Pokutta, Yamada, Nelson, & Weis, 2005; Yamada, Pokutta, Drees, Weis, & Nelson, 2005). α-Catenin can recruit a number of F-actin-binding proteins to the AJ, including vinculin, α-actinin, EPLIN, and afadin (Abe & Takeichi, 2008; le Duc et al., 2010; Rangarajan & Izard, 2012; Sawyer, Harris, Slep, Gaul, & Peifer, 2009). Therefore, it can act as an indirect linker. That said, recent work has suggested that α-catenin exists in different conformations, including ones that may be promoted by tension across the molecule (Drees et al., 2005; Rangarajan & Izard, 2012; Yonemura, Wada, Watanabe, Nagafuchi, & Shibata, 2010). Therefore, under cellular conditions of tension, it is possible that α-catenin can also directly link AJs to F-actin. To date, there is little genetic or cell biological evidence in the epidermis that suggests which of these linkages is most important during development or homeostasis.

3.3. AJs sense and respond to tension

AJs are mechanosensitive and responsive structures. The model that has emerged from studies on cultured cells is that forces on AJs, whether externally applied or generated by myosin II contractility, result in conformational changes in α-catenin that allow it to bind to additional F-actin-binding proteins and perhaps F-actin itself (Ishiyama et al., 2013; Yao et al., 2014; Yonemura et al., 2010). Most notably, vinculin's AJ localization has been demonstrated to be tension dependent in a number of cell lines (le Duc et al., 2010; Miyake et al., 2006; Sumida, Tomita, Shih, & Yamada, 2011; Yonemura et al., 2010). A recent report directly demonstrated that a 5-pN force across α-catenin was sufficient to induce a conformational change to allow vinculin binding (Yao et al., 2014). In this review, we will refer to this tension-induced state as engaged AJs. It is important to point out that while current studies have focused on tension in regulating this AJ state, it is possible that posttranslational modifications could result in a similar change. Also, at present there is no concrete evidence in tissue for these different states of AJs. Below, we will discuss work in keratinocytes and mouse epidermis that suggests a role for AJ engagement in skin physiology.

In polarized columnar simple epithelia, AJs are enriched at the subapical surface, where they form a continuous cortical belt, termed the ZA, which is thought to be under constitutive tension. The ZA lies just basal to the tight junctions and links multiple cells to the circumferential actin belt. In addition to AJ proteins, ZAs contain other peripherally associated proteins, including nonmuscle myosin II, vinculin, α-actinin, and the actin nucleator Arp2/3 (Drenckhahn & Franz, 1986; Vasioukhin, Bauer, Yin, & Fuchs, 2000; Verma et al., 2012).

The epidermis, in contrast, does not have a true ZA. In suprabasal epidermal cells, cell–cell junctions localize to all cell–cell interfaces. However, there is evidence that AJs in the granular layer of the epidermis may exist in an engaged state, similar to what has been shown in cultured cells under tension. In cultured keratinocytes, AJs can be converted into an engaged state through microtubule stabilization, which mimics microtubule organization in the granular layer of the epidermis. These junctions have increased cortical levels of myosin IIA and β-catenin, decreased α-catenin turnover, and exposure of a tension-sensitive epitope of α-catenin epitope (Sumigray et al., 2012). These changes are lost when myosin II is pharmacologically or genetically inhibited. One interpretation of these data is that AJs in the granular layer are under increased tension and that this might be important for the mechanical properties of the epidermis. Indeed, cultured epidermal cells show increased mechanical resilience under conditions where their AJs exist in the engaged state (Sumigray et al., 2012). Whether this change occurs in intact tissue has not yet been addressed, and developing tools to define the tension status of AJs *in vivo* is an important future goal.

There is also evidence that AJ engagement might potentiate tight-junction activity. Tight junctions specifically form in the granular layer of the epidermis, where AJs are expected to be under tension, or at least in a distinct state similar to AJs under tension. In support of the idea that myosin II-induced AJ engagement is important in the epidermis, loss of either E-cadherin or myosin II A/B resulted in similar tight-junction defects (Sumigray et al., 2012; Tunggal et al., 2005). However, it is possible that myosin II acts directly on TJs.

There are also focal tight-junction localization defects in β-catenin-null epidermis. In addition, mechanical perturbation of cultured β-catenin-null keratinocytes resulted in tight-junction defects not seen in wild-type cells (Ray et al., 2013). Cells devoid of β-catenin are unable to engage their AJs. This has been attributed to a failure to recruit vinculin to the junctions in the absence of β-catenin. Therefore, in keratinocytes, β-catenin appears to recruit a local pool of vinculin than can be used to engage AJs under tension. A similar function for β-catenin has been reported in mammary gland-derived MCF10a cells (Peng, Cuff, Lawton, & DeMali, 2010). However, other cell types appear to have low levels of vinculin under resting conditions (Huveneers et al., 2012; le Duc et al., 2010; Twiss et al., 2012; Yonemura et al., 2010). Overall, these data are consistent with a model in which AJ engagement in granular cells is necessary to increase strength of the epidermis and barrier activity of tight junctions.

3.4. AJs in hair follicle morphogenesis

In addition to the interfollicular epidermis, the skin has epidermal appendages, such as hair follicles, sebaceous glands, and sweat glands. Hair is initially specified as thickened regions of the basal epidermal layer, called placodes.

These placodes proliferate and grow into the underlying dermis to eventually form a complex mini-organ with many cell types. A major function of this structure is to produce the hair shaft and to provide it with mechanical scaffolding.

E-cadherin, which is present throughout the basal layer, is downregulated in placodes, while P-cadherin is upregulated (Jamora et al., 2003; Muller-Rover et al., 1999). P-cadherin remains exclusively expressed in the inner matrix of the mature hair, a proliferative compartment that gives rise to differentiated cells that make and surround the hair follicle. While loss of P-cadherin does not cause notable hair follicle defects in mice, it has significant effects in humans (Indelman et al., 2002; Shimomura, Wajid, Shapiro, & Christiano, 2008; Tinkle et al., 2008). A number of families have been identified with germline mutations in the P-cadherin gene. This results in a decrease in hair follicles (hypotrichosis) in addition to eye and limb defects. The mechanism of P-cadherin function is still not clear, and it may act both through AJs and through signaling pathways.

Cadherin switching, which occurs at the placode, is also seen in other cellular contexts. Most notably, E- to N-cadherin switches often correlate with tumorigenesis and invasion (Cavallaro, Schaffhauser, & Christofori, 2002; Christofori, 2003). In the case of the hair follicle, external signals converge on placode cells to cause E-cadherin downregulation through Wnt and Bmp pathways (Gat, DasGupta, Degenstein, & Fuchs, 1998; Jamora et al., 2003). Forced expression of E-cadherin inhibits hair follicle downgrowth (Jamora et al., 2003), though it is still not clear whether this is a gain of function effect or whether maintaining endogenous E-cadherin levels would also prevent invagination.

E-cadherin is clearly required for mouse hair follicle maintenance. Mice with conditional loss of E-cadherin have sparse hair due to degeneration of the differentiated hair follicle cell layers that do not express P-cadherin (Tinkle et al., 2004; Young et al., 2003). It is assumed that this is due to an adhesive role for E-cadherin as these cells are postmitotic and have not been demonstrated to have active Wnt signaling.

In the absence of β-catenin, placodes and hair follicles cannot be specified due to lack of Wnt signaling (Huelsken et al., 2001). In contrast, unregulated β-catenin activation is sufficient to induce ectopic hair growth and hair follicle tumors (Gat et al., 1998). Whether β-catenin's AJ function is also required for hair follicle morphogenesis has been obscured by its signaling role.

3.5. Noncanonical roles for catenins

Perhaps the most surprising findings from genetic analysis of catenins are that they have unexpected phenotypes that were not predicted if catenins had purely adhesive functions. p120-Catenin stabilizes E-cadherin at the membrane and protects it from being endocytosed (Davis et al., 2003; Ishiyama et al., 2010; Miyashita & Ozawa, 2007; Nanes et al., 2012). However, as mentioned, it is not essential for epidermal development. Strikingly though, adult p120-catenin-null epidermis is hyperproliferative and thickened due to chronic inflammation (Perez-Moreno et al., 2006). This is due to a cell autonomous increase in NF-κB activity in epidermal cells. While the underlying mechanism is still not clear, p120-catenin-null epidermis has increased levels of active RhoA (Perez-Moreno et al., 2006), suggesting that p120-catenin negatively regulates RhoA signaling. Increased Rho activity is sufficient to induce NF-κB nuclear translocation in wild-type cells. These data suggest that p120-catenin mediates cell contractility through RhoA, leading to NF-κB activation.

Loss of α-catenin results in a number of epidermal phenotypes, including loss of polarity and hyperproliferation, leading to neonatal lethality (Vasioukhin, Bauer, et al., 2001). However, when mutant skin was grafted onto wild-type mice, tumors closely resembling squamous cell carcinoma developed, demonstrating that α-catenin acts as a tumor suppressor (Kobielak & Fuchs, 2006). In support of this, α-catenin levels were decreased in a panel of human squamous cell carcinomas, suggesting that α-catenin acts similarly in humans and mice. A number of signaling pathways and processes are now known to be disrupted in α-catenin knockout tissue. These include changes in insulin receptor substrate/MAPK signaling, spindle orientation, NF-kB, and YAP1 activity (Kobielak & Fuchs, 2006; Lechler & Fuchs, 2005; Schlegelmilch et al., 2011; Vasioukhin, Bauer, et al., 2001). This last pathway is especially interesting, as forced YAP1 activation is sufficient to drive tumor formation in the epidermis (Schlegelmilch et al., 2011). While YAP1 has clear roles in the Hippo signaling pathway, it also responds to changes in a cell's mechanical status. For example, changes in cell spreading, substrate rigidity, and cell contractility can all alter YAP1 localization and activity (Dupont et al., 2011; Sansores-Garcia et al., 2011; Wada, Itoga, Okano, Yonemura, & Sasaki, 2011). The mechanism underlying YAP1 activation in α-catenin-null epidermis is still unclear. α-Catenin can directly interact with and recruit YAP1 to junctions, suggesting a simple sequestration function (Schlegelmilch et al., 2011). However, it is not clear that this is sufficient to prevent activation, and YAP1 may also be responding to changes in the mechanical status of the cell. Because α-catenin can bind F-actin and recruit additional F-actin binding proteins to the cortex, as well as locally control actin assembly by inhibiting the Arp2/3 complex (Drees et al., 2005), it is likely that cortical tension is also altered in this mutant.

4. DESMOSOMES

Like AJs, desmosomes form around transmembrane cadherin molecules. There are two types of desmosomal cadherins, desmocollins and desmogleins. Mice and humans express three desmocollin genes and four desmoglein genes, which are expressed in cell-type specific manners. The cytoplasmic face of the desmosome contains two distinct electron-dense structures, called the outer-dense plaque and inner-dense plaque. The outer-dense plaque, in which plakoglobin and plakophilins bind the desmosomal cadherin cytoplasmic tails, is proximal to the cell membrane while the inner-dense plaque is the site at which desmoplakin's tail binds to keratins.

4.1. Desmosomal proteins in epidermal integrity and disease

Desmosomes impart the epidermis with the majority of its adhesive strength and resistance to mechanical stress. Unlike the genetics of AJs in which diverse phenotypes are seen, loss of desmosomal proteins results in a spectrum of epidermal integrity defects depending on the severity of disruption. This is most clear in humans, as a large number of mutations covering most desmosomal genes have been described. At the mild end of the spectrum mutations result in thickened skin and woolly hair, while severe blistering and lethality characterize the most severe cases. Excellent reviews on desmosomal mutations have been published (Kottke, Delva, & Kowalczyk, 2006; Lai Cheong, Wessagowit, & McGrath, 2005; Lai-Cheong, Arita, & McGrath, 2007; McGrath & Wessagowit, 2005). We will focus on mouse mutants and cell biological lessons that have been learned from them.

4.1.1 Desmoplakin

Work in cultured cells and with purified proteins demonstrated that desmoplakin forms the connection between the desmosome and the intermediate filament network (Stappenbeck et al., 1993; Stappenbeck & Green, 1992). This is supported by genetic data in humans and mice. In all cell types examined, when desmoplakin is lost, intermediate filament connections to the desmosome are also lost (Gallicano, Bauer, & Fuchs, 2001; Gallicano et al., 1998; Sumigray & Lechler, 2012; Vasioukhin, Bowers, Bauer, Degenstein, & Fuchs, 2001). This is an essential role in the epidermis as well as in the developing embryo (Gallicano et al., 2001, 1998; Vasioukhin, Bowers, et al., 2001). Epidermal-conditional knockout desmoplakin mice die shortly after birth due to barrier defects caused by blistering and peeling skin (Vasioukhin, Bowers, et al., 2001). Much of the peeling is caused by the trauma of birth, and the integrity of the epidermis is severely compromised. Thus, genetics bore out cell biological studies, though additional non-canonical functions of desmoplakin were revealed by these mice and will be discussed later.

Human mutations in the desmoplakin gene result in various diseases ranging in severity from mild hair defects to lethal widespread blistering (Lai Cheong et al., 2005; Lai-Cheong et al., 2007). The mutations fall throughout the length of the gene, with most mutations concentrated within its head domain, which targets it to the desmosome (Kowalczyk et al., 1997), and its C-terminal tail, which binds intermediate filaments (Green, Stappenbeck, Parry, & Virata, 1992; Stappenbeck et al., 1993; Stappenbeck & Green, 1992).

4.1.2 Plakoglobin

Plakoglobin localizes to the desmosome through direct interactions with desmosomal cadherins (Kowalczyk et al., 1997; Mathur, Goodwin, & Cowin, 1994; Troyanovsky et al., 1994). In turn, it binds directly to desmoplakin (Kowalczyk et al., 1997). Two mouse models for plakoglobin have been published—a complete null, which dies between e10.5 and birth (Bierkamp, McLaughlin, Schwarz, Huber, & Kemler, 1996), and an epidermal-conditional knockout, which is viable (Li, Zhang, Liu, Haneline, & Shou, 2012). While the former mirrors the phenotype of the desmoplakin mutant, the latter clearly does not. This may be due, in part, to incomplete recombination and loss of plakoglobin. However, even in cultured keratinocytes cloned from plakoglobin null skin, desmoplakin is still found at significant levels at desmosomes (Acehan et al., 2008). This is likely because plakophilins can also bind to desmoplakin and promote its association with the desmosome (Kowalczyk et al., 1999; Li et al., 2011). Although β-catenin can localize to the desmosome in the absence of plakoglobin by binding to the desmosomal cadherin cytoplasmic tails, it cannot recruit or maintain desmoplakin at the desmosomal plaque (Acehan et al., 2008).

4.1.3 Desmosomal cadherins

Dsg3-null mice exhibit severe oral lesions, causing decreased food intake and runting (Koch et al., 1997). Lesions cause acute local inflammation, and the suprabasal epidermal layers frequently detach from basal cells. Loss of desmocollin 1 in the mouse epidermis results in defects in cell–cell adhesion, but not overt blistering. However, as the mice age, these acantholysis-induced focal lesions progress to chronic dermatitis (Chidgey et al., 2001).

Human mutations in Dsg1 extracellular domains result in striate palmoplantar keratoderma, characterized by skin thickening on the palms of the hand and soles of the feet (Lai-Cheong et al., 2007). Human mutations in Dsg3, Dsc1, and Dsc3 have not been identified. Perhaps mutation in any of these is too severe for development to proceed. This would be consistent with the Dsc3 knockout embryo, which dies early in embryogenesis before implantation into the uterus (Den, Cheng, Merched-Sauvage, & Koch, 2006).

4.1.4 Pemphigus

In pemphigus vulgaris and pemphigus foliaceus, autoantibodies against desmoglein 3 and desmoglein 1 (Amagai & Stanley, 2012), respectively, recognize and bind to desmoglein and inhibit its adhesive function, resulting in skin blistering. The underlying mechanism is still under active research, and both steric hindrance and intracellular-signaling pathway activation are thought to contribute to the disease state (Sharma, Mao, & Payne, 2007). Desmoglein 3 is most highly expressed in the basal layers of the epidermis, while desmoglein 1 is expressed in the differentiated suprabasal layers (Kottke et al., 2006). Because of the differential expression patterns of the target desmogleins, pemphigus foliaceus affects the superficial layers of the epidermis, while pemphigus vulgaris affects the deep layers of the epidermis and oral mucosa to cause widespread blistering and lesions (Amagai, 2010).

Pathogenic pemphigus antibodies activate several signaling pathways, including protein kinase C (PKC), c-myc, RhoA, calmodulin, p38 MAPK, and tyrosine kinase pathways (Berkowitz et al., 2005, 2006; Mao, Sano, Park, & Payne, 2011; Sanchez-Carpintero et al., 2004; Seishima et al., 1999; Spindler & Waschke, 2011; Waschke et al., 2006; Williamson et al., 2006). How these autoantibodies activate signaling pathways is unclear. However, this serves as precedent for desmosomal control of signaling, and it will be important to determine how these same pathways are affected in genetic desmosomal disease and during physiological remodeling of these structures.

4.2. Desmosomes in hair follicle morphogenesis

Desmosomal components are downregulated in the follicle-initiating hair placode (Nanba, Hieda, & Nakanishi, 2000), presumably to allow for cell rearrangement and downgrowth during hair morphogenesis, though this has not yet been tested experimentally. Later, desmosomes are necessary to support the integrity of hair follicles. Individual desmoglein and desmocollin genes are specifically expressed in various hair follicle compartments (Kurzen et al., 1998), and many of these genes have functional roles in hair follicle integrity. Mutations in Dsg4 in both human and mouse lead to hair loss (Kljuic et al., 2003). This is associated with a defective differentiation program of the hair shaft. While other desmogleins have not been implicated in hair function in humans, loss of Dsg3 in mice also leads to a hair loss phenotype (Koch et al., 1998, 1997). In addition, all three desmocollin genes have roles in hair follicle integrity in either human or mouse. Dsc1 loss results in telogen hair loss in mice (Chidgey et al., 2001), Dsc2 mutations in humans leads to a woolly hair phenotype (Simpson et al., 2009), while Dsc3 mutations in both mouse and human result in hypotrichosis (Ayub et al., 2009; Chen, Den, & Koch, 2008).

Similarly, mutations in Pkp1, plakoglobin, and desmoplakin all lead to hair follicle defects in humans (reviewed in McGrath & Wessagowit, 2005). While there is a clear role for desmosomes in providing structural integrity to the hair follicle, it is not yet clear whether or how they regulate differentiation in this compartment.

4.3. Noncanonical roles for desmosomal proteins

4.3.1 Microtubule organization

In addition to its essential role in binding keratins, desmoplakin has an additional function in controlling microtubule reorganization to the cell cortex of suprabasal cells. It does this in part through the recruitment of a group of microtubule-associated proteins, including ninein, Ndel1, Lis1, and CLIP170 (Lechler & Fuchs, 2007; Sumigray, Chen, & Lechler, 2011; Wacker, Rickard, De Mey, & Kreis, 1992). Ninein, Ndel1, and Lis1 are all proteins that localize to the centrosome in fibroblasts and in basal keratinocytes. However, they relocalize to desmosomes upon epidermal differentiation. Trichoplein, initially identified as a desmosomal-associated protein, has also been shown to localize to centrosomes (Ibi et al., 2011; Nishizawa et al., 2005). Loss of Lis1 resulted in desmosome defects in the mouse epidermis, demonstrating that at least a subset of these proteins is important not only for microtubule organization, but also desmosome function (Sumigray et al., 2011). The role of cortical microtubules has been studied in cultured keratinocytes, but its role in intact tissue has not been addressed. In cultured cells, cortical microtubules promote the accumulation of myosin IIA at the cell cortex, the engagement of AJs (which results in epithelial sheet strengthening) and the increased function of tight junctions (Sumigray et al., 2012). These effects are consistent with the physiological roles of mechanical strength and barrier formation that is seen in the granular layer of the epidermis, where cortical microtubules are most robust.

4.3.2 Actin organization

A number of studies in cultured cells have demonstrated changes in F-actin organization upon loss of desmosomal proteins. For example, desmoplakin-null keratinocytes are unable to reorganize their F-actin cytoskeleton into mature cortical networks associated with AJs (Vasioukhin, Bowers, et al., 2001). Despite this, they have normal cortical levels of AJ components. These junctions were distinct from those in WT cells, however, as they are constitutively engaged—they have increased levels of associated myosin IIA and an exposed tension-sensitive epitope of α-catenin (Sumigray, Zhou, & Lechler, 2014). These effects are unlikely due to direct control of F-actin by desmoplakin, but rather a cellular response to desmosome disruption. In support of this, pathogenic pemphigus antibodies can induce similar changes in keratinocytes. Additionally, Dsg3 has been shown to regulate actin dynamics via Rac1 (Tsang et al., 2012). Whether this interaction functions in epidermal development is unknown.

Similarly, loss of Pkp2 resulted in defects in F-actin remodeling upon cell adhesion (Godsel et al., 2010). In this case, RhoA levels are elevated and cells appear more contractile. It is not clear whether Pkp2 can directly control RhoA activity or whether these responses are secondary to desmosome defects.

4.3.3 Signaling/transcription

The strongest evidence for roles of desmosomal proteins outside of adhesion come from both mouse mutant studies and cell culture work. Overexpression of Dsg3 in the suprabasal layers resulted in proliferation, differentiation, and morphological changes converting the interfollicular epidermis into something more similar to the oral mucosa, which normally maintains Dsg3 in suprabasal cells (Elias et al., 2001; Merritt et al., 2002). These results suggest that Dsg3 restriction to the basal layer is important for interfolliclular epidermal identity. Overexpression of Dsg2 in suprabasal epidermis caused a hyperproliferative phenotype with activation of many signaling pathways (Brennan et al., 2007). Desmoglein 1, which is more highly expressed in suprabasal cells, has been shown to play a role in keratinocyte differentiation by inhibiting EGFR signaling and ERK activation (Getsios et al., 2009; Harmon et al., 2013). Thus, while desmosomal cadherins have clear structural roles, they are also involved in cell fate and differentiation through regulation of signaling pathways.

5. TIGHT JUNCTIONS

Unlike AJs and desmosomes, tight junctions are not cadherin-based adhesion structures. Rather, the predominant transmembrane protein components of these junctions are claudins and occludin. Both claudins and occludin have four transmembrane passes, though they are not structurally related beyond that. While occludin is not required for the formation of tight-junction strands, it has complex roles in their function. Claudins, in contrast, are necessary to form the strands of the tight junction (Furuse, Fujita, Hiiragi, Fujimoto, & Tsukita, 1998; Furuse, Sasaki, Fujimoto, & Tsukita, 1998). The claudin family consists of at least 24 members, each encoded by a separate gene. Claudins are expressed in a tissue-specific manner, and the particular combination of claudins within a tight junction alters its paracellular transport abilities and properties (Van Itallie & Anderson, 2004). In addition to generating strands, claudins recruit occludin to the tight junction (Furuse, Fujita, et al., 1998).

On the cytoplasmic side of the tight junction reside three MAGUK family members, Zonula occludens (ZO)-1, -2, and -3. The three ZO proteins can interact with each other, and bind several tight and AJ proteins, as well as F-actin (Fanning, Jameson, Jesaitis, & Anderson, 1998; Wittchen, Haskins, & Stevenson, 2000). Because the ZO proteins can interact with multiple-junctional proteins, they have been proposed to act as scaffolds for complexes associated with the actin cytoskeleton (Van Itallie et al., 2013).

Loss of all three ZO proteins in cultured cells resulted in complete disruption of tight junctions, as shown by loss of cortical staining of claudin and occludin (Umeda et al., 2006). Studies on these cells have shown that tight junctions (or specifically ZO proteins) are required for AJ maturation into belt-like ZAs. ZO proteins seem to regulate cellular responses to tension, though it remains unclear how. Previous studies have suggested that

ZO-1 and -2 are required for recruitment and incorporation of myosin II into the ZA (Yamazaki et al., 2008). In contrast, other studies have shown that loss of ZO-1 and -2 resulted in increased recruitment of myosin II to the ZA, apical actin reorganization, and distorted apical cell shape (Fanning, Van Itallie, & Anderson, 2012). These discrepancies could be due to different cell types used or different culture conditions as cells were cultured on different substrates. Cell morphology and behavior can dramatically change depending on extracellular matrix, and perhaps ZA composition and responses to changes in tension also change.

5.1. Tight junctions in the epidermis

Tight-junction-associated proteins are found in distinctive patterns throughout the epidermis. Some are distributed throughout all the cell layers, such as claudin 1 (Furuse et al., 2002), while others show a tight restriction to granular cells, such as occludin (Morita et al., 1998). Electron microscopic and functional analyses show tight junctions occurring only in the granular cells (Elias & Friend, 1975; Kitajima, Eguchi, Ohno, Mori, & Yaoita, 1983). While initially called maculae occludentes, which were thought to be immature junctions (Elias & Friend, 1975; Elias, McNutt, & Friend, 1977), more recent data suggests that these junctions form a complete network and are functionally active barriers (Brandner et al., 2002; Kirschner, Houdek, Fromm, Moll, & Brandner, 2010; Yuki et al., 2011). The precise localization of functional tight junctions has most clearly been made in human epidermis where a specific layer of granular cells appear to form an effective barrier (Brandner et al., 2002; Pummi et al., 2001; Yoshida et al., 2013). A single cell layer also appears to form tight junctions in the hair follicle. Both immunofluorescence and EM analysis has demonstrated that tight junctions form rather specifically in the Henle layer of the inner root sheath (Brandner, McIntyre, Kief, Wladykowski, & Moll, 2003; Langbein et al., 2002).

At present, we have little understanding of how tight junctions specifically form in the granular and Henle layers. There are likely transcriptional inputs, and a number of claudin genes show upregulation at the mRNA level upon calcium-induced differentiation of keratinocytes (Sen, Reuter, Webster, Zhu, & Khavari, 2010). These changes are consistent with observed changes in tight-junction activity during epidermal development (Celli et al., 2012). That said, claudin 1 is found throughout the epidermis and is clearly not sufficient to form a barrier in the basal and spinous layers

(Furuse et al., 2002). It is likely that the full complement of proteins required for function is only produced in the granular layer. In addition, the assembly and/or function of the tight junctions in the granular layer also require other cell–cell adhesion systems to be functional. For example, loss of E-cadherin results in tight-junction defects even though the tight-junction proteins are expressed. (Tunggal et al., 2005). There were some slight differences in TJ protein localization, which could underlie the defect. Similarly, localization of tight-junction proteins was defective in the footpads of β-catenin conditional null mice (Ray et al., 2013). Finally, loss of myosin IIA/B resulted in similar tight junction function defects with no significant change in the expression/localization of examined components (Sumigray et al., 2012). These data suggest that myosin II-dependent tension may also regulate tight-junction activity. In support of this idea, conditions that promote myosin II accumulation at cell contacts (resulting in engagement of AJs) also increase the transepithelial resistance of epidermal sheets—a function provided by tight junctions (Sumigray et al., 2012, 2014). Therefore, the specific placement of tight junctions is likely to be controlled by the convergence of multiple events—both transcriptional and cell biological (Furuse et al., 2002; Sumigray et al., 2014; Turksen & Troy, 2002).

5.2. Tight junctions in epidermal development and barrier formation

The first clear functional data for tight-junction function in the epidermis came from analysis of claudin 1 knockout mice. Loss of claudin 1 resulted in a defective barrier, and thus, neonatal death due to dehydration (Furuse et al., 2002). In contrast, loss of occludin caused complex phenotypes, but animals survive, suggesting that at least a partial epidermal barrier exists, though this has not been examined in detail (Saitou et al., 2000). The roles of ZO-1,-2, and -3 have also been examined genetically. While both ZO-1 and -2 are embryonic lethal, ZO-3 has no reported gross phenotype (Katsuno et al., 2008; Xu et al., 2008). As embryonic lethality precluded analysis of epidermal function, it is not yet clear whether these proteins have distinct functions in the epidermis or whether they all act redundantly.

5.3. Tight junctions in disease

Genetic studies of claudin 1 have highlighted the importance of tight junctions in barrier formation (Furuse et al., 2002). However the stratum corneum, the highly cross-linked lipid-rich impermeable outermost layer of the epidermis, also contributes to epidermal barrier function. Loss of claudin 1 not only affects tight-junction activity, but also cornified layer composition and morphology (Sugawara et al., 2013). These data, along with reports that expression of claudin 6 can induce epithelial differentiation (Sugimoto et al., 2013), suggest that claudins may also play auxiliary roles in regulating differentiation. This is consistent with their broad expression pattern in the epidermis.

Mutations in claudin 1 have been identified in patients suffering from a syndrome affecting the liver and skin, neonatal ichthyosis-sclerosing cholangitis (Hadj-Rabia et al., 2004). Ichthyotic skin is characterized as dry, thickened, and flaky. Mutations in genes involved in barrier function have also been identified as causing various ichthyosis-related conditions, including transglutaminase 1, filaggrin, fatty acid dehydrogenase, and members of the gap junction family of the connexins (Djalilian et al., 2006; Matsuki et al., 1998; Rizzo, 2014; Rizzo et al., 2010; Thyssen, Godoy-Gijon, & Elias, 2013). Thus, similar to mice, skin thickening is a response of human skin to barrier defects. While we still do not have a good sense of the noncanonical roles of tight-junction proteins in the epidermis, the broad expression pattern of some of them suggest additional roles in cytoskeleton organization, differentiation, and signaling.

6. JUNCTIONAL CROSSTALK IN EPIDERMAL FUNCTION

It is becoming increasingly clear that proper epidermal function requires the integration of all cell–cell adhesion systems and their associated cytoskeletons. This network of cell–cell adhesion systems that sense and respond to stress allows for tissue strengthening and full barrier function. We propose a model for how collaboration between these three elements is required for full barrier and mechanical strength of the granular layer of the epidermis (Fig. 2). The upregulation of cell–cell adhesions (predominantly desmosomes) in differentiated suprabasal cells allows for increased mechanical robustness. At the desmosome, desmoplakin recruits microtubule-associated proteins (Lis1, Ndel1, ninein, CLIP170) to the cell cortex, where they subsequently recruit and help stabilize microtubules. Cortical stabilization of microtubules results in engagement of AJs in a myosin II-dependent manner, thus further strengthening the epidermis. This activity is likely not necessary for basal epidermal integrity, as loss of most AJs does not significantly perturb the tissue, however it may increase integrity in response to mechanical insults. Additionally, AJ engagement enhances

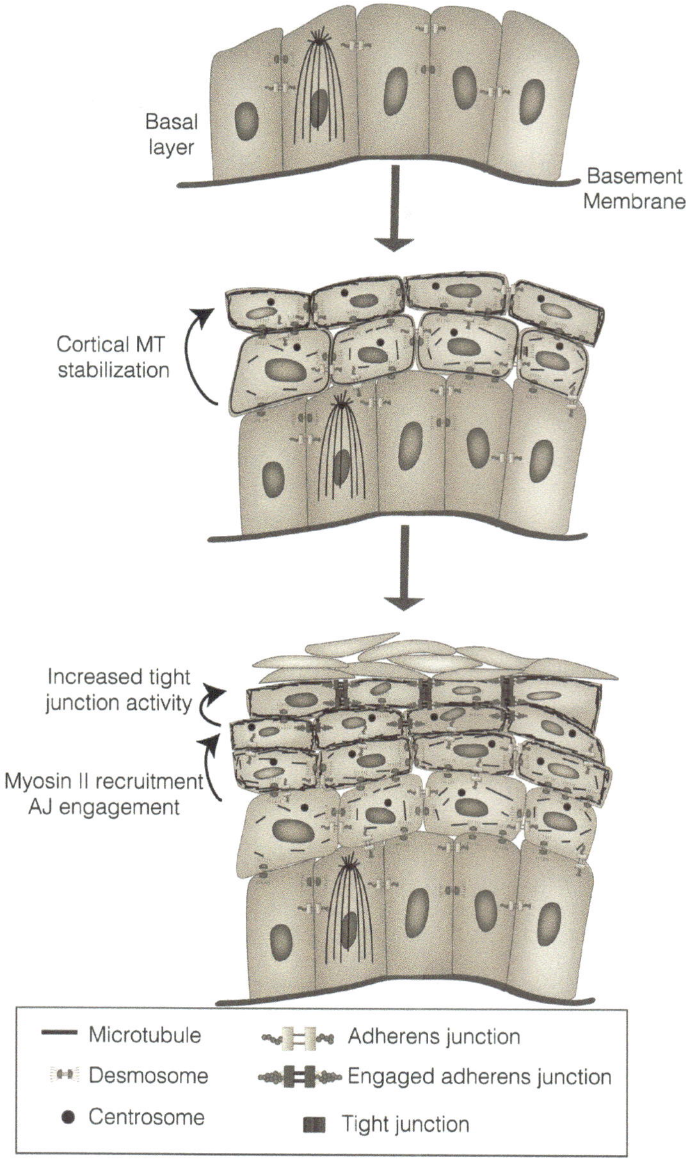

Figure 2 Model for collaboration of cell–cell adhesions to generate a functional barrier. See text for details. (See the color plate.)

tight-junction activity in the granular layer to give the epidermis an effective inside–out barrier. Because the outer surface of the skin is free, there are no forces acting along the proximal-distal axis of the epidermis. Therefore, tension is likely to be planar within the epidermis, and may be important for setting up the precise polygonal assembly of tight junctions on the lateral edges of the granular cells. Many of the mechanisms underlying the integration of these networks are yet to be elucidated, and much of this work has been performed in cultured keratinocytes. Translating this to tissue is an important future goal.

REFERENCES

Abe, K., & Takeichi, M. (2008). EPLIN mediates linkage of the cadherin catenin complex to F-actin and stabilizes the circumferential actin belt. *Proceedings of the National Academy of Sciences of the United States of America, 105*(1), 13–19. http://dx.doi.org/10.1073/pnas.0710504105.

Aberle, H., Butz, S., Stappert, J., Weissig, H., Kemler, R., & Hoschuetzky, H. (1994). Assembly of the cadherin–catenin complex *in vitro* with recombinant proteins. *Journal of Cell Science, 107*(Pt 12), 3655–3663.

Acehan, D., Petzold, C., Gumper, I., Sabatini, D. D., Muller, E. J., Cowin, P., et al. (2008). Plakoglobin is required for effective intermediate filament anchorage to desmosomes. *Journal of Investigative Dermatology, 128*(11), 2665–2675. http://dx.doi.org/10.1038/jid.2008.141.

Amagai, M. (2010). Autoimmune and infectious skin diseases that target desmogleins. *Proceedings of the Japan Academy. Series B, Physical and Biological Sciences, 86*(5), 524–537.

Amagai, M., & Stanley, J. R. (2012). Desmoglein as a target in skin disease and beyond. *Journal of Investigative Dermatology, 132*(3 Pt 2), 776–784. http://dx.doi.org/10.1038/jid.2011.390.

Ayub, M., Basit, S., Jelani, M., Ur Rehman, F., Iqbal, M., Yasinzai, M., et al. (2009). A homozygous nonsense mutation in the human desmocollin-3 (DSC3) gene underlies hereditary hypotrichosis and recurrent skin vesicles. *American Journal of Human Genetics, 85*(4), 515–520. http://dx.doi.org/10.1016/j.ajhg.2009.08.015.

Berkowitz, P., Hu, P., Liu, Z., Diaz, L. A., Enghild, J. J., Chua, M. P., et al. (2005). Desmosome signaling. Inhibition of p38MAPK prevents pemphigus vulgaris IgG-induced cytoskeleton reorganization. *Journal of Biological Chemistry, 280*(25), 23778–23784. http://dx.doi.org/10.1074/jbc.M501365200.

Berkowitz, P., Hu, P., Warren, S., Liu, Z., Diaz, L. A., & Rubenstein, D. S. (2006). p38MAPK inhibition prevents disease in pemphigus vulgaris mice. *Proceedings of the National Academy of Sciences of the United States of America, 103*(34), 12855–12860. http://dx.doi.org/10.1073/pnas.0602973103.

Bierkamp, C., McLaughlin, K. J., Schwarz, H., Huber, O., & Kemler, R. (1996). Embryonic heart and skin defects in mice lacking plakoglobin. *Developmental Biology, 180*(2), 780–785. http://dx.doi.org/10.1006/dbio.1996.0346.

Brandner, J. M., Kief, S., Grund, C., Rendl, M., Houdek, P., Kuhn, C., et al. (2002). Organization and formation of the tight junction system in human epidermis and cultured keratinocytes. *European Journal of Cell Biology, 81*(5), 253–263. http://dx.doi.org/10.1078/0171-9335-00244.

Brandner, J. M., McIntyre, M., Kief, S., Wladykowski, E., & Moll, I. (2003). Expression and localization of tight junction-associated proteins in human hair follicles. *Archives of Dermatological Research, 295*(5), 211–221. http://dx.doi.org/10.1007/s00403-003-0418-3.

Brennan, D., Hu, Y., Joubeh, S., Choi, Y. W., Whitaker-Menezes, D., O'Brien, T., et al. (2007). Suprabasal Dsg2 expression in transgenic mouse skin confers a hyperproliferative and apoptosis-resistant phenotype to keratinocytes. *Journal of Cell Science, 120*(Pt 5), 758–771. http://dx.doi.org/10.1242/jcs.03392.

Brieher, W. M., & Yap, A. S. (2013). Cadherin junctions and their cytoskeleton(s). *Current Opinion in Cell Biology*, *25*(1), 39–46. http://dx.doi.org/10.1016/j.ceb.2012.10.010.
Brieher, W. M., Yap, A. S., & Gumbiner, B. M. (1996). Lateral dimerization is required for the homophilic binding activity of C-cadherin. *Journal of Cell Biology*, *135*(2), 487–496.
Byrne, C., Tainsky, M., & Fuchs, E. (1994). Programming gene expression in developing epidermis. *Development*, *120*(9), 2369–2383.
Cavallaro, U., Schaffhauser, B., & Christofori, G. (2002). Cadherins and the tumour progression: Is it all in a switch? *Cancer Letters*, *176*(2), 123–128.
Celli, A., Zhai, Y., Jiang, Y. J., Crumrine, D., Elias, P. M., Feingold, K. R., et al. (2012). Tight junction properties change during epidermis development. *Experimental Dermatology*, *21*(10), 798–801. http://dx.doi.org/10.1111/j.1600-0625.2012.01573.x.
Chen, J., Den, Z., & Koch, P. J. (2008). Loss of desmocollin 3 in mice leads to epidermal blistering. *Journal of Cell Science*, *121*(Pt 17), 2844–2849. http://dx.doi.org/10.1242/jcs.031518.
Chidgey, M., Brakebusch, C., Gustafsson, E., Cruchley, A., Hail, C., Kirk, S., et al. (2001). Mice lacking desmocollin 1 show epidermal fragility accompanied by barrier defects and abnormal differentiation. *Journal of Cell Biology*, *155*(5), 821–832. http://dx.doi.org/10.1083/jcb.200105009.
Choi, Y. S., Zhang, Y., Xu, M., Yang, Y., Ito, M., Peng, T., et al. (2013). Distinct functions for Wnt/beta-catenin in hair follicle stem cell proliferation and survival and interfollicular epidermal homeostasis. *Cell Stem Cell*, *13*(6), 720–733. http://dx.doi.org/10.1016/j.stem.2013.10.003.
Christofori, G. (2003). Changing neighbours, changing behaviour: Cell adhesion molecule-mediated signalling during tumour progression. *EMBO Journal*, *22*(10), 2318–2323. http://dx.doi.org/10.1093/emboj/cdg228.
Davis, M. A., Ireton, R. C., & Reynolds, A. B. (2003). A core function for p120-catenin in cadherin turnover. *Journal of Cell Biology*, *163*(3), 525–534. http://dx.doi.org/10.1083/jcb.200307111.
Delva, E., & Kowalczyk, A. P. (2009). Regulation of cadherin trafficking. *Traffic*, *10*(3), 259–267. http://dx.doi.org/10.1111/j.1600-0854.2008.00862.x.
Den, Z., Cheng, X., Merched-Sauvage, M., & Koch, P. J. (2006). Desmocollin 3 is required for pre-implantation development of the mouse embryo. *Journal of Cell Science*, *119*(Pt 3), 482–489. http://dx.doi.org/10.1242/jcs.02769.
Djalilian, A. R., McGaughey, D., Patel, S., Seo, E. Y., Yang, C., Cheng, J., et al. (2006). Connexin 26 regulates epidermal barrier and wound remodeling and promotes psoriasiform response. *Journal of Clinical Investigation*, *116*(5), 1243–1253. http://dx.doi.org/10.1172/JCI27186.
Drees, F., Pokutta, S., Yamada, S., Nelson, W. J., & Weis, W. I. (2005). Alpha-catenin is a molecular switch that binds E-cadherin-beta-catenin and regulates actin-filament assembly. *Cell*, *123*(5), 903–915. http://dx.doi.org/10.1016/j.cell.2005.09.021.
Drenckhahn, D., & Franz, H. (1986). Identification of actin-, alpha-actinin-, and vinculin-containing plaques at the lateral membrane of epithelial cells. *Journal of Cell Biology*, *102*(5), 1843–1852.
Dupont, S., Morsut, L., Aragona, M., Enzo, E., Giulitti, S., Cordenonsi, M., et al. (2011). Role of YAP/TAZ in mechanotransduction. *Nature*, *474*(7350), 179–183. http://dx.doi.org/10.1038/nature10137.
Elias, P. M., & Friend, D. S. (1975). The permeability barrier in mammalian epidermis. *Journal of Cell Biology*, *65*(1), 180–191.
Elias, P. M., Matsuyoshi, N., Wu, H., Lin, C., Wang, Z. H., Brown, B. E., et al. (2001). Desmoglein isoform distribution affects stratum corneum structure and function. *Journal of Cell Biology*, *153*(2), 243–249.

Elias, P. M., McNutt, N. S., & Friend, D. S. (1977). Membrane alterations during cornification of mammalian squamous epithelia: A freeze-fracture, tracer, and thin-section study. *Anatomical Record*, *189*(4), 577–594. http://dx.doi.org/10.1002/ar.1091890404.

Fanning, A. S., Jameson, B. J., Jesaitis, L. A., & Anderson, J. M. (1998). The tight junction protein ZO-1 establishes a link between the transmembrane protein occludin and the actin cytoskeleton. *Journal of Biological Chemistry*, *273*(45), 29745–29753.

Fanning, A. S., Van Itallie, C. M., & Anderson, J. M. (2012). Zonula occludens-1 and -2 regulate apical cell structure and the zonula adherens cytoskeleton in polarized epithelia. *Molecular Biology of the Cell*, *23*(4), 577–590. http://dx.doi.org/10.1091/mbc.E11-09-0791.

Foote, H. P., Sumigray, K. D., & Lechler, T. (2013). FRAP analysis reveals stabilization of adhesion structures in the epidermis compared to cultured keratinocytes. *PLoS One*, *8*(8), e71491. http://dx.doi.org/10.1371/journal.pone.0071491.

Furuse, M., Fujita, K., Hiiragi, T., Fujimoto, K., & Tsukita, S. (1998). Claudin-1 and -2: Novel integral membrane proteins localizing at tight junctions with no sequence similarity to occludin. *Journal of Cell Biology*, *141*(7), 1539–1550.

Furuse, M., Hata, M., Furuse, K., Yoshida, Y., Haratake, A., Sugitani, Y., et al. (2002). Claudin-based tight junctions are crucial for the mammalian epidermal barrier: A lesson from claudin-1-deficient mice. *Journal of Cell Biology*, *156*(6), 1099–1111. http://dx.doi.org/10.1083/jcb.200110122.

Furuse, M., Sasaki, H., Fujimoto, K., & Tsukita, S. (1998). A single gene product, claudin-1 or -2, reconstitutes tight junction strands and recruits occludin in fibroblasts. *Journal of Cell Biology*, *143*(2), 391–401.

Gallicano, G. I., Bauer, C., & Fuchs, E. (2001). Rescuing desmoplakin function in extra-embryonic ectoderm reveals the importance of this protein in embryonic heart, neuroepithelium, skin and vasculature. *Development*, *128*(6), 929–941.

Gallicano, G. I., Kouklis, P., Bauer, C., Yin, M., Vasioukhin, V., Degenstein, L., et al. (1998). Desmoplakin is required early in development for assembly of desmosomes and cytoskeletal linkage. *Journal of Cell Biology*, *143*(7), 2009–2022.

Gat, U., DasGupta, R., Degenstein, L., & Fuchs, E. (1998). De Novo hair follicle morphogenesis and hair tumors in mice expressing a truncated beta-catenin in skin. *Cell*, *95*(5), 605–614.

Georgiou, M., Marinari, E., Burden, J., & Baum, B. (2008). Cdc42, Par6, and aPKC regulate Arp2/3-mediated endocytosis to control local adherens junction stability. *Current Biology*, *18*(21), 1631–1638. http://dx.doi.org/10.1016/j.cub.2008.09.029.

Getsios, S., Simpson, C. L., Kojima, S., Harmon, R., Sheu, L. J., Dusek, R. L., et al. (2009). Desmoglein 1-dependent suppression of EGFR signaling promotes epidermal differentiation and morphogenesis. *Journal of Cell Biology*, *185*(7), 1243–1258. http://dx.doi.org/10.1083/jcb.200809044.

Godsel, L. M., Dubash, A. D., Bass-Zubek, A. E., Amargo, E. V., Klessner, J. L., Hobbs, R. P., et al. (2010). Plakophilin 2 couples actomyosin remodeling to desmosomal plaque assembly via RhoA. *Molecular Biology of the Cell*, *21*(16), 2844–2859. http://dx.doi.org/10.1091/mbc.E10-02-0131.

Green, K. J., Stappenbeck, T. S., Parry, D. A., & Virata, M. L. (1992). Structure of desmoplakin and its association with intermediate filaments. *Journal of Dermatology*, *19*(11), 765–769.

Hadj-Rabia, S., Baala, L., Vabres, P., Hamel-Teillac, D., Jacquemin, E., Fabre, M., et al. (2004). Claudin-1 gene mutations in neonatal sclerosing cholangitis associated with ichthyosis: A tight junction disease. *Gastroenterology*, *127*(5), 1386–1390.

Harmon, R. M., Simpson, C. L., Johnson, J. L., Koetsier, J. L., Dubash, A. D., Najor, N. A., et al. (2013). Desmoglein-1/Erbin interaction suppresses ERK activation to support

epidermal differentiation. *Journal of Clinical Investigation, 123*(4), 1556–1570. http://dx.doi.org/10.1172/JCI65220.
Huelsken, J., Vogel, R., Erdmann, B., Cotsarelis, G., & Birchmeier, W. (2001). beta-Catenin controls hair follicle morphogenesis and stem cell differentiation in the skin. *Cell, 105*(4), 533–545.
Huen, A. C., Park, J. K., Godsel, L. M., Chen, X., Bannon, L. J., Amargo, E. V., et al. (2002). Intermediate filament-membrane attachments function synergistically with actin-dependent contacts to regulate intercellular adhesive strength. *Journal of Cell Biology, 159*(6), 1005–1017. http://dx.doi.org/10.1083/jcb.200206098.
Huveneers, S., Oldenburg, J., Spanjaard, E., van der Krogt, G., Grigoriev, I., Akhmanova, A., et al. (2012). Vinculin associates with endothelial VE-cadherin junctions to control force-dependent remodeling. *Journal of Cell Biology, 196*(5), 641–652. http://dx.doi.org/10.1083/jcb.201108120.
Ibi, M., Zou, P., Inoko, A., Shiromizu, T., Matsuyama, M., Hayashi, Y., et al. (2011). Trichoplein controls microtubule anchoring at the centrosome by binding to Odf2 and ninein. *Journal of Cell Science, 124*(Pt 6), 857–864. http://dx.doi.org/10.1242/jcs.075705.
Indelman, M., Bergman, R., Lurie, R., Richard, G., Miller, B., Petronius, D., et al. (2002). A missense mutation in CDH3, encoding P-cadherin, causes hypotrichosis with juvenile macular dystrophy. *Journal of Investigative Dermatology, 119*(5), 1210–1213. http://dx.doi.org/10.1046/j.1523-1747.2002.19528.x.
Ireton, R. C., Davis, M. A., van Hengel, J., Mariner, D. J., Barnes, K., Thoreson, M. A., et al. (2002). A novel role for p120 catenin in E-cadherin function. *Journal of Cell Biology, 159*(3), 465–476. http://dx.doi.org/10.1083/jcb.200205115.
Ishiyama, N., Lee, S. H., Liu, S., Li, G. Y., Smith, M. J., Reichardt, L. F., et al. (2010). Dynamic and static interactions between p120 catenin and E-cadherin regulate the stability of cell–cell adhesion. *Cell, 141*(1), 117–128. http://dx.doi.org/10.1016/j.cell.2010.01.017.
Ishiyama, N., Tanaka, N., Abe, K., Yang, Y. J., Abbas, Y. M., Umitsu, M., et al. (2013). An autoinhibited structure of alpha-catenin and its implications for vinculin recruitment to adherens junctions. *Journal of Biological Chemistry, 288*(22), 15913–15925. http://dx.doi.org/10.1074/jbc.M113.453928.
Jamora, C., DasGupta, R., Kocieniewski, P., & Fuchs, E. (2003). Links between signal transduction, transcription and adhesion in epithelial bud development. *Nature, 422*(6929), 317–322. http://dx.doi.org/10.1038/nature01458.
Katsuno, T., Umeda, K., Matsui, T., Hata, M., Tamura, A., Itoh, M., et al. (2008). Deficiency of zonula occludens-1 causes embryonic lethal phenotype associated with defected yolk sac angiogenesis and apoptosis of embryonic cells. *Molecular Biology of the Cell, 19*(6), 2465–2475. http://dx.doi.org/10.1091/mbc.E07-12-1215.
Kirschner, N., Houdek, P., Fromm, M., Moll, I., & Brandner, J. M. (2010). Tight junctions form a barrier in human epidermis. *European Journal of Cell Biology, 89*(11), 839–842. http://dx.doi.org/10.1016/j.ejcb.2010.07.010.
Kitajima, Y., Eguchi, K., Ohno, T., Mori, S., & Yaoita, H. (1983). Tight junctions of human keratinocytes in primary culture: A freeze-fracture study. *Journal of Ultrastructure Research, 82*(3), 309–313.
Kljuic, A., Bazzi, H., Sundberg, J. P., Martinez-Mir, A., O'Shaughnessy, R., Mahoney, M. G., et al. (2003). Desmoglein 4 in hair follicle differentiation and epidermal adhesion: Evidence from inherited hypotrichosis and acquired pemphigus vulgaris. *Cell, 113*(2), 249–260.
Kobielak, A., & Fuchs, E. (2006). Links between alpha-catenin, NF-kappaB, and squamous cell carcinoma in skin. *Proceedings of the National Academy of Sciences of the United States of America, 103*(7), 2322–2327. http://dx.doi.org/10.1073/pnas.0510422103.

Kobielak, A., Pasolli, H. A., & Fuchs, E. (2004). Mammalian formin-1 participates in adherens junctions and polymerization of linear actin cables. *Nature Cell Biology, 6*(1), 21–30. http://dx.doi.org/10.1038/ncb1075.

Koch, P. J., Mahoney, M. G., Cotsarelis, G., Rothenberger, K., Lavker, R. M., & Stanley, J. R. (1998). Desmoglein 3 anchors telogen hair in the follicle. *Journal of Cell Science, 111*(Pt 17), 2529–2537.

Koch, P. J., Mahoney, M. G., Ishikawa, H., Pulkkinen, L., Uitto, J., Shultz, L., et al. (1997). Targeted disruption of the pemphigus vulgaris antigen (desmoglein 3) gene in mice causes loss of keratinocyte cell adhesion with a phenotype similar to pemphigus vulgaris. *Journal of Cell Biology, 137*(5), 1091–1102.

Kottke, M. D., Delva, E., & Kowalczyk, A. P. (2006). The desmosome: Cell science lessons from human diseases. *Journal of Cell Science, 119*(Pt 5), 797–806. http://dx.doi.org/10.1242/jcs.02888.

Kowalczyk, A. P., Bornslaeger, E. A., Borgwardt, J. E., Palka, H. L., Dhaliwal, A. S., Corcoran, C. M., et al. (1997). The amino-terminal domain of desmoplakin binds to plakoglobin and clusters desmosomal cadherin-plakoglobin complexes. *Journal of Cell Biology, 139*(3), 773–784.

Kowalczyk, A. P., Hatzfeld, M., Bornslaeger, E. A., Kopp, D. S., Borgwardt, J. E., Corcoran, C. M., et al. (1999). The head domain of plakophilin-1 binds to desmoplakin and enhances its recruitment to desmosomes. Implications for cutaneous disease. *Journal of Biological Chemistry, 274*(26), 18145–18148.

Kurzen, H., Moll, I., Moll, R., Schafer, S., Simics, E., Amagai, M., et al. (1998). Compositionally different desmosomes in the various compartments of the human hair follicle. *Differentiation, 63*(5), 295–304. http://dx.doi.org/10.1046/j.1432-0436.1998.6350295.x.

Lai Cheong, J. E., Wessagowit, V., & McGrath, J. A. (2005). Molecular abnormalities of the desmosomal protein desmoplakin in human disease. *Clinical and Experimental Dermatology, 30*(3), 261–266. http://dx.doi.org/10.1111/j.1365-2230.2005.01736.x.

Lai-Cheong, J. E., Arita, K., & McGrath, J. A. (2007). Genetic diseases of junctions. *Journal of Investigative Dermatology, 127*(12), 2713–2725. http://dx.doi.org/10.1038/sj.jid.5700727.

Langbein, L., Grund, C., Kuhn, C., Praetzel, S., Kartenbeck, J., Brandner, J. M., et al. (2002). Tight junctions and compositionally related junctional structures in mammalian stratified epithelia and cell cultures derived therefrom. *European Journal of Cell Biology, 81*(8), 419–435. http://dx.doi.org/10.1078/0171-9335-00270.

Lechler, T., & Fuchs, E. (2005). Asymmetric cell divisions promote stratification and differentiation of mammalian skin. *Nature, 437*(7056), 275–280. http://dx.doi.org/10.1038/nature03922.

Lechler, T., & Fuchs, E. (2007). Desmoplakin: An unexpected regulator of microtubule organization in the epidermis. *Journal of Cell Biology, 176*(2), 147–154. http://dx.doi.org/10.1083/jcb.200609109.

le Duc, Q., Shi, Q., Blonk, I., Sonnenberg, A., Wang, N., Leckband, D., et al. (2010). Vinculin potentiates E-cadherin mechanosensing and is recruited to actin-anchored sites within adherens junctions in a myosin II-dependent manner. *Journal of Cell Biology, 189*(7), 1107–1115. http://dx.doi.org/10.1083/jcb.201001149.

Lee, J., & Tumbar, T. (2012). Hairy tale of signaling in hair follicle development and cycling. *Seminars in Cell & Developmental Biology, 23*(8), 906–916. http://dx.doi.org/10.1016/j.semcdb.2012.08.003.

Lewis, J. E., Wahl, J. K., 3rd., Sass, K. M., Jensen, P. J., Johnson, K. R., & Wheelock, M. J. (1997). Cross-talk between adherens junctions and desmosomes depends on plakoglobin. *Journal of Cell Biology, 136*(4), 919–934.

Li, J., Swope, D., Raess, N., Cheng, L., Muller, E. J., & Radice, G. L. (2011). Cardiac tissue-restricted deletion of plakoglobin results in progressive cardiomyopathy and activation of {beta}-catenin signaling. *Molecular and Cellular Biology*, *31*(6), 1134–1144. http://dx.doi.org/10.1128/MCB.01025-10.

Li, D., Zhang, W., Liu, Y., Haneline, L. S., & Shou, W. (2012). Lack of plakoglobin in epidermis leads to keratoderma. *Journal of Biological Chemistry*, *287*(13), 10435–10443. http://dx.doi.org/10.1074/jbc.M111.299669.

Lim, X., Tan, S. H., Koh, W. L., Chau, R. M., Yan, K. S., Kuo, C. J., et al. (2013). Interfollicular epidermal stem cells self-renew via autocrine Wnt signaling. *Science*, *342*(6163), 1226–1230. http://dx.doi.org/10.1126/science.1239730.

Maconnachie, E. (1979). A study of digit fusion in the mouse embryo. *Journal of Embryology and Experimental Morphology*, *49*, 259–276.

Mao, X., Sano, Y., Park, J. M., & Payne, A. S. (2011). p38 MAPK activation is downstream of the loss of intercellular adhesion in pemphigus vulgaris. *Journal of Biological Chemistry*, *286*(2), 1283–1291. http://dx.doi.org/10.1074/jbc.M110.172874.

Mathur, M., Goodwin, L., & Cowin, P. (1994). Interactions of the cytoplasmic domain of the desmosomal cadherin Dsg1 with plakoglobin. *Journal of Biological Chemistry*, *269*(19), 14075–14080.

Matsuki, M., Yamashita, F., Ishida-Yamamoto, A., Yamada, K., Kinoshita, C., Fushiki, S., et al. (1998). Defective stratum corneum and early neonatal death in mice lacking the gene for transglutaminase 1 (keratinocyte transglutaminase). *Proceedings of the National Academy of Sciences of the United States of America*, *95*(3), 1044–1049.

M'Boneko, V., & Merker, H. J. (1988). Development and morphology of the periderm of mouse embryos (days 9–12 of gestation). *Acta Anatomica (Basel)*, *133*(4), 325–336.

McGrath, J. A., & Wessagowit, V. (2005). Human hair abnormalities resulting from inherited desmosome gene mutations. *Keio Journal of Medicine*, *54*(2), 72–79.

Merritt, A. J., Berika, M. Y., Zhai, W., Kirk, S. E., Ji, B., Hardman, M. J., et al. (2002). Suprabasal desmoglein 3 expression in the epidermis of transgenic mice results in hyperproliferation and abnormal differentiation. *Molecular and Cellular Biology*, *22*(16), 5846–5858.

Millar, S. E. (2002). Molecular mechanisms regulating hair follicle development. *Journal of Investigative Dermatology*, *118*(2), 216–225. http://dx.doi.org/10.1046/j.0022-202x.2001.01670.x.

Mills, A. A., Zheng, B., Wang, X. J., Vogel, H., Roop, D. R., & Bradley, A. (1999). p63 is a p53 homologue required for limb and epidermal morphogenesis. *Nature*, *398*(6729), 708–713. http://dx.doi.org/10.1038/19531.

Miyake, Y., Inoue, N., Nishimura, K., Kinoshita, N., Hosoya, H., & Yonemura, S. (2006). Actomyosin tension is required for correct recruitment of adherens junction components and zonula occludens formation. *Experimental Cell Research*, *312*(9), 1637–1650. http://dx.doi.org/10.1016/j.yexcr.2006.01.031.

Miyashita, Y., & Ozawa, M. (2007). Increased internalization of p120-uncoupled E-cadherin and a requirement for a dileucine motif in the cytoplasmic domain for endocytosis of the protein. *Journal of Biological Chemistry*, *282*(15), 11540–11548. http://dx.doi.org/10.1074/jbc.M608351200.

Morita, K., Itoh, M., Saitou, M., Ando-Akatsuka, Y., Furuse, M., Yoneda, K., et al. (1998). Subcellular distribution of tight junction-associated proteins (occludin, ZO-1, ZO-2) in rodent skin. *Journal of Investigative Dermatology*, *110*(6), 862–866. http://dx.doi.org/10.1046/j.1523-1747.1998.00209.x.

Muller-Rover, S., Tokura, Y., Welker, P., Furukawa, F., Wakita, H., Takigawa, M., et al. (1999). E- and P-cadherin expression during murine hair follicle morphogenesis and cycling. *Experimental Dermatology*, *8*(4), 237–246.

Nanba, D., Hieda, Y., & Nakanishi, Y. (2000). Remodeling of desmosomal and hemidesmosomal adhesion systems during early morphogenesis of mouse pelage hair follicles.

Journal of Investigative Dermatology, 114(1), 171–177. http://dx.doi.org/10.1046/j.1523-1747.2000.00842.x.

Nanes, B. A., Chiasson-MacKenzie, C., Lowery, A. M., Ishiyama, N., Faundez, V., Ikura, M., et al. (2012). p120-catenin binding masks an endocytic signal conserved in classical cadherins. *Journal of Cell Biology, 199*(2), 365–380. http://dx.doi.org/10.1083/jcb.201205029.

Nishizawa, M., Izawa, I., Inoko, A., Hayashi, Y., Nagata, K., Yokoyama, T., et al. (2005). Identification of trichoplein, a novel keratin filament-binding protein. *Journal of Cell Science, 118*(Pt 5), 1081–1090. http://dx.doi.org/10.1242/jcs.01667.

Ozawa, M. (2002). Lateral dimerization of the E-cadherin extracellular domain is necessary but not sufficient for adhesive activity. *Journal of Biological Chemistry, 277*(22), 19600–19608. http://dx.doi.org/10.1074/jbc.M202029200.

Pellegrini, G., Dellambra, E., Golisano, O., Martinelli, E., Fantozzi, I., Bondanza, S., et al. (2001). p63 identifies keratinocyte stem cells. *Proceedings of the National Academy of Sciences of the United States of America, 98*(6), 3156–3161. http://dx.doi.org/10.1073/pnas.061032098.

Peng, X., Cuff, L. E., Lawton, C. D., & DeMali, K. A. (2010). Vinculin regulates cell-surface E-cadherin expression by binding to beta-catenin. *Journal of Cell Science, 123*(Pt 4), 567–577. http://dx.doi.org/10.1242/jcs.056432.

Perez-Moreno, M., Davis, M. A., Wong, E., Pasolli, H. A., Reynolds, A. B., & Fuchs, E. (2006). p120-catenin mediates inflammatory responses in the skin. *Cell, 124*(3), 631–644. http://dx.doi.org/10.1016/j.cell.2005.11.043.

Pummi, K., Malminen, M., Aho, H., Karvonen, S. L., Peltonen, J., & Peltonen, S. (2001). Epidermal tight junctions: ZO-1 and occludin are expressed in mature, developing, and affected skin and *in vitro* differentiating keratinocytes. *Journal of Investigative Dermatology, 117*(5), 1050–1058. http://dx.doi.org/10.1046/j.0022-202x.2001.01493.x.

Rangarajan, E. S., & Izard, T. (2012). The cytoskeletal protein alpha-catenin unfurls upon binding to vinculin. *Journal of Biological Chemistry, 287*(22), 18492–18499. http://dx.doi.org/10.1074/jbc.M112.351023.

Ray, S., Foote, H. P., & Lechler, T. (2013). beta-Catenin protects the epidermis from mechanical stresses. *Journal of Cell Biology, 202*(1), 45–52. http://dx.doi.org/10.1083/jcb.201212140.

Richardson, R., Hammond, N., Coulombe, P. A., Saloranta, C., Nousiainen, H., Salonen, R., et al. (2014). Periderm prevents pathological epithelial adhesions during embryogenesis. *Journal of Clinical Investigation, 124*, 3891–3900.

Rizzo, W. B. (2014). Fatty aldehyde and fatty alcohol metabolism: review and importance for epidermal structure and function. *Biochimica et Biophysica Acta, 1841*(3), 377–389. http://dx.doi.org/10.1016/j.bbalip.2013.09.001.

Rizzo, W. B., S'Aulis, D., Jennings, M. A., Crumrine, D. A., Williams, M. L., & Elias, P. M. (2010). Ichthyosis in Sjogren-Larsson syndrome reflects defective barrier function due to abnormal lamellar body structure and secretion. *Archives of Dermatological Research, 302*(6), 443–451. http://dx.doi.org/10.1007/s00403-009-1022-y.

Saitou, M., Furuse, M., Sasaki, H., Schulzke, J. D., Fromm, M., Takano, H., et al. (2000). Complex phenotype of mice lacking occludin, a component of tight junction strands. *Molecular Biology of the Cell, 11*(12), 4131–4142.

Sanchez-Carpintero, I., Espana, A., Pelacho, B., Lopez Moratalla, N., Rubenstein, D. S., Diaz, L. A., et al. (2004). *In vivo* blockade of pemphigus vulgaris acantholysis by inhibition of intracellular signal transduction cascades. *British Journal of Dermatology, 151*(3), 565–570. http://dx.doi.org/10.1111/j.1365-2133.2004.06147.x.

Sansores-Garcia, L., Bossuyt, W., Wada, K., Yonemura, S., Tao, C., Sasaki, H., et al. (2011). Modulating F-actin organization induces organ growth by affecting the Hippo pathway. *EMBO Journal, 30*(12), 2325–2335. http://dx.doi.org/10.1038/emboj.2011.157.

Sawyer, J. K., Harris, N. J., Slep, K. C., Gaul, U., & Peifer, M. (2009). The Drosophila afadin homologue Canoe regulates linkage of the actin cytoskeleton to adherens junctions during apical constriction. *Journal of Cell Biology*, *186*(1), 57–73. http://dx.doi.org/10.1083/jcb.200904001.

Schlegelmilch, K., Mohseni, M., Kirak, O., Pruszak, J., Rodriguez, J. R., Zhou, D., et al. (2011). Yap1 acts downstream of alpha-catenin to control epidermal proliferation. *Cell*, *144*(5), 782–795. http://dx.doi.org/10.1016/j.cell.2011.02.031.

Schluter, H., Wepf, R., Moll, I., & Franke, W. W. (2004). Sealing the live part of the skin: The integrated meshwork of desmosomes, tight junctions and curvilinear ridge structures in the cells of the uppermost granular layer of the human epidermis. *European Journal of Cell Biology*, *83*(11–12), 655–665. http://dx.doi.org/10.1078/0171-9335-00434.

Seishima, M., Iwasaki-Bessho, Y., Itoh, Y., Nozawa, Y., Amagai, M., & Kitajima, Y. (1999). Phosphatidylcholine-specific phospholipase C, but not phospholipase D, is involved in pemphigus IgG-induced signal transduction. *Archives of Dermatological Research*, *291*(11), 606–613.

Sen, G. L., Reuter, J. A., Webster, D. E., Zhu, L., & Khavari, P. A. (2010). DNMT1 maintains progenitor function in self-renewing somatic tissue. *Nature*, *463*(7280), 563–567. http://dx.doi.org/10.1038/nature08683.

Sharma, P., Mao, X., & Payne, A. S. (2007). Beyond steric hindrance: The role of adhesion signaling pathways in the pathogenesis of pemphigus. *Journal of Dermatological Science*, *48*(1), 1–14. http://dx.doi.org/10.1016/j.jdermsci.2007.05.005.

Shimomura, Y., Wajid, M., Shapiro, L., & Christiano, A. M. (2008). P-cadherin is a p63 target gene with a crucial role in the developing human limb bud and hair follicle. *Development*, *135*(4), 743–753. http://dx.doi.org/10.1242/dev.006718.

Simpson, M. A., Mansour, S., Ahnood, D., Kalidas, K., Patton, M. A., McKenna, W. J., et al. (2009). Homozygous mutation of desmocollin-2 in arrhythmogenic right ventricular cardiomyopathy with mild palmoplantar keratoderma and woolly hair. *Cardiology*, *113*(1), 28–34. http://dx.doi.org/10.1159/000165696.

Spindler, V., & Waschke, J. (2011). Role of Rho GTPases in desmosomal adhesion and pemphigus pathogenesis. *Annals of Anatomy*, *193*(3), 177–180. http://dx.doi.org/10.1016/j.aanat.2011.02.003.

Stappenbeck, T. S., Bornslaeger, E. A., Corcoran, C. M., Luu, H. H., Virata, M. L., & Green, K. J. (1993). Functional analysis of desmoplakin domains: Specification of the interaction with keratin versus vimentin intermediate filament networks. *Journal of Cell Biology*, *123*(3), 691–705.

Stappenbeck, T. S., & Green, K. J. (1992). The desmoplakin carboxyl terminus coaligns with and specifically disrupts intermediate filament networks when expressed in cultured cells. *Journal of Cell Biology*, *116*(5), 1197–1209.

Steven, A. C., & Steinert, P. M. (1994). Protein composition of cornified cell envelopes of epidermal keratinocytes. *Journal of Cell Science*, *107*(Pt 2), 693–700.

Sugawara, T., Iwamoto, N., Akashi, M., Kojima, T., Hisatsune, J., Sugai, M., et al. (2013). Tight junction dysfunction in the stratum granulosum leads to aberrant stratum corneum barrier function in claudin-1-deficient mice. *Journal of Dermatological Science*, *70*(1), 12–18. http://dx.doi.org/10.1016/j.jdermsci.2013.01.002.

Sugimoto, K., Ichikawa-Tomikawa, N., Satohisa, S., Akashi, Y., Kanai, R., Saito, T., et al. (2013). The tight-junction protein claudin-6 induces epithelial differentiation from mouse F9 and embryonic stem cells. *PLoS One*, *8*(10), e75106. http://dx.doi.org/10.1371/journal.pone.0075106.

Sumida, G. M., Tomita, T. M., Shih, W., & Yamada, S. (2011). Myosin II activity dependent and independent vinculin recruitment to the sites of E-cadherin-mediated cell–cell adhesion. *BMC Cell Biology*, *12*, 48. http://dx.doi.org/10.1186/1471-2121-12-48.

Sumigray, K. D., Chen, H., & Lechler, T. (2011). Lis1 is essential for cortical microtubule organization and desmosome stability in the epidermis. *Journal of Cell Biology, 194*(4), 631–642. http://dx.doi.org/10.1083/jcb.201104009.

Sumigray, K. D., Foote, H. P., & Lechler, T. (2012). Noncentrosomal microtubules and type II myosins potentiate epidermal cell adhesion and barrier formation. *Journal of Cell Biology, 199*(3), 513–525. http://dx.doi.org/10.1083/jcb.201206143.

Sumigray, K. D., & Lechler, T. (2012). Desmoplakin controls microvilli length but not cell adhesion or keratin organization in the intestinal epithelium. *Molecular Biology of the Cell, 23*(5), 792–799. http://dx.doi.org/10.1091/mbc.E11-11-0923.

Sumigray, K. D., Zhou, K., & Lechler, T. (2014). Cell–cell adhesions and cell contractility are upregulated upon desmosome disruption. *PLoS One, 9*(7), e101824. http://dx.doi.org/10.1371/journal.pone.0101824.

Tang, V. W., & Brieher, W. M. (2012). alpha-Actinin-4/FSGS1 is required for Arp2/3-dependent actin assembly at the adherens junction. *Journal of Cell Biology, 196*(1), 115–130. http://dx.doi.org/10.1083/jcb.201103116.

Thoreson, M. A., Anastasiadis, P. Z., Daniel, J. M., Ireton, R. C., Wheelock, M. J., Johnson, K. R., et al. (2000). Selective uncoupling of p120(ctn) from E-cadherin disrupts strong adhesion. *Journal of Cell Biology, 148*(1), 189–202.

Thyssen, J. P., Godoy-Gijon, E., & Elias, P. M. (2013). Ichthyosis vulgaris: The filaggrin mutation disease. *British Journal of Dermatology, 168*(6), 1155–1166. http://dx.doi.org/10.1111/bjd.12219.

Tinkle, C. L., Lechler, T., Pasolli, H. A., & Fuchs, E. (2004). Conditional targeting of E-cadherin in skin: Insights into hyperproliferative and degenerative responses. *Proceedings of the National Academy of Sciences of the United States of America, 101*(2), 552–557. http://dx.doi.org/10.1073/pnas.0307437100.

Tinkle, C. L., Pasolli, H. A., Stokes, N., & Fuchs, E. (2008). New insights into cadherin function in epidermal sheet formation and maintenance of tissue integrity. *Proceedings of the National Academy of Sciences of the United States of America, 105*(40), 15405–15410. http://dx.doi.org/10.1073/pnas.0807374105.

Tomschy, A., Fauser, C., Landwehr, R., & Engel, J. (1996). Homophilic adhesion of E-cadherin occurs by a co-operative two-step interaction of N-terminal domains. *EMBO Journal, 15*(14), 3507–3514.

Troyanovsky, S. M., Troyanovsky, R. B., Eshkind, L. G., Krutovskikh, V. A., Leube, R. E., & Franke, W. W. (1994). Identification of the plakoglobin-binding domain in desmoglein and its role in plaque assembly and intermediate filament anchorage. *Journal of Cell Biology, 127*(1), 151–160.

Tsang, S. M., Brown, L., Gadmor, H., Gammon, L., Fortune, F., Wheeler, A., et al. (2012). Desmoglein 3 acting as an upstream regulator of Rho GTPases, Rac-1/Cdc42 in the regulation of actin organisation and dynamics. *Experimental Cell Research, 318*(18), 2269–2283. http://dx.doi.org/10.1016/j.yexcr.2012.07.002.

Tunggal, J. A., Helfrich, I., Schmitz, A., Schwarz, H., Gunzel, D., Fromm, M., et al. (2005). E-cadherin is essential for *in vivo* epidermal barrier function by regulating tight junctions. *EMBO Journal, 24*(6), 1146–1156. http://dx.doi.org/10.1038/sj.emboj.7600605.

Turksen, K., & Troy, T. C. (2002). Permeability barrier dysfunction in transgenic mice overexpressing claudin 6. *Development, 129*(7), 1775–1784.

Twiss, F., Le Duc, Q., Van Der Horst, S., Tabdili, H., Van Der Krogt, G., Wang, N., et al. (2012). Vinculin-dependent Cadherin mechanosensing regulates efficient epithelial barrier formation. *Biology Open, 1*(11), 1128–1140. http://dx.doi.org/10.1242/bio.20122428.

Umeda, K., Ikenouchi, J., Katahira-Tayama, S., Furuse, K., Sasaki, H., Nakayama, M., et al. (2006). ZO-1 and ZO-2 independently determine where claudins are polymerized in tight-junction strand formation. *Cell*, *126*(4), 741–754. http://dx.doi.org/10.1016/j.cell.2006.06.043.

Van Itallie, C. M., & Anderson, J. M. (2004). The role of claudins in determining paracellular charge selectivity. *Proceedings of the American Thoracic Society*, *1*(1), 38–41. http://dx.doi.org/10.1513/pats.2306013.

Van Itallie, C. M., Aponte, A., Tietgens, A. J., Gucek, M., Fredriksson, K., & Anderson, J. M. (2013). The N and C termini of ZO-1 are surrounded by distinct proteins and functional protein networks. *Journal of Biological Chemistry*, *288*(19), 13775–13788. http://dx.doi.org/10.1074/jbc.M113.466193.

Vasioukhin, V., Bauer, C., Degenstein, L., Wise, B., & Fuchs, E. (2001). Hyperproliferation and defects in epithelial polarity upon conditional ablation of alpha-catenin in skin. *Cell*, *104*(4), 605–617.

Vasioukhin, V., Bauer, C., Yin, M., & Fuchs, E. (2000). Directed actin polymerization is the driving force for epithelial cell–cell adhesion. *Cell*, *100*(2), 209–219.

Vasioukhin, V., Bowers, E., Bauer, C., Degenstein, L., & Fuchs, E. (2001). Desmoplakin is essential in epidermal sheet formation. *Nature Cell Biology*, *3*(12), 1076–1085. http://dx.doi.org/10.1038/ncb1201-1076.

Verma, S., Han, S. P., Michael, M., Gomez, G. A., Yang, Z., Teasdale, R. D., et al. (2012). A WAVE2-Arp2/3 actin nucleator apparatus supports junctional tension at the epithelial zonula adherens. *Molecular Biology of the Cell*, *23*(23), 4601–4610. http://dx.doi.org/10.1091/mbc.E12-08-0574.

Wacker, I. U., Rickard, J. E., De Mey, J. R., & Kreis, T. E. (1992). Accumulation of a microtubule-binding protein, pp 170, at desmosomal plaques. *Journal of Cell Biology*, *117*(4), 813–824.

Wada, K., Itoga, K., Okano, T., Yonemura, S., & Sasaki, H. (2011). Hippo pathway regulation by cell morphology and stress fibers. *Development*, *138*(18), 3907–3914. http://dx.doi.org/10.1242/dev.070987.

Waschke, J., Spindler, V., Bruggeman, P., Zillikens, D., Schmidt, G., & Drenckhahn, D. (2006). Inhibition of Rho A activity causes pemphigus skin blistering. *Journal of Cell Biology*, *175*(5), 721–727. http://dx.doi.org/10.1083/jcb.200605125.

Williamson, L., Raess, N. A., Caldelari, R., Zakher, A., de Bruin, A., Posthaus, H., et al. (2006). Pemphigus vulgaris identifies plakoglobin as key suppressor of c-Myc in the skin. *EMBO Journal*, *25*(14), 3298–3309. http://dx.doi.org/10.1038/sj.emboj.7601224.

Wittchen, E. S., Haskins, J., & Stevenson, B. R. (2000). Exogenous expression of the amino-terminal half of the tight junction protein ZO-3 perturbs junctional complex assembly. *Journal of Cell Biology*, *151*(4), 825–836.

Xiao, K., Allison, D. F., Buckley, K. M., Kottke, M. D., Vincent, P. A., Faundez, V., et al. (2003). Cellular levels of p120 catenin function as a set point for cadherin expression levels in microvascular endothelial cells. *Journal of Cell Biology*, *163*(3), 535–545. http://dx.doi.org/10.1083/jcb.200306001.

Xu, J., Kausalya, P. J., Phua, D. C., Ali, S. M., Hossain, Z., & Hunziker, W. (2008). Early embryonic lethality of mice lacking ZO-2, but Not ZO-3, reveals critical and non-redundant roles for individual zonula occludens proteins in mammalian development. *Molecular and Cellular Biology*, *28*(5), 1669–1678. http://dx.doi.org/10.1128/MCB.00891-07.

Yamada, S., Pokutta, S., Drees, F., Weis, W. I., & Nelson, W. J. (2005). Deconstructing the cadherin-catenin-actin complex. *Cell*, *123*(5), 889–901. http://dx.doi.org/10.1016/j.cell.2005.09.020.

Yamazaki, Y., Umeda, K., Wada, M., Nada, S., Okada, M., Tsukita, S., et al. (2008). ZO-1- and ZO-2-dependent integration of myosin-2 to epithelial zonula adherens. *Molecular Biology of the Cell*, *19*(9), 3801–3811. http://dx.doi.org/10.1091/mbc.E08-04-0352.

Yang, A., Schweitzer, R., Sun, D., Kaghad, M., Walker, N., Bronson, R. T., et al. (1999). p63 is essential for regenerative proliferation in limb, craniofacial and epithelial development. *Nature, 398*(6729), 714–718. http://dx.doi.org/10.1038/19539.

Yao, M., Qiu, W., Liu, R., Efremov, A. K., Cong, P., Seddiki, R., et al. (2014). Force-dependent conformational switch of alpha-catenin controls vinculin binding. *Nature Communications, 5*, 4525. http://dx.doi.org/10.1038/ncomms5525.

Yap, A. S., Niessen, C. M., & Gumbiner, B. M. (1998). The juxtamembrane region of the cadherin cytoplasmic tail supports lateral clustering, adhesive strengthening, and interaction with p120ctn. *Journal of Cell Biology, 141*(3), 779–789.

Yonemura, S., Wada, Y., Watanabe, T., Nagafuchi, A., & Shibata, M. (2010). alpha-Catenin as a tension transducer that induces adherens junction development. *Nature Cell Biology, 12*(6), 533–542. http://dx.doi.org/10.1038/ncb2055.

Yoshida, K., Yokouchi, M., Nagao, K., Ishii, K., Amagai, M., & Kubo, A. (2013). Functional tight junction barrier localizes in the second layer of the stratum granulosum of human epidermis. *Journal of Dermatological Science, 71*(2), 89–99. http://dx.doi.org/10.1016/j.jdermsci.2013.04.021.

Young, P., Boussadia, O., Halfter, H., Grose, R., Berger, P., Leone, D. P., et al. (2003). E-cadherin controls adherens junctions in the epidermis and the renewal of hair follicles. *EMBO Journal, 22*(21), 5723–5733. http://dx.doi.org/10.1093/emboj/cdg560.

Yuki, T., Hachiya, A., Kusaka, A., Sriwiriyanont, P., Visscher, M. O., Morita, K., et al. (2011). Characterization of tight junctions and their disruption by UVB in human epidermis and cultured keratinocytes. *Journal of Investigative Dermatology, 131*(3), 744–752. http://dx.doi.org/10.1038/jid.2010.385.

Zhou, K., Muroyama, A., Underwood, J., Leylek, R., Ray, S., Soderling, S. H., et al. (2013). Actin-related protein2/3 complex regulates tight junctions and terminal differentiation to promote epidermal barrier formation. *Proceedings of the National Academy of Sciences of the United States of America, 110*(40), E3820–E3829. http://dx.doi.org/10.1073/pnas.1308419110.

CHAPTER THIRTEEN

Cadherin-Based Transsynaptic Networks in Establishing and Modifying Neural Connectivity

Lauren G. Friedman, Deanna L. Benson, George W. Huntley[1]
Fishberg Department of Neuroscience, Friedman Brain Institute and the Graduate School of Biomedical Sciences, Icahn School of Medicine at Mount Sinai, New York, USA
[1]Corresponding author: e-mail address: george.huntley@mssm.edu

Contents

Current Topics in Developmental Biology, Volume 112
ISSN 0070-2153
http://dx.doi.org/10.1016/bs.ctdb.2014.11.025

Abstract

It is tacitly understood that cell adhesion molecules (CAMs) are critically important for the development of cells, circuits, and synapses in the brain. What is less clear is what CAMs continue to contribute to brain structure and function after the early period of development. Here, we focus on the cadherin family of CAMs to first briefly recap their multidimensional roles in neural development and then to highlight emerging data showing that with maturity, cadherins become largely dispensible for maintaining neuronal and synaptic structure, instead displaying new and narrower roles at mature synapses where they critically regulate dynamic aspects of synaptic signaling, structural plasticity, and cognitive function. At mature synapses, cadherins are an integral component of multiprotein networks, modifying synaptic signaling, morphology, and plasticity through collaborative interactions with other CAM family members as well as a variety of neurotransmitter receptors, scaffolding proteins, and other effector molecules. Such recognition of the ever-evolving functions of synaptic cadherins may yield insight into the pathophysiology of brain disorders in which cadherins have been implicated and that manifest at different times of life.

1. INTRODUCTION

Our behaviors, thoughts, and actions reflect highly organized synaptic networks that are established principally during brain development, when molecular cues and neural activity collaborate to generate neural circuits that are progressively honed by sensory and motor experience during postnatal life (Benson, Colman, & Huntley, 2001). A large number of secreted and cell-surface molecular cues guide all phases of brain development, and key among these are structurally and functionally diverse families of cell adhesion molecules (CAMs). In particular, many different CAMs become concentrated at synapses, which are junctional sites of interneuronal communication, where they compose a transsynaptic adhesive apparatus that bridges rigidly apposed pre- and postsynaptic membranes across the intervening synaptic cleft (Benson & Huntley, 2012; Dalva, McClelland, & Kayser, 2007). There are two emergent features of synaptic CAMs that have modified our understanding of how CAMs contribute to the generation, maintenance, and pathological dysfunction of brain circuits. First, the function of CAMs is ever changing over developmental time, evolving in conformity with the complexities of the circuits they support. While initially providing spot welds of structural adhesion, they become dynamic and multifunctional signaling nodes that are integrated with a vast array of scaffolding, cytoskeletal, and effector proteins on both sides of the synapse, capable of exerting significant influence over neurotransmission. Second, CAMs of one family are unlikely to act in isolation of other CAM families, as recent data suggest surprisingly broad molecular cross talk between CAM families. Together, such emerging recognition of how and when CAMs interact has important implications for how they might contribute to brain disorders that manifest at different times in life. The goal of this chapter is to highlight some of these recent developments focusing on cadherins (*ca*lcium-*d*ependent ad*her*ent prote*ins*) and their varied roles in development and maintenance of synaptic circuits principally in the vertebrate nervous system.

2. THE CADHERIN SUPERFAMILY

Numerous reviews have comprehensively covered the organization and classification of the cadherin superfamily. Here, we outline the basics and refer the reader to other reviews for greater depth and detail (Hulpiau & van Roy, 2009; Yagi & Takeichi, 2000).

2.1. Classification and structure

Cadherins are an evolutionarily conserved family of adhesion proteins having first made their appearance in sponge epithelia where their contributions to cell–cell adhesion likely played a key role in the generation of multicellular animals (Abedin & King, 2008; Sakarya et al., 2007). Over the course of evolution, the cadherin repertoire has become extensive in vertebrates, containing 114 members in humans (Hulpiau, Gul, & van Roy, 2013). The superfamily is composed of Type I to Type IV cadherins, desmocollins, desmogleins, Flamingo/Celsr, clustered and nonclustered protocadherins (Pcdhs), Ret, Dachsous and Fat, calsyntenins, and primitive cadherins (Hulpiau & van Roy, 2009). All are membrane glycoproteins having two or more extracellular cadherin (EC) domains and conserved calcium-binding motifs. Type I and II cadherins, the desmosomal cadherins, Dachsous, Fat, Pcdhs, and calsyntenins engage in or regulate junctional adhesion in mammals. Some superfamily members appear to have evolved highly specialized functions. The desmosomal cadherins, which are specialized for the generation of desmosomes, and Dachsous and Fat bind heterophilically to generate a specialized adhesion site that helps to confer apical identity important for the generation of an organized cerebral cortex in mouse and planar cell polarity in *Drosophila* (Ishiuchi, Misaki, Yonemura, Takeichi, & Tanoue, 2009; Thomas & Strutt, 2012). Others, such as

Flamingo in *Drosophila* and the Celsr family in vertebrates, as well as the calsyntenins, play important and varied roles in brain development that partially overlap with some classic cadherin functions. However, their mechanism(s) of action appears to be mostly distinct from other superfamily members (Berger-Muller & Suzuki, 2011; Boutin, Goffinet, & Tissir, 2012; Pettem et al., 2013; Ster et al., 2014; Um et al., 2014). Here, we will focus principally on those cadherins that participate in and regulate synapse adhesion in mammals: Type I and Type II classic cadherins and Pcdhs. Hereafter, we will refer to classic cadherins simply as "cadherins."

Most cadherins engage in homophilic interactions mediated principally by N-terminal EC domains, but the binding interactions differ between groups. Homophilic adhesion between Type I cadherins is mediated by an N-terminal β-strand containing a conserved tryptophan (Trp2) in the first of five EC domains that engages a hydrophobic pocket in the apposing EC1 to generate an extended, curved *trans* dimer. Because the intermolecular interaction between the two apposing cadherins is the same as would be formed intramolecularly, the domains are "swapped" (Boggon et al., 2002; Shapiro et al., 1995; Shapiro & Weis, 2009). In the presence of physiological levels of calcium, the extracellular domain becomes more rigid (Nagar, Overduin, Ikura, & Rini, 1996; Pokutta, Herrenknecht, Kemler, & Engel, 1994) and *trans* interactions are favored over intramolecular interactions. Individual binding interactions are relatively weak, but when viewed by high-resolution electron microscopy, cadherin-based junctions can form highly ordered arrays, suggesting that strong adhesion is conferred by organized numbers (Al-Amoudi, Diez, Betts, & Frangakis, 2007). The degree to which cadherins interact in *cis in vivo* is not well understood. The binding interface is such that *cis* and *trans* interactions between EC1 domains are mutually exclusive, consistent with the increase in *cis* binding and loss of *trans* binding seen in the absence of calcium (Shapiro & Weis, 2009). At the same time, there may be some *cis* interactions that can coexist with *trans* binding. Crystal structure data suggest that specific *cis* interactions can form between EC1 and a site near the EC2–EC3 linker regions of some classic cadherins, and EM data suggest that there can be *cis* interactions at desmosomes (Shapiro & Weis, 2009; Wu, Vendome, Shapiro, Ben-Shaul, & Honig, 2011). In contrast, homophilic binding between T-cadherin (cadherin-13) occurs only via crossed contacts at a region between EC1 and EC2 (Ciatto et al., 2010), an alternate adhesive conformation utilized in other classic cadherins as well, either as a binding intermediate that facilitates transition into or out of the strand-swapped dimer

(Harrison et al., 2010; Hong, Troyanovsky, & Troyanovsky, 2011; Nagar et al., 1996; Pertz et al., 1999), or as a preferred interaction under certain circumstances. Single molecule force measurements suggest that tensile force can promote the generation of crossed or X-dimers that become stronger when pulled and weaker when released (Manibog, Li, Rakshit, & Sivasankar, 2014; Rakshit, Zhang, Manibog, Shafraz, & Sivasankar, 2012), actions that would be useful for generating traction for cell or growth cone migration. Type II cadherins utilize an adhesion mechanism similar to that of Type I cadherins. They can form the crossed intermediate, but they have two conserved tryptophans in EC1 (Trp2 and Trp4) that bind a larger, apposing hydrophobic pocket (Patel et al., 2006).

Classic cadherins share a highly similar intracellular motif with a conserved juxtacellular domain and an extended β-catenin-binding motif (Huber & Weis, 2001). The juxtamembrane domain binds p120 and δ-catenins, which regulate lateral clustering, surface distribution, and interactions with secondary binding partners (Rubio, Curcio, Chauvet, & Bruses, 2005; Silverman et al., 2007; Yap, Niessen, & Gumbiner, 1998); β-catenin provides an interface with a wide variety of partners, including α-catenin, which provides a dynamic bridge to actin cytoskeleton (Yamada, Pokutta, Drees, Weis, & Nelson, 2005). Thus, different cadherins bind similar intracellular partners and similarly engage in intracellular anchoring and adhesion, and specificity is driven by *trans* engagement with particular cadherin partners as well as by binding preferentially to particular *cis* partners.

Protocadherins have six or seven EC repeats and a short cytoplasmic domain. They are divided into the clustered Pcdhs (α-, β-, and γ-Pcdhs), a term that refers to their genomic organization (Wu & Maniatis, 1999), and the nonclustered δ-Pcdhs (Hulpiau & van Roy, 2009). Multiple versions of α- and γ-Pcdhs are generated from mRNAs having different transcription start sites spliced to shared α or γ constant C-terminal exons (Sugino et al., 2000; Tasic et al., 2002; Wu et al., 2001). β-Pcdhs are generated similarly, but do not have a constant C-terminal domain (Hirano et al., 2012; Wang, Su, & Bradley, 2002). δ-Pcdhs are encoded by separate genes and have two or three conserved cytoplasmic motifs (Vanhalst, Kools, Staes, van Roy, & Redies, 2005).

A study of the molecular basis underlying adhesion between γ-Pcdhs indicates that binding is homophilic, only partially reliant on calcium, and requires the EC1 domain for binding but utilizes EC2 and EC3 (rather than EC1) for recognition (Schreiner & Weiner, 2010). γ-Pcdhs readily form *cis* multimers with themselves and other Pcdhs (Murata, Hamada, Morishita, Mutoh, & Yagi, 2004; Schreiner & Weiner, 2010) in a manner that suggests that combinatorial Pcdh tetramers will bind similar tetramers in *trans* (Schreiner & Weiner, 2010), conferring a rigorous recognition code. Based on the structure of α4-Pcdh (Morishita et al., 2006), Pcdhs have a much shallower hydrophobic pocket in their EC1 domain and lack the conserved Trp residues seen in classic cadherins, but the adhesive interactions shown by γ-Pcdhs strongly suggest that all Pcdhs are likely to be able to use their EC domains to generate adhesion. However, such adhesion must be highly regulated as homophilic adhesion between α-Pcdhs has not yet been demonstrated, and surface expression of α- and γ-Pcdhs is highly regulated. The C-terminal domain of γ-Pcdhs acts to retain most γ-Pcdhs in an intramembranous pool (Fernandez-Monreal, Kang, & Phillips, 2009; O'Leary et al., 2011), and α-Pcdhs appear to use γ-Pcdhs to transit to the membrane surface (Murata et al., 2004). Additionally, some α-Pcdhs can engage in heterophilic interactions with integrins via an RGD motif in EC1 (Morishita et al., 2006; Mutoh, Hamada, Senzaki, Murata, & Yagi, 2004), and many Pcdhs have shown *cis* interactions with classic cadherins (Weiner & Jontes, 2013), suggesting that EC domains may confer *cis* as well as *trans* interaction specificity.

2.2. Cadherin localization in the central nervous system

Cadherins display distinct expression profiles in different regions, circuits, and cell types of the central nervous system (CNS). Their differential spatial and temporal distribution thus influences development and function of these highly structured areas. In general, Type I cadherins, and particularly N-cadherin, which has been the most extensively studied, are broadly expressed in the CNS, whereas Type II cadherins and protocadherins exhibit distinct, and partially overlapping, expression patterns.

The mammalian neocortex contains functionally and structurally distinct areas built upon a highly organized, laminated structure. While gradients of extracellular signaling molecules and transcription factors influence the formation of cortical regions and their laminar structure (reviewed in Sansom & Livesey, 2009), cadherins are also involved in these processes. N-Cadherin is required for development of proper laminar organization, as conditional knockout (KO) of N-cadherin, driven by Cre recombinase under the control of the D6 promoter which inactivates N-cadherin early in embryonic development, causes complete disruption of cortical laminar organization

(Kadowaki et al., 2007). In contrast, extensive *in situ* hybridization studies have revealed that some Type I cadherins (e.g., R-cadherin), Type II cadherins, and Pcdhs mostly exhibit restricted and/or combinatorial expression that is specific to cells, layers and/or cortical regions (Bekirov, Needleman, Zhang, & Benson, 2002; Gil, Needleman, & Huntley, 2002; Hertel & Redies, 2011; Suzuki, Inoue, Kimura, Tanaka, & Takeichi, 1997; Lefebvre et al., 2012; Uemera et al., 2007). Adjacent cortical regions are delineated by unique expression patterns of one or more cadherin(s). For example, cadherin-8 is abundantly expressed in the anterior cingulate cortex (a medial cortical region), while cadherin-6 is concentrated in the adjacent primary motor cortex (Hertel & Redies, 2011). In contrast, cadherin-6, -8, and -10 are all expressed in somatosensory cortex, but exhibit only partially overlapping expression patterns (Gil et al., 2002). Additionally, regions that send projections to other cortical or subcortical areas display similar expression profiles as their targets, suggesting that specific cadherins, or combinations of cadherins, mediate the formation of brain circuits. For example, cadherin-8 is expressed by corticostriatal projection neurons in layer 5 of prefrontal cortex and by medium spiny neurons in striatum and becomes concentrated at striatal excitatory synapses early in postnatal development when this projection is establishing (Fig. 1), suggesting that cadherin-8 is involved in the formation and/or targeting of the corticostriatal projection

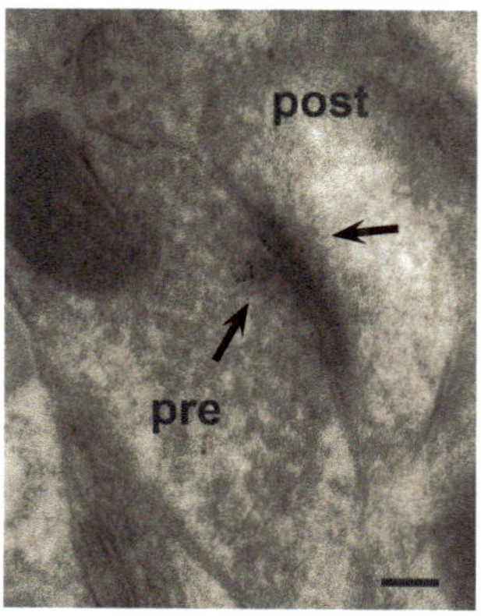

Figure 1 Cadherin-8 is synaptically localized in striatum. This electron micrograph was taken from a section through the dorsal striatum of a P30 mouse that had been processed for immunogold localization of cadherin-8. The gold particles are clustered on both presynaptic (pre) and postsynaptic (post) sides of the asymmetric synapse (arrows), consistent with homophilic recognition, and this is a pattern of localization that is typical for classic cadherins in the brain. Bar, 100 nm.

(Friedman et al., 2014). Similarly, the expression pattern of cadherin-8 in structures of the trigeminal somatosensory system suggests a role in the precise synaptic targeting of specific projections from somatosensory thalamus to the somatosensory (barrel) cortex (Gil et al., 2002). Additionally, several different Type II cadherins are expressed differentially within rat hippocampus. Whereas *in situ* hybridization shows that cadherin-8 is expressed uniformly throughout CA1, CA3, and dentate gyrus, cadherin-10 is most abundantly expressed in CA1, and cadherin-9 in CA3 (Bekirov et al., 2002; Williams et al., 2011).

Although Type I cadherins typically have a rather broad distribution, in the spinal cord, two different Type I cadherins delineate functionally distinct nociceptive synaptic pathways within lamina II of the dorsal horn. E-Cadherin is restricted to synapses formed by nonpeptidergic nociceptive afferents that occupy the inner portion of lamina II (Brock, Elste, & Huntley, 2004; Patil, Brock, Colman, & Huntley, 2009; Shimamura, Takahashi, & Takeichi, 1992), while N-cadherin is found at peptidergic nociceptive afferents that target the outer portion of lamina II (Brock et al., 2004).

Several types of neuroglia, the support cells of the CNS, also express classic cadherins. Astrocytes express N-cadherin and other Type I classic cadherins, but generally do not express Type II cadherins (Gil et al., 2002; Karpowicz et al., 2009; Tomaselli, Neugebauer, Bixby, Lilien, & Reichardt, 1988). During early embryonic development, radial glia, which can divide to form neural and glial progenitors, act as scaffolds to guide newly born neurons during migration to their target layer (Marin & Rubenstein, 2003; Miyata, Kawaguchi, Okano, & Ogawa, 2001). Cadherin-7 is expressed in radial glia and neurons in the basal plate of the spinal cord in chicken (Luo, Ju, & Redies, 2006), suggesting homophilic binding between two cell types aids in early radial migration.

Oligodendrocytes are myelin-producing glia that ensheath axons to facilitate saltatory conduction. Myelination is highly regulated by molecular cues that control specific axon–oligodendrocyte interactions (Bradl & Lassmann, 2010). Several studies suggest that N-cadherin plays an important role in oligodendrocyte targeting and adhesion with axons during the first steps of myelination. N-Cadherin is expressed in cultured oligodendrocytes and is concentrated on major processes in oligodendrocyte progenitors (OPs; Payne, Hemperly, & Lemmon, 1996). Blocking N-cadherin reduces OP migration on astrocyte monolayers (Schnadelbach et al., 2000) and impairs oligodendrocyte-axon adhesion in OPs cocultured on dorsal root ganglia explants or rat cerebellar slices (Schnadelbach, Ozen, Blaschuk,

Meyer, & Fawcett, 2001). Recently, cadherin-19 was identified as a glial-specific cadherin that may be involved in myelination of the avian CNS. Although cadherin-19 expression is absent in rat brain (Takahashi & Osumi, 2005), oligodendrocytes in chicken spinal cord and brain selectively express cadherin-19 (Lin, Luo, & Redies, 2010).

3. FUNCTIONAL UNITS OF ADHESION AND SIGNALING: INTRA- AND INTER-CAM FAMILY CROSS TALK (TABLE 1)

Since classic cadherins and many Pcdhs adhere homophilically in *trans*, some aspects of cadherin-based adhesion are symmetric with the same intracellular partners present on either side of a junction. However, since synapses are polarized, some pre- or postsynaptic-binding partners differ based on local availability or on regulated binding interactions. The complex C-terminal interactions have been reviewed in depth by others (Anderson & Benson, 2006; Arikkath & Reichardt, 2008; Brigidi & Bamji, 2011) and will not be detailed here, but should be noted as a key means by which the strength of cadherin-based adhesion can be adjusted. Fewer intracellular partners have been established for the Pcdhs and those that have been identified appear to be both molecularly and functionally unique to particular Pcdh families (Chen & Gumbiner, 2006; Chen et al., 2009; Nakao, Platek, Hirano, & Takeichi, 2008).

Pre- to postsynaptic adhesion between homophilic pairs gains a layer of specificity by the addition of *cis*- and *trans*-binding partners (Table 1). Classic cadherins commonly engage in *cis* interactions with Pcdhs (Weiner & Jontes, 2013). For example, N-cadherin can bind in *cis* with either Pcdh-19 or -17 and promote homophilic adhesion in *trans* between the Pcdhs (Biswas et al., 2010; Emond et al., 2011). Recent work shows that self-association between phrenic neurons during development requires expression of both cadherin-10 and Pcdh-10, suggesting that the two may be acting as an adhesive complex (Machado et al., 2014). α- and β-Pcdhs can be found in complexes with γ-Pcdhs (Han, Lin, Meng, & Wang, 2010), and γ-Pcdhs can form heteromeric tetramers in *cis* and partner with similar complexes in *trans*, increasing the specificity of interaction (Schreiner & Weiner, 2010). A more limited number of classic cadherins can form *cis* heteromers, but there is no evidence that they prefer heteromers over homomers in *trans* (Shan et al., 2000). Pcdh-8 (δ2 Pcdh, arcadlin) or PAPC serves to reduce adhesion by classic

Table 1 Interactions between cadherins and transmembrane-binding partners

Cadherin	*cis/trans* interactions	Purpose	Citation
N-Cadherin	δ-Pcdh-8/ arcadlin	Activity-mediated induction of δ-Pcdh-8 binding in *trans* promotes N-cadherin internalization in *cis*; *cis* interactions also observed between PAPC and C-cadherin in *Xenopus*	Yasuda et al. (2007) and Chen and Gumbiner (2006)
PAPC (*Xenopus* homolog of δ-Pcdh-8/ arcadlin)	Flrt3 (LRRP) PAPC	*cis* interactions appear to be via C-terminal domains. PAPC attenuates Flrt3 inhibition of C-cadherin adhesion	Chen and Gumbiner (2006) and Chen, Koh, Yoder, and Gumbiner (2009)
N-Cadherin	FGFR	N-Cadherin physically interacts with FGFR1 in breast cancer cells via Ig domains; blocking FGF2 inhibits N-cadherin-mediated growth	Suyama, Shapiro, Guttman, and Hazan (2002) and Williams et al. (2001)
N- and M-cadherin	CDO and BOC	Ig superfamily members CDO and BOC bind N-cadherin and M-cadherin in *cis* and promote muscle cell differentiation	Kang, Feinleib, Knox, Ketteringham, and Krauss (2003)
N-Cadherin	R-cadherin	*cis* or *trans*; does not appear to influence adhesion specificity	Shan et al. (2000)
N-Cadherin	GluA2	GluA2 binds N-cadherin extracellular domain in *cis* or *trans*. Interaction needed to support GluA2 overexpression-mediated increase in dendritic spines and mEPSCs	Saglietti et al. (2007)

Table 1 Interactions between cadherins and transmembrane-binding partners—cont'd

Cadherin	*cis/trans* interactions	Purpose	Citation
N-Cadherin	GluA1	*In vitro* studies in HEK cells are suggestive of *cis* interactions that support surface GluA1 expression	Nuriya and Huganir (2006)
N-Cadherin	δ-Pcdh-19	*Cis* interactions cooperate and promote adhesion mediated by δ-Pcdh-19; δ-Pcdh-17 acts similarly	Biswas, Emond, and Jontes (2010) and Emond, Biswas, Blevins, and Jontes (2011)
γ-Pcdh	Other γ-Pcdhs	Increases binding specificity. Heteromers form in *cis* and bind homomers in *trans*	Schreiner and Weiner (2010)
γ-Pcdhs	α-Pcdhs	*cis* interactions via C-termini as well as extracellular interactions. γ-Pcdhs promote surface distribution of α-Pcdhs	Murata et al. (2004)
Pcdh-α, -γ	Ret	Site of *cis* interaction is not clear. Stimulation by GDNF serves to stabilize both Ret and the Pcdhs	Schalm, Ballif, Buchanan, Phillips, and Maniatis (2010)
Pcdh-γC5	γ2-$GABA_A$ receptor subunit	C-term *cis* interactions that may be unique to γC5. Promotes surface localization of $GABA_A$ receptors and appears to promote inhibitory synapse stability	Li et al. (2012)
Pcdh-α-4 (CNR1)	β1-integrin	*In vitro* studies suggest *trans* binding and adhesion between integrins expressed in HEK cells and the RGD site in Pcdh-α-4	Mutoh et al. (2004)

cadherins. In mouse neurons, Pcdh-8 promotes the internalization of N-cadherin (Yasuda et al., 2007), but interactions between PAPC and Flrt3 outline additional means to modulate adhesion (Chen et al., 2009).

Several additional heterophilic *cis* partners are particularly relevant for cadherin function at synapses. Pcdh-γC5 interacts directly with $GABA_A$-γ receptor subunit via C-terminal domain regions. This interaction is not synaptogenic, but appears to augment the stability of receptors on the surface (Li et al., 2012). N-Cadherin can bind AMPA-type glutamate receptor subunits GluA1, GluA2, or GluA4, and at least for GluA2, such interactions can be in *cis* or in *trans* via an extracellular domain and appear to have a stabilizing effect on the surface levels of GluA subunits (Nuriya & Huganir, 2006; Saglietti et al., 2007). Indirect interactions between N-cadherin and neuroligins, which are probably mediated by the scaffolding protein S-SCAM (MAGI-2), positively modulate synapse formation in developing neurons (Aiga, Levinson, & Bamji, 2011; Jontes, Emond, & Smith, 2004; Stan et al., 2010). Notably, in cultured neurons less than 2 weeks old, N-cadherin is required for neuroligin to have its synaptogenic effect (Stan et al., 2010).

4. DEVELOPMENTAL PHASES OF CIRCUIT ASSEMBLY: EVOLVING ROLES OF THE CADHERIN-BASED ADHESIVE NETWORK

There are phases of circuit formation by which neurons in one brain area (the source) become synaptically coupled to neurons in another brain region (the target; Benson et al., 2001). These phases, which can be overlapping in time, include (1) the generation of source and target neurons (neurogenesis) in germinal zones and migration to their respective terminal locations; (2) the initiation, outgrowth, and targeting of axons from source neurons; (3) elaboration of postsynaptic dendrites and dendritic spines of target neurons; and (4) contact between pre- and postsynaptic membranes and subsequent synaptogenesis between source and target neurons. Cadherins have been implicated in all of these developmental phases, but their contributions at each step are mechanistically distinct (for a comprehensive review of developmental roles, see Hirano & Takeichi, 2012).

4.1. Neurogenesis and migration

A role for cadherin-based adhesive interactions in neurogenesis was speculated upon in early studies showing that neural tissue histogenesis was profoundly disrupted by cadherin function-neutralizing antibodies (Takeichi, Inuzuka, Shimamura, Matsunaga, & Nose, 1990). More recent studies have implicated N-cadherin in regulating proliferation of dopamine neuron progenitors in the ventral brainstem in mice (Sakane & Miyamoto, 2013). Similarly, N-cadherin regulates retinal neurogenesis in zebrafish (Nagashima, Barthel, & Raymond, 2013; Yamaguchi, Imai, Tonou-Fujimori, & Masai, 2010), where N-cadherin and Slit-Robo signaling collaborate to allow detachment of newborn retinal ganglion cells (RGCs) from the apical surface of the neuroepithelium following their terminal mitosis (Wong, Baudet, Norden, Leung, & Harris, 2012). Similar disruptions of N-cadherin-mediated apical attachments of neuroepithelial cells in mice lead to profoundly disorganized cortical structures (Kadowaki et al., 2007). Certain miRNAs (those belonging to the miR379–410 cluster) can bind directly to the 3′-UTR of N-cadherin transcripts and modulate N-cadherin levels in neocortical neural progenitor cells, which in turn controls the proliferation and subsequent migration of young neurons (Rago, Beattie, Taylor, & Winter, 2014). Dynamic trafficking of N-cadherin regulates migration of neocortical glutamatergic neurons along radial glial guides (Kawauchi et al., 2010). Additionally, cortical GABA interneurons require N-cadherin for both tangential migration (from their distant site of origin in the medial ganglionic eminence) to the neocortex and radial migration (which delivers them to their proper cortical layer; Luccardini et al., 2013).

4.2. Axon outgrowth and targeting

There are extensive and compelling data using both *in vitro* and *in vivo* approaches in different model systems showing that classic cadherins promote axon fasciculation and are a potent substrate for axonal outgrowth. Early studies showed that many different types of cultured neurons exhibit robust axonal outgrowth when grown on a purified N-cadherin substrate, on astrocytes, or on various cell lines expressing N-cadherin on their surface (Bixby & Zhang, 1990; Doherty, Skaper, Moore, Leon, & Walsh, 1992; Matsunaga, Hatta, Nagafuchi, & Takeichi, 1988; Tomaselli et al., 1988), effects that are blocked in the presence of N-cadherin-neutralizing antibodies (Doherty, Rowett, Moore, Mann, & Walsh, 1991; Drazba & Lemmon, 1990; Neugebauer, Tomaselli, Lilien, & Reichardt, 1988) or interfering peptides targeting the conserved HAV sequence in Type I cadherins (Blaschuk, Sullivan, David, & Pouliot, 1990). More recent genetic analyses in *Drosophila* and zebrafish also provide strong evidence for N-cadherin and R-cadherin in axon extension, growth cone guidance, and targeting (Babb et al., 2005; Bruses, 2011; Lee, Herman, Clandinin, Lee, & Zipursky, 2001; Nern, Zhu, & Zipursky, 2008; Schwabe, Neuert, & Clandinin, 2013). Cadherin dominant-negative reagents have also been used to investigate cadherin-based outgrowth and targeting. One such reagent, NcadΔE, lacks most of the ectodomain, but retains transmembrane and intracellular (catenin-binding) domains. As such, it inhibits the function of all classic cadherins by competing for β-catenin binding at the membrane (Bozdagi, Valcin, Poskanzer, Tanaka, & Benson, 2004; Kintner, 1992). Hippocampal neurons expressing NcadΔE exhibit significantly impaired axonal growth when grown on N-cadherin-expressing L-cells fibroblasts (Bozdagi et al., 2004), which do not express any classic cadherins endogenously (Nose, Nagafuchi, & Takeichi, 1988; Shan et al., 2000). When NcadΔE was expressed *in vivo* by *Xenopus* RGCs, about 70% of them failed to emit an axon (a failure of axonogenesis), and of those that did, their axon growth was significantly retarded, never reaching the optic tectum, their normal target (Riehl et al., 1996). Some Type II cadherins, such as cadherins 6B, 7, and 11, also promote axonal growth and branching (Barnes, Price, Wentzel, & Guthrie, 2010; Marthiens et al., 2005).

Mechanistically, the axon outgrowth-promoting effects of N-cadherin and other classic cadherins may involve association with coreceptors. Presenting cerebellar neurons with soluble N-cadherin-Fc causes coclustering of endogenous surface N-cadherin and the fibroblast growth factor (FGF) receptor and potently stimulates axon outgrowth in an FGF receptor-dependent manner (Utton, Eickholt, Howell, Wallis, & Doherty, 2001). The FGF receptor may be a common effector of classic cadherin-mediated outgrowth, as studies have shown that cadherin-11 promotes neurite outgrowth by recruiting the FGF receptor through ectodomain binding and triggering downstream effectors involving PLCγ/DAG lipase/CAM kinase and PI3 kinase pathways (Boscher & Mege, 2008; Williams, Mittal, Walsh, & Doherty, 1995; Williams, Walsh, & Doherty, 1994). Other studies support a model in which neurite outgrowth is coordinately regulated by molecular cross talk between N-cadherin and β1-integrin through differential coupling to the protein tyrosine kinase Fer (Arregui, Pathre, Lilien, & Balsamo, 2000). Intracellularly, cadherin-mediated outgrowth is also dependent on distinct signaling cascades initiated by protein tyrosine phosphatase-mu (PTPmu; Burden-Gulley & Brady-Kalnay, 1999; Oblander & Brady-Kalnay, 2010).

While many classic cadherins promote axon extension, others appear to constrain axon growth and targeting. For example, in hippocampal organotypic cultures, the axons of dentate gyrus granule neurons (mossy fibers) normally grow in a tightly fasciculated bundle to innervate CA3 pyramidal cell dendrites principally in stratum lucidum, as they do *in vivo* (Frotscher & Heimrich, 1995). However, treating hippocampal organotypic cultures with an exogenous cadherin-8 peptide inhibitor leads to exuberant mossy fiber growth, loss of fasciculation, and loss of laminar specificity (Bekirov, Nagy, Svoronos, Huntley, & Benson, 2008). These effects are not likely to represent off-target effects of N-cadherin blockade, because the cadherin-8 peptide inhibitor enhances axon outgrowth on a cadherin-8 substrate, but has no effects on outgrowth on an N-cadherin substrate. Similarly, the extension of motor neuron axons expressing T-cadherin is significantly impaired when grown on T-cadherin substrates (Fredette, Miller, & Ranscht, 1996). Together, these studies show that cadherins work coordinately to promote and constrain fasciculation and extension of growing axon tracts (Treubert-Zimmermann, Heyers, & Redies, 2002; Wöhrn et al., 1999).

In addition to affecting axon outgrowth, classic cadherins can also act as a "stop signal" to direct ingrowing axons to their correct terminal target layer or region. For example, N-cadherin becomes concentrated in retinorecipient layers of chick optic tectum contemporaneously with the arrival of retinal axons and subsequent synaptogenesis, but fails to do so in enucleated embryos (Yamagata, Herman, & Sanes, 1995). Treating tectal slices with N-cadherin function-blocking antibodies during this early period of ingrowth impairs laminar targeting of ingrowing retinal axons (Inoue & Sanes, 1997). Similarly, in the mammalian visual system, RGC axons in cadherin-6-deficient mice fail to innervate their appropriate visual targets, but instead project to inappropriate visual nuclei (Osterhout et al., 2011). In rodent barrel cortex—the part of primary somatosensory cortex containing an isomorphic map of the large facial whiskers of the contralateral snout in layer 4—thalamocortical inputs from two distinct nuclei, the ventral posterior medial (VPM) nucleus, and the medial division of the posterior nucleus (POm), both target layer 4 but these two thalamic streams terminate in mutually exclusive, interdigitating zones within layer 4. Studies have shown that N-cadherin becomes concentrated at thalamocortical synapses of one pathway (VPM-to-layer 4 synapses in barrel centers; Huntley & Benson, 1999), while cadherin-8 becomes concentrated at thalamocortical synapses of the other pathway (POm-to-layer 4 synapses in barrel septa; Gil et al., 2002) during the early postnatal period when these projections form. Such cadherin specificity and matching between source (thalamic nucleus) and target (layer 4 neurons) suggest that these cadherins play an important role in the targeting of these two converging, but separated thalamic input streams to layer 4. In partial confirmation of this idea, when N-cadherin function-neutralizing antibodies are applied to organotypic cocultures of thalamus and somatosensory cortex, thalamic axons grow normally through deep layers at the same rate as they grow through deep layers in untreated cocultures, but instead of terminating in their target layer 4, they grow inappropriately through layer 4 to reach the pial surface (Poskanzer, Needleman, Bozdagi, & Huntley, 2003). Direct evidence that Type II cadherins coordinate synaptic targeting and specificity comes from recent studies showing that cadherin-9 is required for selective targeting and synapse formation between hippocampal dentate gyrus (DG) granule neurons and CA3 pyramidal neurons (Williams et al., 2011). This study showed that in culture, DG granule neurons that had equal access to all hippocampal cell types, nonetheless preferentially formed synapses with their natural target, CA3 neurons, rather than indiscriminately with CA1 neurons. The molecular basis for such selective synaptic partnering involves cadherin-9, as loss of this cadherin from either DG neurons or CA3 neurons disrupted mossy fiber targeting and synapse formation.

Some members of the Pcdh family have also been implicated in axon targeting. For example, deletion of the constant region of the α-Pcdh gene cluster in mice leads to abnormal targeting of olfactory axons to their appropriate glomeruli in the olfactory bulb (Hasegawa et al., 2008) and to inhomogeneities in the density of serotonergic innervation in several structures throughout the brain (Katori et al., 2009). Other studies in mice have shown that genetic deletion of Pcdh-10 (a δ-Pcdh) impairs striatal axon outgrowth and impedes the thalamocortical projection through the ventral telencephalon (Uemura, Nakao, Suzuki, Takeichi, & Hirano, 2007). Inactivation of Celsr3 causes hypotrophy of the cortex and impedes tract formation of anterior commissure and corticospinal, corticofugal, and thalamocortical tracts in Celsr3 KO mice (Tissir, Bar, Jossin, De Backer, & Goffinet, 2005).

4.3. Elaboration of dendrites, arborization, and self-avoidance

Classic cadherins and γ-Pcdhs have been implicated in dendritic development (Garrett, Schreiner, Lobas, & Weiner, 2012; Hirano & Takeichi, 2012; Zhu & Luo, 2004). In dissociated hippocampal cultures,

N-cadherin promotes dendritic outgrowth (Esch, Lemmon, & Banker, 2000), while in organotypic hippocampal slice cultures, interfering with N-cadherin leads to impaired dendritic arbors of CA3 pyramidal neurons attributed to both an arrest in initial dendrite growth and an increase in dendrite retraction (Bekirov et al., 2008). In *Drosophila*, DN-cadherin functions in dendrodendritic interactions between olfactory projection neurons that restrict dendritic arbors to a single glomerulus (Zhu & Luo, 2004). Similarly, activity-mediated enhancement of dendritic arborization in cultured hippocampal neurons leads to increased surface expression of N-cadherin/ β-catenin and requires N-cadherin/β-catenin interactions between neurons (Tan, Peng, Song, Zheng, & Yu, 2010). The stabilization of dendritic arbors by classic cadherins may require interactions with other effectors. For example, in chick retina, RGC dendritic arbors are rapidly destabilized and retract following β1-integrin blockade; disrupting the juxtamembrane region of N-cadherin generates a β1-integrin inactivation signal leading to dendritic retraction (Marrs et al., 2006). In cultured cortical neurons, reducing levels of afadin, which is a Ras/Rap effector that associates with and regulates the cadherin/catenin system as well as nectins, Ephs/ephrins, and neurexins (Beaudoin et al., 2012; Buchert et al., 1999; Takai, Ikeda, Ogita, & Rikitake, 2008), leads to impaired dendritic growth (Srivastava et al., 2012). While blocking N-cadherin generally seems to impair dendritic growth, overexpression of N-cadherin, β-catenin, or αN-catenin in hippocampal neurons promotes dendritic growth and complexity (Peng et al., 2009; Yu & Malenka, 2003).

An important binding partner of classic cadherins that also regulates development and maintenance of dendritic arbors is δ-catenin, a member of the p120-catenin family whose localization is restricted to dendrites and spines (Kosik, Donahue, Israely, Liu, & Ochiishi, 2005) and whose hemizygous loss in humans is associated with the mental retardation syndrome Cri-du-Chat (Medina, Marinescu, Overhauser, & Kosik, 2000). In cultured hippocampal neurons, dendritic branching is enhanced by overexpression of δ-catenin; conversely, dendritic morphogenesis is impaired by δ-catenin ablation (Arikkath et al., 2008; Kim et al., 2008). In contrast, in δ-catenin KO mice examined *in vivo*, cortical neuronal dendritic complexity was entirely normal through 5 postnatal weeks, but thereafter, significantly regressed, resulting in late postnatal abnormalities in cortical responsiveness and receptive field properties (Matter, Pribadi, Liu, & Trachtenberg, 2009). Because δ-catenin is a multifunctional protein with a number of interacting signaling and scaffolding partners in addition to cadherins (Brigidi & Bamji, 2011), it is not entirely clear whether or to what extent such regulation of dendritic development or maintenance by δ-catenin reflects interactions with cadherins.

Recent studies have shown that γ-Pcdhs regulate development of cortical pyramidal cell dendritic arbors (Garrett et al., 2012). Using a conditional allele that restricts disruption of the entire gene cluster to cortical pyramidal neurons and astrocytes starting in embryonic life, mutant pyramidal neurons failed to develop their typical complex dendritic arbors, particularly those that ramify in layer 1, but there were no obvious effects of deleting the γ-Pcdh gene cluster on migration, lamination, or synaptogenesis. In contrast to embryonic deletion, using an inducible mutant allele that disrupted the γ-Pcdh cluster after cortical dendritic arbors formed failed to affect dendritic complexity, demonstrating that γ-Pcdhs are required for the elaboration, but not maintenance, of cortical dendritic complexity. Functional analyses indicated that γ-Pcdhs promoted dendritic arborization by regulating a focal adhesion kinase/protein kinase C/MARCKS signal cascade (Garrett et al., 2012).

Dendrite self-avoidance, where dendritic branches of the same neuron repel each other so as to ensure even coverage of receptive territory (Grueber & Sagasti, 2010), is another function that has been attributed to γ-Pcdh. Conditionally ablating the γ-Pcdh gene cluster in retinal starburst amacrine cells or cerebellar Purkinje cells resulted in a cell-autonomous failure in dendritic self-avoidance (Lefebvre, Kostadinov, Chen, Maniatis, & Sanes, 2012). Interestingly, molecular replacement in KO neurons with a single γ-Pcdh isoform restored self-avoidance, but decreased interactions between dendrites of neighboring neurons. Thus, homophilic recognition between sister branches expressing the same γ-Pcdh isoforms mediated a repulsive signal to ensure self-avoidance. However, because neuronal expression of different γ-Pcdh isoforms is stochastic (Toyoda et al., 2014) and isoform interactions are both combinatorial (in *cis*) and strictly homophilic (in *trans*; Schreiner & Weiner, 2010), dendritic branches from neighboring neurons are normally free to intermingle because of the low probability that neighboring neurons would express exactly the same combinations of isoforms.

A potential role for cadherin-8 in dendrite arborization and self-avoidance was recently described (Friedman et al., 2014). Cultured cortical neurons in which levels of cadherin-8 were significantly reduced by siRNA-mediated cadherin-8 knockdown exhibited increased numbers of dendritic branches and an increased number of dendrite self-crossings, suggesting a defect in dendrite self-avoidance.

4.4. Synaptogenesis, dendritic spine formation, and synaptic plasticity in developing neurons

Synaptogenesis is initiated by adhesion between heterophilic binding proteins, some of which are targeted to axons and the others to cell bodies and dendrites. Such synaptogenic proteins are specialized for the generation of excitatory or inhibitory synapses, and heterophilic binding ensures that resulting synapses are polarized, forming mostly between presynaptic axon terminals and postsynaptic cell bodies or dendrites. Neither classic cadherins nor Pcdhs alone can induce the formation of synapses, but strong evidence indicates that in young neurons, classic cadherins are essential partners in the combinatorial process of synapse adhesion and stabilization, and Pcdhs can act as priming factors promoting synaptogenesis (Garrett & Weiner, 2009; Li et al., 2012; Waites, Craig, & Garner, 2005).

Proteins that can induce the formation of synapses have been most commonly identified in an assay in which the protein in question is expressed in a nonneuronal cell line, like COS or HEK cells, that is cocultured with neurons and then assayed for the presence of heterologous "synapses" (Scheiffele, Fan, Choih, Fetter, & Serafini, 2000). While not sufficient for inducing synapses on their own, classic cadherins are required for the inductive process of synaptogenesis because synapses cannot be induced in neurons or cell lines that express synaptogenic proteins, but lack classic cadherins (Aiga et al., 2011; Jungling et al., 2006; Stan et al., 2010) or that overexpress inhibitors of classic cadherin adhesion (Bozdagi et al., 2004; Togashi et al., 2002). Consistent with this early and essential role, N-cadherin and β-catenin are concentrated and distributed evenly at both GABAergic and glutamatergic nascent junctions (Benson & Tanaka, 1998; Elste & Benson, 2006; Jontes et al., 2004; Yamagata et al., 1995), and synapses are lost when cadherins are removed or their adhesion is inhibited (Togashi et al., 2002; Zhang & Benson, 2001). Cadherin actions at synapses can be regulated selectively as β-catenin can be synthesized locally at sites of early contact (Huttelmaier et al., 2005; Taylor, Wu, Tai, & Schuman, 2013), and unprocessed N-cadherin retaining its N-terminal prodomain can compete with processed N-cadherin at synapses to reduce adhesion and decrease synapse number (Reines et al., 2012). Also, N-cadherin may be partially synaptogenic for cholinergic synapses as it can induce the formation of presynaptic terminals with heterologous cells when accompanied by p120-catenin, but such sites do not recruit postsynaptic cholinergic receptors and are not functional (Flannery & Bruses, 2012).

A principal role for the N-cadherin/β-catenin complex at nascent synapses is in the assembly of presynaptic vesicle clusters. Expression of the dominant-negative NcadΔE greatly reduces size, intensity, and function of presynaptic terminals (Bozdagi et al., 2004; Togashi et al., 2002), a finding that resembles cultured neurons expressing defective β-catenin (Bamji et al., 2003). While pan-cadherin manipulations show the strongest effects, stem-cell-derived neurons having a selective deletion of N-cadherin show impaired vesicle recruitment and short-term presynaptic plasticity (Jungling et al., 2006). The mechanism is not fully worked out, but data from cultured neurons suggest that β-catenin tethered by N-cadherin binds to the scaffolding protein, Scribble, and this complex appears to recruit β-pix, a Rac/Cdc42 guanine-exchange factor that can enhance actin polymerization (Sun, Aiga, Yoshida, Humbert, & Bamji, 2009; Sun & Bamji, 2011). Since the generation and maintenance of synaptic vesicle clusters in young neurons are also dependent on an intact F-actin cytoskeleton (Morales, Colicos, & Goda, 2000; Zhang & Benson, 2001), cadherins likely regulate vesicle clusters by promoting the generation of a physical restraint. Additionally, N-cadherin may promote synapse maturation by facilitating calcium influx in developing neurons through an N-cadherin/RhoA/cytoskeletal cascade that affects voltage-gated calcium channels (Marrs, Theisen, & Bruses, 2009).

Classic cadherin adhesion also influences the normal generation and early stability of dendritic spines, highly specialized protrusions receiving the bulk of excitatory synaptic input. Neurons expressing a pan-cadherin dominant-negative construct produce thin and pointed dendritic protrusions that nevertheless retain synaptic contact (Togashi et al., 2002). The phenotype closely resembles that seen in neurons in which F-actin has been depolymerized (Zhang & Benson, 2001) and is reproduced in neurons lacking αN-catenin (Abe, Chisaka, Van Roy, & Takeichi, 2004; Togashi et al., 2002), which links classic cadherin tails to F-actin cytoskeleton. Collectively, the data support the view that classic cadherin adhesion recruits and probably stabilizes F-actin in developing dendritic spines.

Pcdhs have a pleiomorphic and less well understood impact on synapse formation. Based on their molecular diversity, timing of expression, and unique enrichment in vertebrate brain, most early work on Pcdhs conjectured that family members would drive synapse recognition and adhesion. There are few data that support this idea, but Pcdhs clearly regulate synaptogenesis (Weiner & Jontes, 2013). Mice lacking all γ-Pcdhs have a decreased density of excitatory and inhibitory synapses in spinal cord. While it was initially suspected that the decreased density could be due to the higher rates of cell death that are also seen in the absence of γ-Pcdhs (Wang, Weiner, et al., 2002), these two effects are separable as synapse density remains diminished even when the effects on cell death are bypassed by crossing the γ-Pcdhs KO mice to mice lacking the proapoptotic protein Bax (Weiner, Wang, Tapia, & Sanes, 2005). The synapse promoting effect appears to arise, at least in part, from γ-Pcdh expression in astrocytes and binding to neuronal γ-Pcdh. Astrocytes produce a variety of soluble priming factors that promote synaptogenesis (Ullian, Christopherson, & Barres, 2004; Waites et al., 2005), but this priming effect is substantially (albeit, transiently) inhibited when astrocytes lacking γ-Pcdhs are cocultured with wild-type spinal cord neurons, presumably by limiting neuron-astrocyte contact (Garrett & Weiner, 2009). Cultured hippocampal neurons lacking a subgroup of γ-Pcdhs (γ-Pcdh-C3, -C4, and -C5) have fewer GABAergic synapses. Here, however, the action is not on priming new synapse formation, but rather is on postsynaptic *cis* interactions between this group of γ-Pcdhs and $GABA_A$ receptor γ2 subunits, which appear to be important for stabilizing inhibitory synapses (Li et al., 2012). In an entirely different role, in dorsal striatum, δ-Pcdh-17 is dispensable for the normal targeting and generation of synapses, but appears to limit presynaptic vesicle number and distribution in adolescent mice (Hoshina et al., 2013). In all cases, the actions of particular Pcdhs appear to be restricted to particular neuron groups or areas.

The actions of classic and Pcdhs at the synapse appear to change over the course of synapse development. This has been best studied with classic cadherins. Once synapses have formed, classic cadherin localization assumes a more clustered distribution either within synapses, where they form discrete sites of adhesion, or sometimes within adjacent and extrasynaptic adherens junctions (Elste & Benson, 2006; Uchida, Honjo, Johnson, Wheelock, & Takeichi, 1996). At this stage of synapse development, classic cadherins can be removed from synapses and the basic structure of synapses is still retained (Zhang & Benson, 2001). However, interfering with cadherin-based adhesion at this time alters the shape and/or stability of dendritic spines (Abe et al., 2004; Arikkath et al., 2009; Israely et al., 2004; Mendez, De Roo, Poglia, Klauser, & Muller, 2010) in the absence of synapse loss. Additionally, presynaptic N-cadherin is essential for maintaining vesicle availability during high-frequency stimulation (Jungling et al., 2006).

In adolescent rodent hippocampus and in cultured hippocampal neurons older than about 3 weeks ("mature" cultures), N-cadherin regulates lasting forms of synapse plasticity. Lasting changes in synapses involve both changes in size (bigger synapses are stronger synapses) and changes in the efficacy of pre- to postsynaptic communication. In cultured neurons, strong stimulation of glutamate receptors induces the expansion of dendritic spine heads and a temporary relocalization of N-cadherin (Okamura et al., 2004; Tanaka et al., 2000; Yam et al., 2013). While the precise mechanism is not understood, maintained spine enlargement in cultured neurons requires N-cadherin and an intact actin cytoskeleton (Okamura et al., 2004). In what may be a related process, N-cadherin is also required for the stabilization of newly formed dendritic spines induced by long-term potentiation (LTP) in organotypic hippocampal slice cultures (Mendez et al., 2010). Data from cultured neurons suggest that such activity-driven changes in N-cadherin adhesion are mediated by regulated recruitment and stabilization of β-catenin to active synapses (Murase, Mosser, & Schuman, 2002), as well as by a more stable form of cadherin adhesion (Tanaka et al., 2000). In acute hippocampal slices, N-cadherin adhesion is required for the induction (Tang, Hung, & Schuman, 1998) and maintenance (Bozdagi, Shan, Tanaka, Benson, & Huntley, 2000; Bozdagi et al., 2010) of LTP at CA1 synapses. As mentioned, N-cadherin can interact in *cis* with AMPA receptor subunits GluA1 and GluA2, providing a direct mechanism to stabilize AMPA-type glutamate receptors at sites of synaptic adhesion. Additionally, the C-terminal domain of N-cadherin can bind directly to the scaffolding protein GRIP1, which can independently bind GluA2 (and probably GluA3) and motor protein KIF5. All of these proteins can be found in trafficking vesicles, suggesting a means by which increases in synapse size and activity could be coupled (Heisler et al., 2014).

The converse of synaptic LTP is synaptic long-term depression (LTD). Brief stimulation of cultured neurons with NMDA, which can induce the internalization of AMPA-type glutamate receptors and produce LTD, alters the regulated endocytosis of N-cadherin (Tai, Mysore, Chiu, & Schuman, 2007) and severs links between AKAP79 (A Kinase-Anchoring Protein) and an N-cadherin/β-catenin complex in a process likely to diminish phosphorylation and promote internalization of AMPA receptors (Dell'Acqua et al., 2006; Gorski, Gomez, Scott, & Dell'Acqua, 2005). Over a longer time scale (hours vs. minutes), strong stimulation in cultured hippocampal neurons promotes homophilic Pcdh-8 binding, activation of TAO2β kinase, a C-terminal Pcdh-8-binding partner, and MEK3 activator and increases p38 MAP kinase activity, all of which serves to promote N-cadherin recycling and decrease adhesion (Yasuda et al., 2007). Collectively, the data obtained from young to adolescent neurons suggest that strong, but dynamic N-cadherin-based synaptic adhesion stabilizes dendritic spines and can increase the strength of glutamatergic synapses. However, one caveat is that the effects of this ubiquitous glutamatergic synaptic cadherin are commonly used to make generalizations about the actions of the family, but they may be distinct. For example, mice lacking cadherin-11 show enhanced, rather than diminished, hippocampal LTP (Manabe et al., 2000).

5. BEYOND DEVELOPMENT: WHAT DOES THE CADHERIN–CATENIN TRANSSYNAPTIC NETWORK CONTRIBUTE TO BRAIN FUNCTION THROUGHOUT LIFE?

A great deal is now known about the multifaceted ways in which cadherins contribute to structural and functional development of brain circuits, particularly stages encompassing generation of synaptic connectivity, as attested to above. However, CAMs generally and cadherins particularly remain prominent components of synapses into adulthood, well beyond the developmental period of circuit and synapse formation. This raises some important questions: What do cadherins contribute to mature synapse and brain function? Are their developmental roles merely extended into maturity or do they take on different functions? If so, how are these functions achieved molecularly? It has generally not been possible to address these questions in the types of cell culture models that dominate studies of cadherin function during development because neurons grown in culture never achieve the level of maturity and functional complexity equivalent to neurons in an adult brain. Additionally, cell culture models cannot be used to assess behavioral consequences of altering the function of the cadherin–catenin transsynaptic apparatus. Accordingly, several recent studies have tackled these and related questions using similar Cre–LoxP-based strategies to conditionally manipulate cadherin and/or catenin levels, stability and function in the hippocampus of mice after the period of early brain development, and assessing connectivity, synaptic plasticity, and cognitive and behavioral performance in adulthood. The postnatal timing of the genetic manipulation in these studies ensures that the cadherin/catenin system is operational during early stages of brain development.

5.1. Cadherins become dispensible for maintaining hippocampal connectivity but regulate dynamic aspects of synaptic function and plasticity

Several recent studies used Cre–LoxP technology to postnatally manipulate cadherins/β-catenin with a Cre-driver line of mice in which Cre recombinase was driven by the α-calcium–calmodulin-dependent protein kinase type-II (αCaMKII) promoter (Tsien et al., 1996). This Cre-driver line ensures conditional gene manipulation starting around the third postnatal week in hippocampal and other forebrain excitatory neurons, but does not affect GABA neurons or glial cells, which do not express αCaMKII. In one set of studies (Bozdagi et al., 2010; Nikitczuk et al., 2014), synaptic N-cadherin was ablated and β-catenin levels were lowered in excitatory neurons of hippocampus. In a second study (Mills et al., 2014), synaptic cadherins in hippocampal excitatory neurons were surface stabilized by a heterozygous mutation of β-catenin that rendered the molecule resistant to degradation. This mutant β-catenin showed enhanced association with synaptic cadherins, but otherwise retained all of its natural binding sites and functional abilities, including interactions with αN-catenin and components of the Wnt-signaling pathway. Because a pan-cadherin antibody was used for synaptic localization, it is not possible to know definitively which cadherin(s) were stabilized at the synapse by this particular β-catenin mutation. Remarkably, both of these studies showed that regardless of whether N-cadherin was eliminated (and β-catenin reduced) at synapses or cadherins (and β-catenin) were stabilized at synapses, adult hippocampal structure, dendritic organization, synapse and spine density, and baseline properties of synaptic neurotransmission were entirely indistinguishable from adult control (floxed) mice. Additionally, there were no (Bozdagi et al., 2010) or only very subtle (Mills et al., 2014) effects on presynaptic release. The lack of effects of two very different manipulations of synaptic cadherin/β-catenin levels and function on any of these parameters of hippocampal architecture, baseline neurotransmission, and presynaptic release was surprising, given the broad developmental roles the cadherin–catenin system has in establishing all of them. Instead, both sets of studies showed profound deficits in persistence of long-lasting forms of synaptic plasticity at mutant hippocampal synapses, as well as impairments in cognitive flexibility, and both point to the idea that deficits in trafficking or localization of particular glutamate receptor subunits underlie such synaptic plasticity and cognitive deficits. Previous studies had shown that through ectodomain interactions, N-cadherin could bind in *cis* with GluA1 to control its surface stability (Nuriya & Huganir, 2006), as well as in *cis* or in *trans* with GluA2 (Saglietti et al., 2007), but the significance of these interactions to adult synaptic and behavioral function *in vivo* had not been clear.

In the case of the conditional N-cadherin KO mice (Bozdagi et al., 2010; Nikitczuk et al., 2014), LTP and concurrent spine enlargement at adult CA1 synapses were both induced normally, but both became unstable, collapsing rapidly to baseline levels of synaptic strength or spine size, indicating a deficit in maintenance. In contrast, LTD of mutant CA1 synapses was unaffected. Normally, at wild-type CA1 synapses, both LTP and LTD require NMDA receptors (NMDARs) for induction (Collingridge & Bliss, 1987), but depend on modifications in the function and trafficking of AMPAR subunits for persistence (Malinow & Malenka, 2002). In particular, the insertion of GluA1 into the synaptic membrane, which is enabled by the scaffolding protein PSD95 as part of an activity-regulated delivery pathway (Ehlers, Heine, Groc, Lee, & Choquet, 2007; Opazo & Choquet, 2011; Passafaro, Piech, & Sheng, 2001; Schnell et al., 2002; Stein, House, Bredt, & Nicoll, 2003; Yang, Wang, Frerking, & Zhou, 2008; Yudowski et al., 2007), is required for, and links, the maintenance of both LTP and spine enlargement (Ehrlich & Malinow, 2004; Kopec, Li, Wei, Boehm, & Malinow, 2006; Kopec, Real, Kessels, & Malinow, 2007; Park, Penick, Edwards, Kauer, & Ehlers, 2004; Shi et al., 1999; Yang et al., 2008). Consistent with the plasticity deficits observed in the mutant mice, further analysis revealed that N-cadherin-deficient excitatory synapses contained normal levels and distribution of NMDAR subunits (hence, normal LTP/LTD induction), but significantly lower levels and altered distribution of both GluA1 and PSD95. This suggests that without N-cadherin, delivery, insertion, and/or trapping of GluA1 at the synaptic membrane is deficient, thus rendering both structural and functional forms of plasticity unstable. Additionally, one recent model proposed that N-cadherin is part of a transsynaptic molecular complex in which postsynaptic δ-catenin links N-cadherin to both the AMPARs and the NMDAR–CamKII complex, thus acting as a molecular "tag" at potentiated synapses to coordinately regulate LTP-associated changes in synaptic strength with pre- to postsynaptic synapse growth (Sanhueza & Lisman, 2013).

In contrast to changes in GluA1 and PSD95, levels and localization of GluA2 and its scaffolding protein S-SCAM/MAGI-2 were unaffected at N-cadherin-deficient synapses, which may have accounted for normal LTD and baseline properties of neurotransmission. LTD requires endocytocis of GluRs, particularly GluA2, from the synaptic membrane (Malinow & Malenka, 2002), while baseline neurotransmission at CA1 synapses is mediated mostly by non-GluA1-containing AMPARs (Danielson et al., 2012; Mainen, Jia, Roder, & Malinow, 1998; Zamanillo et al., 1999).

In the case of the cadherin/β-catenin-stabilized mutant mice (Mills et al., 2014), there were different plasticity deficits that were observed in comparison with the N-cadherin-deficient mice. Nevertheless, such deficits could also be accounted for by defective AMPAR trafficking that was largely complementary to that suggested by the N-cadherin-deficient mice. In the cadherin/β-catenin-stabilized mice, both the induction and the persistence of LTP at CA1 synapses were unaffected. In contrast, the persistence of NMDAR-dependent LTD was abolished. Unlike the N-cadherin-deficient mice, however, there were no effects of the β-catenin-stabilizing mutation on overall levels of GluA1, GluA2, NMDAR subunits, or PSD95. However, immunoelectron microscopy of CA1 synapses showed LTD-related effects on local synaptic trafficking of both cadherin and GluA1. Under basal conditions, mutant synapses had a greater accumulation of cadherin (as expected from stabilization) in comparison with floxed control slices, but a normal accumulation of GluA1. In contrast, inducing LTD in control slices (by bath-applied NMDA, a potent chemical LTD induction protocol) resulted in a reduction of both synaptic N-cadherin and GluA1, indicating that both are endocytosed from the synaptic membrane in response to NMDA. This is somewhat at odds with previous studies that have shown that NMDA treatment *stabilizes* surface N-cadherin via heightened β-catenin binding (Tai et al., 2007), while simultaneously accelerating GluR internalization to produce LTD (Beattie et al., 2000; Ehlers, 2000), a discrepancy likely attributable to vastly different experimental conditions between studies [dissociated neurons in culture, (Tai et al., 2007) vs acute slices from 1-year-old mice (Mills et al., 2014)]. Nevertheless, in mutant slices, NMDA treatment failed to change the synaptic accumulation of both cadherin and GluA1, suggesting that stabilized cadherins trapped GluRs in the synaptic membrane, thereby preventing GluR endocytocis and thus the LTD-associated decrease in synaptic signal strength. The fact that LTP was normal in these mutant mice presumably indicates that insertion/trapping of GluA1 in response to LTP is not affected by stabilized N-cadherin. N-cadherin-GluA2 interactions may also be important for regulating a form of LTD that requires metabotropic GluRs (Zhou, Hu, Passafaro, Xie, & Jia, 2011). Figure 2 highlights a model of how the cadherin/catenin system can differentially regulate trafficking and stability of GluRs at adult synapses to bidirectionally control persistent increases or decreases in synaptic strength all within the framework of existing connectivity.

In addition to apparent roles in trafficking or trapping of AMPA-type receptors during LTP or LTD, the N-cadherin/β-catenin system has also been shown to regulate trafficking and surface stabilization of kainate receptor subunits (Coussen et al., 2002) and nicotinic acetylcholine receptor

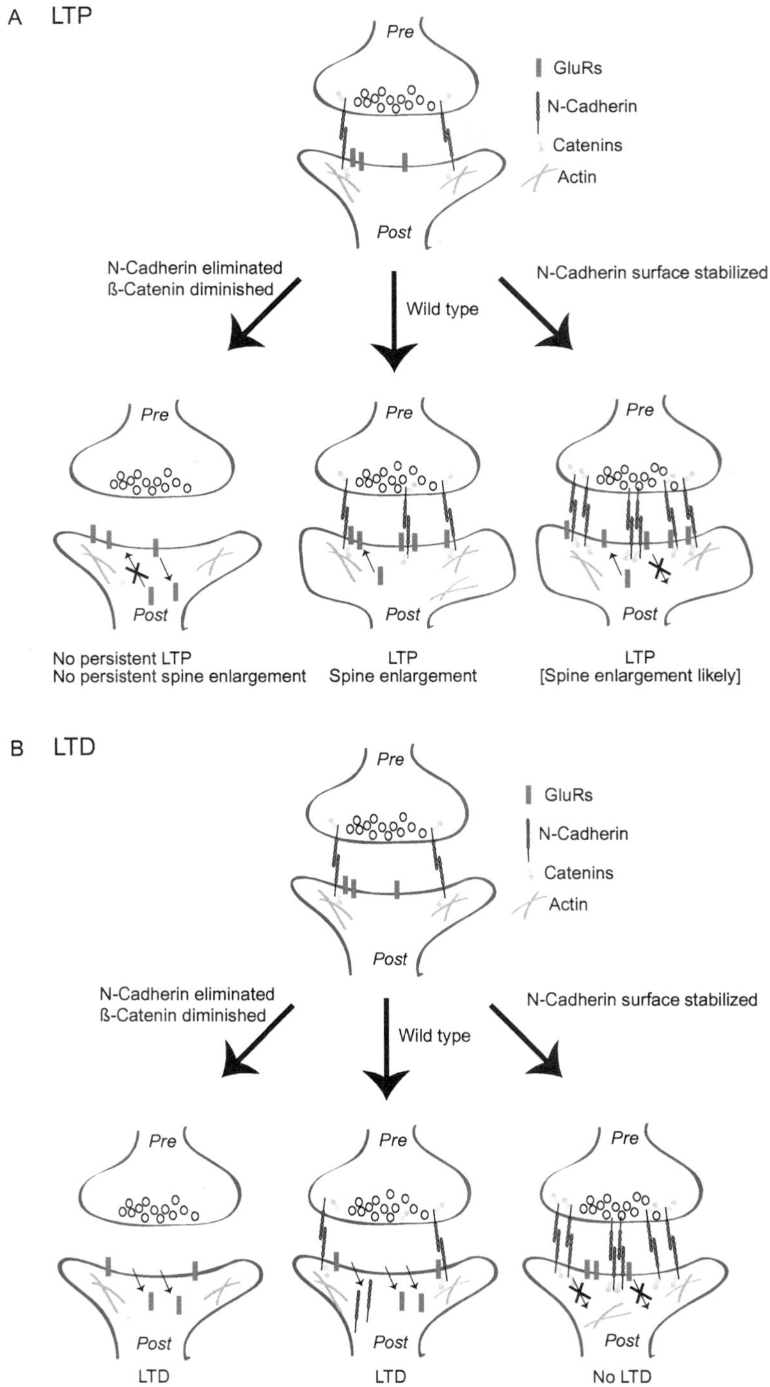

Figure 2 See legend on next page.

subunits (Bruses, 2013). Thus, cadherins may function broadly to promote trafficking or stabilization of a variety of neurotransmitter receptor types.

Interestingly, not all classic cadherins respond similarly to plasticity-inducing synaptic activity in adulthood. For example, while both cadherin-8 and N-cadherin are localized to mature synapses formed by perforant path axons and DG granule neurons, an electrical stimulation paradigm that produces robust LTP and synaptic morphological remodeling at such synapses leads to the selective loss of cadherin-8 but the retention of N-cadherin (Huntley et al., 2012). Although the significance of such differential responses is unknown, the data suggest that activity-mediated loss and retention of different cadherins could promote both flexibility (enabling morphological plasticity) and stability (to maintain synaptic apposition).

5.2. Postnatal deletion of N-cadherin may alter the balance of excitation and inhibition

The studies discussed above show that altering N-cadherin/β-catenin function in mature hippocampal neurons has little or no effect on baseline properties of excitatory neurotransmission, but produces profound changes in plasticity of the excitatory synapses at which these molecules are normally

Figure 2 Schematic diagram showing how the cadherin/catenin system can bidirectionally regulate synaptic plasticity by differentially controlling GluR subunit trafficking and/or stability at the surface. (A) During LTP, a synapse (top) becomes strengthened (wild-type synapse, bottom middle) by the insertion of GluRs into the membrane (small arrow in the spine head) and also undergoes a coordinated increase in the size of the spine head. N-Cadherin facilitates both processes by trapping and/or promoting GluR positioning through direct ectodomain interactions as well as through linkage to the actin cytoskeleton. In the case of the conditional N-cadherin deletion mutant (bottom left synapse), GluRs fail to insert and/or fail to become trapped at the synapse in the absence of N-cadherin (blocked arrow), thus persistent LTP and spine enlargement are both abolished. GluR endocytocis is presumably unaffected (small arrow), thus accounting for normal LTD in these mice. In the case of the cadherin/β-catenin stabilized mice (bottom right synapse), excessively stabilized surface N-cadherin has no effect on GluR insertion (small arrow), thus accounting for normal LTP in these mice. However, the surface-stabilized N-cadherin abnormally traps synaptic GluRs, thus preventing GluR endocytocis (blocked arrow) and blocking LTD. (B) During LTD, wild-type synapses (middle bottom) undergo endocytocis of both N-cadherin and GluRs from the synaptic surface (small arrows). In the N-cadherin conditional KO synapse (left synapse), GluR endocytocis is presumably not affected, thus yielding normal LTD. In contrast, when N-cadherin is surface stabilized (right synapse), GluRs are abnormally trapped, preventing endocytocis, thus these synapses fail to exhibit LTD. *This cartoon is based on the studies of Bozdagi et al. (2010), Mills et al. (2014), and Nikitczuk et al. (2014).* (See the color plate.)

located. However, further examination of the conditional N-cadherin KO mice showed that postnatal deletion of N-cadherin from excitatory synapses also had significant effects on GABAergic inhibitory synapses (Nikitczuk et al., 2014). This was surprising because N-cadherin is not normally found at hippocampal or other cortical GABA synapses in maturity (Benson & Tanaka, 1998; Huntley & Benson, 1999). When examined in adulthood, the density of puncta immunolabeled for pre- and postsynaptic molecular markers of GABA synapses was significantly elevated in the hippocampus of conditional N-cadherin KO mice, suggesting functionally enhanced inhibition. This possibility was indirectly confirmed by examining the severity of kainic acid-induced seizures in these adult mice as a proxy for overall levels of GABA inhibition, as GABA inhibition generally can regulate seizure threshold (Ben-Ari & Cossart, 2000). As expected from the anatomical results, adult conditional N-cadherin KO mice displayed a significantly diminished severity of kainic acid-induced seizures. The mechanism(s) behind such apparent changes in inhibitory synapses following the deletion of N-cadherin from excitatory synapses is not yet known, but there are several possibilities. One is that compensatory changes in levels or localization of other classic cadherins could have affected numbers of GABA synapses, as GABA synapse numbers are reduced in cultured hippocampal neurons when levels of E-cadherin, cadherin-11, or cadherin-13 are reduced (Fiederling, Ewert, Andreyeva, Jungling, & Gottmann, 2011; Paradis et al., 2007). Alternatively, PSD95 collaborates with different neuroligins to specify excitatory and inhibitory synapses (Graf, Zhang, Jin, Linhoff, & Craig, 2004; Levinson et al., 2005, 2010; Prange, Wong, Gerrow, Wang, & El-Husseini, 2004). The diminished levels of PSD95 in the N-cadherin KO mice could have altered neuroligin specificity resulting in the promotion of GABA synapses. Finally, since the conditional deletion of N-cadherin did not affect N-cadherin expression by GABAergic interneurons, it is possible that abnormally asymmetric (non-homophilic) partnering of N-cadherin presented on the surface of GABA neurons with other molecules, either in *cis* or in *trans*, could have promoted GABA synapse proliferation (Pielarski et al., 2013).

5.3. Deficits in cognitive flexibility in adult cadherin-mutant mice

Learning, memory, and other forms of information processing require various forms of persistent synaptic plasticity as well as a normal balance of excitation and inhibition (Silva, 2003; Yizhar et al., 2011) which are the very aspects of synaptic and network function that are altered in adult hippocampus when levels, localization, and function of the cadherin/catenin system are perturbed. As expected from such cellular and molecular alterations, disrupting the cadherin/catenin system in adulthood produces selective deficits in the ability of mutant mice to adapt quickly to unexpectedly changing contingencies in several hippocampal-based behavioral tasks, an ability that wild-type mice readily display and which is called here "cognitive flexibility." Such deficits in cognitive flexibility occur without overt affects on the ability of the mice to learn task rules initially.

In one set of studies using the adult conditional N-cadherin KO mice in which persistent LTP was impaired (Nikitczuk et al., 2014), animals were trained in a six-arm radial water maze to find the location of a submerged escape platform located at the end of one arm (the goal arm). Animals were placed randomly in one of the other five start arms at the beginning of each trial and were given one initial sample trial to find the escape platform by trial-and-error, then tested on subsequent match trials for memory of the position of the hidden platform in the goal arm, which remained constant during each day of testing, but changed across days. Performance measurements included entry into both correct and incorrect goal arms, swim-path lengths, and time to reach the platform (escape latency). Both mutant and control mice learned the general rules of the task at the same rates and performed similarly in the sample trials, where memory could not guide performance. However, in subsequent match trials, mutant mice were significantly impaired in their memory for spatial episodes. Interestingly, the pattern of errors made by the control mice suggested memory for the previous day's goal arm (perseverative errors), while the mutant mice showed no evidence for perseverative errors. Control experiments ruled out any deficits in motivation, motor abilities, or sensory perceptual abilities. Interestingly, when the "cognitive load" of the task was increased by changing the position of the goal arm three times during each daily block of testing, both mutant and control mice performed similarly during the first 2 days of testing, but by the last 2 days of testing, while the control mice had adapted to the changing contingencies and performed significantly better across all measures than during the first 2 days, the mutant mice failed to adapt, revealing a deficit in the cognitive flexibility necessary to adapt to the changing task requirements.

A complimentary set of deficits in cognitive flexibility was observed in the cadherin/β-catenin-stabilized mutant mice in which LTD was deficient (Mills et al., 2014). Mutant and control mice were trained to find the location of a hidden escape platform using a standard Morris water maze. Like the

N-cadherin conditional KO mice, the cadherin/β-catenin-stabilized mutant mice learned the initial rules of the task at similar rates to the control mice. However, after 5 days of testing where the platform remained in the same position, the escape platform was unexpectedly shifted to a new location. Control mice adapted quickly, finding the new location and eliminating entry into the previous goal location. In contrast, the mutant mice largely failed to learn the new position, instead displaying repeated entry into the previous goal position (perseverative errors). The ability to perform successfully in this task requires the extinction of the memory for the position of the initial goal platform, which is thought to require LTD (Dong et al., 2013). Thus, the deficient LTD displayed by these mutant mice likely contributed to their particular deficits in cognitive flexibility.

Other studies have shown that conditional deletion of β-catenin disrupted persistent storage of newly learned fear memories (Maguschak & Ressler, 2008), while application of exogenous HAV peptide blockers of Type I cadherins into dorsal hippocampus of adult mice impaired contextual fear conditioning (Schrick et al., 2007). Furthermore, mutations in δ-catenin have been associated with significant deficits in cognitive function (Israely et al., 2004), which may involve activity-dependent interactions between palmitoylated δ-catenin and N-cadherin that stabilizes N-cadherin at synapses and regulates GluR insertion (Brigidi et al., 2014).

6. CADHERIN-BASED TRANSSYNAPTIC NETWORKS IN PATHOPHYSIOLOGY OF BRAIN CIRCUITS

Various neurodevelopmental and psychiatric disorders are associated with genetic variations in proteins that mediate brain connectivity and synaptic function. Accumulating evidence implicates defects in cadherin-based adhesion and signal transduction in many of these disorders (Redies, Hertel, & Hubner, 2012). Aberrant function of Type I cadherins is also involved in some neurological disorders, as well as neuroprotective repair mechanisms following CNS injury. Finally, recent studies suggest that dysfunctional cadherin signaling may contribute to some types of neurodegenerative disease.

6.1. Neurodevelopmental disorders

Autism spectrum disorders (ASDs) are a clinically heterogeneous group of neurodevelopmental conditions defined by the following core characteristics: impaired reciprocal social interactions, communication deficits, and repetitive restrictive behaviors (Pardo & Eberhart, 2007). Additionally, affected individuals commonly have comorbid anxiety disorders, attention deficit/hyperactivity disorder (ADHD), obsessive compulsive disorder (OCD), Tourette syndrome, or schizophrenia (Geurts et al., 2008; Rapoport, Chavez, Greenstein, Addington, & Gogtay, 2009; Simonoff et al., 2008). Extensive genetic studies have identified inherited and *de novo* copy number variants (CNVs) and single nucleotide polymorphisms (SNPs) in classic cadherins and Pcdhs that are associated with ASDs and comorbid disorders.

Data from the Autism Genome Project Consortium implicated a rare familial microdeletion on chromosome 16q21, containing only the gene encoding cadherin-8, in two families affected by autism and intellectual disability (ID; Pagnamenta et al., 2011). Genetic variations in nongenic regions that alter cadherin gene expression represent risk factors for ASD. Rare inherited CNVs in the intergenic region immediately upstream of *cadherin-8* have been identified as new ASD susceptibility loci (Walker & Scherer, 2013). Genome-wide association studies have also uncovered autism-linked SNPs in the intergenic region between *cadherin-9* and *cadherin-10* genes on chromosome 5p14.1 (Wang et al., 2009). Homozygosity mapping of individuals with related parents revealed a large deletion within a protocadherin cluster proximal to the gene that encodes Pcdh-10 (Morrow et al., 2008), which promotes activity-dependent synapse elimination (Tsai et al., 2012). ASDs have a very high rate of *de novo* mutations in individuals with unaffected parents. A study of rare *de novo* CNVs found a microdeletion on 16q23, which disrupts *cadherin-13* (Sanders et al., 2011).

Children with autism often exhibit ID (Matson & Shoemaker, 2009). Gene variants of *cadherin-15* that disrupt its cell–cell adhesion function were found in patients with ID (Bhalla et al., 2008). Mutations in proteins that directly interact with cadherins have also been shown to be associated with ID. *CTNND2* encodes δ-catenin which is deleted in Cri-du-Chat syndrome, a chromosomal disorder involving 5p chromosomal deletions that cause severe ID and speech impairment (Israely et al., 2004).

N-cadherin has numerous developmental and synaptic functions (discussed in the earlier sections), and its dysfunction is associated with multiple disorders; however until recently, no neurodevelopmental disease-linked N-cadherin mutation had been identified. Exon sequencing in a subset of OCD and Tourette's patients identified rare single nucleotide missense variants in *cadherin-2*, which encodes N-cadherin (Moya et al., 2013). Although this study identified the domains that contain the genetic variants, it remains unknown how these SNPs affect N-cadherin function (Moya et al., 2013).

While several independent genome-wide association analyses have identified SNPs and CNVs in *cadherin-13* that are associated with ADHD (Rivero et al., 2013), multiple studies have implicated cadherin-13 in a variety of neuropsychiatric disorders, including major depressive disorder (Sibille et al., 2009), substance abuse (Hart et al., 2012), schizophrenia (Borglum et al., 2014), and autism (Sanders et al., 2011). Although cadherin-13 contains the conserved extracellular domains of classic cadherins, it is linked to the cell membrane by a glycosylphosphatidylinositol anchor and lacks transmembrane and cytoplasmic domains (Ranscht & Dours-Zimmermann, 1991). Unlike other classic cadherins and Pcdhs, mutations in *cadherin-13* may have pleiotropic effects due to its atypical structure and downstream signaling.

6.2. Seizure/epilepsy

A seizure-induced injury in the hippocampus promotes the loss of CA3 neurons and synaptic reorganization of mossy fibers that leads to *de novo* synapse formation and recurrent excitatory firing (Wenzel, Woolley, Robbins, & Schwartzkroin, 2000). N-Cadherin expression markedly increases 1–4 weeks following seizure induction in hippocampus, which is consistent with the time course of mossy fiber axon sprouting (Shan, Yoshida, Wu, Huntley, & Colman, 2002). Multiple familial and *de novo* mutations or whole-gene deletions of *PCDH-19* (δ2 nonclustered Pcdh) have been linked to epilepsy and mental retardation limited to females (EFMR; Depienne et al., 2011; Dibbens et al., 2008; Jamal, Basran, Newton, Wang, & Milunsky, 2010). Affected EFMR individuals experience recurrent seizures that begin in infancy and mild to moderate ID and autism (Ryan et al., 1997). Due to the multiple mutations associated with Pcdh-19 loss of function, it is not surprising that EFMR patients have highly variable clinical features. However, unlike other cadherins associated with psychiatric disorders, *PCDH-19* is located on the X-chromosome. Monozygotic twins carrying the same Pcdh-19 mutation can exhibit different clinical features (Higurashi et al., 2012), suggesting that random X inactivation of EFMR-linked *PCDH-19* may lead to varying severity of symptoms (Depienne & LeGuern, 2012). Male carriers of the *Pcdh-19* mutations do not show signs of epilepsy, but are susceptible to behaviors associated with ASDs (van Harssel et al., 2013).

6.3. CNS lesions, neuropathic pain, astrogliosis, and remyelination

Neuropathic pain is defined as abnormal pain sensations resulting from injuries to the peripheral or segmental spinal nerves, in part, stemming from the reorganization of microcircuits in the dorsal horn and dorsal root ganglia (Finnerup, Sindrup, & Jensen, 2007). In a mouse model of neuropathic pain, spinal nerve ligation and transection produce a persistent and abnormal responsiveness to nonnoxious mechanical stimuli that are otherwise normally tolerated and result in loss of E-cadherin from synapses formed by small-diameter primary afferents to spinal lamina II as well as a transient elevation of N-cadherin throughout lamina I and II (Brock et al., 2004; Patil et al., 2009). These changes together may mediate synaptic loss and reactive axon sprouting at the site of injury, or represent astrogliosis at the lesion site, leading to abnormal transmission of pain sensation.

N-cadherin expression is also upregulated following neocortical stab wound injury, which induces astrogliosis at the lesion site. In an inducible astrocyte-specific N-cadherin KO, there were fewer reactive astrocytes, but increased reactive microglia and neuronal cell death. These data indicate that N-cadherin is required for neuroprotection following brain injury through the maintenance of reactive astrogliosis (Kanemaru et al., 2013). Enhanced N-cadherin immunoreactivity has also been observed in early remyelinating lesions in mouse models of multiple sclerosis, with the highest expression in oligodendrocytes at lesion sites. N-Cadherin may confer this effect through homophilic adhesion between oligodendrocytes or between axons and oligodendrocytes (Hochmeister et al., 2012).

6.4. Alzheimer's disease

Although current evidence is limited, several studies have implicated dysfunctional N-cadherin/β-catenin signaling in Alzheimer's disease (AD). AD is characterized pathologically by the formation of extracellular β-amyloid (Aβ)-containing plaques, intracellular neurofibrillary tangles consisting of hyperphosphorylated tau, and subsequent synaptic degeneration. Compared to age-matched controls, human AD brains exhibit reduced N-cadherin protein levels and increased p38 MAPK signaling (Ando et al., 2011). Extracellular Aβ oligomers promote the marked loss of synapses (Shankar et al., 2007) and lead to reduced N-cadherin expression and concomitant activation of p38 MAPK (Ando et al., 2011). Recent studies suggest that a γ-secretase-dependent, proteolytically derived C-terminal fragment of N-cadherin accelerates synaptic dysfunction by β-amyloid (Andreyeva et al., 2012).

In contrast, other studies report impaired N-cadherin metabolism that may be associated with AD. N-Cadherin undergoes sequential cleavage by the extracellular protease ADAM10 (A Disintegrin And Metalloprotease-10) following NMDA receptor stimulation (Reiss et al., 2005), and presenilin 1 (PS1), a γ-secretase complex component (Marambaud et al., 2003), to mediate cell–cell adhesion and downstream gene expression via β-catenin. In CHO cells, N-cadherin cleavage by ADAM10 is impaired by extracellular Aβ. Familial AD-linked mutations of PS1 (which also cleaves amyloid precursor protein to generate Aβ peptides) impede cleavage of N-cadherin (Uemura, Kuzuya, et al., 2007). To date, no other cadherins have been associated with AD. Although SNPs in *Pcdh-11x* were initially linked to AD susceptibility (Carrasquillo et al., 2009), recent studies have failed to replicate the results (Miar et al., 2011; Wu et al., 2010).

7. CONCLUSIONS

Cadherin-based adhesion and signaling regulate the development of neurons and their intricate, highly specific connections within the CNS through varied types of interactions and defined signaling cascades. With the development of genome-wide microarray technologies, the number of cadherins and associated proteins linked to neurodevelopmental disorders is growing, as multiple genetic variants in both coding and proximal nongenic regions are being identified, but we are only beginning to understand exactly how these variants affect synaptic function and contribute to disease pathophysiology. The emerging view that cadherins play an indispensable role in plasticity of mature synapses in hippocampus and elsewhere, and thus significantly influence cognitive abilities, suggests that cadherin/catenin dysfunction later in life can have profound effects on memory and other aspects of higher-order brain function and support recent reports of dysfunctional N-cadherin signaling in AD. It is also critical to bear in mind that CAMs from different families are all part of a highly integrated network of synaptic proteins, probably rarely, if ever, operating in isolation of each other and other effectors. For example, the postnatal conditional deletion of synaptic β1-integrin leads to compensatory changes in both N-cadherin and neuroligins, but does not affect synaptic levels of SynCAM (Mortillo et al., 2012). The significance of this for human health is that the variable and complex phenotypes of certain disorders (e.g., autism) that are attributable to a particular genetic variant in one particular CAM (e.g., neuroligin) may in fact reflect a number of compensatory readjustments in synaptic adhesion and signaling molecules that extend across multiple CAM families. Future studies should focus specifically on how different CAM family members influence the total repertoire of synaptic CAMs in order to uncover how these multifaceted proteins collaborate to mediate complex behaviors and pathological dysfunction in vertebrates.

ACKNOWLEDGMENTS

We thank members of the Huntley and Benson laboratories for their contributions to the personal work discussed in this review. During the writing of this chapter, we were supported by grants from the National Institute of Mental Health (MH100600, MH095229, and MH099778); the Michael J. Fox Foundation; and an NIMH T32 Training Program in Mental Health grant (T32 MH096678).

REFERENCES

Abe, K., Chisaka, O., Van Roy, F., & Takeichi, M. (2004). Stability of dendritic spines and synaptic contacts is controlled by alpha N-catenin. *Nature Neuroscience*, *7*, 357–363.

Abedin, M., & King, N. (2008). The premetazoan ancestry of cadherins. *Science*, *319*, 946–948.

Aiga, M., Levinson, J. N., & Bamji, S. X. (2011). N-cadherin and neuroligins cooperate to regulate synapse formation in hippocampal cultures. *The Journal of Biological Chemistry*, *286*, 851–858.

Al-Amoudi, A., Diez, D. C., Betts, M. J., & Frangakis, A. S. (2007). The molecular architecture of cadherins in native epidermal desmosomes. *Nature*, *450*, 832–837.

Anderson, T. R., & Benson, D. L. (2006). Cadherin-mediated adhesion and signaling during vertebrate central synapse formation. In A. Dityatev, & A. El-Husseini (Eds.), *Molecular mechanisms of synaptogenesis* (pp. 83–93). New York, NY: Springer.

Ando, K., Uemura, K., Kuzuya, A., Maesako, M., Asada-Utsugi, M., Kubota, M., et al. (2011). N-cadherin regulates p38 MAPK signaling via association with JNK-associated leucine zipper protein: Implications for neurodegeneration in Alzheimer disease. *The Journal of Biological Chemistry*, *286*, 7619–7628.

Andreyeva, A., Nieweg, K., Horstmann, K., Klapper, S., Müller-Schiffmann, A., Korth, C., et al. (2012). C-terminal fragment of N-cadherin accelerates synapse destabilization by amyloid-β. *Brain*, *135*, 2140–2154.

Arikkath, J., Israely, I., Tao, Y., Mei, L., Liu, X., & Reichardt, L. F. (2008). Erbin controls dendritic morphogenesis by regulating localization of delta-catenin. *The Journal of Neuroscience*, *28*, 7047–7056.

Arikkath, J., Peng, I. F., Ng, Y. G., Israely, I., Liu, X., Ullian, E. M., et al. (2009). Delta-catenin regulates spine and synapse morphogenesis and function in hippocampal neurons during development. *The Journal of Neuroscience*, *29*, 5435–5442.

Arikkath, J., & Reichardt, L. F. (2008). Cadherins and catenins at synapses: Roles in synaptogenesis and synaptic plasticity. *Trends in Neurosciences*, *31*, 487–494.

Arregui, C., Pathre, P., Lilien, J., & Balsamo, J. (2000). The nonreceptor tyrosine kinase fer mediates cross-talk between N-cadherin and beta1-integrins. *The Journal of Cell Biology*, *149*, 1263–1274.

Babb, S. G., Kotradi, S. M., Shah, B., Chiappini-Williamson, C., Bell, L. N., Schmeiser, G., et al. (2005). Zebrafish R-cadherin (Cdh4) controls visual system development and differentiation. *Developmental Dynamics, 233*, 930–945.

Bamji, S. X., Shimazu, K., Kimes, N., Huelsken, J., Birchmeier, W., Lu, B., et al. (2003). Role of beta-catenin in synaptic vesicle localization and presynaptic assembly. *Neuron, 40*, 719–731.

Barnes, S. H., Price, S. R., Wentzel, C., & Guthrie, S. C. (2010). Cadherin-7 and cadherin-6B differentially regulate the growth, branching and guidance of cranial motor axons. *Development, 137*, 805–814.

Beattie, E. C., Carroll, R. C., Yu, X., Morishita, W., Yasuda, H., von Zastrow, M., et al. (2000). Regulation of AMPA receptor endocytosis by a signaling mechanism shared with LTD. *Nature Neuroscience, 3*, 1291–1300.

Beaudoin, G. M., 3rd, Schofield, C. M., Nuwal, T., Zang, K., Ullian, E. M., Huang, B., et al. (2012). Afadin, a Ras/Rap effector that controls cadherin function, promotes spine and excitatory synapse density in the hippocampus. *The Journal of Neuroscience, 32*, 99–110.

Bekirov, I. H., Nagy, V., Svoronos, A., Huntley, G. W., & Benson, D. L. (2008). Cadherin-8 and N-cadherin differentially regulate pre- and postsynaptic development of the hippocampal mossy fiber pathway. *Hippocampus, 18*, 349–363.

Bekirov, I. H., Needleman, L. A., Zhang, W., & Benson, D. L. (2002). Identification and localization of multiple classic cadherins in developing rat limbic system. *Neuroscience, 115*, 213–227.

Ben-Ari, Y., & Cossart, R. (2000). Kainate, a double agent that generates seizures: Two decades of progress. *Trends in Neurosciences, 23*, 580–587.

Benson, D. L., Colman, D. R., & Huntley, G. W. (2001). Molecules, maps and synapse specificity. *Nature Reviews. Neuroscience, 2*, 899–909.

Benson, D. L., & Huntley, G. W. (2012). Building and remodeling synapses. *Hippocampus, 22*, 954–968.

Benson, D. L., & Tanaka, H. (1998). N-cadherin redistribution during synaptogenesis in hippocampal neurons. *The Journal of Neuroscience, 18*, 6892–6904.

Berger-Muller, S., & Suzuki, T. (2011). Seven-pass transmembrane cadherins: Roles and emerging mechanisms in axonal and dendritic patterning. *Molecular Neurobiology, 44*, 313–320.

Bhalla, K., Luo, Y., Buchan, T., Beachem, M. A., Guzauskas, G. F., Ladd, S., et al. (2008). Alterations in CDH15 and KIRREL3 in patients with mild to severe intellectual disability. *American Journal of Human Genetics, 83*, 703–713.

Biswas, S., Emond, M. R., & Jontes, J. D. (2010). Protocadherin-19 and N-cadherin interact to control cell movements during anterior neurulation. *The Journal of Cell Biology, 191*, 1029–1041.

Bixby, J. L., & Zhang, R. (1990). Purified N-cadherin is a potent substrate for the rapid induction of neurite outgrowth. *The Journal of Cell Biology, 110*, 1253–1260.

Blaschuk, O. W., Sullivan, R., David, S., & Pouliot, Y. (1990). Identification of a cadherin cell adhesion recognition sequence. *Developmental Biology, 139*, 227–229.

Boggon, T. J., Murray, J., Chappuis-Flament, S., Wong, E., Gumbiner, B. M., & Shapiro, L. (2002). C-cadherin ectodomain structure and implications for cell adhesion mechanisms. *Science, 296*, 1308–1313.

Borglum, A. D., Demontis, D., Grove, J., Pallesen, J., Hollegaard, M. V., Pedersen, C. B., et al. (2014). Genome-wide study of association and interaction with maternal cytomegalovirus infection suggests new schizophrenia loci. *Molecular Psychiatry, 19*, 325–333.

Boscher, C., & Mege, R. M. (2008). Cadherin-11 interacts with the FGF receptor and induces neurite outgrowth through associated downstream signalling. *Cellular Signalling, 20*, 1061–1072.

Boutin, C., Goffinet, A. M., & Tissir, F. (2012). Celsr1-3 cadherins in PCP and brain development. *Current Topics in Developmental Biology, 101*, 161–183.
Bozdagi, O., Shan, W., Tanaka, H., Benson, D. L., & Huntley, G. W. (2000). Increasing numbers of synaptic puncta during late-phase LTP: N-cadherin is synthesized, recruited to synaptic sites, and required for potentiation. *Neuron, 28*, 245–259.
Bozdagi, O., Valcin, M., Poskanzer, K., Tanaka, H., & Benson, D. L. (2004). Temporally distinct demands for classic cadherins in synapse formation and maturation. *Molecular and Cellular Neurosciences, 27*, 509–521.
Bozdagi, O., Wang, X. B., Nikitczuk, J. S., Anderson, T. R., Bloss, E. B., Radice, G. L., et al. (2010). Persistence of coordinated long-term potentiation and dendritic spine enlargement at mature hippocampal CA1 synapses requires N-cadherin. *The Journal of Neuroscience, 30*, 9984–9989.
Bradl, M., & Lassmann, H. (2010). Oligodendrocytes: Biology and pathology. *Acta Neuropathologica, 119*, 37–53.
Brigidi, G. S., & Bamji, S. X. (2011). Cadherin-catenin adhesion complexes at the synapse. *Current Opinion in Neurobiology, 21*, 208–214.
Brigidi, G. S., Sun, Y., Beccano-Kelly, D., Pitman, K., Mobasser, M., Borgland, S. L., et al. (2014). Palmitoylation of delta-catenin by DHHC5 mediates activity-induced synapse plasticity. *Nature Neuroscience, 17*, 522–532.
Brock, J. H., Elste, A., & Huntley, G. W. (2004). Distribution and injury-induced plasticity of cadherins in relationship to identified synaptic circuitry in adult rat spinal cord. *The Journal of Neuroscience, 24*, 8806–8817.
Bruses, J. L. (2011). N-cadherin regulates primary motor axon growth and branching during zebrafish embryonic development. *The Journal of Comparative Neurology, 519*, 1797–1815.
Bruses, J. L. (2013). Cell surface localization of alpha3beta4 nicotinic acetylcholine receptors is regulated by N-cadherin homotypic binding and actomyosin contractility. *PLoS One, 8*, e62435. http://dx.doi.org/10.1371/journal.pone.0062435.
Buchert, M., Schneider, S., Meskenaite, V., Adams, M. T., Canaani, E., Baechi, T., et al. (1999). The junction-associated protein AF-6 interacts and clusters with specific Eph receptor tyrosine kinases at specialized sites of cell-cell contact in the brain. *The Journal of Cell Biology, 144*, 361–371.
Burden-Gulley, S. M., & Brady-Kalnay, S. M. (1999). PTPmu regulates N-cadherin-dependent neurite outgrowth. *The Journal of Cell Biology, 144*, 1323–1336.
Carrasquillo, M. M., Zou, F., Pankratz, V. S., Wilcox, S. L., Ma, L., Walker, L. P., et al. (2009). Genetic variation in PCDH11X is associated with susceptibility to late-onset Alzheimer's disease. *Nature Genetics, 41*, 192–198.
Chen, X., & Gumbiner, B. M. (2006). Paraxial protocadherin mediates cell sorting and tissue morphogenesis by regulating C-cadherin adhesion activity. *The Journal of Cell Biology, 174*, 301–313.
Chen, X., Koh, E., Yoder, M., & Gumbiner, B. M. (2009). A protocadherin-cadherin-FLRT3 complex controls cell adhesion and morphogenesis. *PLoS One, 4*, e8411. http://dx.doi.org/10.1371/journal.pone.0008411.
Ciatto, C., Bahna, F., Zampieri, N., VanSteenhouse, H. C., Katsamba, P. S., Ahlsen, G., et al. (2010). T-cadherin structures reveal a novel adhesive binding mechanism. *Nature Structural & Molecular Biology, 17*, 339–347.
Collingridge, G. L., & Bliss, T. V. P. (1987). NMDA receptors-their role in long-term potentiation. *Trends in Neurosciences, 10*, 288–293.
Coussen, F., Normand, E., Marchal, C., Costet, P., Choquet, D., Lambert, M., et al. (2002). Recruitment of the kainate receptor subunit glutamate receptor 6 by cadherin/catenin complexes. *The Journal of Neuroscience, 22*, 6426–6436.

Dalva, M. B., McClelland, A. C., & Kayser, M. S. (2007). Cell adhesion molecules: Signalling functions at the synapse. *Nature Reviews Neuroscience*, *8*, 206–220.

Danielson, E., Zhang, N., Metallo, J., Kaleka, K., Shin, S. M., Gerges, N., et al. (2012). S-SCAM/MAGI-2 is an essential synaptic scaffolding molecule for the GluA2-containing maintenance pool of AMPA receptors. *The Journal of Neuroscience*, *32*, 6967–6980.

Dell'Acqua, M. L., Smith, K. E., Gorski, J. A., Horne, E. A., Gibson, E. S., & Gomez, L. L. (2006). Regulation of neuronal PKA signaling through AKAP targeting dynamics. *European Journal of Cell Biology*, *85*, 627–633.

Depienne, C., & LeGuern, E. (2012). PCDH19-related infantile epileptic encephalopathy: An unusual X-linked inheritance disorder. *Human Mutation*, *33*, 627–634.

Depienne, C., Trouillard, O., Bouteiller, D., Gourfinkel-An, I., Poirier, K., Rivier, F., et al. (2011). Mutations and deletions in PCDH19 account for various familial or isolated epilepsies in females. *Human Mutation*, *32*, E1959–E1975.

Dibbens, L. M., Tarpey, P. S., Hynes, K., Bayly, M. A., Scheffer, I. E., Smith, R., et al. (2008). X-linked protocadherin 19 mutations cause female-limited epilepsy and cognitive impairment. *Nature Genetics*, *40*, 776–781. http://dx.doi.org/10.1002/humu.21373.

Doherty, P., Rowett, L. H., Moore, S. E., Mann, D. A., & Walsh, F. S. (1991). Neurite outgrowth in response to transfected N-CAM and N-cadherin reveals fundamental differences in neuronal responsiveness to CAMs. *Neuron*, *6*, 247–258.

Doherty, P., Skaper, S. D., Moore, S. E., Leon, A., & Walsh, F. S. (1992). A developmentally regulated switch in neuronal responsiveness to NCAM and N-cadherin in the rat hippocampus. *Development*, *115*, 885–892.

Dong, Z., Bai, Y., Wu, X., Li, H., Gong, B., Howland, J. G., et al. (2013). Hippocampal long-term depression mediates spatial reversal learning in the Morris water maze. *Neuropharmacology*, *64*, 65–73.

Drazba, J., & Lemmon, V. (1990). The role of cell adhesion molecules in neurite outgrowth on Muller cells. *Developmental Biology*, *138*, 82–93.

Ehlers, M. D. (2000). Reinsertion or degradation of AMPA receptors determined by activity-dependent endocytic sorting. *Neuron*, *28*, 511–525.

Ehlers, M. D., Heine, M., Groc, L., Lee, M. C., & Choquet, D. (2007). Diffusional trapping of GluR1 AMPA receptors by input-specific synaptic activity. *Neuron*, *54*, 447–460.

Ehrlich, I., & Malinow, R. (2004). Postsynaptic density 95 controls AMPA receptor incorporation during long-term potentiation and experience-driven synaptic plasticity. *The Journal of Neuroscience*, *24*, 916–927.

Elste, A. M., & Benson, D. L. (2006). Structural basis for developmentally regulated changes in cadherin function at synapses. *The Journal of Comparative Neurology*, *495*, 324–335.

Emond, M. R., Biswas, S., Blevins, C. J., & Jontes, J. D. (2011). A complex of Protocadherin-19 and N-cadherin mediates a novel mechanism of cell adhesion. *The Journal of Cell Biology*, *195*, 1115–1121.

Esch, T., Lemmon, V., & Banker, G. (2000). Differential effects of NgCAM and N-cadherin on the development of axons and dendrites by cultured hippocampal neurons. *Journal of Neurocytology*, *29*, 215–223.

Fernandez-Monreal, M., Kang, S., & Phillips, G. R. (2009). Gamma-protocadherin homophilic interaction and intracellular trafficking is controlled by the cytoplasmic domain in neurons. *Molecular and Cellular Neurosciences*, *40*, 344–353.

Fiederling, A., Ewert, R., Andreyeva, A., Jungling, K., & Gottmann, K. (2011). E-cadherin is required at GABAergic synapses in cultured cortical neurons. *Neuroscience Letters*, *501*, 167–172.

Finnerup, N. B., Sindrup, S. H., & Jensen, T. S. (2007). Chronic neuropathic pain: Mechanisms, drug targets and measurement. *Fundamental & Clinical Pharmacology*, *21*, 129–136.

Flannery, R. J., & Bruses, J. L. (2012). N-cadherin induces partial differentiation of cholinergic presynaptic terminals in heterologous cultures of brainstem neurons and CHO cells. *Frontiers in Synaptic Neuroscience*, *4*, 1–11.
Fredette, B. J., Miller, J., & Ranscht, B. (1996). Inhibition of motor axon growth by T-cadherin substrata. *Development*, *122*, 3163–3171.
Friedman, L. G., Riemslagh, F. G., Sullivan, J. M., Mesias, R., Williams, F. M., Huntley, G. W., et al. (2014). Cadherin-8 expression, synaptic localization and molecular control of neuronal form in prefrontal cortico-striatal circuits. *The Journal of Comparative Neurology*, *523*, 75–92.
Frotscher, M., & Heimrich, B. (1995). Lamina-specific synaptic connections of hippocampal neurons in vitro. *Journal of Neurobiology*, *26*, 350–359.
Garrett, A. M., Schreiner, D., Lobas, M. A., & Weiner, J. A. (2012). Gamma-protocadherins control cortical dendrite arborization by regulating the activity of a FAK/PKC/MARCKS signaling pathway. *Neuron*, *74*, 269–276.
Garrett, A. M., & Weiner, J. A. (2009). Control of CNS synapse development by {gamma}-protocadherin-mediated astrocyte-neuron contact. *The Journal of Neuroscience*, *29*, 11723–11731.
Geurts, H. M., Grasman, R. P., Verte, S., Oosterlaan, J., Roeyers, H., van Kammen, S. M., et al. (2008). Intra-individual variability in ADHD, autism spectrum disorders and Tourette's syndrome. *Neuropsychologia*, *46*, 3030–3041.
Gil, O. D., Needleman, L., & Huntley, G. W. (2002). Developmental patterns of cadherin expression and localization in relation to compartmentalized thalamocortical terminations in rat barrel cortex. *The Journal of Comparative Neurology*, *453*, 372–388.
Gorski, J. A., Gomez, L. L., Scott, J. D., & Dell'Acqua, M. L. (2005). Association of an A-kinase-anchoring protein signaling scaffold with cadherin adhesion molecules in neurons and epithelial cells. *Molecular Biology of the Cell*, *16*, 3574–3590.
Graf, E. R., Zhang, X., Jin, S. X., Linhoff, M. W., & Craig, A. M. (2004). Neurexins induce differentiation of GABA and glutamate postsynaptic specializations via neuroligins. *Cell*, *119*, 1013–1026.
Grueber, W. B., & Sagasti, A. (2010). Self-avoidance and tiling: Mechanisms of dendrite and axon spacing. *Cold Spring Harbor Perspectives in Biology*, *2*, a001750.
Han, M. H., Lin, C., Meng, S., & Wang, X. (2010). Proteomics analysis reveals overlapping functions of clustered protocadherins. *Molecular & Cellular Proteomics*, *9*, 71–83.
Harrison, O. J., Bahna, F., Katsamba, P. S., Jin, X., Brasch, J., Vendome, J., et al. (2010). Two-step adhesive binding by classical cadherins. *Nature Structural & Molecular Biology*, *17*, 348–357.
Hart, A. B., Engelhardt, B. E., Wardle, M. C., Sokoloff, G., Stephens, M., de Wit, H., et al. (2012). Genome-wide association study of D-amphetamine response in healthy volunteers identifies putative associations, including cadherin 13 (CDH13). *PLoS One*, *7*, e42646. http://dx.doi.org/10.1371/journal.pone.0042646.
Hasegawa, S., Hamada, S., Kumode, Y., Esumi, S., Katori, S., Fukuda, E., et al. (2008). The protocadherin-alpha family is involved in axonal coalescence of olfactory sensory neurons into glomeruli of the olfactory bulb in mouse. *Molecular and Cellular Neurosciences*, *38*, 66–79.
Heisler, F. F., Lee, H. K., Gromova, K. V., Pechmann, Y., Schurek, B., Ruschkies, L., et al. (2014). GRIP1 interlinks N-cadherin and AMPA receptors at vesicles to promote combined cargo transport into dendrites. *Proceedings of the National Academy of Sciences of the United States of America*, *111*, 5030–5035.
Hertel, N., & Redies, C. (2011). Absence of layer-specific cadherin expression profiles in the neocortex of the reeler mutant mouse. *Cerebral Cortex*, *21*, 1105–1117.
Higurashi, N., Shi, X., Yasumoto, S., Oguni, H., Sakauchi, M., Itomi, K., et al. (2012). PCDH19 mutation in Japanese females with epilepsy. *Epilepsy Research*, *99*, 28–37.

Hirano, K., Kaneko, R., Izawa, T., Kawaguchi, M., Kitsukawa, T., & Yagi, T. (2012). Single-neuron diversity generated by Protocadherin-beta cluster in mouse central and peripheral nervous systems. *Frontiers in Molecular Neuroscience*, *5*, 1–13.
Hirano, S., & Takeichi, M. (2012). Cadherins in brain morphogenesis and wiring. *Physiological Reviews*, *92*, 597–634.
Hochmeister, S., Romauch, M., Bauer, J., Seifert-Held, T., Weissert, R., Linington, C., et al. (2012). Re-expression of N-cadherin in remyelinating lesions of experimental inflammatory demyelination. *Experimental Neurology*, *237*, 70–77.
Hong, S., Troyanovsky, R. B., & Troyanovsky, S. M. (2011). Cadherin exits the junction by switching its adhesive bond. *The Journal of Cell Biology*, *192*, 1073–1083.
Hoshina, N., Tanimura, A., Yamasaki, M., Inoue, T., Fukabori, R., Kuroda, T., et al. (2013). Protocadherin 17 regulates presynaptic assembly in topographic corticobasal Ganglia circuits. *Neuron*, *78*, 839–854.
Huber, A. H., & Weis, W. I. (2001). The structure of the beta-catenin/E-cadherin complex and the molecular basis of diverse ligand recognition by beta-catenin. *Cell*, *105*, 391–402.
Hulpiau, P., Gul, I. S., & van Roy, F. (2013). New insights into the evolution of metazoan cadherins and catenins. *Progress in Molecular Biology and Translational Science*, *116*, 71–94.
Hulpiau, P., & van Roy, F. (2009). Molecular evolution of the cadherin superfamily. *The International Journal of Biochemistry & Cell Biology*, *41*, 349–369.
Huntley, G. W., & Benson, D. L. (1999). Neural (N)-cadherin at developing thalamocortical synapses provides an adhesion mechanism for the formation of somatopically organized connections. *The Journal of Comparative Neurology*, *407*, 453–471.
Huntley, G. W., Elste, A. M., Patil, S. B., Bozdagi, O., Benson, D. L., & Steward, O. (2012). Synaptic loss and retention of different classic cadherins with LTP-associated synaptic structural remodeling in vivo. *Hippocampus*, *22*, 17–28.
Huttelmaier, S., Zenklusen, D., Lederer, M., Dictenberg, J., Lorenz, M., Meng, X., et al. (2005). Spatial regulation of beta-actin translation by Src-dependent phosphorylation of ZBP1. *Nature*, *438*, 512–515.
Inoue, A., & Sanes, J. R. (1997). Lamina-specific connectivity in the brain: Regulation by N-cadherin, neurotrophins, and glycoconjugates. *Science*, *276*, 1428–1431.
Ishiuchi, T., Misaki, K., Yonemura, S., Takeichi, M., & Tanoue, T. (2009). Mammalian Fat and dachsous cadherins regulate apical membrane organization in the embryonic cerebral cortex. *The Journal of Cell Biology*, *185*, 959–967.
Israely, I., Costa, R. M., Xie, C. W., Silva, A. J., Kosik, K. S., & Liu, X. (2004). Deletion of the neuron-specific protein delta-catenin leads to severe cognitive and synaptic dysfunction. *Current Biology*, *14*, 1657–1663.
Jamal, S. M., Basran, R. K., Newton, S., Wang, Z., & Milunsky, J. M. (2010). Novel de novo PCDH19 mutations in three unrelated females with epilepsy female restricted mental retardation syndrome. *American Journal of Medical Genetics. Part A*, *152A*, 2475–2481.
Jontes, J. D., Emond, M. R., & Smith, S. J. (2004). In vivo trafficking and targeting of N-cadherin to nascent presynaptic terminals. *The Journal of Neuroscience*, *24*, 9027–9034.
Jungling, K., Eulenburg, V., Moore, R., Kemler, R., Lessmann, V., & Gottmann, K. (2006). N-cadherin transsynaptically regulates short-term plasticity at glutamatergic synapses in embryonic stem cell-derived neurons. *The Journal of Neuroscience*, *26*, 6968–6978.
Kadowaki, M., Nakamura, S., Machon, O., Krauss, S., Radice, G. L., & Takeichi, M. (2007). N-Cadherin mediates cortical organization in the mouse brain. *Developmental Biology*, *304*, 22–33.
Kanemaru, K., Kubota, J., Sekiya, H., Hirose, K., Okubo, Y., & Iino, M. (2013). Calcium-dependent N-cadherin up-regulation mediates reactive astrogliosis and neuroprotection after brain injury. *Proceedings of the National Academy of Sciences of the United States of America*, *110*, 11612–11617.

Kang, J. S., Feinleib, J. L., Knox, S., Ketteringham, M. A., & Krauss, R. S. (2003). Promyogenic members of the Ig and cadherin families associate to positively regulate differentiation. *Proceedings of the National Academy of Sciences of the United States of America*, *100*, 3989–3994.

Karpowicz, P., Willaime-Morawek, S., Balenci, L., DeVeale, B., Inoue, T., & van der Kooy, D. (2009). E-Cadherin regulates neural stem cell self-renewal. *The Journal of Neuroscience*, *29*, 3885–3896.

Katori, S., Hamada, S., Noguchi, Y., Fukuda, E., Yamamoto, T., Yamamoto, H., et al. (2009). Protocadherin-alpha family is required for serotonergic projections to appropriately innervate target brain areas. *The Journal of Neuroscience*, *29*, 9137–9147.

Kawauchi, T., Sekine, K., Shikanai, M., Chihama, K., Tomita, K., Kubo, K., et al. (2010). Rab GTPases-dependent endocytic pathways regulate neuronal migration and maturation through N-cadherin trafficking. *Neuron*, *67*, 588–602.

Kim, H., Han, J. R., Park, J., Oh, M., James, S. E., Chang, S., et al. (2008). Delta-catenin-induced dendritic morphogenesis. An essential role of p190RhoGEF interaction through Akt1-mediated phosphorylation. *The Journal of Biological Chemistry*, *283*, 977–987.

Kintner, C. (1992). Regulation of embryonic cell adhesion by the cadherin cytoplasmic domain. *Cell*, *69*, 225–236.

Kopec, C. D., Li, B., Wei, W., Boehm, J., & Malinow, R. (2006). Glutamate receptor exocytosis and spine enlargement during chemically induced long-term potentiation. *The Journal of Neuroscience*, *26*, 2000–2009.

Kopec, C. D., Real, E., Kessels, H. W., & Malinow, R. (2007). GluR1 links structural and functional plasticity at excitatory synapses. *The Journal of Neuroscience*, *27*, 13706–13718.

Kosik, K. S., Donahue, C. P., Israely, I., Liu, X., & Ochiishi, T. (2005). Delta-catenin at the synaptic-adherens junction. *Trends in Cell Biology*, *15*, 172–178.

Lee, C. H., Herman, T., Clandinin, T. R., Lee, R., & Zipursky, S. L. (2001). N-Cadherin regulates target specificity in the Drosophila visual system. *Neuron*, *30*, 437–450.

Lefebvre, J. L., Kostadinov, D., Chen, W. V., Maniatis, T., & Sanes, J. R. (2012). Protocadherins mediate dendritic self-avoidance in the mammalian nervous system. *Nature*, *488*, 517–521.

Levinson, J. N., Chery, N., Huang, K., Wong, T. P., Gerrow, K., Kang, R., et al. (2005). Neuroligins mediate excitatory and inhibitory synapse formation: Involvement of PSD-95 and neurexin-1beta in neuroligin-induced synaptic specificity. *The Journal of Biological Chemistry*, *280*, 17312–17319.

Levinson, J. N., Li, R., Kang, R., Moukhles, H., El-Husseini, A., & Bamji, S. X. (2010). Postsynaptic scaffolding molecules modulate the localization of neuroligins. *Neuroscience*, *165*, 782–793.

Li, Y., Xiao, H., Chiou, T. T., Jin, H., Bonhomme, B., Miralles, C. P., et al. (2012). Molecular and functional interaction between protocadherin-gammaC5 and GABAA receptors. *The Journal of Neuroscience*, *32*, 11780–11797.

Lin, J., Luo, J., & Redies, C. (2010). Cadherin-19 expression is restricted to myelin-forming cells in the chicken embryo. *Neuroscience*, *165*, 168–178.

Luccardini, C., Hennekinne, L., Viou, L., Yanagida, M., Murakami, F., Kessaris, N., et al. (2013). N-cadherin sustains motility and polarity of future cortical interneurons during tangential migration. *The Journal of Neuroscience*, *33*, 18149–18160.

Luo, J., Ju, M. J., & Redies, C. (2006). Regionalized cadherin-7 expression by radial glia is regulated by Shh and Pax7 during chicken spinal cord development. *Neuroscience*, *142*, 1133–1143.

Machado, C. B., Kanning, K. C., Kreis, P., Stevenson, D., Crossley, M., Nowak, M., et al. (2014). Reconstruction of phrenic neuron identity in embryonic stem cell-derived motor neurons. *Development*, *141*, 784–794.

Maguschak, K. A., & Ressler, K. J. (2008). Beta-catenin is required for memory consolidation. *Nature Neuroscience, 11*, 1319–1326.
Mainen, Z. F., Jia, Z., Roder, J., & Malinow, R. (1998). Use-dependent AMPA receptor block in mice lacking GluR2 suggests postsynaptic site for LTP expression. *Nature Neuroscience, 1*, 579–586.
Malinow, R., & Malenka, R. C. (2002). AMPA receptor trafficking and synaptic plasticity. *Annual Review of Neuroscience, 25*, 103–126.
Manabe, T., Togashi, H., Uchida, N., Suzuki, S. C., Hayakawa, Y., Yamamoto, M., et al. (2000). Loss of cadherin-11 adhesion receptor enhances plastic changes in hippocampal synapses and modifies behavioral responses. *Molecular and Cellular Neurosciences, 15*, 534–546.
Manibog, K., Li, H., Rakshit, S., & Sivasankar, S. (2014). Resolving the molecular mechanism of cadherin catch bond formation. *Nature Communications, 5*, 3941.
Marambaud, P., Wen, P. H., Dutt, A., Shioi, J., Takashima, A., Siman, R., et al. (2003). A CBP binding transcriptional repressor produced by the PS1/epsilon-cleavage of N-cadherin is inhibited by PS1 FAD mutations. *Cell, 114*, 635–645.
Marin, O., & Rubenstein, J. L. (2003). Cell migration in the forebrain. *Annual Review of Neuroscience, 26*, 441–483.
Marrs, G. S., Honda, T., Fuller, L., Thangavel, R., Balsamo, J., Lilien, J., et al. (2006). Dendritic arbors of developing retinal ganglion cells are stabilized by beta 1-integrins. *Molecular and Cellular Neurosciences, 32*, 230–241.
Marrs, G. S., Theisen, C. S., & Bruses, J. L. (2009). N-cadherin modulates voltage activated calcium influx via RhoA, p120-catenin, and myosin-actin interaction. *Molecular and Cellular Neurosciences, 40*, 390–400.
Marthiens, V., Gavard, J., Padilla, F., Monnet, C., Castellani, V., Lambert, M., et al. (2005). A novel function for cadherin-11 in the regulation of motor axon elongation and fasciculation. *Molecular and Cellular Neurosciences, 28*, 715–726.
Matson, J. L., & Shoemaker, M. (2009). Intellectual disability and its relationship to autism spectrum disorders. *Research in Developmental Disabilities, 30*, 1107–1114.
Matsunaga, M., Hatta, K., Nagafuchi, A., & Takeichi, M. (1988). Guidance of optic nerve fibres by N-cadherin adhesion molecules. *Nature, 334*, 62–64.
Matter, C., Pribadi, M., Liu, X., & Trachtenberg, J. T. (2009). Delta-catenin is required for the maintenance of neural structure and function in mature cortex in vivo. *Neuron, 64*, 320–327.
Medina, M., Marinescu, R. C., Overhauser, J., & Kosik, K. S. (2000). Hemizygosity of delta-catenin (CTNND2) is associated with severe mental retardation in cri-du-chat syndrome. *Genomics, 63*, 157–164.
Mendez, P., De Roo, M., Poglia, L., Klauser, P., & Muller, D. (2010). N-Cadherin mediates plasticity-induced long-term spine stabilization. *The Journal of Cell Biology, 189*, 589–600.
Miar, A., Alvarez, V., Corao, A. I., Alonso, B., Diaz, M., Menendez, M., et al. (2011). Lack of association between protocadherin 11-X/Y (PCDH11X and PCDH11Y) polymorphisms and late onset Alzheimer's disease. *Brain Research, 1383*, 252–256.
Mills, F., Bartlett, T. E., Dissing-Olesen, L., Wisniewska, M. B., Kuznicki, J., Macvicar, B. A., et al. (2014). Cognitive flexibility and long-term depression (LTD) are impaired following beta-catenin stabilization in vivo. *Proceedings of the National Academy of Sciences of the United States of America, 111*, 8631–8636.
Miyata, T., Kawaguchi, A., Okano, H., & Ogawa, M. (2001). Asymmetric inheritance of radial glial fibers by cortical neurons. *Neuron, 31*, 727–741.
Morales, M., Colicos, M. A., & Goda, Y. (2000). Actin-dependent regulation of neurotransmitter release at central synapses. *Neuron, 27*, 539–550.
Morishita, H., Umitsu, M., Murata, Y., Shibata, N., Udaka, K., Higuchi, Y., et al. (2006). Structure of the cadherin-related neuronal receptor/protocadherin-alpha first

extracellular cadherin domain reveals diversity across cadherin families. *The Journal of Biological Chemistry*, *281*, 33650–33663.
Morrow, E. M., Yoo, S. Y., Flavell, S. W., Kim, T. K., Lin, Y., Hill, R. S., et al. (2008). Identifying autism loci and genes by tracing recent shared ancestry. *Science*, *321*, 218–223.
Mortillo, S., Elste, A., Ge, Y., Patil, S. B., Hsiao, K., Huntley, G. W., et al. (2012). Compensatory redistribution of neuroligins and N-cadherin following deletion of synaptic beta1-integrin. *The Journal of Comparative Neurology*, *520*, 2041–2052.
Moya, P. R., Dodman, N. H., Timpano, K. R., Rubenstein, L. M., Rana, Z., Fried, R. L., et al. (2013). Rare missense neuronal cadherin gene (CDH2) variants in specific obsessive-compulsive disorder and Tourette disorder phenotypes. *European Journal of Human Genetics*, *21*, 850–854.
Murase, S., Mosser, E., & Schuman, E. M. (2002). Depolarization drives beta-catenin into neuronal spines promoting changes in synaptic structure and function. *Neuron*, *35*, 91–105.
Murata, Y., Hamada, S., Morishita, H., Mutoh, T., & Yagi, T. (2004). Interaction with protocadherin-gamma regulates the cell surface expression of protocadherin-alpha. *The Journal of Biological Chemistry*, *279*, 49508–49516.
Mutoh, T., Hamada, S., Senzaki, K., Murata, Y., & Yagi, T. (2004). Cadherin-related neuronal receptor 1 (CNR1) has cell adhesion activity with beta1 integrin mediated through the RGD site of CNR1. *Experimental Cell Research*, *294*, 494–508.
Nagar, B., Overduin, M., Ikura, M., & Rini, J. M. (1996). Structural basis of calcium-induced E-cadherin rigidification and dimerization. *Nature*, *380*, 360–364.
Nagashima, M., Barthel, L. K., & Raymond, P. A. (2013). A self-renewing division of zebrafish Muller glial cells generates neuronal progenitors that require N-cadherin to regenerate retinal neurons. *Development*, *140*, 4510–4521.
Nakao, S., Platek, A., Hirano, S., & Takeichi, M. (2008). Contact-dependent promotion of cell migration by the OL-protocadherin-Nap1 interaction. *The Journal of Cell Biology*, *182*, 395–410.
Nern, A., Zhu, Y., & Zipursky, S. L. (2008). Local N-cadherin interactions mediate distinct steps in the targeting of lamina neurons. *Neuron*, *58*, 34–41.
Neugebauer, K. M., Tomaselli, K. J., Lilien, J., & Reichardt, L. F. (1988). N-Cadherin, NCAM, and integrins promote retinal neurite outgrowth on astrocytes in vitro. *The Journal of Cell Biology*, *107*, 1177–1187.
Nikitczuk, J. S., Patil, S. B., Matikainen-Ankney, B. A., Scarpa, J., Shapiro, M. L., Benson, D. L., et al. (2014). N-Cadherin regulates molecular organization of excitatory and inhibitory synaptic circuits in adult hippocampus in vivo. *Hippocampus*, *24*, 943–962. http://dx.doi.org/10.1002/hipo.22282.
Nose, A., Nagafuchi, A., & Takeichi, M. (1988). Expressed recombinant cadherins mediate cell sorting in model systems. *Cell*, *54*, 993–1001.
Nuriya, M., & Huganir, R. L. (2006). Regulation of AMPA receptor trafficking by N-cadherin. *Journal of Neurochemistry*, *97*, 652–661.
Oblander, S. A., & Brady-Kalnay, S. M. (2010). Distinct PTPmu-associated signaling molecules differentially regulate neurite outgrowth on E-, N-, and R-cadherin. *Molecular and Cellular Neurosciences*, *44*, 78–93.
Okamura, K., Tanaka, H., Yagita, Y., Saeki, Y., Taguchi, A., Hiraoka, Y., et al. (2004). Cadherin activity is required for activity-induced spine remodeling. *The Journal of Cell Biology*, *167*, 961–972.
O'Leary, R., Reilly, J. E., Hanson, H. H., Kang, S., Lou, N., & Phillips, G. R. (2011). A variable cytoplasmic domain segment is necessary for gamma-protocadherin trafficking and tubulation in the endosome/lysosome pathway. *Molecular Biology of the Cell*, *22*, 4362–4372.

Opazo, P., & Choquet, D. (2011). A three-step model for the synaptic recruitment of AMPA receptors. *Molecular and Cellular Neurosciences*, *46*, 1–8.
Osterhout, J. A., Josten, N., Yamada, J., Pan, F., Wu, S. W., Nguyen, P. L., et al. (2011). Cadherin-6 mediates axon-target matching in a non-image-forming visual circuit. *Neuron*, *71*, 632–639.
Pagnamenta, A. T., Khan, H., Walker, S., Gerrelli, D., Wing, K., Bonaglia, M. C., et al. (2011). Rare familial 16q21 microdeletions under a linkage peak implicate cadherin 8 (CDH8) in susceptibility to autism and learning disability. *Journal of Medical Genetics*, *48*, 48–54.
Paradis, S., Harrar, D. B., Lin, Y., Koon, A. C., Hauser, J. L., Griffith, E. C., et al. (2007). An RNAi-based approach identifies molecules required for glutamatergic and GABAergic synapse development. *Neuron*, *53*, 217–232.
Pardo, C. A., & Eberhart, C. G. (2007). The neurobiology of autism. *Brain Pathology*, *17*, 434–447.
Park, M., Penick, E. C., Edwards, J. G., Kauer, J. A., & Ehlers, M. D. (2004). Recycling endosomes supply AMPA receptors for LTP. *Science*, *305*, 1972–1975.
Passafaro, M., Piech, V., & Sheng, M. (2001). Subunit-specific temporal and spatial patterns of AMPA receptor exocytosis in hippocampal neurons. *Nature Neuroscience*, *4*, 917–926.
Patel, S. D., Ciatto, C., Chen, C. P., Bahna, F., Rajebhosale, M., Arkus, N., et al. (2006). Type II cadherin ectodomain structures: Implications for classical cadherin specificity. *Cell*, *124*, 1255–1268.
Patil, S., Brock, J. H., Colman, D. R., & Huntley, G. W. (2009). GDNF-dependent modulation of cadherins in spinal dorsal horn in relationship to neuropathic pain. *Pain*, *152*, 924–935.
Payne, H. R., Hemperly, J. J., & Lemmon, V. (1996). N-Cadherin expression and function in cultured oligodendrocytes. *Brain Research. Developmental Brain Research*, *97*, 9–15.
Peng, Y. R., He, S., Marie, H., Zeng, S. Y., Ma, J., Tan, Z. J., et al. (2009). Coordinated changes in dendritic arborization and synaptic strength during neural circuit development. *Neuron*, *61*, 71–84.
Pertz, O., Bozic, D., Koch, A. W., Fauser, C., Brancaccio, A., & Engel, J. (1999). A new crystal structure, Ca2+ dependence and mutational analysis reveal molecular details of E-cadherin homoassociation. *The EMBO Journal*, *18*, 1738–1747.
Pettem, K. L., Yokomaku, D., Luo, L., Linhoff, M. W., Prasad, T., Connor, S. A., et al. (2013). The specific alpha-neurexin interactor calsyntenin-3 promotes excitatory and inhibitory synapse development. *Neuron*, *80*, 113–128.
Pielarski, K. N., van Stegen, B., Andreyeva, A., Nieweg, K., Jungling, K., Redies, C., et al. (2013). Asymmetric N-cadherin expression results in synapse dysfunction, synapse elimination, and axon retraction in cultured mouse neurons. *PLoS One*, *8*, e54105. http://dx.doi.org/10.1371/journal.pone.0054105.
Pokutta, S., Herrenknecht, K., Kemler, R., & Engel, J. (1994). Conformational changes of the recombinant extracellular domain of E-cadherin upon calcium binding. *European Journal of Biochemistry*, *223*, 1019–1026.
Poskanzer, K., Needleman, L. A., Bozdagi, O., & Huntley, G. W. (2003). N-Cadherin regulates ingrowth and laminar targeting of thalamocortical axons. *The Journal of Neuroscience*, *23*, 2294–2305.
Prange, O., Wong, T. P., Gerrow, K., Wang, Y. T., & El-Husseini, A. (2004). A balance between excitatory and inhibitory synapses is controlled by PSD-95 and neuroligin. *Proceedings of the National Academy of Sciences of the United States of America*, *101*, 13915–13920.
Rago, L., Beattie, R., Taylor, V., & Winter, J. (2014). miR379-410 cluster miRNAs regulate neurogenesis and neuronal migration by fine-tuning N-cadherin. *The EMBO Journal*, *33*, 906–920.

Rakshit, S., Zhang, Y., Manibog, K., Shafraz, O., & Sivasankar, S. (2012). Ideal, catch, and slip bonds in cadherin adhesion. *Proceedings of the National Academy of Sciences of the United States of America, 109*, 18815–18820.

Ranscht, B., & Dours-Zimmermann, M. T. (1991). T-Cadherin, a novel cadherin cell adhesion molecule in the nervous system lacks the conserved cytoplasmic region. *Neuron, 7*, 391–402.

Rapoport, J., Chavez, A., Greenstein, D., Addington, A., & Gogtay, N. (2009). Autism spectrum disorders and childhood-onset schizophrenia: Clinical and biological contributions to a relation revisited. *Journal of the American Academy of Child and Adolescent Psychiatry, 48*, 10–18.

Redies, C., Hertel, N., & Hubner, C. A. (2012). Cadherins and neuropsychiatric disorders. *Brain Research, 1470*, 130–144.

Reines, A., Bernier, L. P., McAdam, R., Belkaid, W., Shan, W., Koch, A. W., et al. (2012). N-Cadherin prodomain processing regulates synaptogenesis. *The Journal of Neuroscience, 32*, 6323–6334.

Reiss, K., Maretzky, T., Ludwig, A., Tousseyn, T., de Strooper, B., Hartmann, D., et al. (2005). ADAM10 cleavage of N-cadherin and regulation of cell-cell adhesion and beta-catenin nuclear signalling. *The EMBO Journal, 24*, 742–752.

Riehl, R., Johnson, K., Bradley, R., Grunwald, G. B., Cornel, E., Lilienbaum, A., et al. (1996). Cadherin function is required for axon outgrowth in retinal ganglion cells in vivo. *Neuron, 17*, 837–848.

Rivero, O., Sich, S., Popp, S., Schmitt, A., Franke, B., & Lesch, K. P. (2013). Impact of the ADHD-susceptibility gene CDH13 on development and function of brain networks. *European Neuropsychopharmacology, 23*, 492–507.

Rubio, M. E., Curcio, C., Chauvet, N., & Bruses, J. L. (2005). Assembly of the N-cadherin complex during synapse formation involves uncoupling of p120-catenin and association with presenilin 1. *Molecular and Cellular Neurosciences, 30*, 611–623.

Ryan, S. G., Chance, P. F., Zou, C. H., Spinner, N. B., Golden, J. A., & Smietana, S. (1997). Epilepsy and mental retardation limited to females: An X-linked dominant disorder with male sparing. *Nature Genetics, 17*, 92–95.

Saglietti, L., Dequidt, C., Kamieniarz, K., Rousset, M. C., Valnegri, P., Thoumine, O., et al. (2007). Extracellular interactions between GluR2 and N-cadherin in spine regulation. *Neuron, 54*, 461–477.

Sakane, F., & Miyamoto, Y. (2013). N-Cadherin regulates the proliferation and differentiation of ventral midbrain dopaminergic progenitors. *Developmental Neurobiology, 73*, 518–529.

Sakarya, O., Armstrong, K. A., Adamska, M., Adamski, M., Wang, I. F., Tidor, B., et al. (2007). A post-synaptic scaffold at the origin of the animal kingdom. *PLoS One, 2*, e506. http://dx.doi.org/10.1371/journal.pone.0000506.

Sanders, S. J., Ercan-Sencicek, A. G., Hus, V., Luo, R., Murtha, M. T., Moreno-De-Luca, D., et al. (2011). Multiple recurrent de novo CNVs, including duplications of the 7q11.23 Williams syndrome region, are strongly associated with autism. *Neuron, 70*, 863–885.

Sanhueza, M., & Lisman, J. (2013). The CaMKII/NMDAR complex as a molecular memory. *Molecular Brain, 6*, 10.

Sansom, S. N., & Livesey, F. J. (2009). Gradients in the brain: The control of the development of form and function in the cerebral cortex. *Cold Spring Harbor Perspectives in Biology, 1*, a002519.

Schalm, S. S., Ballif, B. A., Buchanan, S. M., Phillips, G. R., & Maniatis, T. (2010). Phosphorylation of protocadherin proteins by the receptor tyrosine kinase Ret. *Proceedings of the National Academy of Sciences of the United States of America, 107*, 13894–13899.

Scheiffele, P., Fan, J., Choih, J., Fetter, R., & Serafini, T. (2000). Neuroligin expressed in nonneuronal cells triggers presynaptic development in contacting axons. *Cell, 101*, 657–669.

Schnadelbach, O., Blaschuk, O. W., Symonds, M., Gour, B. J., Doherty, P., & Fawcett, J. W. (2000). N-Cadherin influences migration of oligodendrocytes on astrocyte monolayers. *Molecular and Cellular Neurosciences*, *15*, 288–302.

Schnadelbach, O., Ozen, I., Blaschuk, O. W., Meyer, R. L., & Fawcett, J. W. (2001). N-Cadherin is involved in axon-oligodendrocyte contact and myelination. *Molecular and Cellular Neurosciences*, *17*, 1084–1093.

Schnell, E., Sizemore, M., Karimzadegan, S., Chen, L., Bredt, D. S., & Nicoll, R. A. (2002). Direct interactions between PSD-95 and stargazin control synaptic AMPA receptor number. *Proceedings of the National Academy of Sciences of the United States of America*, *99*, 13902–13907.

Schreiner, D., & Weiner, J. A. (2010). Combinatorial homophilic interaction between gamma-protocadherin multimers greatly expands the molecular diversity of cell adhesion. *Proceedings of the National Academy of Sciences of the United States of America*, *107*, 14893–14898.

Schrick, C., Fischer, A., Srivastava, D. P., Tronson, N. C., Penzes, P., & Radulovic, J. (2007). N-cadherin regulates cytoskeletally associated IQGAP1/ERK signaling and memory formation. *Neuron*, *55*, 786–798.

Schwabe, T., Neuert, H., & Clandinin, T. R. (2013). A network of cadherin-mediated interactions polarizes growth cones to determine targeting specificity. *Cell*, *154*, 351–364.

Shan, W. S., Tanaka, H., Phillips, G. R., Arndt, K., Yoshida, M., Colman, D. R., et al. (2000). Functional cis-heterodimers of N- and R-cadherins. *The Journal of Cell Biology*, *148*, 579–590.

Shan, W., Yoshida, M., Wu, X. R., Huntley, G. W., & Colman, D. R. (2002). Neural (N-) cadherin, a synaptic adhesion molecule, is induced in hippocampal mossy fiber axonal sprouts by seizure. *Journal of Neuroscience Research*, *69*, 292–304.

Shankar, G. M., Bloodgood, B. L., Townsend, M., Walsh, D. M., Selkoe, D. J., & Sabatini, B. L. (2007). Natural oligomers of the Alzheimer amyloid-beta protein induce reversible synapse loss by modulating an NMDA-type glutamate receptor-dependent signaling pathway. *The Journal of Neuroscience*, *27*, 2866–2875.

Shapiro, L., Fannon, A. M., Kwong, P. D., Thompson, A., Lehmann, M. S., Grubel, G., et al. (1995). Structural basis of cell-cell adhesion by cadherins. *Nature*, *374*, 327–337.

Shapiro, L., & Weis, W. I. (2009). Structure and biochemistry of cadherins and catenins. *Cold Spring Harbor Perspectives in Biology*, *1*, a003053.

Shi, S. H., Hayashi, Y., Petralia, R. S., Zaman, S. H., Wenthold, R. J., Svoboda, K., et al. (1999). Rapid spine delivery and redistribution of AMPA receptors after synaptic NMDA receptor activation. *Science*, *284*, 1811–1816.

Shimamura, K., Takahashi, T., & Takeichi, M. (1992). E-cadherin expression in a particular subset of sensory neurons. *Developmental Biology*, *152*, 242–254.

Sibille, E., Wang, Y., Joeyen-Waldorf, J., Gaiteri, C., Surget, A., Oh, S., et al. (2009). A molecular signature of depression in the amygdala. *The American Journal of Psychiatry*, *166*, 1011–1024.

Silva, A. J. (2003). Molecular and cellular cognitive studies of the role of synaptic plasticity in memory. *Journal of Neurobiology*, *54*, 224–237.

Silverman, J. B., Restituito, S., Lu, W., Lee-Edwards, L., Khatri, L., & Ziff, E. B. (2007). Synaptic anchorage of AMPA receptors by cadherins through neural plakophilin-related arm protein AMPA receptor-binding protein complexes. *The Journal of Neuroscience*, *27*, 8505–8516.

Simonoff, E., Pickles, A., Charman, T., Chandler, S., Loucas, T., & Baird, G. (2008). Psychiatric disorders in children with autism spectrum disorders: Prevalence, comorbidity, and associated factors in a population-derived sample. *Journal of the American Academy of Child and Adolescent Psychiatry*, *47*, 921–929.

Srivastava, D. P., Copits, B. A., Xie, Z., Huda, R., Jones, K. A., Mukherji, S., et al. (2012). Afadin is required for maintenance of dendritic structure and excitatory tone. *The Journal of Biological Chemistry*, *287*, 35964–35974.

Stan, A., Pielarski, K. N., Brigadski, T., Wittenmayer, N., Fedorchenko, O., Gohla, A., et al. (2010). Essential cooperation of N-cadherin and neuroligin-1 in the transsynaptic control of vesicle accumulation. *Proceedings of the National Academy of Sciences of the United States of America*, *107*, 11116–11121.

Stein, V., House, D. R., Bredt, D. S., & Nicoll, R. A. (2003). Postsynaptic density-95 mimics and occludes hippocampal long-term potentiation and enhances long-term depression. *The Journal of Neuroscience*, *23*, 5503–5506.

Ster, J., Steuble, M., Orlando, C., Diep, T. M., Akhmedov, A., Raineteau, O., et al. (2014). Calsyntenin-1 regulates targeting of dendritic NMDA receptors and dendritic spine maturation in CA1 hippocampal pyramidal cells during postnatal development. *The Journal of Neuroscience*, *34*, 8716–8727.

Sugino, H., Hamada, S., Yasuda, R., Tuji, A., Matsuda, Y., Fujita, M., et al. (2000). Genomic organization of the family of CNR cadherin genes in mice and humans. *Genomics*, *63*, 75–87.

Sun, Y., Aiga, M., Yoshida, E., Humbert, P. O., & Bamji, S. X. (2009). Scribble interacts with beta-catenin to localize synaptic vesicles to synapses. *Molecular Biology of the Cell*, *20*, 3390–3400.

Sun, Y., & Bamji, S. X. (2011). beta-Pix modulates actin-mediated recruitment of synaptic vesicles to synapses. *The Journal of Neuroscience*, *31*, 17123–17133.

Suyama, K., Shapiro, I., Guttman, M., & Hazan, R. B. (2002). A signaling pathway leading to metastasis is controlled by N-cadherin and the FGF receptor. *Cancer Cell*, *2*, 301–314.

Suzuki, S. C., Inoue, T., Kimura, Y., Tanaka, T., & Takeichi, M. (1997). Neuronal circuits are subdivided by differential expression of Type-II classic cadherins in postnatal mouse brains. *Molecular and Cellular Neurosciences*, *9*, 433–447.

Tai, C. Y., Mysore, S. P., Chiu, C., & Schuman, E. M. (2007). Activity-regulated N-cadherin endocytosis. *Neuron*, *54*, 771–785.

Takahashi, M., & Osumi, N. (2005). Identification of a novel type II classical cadherin: Rat cadherin19 is expressed in the cranial ganglia and Schwann cell precursors during development. *Developmental Dynamics*, *232*, 200–208.

Takai, Y., Ikeda, W., Ogita, H., & Rikitake, Y. (2008). The immunoglobulin-like cell adhesion molecule nectin and its associated protein afadin. *Annual Review of Cell and Developmental Biology*, *24*, 309–342.

Takeichi, M., Inuzuka, H., Shimamura, K., Matsunaga, M., & Nose, A. (1990). Cadherin-mediated cell-cell adhesion and neurogenesis. *Neuroscience Research. Supplement*, *13*, S92–S96.

Tan, Z. J., Peng, Y., Song, H. L., Zheng, J. J., & Yu, X. (2010). N-Cadherin-dependent neuron-neuron interaction is required for the maintenance of activity-induced dendrite growth. *Proceedings of the National Academy of Sciences of the United States of America*, *107*, 9873–9878.

Tanaka, H., Shan, W., Phillips, G. R., Arndt, K., Bozdagi, O., Shapiro, L., et al. (2000). Molecular modification of N-cadherin in response to synaptic activity. *Neuron*, *25*, 93–107.

Tang, L., Hung, C. P., & Schuman, E. M. (1998). A role for the cadherin family of cell adhesion molecules in hippocampal long-term potentiation. *Neuron*, *20*, 1165–1175.

Tasic, B., Nabholz, C. E., Baldwin, K. K., Kim, Y., Rueckert, E. H., Ribich, S. A., et al. (2002). Promoter choice determines splice site selection in protocadherin alpha and gamma pre-mRNA splicing. *Molecular Cell*, *10*, 21–33.

Taylor, A. M., Wu, J., Tai, H. C., & Schuman, E. M. (2013). Axonal translation of beta-catenin regulates synaptic vesicle dynamics. *The Journal of Neuroscience*, *33*, 5584–5589.
Thomas, C., & Strutt, D. (2012). The roles of the cadherins Fat and dachsous in planar polarity specification in drosophila. *Developmental Dynamics*, *241*, 27–39.
Tissir, F., Bar, I., Jossin, Y., De Backer, O., & Goffinet, A. M. (2005). Protocadherin Celsr3 is crucial in axonal tract development. *Nature Neuroscience*, *8*, 451–457.
Togashi, H., Abe, K., Mizoguchi, A., Takaoka, K., Chisaka, O., & Takeichi, M. (2002). Cadherin regulates dendritic spine morphogenesis. *Neuron*, *35*, 77–89.
Tomaselli, K. J., Neugebauer, K. M., Bixby, J. L., Lilien, J., & Reichardt, L. F. (1988). N-Cadherin and integrins: Two receptor systems that mediate neuronal process outgrowth on astrocyte surfaces. *Neuron*, *1*, 33–43.
Toyoda, S., Kawaguchi, M., Kobayashi, T., Tarusawa, E., Toyama, T., Okano, M., et al. (2014). Developmental epigenetic modification regulates stochastic expression of clustered protocadherin genes, generating single neuron diversity. *Neuron*, *82*, 94–108.
Treubert-Zimmermann, U., Heyers, D., & Redies, C. (2002). Targeting axons to specific fiber tracts in vivo by altering cadherin expression. *The Journal of Neuroscience*, *22*, 7617–7626.
Tsai, N. P., Wilkerson, J. R., Guo, W., Maksimova, M. A., DeMartino, G. N., Cowan, C. W., et al. (2012). Multiple autism-linked genes mediate synapse elimination via proteasomal degradation of a synaptic scaffold PSD-95. *Cell*, *151*, 1581–1594.
Tsien, J. Z., Chen, D. F., Gerber, D., Tom, C., Mercer, E. H., Anderson, D. J., et al. (1996). Subregion- and cell type-restricted gene knockout in mouse brain. *Cell*, *87*, 1317–1326.
Uchida, N., Honjo, Y., Johnson, K. R., Wheelock, M. J., & Takeichi, M. (1996). The catenin/cadherin adhesion system is localized in synaptic junctions bordering transmitter release zones. *Journal of Cell Biology*, *135*, 767–779.
Uemura, K., Kuzuya, A., Aoyagi, N., Ando, K., Shimozono, Y., Ninomiya, H., et al. (2007). Amyloid beta inhibits ectodomain shedding of N-cadherin via down-regulation of cell-surface NMDA receptor. *Neuroscience*, *145*, 5–10.
Uemura, M., Nakao, S., Suzuki, S. T., Takeichi, M., & Hirano, S. (2007). OL-protocadherin is essential for growth of striatal axons and thalamocortical projections. *Nature Neuroscience*, *10*, 1151–1159.
Ullian, E. M., Christopherson, K. S., & Barres, B. A. (2004). Role for glia in synaptogenesis. *Glia*, *47*, 209–216.
Um, J. W., Pramanik, G., Ko, J. S., Song, M. Y., Lee, D., Kim, H., et al. (2014). Calsyntenins function as synaptogenic adhesion molecules in concert with neurexins. *Cell Reports*, *6*, 1096–1109.
Utton, M. A., Eickholt, B., Howell, F. V., Wallis, J., & Doherty, P. (2001). Soluble N-cadherin stimulates fibroblast growth factor receptor dependent neurite outgrowth and N-cadherin and the fibroblast growth factor receptor co-cluster in cells. *Journal of Neurochemistry*, *76*, 1421–1430.
Vanhalst, K., Kools, P., Staes, K., van Roy, F., & Redies, C. (2005). Delta-protocadherins: A gene family expressed differentially in the mouse brain. *Cellular and Molecular Life Sciences*, *62*, 1247–1259.
van Harssel, J. J., Weckhuysen, S., van Kempen, M. J., Hardies, K., Verbeek, N. E., de Kovel, C. G., et al. (2013). Clinical and genetic aspects of PCDH19-related epilepsy syndromes and the possible role of PCDH19 mutations in males with autism spectrum disorders. *Neurogenetics*, *14*, 23–34.
Waites, C. L., Craig, A. M., & Garner, C. C. (2005). Mechanisms of vertebrate synaptogenesis. *Annual Review of Neuroscience*, *28*, 251–274.
Walker, S., & Scherer, S. W. (2013). Identification of candidate intergenic risk loci in autism spectrum disorder. *BMC Genomics*, *14*, 499.

Wang, X., Su, H., & Bradley, A. (2002). Molecular mechanisms governing pcdh-gamma gene expression: Evidence for a multiple promoter and cis-alternative splicing model. *Genes & Development, 16*, 1890–1905.
Wang, X., Weiner, J. A., Levi, S., Craig, A. M., Bradley, A., & Sanes, J. R. (2002). Gamma protocadherins are required for survival of spinal interneurons. *Neuron, 36*, 843–854.
Wang, K., Zhang, H., Ma, D., Bucan, M., Glessner, J. T., Abrahams, B. S., et al. (2009). Common genetic variants on 5p14.1 associate with autism spectrum disorders. *Nature, 459*, 528–533.
Weiner, J. A., & Jontes, J. D. (2013). Protocadherins, not prototypical: A complex tale of their interactions, expression, and functions. *Frontiers in Molecular Neuroscience, 6*, 4.
Weiner, J. A., Wang, X., Tapia, J. C., & Sanes, J. R. (2005). Gamma protocadherins are required for synaptic development in the spinal cord. *Proceedings of the National Academy of Sciences of the United States of America, 102*, 8–14.
Wenzel, H. J., Woolley, C. S., Robbins, C. A., & Schwartzkroin, P. A. (2000). Kainic acid-induced mossy fiber sprouting and synapse formation in the dentate gyrus of rats. *Hippocampus, 10*, 244–260.
Williams, E. J., Mittal, B., Walsh, F. S., & Doherty, P. (1995). A Ca2+/calmodulin kinase inhibitor, KN-62, inhibits neurite outgrowth stimulated by CAMs and FGF. *Molecular and Cellular Neurosciences, 6*, 69–79.
Williams, E. J., Walsh, F. S., & Doherty, P. (1994). The production of arachidonic acid can account for calcium channel activation in the second messenger pathway underlying neurite outgrowth stimulated by NCAM, N-cadherin, and L1. *Journal of Neurochemistry, 62*, 1231–1234.
Williams, M. E., Wilke, S. A., Daggett, A., Davis, E., Otto, S., Ravi, D., et al. (2011). Cadherin-9 regulates synapse-specific differentiation in the developing hippocampus. *Neuron, 71*, 640–655.
Williams, E. J., Williams, G., Howell, F. V., Skaper, S. D., Walsh, F. S., & Doherty, P. (2001). Identification of an N-cadherin motif that can interact with the fibroblast growth factor receptor and is required for axonal growth. *The Journal of Biological Chemistry, 276*, 43879–43886.
Wöhrn, J. C., Nakagawa, S., Ast, M., Takeichi, M., & Redies, C. (1999). Combinatorial expression of cadherins in the tectum and the sorting of neurites in the tectofugal pathways of the chicken embryo. *Neuroscience, 90*, 985–1000.
Wong, G. K., Baudet, M. L., Norden, C., Leung, L., & Harris, W. A. (2012). Slit1b-Robo3 signaling and N-cadherin regulate apical process retraction in developing retinal ganglion cells. *The Journal of Neuroscience, 32*, 223–228.
Wu, Q., & Maniatis, T. (1999). A striking organization of a large family of human neural cadherin-like cell adhesion genes. *Cell, 97*, 779–790.
Wu, Y., Vendome, J., Shapiro, L., Ben-Shaul, A., & Honig, B. (2011). Transforming binding affinities from three dimensions to two with application to cadherin clustering. *Nature, 475*, 510–513.
Wu, Z. C., Yu, J. T., Wang, N. D., Yu, N. N., Zhang, Q., Chen, W., et al. (2010). Lack of association between PCDH11X genetic variation and late-onset Alzheimer's disease in a Han Chinese population. *Brain Research, 1357*, 152–156.
Wu, Q., Zhang, T., Cheng, J. F., Kim, Y., Grimwood, J., Schmutz, J., et al. (2001). Comparative DNA sequence analysis of mouse and human protocadherin gene clusters. *Genome Research, 11*, 389–404.
Yagi, T., & Takeichi, M. (2000). Cadherin superfamily genes: Functions, genomic organization, and neurologic diversity. *Genes & Development, 14*, 1169–1180.
Yam, P. T., Pincus, Z., Gupta, G. D., Bashkurov, M., Charron, F., Pelletier, L., et al. (2013). N-Cadherin relocalizes from the periphery to the center of the synapse after transient synaptic stimulation in hippocampal neurons. *PLoS One, 8*, e79679.

Yamada, S., Pokutta, S., Drees, F., Weis, W. I., & Nelson, W. J. (2005). Deconstructing the cadherin-catenin-actin complex. *Cell*, *123*, 889–901.
Yamagata, M., Herman, J. P., & Sanes, J. R. (1995). Lamina-specific expression of adhesion molecules in developing chick optic tectum. *The Journal of Neuroscience*, *15*, 4556–4571.
Yamaguchi, M., Imai, F., Tonou-Fujimori, N., & Masai, I. (2010). Mutations in N-cadherin and a Stardust homolog, Nagie oko, affect cell-cycle exit in zebrafish retina. *Mechanisms of Development*, *127*, 247–264.
Yang, Y., Wang, X. B., Frerking, M., & Zhou, Q. (2008). Delivery of AMPA receptors to perisynaptic sites precedes the full expression of long-term potentiation. *Proceedings of the National Academy of Sciences of the United States of America*, *105*, 11388–11393.
Yap, A. S., Niessen, C. M., & Gumbiner, B. M. (1998). The juxtamembrane region of the cadherin cytoplasmic tail supports lateral clustering, adhesive strengthening, and interaction with p120ctn. *The Journal of Cell Biology*, *141*, 779–789.
Yasuda, S., Tanaka, H., Sugiura, H., Okamura, K., Sakaguchi, T., Tran, U., et al. (2007). Activity-induced protocadherin arcadlin regulates dendritic spine number by triggering N-cadherin endocytosis via TAO2beta and p38 MAP kinases. *Neuron*, *56*, 456–471.
Yizhar, O., Fenno, L. E., Prigge, M., Schneider, F., Davidson, T. J., O'Shea, D. J., et al. (2011). Neocortical excitation/inhibition balance in information processing and social dysfunction. *Nature*, *477*, 171–178.
Yu, X., & Malenka, R. C. (2003). Beta-catenin is critical for dendritic morphogenesis. *Nature Neuroscience*, *6*, 1169–1177.
Yudowski, G. A., Puthenveedu, M. A., Leonoudakis, D., Panicker, S., Thorn, K. S., Beattie, E. C., et al. (2007). Real-time imaging of discrete exocytic events mediating surface delivery of AMPA receptors. *The Journal of Neuroscience*, *27*, 11112–11121.
Zamanillo, D., Sprengel, R., Hvalby, O., Jensen, V., Burnashev, N., Rozov, A., et al. (1999). Importance of AMPA receptors for hippocampal synaptic plasticity but not for spatial learning. *Science*, *284*, 1805–1811.
Zhang, W., & Benson, D. L. (2001). Stages of synapse development defined by dependence on F-actin. *The Journal of Neuroscience*, *21*, 5169–5181.
Zhou, Z., Hu, J., Passafaro, M., Xie, W., & Jia, Z. (2011). GluA2 (GluR2) regulates metabotropic glutamate receptor-dependent long-term depression through N-cadherin-dependent and cofilin-mediated actin reorganization. *The Journal of Neuroscience*, *31*, 819–833.
Zhu, H., & Luo, L. (2004). Diverse functions of N-cadherin in dendritic and axonal terminal arborization of olfactory projection neurons. *Neuron*, *42*, 63–75.

CHAPTER FOURTEEN

Cell–Cell Interactions Driving Kidney Morphogenesis

Alexander N. Combes*, Jamie A. Davies†, Melissa H. Little*,[1]
*Institute for Molecular Bioscience, The University of Queensland, St. Lucia, Brisbane, Queensland, Australia
†Centre for Integrative Physiology, University of Edinburgh, Edinburgh, United Kingdom
[1]Corresponding author: e-mail address: m.little@imb.uq.edu.au

Contents

Abstract

The mammalian kidney forms via cell–cell interactions between an epithelial outgrowth of the nephric duct and the surrounding nephrogenic mesenchyme. Initial morphogenetic events include ureteric bud branching to form the collecting duct (CD) tree and mesenchymal-to-epithelial transitions to form the nephrons, requiring reciprocal induction between adjacent mesenchyme and epithelial cells. Within the tips of the branching ureteric epithelium, cells respond to mesenchyme-derived trophic factors by proliferation, migration, and mitosis-associated cell dispersal. Self-inhibition signals from one tip to another play a role in branch patterning. The position, survival, and fate of the nephrogenic mesenchyme are regulated by ECM and secreted signals from adjacent tip and stroma. Signals from the ureteric tip promote mesenchyme self-renewal and trigger nephron formation. Subsequent fusion to the CDs, nephron segmentation and maturation, and formation of a patent glomerular basement membrane also require specialized cell–cell interactions. Differential cadherin, laminin, nectin, and integrin expression, as well as intracellular kinesin and actin-mediated regulation of cell shape and adhesion, underlies these cell–cell interactions. Indeed, the capacity for the

ISSN 0070-2153
http://dx.doi.org/10.1016/bs.ctdb.2014.12.002

kidney to form via self-organization has now been established both via the recapitulation of expected morphogenetic interactions after complete dissociation and reassociation of cellular components during development as well as the *in vitro* formation of 3D kidney organoids from human pluripotent stem cells. As we understand more about how the many cell–cell interactions required for kidney formation operate, this enables the prospect of bioengineering replacement structures based on these self-organizing properties.

1. INTRODUCTION

Tissue development relies upon local cellular interactions mediated by secreted morphogens, cell–extracellular matrix, or direct cell–cell interactions. Whether applied to invertebrate or vertebrate development, the final arrangement of differentiated cells relies upon locally acting signals driving competence, induction, cellular identity, and ultimately functional specialization. This combination of inherent locally acting interactions between cells enables the formation of complex structures without a template or scaffold. Such an internally regulated process is referred to as self-organization. As in other tissues, organogenesis of the mammalian permanent kidney, the metanephros, involves such self-organizing cell–cell interactions.

The mammalian kidney is mesodermal in origin, positioned at a specific rostrocaudal and mediolateral position of the embryo and arises from the intermediate mesoderm (reviewed by Kopan, Chen, & Little, 2014; Little & McMahon, 2012). Indeed, three paired excretory organs arise in a rostrocaudal sequence during mammalian embryogenesis: the pronephros, mesonephros, and finally the metanephros. The formation of all these structures is preceded by the initiation of nephric duct formation from the intermediate mesoderm at the level of the forelimb at around E8.5 in the mouse (reviewed by Dressler, 2009; Saxen, 1987). The nephric duct (also called the mesonephric duct or the Wolffian duct) extends in a caudal direction as the embryo develops, with the tubular components of the pronephros, mesonephros, and finally the metanephros forming from an adjacent mesodermal population referred to as the nephric cord. The permanent mammalian excretory organ, the metanephros, is this third set of paired structures and we focus on the development of this organ in this chapter. Until recently, the origin of the metanephric mesenchyme (MM) was regarded to be the same as that of the temporally earlier mesonephros. However, distinctions in gene expression in this more caudal mesenchymal population (Challen et al., 2004;

Wellik, Hawkes, & Capecchi, 2002), as well as lineage analyses (Taguchi et al., 2014; Xu et al., 2014), suggest that the mesenchyme giving rise to the caudal mesonephric tubules and the metanephros represents a more caudal or temporally distinct region of the intermediate mesoderm.

In the E10–10.5 mouse, the nephric duct at the level of the hindlimb begins to swell and a single side-branch, the ureteric bud (UB), arises from the duct and migrates toward the MM (reviewed by Costantini & Kopan, 2010). This outbudding is a response to the production of glial-derived neurotrophic factor (Gdnf) by the mesenchyme. Gdnf acts as a long-distance diffusible inducer, being detected within the nephric duct population via the Ret tyrosine kinase receptor (Ret) and a coreceptor, GFRα1. The UB elongates as an unbranched epithelial duct until it reaches the MM, where branching is initiated. The positioning of UB outgrowth and suppression of premature branching is regulated via the balance between bone morphogenic protein (BMP4) and Gremlin signaling along the nephric duct and extending UB (Michos et al., 2007). UB branching has been regarded as a simple dichotomous branching event lacking the apparent stereotypic patterning of organs such as the lung (Metzger, Klein, Martin, & Krasnow, 2008). However, the kidney achieves a specific shape and orientation, highlighting the fact that branching is not spatially uniform or synchronized. Recent comprehensive imaging of the branching ureteric tree has revealed an underlying initial lobe patterning and the frequent presence of trifurcations early in organogenesis when the proliferation rate is highest, with each of these representing an asynchronous branching event that resolves to bifurcations with time (Short et al., 2014; Fig. 1A). Time-lapse imaging and population modeling show that cells within the ureteric tip contribute both to subsequent tips as well as the intervening branches (Chi et al., 2009; Short et al., 2014).

Once the UB reaches the MM, a subpopulation of MM cells closest to the ureteric epithelium condenses to form the Gdnf^{+}Six2^{+} cap mesenchyme (CM). The CM is a self-renewing progenitor population (Boyle et al., 2008; Kobayashi et al., 2008) also competent to form the second epithelial compartment of the kidney, the nephrons (Little, Georgas, Pennisi, & Wilkinson, 2010). The interaction between the ureteric tip and the adjacent CM is classically regarded as a reciprocal inductive event (reviewed by Kopan et al., 2014). While the CM requires support from the UB to survive and to form nephrons, conversely the identity, proliferation rate, and branching of the UB requires continued Gdnf production by the CM. Ret signaling in response to ligand binding induces the expression of

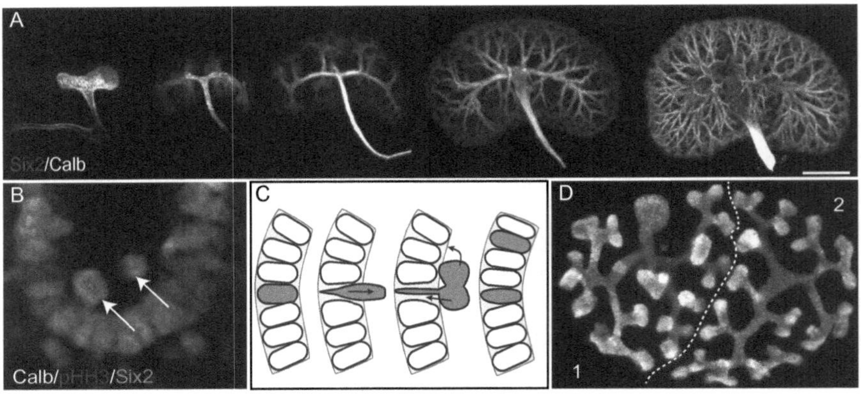

Figure 1 Critical processes in ureteric branching. (A) Optical projection tomograpy series of the developing mouse kidney from 11.5 to 15.5 dpc with Calbindin (white) marking the ureteric epithelium and Six2 (red) marking the surrounding cap mesenchyme (CM). (B). Cells of the ureteric tip undergo mitosis in the lumen arrows) in a process termed mitosis-associated cell dispersal (MACD; Packard et al., 2013). Ureteric tip (Calbindin, white), CM (Six2, green), and mitotic nuclei (pHH3, red). (C) Diagram illustrating the process of MACD (dividing cell in green) with the mother cell partially delaminating into the lumen, dividing, then reinserting into the epithelium while the daughter cell inserts 1–3 cell diameters away. (D) Self-avoidance during branching. The ureteric bud/collecting duct trees of two E11.5 mouse kidney rudiments (1 and 2), grown close to one another, produce distorted shapes as they avoid mutual collisions and leave a fairly straight border (dotted line). This mutual avoidance, which seems to be mediated by BMP7, can account for the automatic spacing out of the tips of the tree, and for the change of branch angle with developmental stage. (See the color plate.)

Wnt11 by the ureteric tip which in turn is thought to increase *Gdnf* expression within the MM (Majumdar, Vainio, Kispert, McMahon, & McMahon, 2003).

As noted above, the CM can be induced by signals from the ureteric tip to form the nephrons. This induction event represents a mesenchyme-to-epithelial transition (MET), a relatively rare event in organogenesis. Both the survival of the CM as a progenitor population and its induction to form nephrons via MET are directed by signals from the adjacent ureteric tip (Barak et al., 2012; Carroll, Park, Hayashi, Majumdar, & McMahon, 2005; Karner et al., 2011). The first visible evidence of nephron induction is the formation of a pretubular aggregate, an event triggered by canonical Wnt signaling, which begins to express markers such as *Wnt4*, *Fgfrl1*, and *Fgf8*. This subsequently undergoes MET as a response to Wnt4-mediated noncanonical Wnt signaling (Burn et al., 2011; Tanigawa et al., 2011) to form an epithelial renal vesicle (RV). The RV is defined as the first stage of nephron formation (defined as such in Georgas et al., 2009; Little et al., 2007). As soon as the RV forms, the cells within the RV show evidence of positional identity with distinct molecular profiles defining the proximal pole, furthest from the adjacent UB tip, and the distal pole, adjacent to the tip. By late RV, the distal pole invades and fuses with the adjacent ureteric tip to form a contiguous tubular lumen (Georgas et al., 2009). After substantial elongation and segmentation, the proximal pole will ultimately vascularize to form the filtering glomerulus.

RV polarization represents the beginning of the process of nephron segmentation that results in a precisely patterned tube comprising more than a dozen distinct functional cell types, including components of the glomerulus, proximal tubule (PT), loop of Henle (LoH), and distal tubule (Costantini & Kopan, 2010; Little et al., 2010). This early patterning is accompanied by differential expression of a variety of cell–cell adhesion molecules (Cho et al., 1998; Goto et al., 1998; Mah, Saueressig, Goulding, Kintner, & Dressler, 2000). The precise shape (convolution) and alignment of the forming nephrons with respect to the advancing and elongating collecting duct (CD) and the surrounding interstitial elements, including vasculature, are essential for ultimate renal function. While the signals directing this patterning remain imprecisely understood, this again involves long- and short-range signals between adjacent cell populations, as well as the formation of specialized cell–cell and cell–matrix interactions. Specification of the PT involves notch signaling (Cheng et al., 2007), while nephron and CD elongation requires canonical and noncanonical Wnt signaling between interstititum and epithelium as well as between one epithelial element and another (Karner et al., 2009; Lienkamp et al., 2012; Yu et al., 2009). At the cell–cell interface, differential cadherin expression across space as time is involved in nephron segmentation and specific integrin and laminin subunit interactions regulate appropriate CM survival and glomerular formation and maturation (Chen et al., 2004; Goto et al., 1998; Kanwar et al., 2004; Mathew, Chen, Pozzi, & Zent, 2012; Miner, 2012; Yang, Zimmerman, et al., 2013).

While the identity and lineage relationships of most of the cell types involved in kidney development have now been defined, along with a number of the critical growth factor/receptor interactions and their downstream subcellular consequences, much remain a mystery. At the subcellular level, what produces the spatially constrained branching CD or the precise formation and fusion of RVs around the ureteric tips? How does a patent glomerular basement membrane (GBM) result from the cell–cell interactions between the podocytes and glomerular endothelial cells? How is the balance between self-renewal and nephron induction regulated to produce a final organ? In this chapter, we will focus on what we know about the nature of critical cell–cell interactions during the formation of key components of the developing kidney and how we may ultimately be able to use this understanding to recapitulate organogenesis *in vitro* for the purposes of regenerative medicine.

2. CELL–CELL INTERACTIONS WITHIN THE DEVELOPING URETERIC EPITHELIUM

In vitro models of epithelial biology have been used to great effect to study the molecular mechanisms underlying cell–cell interactions, polarization, lumen formation, and disease (Debnath & Brugge, 2005; Rodriguez-Fraticelli & Martin-Belmonte, 2014; Roignot, Peng, & Mostov, 2013; Shewan, Eastburn, & Mostov, 2011). In parallel, developmental biologists have been investigating the formation of branched epithelial structures *in vivo* with a focus on the mechanisms regulating the broader morphogenic program (Affolter, Zeller, & Caussinus, 2009; Costantini & Kopan, 2010). With a few exceptions (Kuure et al., 2010), the details of cell–cell interactions in this system remain poorly understood. However, studies of the developing ureteric epithelium in mice have increased our understanding of the genetic regulation of branching morphogenesis and the establishment of regional molecular identity and cell behavior within tubular epithelia (Costantini & Kopan, 2010). Live imaging of organ explants has revealed a surprising amount of lateral cell movement within the epithelium and a novel mitotic behavior that involves dispersal of daughter cells (Packard et al., 2013; Shakya, Watanabe, & Costantini, 2005). Together these organ-based studies build a picture of a rich and dynamic epithelial environment.

The CD network of the metanephric kidney serves to conduct renal filtrate from tens of thousands to millions of nephrons to the bladder (Hughson, Farris, Douglas-Denton, Hoy, & Bertram, 2003; Merlet-Benichou et al., 1999). This tubular network is organized into a tree structure, with each branch tip connected to multiple nephrons (Rumballe et al., 2011; Short et al., 2014). CD architecture forms through a process of branching morphogenesis during development. Initial outgrowth of the UB is regulated by a gene regulatory network centered around GDNF–Ret signaling between the MM and the adjacent nephric duct (reviewed in Costantini, 2012; Costantini & Kopan, 2010). The nephric duct is a tube comprised of a simple cuboidal epithelium that responds to regional mesenchyme-derived signals by undergoing a cellular reorganization to a pseudostratified epithelium before cell proliferation drives outgrowth of the UB (Chi et al., 2009). This cellular reorganization is dependent on Ret signaling. In chimeric experiments, cells without Ret are excluded from the forming UB domain and cells with increased Ret signaling are preferentially included (Chi et al., 2009; Shakya et al., 2005).

Cells in the tip of the ingrowing bud maintain high levels of Ret signaling (Chi et al., 2009). This tip domain then swells to form an ampulla before breaking symmetry to undergo a branching event where the ampulla splits into two new tips. Recent mathematical modeling suggests that a ligand–receptor-based Turing mechanism-based around GDNF–Ret signaling is able to explain symmetry breaking and branching morphogenesis in kidney (Menshykau & Iber, 2013). As this first branching event occurs, the MM condenses around the site of UB ingrowth and nephron progenitor cells marked by transcription factor Six2 (Self et al., 2006) cluster around the tips to eventually form "cap" domains at the end of each tip. These cap domains are collectively referred to as the CM. Ureteric epithelium underlying the mesenchyme in the tip domain has a distinct molecular identity to trunk despite the two domains being part of a continuous tubular epithelium (Caruana et al., 2006; Kuure et al., 2010; Lu et al., 2009; Schmidt-Ott et al., 2005; Thiagarajan et al., 2011). Tip identity is defined by the response to mesenchyme-derived signals including GDNF and fibroblast growth factors (Fgfs) (Costantini, 2012; Kuure et al., 2010). Cells in the tips represent a population of progenitor cells that can self-renew within the tip or be left behind to give rise to trunk cells (Shakya et al., 2005).

The tips and trunk both express classic markers of polarized epithelia including E-cadherin at lateral and apical cell junctions; apical junctions are also enriched for ZO-1 and aPKC (Cerruti et al., 2013; Packard et al., 2013). Despite the expression of these cell adhesion and polarity markers, some cells within the epithelium are not locked into place but appear to have the capacity to move independently of their neighbors (Shakya et al., 2005). Perhaps, the most striking example of this is during tip cell division. In most epithelia, mitotic cell division results in the production of two adjacent daughter cells at the site of the original cell. In the tips of the ureteric epithelium, a very different sequence of events unfolds. The majority of proliferating cells become rounded, then traverse the apical surface of the epithelium into the luminal space where they then divide (Packard et al., 2013; Fig. 1B). One daughter cell retains a basal process and reinserts into the original location, while the other cell reinserts 1–3 cell diameters away (Packard et al., 2013; Fig. 1B and C). This behavior has not been previously been observed in a physiological context, but an analogous behavior has been seen in single oncogene-expressing cells in 3D cyst models (Leung & Brugge, 2012). In this context, the epithelial environment has been proposed to suppress proliferation (Leung & Brugge, 2012). The requirement for mitosis-associated cell dispersal in tip growth and branching is unclear (Davies, 2013a). However, further insight from 3D cyst culture suggests that high rates of in-plane proliferation can result in aberrant multilumen phenotypes (Cerruti et al., 2013). Rates of proliferation in the tip compartment during morphogenesis are high (Michael & Davies, 2004; Short et al., 2014). Perhaps, the out-of-plane division and dispersal of daughter cells in the ureteric tip relieves this compartment from destructive mechanical forces that would be generated by in-plane division. Whether cell division is oriented in the tip is unclear but Wnt9b plays a role in mediating tubule elongation in the branches and trunk of the ureteric tree through convergent extension during embryogenesis then planar cell polarity in the postnatal kidney (Karner et al., 2009).

The broader program of branching morphogenesis is dynamically regulated during development (Cebrian, Borodo, Charles, & Herzlinger, 2014; Short et al., 2014). After the initial budding event, the ureteric tree rapidly elaborates through asynchronous bifurcations and trifurcations that conform to a core stereotypical lobe structure (Lamberton, Lefevre, Short, Smyth, & Hamilton, 2014; Short et al., 2014). The rate of branching slows over time before ceasing shortly after birth with a total of ~3400 tips in C57Bl/6 mice (Short et al., 2014). Across development the number of tip cells and the overall size and volume of this compartment decreases (Short et al., 2014; Fig. 1A). Trifurcations are only observed during earlier stages when tip size is large. Later branching events are exclusively bifurcations. Branching in a wild-type context *in vivo* occurs exclusively in tip domains (Short et al., 2014). However, lateral branching—the formation of a new tip domain from a branch or trunk segment—has been observed *in vitro* (Sweeney, Lindstrom, & Davies, 2008; Watanabe & Costantini, 2004). This could be due to mislocalization of the CM in this 2D culture method, yet it demonstrates that trunk cells are capable of reacquiring tip cell fate.

Little is known about the regulation of tree architecture; however, the rotational angle between one set of branches and the next is relatively fixed at ~65°, implying a constant mechanism (Short et al., 2014). A recent series of culture experiments has experimentally addressed the mechanism of tip spacing (Davies, Hohenstein, Chang, & Berry, 2014), revealing a self-avoidance mechanism previously unrecognized (Fig. 1D). Mechanisms controlling branch patterns are difficult to test *in vivo* due to the requirement for time-consuming mouse genetics, and so culture of isolated embryonic mouse kidney has been used extensively to study branching in the ureteric epithelium. In culture, embryonic kidneys lose their three-dimensional form but continue to branch in two dimensions. Growth rates, branch patterns, and quantitative assessments of branching *in vitro* do not always reflect equivalent measures *in vivo* (Short et al., 2014; Watanabe & Costantini, 2004). However, fundamental properties of the epithelium and branching program are maintained (Chi et al., 2009; Packard et al., 2013; Shakya et al., 2005; Watanabe & Costantini, 2004). During development, tips and associated mesenchymal domains are arranged in distinct domains in the periphery of the kidney. *In vivo*, angles between adjacent branches are initially broad, but decrease as the tree elaborates (Short et al., 2014). The mechanism underlying this change was unclear until a recent series of culture experiments was used to test whether branch angle and spacing is regulated by an intrinsic mechanism or influenced by the surrounding environment. In culture, branch angles were carefully measured for the first and subsequent generations of branches. Like the native branching pattern, angles were initially broad, then decreased as the tree elaborated (Davies et al., 2014). Branch angles were then compared in isolated culture fragments containing a tip and its surrounding mesenchyme. Cultured tips that had not previously branched maintained the broad branch angles characteristic of early branching events, as did tips that had already undergone a previous branching event before being isolated; subsequent branching events in both cases showed a progression toward decreased angles. These data suggest that branch angles are regulated by the proximity of other tips. Computational modeling of a tip collision avoidance scenario faithfully simulated analogous experimental results (Fig. 1D). Finally, culture experiments involving disruption of TGFβ-superfamily signaling identified this pathway, specifically Bmp7, as a likely candidate for regulating this avoidance behavior (Davies et al., 2014).

The recent application of multiscale imaging and custom image analysis to kidney development now makes it possible to quantitatively assess morphogenesis in whole fixed organs at the cell and tissue level (Short et al., 2014; Figs. 1A, 2A and B). This approach is likely to build an integrated

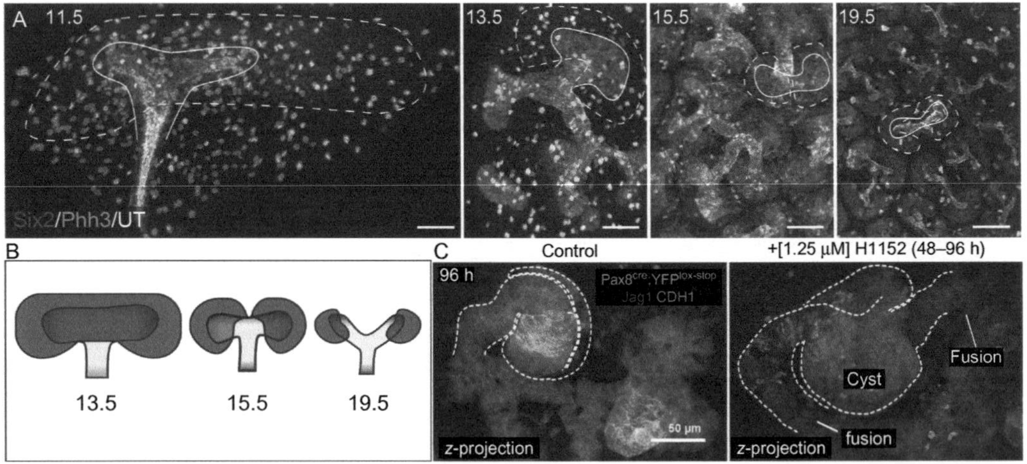

Figure 2 Critical processes in nephron formation and maturation. (A) Confocal imaging of developing mouse kidneys from 11.5 to 19.5 (birth) dpc with labeling of the cap mesenchyme (CM; Six2; red), ureteric epithelium (cytokeratin (11.5 and 19.5) or calbindin (13.5 and 15.5; white), and cells undergoing mitosis (pHH3; green). This series serves to illustrate changes in the cap/tip relationship within the nephrogenic niche as branching proceeds. Dashed lines in each image indicate the boundaries of representative CM domain/s. Solid outlines indicate the boundary of the underlying epithelium. (B) Diagrammatic representations of niche morphology across murine kidney development showing the changing relationship between the CM (red) and ureteric epithelium (white). (C) Rho kinase in fusion and patterning. The left panel shows a nephron (white dotted lines) developing in a cultured kidney, displaying a normal anatomy with a developing proximal glomerular cleft and one point of connection to the ureteric bud system. The right panel shows a nephron growing in the presence of the ROCK inhibitor H1152: the center of the nephron is cystic, and it has connected to the ureteric bud system at both ends. *This picture is CC from Lindström, Hohenstein, and Davies (2013).* (See the color plate.)

picture of how cellular-level events affect the broader organ structure. Continued progress in microscopy and image analysis, culture methods, and the application of tissue-specific gene ablation and labeling tools will drive further insight into the biology of the dynamic and often surprising ureteric epithelium.

3. THE NEPHROGENIC NICHE—BALANCING SELF-RENEWAL AND DIFFERENTIATION

The CM, which represents the nephron progenitor population of the kidney, exists in a histologically distinct domain around each ureteric tip. The CM, together with the underlying tip epithelium and surrounding stroma, comprises the nephrogenic niche (Fig. 2A and B). For quantification purposes, a niche can be defined as the region surrounding or underlying a spatially distinct cluster of CM cells. This definition differs from tip number as, before 14.5 dpc, cap domains are broad and can overlay more than one tip. After 14.5 dpc, the number of niches is essentially equivalent to the number of tips and niche number can be quantified to determine the extent of branching morphogenesis (Short et al., 2014). The number of cells per niche is not stable across development. Instead, the number of cells in both CM and ureteric tip populations decreases over time (Short et al., 2014). This is driven by high rates of branching morphogenesis between 11.5 and 15.5 dpc. Each branching event subdivides the cap and tip cells of a single niche into two new niches (Short et al., 2014). Proliferation rates in CM and tip are not high enough to recover niche size prior to the next branching event. There are initially more CM cells than tip cells in each niche, but the ratio of CM cells to tip cells per niche decreases over time. This corresponds with slowing rates of branching and an increase in cell cycle length in both compartments (Short et al., 2014), possibly reflecting changes in the signaling relationships between these two compartments. Signals within and between cellular compartments within this niche are critical for kidney morphogenesis.

Since the initial concept of induction, proposed in the context of Spemann's organizer (Spemann, 1938) in early embryogenesis, it has been appreciated that tissues can pattern as a result of signals from one cell type influencing the fate of another. A role for induction in kidney morphogenesis was first established by Grobstein (1953). Using careful microdissection to separate early epithelial structures from their surrounding mesenchyme,

Grobstein illustrated the requirement by the mesenchyme of an inductive signal from the epithelium in both the salivary gland and the kidney. The inductive signal in the kidney was a signal from the tip to induce nephron formation by the mesenchyme. Subsequent studies by Saxen (1987) investigated the interaction between these inducing populations and revealed a requirement for cell–cell contact via the use of intervening porous membranes. While the molecular nature of the inductive signal from the epithelium driving nephron formation was subsequently identified as Wnt9b (Carroll et al., 2005), the precise mode of transduction is still to be understood. Indeed, this is better described as a reciprocal induction between the nephron progenitors of the CM and the ureteric tip and is more complex than a simple signal for nephron initiation. Signals from the UB must balance self-renewal versus differentiation of the CM to ensure optimal organ growth and maximize functional capacity. A signal to differentiate competes with a signal for survival as morphogenesis relies upon a balance between the self-renewal of the CM and its depletion via differentiation. The net result of an imbalance is seen in the loss of Six2, a critical transcription factor driving CM self-renewal. As the first UB branches occur within the mesenchyme of the *Six2* KO kidney, the entire CM is induced to form nephrons, preventing all onward nephron formation (Self et al., 2006). Importantly, this also appears to prevent all subsequent branching as well due to the loss of a Gdnf source, usually supplied by the CM.

One can envisage that any cell receiving an inductive signal will respond. Induction by morphogens is rarely that simple, with secreted morphogens often available in a gradient of concentration from the source with that gradient allowing several possible responses depending upon distance (Wolpert, 1969). In addition, the gradient can vary across developmental time with the response of the receiving cells determined by multiple parameters including distance from the source, the presence of a local sink effectively reducing the signal, and/or the possibility of a relay system where one inductive response triggers another. In the case of the CM, while it can respond to Wnt9b via nephron induction, it also requires canonical Wnt signaling (presumably also Wnt9b) for survival as a self-renewing progenitor population (Karner et al., 2011). A gradient of morphogen may be envisaged to explain this. Paradoxically, Wnt9b is expressed along the length of the UB, whereas the CM only surrounds the ureteric tip (Kobayashi et al., 2008) and nephrons only form in the "armpit" between the trunk and the tip (Little & McMahon, 2012). One simple model to explain how both outcomes (self-renewal and nephron formation) can arise is additional spatial regulation of CM competence by other signals/signaling centers, such as the surrounding stroma (Das et al., 2013; discussed later) or existing nascent nephrons. Another would be a stochastic response by the CM in which an individual cell, once induced, migrated to form a pretubular aggregate. This suggests the generation of heterogeneity within the CM. There is evidence from gene expression analyses of differential gene expression within the CM domain, with genes such as *Cited1* apparently marking the less committed self-renewing progenitors (Mugford, Yu, Kobayashi, & McMahon, 2009). This suggests a variation in CM response based on position. This view is supported by comprehensive confocal analyses suggesting a higher level of the transcription factor, Six2, within the more peripheral domain of CM (Short et al., 2014). This study also identified two distinct cellular populations within the CM based upon cell cycle length, which could also be interpreted as evidence of a stochastic induction process sensitive to a threshold at the level of the individual cell (Short et al., 2014).

In addition to the effects of tip-produced signals, nephron progenitor identity and fate is known to also depend on signals originating from the CM itself and the surrounding stroma. Few of these signals have been identified, with these hampering efforts to isolate and culture the CM. However, aside from a role for low canonical Wnt signaling, it has also been shown that Fgf9 and Fgf20 signals act redundantly in the CM through Fgf receptors 1 and 2 (Fgfr1/2) to promote CM proliferation and self-renewal (Barak et al., 2012; Poladia et al., 2006). BMP signaling is also required for nephron progenitor maintenance and induction (Brown et al., 2013; Dudley, Godin, & Robertson, 1999; Dudley, Lyons, & Robertson, 1995; Oxburgh, Brown, Muthukrishnan, & Fetting, 2014).

4. MEDIATORS OF CM INTEGRITY, IDENTITY, AND MORPHOLOGY

The CM is a derivative of the MM. While it is described as a mesenchyme and does not show evidence of classical epithelial morphology, these cells do show some anatomical signs of alignment around the ureteric tip, suggesting an integrity requiring cell–cell interactions both within the CM and between the CM and the ureteric tip.

A number of cell adhesion components or proteins that associate with cell–cell junctions contribute to CM morphology or maintenance. The

CM produces neural cell adhesion molecule (NCAM), OB-cadherin (Cdh11), and N-cadherin (Klein, Langegger, Goridis, & Ekblom, 1988; Nouwen, Dauwe, van der Biest, & De Broe, 1993), although expression of these proteins persists into early nephron formation. The CM also expresses the cell-membrane-associated protein, integrin α8, which is required for maintenance and differentiation of the nephron progenitor population. Integrins are a well-characterized family of receptor proteins that interact with various extracellular matrix components to regulate cell migration, cell adhesion, proliferation, and cancer biology (Bokel & Brown, 2002; Desgrosellier & Cheresh, 2010). Integrins α8 and β1 form a protein complex (Itgα8β1) that serves as a receptor for various extracellular matrix proteins. Itgα8 is reported to be induced in the MM by UB outgrowth and is maintained in the CM throughout development (Müller et al., 1997). Polarized accumulation of Itgα8 protein toward the ureteric epithelium suggests a role in mediating interactions between the CM and UB (Muller et al., 1997; Uchiyama et al., 2010). Knockout of the *Itgα8* resulted in a failure of one or both kidneys to form due to a defect in UB outgrowth. As a result of this agenesis, many pups die within a day or two of birth (Muller et al., 1997). Recessive *Itgα8* mutations have also been linked renal agenesis in humans (Humbert et al., 2014). Further investigation of proteins that potentially interact with Itgα8β1, including nephronectin in the UB and Kif26b in the CM, suggest that such interactions regulate *Gdnf* expression, consistent with the delay or failure of UB outgrowth in loss-of-function studies (Linton, Martin, & Reichardt, 2007; Uchiyama et al., 2010). It is unclear whether these proteins are directly involved in Gdnf–Ret signaling, or whether reduced *Gdnf* levels are a secondary outcome of a loss of cell integrity and intercellular communication. Kif26b is a direct target of CM transcriptional regulator Sall1, and may be the main downstream effector as *Sall1* knockout results in a similar UB outgrowth defect (Nishinakamura et al., 2001). Careful analysis of cell morphology in *Kif26b*-null mice revealed a loss of polarized Itgα8 expression and disturbed lateral cell–cell junctions marked by N-cadherin (Uchiyama et al., 2010). Cell culture-based assays demonstrated a role for Kif26b in mediating N-cadherin-dependent cell–cell adhesion, likely through an interaction with nonmuscle myosin heavy chain type IIB (Myh10; Uchiyama et al., 2010). Subsequent analysis of CM-specific loss of Myh9 and Myh10 resulted in reduced N-cadherin protein levels, lower expression of *Itga8*, the nephron progenitor marker *Cited1*, and less nephron progenitors at birth (Recuenco et al., 2014).

Disks-large homolog 1 (Dlg1) and calcium-/calmodulin-dependent serine protein kinase (Cask) are two proteins that interact at cell membranes and function to link structural proteins and components of signaling pathways to cell–cell junctions. Targeted genetic disruption of *Dlg1* in the mouse results in a range of renal phenotypes from mild disruptions in branching and nephron formation to hydronephrosis and infrequent unilateral renal agenesis (Iizuka-Kogo, Ishidao, Akiyama, & Senda, 2007; Mahoney et al., 2006; Naim, Bernstein, Bertram, & Caruana, 2005). *Cask*-null mice die at birth with apparently normal kidneys (Ahn, Kim, Kim, Swat, & Miner, 2013). Deletion of both *Dlg1* and *Cask* results in highly penetrant hypoplastic and dysplastic kidneys, with a severe reduction in the number of nephron progenitors (Ahn et al., 2013). Although *Dlg1* and *Cask* are expressed in the CM and UB, compartment-specific knockouts revealed that the phenotypic consequences of the double knockout are due to a requirement for these genes in the CM (Ahn et al., 2013). The underlying cause of the phenotype is unclear. However, there was some evidence for a reduction in native and phosphorylated forms of signaling components associated with Fgf signaling (ERK, JNK, c-Raf) and reduced *Bmp7*, *Gdnf*, and *Fgf8* mRNA levels (Ahn et al., 2013). Further insight would be gained from more comprehensive profiling of the CM in the double knockout state to determine the extent of disruption to Fgf signaling and whether other pathways are also affected.

While the kidney development field has focused on the interactions between the CM and the ureteric tip, recent studies have demonstrated a significant role for the stromal population that surrounds the CM in regulating nephron progenitor expansion and differentiation. Foxd1 is a transcriptional regulator expressed in the cortical stroma. Knockout of *Foxd1* or ablation of the *Foxd1*-expressing population using diphtheria toxin ("*stromaless*" mouse) results in a surprising expansion of the nephron progenitor population and a reduction in markers of early nephron formation (Das et al., 2013; Hatini, Huh, Herzlinger, Soares, & Lai, 1996; Levinson & Mendelsohn, 2003). Expression profiling of *Foxd1*$^{-/-}$ kidneys identified a number of dysregulated genes including a stromal-expressed proteoglycan, Decorin (Fetting et al., 2014). Deletion of Decorin (*Dcn*) on a *Foxd1* null background was sufficient to partially rescue the defect in nephron progenitor differentiation (Fetting et al., 2014). Decorin has been proposed to bind Bmp7 and repress a Bmp/Smad-dependent sensitization of nephron progenitors to inductive Wnt signals (Brown et al., 2013; Fetting et al.,

2014). Under normal conditions, *Dcn* is repressed by Foxd1, and while it clearly plays a role in mediating the progenitor expansion phenotype in the absence of *Foxd1*, the role of this protein in a physiological context remains unclear. As discussed earlier, nephron induction is dependent on Wnt9b production within the UB, which is proposed to stabilize β-catenin in the CM to maintain progenitor identity at low levels and promote differentiation at high levels (Karner et al., 2011). Culture experiments have shown that in the absence of stroma, CM survival still requires β-catenin but is less sensitive to inhibition of Wnt secretion from the UB (Das et al., 2013). Drawing on studies linking the Hippo/Warts signaling pathway with β-catenin regulation, Das and colleagues investigated whether the localization or activation of the Hippo/Warts transcriptional regulator, Yap, was altered in the *Stromaless* mouse model. These mice do show an increase in nuclear Yap throughout the cortical region and this is also seen, although to a lesser extent, in mice lacking Fat4, a stroma-expressed protocadherin that mediates intercellular Hippo signaling. Knockdown of *Yap* or the related Hippo transcriptional activator, *Taz*, in isolated progenitor cells caused a reduction in some nephron progenitor markers. Knockout of *Yap* or *Taz* in the CM was inefficient, and no overt phenotype was observed early in development yet double-mutant mice died shortly after birth. Just prior to birth, a reduction in progenitor markers and precocious CM loss was seen (Das et al., 2013). These data suggest that stromal Fat4 promotes nephron progenitor differentiation by activating Hippo signaling in the CM through an intercellular interaction that suppresses nuclear Yap/Taz localization in the CM. In an independent study, Reginensi and colleagues removed *Yap* and/or *Taz* from the CM and observed a defect in nephron induction and patterning (Reginensi et al., 2013). Kidneys lacking *Yap* in the CM were hypoplastic and the mice died shortly after birth, with indications of a failure to produce urine (Reginensi et al., 2013). Mice lacking *Taz* in the CM possess cystic nephrons, as previously described, but did not appear to have any deficit in nephron progenitor number or impaired nephron induction (Reginensi et al., 2013). Both studies noted a reduction in markers of uninduced nephron progenitors, supporting a role for *Yap* in nephron progenitor maintenance. The partial recovery of *Foxd1* loss of function by ablation of *decorin* and moderate progenitor phenotypes in *Yap* and *Taz* mutants suggest that we have more to learn about stromal contributions to nephron progenitor regulation.

5. DIFFERENTIAL CELL–CELL ADHESION IN NEPHRON FORMATION, FUSION, PATTERNING, AND SEGMENTATION

The transition from pretubular aggregate to RV includes a profound alteration in the nature of the tissue, from a mesenchyme to an epithelium. This involves a substantial change in both the identity and the placement of the adhesion molecules involved. Before MET has begun, cell–cell adhesion molecule expression is dominated by NCAM and by three cadherins, OB-cadherin (Cdh11), N-cadherin (Cdh2), and R-cadherin (Chd4; Goto et al., 1998; Klein et al., 1998). OB-cadherin is typically mesenchymal and cells expressing it do not adhere to epithelia (Kimura et al., 1995); N-cadherin is neural and mesenchymal, while R-cadherin is associated mainly with retina, brain, and muscle.

MET is immediately preceded by expression of K-cadherin (Cdh6), which begins to be expressed at about the same time as other classical epithelial markers, such as the tight junction protein, ZO1, but before the transition to a morphologically visible epithelial state (Mah et al., 2000). Adhesion by this K-cadherin is important for formation of the epithelium: function-blocking antibodies prevent the formation of a tight condensate and prevent epithelialization in organ culture (Cho et al., 1998), while knockouts show delayed and reduced epitheliogenesis (Mah et al., 2000).

Shortly after K-cadherin is expressed, cells activate expression of E-cadherin (Cdh1); surprisingly, perhaps, E-cadherin itself is not required for epithelialization (Vestweber, Kemler, & Ekblom, 1985). Desmosomal cadherins appear at around the same time as E-cadherin (Davies & Garrod, 1995; Garrod & Fleming, 1990), though electron microscopy suggests that mature desmosomes develop only later (Garrod & Fleming, 1990). Cytokeratins, the cytoskeletal elements with which desmosomes associate, also appear a little later than E-cadherin (Davies & Garrod, 1995). Formation of a lumen results as a consequence of cellular polarization. While little is known about the regulation of polarization in the RV, this has been shown to require afadin with some evidence for the sequential formation and fusion of lumens as the RV elongates (Yang, Zimmerman, et al., 2013). Defects in lumen formation in nascent nephrons are also observed in mice lacking *Myh9* and *Myh10*, which encode the genes for nonmuscle myosin IIA and IIB, respectively (Recuenco et al., 2014).

Even in the earliest epithelial cyst, K-cadherin and E-cadherin show a reciprocal gradient of expression so that E-cadherin expression dominates on the side of the vesicle closest to the nearest developing CD and K-cadherin dominates on the side furthest away (Cho et al., 1998; Georgas et al., 2009). Catenins, which associate with the cytoplasmic tails of classical cadherins, are present throughout (Piepenhagen & Nelson, 1995). This is an early manifestation of the proximo-distal polarity of the nephron and, as development proceeds, the region of overlap diminishes so that E-cadherin is essentially restricted to the future distal domains of the nephron, K-cadherin to the future proximal domains, and they overlap only in the intermediate domain that forms the LoH. As the glomerular cleft forms, cells that will go on to form the podocytes lose their K-cadherin expression and gain expression of P-cadherin (Cdh3; Cho et al., 1998; Tassin, Beziau, Gubler, & Boyer, 1994). It is not clear whether this change of cadherin expression is a cause or a consequence of the cell separation that produces the glomerular cleft. The formation of the glomerular filter itself, which involves formation of modified adherens junctions (Reiser, Kriz, Kretzler, & Mundel, 2000), is described below.

K-cadherin seems to be important in achieving a connection between the distal pole of a developing nephron and a nearby CD, based on the observation that this process becomes unreliable in K-cadherin knockout mice (Mah et al., 2000). This function of K-cadherin seems surprising, given that it is expressed most strongly at the opposite (proximal) end of the RV. Connection is preceded by enhanced cell proliferation in the distal region of the RV, and by a disappearance of the basement membranes of the distal RV and apposing CD epithelium (Georgas et al., 2009). Actual connection seems to take place by distal RV cells infiltrating the epithelium of the CD, which accounts for the connecting tubule (the short tubule that runs between distal nephron and duct) having a nephron-like character, for example, in not showing the CD's vitamin D-independent expression of calbindin-D28k (Davies, 1994). Connection is not necessary for subsequent development of the nephron, a point illustrated by methods of organ culture that induce nephron formation with other tissues, such as spinal cord *in lieu* of the CD system. Connection is, however, necessary for function as it provides the only route for urine drainage, and unconnected nephrons are lost *in vivo* (Mah et al., 2000).

After the formation of the glomerular cleft (the "S"-shape stage of nephrogenesis), the developing nephron begins to express Ksp-cadherin (Cdh16; Shao, Johnson, Richardson, Hiesberger, & Igarashi, 2002). As the nephron matures, expression of cadherins changes so that the previously K-cadherin-dominated PTs express E-cadherin, and K-cadherin is seen only in the LoH (with some overlap between the two at the entrance and exit of the loop); the distal tubules continue to express E-cadherin and the glomerulus continues to express P-cadherin (Cho et al., 1998). N-cadherin is found mainly in the PTs (Prozialeck, Lamar, & Appelt, 2004), but plakoglobin is restricted to the distal tubules (Piepenhagen & Nelson, 1995). Ksp-cadherin is expressed most strongly in the thick ascending limb of the LoH and to a lesser extent in the PT (Shao et al., 2002).

The association of different patterns of cadherin expression in different segments of the nephron raises the question of whether the distinct identity of the segments is maintained through cadherin-mediated cell sorting. It has been observed that cells expressing different cadherins, or different amounts of the same cadherin, sort out into two populations rather than mixing evenly. This was first explained on thermodynamic grounds (minimization of free energy by maximizing adhesion; Steinberg, 1970) and more recently by suggesting, with support from a few experimental systems, that heterophilic contacts encourage the formation of actin–myosin stress fibers along the boundary (perhaps via Rho-ROCK signaling—see below) and that the "surface tension" minimizes the area of contact, thus separating cells (Davies, 2013a, 2013b).

Catenin p120 (p120ctn) is a regulator of cadherin stability and is expressed widely in the kidney. Kidney-specific knockout of p120ctn results in marked dilation of PTs, leading to cyst formation; there is also some dilation of distal tubules (Marciano et al., 2011). This dilation effect seems secondary to a loss of cadherin stability and consequent loss of actin filament organization and an increase in cell proliferation (at least in the PT). S-shaped tubules appear disorganized, and it is possible that segment boundaries are affected as would be expected if boundary formation used the mechanisms outlined in the paragraph above.

Cadherin function is required not just for development of nephrons but for their continuing health and function. Aging tubules show loss of N-cadherin in the PT, alterations in cell polarization, and eventual tubule failure (Jung et al., 2004). The order of cause-and-effect is not clear but, at least in cell culture experiments, cadherin-mediated cell–cell adhesion is critical to prevent PT cell apoptosis (Bergin, Levine, Koh, & Lieberthal, 2000). Many toxins that damage the PT (e.g., cadmium, bismuth) affect cadherin expression or localization (Leussink et al., 2001; Prozialeck, Lamar, & Lynch, 2003).

In addition to cadherin-type (adherens) junctions, renal epithelial cells possess tight junctions to control flow via the paracellular route. Different segments of the tubule show very different paracellular permeabilities, the PT being relatively "leaky," for example, to water and ions, and the LoH being watertight. This variation in tightness correlates with the expression of different claudins (Kirk, Campbell, Bass, Mason, & Collins, 2010): Claudin 1 is present and associated with tight junctions of Bowman's capsule. Claudins 2, 10, and 11 are expressed in PTs, Claudin 10 being basolateral, Claudin 11 apical, and Claudin 2 apparently associated with tight junctions. Claudins 10 and 11 are also present in the LoH, with Claudin 16 also present and associated with tight junctions. There is some Claudin 3 in the thick ascending loop. In the distal tubule, Claudins 3, 4, 8, and 14 dominate, Claudin 8 being associated with tight junctions.

6. THE ADHESION–CYTOSKELETON–SIGNALING AXIS IN KIDNEY TUBULOGENESIS

Whether within the CD epithelium or the many mesenchyme-derived segments of the nephron, final morphology results from the intracellular transduction of external signals. External cell–cell and cell–substrate junctions link via the inner face of the plasma membrane with the microfilament or intermediate filament cytoskeletal systems of the cell. Structurally, this creates a network of cytoskeletal and junctional elements that enables tissues to withstand mechanical loads. The junctional and cytoskeletal elements of this network are interdependent, the survival of each depending on mechanical engagement and tension in the whole (de Rooij, 2014). At the interfaces between the junctions and the cytoskeleton, there are signaling molecules that allow the mechanical state of the cell to be communicated chemically to the rest of the cytoplasm and to the nucleus (Twiss & de Rooij, 2013). Adhesive junctions are therefore highly significant mechanical and signaling centers within the cell.

The integrity of the junction–cytoskeletal system has been known for some years to be critical to renal morphogenesis. In the developing UB/CD system, actin microfilaments are expressed particularly strongly at the branching tips, as long as the tips are actively branching. Disruption of the actin cytoskeleton, using cytochalasins, blocks normal branching and instead results in radial expansion of the epithelium to produce an aberrant, bloated tube showing a higher mitotic index than normal (Meyer et al., 2006; Michael, Sweeney, & Davies, 2005). Inhibition of either ROCK, which activates myosin-mediated contraction of actin–myosin filaments, or of myosin itself, has the same result, suggesting that mechanical tension of the cytoskeleton is important in maintaining a normal tubule diameter and in tubule bifurcation. Inhibition of these systems also results in cells apparently leaving the tips of the epithelium to move freely through the mesenchyme, suggesting that a loss of cytoskeletal tension may result in failure of cell–cell junctions. It is also possible, though, that these cells are dying remnants left behind by a contracting tip.

In nephron formation, activity of the Rho kinase ROCK is important both for the initial formation of RVs, and for their subsequent morphogenesis into mature nephrons (Lindström et al., 2013; Fig. 2C). In the presence of ROCK-inhibiting drugs, fewer nephrons form and those that do form show seriously disrupted morphology: MET still occurs but many of the nephrons that form remain as rounded cysts and fail to elongate, or display a partially elongated and partially cystic nature (Fig. 2C). This seems to reflect a defect in planar cell polarity, detectable by a failure of mitotic axes to align normally with the long axis of the tubule (Carroll & Yu, 2012). Cell volumes are also increased. These two effects probably reflect the role of ROCK as an effector of the canonical planar cell polarity pathway (Strutt, 2001). Strikingly, while segment-specific markers are expressed at the normal time, cells expressing them do not segregate properly so that there are regions of nephrons that show podocyte and PT markers at the same time, and some tubules connect to the CD at proximal rather than distal domains. Occasionally, nephrons even connected to one another. These defects suggest a failure of boundary formation so that cells are free to mix inappropriately and perhaps to interact inappropriately. In fruitflies, actin–myosin contraction is needed to maintain boundaries in the plane of an epithelium (Monier, Pélissier-Monier, & Sanson, 2011). One effect of ROCK activity is activation of myosin, so it may be that boundary formation in ROCK-inhibited nephrons fails because of a failure of tension.

The segmentation defects in ROCK-inhibited nephrons may have a more complex and nonmechanical explanation. Cell–cell junctions are associated with the signaling molecule β-catenin, which is involved in "canonical" Wnt signaling (Valenta, Hausmann, & Basler, 2012). There is a gradient of β-catenin activation along a normal nephron, and perturbation of β-catenin activation levels leads to a predictable proximalization or distalization of the nephron (Lindstrom N, Hohenstein P, Davies JA, unpublished). Mechanical force at cell junctions regulates α-catenin and this molecule interacts with β-catenin (Twiss & de Rooij, 2013). In at least some systems, cadherin activities regulate the gene-regulating activity of β-catenin by determining how much is in the junctions and how much in the cytoplasm (Samuelov et al., 2012; Valenta et al., 2012), so it is possible that the gradient of β-catenin in the nephron is a consequence of differential cadherin expression. Cadherin expression can also be regulated by β-catenin activity (for example, via Twist and Slug; Heuberger & Birchmeier, 2010), making the deep mechanisms of nephron segmentation and specification potentially rich in feedback loops and consequently difficult to analyze.

In glomeruli, activation of ROCK is associated with retraction of foot processes and inhibition of ROCK in cultured podocytes increases the formation of foot processes in a mechanism connected (though not necessarily directly) with increased microtubule stability (Gao et al., 2004). ROCK activity may also be important in pathological retraction of foot processes as a response to injury or high glucose: in experimental animals treated with renotoxins or featuring high glucose, treatment with ROCK-inhibiting drugs helps to conserve podocyte architecture (Komers, 2013; Wang, Ellis, Fields, Howell, & Spurney, 2008; Yang, Zhao, et al., 2013). In the case of diabetes, the skeleton of a causative pathway has been determined, glucose increasing the concentration of reactive oxygen species, thus enhancing TRPC6-mediated cellular Ca^{2+} influx, which activates RhoA and hence ROCK (Yang, Zhao, et al., 2013). Mechanical stretch can also activate RhoA, and this may be the basis of hypertension-mediated podocyte damage (Babelova et al., 2013).

Vascular cells of the glomerulus also show regulation by small GTPases. Rac1, a member of the same family as RhoA, is activated by high glucose and results in changes in tight junctions and vascular hyperpermeability (Peng et al., 2010). In an endothelium that anyway has fenestrae, permeability itself is presumably not of great importance, but degradation of endothelial cell–cell junctions may place the podocytes that overlie endothelial cells under greater mechanical strain. There is much to be explored in this area, but it is already clear that cell adhesion molecules, signaling molecules, and components of the cytoskeleton operate as an integrated whole in the kidney and a focus on just one of them will omit important parts of the overall picture.

7. FORMATION OF THE GLOMERULAR FILTER

The filter of the glomerulus has three main components: (i) the fenestrated endothelium of the glomerular capillaries, the 60–80 nm pores through which molecules can pass but not cells; (ii) the three-layered (or perhaps five-layered: Salmon, Neal, & Harper, 2009) GBM, which provides the modest filtration function; and (iii) the slit diaphragm between podocytes of the nephron, which provides the finest filtering.

Endothelial cells, most of which differentiate from renal stroma (Hyink et al., 1996), are attracted to the developing glomerular cleft by a gradient of VEGF, which is synthesized by developing podocytes. Having formed a dense aggregate in the cleft, endothelial cells create hollow tubes apparently by TGFβ-mediated apoptosis, a process that may also set the final population to be the right size for the glomerulus (Fierlbeck, Liu, Coyle, & Ballermann, 2003). Once capillaries have begun to form, the walls of endothelial cells become punctuated by 60–80 nm fenestrae (holes). This process depends on podocytes secreting VEGF-A (Eremina et al., 2003) and, while the cellular mechanism for pore formation is unclear, it is remarkable that when even skin capillaries and venules are treated with VEGF, they become fenestrated within 10 min (Roberts & Palade, 1995). It may therefore be that glomeruli make permanent use of an effect that is normally transient in the rest of the body; healthy podocytes produce VEGF-A throughout life (Eremina et al., 2003).

The GBM is a planar sheet composed mainly of type IV collagen (initially α1α2α1, later α3α4α5), laminin (initially LM-111, then LM-511, and finally LM-521), and nitrogen and agrin (Miner, 2012). It lies between endothelial cells and podocytes and is product of them both (Droz, Diebold, Jan, & Jaubert, 1990; Abrahamson, 2012), which may account for its multilayered structure (Miner, 2011). During the earliest stages of GBM synthesis, as endothelial cells are invading the glomerular slit of the S-shaped tubule, podocytes and endothelia begin to synthesize GBM separately and only when the cells achieve close apposition to the independent layers of the GBM fuse (Abrahamson, 2012). Cells bind to the GBM mainly by integrins, using α1β1 and α2β1 to bind collagen IV and α3β1 to bind laminin (Abrahamson, 2012). Podocytes also use Dystroglycan to bind laminin.

Podocytes are derived from epithelial cells of the proximal-most pole of the RV: they can be considered to have a part epithelial, part mesenchymal character, expressing ZO-1 (normally associated with epithelia), and vimentin (normally associated with mesenchyme; Fujimoto & Singer, 1987). The part epithelial/part mesenchymal nature of podocytes is supported by their lifelong expression of the zinc finger protein WT1, expression of which is generally associated with transitions between epithelial and mesenchymal states (Miller-Hodges & Hohenstein, 2012). Mutual adhesion of developing podocytes is at first achieved by conventional apical junctional complexes between cells that, at the comma-shaped body stage of nephron development, still look like normal epithelia. Through the S-shaped body and capillary loop stage of nephron development, the junctional zone of presumptive podocytes moves in a basal direction so that most of the cell's volume is in the apical domain and the junctions are located only just above the basement membrane (Schell, Wanner, & Huber, 2014). As this translocation of junctions takes place, the junctional components change in composition and morphology to become the slit diaphragm. Slit diaphragms are junction-like complexes with some specialized components and some components common in other types of junctions. The common components include P-cadherin, the protocadherin FAT, the tight junction protein ZO1, β-catenin, and actin filaments (desmocollins, though, are absent; Garrod & Fleming, 1990). The specialized components include Nephrin, which can bind homophilically between cells or bind Neph1–3, also present (Ristola & Lehtonen, 2014), and podocin, which associates with Nephrin (Ha, 2013). These molecules also mediate presynaptic to postsynaptic cell–cell adhesion in neuronal synapses. As the junctions change, the podocyte surface expresses large quantities of the negatively charged glycoprotein podocalyxin, which interacts internally with the cytoskeleton.

Podocytes have numerous foot processes (hence the name: podo (foot) cyte (cell)), and the processes of adjacent cells interdigitate rather like the teeth on a zip fastener. How cells organize these interdigitations remain to be understood. It is not even clear to what extent process formation is driven by protrusive cytoskeletal activity of the kind that supports process in other cell types, or to what extent it may be driven by an energetic favoring of maximization of cell–cell contacts due to the slit diaphragm adhesive systems (Schell et al., 2014). Podocalyxin is required for foot process formation (Doyonnas et al., 2001), arguing for cytoskeletal involvement, but mutants for various actin-organizing proteins can form foot processes, albeit with problems in stability (Blattner et al., 2013; Schell et al., 2013). The small GTPase, cdc42, which is involved in filopodia formation in neuronal growth cones, is needed for foot process formation (Scott et al., 2012), but cdc42 is also involved in the Par3/Par6/aPKC complex that is known, in other systems, to be necessary for cell polarity and the placement of junctions. The phenotype of cdc42 knockout podocytes does not, therefore, discriminate between the cytoskeleton-led or adhesion-led hypothetical mechanisms.

Filtration function depends on both the specialized and the common components of slit diaphragms and is lost on the loss of Tjp1 and Z01 (Itoh et al., 2014). Most of these losses cause loss of slit diaphragm architecture, detectable under electron microscopy, as well as loss of function. As in the rest of the tubule, the cadherin regulator p120 catenin is needed for proper assembly of cell–cell junctions, and the podocytes of p120-knockout mice show only minimal mutual adhesion although they do express slit diaphragm-type proteins such as podocin (Marciano et al., 2011).

Like conventional adherens junctions, slit diaphragms are involved in signaling as well as simple mechanics (Grahammer, Schell, & Huber, 2013). Nephrin and Neph1 are targets of phosphorylation by the Src family, the susceptibility depending on their extracellular interactions and probably mechanical stress; phosphorylation leads to intracellular engagement with adaptor proteins, PI-3-kinase, etc. (Ha, 2013), and major consequent intracellular events such as actin reorganization (Arif et al., 2014). In addition to this, P-cadherin–β-catenin signaling takes place as in normal adherens junctions.

Injury to podocytes typically promotes change toward the mesenchymal side of the epithelial–mesenchymal balance. This takes place partially by the canonical Wnt pathway inhibiting GSK3β-mediated phosphorylation and consequent destruction of the DNA-binding molecule Snail. Allowed to accumulate, Snail inhibits transcription of genes with E-box-containing promoters, including cadherins and nephrin (Matsui et al., 2007). High blood glucose (as found in diabetes) triggers the canonical Wnt pathway via Rac1/PAK1 and triggers snail-mediated mesenchymal transition. The loss of nephrin causes slit diaphragms, and consequently normal glomerular function, to be lost.

8. *IN VITRO* SELF-ORGANIZATION GENERATES KIDNEY ORGANOIDS

As described above, kidney organogenesis involves temporospatially defined reciprocal induction events, cell migration, and proliferation in response to secreted and cell adhesion-based signals. As such, the component cells involved in kidney organogenesis appear capable of elaborating a highly complex architecture without a scaffold or template, suggestive of a self-organizing structure (Camazine et al., 2001; Sasai, Eiraku, & Suga, 2012). The formation of functional tissues and organisms by self-organization of dispersed cells was first reported over a century ago by Wilson (1907), who worked on sponges. The following 50 years identified examples in higher animals and also brought some mechanistic understanding in terms of chemotaxis (Bonner, 1947) and differential adhesiveness (Holtfreter, 1939, 1943). The growing dominance of embryology by paradigms of "genetic programs" and genetic determinism drew mainstream developmental biologists' attention away from self-organization, although a few researchers pressed and connected the adhesive properties with specific behaviors and molecules (Duguay, Foty, & Steinberg, 2003; Steinberg, 1970).

It was observed many years ago that disaggregated cells of some organ primordia (e.g., embryonic lung) will reorganize organotypically when allowed to reaggregate (Grover, 1963). The self-organization capacity of embryonic kidney was illustrated in the studies of Lusis et al. (2010) and Unbekandt and Davies (2010) (Fig. 3A–C). The approaches used in these two studies varied slightly, with one reaggregating embryonic kidneys from E12.5 kidneys and plating on collagen-coated filters above a Wnt4-producing feeder layer (Lusis et al., 2010), while the other plated reaggregated cell suspensions on either polycarbonate filters or glass with no feeder layer in the presence of ROCK-inhibiting drugs (24 h) to inhibit cell death (Unbekandt & Davies, 2010). Both studies showed that the embryonic mouse kidney could be completely dissociated to single cells, reaggregated as a pellet, and cultured *in vitro* at an air–media interface for several days (Fig. 3A). Despite the complete disruption on any existing spatial relationships between the component cells, "like" cells could reassociate with appropriate spatial relationships between themselves and surrounding populations. Hence, within approximately 8 h (Little MH & Combes AN, unpublished), the recreation of Pax2$^+$Calbindin$^+$ ureteric epithelial structures is seen and by 48 h these become surrounded by a Six2$^+$ CM. This is followed by nephron induction, including the formation of Aqp1$^+$ PTs and WT1$^+$ parietal epithelial cells and as would have occurred *in vivo* (Hendry et al., 2013).

From a mechanistic point of view, the observed self-organization seen in reaggregations of embryonic kidney may be regarded as representing a simple adhesion-based reassortment. As described, the early kidney consists of at least two very different cell types: epithelia (expressing E-cadherin and other typical epithelial adhesion systems) and mesenchyme (expressing mesenchymal cell–cell and cell–matrix adhesion systems). The early work of Steinberg (1962, 1970) argued that cells expressing different adhesion systems or the same system to different extents (Steinberg & Takeichi, 1994) will sort out spontaneously. Steinberg believed that this sorting was a simple

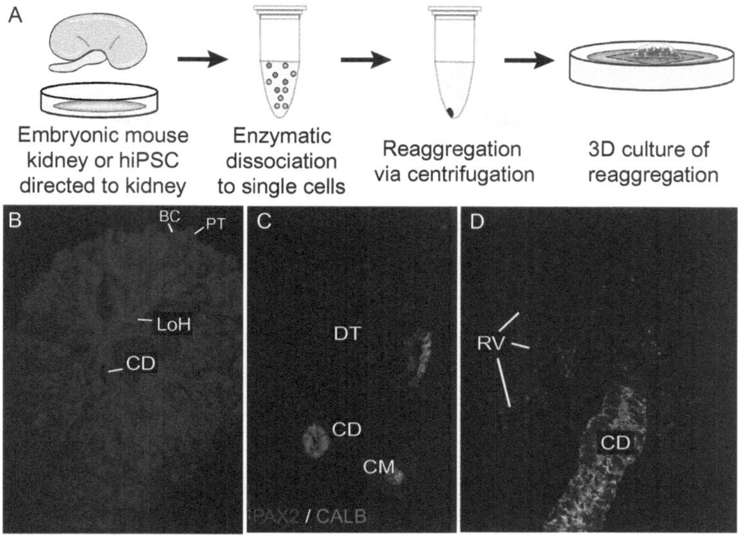

Figure 3 Self-organization in kidney organoids. (A) Diagram of the methodology for dissociation and reaggregation cultures in which total organ primordia or differentiated human pluripotent cells are used. This can also be achieved using serial culture in which an intact tree is reaggregated with dissociated cells (Davies, Unbekandt, Ineson, Lusis, & Little, 2012). (B) A low-power view of a serial reaggregate using the method of Chang and Davies (2012), showing a branched collecting duct (CD) system, nephrons with Bowman's capsules (BC) and proximal tubules (PT) in the cortex, and loops of Henle (LoH) descending into the medulla. (C) Section through a reaggreation generated from dissociated 12.5 dpc mouse kidney illustrating immunofluroescence for Pax2 (red), Calbindin (green), and DAPI (blue). This shows the presence of CD epithelium ($Pax2^{+}Calb^{+}$), cap mesenchyme (CM; $Pax2^{+}$ surrounding CD), and distal tubule ($Pax2^{+}$ epithelial structures). (D) Whole-mount IF of a kidney organoid generated after 2D culture of human pluripotent stem cells subjected to directed differentiation as described in Takasato et al. (2014). Markers used are PAX2 (red) and cytokeratin (CYTOK; green). The organoid shows the presence of CD ($PAX2^{+}CYTOK^{+}$) and early renal vesicles (RV) $PAX2^{+}$ polarizing structures. (See the color plate.)

thermodynamic consequence of different binding energies resulting from different affinities, the system naturally trying to minimize free energy by maximizing adhesion. Recent examination of the actual binding strength of different cadherins (Shi, Chien, & Leckband, 2008) has cast doubt on this simple view, and active responses by cells, for example, in organizing contractile actin–myosin complexes below adhesions, may play a major role (for discussion, see Davies, 2013b, pp. 275–279). What is observed is that this organotypic self-organization does not occur after dissociation and reaggregation of more mature embryonic structures, suggesting a window of competence before more refined cell–cell interactions are established that fail to reform when disrupted.

This dissociation/reaggregation assay has already been useful for evaluating the effect of RNA interference on kidney development (Davies & Unbekandt, 2012; Unbekandt & Davies, 2010), a process that is problematic using whole embryonic kidney cultures (Lee, Berry, Hohenstein, & Davies, 2008), and in testing the lineage commitment or renal potential of exogenous cells (Davies et al., 2012; Hendry et al., 2013; Lusis et al., 2010; Siegel et al., 2010). The stringency of this assay with respect to the capacity of any cell type to participate was thoroughly investigated in Hendry et al. (2013). This study focused on using the reaggregation assay as a means to test the renal potential of exogenous cell types. For this reason, it was important to demonstrate that a passenger cell not competent to respond to the signals required for kidney self-organization would not contribute productively to morphogenesis by chance. Initial studies were performed mixing disaggregated E12.5 embryonic kidneys from wild type versus constitutionally GFP^+ mice. This showed clear evidence of admixing between cells derived from different embryonic kidneys. Fluorescence activated cell sorting (FACS) was then used to isolate $Sall1GFP^+$ nephric mesenchyme from E12.5 embryonic kidneys. This population was mixed with disaggregated wild-type E12.5 embryonic kidneys, revealing the selective integration of GFP^+ mesenchyme into the cap mesenchymal compartments aggregating around the newly formed ureteric epithelium. A number of cell types not derived from the embryonic kidney were then tested for their capacity to contribute to renal structures within such aggregations. These included bone-marrow-derived mesenchymal stem cells (bmMSCs). Previous studies had reported that human bmMSCs could contribute to the final kidney of a rat after whole embryo culture if injected into the rat renal field at E11.5 (Yokoo et al., 2005). However, in the reaggregation assays of Hendry et al. (2013), MSCs showed no capacity to contribute to kidney morphogenesis. Indeed, the presence of as few as 10% MSC in the reaggregations resulted in a structural perturbation that prevented the formation of the normally anticipated structures. In addition, no integration into the CM or forming nephrons was seen when the human immortalized cell lines HEK293T (derived from embryonic kidney; Graham, Smiley, Russell, &Nairn, 1977), HK2 (adult PT; Ryan et al., 1994) or the mouse metanephric mesenchymal cell line, M15 (Larsson et al., 1995), were included in reaggregations. Similar studies by Rak-Raszewska et al. (2012) showed that while mesenchyme derived from pluripotent stem cells was able to contribute to renal structures within reaggregation cultures, undifferentiated ES cells, or ES cells differentiated to nonmesodermal elements did not. Siegel et al. (2010) established the presence of human amniotic stem cells within similar aggregation cultures. What is clear from these studies is that appropriate cell identity and competence to respond are both requirements for contribution to successful self-organization in a kidney organoid.

While substantial recreation of reciprocally interacting cellular domains does occur in these reaggregated organoids, the presence of a dichotomously branching ureteric tree is lost. Instead, clusters of ureteric epithelium reform initiating a surrounding mesenchymal niche. Some of these ureteric epithelial structures do branch, but the prospect of a single exit path for urinary filtrate, were these organoids vascularized, is not likely (Davies et al., 2012). More sophisticated serial culture versions of this assay have been developed in which one UB cyst formed in a reaggregate is removed and combined with a fresh suspension of mesenchyme. This approach produces an anatomically realistic kidney with nephrons arranged around a single UB/CD tree, and with distinct cortical and medullary zones with LoH dipping into the medulla in the usual way (Chang & Davies, 2012; Davies et al., 2012; Ganeva, Unbekandt, & Davies, 2011). Further refinements have also employed low-volume culture techniques to promote maturation of nephron segments (Chang & Davies, 2012; Fig. 3B). As well as being anatomically realistic, the PTs of the nephrons are polarized and physiologically active, in terms of organic anion and cation transport (Lawrence LM & Davies JA, unpublished). Hence, with the appropriate components, this assay may evolve for the production of accurate models of the normal kidney.

9. SELF-ORGANIZATION IN DIRECTED DIFFERENTIATION TO KIDNEY

In the twenty-first century, self-assembly or self-organization is being seen by some tissue engineers as a powerful method for construction of complex tissues without a scaffold. The power of morphogenesis via self-organization has recently been highlighted by observations in studies of the directed differentiation of human pluripotent stem cells. The approach commonly taken to direct a pluripotent cell source to a specific mature cell type is to recapitulate in a stepwise fashion the process of normal organogenesis. However, rather than being able to direct toward a uniform differentiated cell type, many studies have now observed for formation of complex self-organizing organoids arising *in vitro* without the supply of any scaffold. The first report of such self-organization was after the *in vitro* induction of human ES cells to adopt a posterior primitive streak fate in response to activin A. Here, the addition of Wnt3a and Fgf4 was shown not only to cause the induction of cells expressing the appropriate genes but also the formation of complex 3D structures similar to the developing pituitary or optic cup (Eiraku et al., 2011; Nakano et al., 2012; Suga et al., 2011). This self-organization is likely to stem from the simultaneous induction of multiple required cell types that then drive self-organization via interacting growth factor–receptor relationships. Self-organization toward liver, brain, thymus, pituitary, and intestine have now been reported to occur after directed differentiation of human pluripotent cells (Lancaster & Knoblich, 2014; Sasai et al., 2012; Takebe et al., 2014). This is true, also, for the kidney (Takasato et al., 2014). Our own studies into the directed differentiation of human ES cells toward a renal fate established the simultaneous induction of both $PAX2^{+}GATA3^{+}ECAD^{+}$ ureteric epithelium surrounded by $WT1^{+}SIX2^{+}ECAD^{-}$ mesenchyme (Takasato et al., 2014). This was observed after the sequential induction of posterior primitive streak and intermediate mesoderm via the addition of recombinant proteins and small molecule agonists in a monolayer culture of human pluripotent stem cells (Fig. 3D). To test whether the two populations present represented true renal progenitors able to undergo appropriate reciprocal self-organization to form a kidney, cells from these differentiation cultures were harvested and aggregated for culture in 3D, as for the embryonic reaggregation assay described above. Analysis of the resulting 3D organoids showed evidence for all of the components of the developing organ one would anticipate from an embryonic mouse kidney reaggregation culture (Takasato et al., 2014). This included the presence of epithelial CDs ($ECAD^{+}AQP2^{+}PAX2^{+}$) surrounded by $SIX2^{+}WT1^{+}$ CM as well as the presence of RVs ($CDH6^{+}JAG1^{+}$) and PTs ($SLC3A1^{+}$), as seen in mouse reaggregations (Takasato et al., 2014).

10. APPLICATION OF CELL–CELL AND CELL–MATRIX INTERACTIONS KIDNEY TISSUE ENGINEERING

The power of self-organization raises the prospect of the generation of replacement organs based around the spontaneous recapitulation of organogenesis without the need for positional direction or scaffolding. The capacity to direct the differentiation of pluripotent human cells toward kidney by capitalizing on the critical cell–cell communication that normally occurs during development also raises the prospect of the generation of large amounts of starting material for template-based bioengineering approaches, such as the recellularization of organ scaffolds. The capacity for an entire human organ to be completely decellularized, leaving behind a complete extracellular scaffold, has now been demonstrated for a variety of organs, including the human kidney (Bonandrini et al., 2014; Orlando et al., 2013; Song et al., 2013). The source material has been donor organs that were unfit for transplantation (Orlando et al., 2013). The concept is that by preserving the appropriate extracellular matrix, that this can act as an instructive template able to induce cells redelivered into the scaffold to readopt appropriate identity and functional capacity, so as to reconstruct a functional organ. The challenge has been a source of cells and a means by which to efficiently deliver these to all sites of the scaffold. The latter is a formidable challenge in an organ such as the kidney. Two obvious anatomical ports exist for this: the ureter/pelvis and the renal artery/vein. However, the delivery of cells along the length of the nephrons, including right up into the glomerulus, from either of these access points will be very difficult. More importantly, the instructive properties of a decellularized ECM are yet to be proved. The presence of the ECM itself is likely to hamper the early embryonic inductive events driving self-organization in reaggregation cultures; however, further studies into the instructive interactions between more differentiated nephron components and their neighbors are likely to prove fruitful in guiding this approach to regeneration. Alternatively, the development of suitable three-dimensional culture systems able to overcome the oxygen supply challenges of a multicellular structure may advance kidney tissue engineering based on the fundamental self-organization properties used by the embryo.

11. CONCLUSION

In conclusion, we have presented here the existing understanding of how kidney morphogenesis occurs at the level of cell–cell interactions. This represents the intersection between what observational anatomical developmental biology and mouse genetics has revealed at the organ level for many decades and what systems biology and gene expression studies tell us are the key components at the molecular level. With the advent of reporters of intracellular signaling, cytoskeletal structure and tension, remaining mysteries, such as the mode of tubular convolution, nephron arcading (which while described only in humans may be pertinent in mice), and the formation of nerves, lymphatics, and vasculature will be elucidated and contribute further to advances in tissue engineering approaches based on development.

ACKNOWLEDGMENTS

M. H. L. is a National Health and Medical Research Council of Australia Senior Principal Research Fellow. M. H. Little's research is supported by the National Health and Medical Research Council of Australia, the Australian Research Council, the Human Frontiers Science Program and Organovo Inc. J. A. Davies' research is supported by BBSRC, British Heart Foundation, European Union, Leverhulme Trust, MRC, NIH/NIDDK, and The Wellcome Trust. A. N. C. is a DECRA Fellow of the Australian Research Council, Australia.

REFERENCES

Abrahamson, D. R. (2012). Role of the podocyte (and glomerular endothelium) in building the GBM. *Seminars in Nephrology*, *32*(4), 342–349.

Affolter, M., Zeller, R., & Caussinus, E. (2009). Tissue remodelling through branching morphogenesis. *Nature Reviews. Molecular Cell Biology*, *10*(12), 831–842.

Ahn, S. Y., Kim, Y., Kim, S. T., Swat, W., & Miner, J. H. (2013). Scaffolding proteins DLG1 and CASK cooperate to maintain the nephron progenitor population during kidney development. *Journal of the American Society of Nephrology*, *24*(7), 1127–1138.

Arif, E., Rathore, Y. S., Kumari, B., Ashish, F., Wong, H. N., Holzman, L. B., et al. (2014). Slit diaphragm protein Neph1 and its signaling: A novel therapeutic target for protection of podocytes against glomerular injury. *Journal of Biological Chemistry*, *289*(14), 9502–9518.

Babelova, A., Jansen, F., Sander, K., Löhn, M., Schäfer, L., Fork, C., et al. (2013). Activation of Rac-1 and RhoA contributes to podocyte injury in chronic kidney disease. *PLoS One*, *8*(11), e80328.

Barak, H., Huh, S. H., Chen, S., Jeanpierre, C., Martinovic, J., Parisot, M., et al. (2012). FGF9 and FGF20 maintain the stemness of nephron progenitors in mice and man. *Developmental Cell*, *22*(6), 1191–1207.

Bergin, E., Levine, J. S., Koh, J. S., & Lieberthal, W. (2000). Mouse proximal tubular cell–cell adhesion inhibits apoptosis by a cadherin-dependent mechanism. *American Journal of Physiology. Renal Physiology*, *278*(5), F758–F768.

Blattner, S. M., Hodgin, J. B., Nishio, M., Wylie, S. A., Saha, J., Soofi, A. A., et al. (2013). Divergent functions of the Rho GTPases Rac1 and Cdc42 in podocyte injury. *Kidney International*, *84*(5), 920–930.

Bokel, C., & Brown, N. H. (2002). Integrins in development: Moving on, responding to, and sticking to the extracellular matrix. *Developmental Cell*, *3*(3), 311–321.

Bonandrini, B., Figliuzzi, M., Papadimou, E., Morigi, M., Perico, N., et al. (2014). Recellularization of well-preserved acellular kidney scaffold using embryonic stem cells. *Tissue Engineering. Part A*, *20*(9–10), 1486–1498.

Bonner, J. T. (1947). Evidence for the formation of cell aggregates by chemotaxis in the development of the slime mould Dictyostelium discoideum. *Journal of Experimental Zoology*, *106*, 1.

Boyle, S., Misfeldt, A., Chandler, K. J., Deal, K. K., Southard-Smith, E. M., Mortlock, D. P., et al. (2008). Fate mapping using Cited1-CreERT2 mice demonstrates that the cap mesenchyme contains self-renewing progenitor cells and gives rise exclusively to nephronic epithelia. *Developmental Biology*, *313*(1), 234–245.

Brown, A. C., Muthukrishnan, S. D., Guay, J. A., Adams, D. C., Schafer, D. A., Fetting, J. L., et al. (2013). Role for compartmentalization in nephron progenitor differentiation. *Proceedings of the National Academy of Sciences of the United States of America*, *110*(12), 4640–4645.

Burn, S. F., Webb, A., Berry, R. L., Davies, J. A., Ferrer-Vaquer, A., Hadjantonakis, A. K., et al. (2011). Calcium/NFAT signalling promotes early nephrogenesis. *Developmental Biology*, *352*(2), 288–298.

Camazine, S., Deneubourg, J.-L., Franks, N. R., Sneyd, J., Theraulaz, G., & Bonabeau, E. (2001). *Self-organization in biological systems*. Princeton, NJ: Princeton University Press.

Carroll, T. J., Park, J. S., Hayashi, S., Majumdar, A., & McMahon, A. P. (2005). Wnt9b plays a central role in the regulation of mesenchymal to epithelial transitions underlying organogenesis of the mammalian urogenital system. *Developmental Cell*, *9*(2), 283–292.

Carroll, T. J., & Yu, J. (2012). The kidney and planar cell polarity. *Current Topics in Developmental Biology*, *101*, 185–212.

Caruana, G., Cullen-McEwen, L., Nelson, A. L., Kostoulias, X., Woods, K., Gardiner, B., et al. (2006). Spatial gene expression in the T-stage mouse metanephros. *Gene Expression Patterns*, *6*(8), 807–825.

Cebrian, C., Asai, N., D'Agati, V., & Costantini, F. (2014). The number of fetal nephron progenitor cells limits ureteric branching and adult nephron endowment. *Cell Reports*, 7(1), 127–137.

Cerruti, B., Puliafito, A., Shewan, A. M., Yu, W., Combes, A. N., Little, M. H., et al. (2013). Polarity, cell division, and out-of-equilibrium dynamics control the growth of epithelial structures. *Journal of Cell Biology*, *203*(2), 359–372.

Challen, G. A., Martinez, G., Davis, M. J., Taylor, D. F., Crowe, M., Teasdale, R. D., et al. (2004). Identifying the molecular phenotype of renal progenitor cells. *Journal of the American Society of Nephrology*, *15*(9), 2344–2357.

Chang, C. H., & Davies, J. A. (2012). An improved method of renal tissue engineering, by combining renal dissociation and reaggregation with a low-volume culture technique, results in development of engineered kidneys complete with loops of Henle. *Nephron. Experimental Nephrology*, *121*(3–4), e79–e85.

Chen, D., Roberts, R., Pohl, M., Nigam, S., Kreidberg, J., Wang, Z., et al. (2004). Differential expression of collagen- and laminin-binding integrins mediates ureteric bud and inner medullary collecting duct cell tubulogenesis. *American Journal of Physiology. Renal Physiology*, *287*(4), F602–F611.

Cheng, H. T., Kim, M., Valerius, M. T., Surendran, K., Schuster-Gossler, K., Gossler, A., et al. (2007). Notch2, but not Notch1, is required for proximal fate acquisition in the mammalian nephron. *Development*, *134*(4), 801–811.

Chi, X., Michos, O., Shakya, R., Riccio, P., Enomoto, H., Licht, J. D., et al. (2009). Ret-dependent cell rearrangements in the Wolffian duct epithelium initiate ureteric bud morphogenesis. *Developmental Cell*, *17*(2), 199–209.

Cho, E. A., Patterson, L. T., Brookhiser, W. T., Mah, S., Kintner, C., & Dressler, G. R. (1998). Differential expression and function of cadherin-6 during renal epithelium development. *Development*, *125*(5), 803–812.

Costantini, F. (2012). Genetic controls and cellular behaviors in branching morphogenesis of the renal collecting system. *Wiley Interdisciplinary Reviews: Developmental Biology*, *1*(5), 693–713.

Costantini, F., & Kopan, R. (2010). Patterning a complex organ: Branching morphogenesis and nephron segmentation in kidney development. *Developmental Cell*, *18*(5), 698–712.

Das, A., Tanigawa, S., Karner, C. M., Xin, M., Lum, L., Chen, C., et al. (2013). Stromal–epithelial crosstalk regulates kidney progenitor cell differentiation. *Nature Cell Biology*, *15*(9), 1035–1044.

Davies, J. (1994). Control of calbindin-D28K expression in developing mouse kidney. *Developmental Dynamics*, *199*(1), 45–51.

Davies, J. A. (2013a). Disinherited daughters travel by tube. *Developmental Cell*, *27*(3), 245–246.

Davies, J. A. (2013b). *Mechanisms of morphogenesis* (2nd ed.). London: Academic Press.

Davies, J. A., & Garrod, D. R. (1995). Induction of early stages of kidney tubule differentiation by lithium ions. *Developmental Biology*, *167*(1), 50–60.

Davies, J. A., Hohenstein, P., Chang, C. H., & Berry, R. (2014). A self-avoidance mechanism in patterning of the urinary collecting duct tree. *BMC Developmental Biology*, *14*(1), 35.

Davies, J. A., & Unbekandt, M. (2012). siRNA-mediated RNA interference in embryonic kidney organ culture. *Methods in Molecular Biology*, *886*, 295–303.

Davies, J. A., Unbekandt, M., Ineson, J., Lusis, M., & Little, M. H. (2012). Dissociation of embryonic kidney followed by re-aggregation as a method for chimeric analysis. *Methods in Molecular Biology*, *886*, 135–146.

de Rooij, J. (2014). Cadherin adhesion controlled by cortical actin dynamics. *Nature Cell Biology*, *16*(6), 508–510.

Debnath, J., & Brugge, J. S. (2005). Modelling glandular epithelial cancers in three-dimensional cultures. *Nature Reviews. Cancer*, *5*(9), 675–688.

Desgrosellier, J. S., & Cheresh, D. A. (2010). Integrins in cancer: Biological implications and therapeutic opportunities. *Nature Reviews. Cancer*, *10*(1), 9–22.

Doyonnas, R., Kershaw, D. B., Duhme, C., Merkens, H., Chelliah, S., Graf, T., et al. (2001). Anuria, omphalocele, and perinatal lethality in mice lacking the CD34-related protein podocalyxin. *Journal of Experimental Medicine*, *194*(1), 13–27.

Dressler, G. R. (2009). Advances in early kidney specification, development and patterning. *Development*, *136*(23), 3863–3874.

Droz, D., Diebold, N., Jan, D., & Jaubert, F. (1990). Deduction from Wilms' tumour that glomerular podocytes produce the basement membrane material bearing Goodpasture determinants. *Journal of Pathology*, *162*(4), 323–327.

Dudley, A. T., Godin, R. E., & Robertson, E. J. (1999). Interaction between FGF and BMP signaling pathways regulates development of metanephric mesenchyme. *Genes and Development*, *13*(12), 1601–1613.

Dudley, A. T., Lyons, K. M., & Robertson, E. J. (1995). A requirement for bone morphogenetic protein-7 during development of the mammalian kidney and eye. *Genes and Development*, *9*(22), 2795–2807.

Duguay, D., Foty, R. A., & Steinberg, M. S. (2003). Cadherin-mediated cell adhesion and tissue segregation: Qualitative and quantitative determinants. *Developmental Biology*, *253*(2), 309–323.

Eiraku, M., Takata, N., Ishibashi, H., Kawada, M., Sakakura, E., Okuda, S., et al. (2011). Self-organizing optic-cup morphogenesis in three-dimensional culture. *Nature*, *472*(7341), 51–56.

Eremina, V., Sood, M., Haigh, J., Nagy, A., Lajoie, G., Ferrara, N., et al. (2003). Glomerular-specific alterations of VEGF-A expression lead to distinct congenital and acquired renal diseases. *Journal of Clinical Investigation*, *111*(5), 707–716.

Fetting, J. L., Guay, J. A., Karolak, M. J., Iozzo, R. V., Adams, D. C., Maridas, D. E., et al. (2014). FOXD1 promotes nephron progenitor differentiation by repressing decorin in the embryonic kidney. *Development*, *141*(1), 17–27.

Fierlbeck, W., Liu, A., Coyle, R., & Ballermann, B. J. (2003). Endothelial cell apoptosis during glomerular capillary lumen formation in vivo. *Journal of the American Society of Nephrology*, *14*(5), 1349–1354.

Fujimoto, T., & Singer, S. J. (1987). Immunocytochemical studies of desmin and vimentin in pericapillary cells of chicken. *Journal of Histochemistry and Cytochemistry*, *35*(10), 1105–1115.

Ganeva, V., Unbekandt, M., & Davies, J. A. (2011). An improved kidney dissociation and reaggregation culture system results in nephrons arranged organotypically around a single collecting duct system. *Organogenesis*, 7(2), 83–87.

Gao, S. Y., Li, C. Y., Chen, J., Pan, L., Saito, S., Terashita, T., et al. (2004). Rho-ROCK signal pathway regulates microtubule-based process formation of cultured podocytes—Inhibition of ROCK promoted process elongation. *Nephron. Experimental Nephrology*, *97*(2), e49–e61.

Garrod, D. R., & Fleming, S. (1990). Early expression of desmosomal components during kidney tubule morphogenesis in human and murine embryos. *Development*, *108*(2), 313–321.

Georgas, K., Rumballe, B., Valerius, M. T., Chiu, H. S., Thiagarajan, R. D., Lesieur, E., et al. (2009). Analysis of early nephron patterning reveals a role for distal RV proliferation in fusion to the ureteric tip via a cap mesenchyme-derived connecting segment. *Developmental Biology*, *332*(2), 273–286.

Goto, S., Yaoita, E., Matsunami, H., Kondo, D., Yamamoto, T., Kawasaki, K., et al. (1998). Involvement of R-cadherin in the early stage of glomerulogenesis. *Journal of the American Society of Nephrology*, *9*(7), 1234–1241.

Graham, F. L., Smiley, J., Russell, W. C., & Nairn, R. (1977). Characteristics of a human cell line transformed by DNA from human adenovirustype 5. *Journal of General Virology*, *36*(1), 59–74.

Grahammer, F., Schell, C., & Huber, T. B. (2013). The podocyte slit diaphragm—From a thin grey line to a complex signalling hub. *Nature Reviews. Nephrology*, *9*(10), 587–598.

Grobstein, C. (1953). Inductive epitheliomesenchymal interaction in cultured organ rudiments of the mouse. *Science*, *118*(3054), 52–55.

Grover, J. W. (1963). Reaggregation and organotypic redevelopment of dissociated embryonic chick lung cells in short-term culture. *National Cancer Institute Monograph*, *11*, 35–50.

Ha, T. S. (2013). Roles of adaptor proteins in podocyte biology. *World Journal of Nephrology*, *2*(1), 1–10.

Hatini, V., Huh, S. O., Herzlinger, D., Soares, V. C., & Lai, E. (1996). Essential role of stromal mesenchyme in kidney morphogenesis revealed by targeted disruption of Winged Helix transcription factor BF-2. *Genes and Development*, *10*(12), 1467–1478.

Hendry, C. E., Vanslambrouck, J. M., Ineson, J., Suhaimi, N., Takasato, M., Rae, F., et al. (2013). Direct transcriptional reprogramming of adult cells to embryonic nephron progenitors. *Journal of the American Society of Nephrology*, *24*(9), 1424–1434.

Heuberger, J., & Birchmeier, W. (2010). Interplay of cadherin-mediated cell adhesion and canonical Wnt signaling. *Cold Spring Harbor Perspectives in Biology*, *2*(2), a002915.

Holtfreter, J. (1939). Gewebeaffinitat, ein Mittel der embryolnalen Formbilding. *Archiv Fur Experimentelle Zellforschung*, *23*, 169.

Holtfreter, J. (1943). A study of the mechanics of gastrulation. *Journal of Experimental Zoology*, *94*, 261.

Hughson, M., Farris, A. B., III, Douglas-Denton, R., Hoy, W. E., & Bertram, J. F. (2003). Glomerular number and size in autopsy kidneys: The relationship to birth weight. *Kidney International*, *63*(6), 2113–2122.

Humbert, C., Silbermann, F., Morar, B., Parisot, M., Zarhrate, M., Masson, C., et al. (2014). Integrin alpha 8 recessive mutations are responsible for bilateral renal agenesis in humans. *The American Journal of Human Genetics, 94*(2), 288–294.

Hyink, D. P., Tucker, D. C., St John, P. L., Leardkamolkarn, V., Accavitti, M. A., Abrass, C. K., et al. (1996). Endogenous origin of glomerular endothelial and mesangial cells in grafts of embryonic kidneys. *American Journal of Physiology, 270*(5 Pt. 2), F886–F899.

Iizuka-Kogo, A., Ishidao, T., Akiyama, T., & Senda, T. (2007). Abnormal development of urogenital organs in Dlgh1-deficient mice. *Development, 134*(9), 1799–1807.

Itoh, M., Nakadate, K., Horibata, Y., Matsusaka, T., Xu, J., Hunziker, W., et al. (2014). The structural and functional organization of the podocyte filtration slits is regulated by Tjp1/ZO-1. *PLoS One, 9*(9), e106621.

Jung, K. Y., Dean, D., Jiang, J., Gaylor, S., Griffith, W. H., Burghardt, R. C., et al. (2004). Loss of N-cadherin and alpha-catenin in the proximal tubules of aging male Fischer 344 rats. *Mechanisms of Ageing and Development, 125*(6), 45–53.

Kanwar, Y. S., Wada, J., Lin, S., Danesh, F. R., Chugh, S. S., Yang, Q., et al. (2004). Update of extracellular matrix, its receptors, and cell adhesion molecules in mammalian nephrogenesis. *American Journal of Physiology. Renal Physiology, 286*(2), F202–F215.

Karner, C. M., Chirumamilla, R., Aoki, S., Igarashi, P., Wallingford, J. B., & Carroll, T. J. (2009). Wnt9b signaling regulates planar cell polarity and kidney tubule morphogenesis. *Nature Genetics, 41*(7), 793–799.

Karner, C. M., Das, A., Ma, Z., Self, M., Chen, C., Lum, L., et al. (2011). Canonical Wnt9b signaling balances progenitor cell expansion and differentiation during kidney development. *Development, 138*(7), 1247–1257.

Kimura, Y., Matsunami, H., Inoue, T., Shimamura, K., Uchida, N., Ueno, T., et al. (1995). Cadherin-11 expressed in association with mesenchymal morphogenesis in the head, somite, and limb bud of early mouse embryos. *Developmental Biology, 169*(1), 347–358.

Kirk, A., Campbell, S., Bass, P., Mason, J., & Collins, J. (2010). Differential expression of claudin tight junction proteins in the human cortical nephron. *Nephrology, Dialysis, Transplantation, 25*(7), 2107–2119.

Klein, G., Langegger, M., Goridis, C., & Ekblom, P. (1998). Neural cell adhesion molecules during embryonic induction and development of the kidney. *Development, 102*(4), 749–761.

Kobayashi, A., Valerius, M. T., Mugford, J. W., Carroll, T. J., Self, M., Oliver, G., et al. (2008). Six2 defines and regulates a multipotent self-renewing nephron progenitor population throughout mammalian kidney development. *Cell Stem Cell, 3*(2), 169–181.

Komers, R. (2013). Rho kinase inhibition in diabetic kidney disease. *British Journal of Clinical Pharmacology, 276*(4), 551–559.

Kopan, R., Chen, S., & Little, M. (2014). Nephron progenitor cells: Shifting the balance of self-renewal and differentiation. *Current Topics in Developmental Biology, 107*, 293–331.

Kuure, S., Cebrian, C., Machingo, Q., Lu, B. C., Chi, X., Hyink, D., et al. (2010). Actin depolymerizing factors cofilin1 and destrin are required for ureteric bud branching morphogenesis. *PLoS Genetics, 6*(10), e1001176.

Lamberton, T. O., Lefevre, J., Short, K. M., Smyth, I. M., & Hamilton, N. A. (2014). Comparing and distinguishing the structure of biological branching. *Journal of Theoretical Biology, 365*, 226–237.

Lancaster, M. A., & Knoblich, J. A. (2014). Generation of cerebral organoids from human pluripotent stem cells. *Nature Protocols, 9*(10), 2329–2340.

Larsson, S. H., Charlieu, J.-P., Miyagawa, K., Engelkamp, D., Rassoulzadegan, M., Ross, A., et al. (1995). Subnuclear localization of WT1 in splicing or transcription factor domains is regulated by alternative splicing. *Cell, 81*, 391–401.

Lee, W. C., Berry, R., Hohenstein, P., & Davies, J. (2008). siRNA as a tool for investigating organogenesis: The pitfalls and the promises. *Organogenesis*, *4*(3), 176–181.
Leung, C. T., & Brugge, J. S. (2012). Outgrowth of single oncogene-expressing cells from suppressive epithelial environments. *Nature*, *482*(7385), 410–413.
Leussink, B. T., Litvinov, S. V., de Heer, E., Slikkerveer, A., van der Voet, G. B., Bruijn, J. A., et al. (2001). Loss of homotypic epithelial cell adhesion by selective N-cadherin displacement in bismuth nephrotoxicity. *Toxicology and Applied Pharmacology*, *175*(1), 54–59.
Levinson, R., & Mendelsohn, C. (2003). Stromal progenitors are important for patterning epithelial and mesenchymal cell types in the embryonic kidney. *Seminars in Cell and Developmental Biology*, *14*(4), 225–231.
Lienkamp, S. S., Liu, K., Karner, C. M., Carroll, T. J., Ronneberger, O., Wallingford, J. B., et al. (2012). Vertebrate kidney tubules elongate using a planar cell polarity-dependent, rosette-based mechanism of convergent extension. *Nature Genetics*, *44*(12), 1382–1387.
Lindström, N. O., Hohenstein, P., & Davies, J. A. (2013). Nephrons require Rho-kinase for proximal-distal polarity development. *Scientific Reports*, *3*, 2692.
Linton, J. M., Martin, G. R., & Reichardt, L. F. (2007). The ECM protein nephronectin promotes kidney development via integrin alpha8beta1-mediated stimulation of Gdnf expression. *Development*, *134*(13), 2501–2509.
Little, M. H., Brennan, J., Georgas, K., Davies, J. A., Davidson, D. R., Baldock, R. A., et al. (2007). A high-resolution anatomical ontology of the developing murine genitourinary tract. *Gene Expression Patterns*, *7*(6), 680–699.
Little, M., Georgas, K., Pennisi, D., & Wilkinson, L. (2010). Kidney development: Two tales of tubulogenesis. *Current Topics in Developmental Biology*, *90*, 193–229.
Little, M. H., & McMahon, A. P. (2012). Mammalian kidney development: Principles, progress, and projections. *Cold Spring Harbor Perspectives in Biology*, *4*(5).
Lu, B. C., Cebrian, C., Chi, X., Kuure, S., Kuo, R., Bates, C. M., et al. (2009). Etv4 and Etv5 are required downstream of GDNF and Ret for kidney branching morphogenesis. *Nature Genetics*, *41*(12), 1295–1302.
Lusis, M., Li, J., Ineson, J., Christensen, M. E., Rice, A., & Little, M. H. (2010). Isolation of clonogenic, long-term self renewing embryonic renal stem cells. *Stem Cell Research*, *5*(1), 23–39.
Mah, S. P., Saueressig, H., Goulding, M., Kintner, C., & Dressler, G. R. (2000). Kidney development in cadherin-6 mutants: Delayed mesenchyme-to-epithelial conversion and loss of nephrons. *Developmental Biology*, *223*(1), 38–53.
Mahoney, Z. X., Sammut, B., Xavier, R. J., Cunningham, J., Go, G., Brim, K. L., et al. (2006). Discs-large homolog 1 regulates smooth muscle orientation in the mouse ureter. *Proceedings of the National Academy of Sciences of the United States of America*, *103*(52), 19872–19877.
Majumdar, A., Vainio, S., Kispert, A., McMahon, J., & McMahon, A. P. (2003). Wnt11 and Ret/Gdnf pathways cooperate in regulating ureteric branching during metanephric kidney development. *Development*, *130*(14), 3175–3185.
Marciano, D. K., Brakeman, P. R., Lee, C. Z., Spivak, N., Eastburn, D. J., Bryant, D. M., et al. (2011). p120 catenin is required for normal renal tubulogenesis and glomerulogenesis. *Development*, *138*(10), 2099–2109.
Mathew, S., Chen, X., Pozzi, A., & Zent, R. (2012). Integrins in renal development. *Pediatric Nephrology*, *27*(6), 891–900.
Matsui, I., Ito, T., Kurihara, H., Imai, E., Ogihara, T., & Hori, M. (2007). Snail, a transcriptional regulator, represses nephrin expression in glomerular epithelial cells of nephrotic rats. *Laboratory Investigation*, *87*(3), 273–283.
Menshykau, D., & Iber, D. (2013). Kidney branching morphogenesis under the control of a ligand–receptor-based Turing mechanism. *Physical Biology*, *10*(4), 046003.

Merlet-Benichou, C., Gilbert, T., Vilar, J., Moreau, E., Freund, N., & Lelievre-Pegorier, M. (1999). Nephron number: Variability is the rule. Causes and consequences. *Laboratory Investigation*, *79*(5), 515–527.

Metzger, R. J., Klein, O. D., Martin, G. R., & Krasnow, M. A. (2008). The branching programme of mouse lung development. *Nature*, *453*(7196), 745–750.

Meyer, T. N., Schwesinger, C., Sampogna, R. V., Vaughn, D. A., Stuart, R. O., Steer, D. L., et al. (2006). Rho kinase acts at separate steps in ureteric bud and metanephric mesenchyme morphogenesis during kidney development. *Differentiation*, *74*(9–10), 638–647.

Michael, L., & Davies, J. A. (2004). Pattern and regulation of cell proliferation during murine ureteric bud development. *Journal of Anatomy*, *204*(4), 241–255.

Michael, L., Sweeney, D. E., & Davies, J. A. (2005). A role for microfilament-based contraction in branching morphogenesis of the ureteric bud. *Kidney International*, *68*(5), 2010–2018.

Michos, O., Gonçalves, A., Lopez-Rios, J., Tiecke, E., Naillat, F., Beier, K., et al. (2007). Reduction of BMP4 activity by gremlin 1 enables ureteric bud outgrowth and GDNF/WNT11 feedback signalling during kidney branching morphogenesis. *Development*, *134*(13), 2397–2405.

Miller-Hodges, E., & Hohenstein, P. (2012). WT1 in disease: Shifting the epithelial–mesenchymal balance. *Journal of Pathology*, *226*(2), 229–240.

Miner, J. H. (2011). Organogenesis of the kidney glomerulus: Focus on the glomerular basement membrane. *Organogenesis*, 7(2), 75–82.

Miner, J. H. (2012). The glomerular basement membrane. *Experimental Cell Research*, *318*(9), 973–978.

Monier, B., Pélissier-Monier, A., & Sanson, B. (2011). Establishment and maintenance of compartmental boundaries: Role of contractile actomyosin barriers. *Cellular and Molecular Life Sciences*, *68*(11), 1897–1910.

Mugford, J. W., Yu, J., Kobayashi, A., & McMahon, A. P. (2009). High-resolution gene expression analysis of the developing mouse kidney defines novel cellular compartments within the nephron progenitor population. *Developmental Biology*, *333*(2), 312–323.

Müller, U., Wang, D., Denda, S., Meneses, J. J., Pedersen, R. A., & Reichardt, L. F. (1997). Integrin alpha8beta1 is critically important for epithelial-mesenchymal interactions during kidney morphogenesis. *Cell*, *88*(5), 603–613.

Naim, E., Bernstein, A., Bertram, J. F., & Caruana, G. (2005). Mutagenesis of the epithelial polarity gene, discs large 1, perturbs nephrogenesis in the developing mouse kidney. *Kidney International*, *68*(3), 955–965.

Nakano, T., Ando, S., Takata, N., Kawada, M., Muguruma, K., Sekiguchi, K., et al. (2012). Self-formation of optic cups and storable stratified neural retina from human ESCs. *Cell Stem Cell*, *10*(6), 771–785.

Nishinakamura, R., Matsumoto, Y., Nakao, K., Nakamura, K., Sato, A., Copeland, N. G., et al. (2001). Murine homolog of SALL1 is essential for ureteric bud invasion in kidney development. *Development*, *128*(16), 3105–3115.

Nouwen, E. J., Dauwe, S., van der Biest, I., & De Broe, M. E. (1993). Stage- and segment-specific expression of cell-adhesion molecules N-CAM, A-CAM, and L-CAM in the kidney. *Kidney International*, *44*(1), 147–158.

Orlando, G., Booth, C., Wang, Z., Totonelli, G., Ross, C. L., Moran, E., et al. (2013). Discarded human kidneys as a source of ECM scaffold for kidney regeneration technologies. *Biomaterials*, *34*(24), 5915–5925.

Oxburgh, L., Brown, A. C., Muthukrishnan, S. D., & Fetting, J. L. (2014). Bone morphogenetic protein signaling in nephron progenitor cells. *Pediatric Nephrology*, *29*(4), 531–536.

Packard, A., Georgas, K., Michos, O., Riccio, P., Cebrian, C., Combes, A. N., et al. (2013). Luminal mitosis drives epithelial cell dispersal within the branching ureteric bud. *Developmental Cell*, *27*(3), 319–330.

Peng, H., Wang, C., Ye, Z. C., Chen, Y. R., Zhang, J., Chen, Z. J., et al. (2010). How increased VEGF induces glomerular hyperpermeability: A potential signaling pathway of Rac1 activation. *Acta Diabetologica, 47*(Suppl. 1), 57–63.

Piepenhagen, P. A., & Nelson, W. J. (1995). Differential expression of cell–cell and cell–substratum adhesion proteins along the kidney nephron. *American Journal of Physiology, 269*(6 Pt. 1), C1433–C1449.

Poladia, D. P., Kish, K., Kutay, B., Hains, D., Kegg, H., Zhao, H., et al. (2006). Role of fibroblast growth factor receptors 1 and 2 in the metanephric mesenchyme. *Developmental Biology, 291*(2), 325–339.

Prozialeck, W. C., Lamar, P. C., & Appelt, D. M. (2004). Differential expression of E-cadherin, N-cadherin and beta-catenin in proximal and distal segments of the rat nephron. *BMC Physiology, 4*, 10.

Prozialeck, W. C., Lamar, P. C., & Lynch, S. M. (2003). Cadmium alters the localization of N-cadherin, E-cadherin, and beta-catenin in the proximal tubule epithelium. *Toxicology and Applied Pharmacology, 189*(3), 180–195.

Rak-Raszewska, A., Wilm, B., Edgar, D., Kenny, S., Woolf, A. S., & Murray, P. (2012). Development of embryonic stem cells in recombinant kidneys. *Organogenesis, 8*(4), 125–136.

Recuenco, M. C., Ohmori, T., Tanigawa, S., Taguchi, A., Fujimura, S., Conti, M. A., et al. (2014). Nonmuscle myosin II regulates the morphogenesis of metanephric mesenchyme-derived immature nephrons. *Journal of the American Society of Nephrology*. http://dx.doi.org/10.1681/ASN.2014030281 (in press). [Epub ahead of print].

Reginensi, A., Scott, R. P., Gregorieff, A., Bagherie-Lachidan, M., Chung, C., Lim, D. S., et al. (2013). Yap- and Cdc42-dependent nephrogenesis and morphogenesis during mouse kidney development. *PLoS Genetics, 9*(3), e1003380.

Reiser, J., Kriz, W., Kretzler, M., & Mundel, P. (2000). The glomerular slit diaphragm is a modified adherens junction. *Journal of the American Society of Nephrology, 11*(1), 1–8.

Ristola, M., & Lehtonen, S. (2014). Functions of the podocyte proteins nephrin and Neph3 and the transcriptional regulation of their genes. *Clinical Science (London), 126*(5), 315–328.

Roberts, W. G., & Palade, G. E. (1995). Increased microvascular permeability and endothelial fenestration induced by vascular endothelial growth factor. *Journal of Cell Science, 108*(Pt. 6), 2369–2379.

Rodriguez-Fraticelli, A. E., & Martin-Belmonte, F. (2014). Picking up the threads: Extracellular matrix signals in epithelial morphogenesis. *Current Opinion in Cell Biology, 30C*, 83–90.

Roignot, J., Peng, X., & Mostov, K. (2013). Polarity in mammalian epithelial morphogenesis. *Cold Spring Harbor Perspectives in Biology, 5*(2).

Rumballe, B. A., Georgas, K. M., Combes, A. N., Ju, A. L., Gilbert, T., & Little, M. H. (2011). Nephron formation adopts a novel spatial topology at cessation of nephrogenesis. *Developmental Biology, 360*(1), 110–122.

Ryan, M. J., Johnson, G., Kirk, J., Fuerstenberg, S. M., Zager, R. A., & Torok-Storb, B. (1994). HK-2: An immortalized proximal tubule epithelial cell line from normal adult human kidney. *Kidney International, 45*(1), 48–57.

Salmon, A. H., Neal, C. R., & Harper, S. J. (2009). New aspects of glomerular filtration barrier structure and function: Five layers (at least) not three. *Current Opinion in Nephrology and Hypertension, 18*(3), 197–205.

Samuelov, L., Sprecher, E., Tsuruta, D., Bíró, T., Kloepper, J. E., & Paus, R. (2012). P-cadherin regulates human hair growth and cycling via canonical Wnt signaling and transforming growth factor-β2. *Journal of Investigative Dermatology, 132*(10), 2332–2341.

Sasai, Y., Eiraku, M., & Suga, H. (2012). In vitro organogenesis in three dimensions: Self-organising stem cells. *Development*, *139*(22), 4111–4121.

Saxen, L. (1987). Organogenesis of the kidney. In *Developmental and cell biology series: Vol. 19*. Cambridge: Cambridge University Press.

Schell, C., Baumhakl, L., Salou, S., Conzelmann, A. C., Meyer, C., Helmstädter, M., et al. (2013). N-wasp is required for stabilization of podocyte foot processes. *Journal of the American Society of Nephrology*, *24*(5), 713–721.

Schell, C., Wanner, N., & Huber, T. B. (2014). Glomerular development—Shaping the multi-cellular filtration unit. *Seminars in Cell and Developmental Biology*, *36C*, 39–49, pii: S1084-9521(14)00239-0.

Schmidt-Ott, K. M., Yang, J., Chen, X., Wang, H., Paragas, N., Mori, K., et al. (2005). Novel regulators of kidney development from the tips of the ureteric bud. *Journal of the American Society of Nephrology*, *16*(7), 1993–2002.

Scott, R. P., Hawley, S. P., Ruston, J., Du, J., Brakebusch, C., Jones, N., et al. (2012). Podocyte-specific loss of Cdc42 leads to congenital nephropathy. *Journal of the American Society of Nephrology*, *23*(7), 1149–1154.

Self, M., Lagutin, O. V., Bowling, B., Hendrix, J., Cai, Y., Dressler, G. R., et al. (2006). Six2 is required for suppression of nephrogenesis and progenitor renewal in the developing kidney. *EMBO Journal*, *25*(21), 5214–5228.

Shakya, R., Watanabe, T., & Costantini, F. (2005). The role of GDNF/Ret signaling in ureteric bud cell fate and branching morphogenesis. *Developmental Cell*, *8*(1), 65–74.

Shao, X., Johnson, J. E., Richardson, J. A., Hiesberger, T., & Igarashi, P. (2002). A minimal Ksp-cadherin promoter linked to a green fluorescent protein reporter gene exhibits tissue-specific expression in the developing kidney and genitourinary tract. *Journal of the American Society of Nephrology*, *13*(7), 1824–1836.

Shewan, A., Eastburn, D. J., & Mostov, K. (2011). Phosphoinositides in cell architecture. *Cold Spring Harbor Perspectives in Biology*, *3*(8), a004796.

Shi, Q., Chien, Y. H., & Leckband, D. (2008). Biophysical properties of cadherin bonds do not predict cell sorting. *Journal of Biological Chemistry*, *283*(42), 28454–28463.

Short, K. M., Combes, A. N., Lefevre, J., Ju, A. L., Georgas, K. M., Lamberton, T., et al. (2014). Global quantification of tissue dynamics in the developing mouse kidney. *Developmental Cell*, *29*(2), 188–202.

Siegel, N., Rosner, M., Unbekandt, M., Fuchs, C., Slabina, N., Dolznig, H., et al. (2010). Contribution of human amniotic fluid stem cells to renal tissue formation depends on mTOR. *Human Molecular Genetics*, *19*(17), 3320–3331.

Song, J. J., Guyette, J. P., Gilpin, S. E., Gonzalez, G., Vacanti, J. P., & Ott, H. C. (2013). Regeneration and experimental orthotopic transplantation of a bioengineered kidney. *Nature Medicine*, *19*(5), 646–651.

Spemann, H. (1938). *Embryonic development and induction*. London: Yale University Press, New Haven.

Steinberg, M. S. (1962). On the mechanism of tissue reconstruction by dissociated cells, III. Free energy relations and the reorganisation of fused, heternomic tissue fragments. *Proceedings of the National Academy of Sciences of the United States of America*, *48*(10), 1769–1776.

Steinberg, M. S. (1970). Does differential adhesion govern self-assembly processes in histogenesis? Equilibrium configurations and the emergence of a hierarchy among populations of embryonic cells. *Journal of Experimental Zoology*, *173*(4), 395–433.

Steinberg, M. S., & Takeichi, M. (1994). Experimental specification of cell sorting, tissue spreading, and specific spatial patterning by quantitative differences in cadherin expression. *Proceedings of the National Academy of Sciences of the United States of America*, *91*(1), 206–209.

Strutt, D. (2001). Planar polarity: Getting ready to ROCK. *Current Biology, 11*(13), R506–R509.

Suga, H., Kadoshima, T., Minaguchi, M., Ohgushi, M., Soen, M., & Nakano, T. (2011). Self-formation of functional adenohypophysis in three-dimensional culture. *Nature, 480*(7375), 57–62.

Sweeney, D., Lindstrom, N., & Davies, J. A. (2008). Developmental plasticity and regenerative capacity in the renal ureteric bud/collecting duct system. *Development, 135*(15), 2505–2510.

Taguchi, A., Kaku, Y., Ohmori, T., Sharmin, S., Ogawa, M., Sasaki, H., et al. (2014). Redefining the in vivo origin of metanephric nephron progenitors enables generation of complex kidney structures from pluripotent stem cells. *Cell Stem Cell, 14*(1), 53–67.

Takasato, M., Er, P. X., Becroft, M., Vanslambrouck, J. M., Stanley, E. G., Elefanty, A. G., et al. (2014). Directing human embryonic stem cell differentiation towards a renal lineage generates a self-organizing kidney. *Nature Cell Biology, 16*(1), 118–126.

Takebe, T., Zhang, R. R., Koike, H., Kimura, M., Yoshizawa, E., Enomura, M., et al. (2014). Generation of a vascularized and functional human liver from an iPSC-derived organ bud transplant. *Nature Protocols, 9*(2), 396–409.

Tanigawa, S., Wang, H., Yang, Y., Sharma, N., Tarasova, N., Ajima, R., et al. (2011). Wnt4 induces nephronic tubules in metanephric mesenchyme by a non-canonical mechanism. *Developmental Biology, 352*(1), 58–69.

Tassin, M. T., Beziau, A., Gubler, M. C., & Boyer, B. (1994). Spatiotemporal expression of molecules associated with junctional complexes during the in vivo maturation of renal podocytes. *International Journal of Developmental Biology, 38*(1), 45–54.

Thiagarajan, R. D., Cloonan, N., Gardiner, B. B., Mercer, T. R., Kolle, G., Nourbakhsh, E., et al. (2011). Refining transcriptional programs in kidney development by integration of deep RNA-sequencing and array-based spatial profiling. *BMC Genomics, 12*, 441.

Twiss, F., & de Rooij, J. (2013). Cadherin mechanotransduction in tissue remodeling. *Cellular and Molecular Life Sciences, 70*(21), 4101–4116.

Uchiyama, Y., Sakaguchi, M., Terabayashi, T., Inenaga, T., Inoue, S., Kobayashi, C., et al. (2010). Kif26b, a kinesin family gene, regulates adhesion of the embryonic kidney mesenchyme. *Proceedings of the National Academy of Sciences of the United States of America, 107*(20), 9240–9245.

Unbekandt, M., & Davies, J. A. (2010). Dissociation of embryonic kidneys followed by reaggregation allows the formation of renal tissues. *Kidney International*, 77(5), 407–416.

Valenta, T., Hausmann, G., & Basler, K. (2012). The many faces and functions of β-catenin. *EMBO Journal, 31*(12), 2714–2736.

Vestweber, D., Kemler, R., & Ekblom, P. (1985). Cell-adhesion molecule uvomorulin during kidney development. *Developmental Biology, 112*(1), 213–221.

Wang, L., Ellis, M. J., Fields, T. A., Howell, D. N., & Spurney, R. F. (2008). Beneficial effects of the Rho kinase inhibitor Y27632 in murine puromycin aminonucleoside nephrosis. *Kidney and Blood Pressure Research, 31*(2), 111–121.

Watanabe, T., & Costantini, F. (2004). Real-time analysis of ureteric bud branching morphogenesis in vitro. *Developmental Biology, 271*(1), 98–108.

Wellik, D. M., Hawkes, P. J., & Capecchi, M. R. (2002). Hox11 paralogous genes are essential for metanephric kidney induction. *Genes and Development, 16*(11), 1423–1432.

Wilson, H. V. (1907). On some phenomena of coalescence and regeneration in sponges. *The Journal of Experimental Zoology, 5*, 245–258.

Wolpert, L. (1969). Positional information and the spatial pattern of cellular differentiation. *Journal of Theoretical Biology, 25*(1), 1–47.

Xu, J., Wong, E. Y., Cheng, C., Li, J., Sharkar, M. T., Xu, C. Y., et al. (2014). Eya1 Interacts with Six2 and Myc to Regulate Expansion of the Nephron

Progenitor Pool during Nephrogenesis. *Developmental Cell*, *31*(4), 434–447. http://dx.doi.org/10.1016/j.devcel.2014.10.015. Epub 2014 Nov 24.

Yang, H., Zhao, B., Liao, C., Zhang, R., Meng, K., Xu, J., et al. (2013). High glucose-induced apoptosis in cultured podocytes involves TRPC6-dependent calcium entry via the RhoA/ROCK pathway. *Biochemical and Biophysical Research Communications*, *434*(2), 394–400.

Yang, Z., Zimmerman, S., Brakeman, P. R., Beaudoin, G. M., III, Reichardt, L. F., & Marciano, D. K. (2013). De novo lumen formation and elongation in the developing nephron: A central role for afadin in apical polarity. *Development*, *140*(8), 1774–1784.

Yokoo, T., Ohashi, T., Shen, J. S., Sakurai, K., Miyazaki, Y., Utsunomiya, Y., et al. (2005). Human mesenchymal stem cells in rodent whole-embryo culture are reprogrammed to contribute to kidney tissues. *Proceedings of the National Academy of Sciences of the United States of America*, *102*(9), 3296–3300.

Yu, J., Carroll, T. J., Rajagopal, J., Kobayashi, A., Ren, Q., & McMahon, A. P. (2009). A Wnt7b-dependent pathway regulates the orientation of epithelial cell division and establishes the cortico-medullary axis of the mammalian kidney. *Development*, *136*(1), 161–171.

INDEX

Note: Page numbers followed by "*f*" indicate figures and "*t*" indicate tables.

C

D

E

F

G

H

I

K

L

M

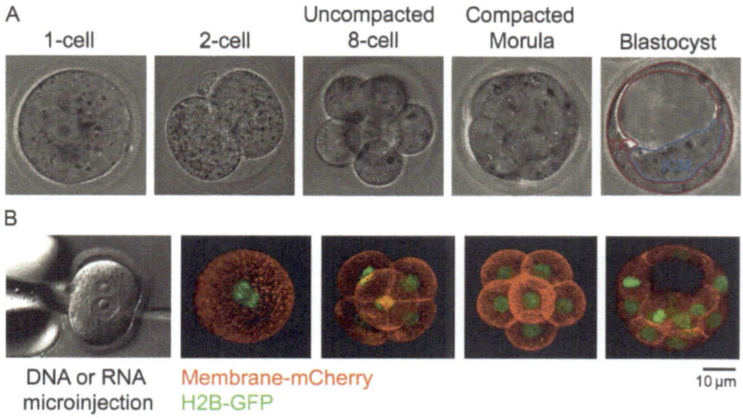

Melanie D. White and Nicolas Plachta, Figure 1 Imaging preimplantation development in the mouse embryo. (A) DIC images showing development of mouse embryo from 1-cell to blastocyst stage. (B) Microinjection of mRNA or DNA into the pronucleus allows visualization of proteins of interest throughout preimplantation development. In the example shown, the membrane is labeled with mCherry and the nucleus is labeled with H2B-GFP. ICM, inner cell mass; TE, trophectoderm.

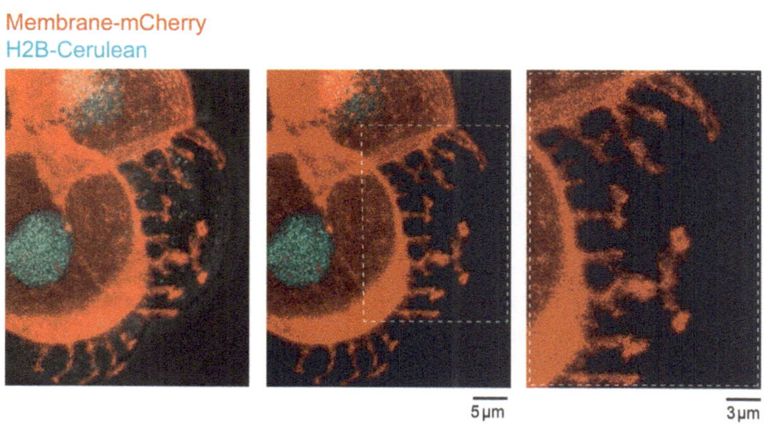

Melanie D. White and Nicolas Plachta, Figure 4 E-cadherin-dependent filopodia control embryo compaction. Two cells labeled with a mCherry protein targeted to the cell membrane extend filopodia over their nonlabeled neighbors during compaction at the late 8-cell stage. Dashed box shows filopodia at higher magnification.

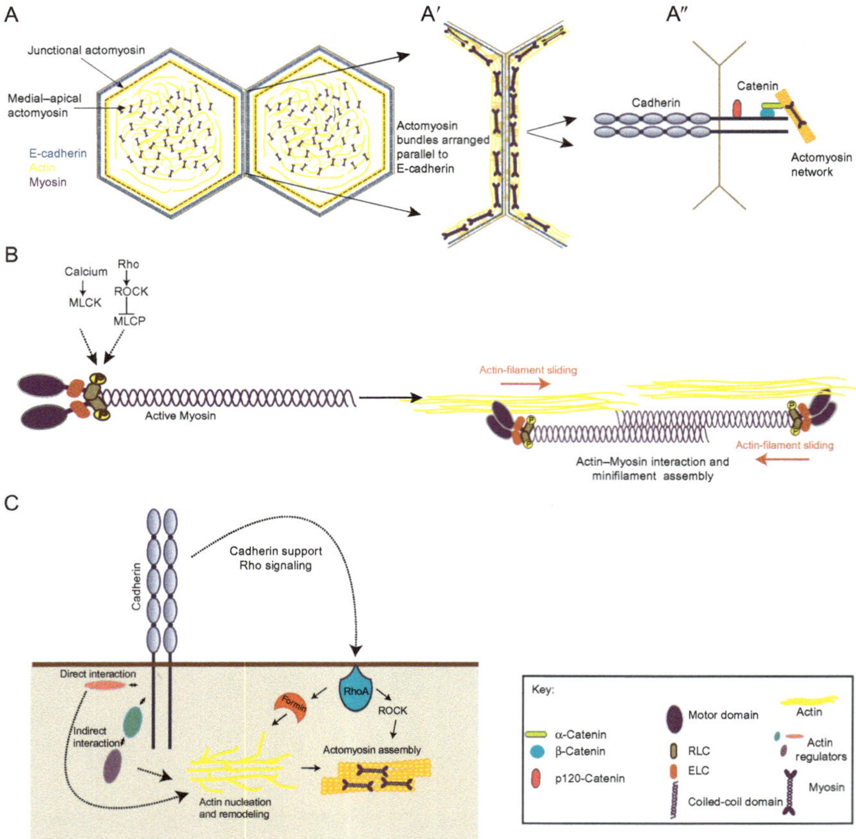

Rashmi Priya and Alpha S. Yap, Figure 1 The contractile apparatus at junctions and its biogenesis. (A) Spatial relation of actomyosin network with cadherin junctions. F-actin and myosin form a mesh at the apical pole of the cells (medial–apical network) and also orient themselves parallel to the junctions (junctional actomyosin). (A′) The contractile cortices of neighboring cells are mechanically coupled through association with the cadherin–catenin complex (A″). (For clarity, this coupling is illustrated only for junctional actomyosin.) (B) Kinases (ROCK and MLCK) activate myosin by phosphorylating its RLC. This leads to association of myosin with F-actin and minifilament assembly (which is also controlled by heavy chain phosphorylation). (C) Cadherin ligation recruits actin regulators directly or indirectly via adaptor proteins to modulate local actin assembly and remodeling. Cadherin can also scaffold upstream Rho activators and thus support Rho signaling, which further mediates actomyosin assembly via its effectors.

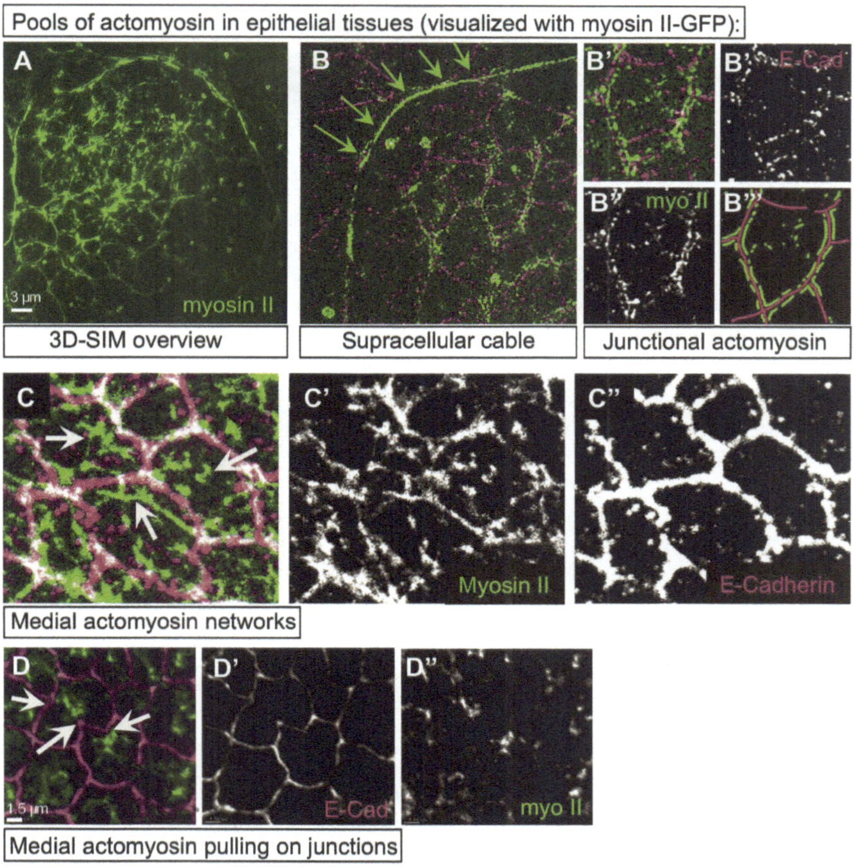

Katja Röper, Figure 2 Apical actomyosin assemblies in epithelial cells. (A) Overview of apical myosin II organization in a patch of epithelial cells in a *Drosophila* embryo. The patch of cells shown here form the so-called salivary gland placode that will undergo the morphogenetic process of tubulogenesis through a process of regulated apical constriction and convergent extension (for review, see Girdler & Röper, 2014). This image was acquired using structured illumination microscopy (3D-SIM) on an embryo expressing myosin II-GFP. Extensive networks of myosin II, some of which are spanning many cell diameters can be seen (the cells in the placode have an apical diameter of 2–5 µm). (B) 3D-SIM image of myosin II incomparison to E-Cadherin labeling at the boundary of the placode shown in (A). A supracellular actomyosin cable surrounds the placode (arrows). The cable is perfectly aligned between many cells. (B′)–(B″) Junctional myosin II is in close apposition with E-Cadherin on either side of the junction. (B″) is a schematic of (B′). (C)–(C‴). In many tissues undergoing morphogenesis, actomyosin forms an apical–medial meshwork that connects between neighboring cells through spot AJ. Arrows point to medial pools of myosin II. (D)–(D‴) AJ labeled by E-Cadherin appear deformed near clusters of apical–medial actomyosin, indicating pulling forces (arrows). Myosin II is in green and E-Cadherin in magenta. Scale bar in (A) is 3 µm, in (D) 1.5 µm. *Panels (A) and (B) were adapted from Fig. 3B and H in Röper (2012).*

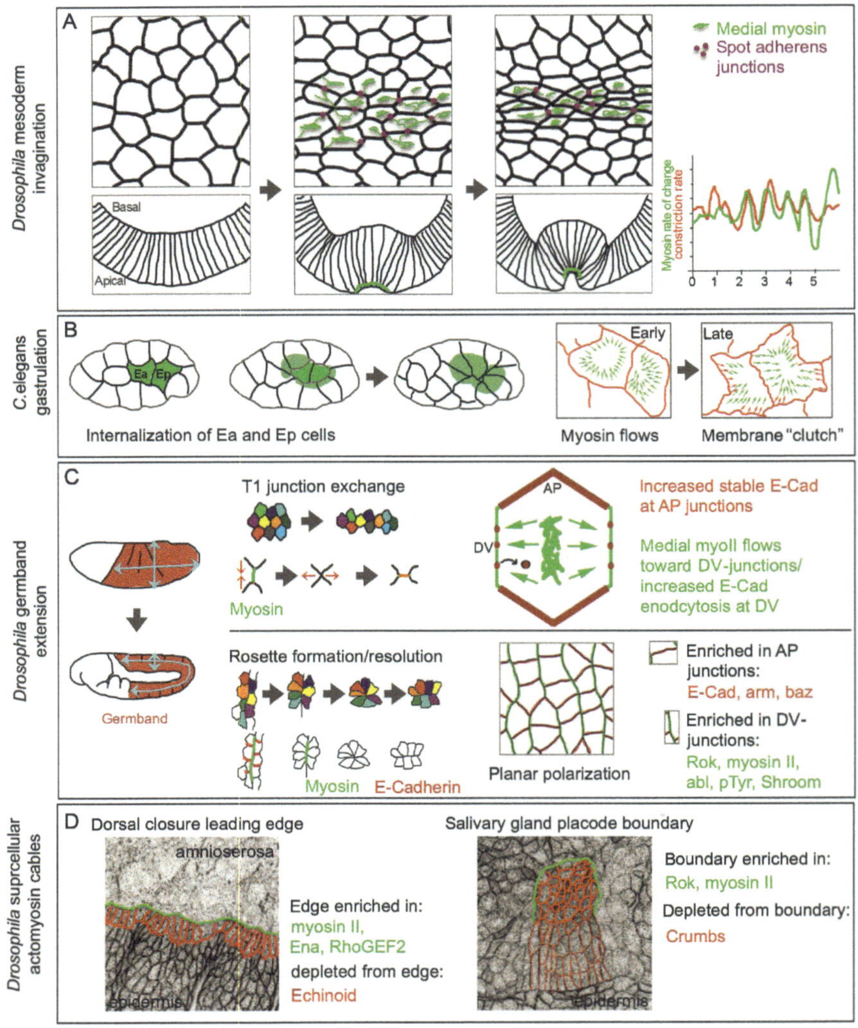

Katja Röper, Figure 3 Adherens junction–actomyosin interactions in morphogenesis. (A) During mesoderm invagination in the fly embryo, apical–medial actomyosin networks develop that show a pulsatile increase/decrease in intensity (see graph on right). These myosin II assemblies (green) pull on spot AJ (magenta) to decrease apical area (Martin et al., 2009; Martin & Wieschaus, 2010). Top panels show surface views, bottom panels show cross sections of the presumptive mesoderm. (B) During gastrulation and internalization of the Ea and Ep cells (green) on the ventral side of the *C. elegans* embryo, pulsatile centripetal myosin II flows can already be observed prior to apical constriction. A clutch mechanism that depends on Cadherin and Rac activity allows the flows to be harnessed for productive constriction at a later stage (Roh-Johnson et al., 2012). (C) During *Drosophila* germband extension, the germband shrinks in width by half and

(Continued)

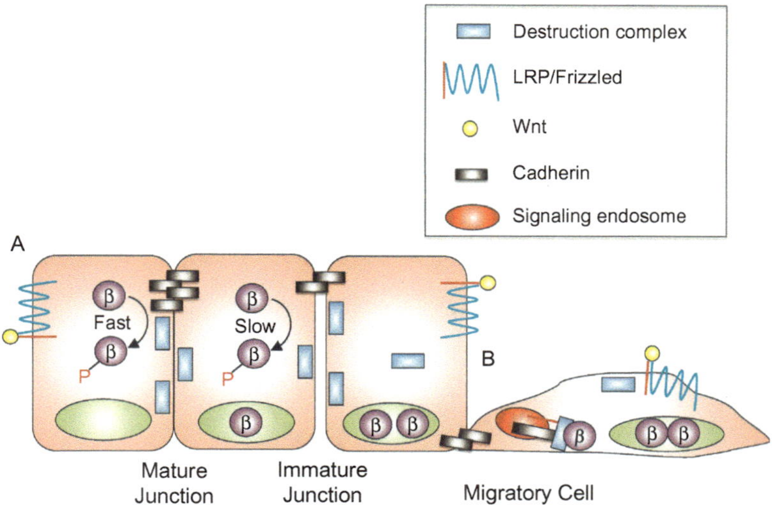

Pierre D. McCrea *et al.*, Figure 2 Cadherin-based adhesion as an enhancer of β-catenin destruction. In densely confluent cells with mature junctions, cadherins promote a faster turnover of β-catenin than in less adhesive (subconfluent) cells with immature junctions (A). This may explain why cells migrating adjacent to a wound appear sensitized to Wnt signals (Howard, Deroo, Fujita, & Itasaki, 2011; Maher et al., 2009). Mechanistically, the latter study suggests that enhanced β-catenin signaling is due to an activation step that depends on cadherins and a signaling endosome (B), while the former suggests that these cells show reduced capacity to degrade β-catenin (A).

Figure 3—Cont'd doubles in length within the confinement of the egg (schematic on the left). This is achieved through a process termed convergent extension, that in the germband seems to be fueled by two main mechanism. T1 junctional exchanges involve groups of four cells where a vertical or DV junction shrinks, followed by expansion of a perpendicular horizontal or AP junction, leading to exchange in neighbors (Bertet, Sulak, & Lecuit, 2004). Flows of medial myosin II clusters flow toward DV-junctions and help shrink them (tope panel diagram), with the flows being triggered by oscillating amounts of E-Cadherin-mediated tension from those junction. There is also an increase in E-Cadherin endocytosis from DV-junctions (Levayer & Lecuit, 2013; Rauzi et al., 2010). The second mechanisms affect larger groups of cells. Myosin II localization and activity is overall planar polarized within the germband, with myosin II being enriched in DV-junctions together with Rok, phospho-Tyrosine (pTyr), abelson kinase (abl), and Shroom (that can recruit Rok) (Blankenship, Backovic, Sanny, Weitz, & Zallen, 2006; Simoes Sde et al., 2010; Simoes Sde, Mainieri, & Zallen, 2014; Tamada, Farrell, & Zallen, 2012). Myosin II in DV-junctions leads to the pulling of larger groups of cells into a single vertex, forming a rosette structure than then resolves in a perpendicular fashion (schematic in lower panel). (D) During dorsal closure in the fly embryo, an actomyosin cable form at the dorsal-most edge of the leading edge cells, where Ena and RhoGEF2 are also enriched. This enrichment depends on the anisotropic localization of Echinoid in the leading edge cells, with loss of Echinoid from the edge surface triggered by the absence of Echinoid expression in the amnioserosa (Laplante & Nilson, 2011). An actomyosin cable also forms around the salivary gland placode in the fly embryo, where instead of Echinoid the anisotropic localization of Crumbs triggers Rok and myosin II enrichment at the boundary (Röper, 2012).

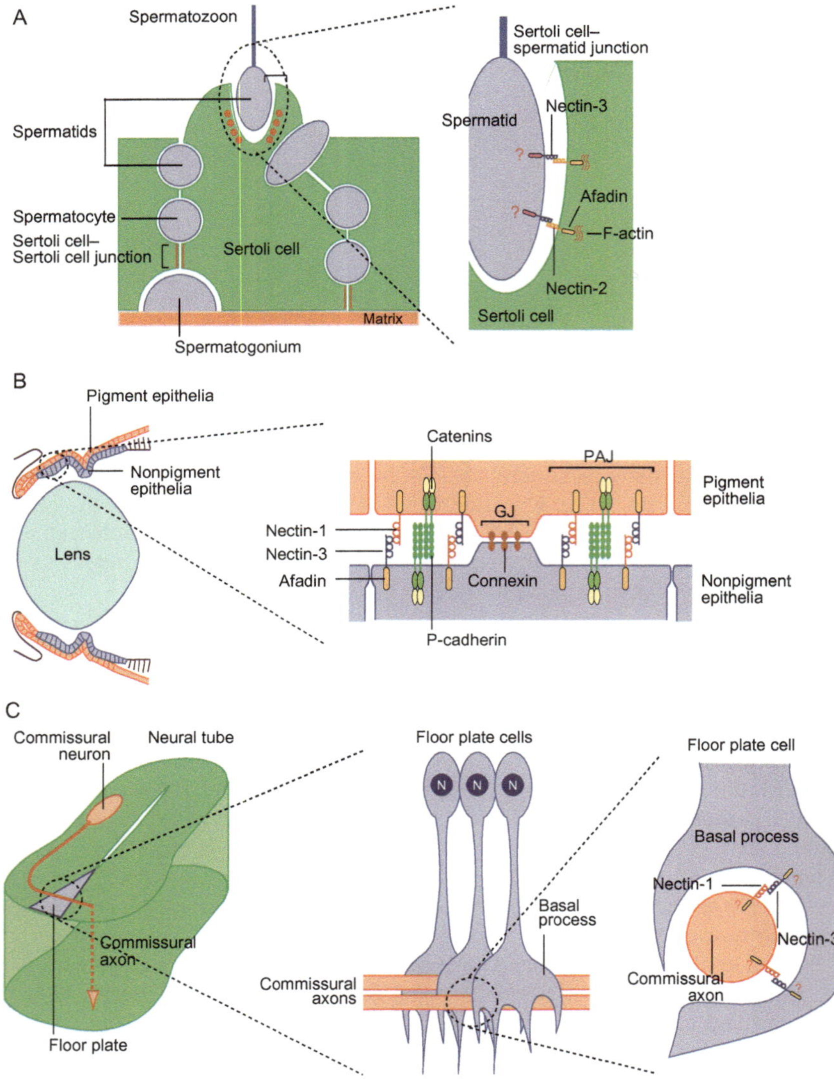

Kenji Mandai *et al.*, Figure 4 Various types of heterotypic cell–cell adhesion mediated by nectins. (A) Localization of nectin-2 and -3 at Sertoli cell–spermatid junctions in the testis. (B) Localization of nectin-1 and -3 at the contact site between the pigment cell and nonpigment cell layers of the ciliary epithelium in the eye. GJ, gap junction; puncta adherentia junction. (C) Localization of nectin-1 and -3 at the contact site between commissural axons and floor plate cells in the neural tube. F-actin is not shown in (B) and (C).

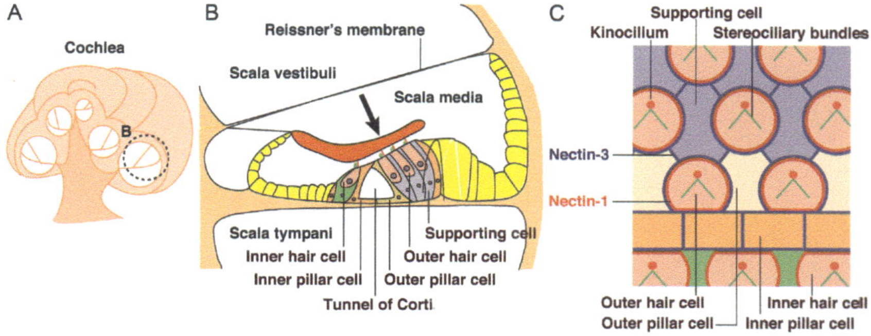

Kenji Mandai *et al.*, Figure 5 Formation of the checkerboard-like cellular pattern in the organ of Corti. (A) Schematic illustration of the cochlea. The area encircled by dashed dots is enlarged in (B). (B) Cross-sectional illustration of the organ of Corti. The cochlea includes three chambers, and the organ of Corti is located in the scala media. A single row of inner hair cells is located on the medial side of the epithelium, whereas three rows of outer hair cells are located more laterally. The inner and outer hair cell regions are separated by the tunnel of Corti, which is surrounded by single rows of inner and outer pillar cells. An arrow indicates the direction of view shown in (C). (C) Luminal surface illustration of the auditory epithelium in the organ of Corti. Outer hair cells, inner hair cells, and various types of supporting cells are arranged in a checkerboard-like cellular pattern in the auditory epithelium. Nectin-1 and -3 are expressed exclusively by hair cells and supporting cells, respectively. Therefore, the *trans*-interaction between nectin-1 and -3 occurs at the boundaries between these cells.

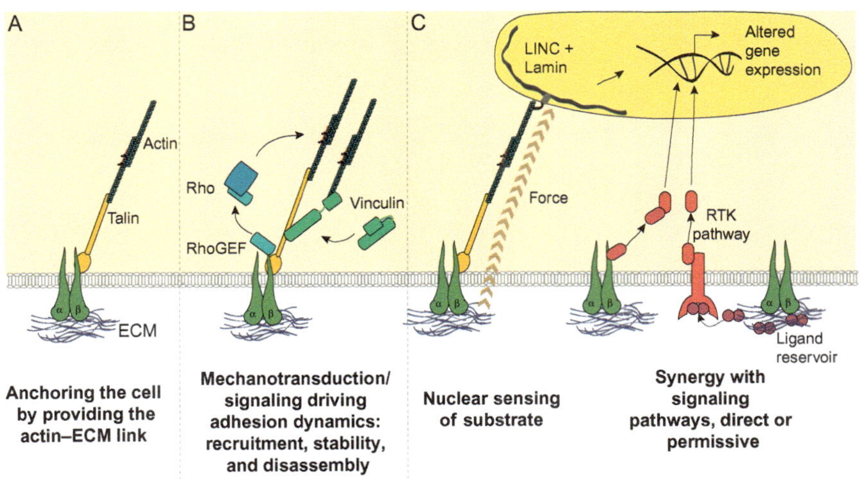

Aidan P. Maartens and Nicholas H. Brown, Figure 1 Adhesion architecture, mechanics, and signaling. (A) Basic architecture of an integrin adhesion. (B) Signaling interactions that feed back into adhesion dynamics. RhoGEF recruitment promotes RhoGTPase activation, and a key target of Rho activity is the actin cytoskeleton. Force on the adhesion can also stretch talin, leading to recruitment of vinculin and providing additional links to actin. (C) Integrin influence on gene expression. Actin can connect to the LINC/lamin complex, allowing force transduced through integrins to translate into nuclear deformation and changes in gene expression. Integrins may also recruit proteins that influence signaling pathways, or play a permissive role by maintaining proximity between the cell and an ECM rich in ligands. Abbreviations: ECM, extracellular matrix; LINC, linker of nucleoskeleton and cytoskeleton complex; RTK, receptor tyrosine kinase.

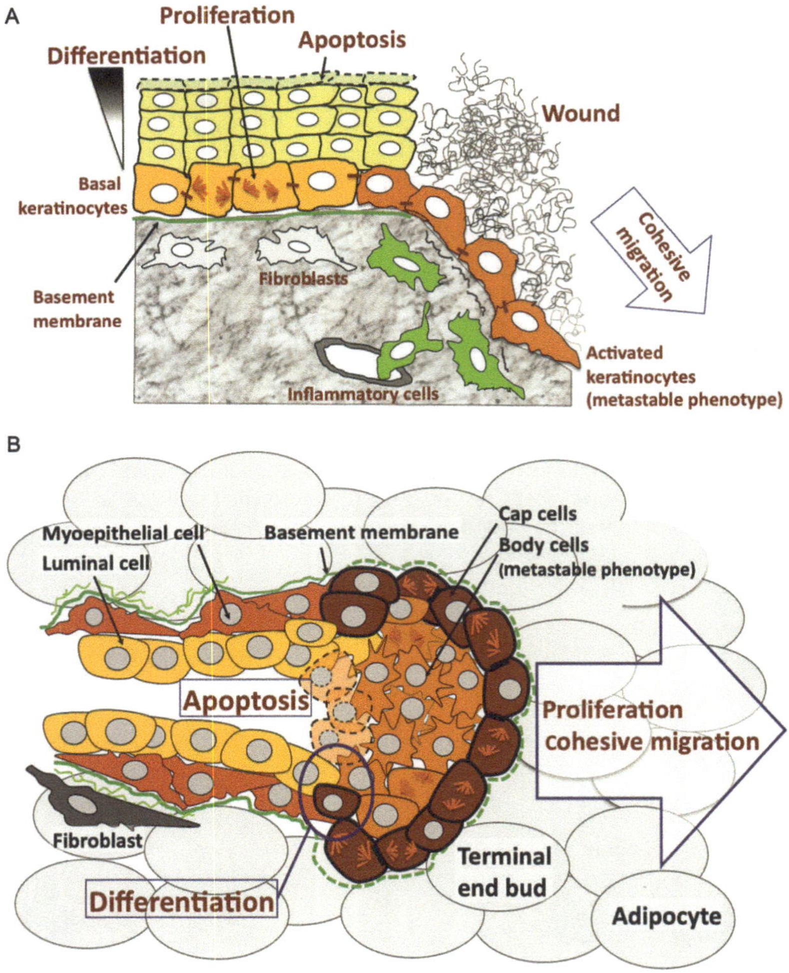

Pierre Savagner, Figure 2 Physiological morphogenetic processes can share some functional modules with EMT without involving cell individualization. These processes produce cells with a metastable phenotype involving transient new properties including motility and phenotype plasticity. (A) During cutaneous wound healing re-epithelialization, activation of basal keratinocytes induces a cohort migration characterizing a metastable phenotype. (B) During mammary tubulogenesis, the terminal end bud is a transient structure composed of cap and body cells expressing cell–cell adhesion structures. Body cells are poorly polarized, motile, and progressively reach a differentiation stage as they reverse from a metastable phenotype.

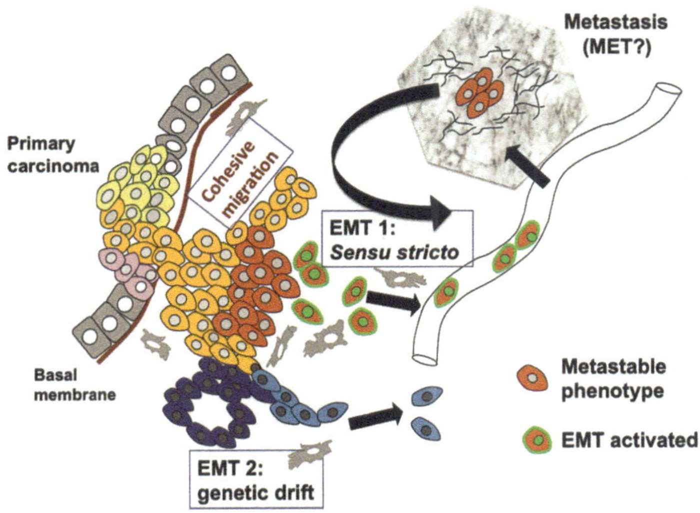

Pierre Savagner, Figure 3 During carcinoma progression, a genetic drift engender transformed cells with new properties. Among them, the acquisition of a metastable phenotype confers to the cells a plasticity necessary to the metastatic process.

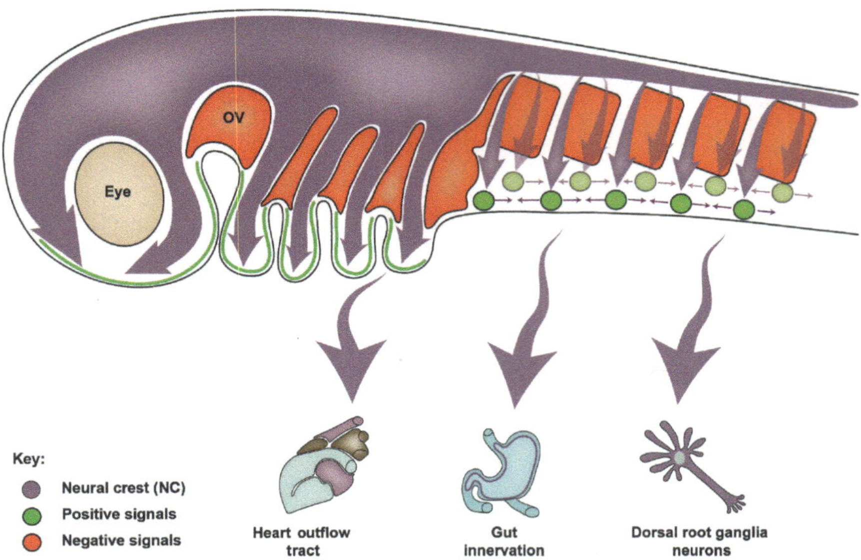

Elias H. Barriga and Roberto Mayor, Figure 1 Neural crest cell directional migration. Illustration of a hypothetical vertebrate embryo. Neural crest cells are shown in purple, restrictive signals in red, and attractant cues are represented in green. Thick arrows represent migratory stream of cephalic neural crest cell, migrating around the eye and the otic vesicle (OV). Neural crest cells streams follow chemotactic signals (green) and avoid chemorepulsive cues (red). Trunk neural crest cells are shown as thinner streams that avoid restrictive signals secreted by the somite (red). Migratory neural crest cells reach their target tissues and differentiate to distinct cell types. Among other tissues, neural crest cells contribute to the heart outflow tract, gut innervations, and dorsal root ganglia neurons.

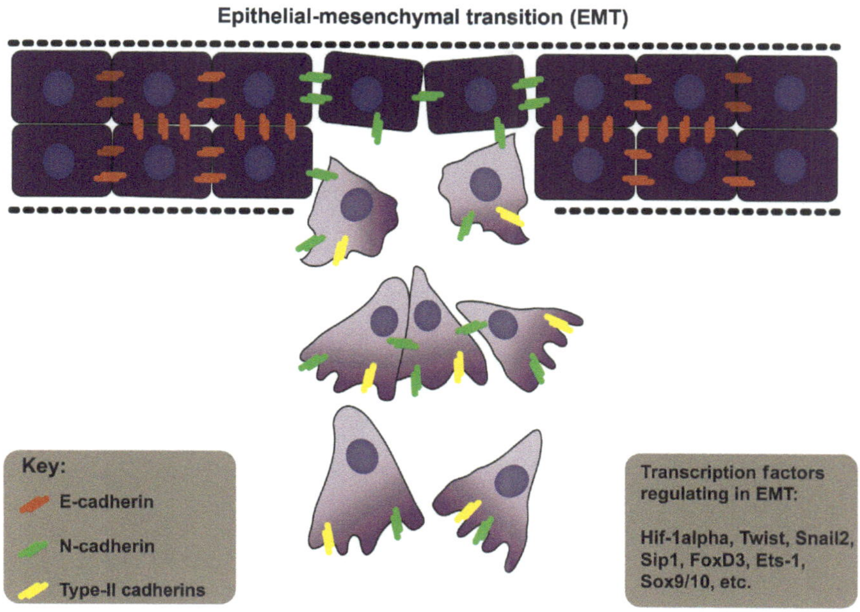

Elias H. Barriga and Roberto Mayor, Figure 2 Cadherins switch during neural crest cells epithelial-to-mesenchymal transition (EMT). Events occurring during neural crest cell EMT. Presumptive neural crest cells form an epithelium-like group of cells along the lateral edge of the neural tube. Before migration, neural crest cells express E-cadherin (red parallel lines) and N-cadherin (green parallel lines) as cell–cell adhesion molecules. Prior to migration, neural crest cells undergo EMT. This process involves transcriptional repression of E-cadherin, to allow cells to separate from the neuroectoderm and ectoderm, via delamination. The expression levels of N-cadherin are slightly but not completely reduced, allowing cells to form transient contacts that mediate signaling during collective migration. Once migrating, neural crest cells start to express type-II cadherins such as Cadherin-11, -7, -19, and -20 (yellow parallel lines), this proteins are described to promote protrusive activity, and to be involved in modulating contact inhibition of locomotion (CIL). In the lower right corner are listed some of the master transcription factors that control neural crest cells EMT in different animal models.

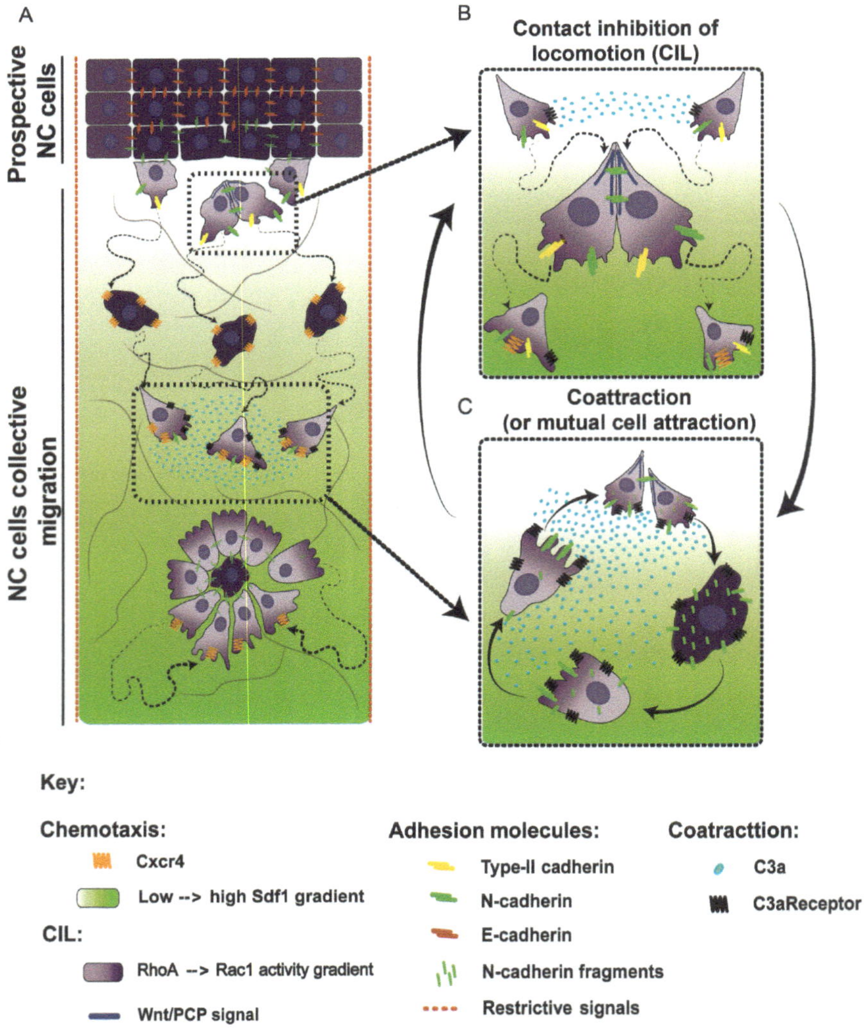

Elias H. Barriga and Roberto Mayor, Figure 4 Integration of chemotaxis, contact inhibition of locomotion (CIL), and coattraction along neural crest cells collective migration. (A) Events occurring within a stream of migratory neural crest cells. Restrictive signals are shown as parallel dotted red lines; chemotactic cues (Sdf1) are represented as a green gradient (with high concentration of Sdf1 in green). After prospective neural crest cells undergo EMT, cells experience CIL, repolarize, and disperse. If Sdf1 levels are low neural crest cells disperse and become polarized single cells. Sdf1 can guide polarized cells and, by coattraction, these cells come together to form transient cell–cell contacts. Upon contact, cells undergo CIL, repolarize, and move away from each other again. Repetitive cycle of CIL and coattraction are proposed to maintain cells migrating in a collective and directional manner. (B) Contact inhibition of locomotion (CIL). CIL is a cellular behavior by which upon contact, two migratory cells stop their migration and

(*Continued*)

Figure 4—Cont'd repolarize in opposite directions. At the molecular level, two neural crest cells collide and upon contact they form very transient N-cadherin-based cell–cell adhesions (green lines). N-cadherin generates local Wnt/PCP signals (blue), which in turn activate RhoA and repress Rac1 at the contact point to repolarize the cells in opposite directions. Type-II cadherins (yellow lines) have also been shown to be involved in CIL. (C) Mutual cell attraction or coattraction. During coattraction, neural crest cells mutually attract each other by expressing and secreting the complement fragment C3a (light blue dots) and the cognate receptor C3aR (black barrels). After experiencing CIL, a neural crest cell moves away from the group. Soon after, the lost cell repolarizes and migrates back toward the group following a C3a gradient. When the cell is docked back to the group, it establishes transient N-cadherin-dependent cell contacts again and the cycle is repeated.

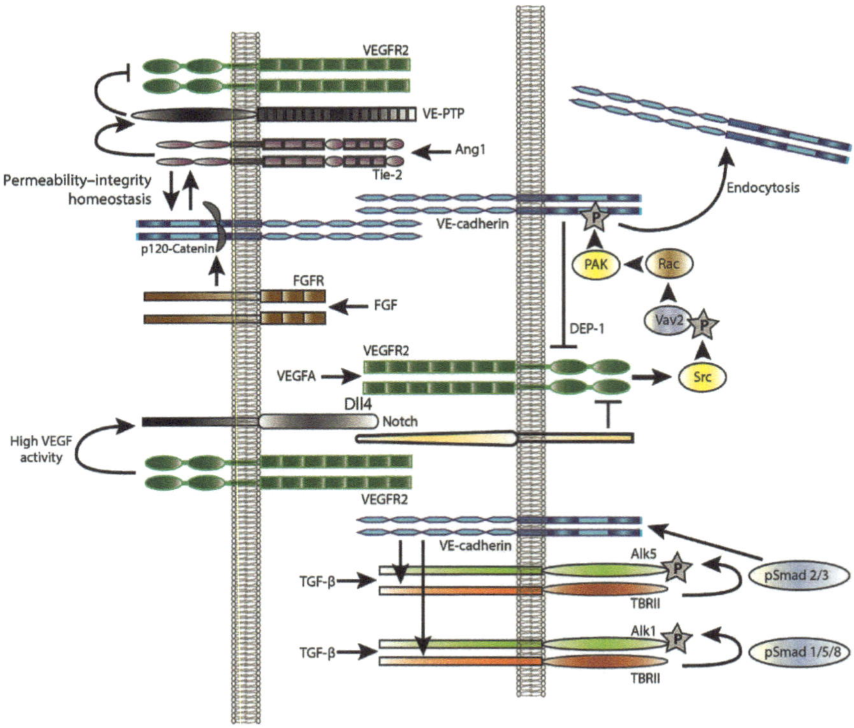

Anne Karine Lagendijk and Benjamin M. Hogan, Figure 3 *Schematic representation of relationships between major vascular signaling pathways and VE-cadherin.* VEGFR2 activation by VEGFA results in an activation of Src kinase, which subsequently phosphorylates Vav2, ultimately activating PAK (via Rac). PAK phosphorylation of the VE-cadherin cytoplasmic tail triggers endocytosis of the protein and thus enhances junctional permeability. Reciprocally VE-cadherin can dampen VEGFR2 activity via activation of the phosphatase DEP-1. Another mechanism to inhibit VEGF signaling activity is via active Dll4–Notch signaling which dampens the VEGF-induced effect at VE-cadherin. Ang–Tie2 signaling can also inhibit VEGF-induced signaling via binding to VE-PTP. Levels of active Tie2 and VE-cadherin at the junctions were shown to maintain a balance in permeability and integrity of the junctions. FGF signals enhance VE-cadherin stability by promoting VE-cadherin presence at the junctions and also by enhancing the interaction between VE-cadherin and p120-catenin. Finally, TGF-β ligands in endothelial cells activate TBRII receptors that form active receptor complexes with both Alk1 and Alk5 (by phosphorylation) resulting in active pSmad2/3 and pSmad 1/5/8 transcriptional complexes. pSmad 2/3 activation has been shown to be required for VE-cadherin junctional stability and VE-cadherin in return enhances the formation of both TBRII–TBRI complexes upon TGF-β activation.

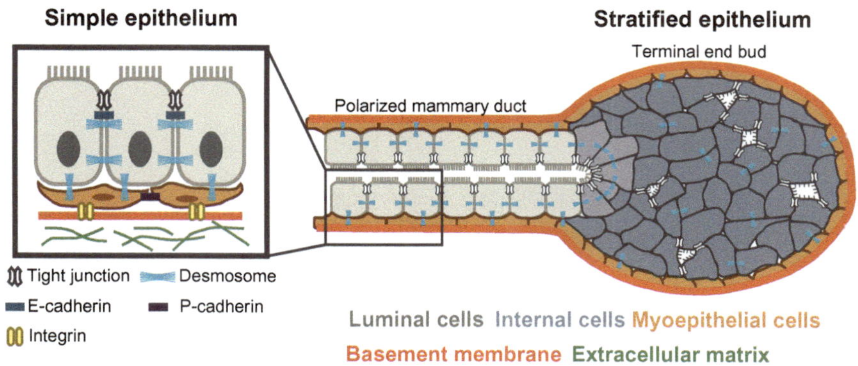

Eliah R. Shamir and Andrew J. Ewald, Figure 1 Normal transitions in adhesion during epithelial branching morphogenesis. Mammary epithelium initiates branching morphogenesis postnatally. Tube elongation is accomplished by a stratified terminal end bud, which contains many internal luminal cells that lack apicobasal polarity and display reduced numbers of intercellular junctions. The epithelium at the rear polarizes to a bilayered, simple ductal architecture consisting of an inner layer of luminal cells and a basal layer of myoepithelial cells. Epithelial cells in the ducts are connected by many cell–cell junctions. *Schematic adapted from original by Robert Huebner, with permission.*

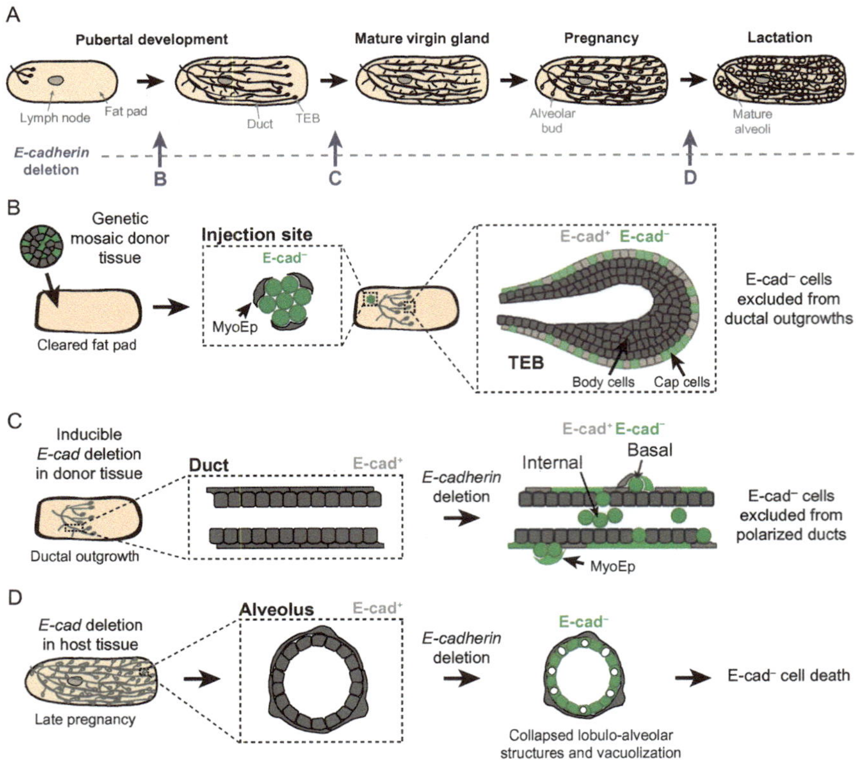

Eliah R. Shamir and Andrew J. Ewald, Figure 2 E-cadherin is required at multiple stages of mammary epithelial development. (A) Mammary pubertal branching morphogenesis initiates at approximately 3 weeks postnatal. A stratified terminal end bud (TEB) leads the elongation front, and a full ductal network forms over 7 weeks. During pregnancy, lactogenic hormones induce the ductal epithelium to form specialized structures called alveoli that mature to secrete milk during lactation. Different genetic approaches have been used to test the consequences of E-cadherin loss during distinct stages of development. (B) E-cadherin is required for branching morphogenesis (Shamir et al., 2014). E-cadherin$^-$ cells (green) within transplanted genetic mosaic epithelium are excluded from a ductal network elaborated exclusively by E-cadherin$^+$ cells. (C) E-cadherin is required for maintenance of epithelial architecture (Shamir et al., 2014). E-cadherin$^-$ cells (green) are extruded apically and basally from mature E-cadherin$^+$ epithelial ducts. (D) E-cadherin is required for terminal differentiation and cell survival of alveoli in the lactating gland (Boussadia et al., 2002). E-cadherin loss induces lobuloalveolar collapse and massive apoptosis.

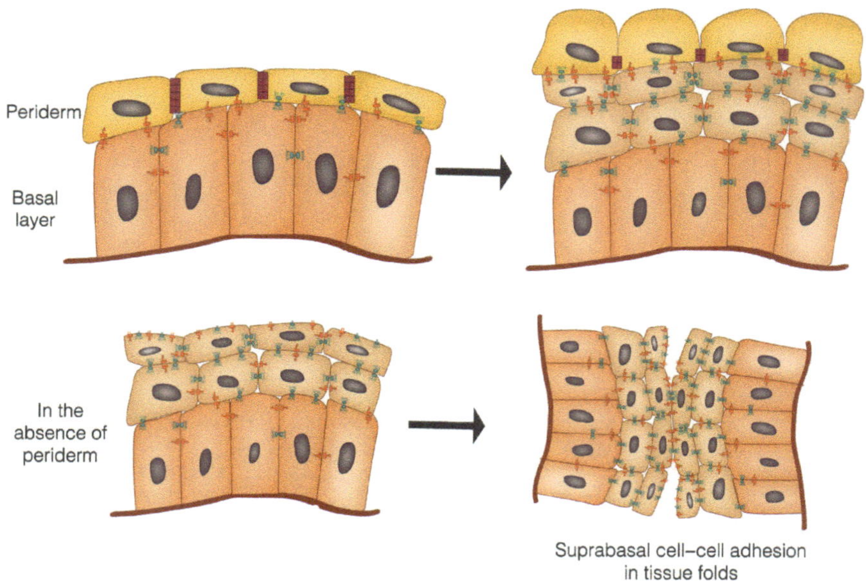

Kaelyn D. Sumigray and Terry Lechler, Figure 1 Periderm function in the embryo. The periderm is found on top of the embryonic epidermis where it forms tight junctions. In the absence of periderm, suprabasal epithelial cells are competent to interact with other cells, resulting in epidermal sheet fusions.

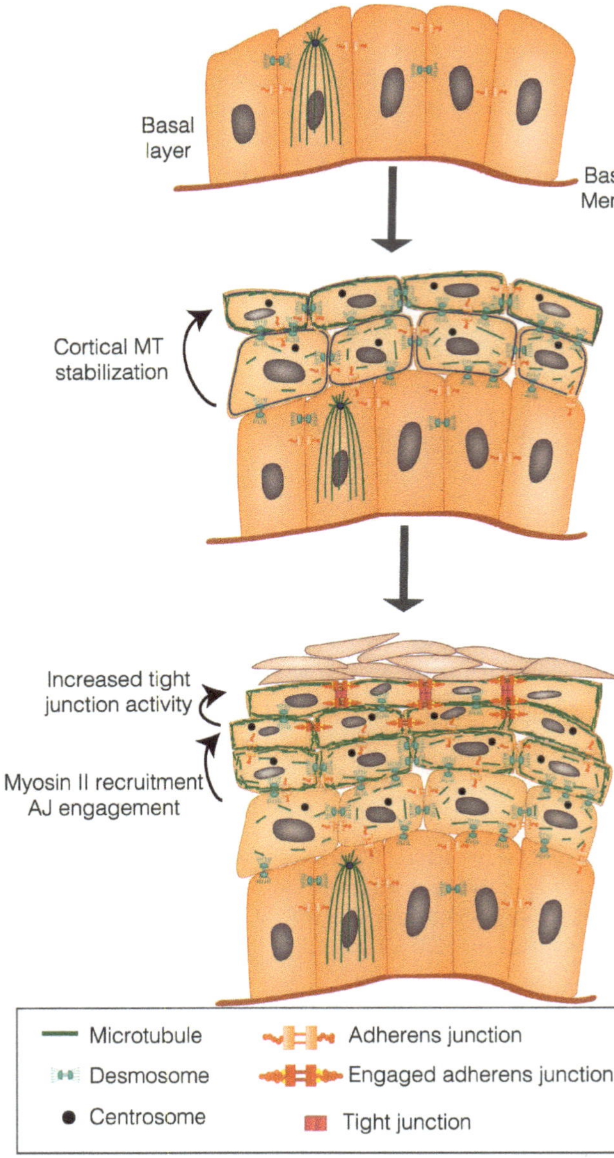

Kaelyn D. Sumigray and Terry Lechler, Figure 2 Model for collaboration of cell–cell adhesions to generate a functional barrier. See text for details.

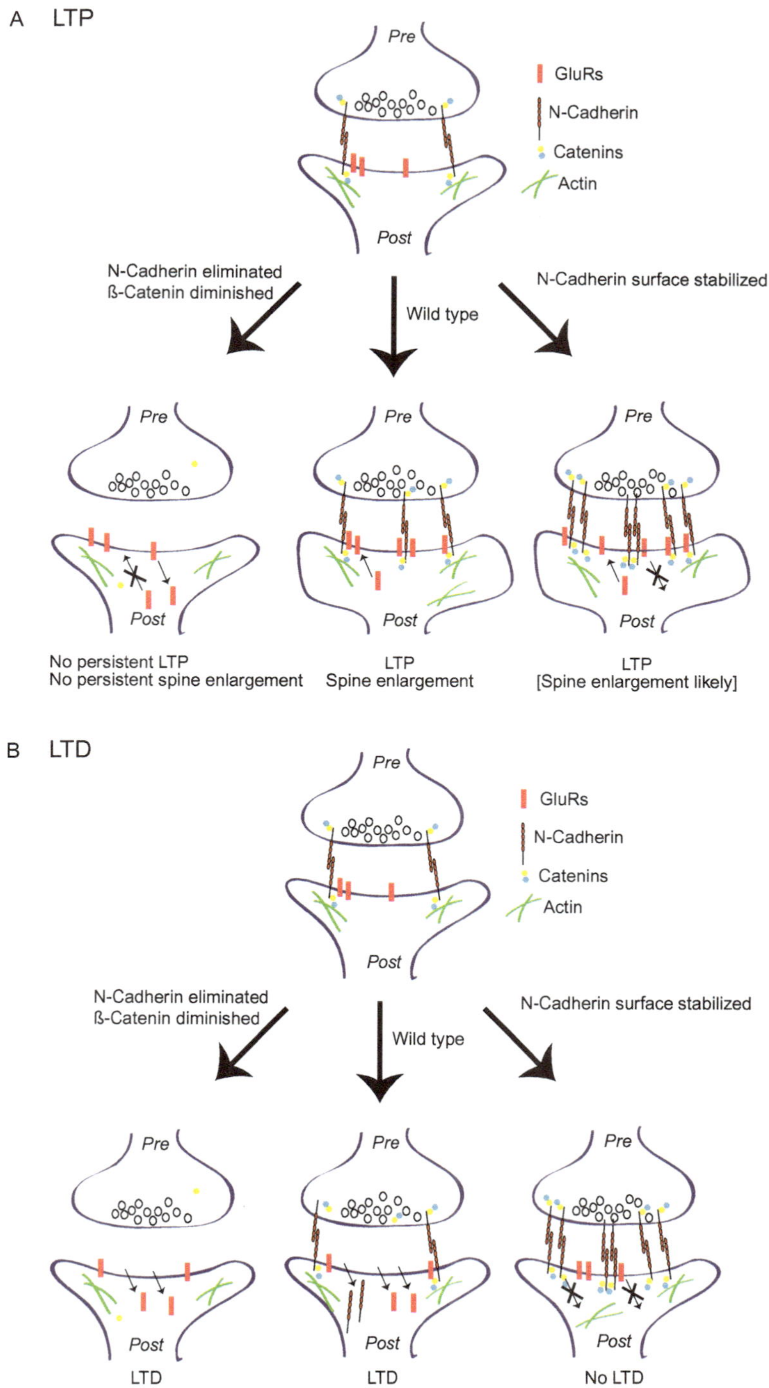

Lauren G. Friedman *et al.*, Figure 2 Schematic diagram showing how the cadherin/catenin system can bidirectionally regulate synaptic plasticity by differentially controlling GluR subunit trafficking and/or stability at the surface. (A) During LTP, a synapse (top) becomes strengthened (wild-type synapse, bottom middle) by the insertion of GluRs into the membrane (small arrow in the spine head) and also undergoes a coordinated increase in the size of the spine head. N-Cadherin facilitates both processes by

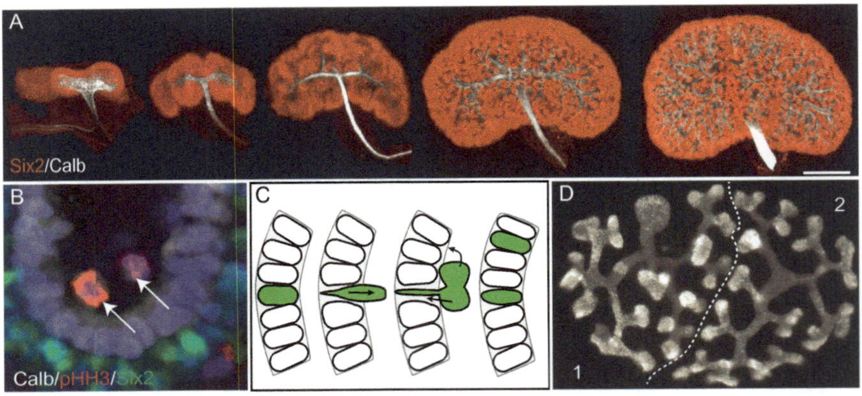

Alexander N. Combes *et al.*, Figure 1 Critical processes in ureteric branching. (A) Optical projection tomograpy series of the developing mouse kidney from 11.5 to 15.5 dpc with Calbindin (white) marking the ureteric epithelium and Six2 (red) marking the surrounding cap mesenchyme (CM). (B). Cells of the ureteric tip undergo mitosis in the lumen arrows) in a process termed mitosis-associated cell dispersal (MACD; Packard et al., 2013). Ureteric tip (Calbindin, white), CM (Six2, green), and mitotic nuclei (pHH3, red). (C) Diagram illustrating the process of MACD (dividing cell in green) with the mother cell partially delaminating into the lumen, dividing, then reinserting into the epithelium while the daughter cell inserts 1–3 cell diameters away. (D) Self-avoidance during branching. The ureteric bud/collecting duct trees of two E11.5 mouse kidney rudiments (1 and 2), grown close to one another, produce distorted shapes as they avoid mutual collisions and leave a fairly straight border (dotted line). This mutual avoidance, which seems to be mediated by BMP7, can account for the automatic spacing out of the tips of the tree, and for the change of branch angle with developmental stage.

---

trapping and/or promoting GluR positioning through direct ectodomain interactions as well as through linkage to the actin cytoskeleton. In the case of the conditional N-cadherin deletion mutant (bottom left synapse), GluRs fail to insert and/or fail to become trapped at the synapse in the absence of N-cadherin (blocked arrow), thus persistent LTP and spine enlargement are both abolished. GluR endocytocis is presumably unaffected (small arrow), thus accounting for normal LTD in these mice. In the case of the cadherin/β-catenin stabilized mice (bottom right synapse), excessively stabilized surface N-cadherin has no effect on GluR insertion (small arrow), thus accounting for normal LTP in these mice. However, the surface-stabilized N-cadherin abnormally traps synaptic GluRs, thus preventing GluR endocytocis (blocked arrow) and blocking LTD. (B) During LTD, wild-type synapses (middle bottom) undergo endocytocis of both N-cadherin and GluRs from the synaptic surface (small arrows). In the N-cadherin conditional KO synapse (left synapse), GluR endocytocis is presumably not affected, thus yielding normal LTD. In contrast, when N-cadherin is surface stabilized (right synapse), GluRs are abnormally trapped, preventing endocytocis, thus these synapses fail to exhibit LTD. *This cartoon is based on the studies of Bozdagi et al. (2010), Mills et al. (2014), and Nikitczuk et al. (2014).*

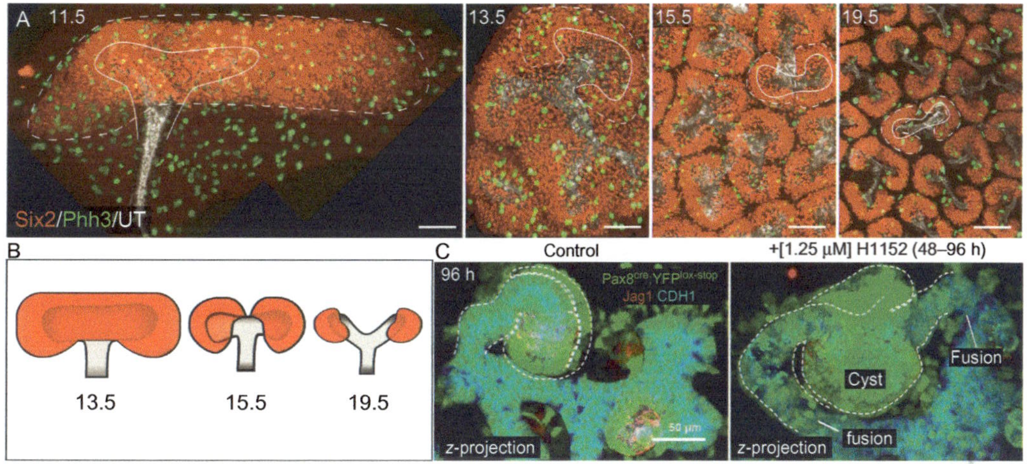

Alexander N. Combes *et al.*, Figure 2 Critical processes in nephron formation and maturation. (A) Confocal imaging of developing mouse kidneys from 11.5 to 19.5 (birth) dpc with labeling of the cap mesenchyme (CM; Six2; red), ureteric epithelium (cytokeratin (11.5 and 19.5) or calbindin (13.5 and 15.5; white), and cells undergoing mitosis (pHH3; green). This series serves to illustrate changes in the cap/tip relationship within the nephrogenic niche as branching proceeds. Dashed lines in each image indicate the boundaries of representative CM domain/s. Solid outlines indicate the boundary of the underlying epithelium. (B) Diagrammatic representations of niche morphology across murine kidney development showing the changing relationship between the CM (red) and ureteric epithelium (white). (C) Rho kinase in fusion and patterning. The left panel shows a nephron (white dotted lines) developing in a cultured kidney, displaying a normal anatomy with a developing proximal glomerular cleft and one point of connection to the ureteric bud system. The right panel shows a nephron growing in the presence of the ROCK inhibitor H1152: the center of the nephron is cystic, and it has connected to the ureteric bud system at both ends. *This picture is CC from Lindström, Hohenstein, and Davies (2013).*

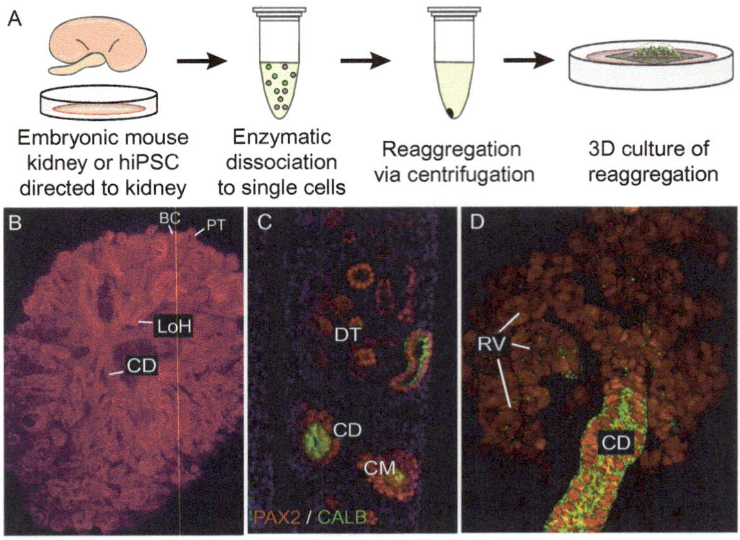

Alexander N. Combes *et al.*, Figure 3 Self-organization in kidney organoids. (A) Diagram of the methodology for dissociation and reaggregation cultures in which total organ primordia or differentiated human pluripotent cells are used. This can also be achieved using serial culture in which an intact tree is reaggregated with dissociated cells (Davies, Unbekandt, Ineson, Lusis, & Little, 2012). (B) A low-power view of a serial reaggregate using the method of Chang and Davies (2012), showing a branched collecting duct (CD) system, nephrons with Bowman's capsules (BC) and proximal tubules (PT) in the cortex, and loops of Henle (LoH) descending into the medulla. (C) Section through a reaggreation generated from dissociated 12.5 dpc mouse kidney illustrating immunofluroescence for Pax2 (red), Calbindin (green), and DAPI (blue). This shows the presence of CD epithelium ($Pax2^{+}Calb^{+}$), cap mesenchyme (CM; $Pax2^{+}$ surrounding CD), and distal tubule ($Pax2^{+}$ epithelial structures). (D) Whole-mount IF of a kidney organoid generated after 2D culture of human pluripotent stem cells subjected to directed differentiation as described in Takasato et al. (2014). Markers used are PAX2 (red) and cytokeratin (CYTOK; green). The organoid shows the presence of CD ($PAX2^{+}CYTOK^{+}$) and early renal vesicles (RV) $PAX2^{+}$ polarizing structures.